Solution Key

Geometry

Ray C. Jurgensen
Richard G. Brown
John W. Jurgensen

Evanston, Illinois • Boston • Dallas

2004 Impression
Copyright © 1994, 1992, 1990 by Houghton Mifflin Company.
All rights reserved.

No part of this work may be reproduced or transmitted in any form or by any means, electronic or mechanical, including photocopying and recording, or by any information storage or retrieval system without the prior written permission of McDougal Littell, a division of Houghton Mifflin Company, unless such copying is expressly permitted by federal copyright law. Address inquiries to Manager, Rights and Permissions, McDougal Littell, a division of Houghton Mifflin Company, P.O. Box 1667, Evanston, IL 60204.

Printed in U.S.A.

ISBN: 0-395-67766-1

11 12-DCI-07

Contents

Chapter

1.	Points, Lines, Planes, and Angles	1
2.	Deductive Reasoning	10
3.	Parallel Lines and Planes	23
4.	Congruent Triangles	42
5.	Quadrilaterals	79
6.	Inequalities in Geometry	110
7.	Similar Polygons	126
8.	Right Triangles	153
9.	Circles	176
10.	Constructions and Loci	208
11.	Areas of Plane Figures	239
12.	Areas and Volumes of Solids	274
13.	Coordinate Geometry	297
14.	Transformations	347
	Examinations	377
	Logic	379
	Flow Proofs	389
	Handbook for Integrating Coordinate and Transformational Geometry	393
	Discrete Mathematics	412
	Fractal Geometry	416
	Portfolio Projects	421

CHAPTER 1 • Points, Lines, Planes, and Angles

Pages 2–3 • CLASSROOM EXERCISES

1. a circle **2.** Yes; 2 **3.** No **4.** No

5. one line with two different names **6.** No

7. No **8.** Yes

Pages 3–4 • WRITTEN EXERCISES

A **1.**

Distance between	Diagram distance	Ground distance
X and P	5 cm	10 m
X and F	7 cm	14 m
X and T	7 cm	14 m
Y and F	9.5 cm	19 m
F and T	12 cm	24 m

2. a, b.

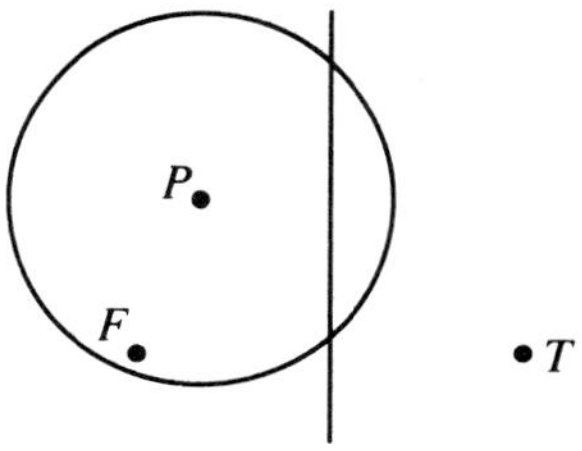

P
F
T

c. 2 **3. a, b.**

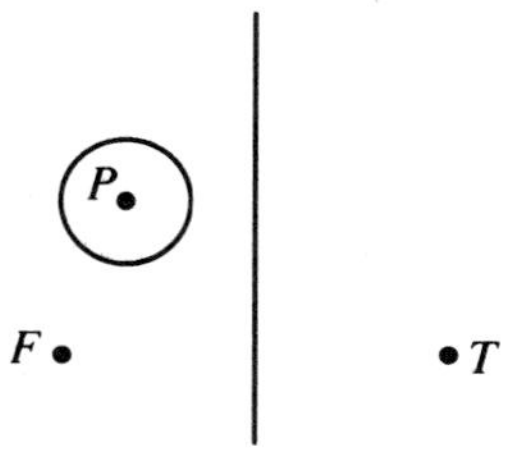

P
F
T

c. none

4. about 4.7 cm **5.** the distance from R to S

6. the distance from A to B = the distance from A to C

B **7.** It is twice the area of the inner square. **8.** The areas are equal. **9.** $ab = cd$

10. 34

Page 7 • CLASSROOM EXERCISES

1. True **2.** False **3.** False **4.** True **5.** True **6.** False **7.** True **8.** No

9. Yes; yes **10.** Yes; yes **11.** D **12.** G **13.** G **14.** H **15.** A **16.** F

17. Yes **18.** Yes **19.** $\overleftrightarrow{BF}$

20. $ABCD$ and $EFGH$, or $ADHE$ and $BCGF$, or $ABFE$ and $DCGH$

Pages 7–9 • WRITTEN EXERCISES

A **1.** True **2.** True **3.** True **4.** True **5.** True **6.** False **7.** True **8.** False

9. False **10.** True

11.

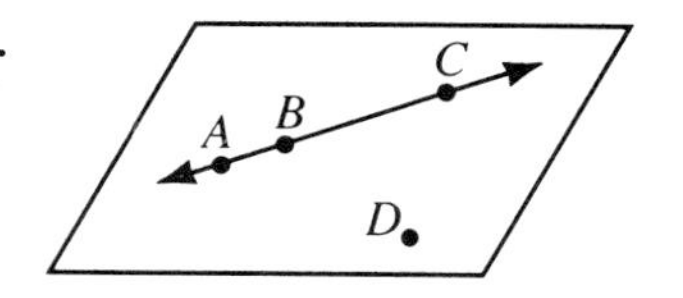

12.

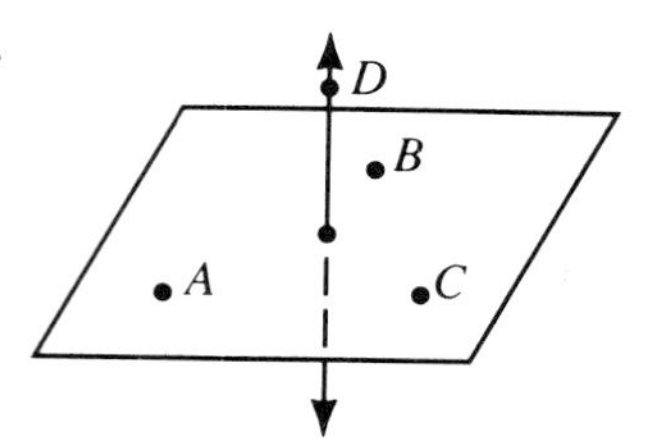

13. *VWT*, *VST*, *VRS*, *VWR*, *WRST* **14.** No **15.** $\overleftrightarrow{RW}$, $\overleftrightarrow{RV}$, $\overleftrightarrow{RS}$ **16.** *VST*, *WRST*

17. *VRS*, *VST*, *WRST* **18.** Answers may vary; for example, $\overleftrightarrow{VS}$ and *WRST*

19, 20. Check students' drawings.

21. *RSGF* and *FGCB* **22.** $\overleftrightarrow{EA}$, $\overleftrightarrow{ER}$, $\overleftrightarrow{EH}$ **23.** *ABCD*, *REABF*, *GFBC*

24. a. No **b.** Yes **25. a.** Yes **b.** No

26. Answers may vary; for example, **a.** *REABF* and *SHDCG* **b.** *AEHD* and *BFGC*

B **27.** No **28. a.** Yes **b.** No **29–36.** Sketches may vary.

29.

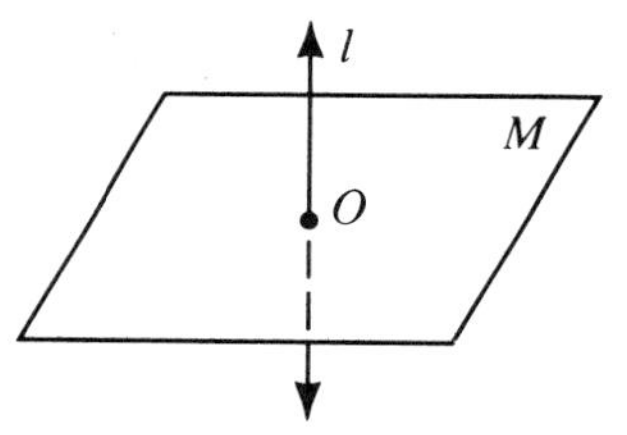

30.

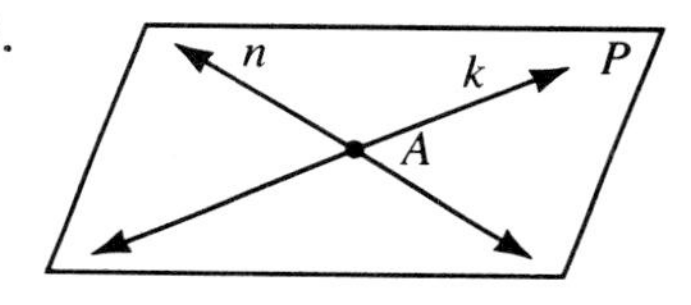

31.

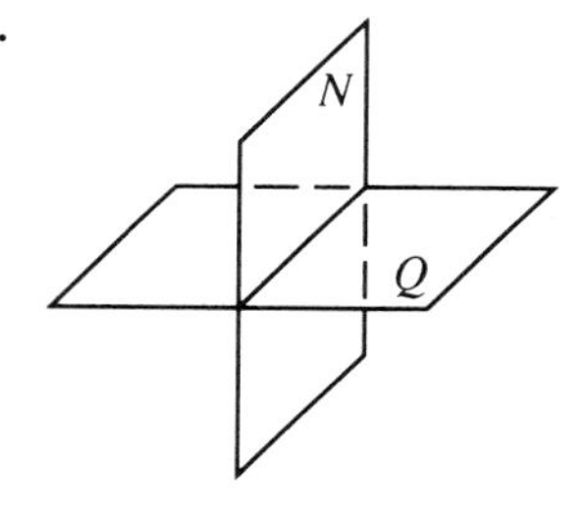

32.

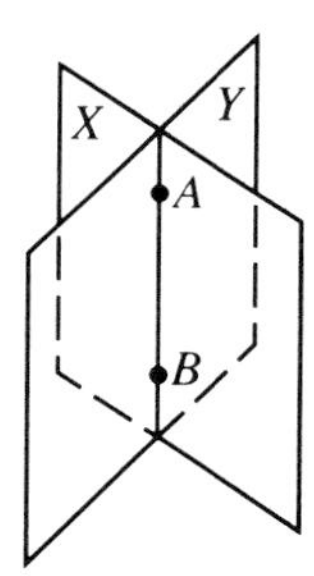

33.

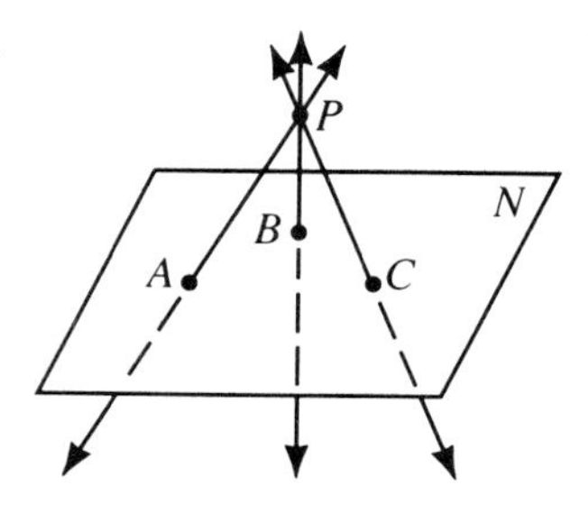

C **34.**

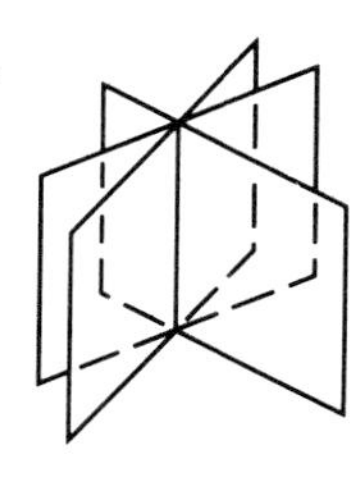

35.

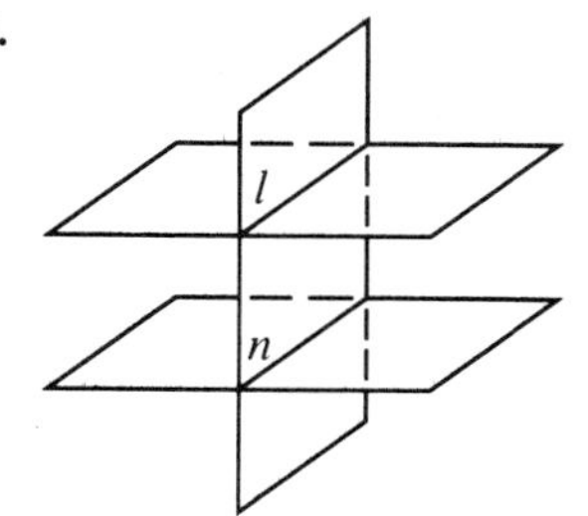

36.

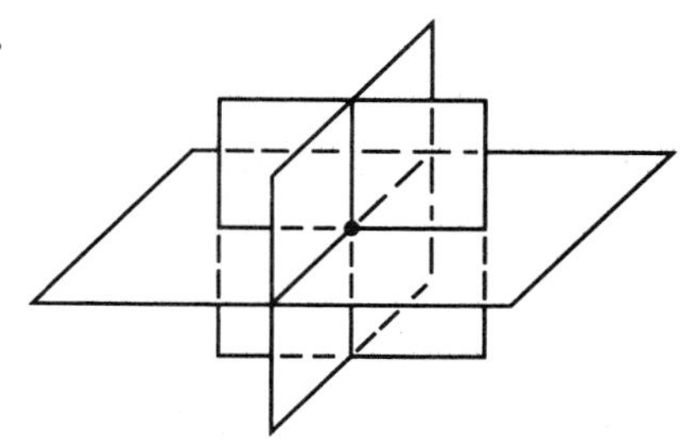

Page 9 • CHALLENGE

Since one half the area of the red square equals one fourth the area of the blue one, the area of the blue square is 2 square units.

Page 10 • SELF-TEST 1

1. T **2.** T **3.** V **4.** True **5.** True **6.** False **7.**

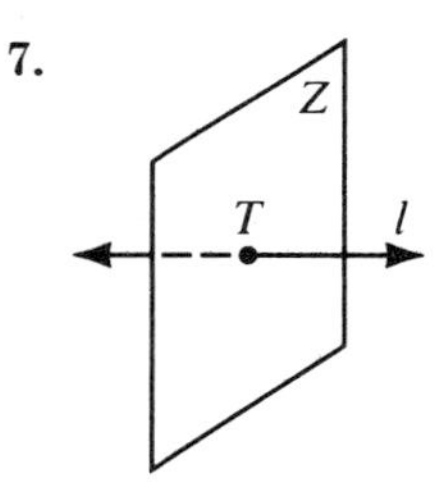

Page 10 • ALGEBRA REVIEW

1. $c + 5 = 12; c = 7$

2. $8 + c = 13; c = 5$

3. $c - 5 = 12; c = 17$

4. $7 - z = 13; z = -6$

5. $15 - z = 0; z = 15$

6. $4x = 28; x = 7$

7. $3x = 15; x = 5$

8. $7x = -35; x = -5$

9. $-5x = -5; x = 1$

10. $\frac{1}{3}a = 2; a = 6$

11. $\frac{3}{4}a = 9; a = 12$

12. $\frac{4}{5}a = -20; a = -25$

13. $-2b = 6; b = -3$

14. $-3b = -9; b = 3$

15. $-9b = 2; b = -\frac{2}{9}$

16. $42 = 6k; k = 7$

17. $5 = 10k; k = \frac{1}{2}$

18. $-16 = -4k; k = 4$

19. $12 = \frac{e}{2}; e = 24$

20. $-9 = \frac{e}{3}; e = -27$

21. $5 = -\frac{e}{3}; e = -15$

22. $2p + 5 = 13; 2p = 8; p = 4$

23. $3p - 5 = 13; 3p = 18; p = 6$
24. $4p + 2 = 22; 4p = 20; p = 5$
25. $60 = 6t + 12; 48 = 6t; t = 8$
26. $12 = 3r - 9; 21 = 3r; r = 7$
27. $55 = 7s - 8; 63 = 7s; s = 9$
28. $8x + 2x = 90; 10x = 90; x = 9$
29. $8x - 2x = 90; 6x = 90; x = 15$
30. $x + 9x = 5; 10x = 5; x = \frac{1}{2}$
31. $(2g - 15) + g = 9; 3g - 15 = 9; 3g = 24; g = 8$
32. $3u + (u - 2) = 10; 4u - 2 = 10; 4u = 12; u = 3$
33. $(w - 20) + 5w = 28; 6w - 20 = 28; 6w = 48; w = 8$
34. $3x = 2x - 17; x = -17$
35. $5y = 3y + 26; 2y = 26; y = 13$
36. $7z = 180 - 2z; 9z = 180; z = 20$
37. $12 + 3b = 2 + 5b; 10 = 2b; b = 5$
38. $4c + 23 = 9c - 7; 30 = 5c; c = 6$
39. $7h + (90 - h) = 210; 6h + 90 = 210; 6h = 120; h = 20$
40. $5x + (180 - x) = 300; 4x + 180 = 300; 4x = 120; x = 30$
41. $(4f + 5) + (5f + 40) = 180; 9f + 45 = 180; 9f = 135; f = 15$
42. $(3g - 4) + (4g + 10) = 90; 7g + 6 = 90; 7g = 84; g = 12$
43. $2(4d + 4) = d + 1; 8d + 8 = d + 1; 7d = -7; d = -1$
44. $2(d + 5) = 3(d - 2); 2d + 10 = 3d - 6; d = 16$
45. $180 - x = 3(90 - x); 180 - x = 270 - 3x; 2x = 90; x = 45$
46. $3(180 - y) = 2(90 - y); 540 - 3y = 180 - 2y; y = 360$

Page 14 • CLASSROOM EXERCISES

1. a. segment **b.** ray **c.** line **d.** length **2.** 2; 1; 0 **3.** Yes **4.** No **5.** Yes **6.** Yes **7.** $-2; 0$ **8.** T **9. a.** 1 **b.** 1 **c.** 4 **10.** $\overline{QR}, \overline{RS}, \overline{ST}$ **11.** $\overrightarrow{ST}$ **12.** R **13. a.** 1.5 **b.** 1.5 **14. a.** No **b.** Yes **c.** No **15.** Yes **16.** Yes **17.** No **18.** No **19.** Yes **20.** Yes

21. X is the midpoint of $\overline{PQ}$ because it lies on $\overline{PQ}$ and $PX = XQ$.

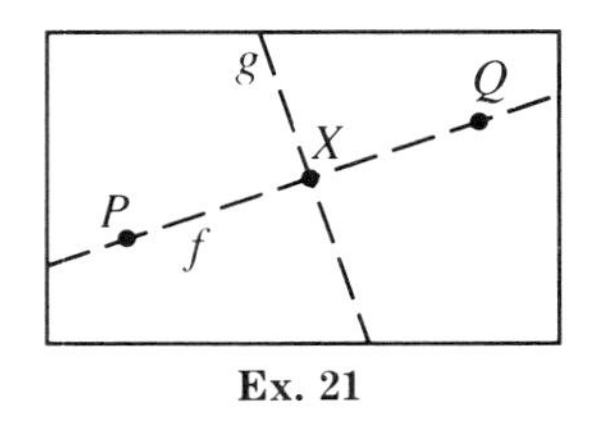

Ex. 21

22. No; B does not have to be on $\overline{AC}$.

23. 8 **24.** 4 **25.** 8 **26.** 8

Pages 15–16 • WRITTEN EXERCISES

A **1.** 15 **2.** 14 **3.** 4.5 **4.** 7.1 **5.** True **6.** False **7.** True **8.** True **9.** False **10.** True **11.** False **12.** True **13.** True **14.** False **15.** True **16.** False **17.** False **18.** True **19.** B **20.** F **21.** C, G **22.** $\overrightarrow{BA}$ **23.** D **24.** -1 **25.** -1 **26.** $\overline{BG}$

27–30. Sketches may vary.

27.

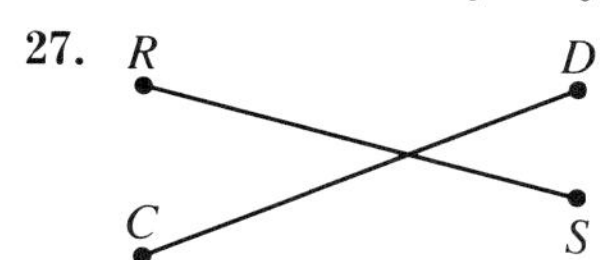

28.

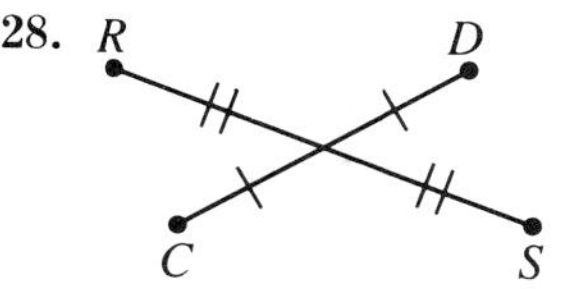

29.

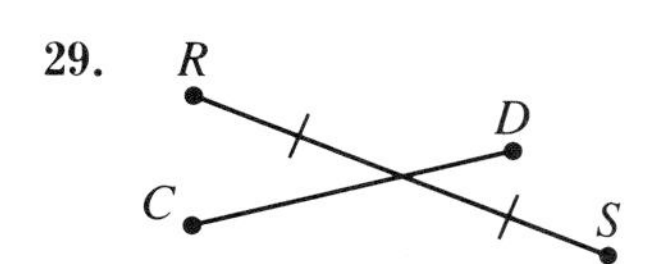

30.

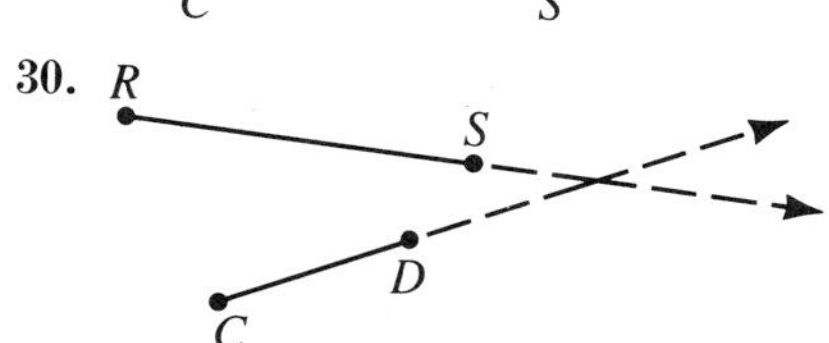

B 31. a. 5 **b.** 10 **c.** 10 **d.** 6 **32.** $-7, 5, 8$ **33.** $x = 6$ **34.** $x = 11$

35. $x = 3$ **36.** $x = 7$ **37.** $y = 6$ **38.** $y = 7$

39. $z = 8$; $GE = 10$; $EH = 10$; yes **40.** $z = 5$; $GE = 5$; $EH = 6$; no

41. $\overline{HN}$ **42.** $\overrightarrow{MH}$ or $\overrightarrow{MG}$ **43.** $\overline{GT}$

44. Answers may vary; for example, $\overleftrightarrow{GT}$ **45.** M

46. a. 1

P
A B
⊢ 3 cm ⊣

b. 2

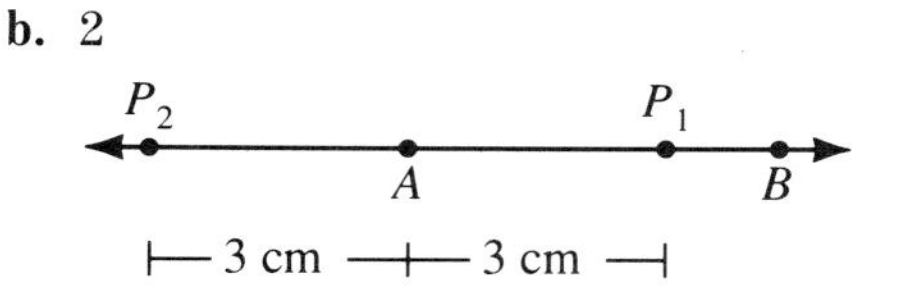

C 47. 2 if $AB \geq 3$ cm

P2 P1
A B
⊢ 3 cm ⊢ 3 cm ⊣

1 if $AB < 3$ cm

P
A B
⊢ 2 cm ⊢ 3 cm ⊣

48. a. $C = \frac{5}{9}(212 - 32) = 100$, $212°\text{F} = 100°\text{C}$; $C = \frac{5}{9}(98.6 - 32) = 37$, $98.6°\text{F} = 37°\text{C}$ **b.** $C = \frac{5}{9}(F - 32)$; $F - 32 = \frac{9}{5}C$; $F = \frac{9}{5}C + 32$

c. $F = \frac{9}{5}(-40) + 32 = -40$, $-40°\text{C} = -40°\text{F}$; $F = \frac{9}{5}(2000) + 32 = 3632$, $2000°\text{C} = 3632°\text{F}$

Page 20 • CLASSROOM EXERCISES

1. C; $\overrightarrow{CD}$, $\overrightarrow{CB}$ **2.** A; $\overrightarrow{AD}$, $\overrightarrow{AB}$ **3.** D; $\overrightarrow{DC}$, $\overrightarrow{DB}$ **4.** $\angle 7$, $\angle 5$

5. $\angle ABC$, $\angle DBA$, $\angle DBC$ **6.** 6 **7–12.** Accept reasonable estimates.

7. acute; about 35 **8.** right; 90 **9.** obtuse; about 125 **10.** acute; about 65

11. obtuse; about 120 **12.** straight; 180 **13.** EDB **14.** ADC **15.** 180

16. 5, 6 **17.** 70 **18.** 20 **19.** 40 **20.** 30 **21.** 140 **22.** 180

23. $\angle GOH$, $\angle COF$, $\angle FOB$, $\angle FOA$ **24.** $\overrightarrow{OC}$ bisects $\angle HOA$ and $\angle GOB$.

25. a. $\angle GOH$ and $\angle BOA$, or $\angle GOC$ and $\angle COB$ **b.** $\angle HOC$ and $\angle COA$

c. $\angle HOB$ and $\angle GOA$

26. 3

27. $\overrightarrow{OK}$ bisects $\angle AOB$ because $\angle AOK \cong \angle BOK$.

28. Yes **29.** No **30.** Yes **31.** No

32. No **33.** Yes **34.** Yes **35.** Yes

36. No **37.** Yes **38.** Yes

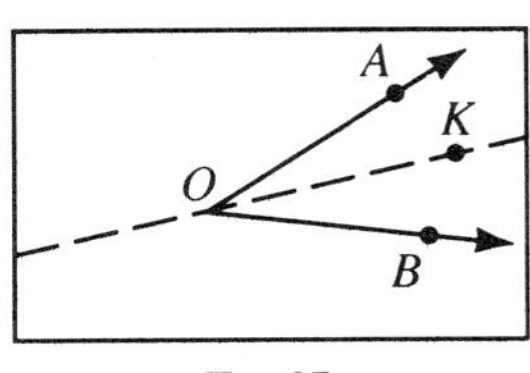

Ex. 27

Pages 21–22 • WRITTEN EXERCISES

A **1.** E; $\overrightarrow{EL}$ and $\overrightarrow{EA}$ **2.** $\angle ADL$, $\angle EDT$ **3–8.** Answers may vary. **3.** $\angle DLT$

4. $\angle LAT$ **5.** $\angle AEL$ **6.** $\angle 2$ **7.** $\angle 7$ **8.** $\angle 6$ **9.** acute **10.** acute **11.** right

12. acute **13.** straight **14.** obtuse **15.** LAS **16.** 1 **17.** $\overrightarrow{LE}$, $\angle ALS$

18. 180 **19–22.** Accept reasonable sketches.

19. **20.** **21.** **22.**

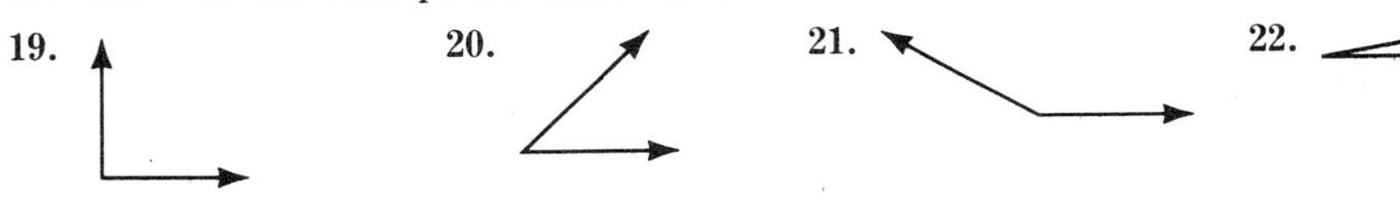

23. 2 **24.** 1

B **25.** The sum of the measures of the angles of a triangle is 180. Yes, the result is the same for triangles with different shapes.

26. a. 90, 90, 90 **b.** 87, 93, 87 **27.** $180 - t$, t, $180 - t$

28. B must be in the interior of $\angle AOC$. **29.** $x = 18$ **30.** $x = 4$ **31.** $x = 9$

32. $x = 30$ **33.** $x = 20$ **34.** $x = 22$

C **35. a.** 6; 10 **b.** 15 **c.** $\frac{n(n-1)}{2}$ **36. a.** 16 **b.** 25

Page 24 • CLASSROOM EXERCISES

1. Through any 2 points there is exactly one line.

2. Through any 2 points there is one and only one line.

3. Every segment has at least one midpoint. A segment has no more than one midpoint.

4. A line and a point not in the line determine a plane.

5. Yes **6.** Yes; at least one line **7.** No; unless the points are noncollinear.

8. C, A; through any 2 points there is exactly one line.

9. Answers may vary; any 3 of A, B, C, and D. Through any 3 noncollinear points there is exactly one plane.

10. $\overleftrightarrow{AB}$; if 2 planes intersect, then their intersection is a line.

11. Yes; if 2 points are in a plane, then the line that contains the points is in that plane.

12. Yes; a plane contains at least 3 points not all in one line.

13. The ends of three legs determine a plane (the floor); the end of a fourth leg might not be in that plane.

14. The legs may not be the same length, and their ends may not be coplanar.

15. If 2 points are in a plane, then the line that contains the points is in that plane.

16. **a.** Yes **b.** Through a line and a point not in the line, there is exactly one plane.

Pages 25–26 • WRITTEN EXERCISES

A **1.** If there is a line and a point not on the line, then one and only one plane contains them.

2. If 2 lines intersect, then at least one plane contains the lines. If 2 lines intersect, then no more than one plane contains the lines.

3. **a.** a line **b.** If 2 planes intersect, then their intersection is a line.

4. **a.** $\overleftrightarrow{AB}$ is in the plane. **b.** If 2 points are in a plane, then the line that contains the points is in that plane.

5. Through any 2 points there is exactly one line. **6.** $ABCD$ **7.** $ACGE$ **8.** $\overleftrightarrow{CD}$

9. $\overleftrightarrow{AB}$, $\overleftrightarrow{CD}$, $\overleftrightarrow{AD}$, $\overleftrightarrow{BC}$ **10.** $\overleftrightarrow{AC}$, $\overleftrightarrow{BD}$ **11.** $ABCD$, $DCGH$, $ABGH$

12.

	$\angle EFG$	$\angle AEF$	$\angle DCB$	$\angle FBC$
In the diagram	obtuse	right	acute	obtuse
In the box	right	right	right	right

B **13.** No **14.** Yes **15.** No **16.** Yes

17. **a.** Through any 3 points there is at least one plane. **b.**

c. Yes; if 2 points are in a plane, then the line that contains the points is in that plane.

d. Through any 2 points there is exactly one line.

e. If 2 points are in a plane, then the line that contains the points is in that plane.

18. **a.** Through any 3 points there is at least one plane.

b. According to the Ruler Postulate, the points on $\overleftrightarrow{AD}$ can be paired with the real numbers. Between any 2 real numbers there is another real number. So there must exist a point P between A and D. **c.** Through any 3 points there is at least one plane. **d.** There are an infinite number of points P on $\overleftrightarrow{AD}$. For each P there exists a plane BCP.

C **19.** **a.** 3 **b.** 6 **c.** 10 **d.** 15 **e.** 21 **f.** $\frac{n(n-1)}{2}$

20. **a.** A line contains at least 2 points. **b.** Through any 3 points there is at least one plane. **c.** If 2 points are in a plane, then the line that contains the points is in that plane. **d.** Through any 3 noncollinear points there is exactly one plane.

Page 28 • APPLICATION

1. **a–c.**

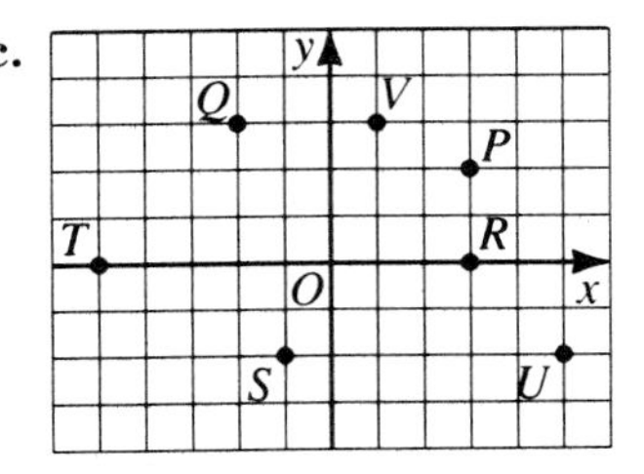

2. **a.** P **b.** I **c.** G

3. **a.** (2, 210°) or (2, −150°) **b.** (1.5, 150°) or (1.5, −210°) **c.** (1, 0°)

4. **a.** (1, 240°) **b.** (2, −60°) **c.** (2.5, 180°)

5–6. Accept reasonable answers.

5. **a.** (5, 53°) **b.** (5.4, 112°) **c.** (4, 0°) **d.** (10, −37°)

6. **a.** (1.3, 1.5) **b.** (0.5, −1.4) **c.** (0, 3) **d.** (−0.5, 0.9)

Page 29 • SELF-TEST 2

1. $\overleftrightarrow{RN}$, $\overleftrightarrow{RC}$, $\overleftrightarrow{NC}$ **2.** $\overrightarrow{NR}$ **3.** No **4.** $x = 2$ **5.** JOT **6.** $\overrightarrow{OK}$; JOT

7. 180; straight **8.** c **9.** there is exactly one line **10.** then $\overleftrightarrow{AB}$ is in Z

11. their intersection is a line

12. there is exactly one plane that contains j and P

Page 30 • CHAPTER REVIEW

1. infinitely many **2.** 2 **3.** 2 **4.** equidistant

5.

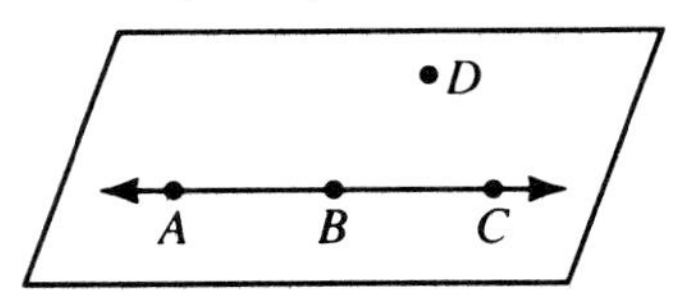

6.

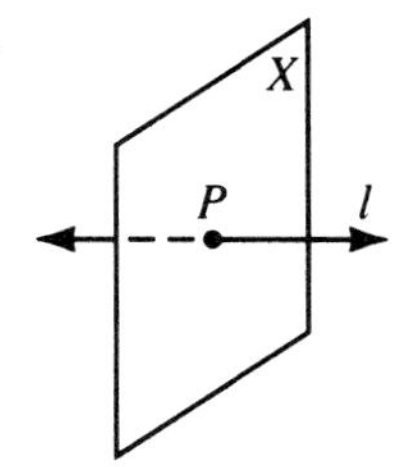

7. 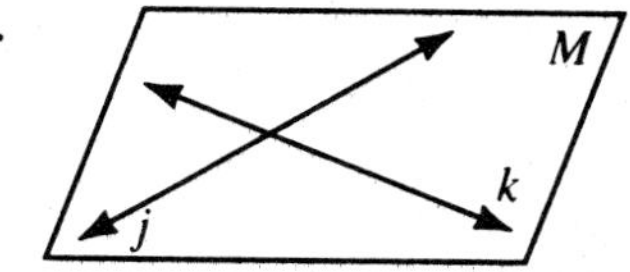

8.

9. U or V **10.** 3, 3 **11.** congruent **12.** $x = 9$
13. $\angle 1$, $\angle 2$, $\angle ADC$; $\angle 1$ and $\angle 2$ are adjacent. **14. a.** 92 **b.** $\angle$ Add. Post.
15. obtuse **16.** $x = 7$ **17.** True **18.** False **19.** False **20.** True

Page 31 • CHAPTER TEST

1. an unlimited number **2.** none **3.** Yes **4.** Yes **5.** No **6.** No **7.** No
8. A, X, and C; or D, X, and B **9.** A **10.** Seg. Add. Post. **11.** $\overline{DX}$, $\overline{XB}$
12. X; $\overrightarrow{XD}$, $\overrightarrow{XA}$ **13.** $\angle DAB$ **14.** 134; 46 **15.** 20 **16.** 3 **17.** 3.5 **18.** -1
19. 9 **20.** 1 **21.** No **22.** No **23.** a basic assumption
24. the line that contains the points is in that plane

CHAPTER 2 • Deductive Reasoning

Page 34 • CLASSROOM EXERCISES

1. H: $2x - 1 = 5$, C: $x = 3$ **2.** H: she's smart, C: I'm a genius

3. H: $8y = 40$, C: $y = 5$ **4.** H: S is the midpoint of $\overline{RT}$, C: $RS = \frac{1}{2}RT$

5. H: $m\angle 1 = m\angle 2$, C: $\angle 1 \cong \angle 2$ **6.** H: $\angle 1 \cong \angle 2$, C: $m\angle 1 = m\angle 2$

7. $\angle 1 \cong \angle 2$ if and only if $m\angle 1 = m\angle 2$, or $m\angle 1 = m\angle 2$ if and only if $\angle 1 \cong \angle 2$.

8–11. Answers will vary. Examples are given.

8.

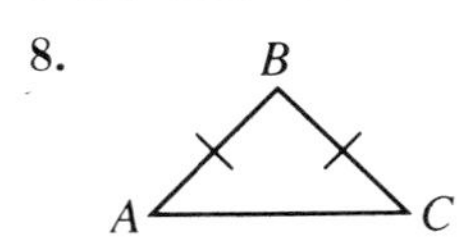

9.

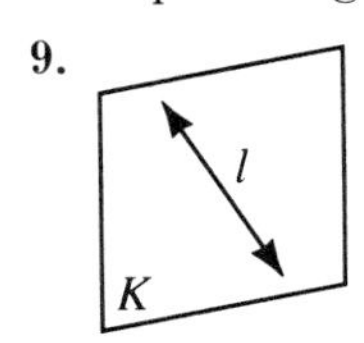

10. $n = 16$ **11.** $x = -7$

12. If tomorrow is Saturday, then today is Friday; true.

13. If $x^2 > 0$, then $x > 0$; false.

14. If a number is divisible by 3, then it is divisible by 6; false.

15. If $x = 3$, then $6x = 18$; true.

16. Answers will vary; for example, if $\overline{AB} \cong \overline{BC}$, then B is the midpoint of $\overline{AC}$.

Page 35 • WRITTEN EXERCISES

A **1.** H: $3x - 7 = 32$, C: $x = 13$ **2.** H: I'm not tired, C: I can't sleep

3. H: you will, C: I'll try **4.** H: $m\angle 1 = 90$, C: $\angle 1$ is a right angle

5. H: $a + b = a$, C: $b = 0$ **6.** H: $x = -5$, C: $x^2 = 25$

7. B is between A and C if and only if $AB + BC = AC$, or $AB + BC = AC$ if and only if B is between A and C.

8. $m\angle AOC = 180$ if and only if $\angle AOC$ is a straight angle, or $\angle AOC$ is a straight angle if and only if $m\angle AOC = 180$.

9. If pts. are collinear, then they all lie in one line. If pts. all lie in one line, then they are collinear.

10. If pts. lie in one plane, then they are coplanar. If pts. are coplanar, then they lie in one plane.

11–16. Answers will vary. Examples are given.

11. $a = 1$, $b = -1$ **12.** $n = 0$ **13.**

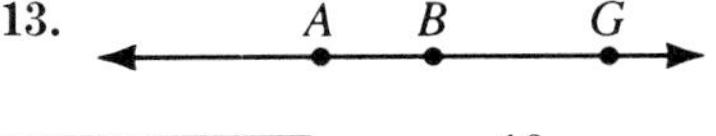

14. $x = -1$, $y = -1$ **15.**

16.

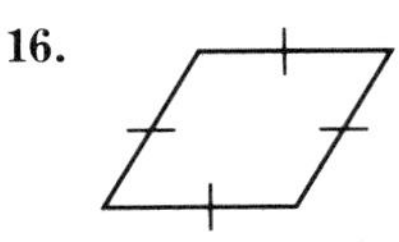

17. True. If $|x| = 6$, then $x = -6$; false. **18.** False. If $x = -2$, then $x^2 = 4$; true.

19. True. If $5b > 20$, then $b > 4$; true.

20. True. If $\angle T$ is not obtuse, then $m\angle T = 40$; false.

21. True. If Pam lives in Illinois, then she lives in Chicago; false.

22. True. If $m\angle A = m\angle B$, then $\angle A \cong \angle B$; true.

B 23. True. If $a^2 > 9$, then $a > 3$; false. **24.** True. If $x^2 = x$, then $x = 1$; false.

25. False. If $n > 7$, then $n > 5$; true.

26. True. If $a = 0$ or $b = 0$, then $ab = 0$; true.

27. False. If $DE + EF = DF$, then points D, E, and F are collinear; true.

28. True. If $GH = 2PG$, then P is the midpoint of $\overline{GH}$; false.

29. Two $\angle$s are $\cong$ if and only if their measures are equal.

30. An $\angle$ is a rt. $\angle$ if and only if its measure is 90.

C 31. Possible conclusions are: q, not r, and s.

Page 37 • MIXED REVIEW EXERCISES

1. $\overline{AM}$; $\overline{MB}$ **2.** $\angle ABX$; $\angle XBC$ **3.** AOB; BOC; AOC **4.** POR; ROQ; 180

Page 40 • CLASSROOM EXERCISES

1. Refl. Prop. **2.** Trans. Prop. **3.** Sym. Prop. **4.** Subtr. Prop. of =

5. Div. Prop. of = **6.** Mult. Prop. of = **7.** Distributive Prop.

8. Add. Prop. of = **9.** Substitution Prop. **10.** Trans. Prop.

11. 1. Given 2. Add. Prop. of = 3. $\angle$ Add. Post. 4. Substitution Prop.

12. 1. Given 2. ST, RN; Seg. Add. Post. 3. Substitution Prop. 4. Given
5. $SI = UN$; Subtr. Prop. of =

Pages 41–43 • WRITTEN EXERCISES

A 1. Given; Add. Prop. of =; Div. Prop. of =

2. Given; Mult. Prop. of =; Div. Prop. of =

3. Given; Mult. Prop. of =; Subtr. Prop. of =

4. Given; Add. Prop. of =; Subtr. Prop. of =; Div. Prop. of =

5. Given; Mult. Prop. of =; Dist. Prop.; Add. Prop. of =; Div. Prop. of =

6. Given; Mult. Prop. of =; Dist. Prop.; Subtr. Prop. of =; Add. Prop. of =;
Div. Prop. of =

7. 1. $\angle$ Add. Post. 2. $\angle$ Add. Post. 3. $m\angle AOD = m\angle 1 + m\angle 2 + m\angle 3$;
Substitution Prop.

8. 1. $FL = AT$ 2. Refl. Prop. 3. Add. Prop. of = 4. Seg. Add. Post.
5. $FA = LT$

9. 1. Given 2. OW, WN; Seg. Add. Post. 3. $DO + OW = OW + WN$
4. Refl. Prop. 5. $DO = WN$; Subtr. Prop. of =

10. 1. Given 2. ∠ Add. Post. 3. Substitution Prop. 4. Refl. Prop.
5. $m\angle 5 = m\angle 6$; Subtr. Prop. of =

B 11.

Statements	Reasons
1. $m\angle 1 = m\angle 2$; $m\angle 3 = m\angle 4$	1. Given
2. $m\angle 1 + m\angle 3 = m\angle 2 + m\angle 4$	2. Add. Prop. of =
3. $m\angle 1 + m\angle 3 = m\angle SRT$; $m\angle 2 + m\angle 4 = m\angle STR$	3. ∠ Add. Post.
4. $m\angle SRT = m\angle STR$	4. Substitution Prop.

12.

Statements	Reasons
1. $RP = TQ$; $PS = QS$	1. Given
2. $RP + PS = TQ + QS$	2. Add. Prop. of =
3. $RP + PS = RS$; $TQ + QS = TS$	3. Seg. Add. Post.
4. $RS = TS$	4. Substitution Prop.

13.

Statements	Reasons
1. $RQ = TP$	1. Given
2. $RZ + ZQ = RQ$; $TZ + ZP = TP$	2. Seg. Add. Post.
3. $RZ + ZQ = TZ + ZP$	3. Substitution Prop.
4. $ZQ = ZP$	4. Given
5. $RZ = TZ$	5. Subtr. Prop. of =

14.

Statements	Reasons
1. $m\angle SRT = m\angle STR$	1. Given
2. $m\angle 1 + m\angle 3 = m\angle SRT$; $m\angle 2 + m\angle 4 = m\angle STR$	2. ∠ Add. Post.
3. $m\angle 1 + m\angle 3 = m\angle 2 + m\angle 4$	3. Substitution Prop.
4. $m\angle 3 = m\angle 4$	4. Given
5. $m\angle 1 = m\angle 2$	5. Subtr. Prop. of =

C 15. b

Page 45 • CLASSROOM EXERCISES

1. ∠ Add. Post. **2.** Seg. Add. Post. **3.** ∠ Add. Post. **4.** Def. of midpt.
5. Midpt. Thm. **6.** Def. of seg. bis. **7.** Def. of seg. bis. **8.** ∠ Bis. Thm.
9. Def. of ∠ bis.

10. 1. Given 2. $\angle XBC$, $m\angle XBC$; Def. of ∠ bis. 3. ∠ Add. Post.
4. Substitution Prop. 5. Mult. Prop. of = 6. 2, 5

Pages 46–47 • WRITTEN EXERCISES

A **1.** Def. of midpt. **2.** Def. of ∠ bis. **3.** Def. of ∠ bis. **4.** ∠ Add. Post.
5. Def. of midpt. **6.** Midpt. Thm. **7.** ∠ Add. Post. **8.** Seg. Add. Post.
9. 60 **10.** 75 **11.** 70

12. a, b.

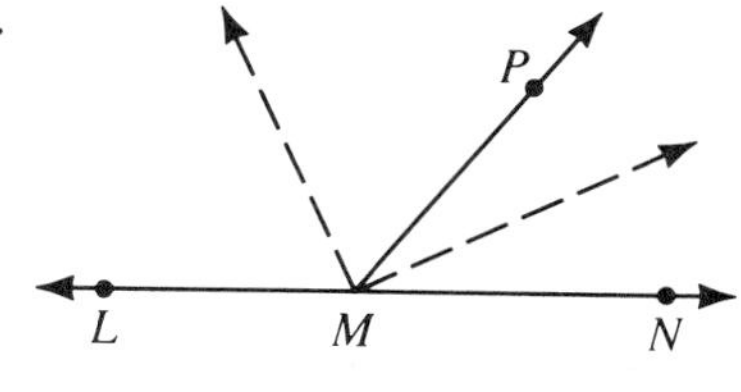

c. 90
d. Since $m\angle LMP + m\angle PMN = 180$, then $\frac{1}{2} m\angle LMP + \frac{1}{2} m\angle PMN = \frac{1}{2}(m\angle LMP + m\angle PMN) = \frac{1}{2}(180) = 90$

B **13.** L Y N X (16, 40) **a.** 12 **b.** 28 **c.** 6 **d.** 22

14. $m\angle RSZ = 18$, $m\angle NSZ = 54$.

15. a. $\overline{LM} \cong \overline{MK}$, $\overline{GN} \cong \overline{NH}$ **b.** Answers may vary; for example, $\overline{LK} \cong \overline{GH}$.
16. a. $\angle RSV \cong \angle VST$, $\angle SRU \cong \angle URT$ **b.** Answers may vary; for example, $\angle RST \cong \angle SRT$.
17. $AC = BD$ **18.** $AC = BD$ and $AE = DE = CE = BE$
19. 1. Given 2. Ruler Post. 3. Given 4. Def. of midpt. 5. Substitution Prop. 6. $a + b$; Add. Prop. of = 7. Div. Prop. of =

C **20.** $m\angle 1 + m\angle 2 + m\angle 3 + m\angle 4 = 180$; $m\angle 1 = m\angle 2$; $m\angle 3 = m\angle 4$; $2m\angle 2 + 2m\angle 3 = 180$; $m\angle 2 + m\angle 3 = 90$

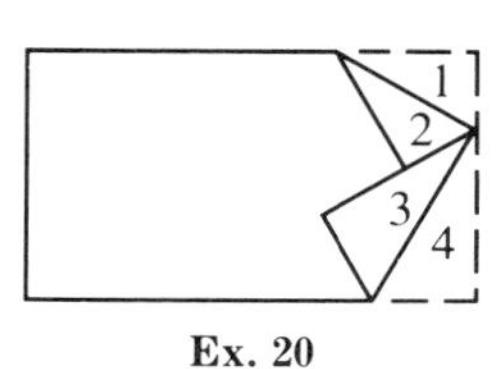

Ex. 20

21. Q: $\frac{3a + b}{4}$; T: $\frac{5a + 3b}{8}$
22. If $WT = y$, $ZS = x + y$ and $TS = 2x + y$. $RT = 2y$, so $2y = 2x + y$ or $y = 2x$.
a. $2x$ **b.** $3x$ **c.** $8x$ **d.** $3x$

Page 48 • COMPUTER KEY-IN

1. 40 FOR N = 2 to 40; $\frac{2}{3}$

3. $1 - \frac{1}{3} + \frac{1}{9} - \frac{1}{27} + \cdots + \left(-\frac{1}{3}\right)^{n-1}$; P(N) = P(N − 1) + $\left(-\frac{1}{3}\right)$ ↑ (N − 1); $\frac{3}{4}$

Page 49 • SELF-TEST 1

1. H: $\overline{AB}$ and $\overline{CD}$ intersect; C: $\overrightarrow{AB}$ and $\overrightarrow{CD}$ intersect
2. If $\overrightarrow{AB}$ and $\overrightarrow{CD}$ intersect, then $\overline{AB}$ and $\overline{CD}$ intersect; false.
3. $\overline{AB} \cong \overline{CD}$ if and only if $AB = CD$.
4. Answers may vary; for example, $m\angle A = 95$.
5. Substitution Prop. **6.** $x = 3$ **7.** 81 **8.** definitions, postulates

Pages 51–52 • CLASSROOM EXERCISES

1. 80, 170 **2.** 15, 105 **3.** 1, 91 **4.** $90 - y$, $180 - y$ **5.** $\angle M$; $\angle QPM$
6. $\angle LPQ$; $\angle LPM$ **7.** $\angle LPM$ and $\angle MLP$
8. a. $\angle LQP$ **b.** $\angle M$ and $\angle MPQ$
9. a. $\angle AXB$ and $\angle BXD$ **b.** $\angle AXC$ and $\angle CXD$ **c.** $\angle BXC$ and $\angle CXD$
d. $\angle AXD$
10. $\angle EOD$ **11.** $\angle BOD$ **12.** $\angle EOC$ **13.** $\angle FOD$ **14.** 60 **15.** 40
16. 100 **17.** 80 **18.** 120 **19.** 120 **20. a.** 90 **b.** 60 **c.** 120 **d.** 90
21. a. $\angle 1 \cong \angle 4$ **b.** $\angle 1 \cong \angle 2$ and $\angle 3 \cong \angle 4$ because vertical angles are $\cong$. $\angle 2 \cong \angle 3$ is given. So $\angle 1 \cong \angle 4$ by the Trans. Prop.

Pages 52–54 • WRITTEN EXERCISES

A **1.** 70, 160 **2.** $17\frac{1}{2}$, $107\frac{1}{2}$ **3.** $90 - x$, $180 - x$ **4.** $90 - 2y$, $180 - 2y$
5. $x = 90 - x$; $2x = 90$; $x = 45$; $90 - x = 45$
6. $x = 180 - x$; $2x = 180$; $x = 90$; $180 - x = 90$ **7.** $\angle AFD$
8. $\angle AFE$ and $\angle EFD$, or $\angle AFE$ and $\angle BFC$ **9.** $\angle AFD$ and $\angle AFB$
10. $\angle BFE$ and $\angle EFD$, $\angle CFA$ and $\angle AFE$, $\angle DFC$ and $\angle CFB$, $\angle CFB$ and $\angle BFE$, or $\angle CFD$ and $\angle DFE$
11. $\angle BFC$ and $\angle EFD$ **12.** $\angle BFE$ and $\angle CFD$
13. 35 **14.** 155 **15.** 25 **16.** 120 **17.** 60 **18.** 85
19. $3x - 5 = 70$; $3x = 75$; $x = 25$ **20.** $3x + 8 = 6x - 22$; $3x = 30$; $x = 10$
21. $4x = 100$; $x = 25$
22. a. $m\angle 1 + m\angle 2 = 180$; $m\angle 2 + 27 = 180$; $m\angle 2 = 153$. Similarly, $m\angle 4 = 153$.
b. $m\angle 1 + m\angle 2 = 180$; $m\angle 2 + x = 180$; $m\angle 2 = 180 - x$. Similarly, $m\angle 4 = 180 - x$. **c.** Yes
23. 1. Vert. $\angle$s are $\cong$. 2. Given 3. Vert. $\angle$s are $\cong$. 4. $\angle 1 \cong \angle 4$
B **24.** $2x + (x - 15) = 180$, $3x = 195$, $x = 65$; $m\angle A = 2x = 130$; $m\angle B = x - 15 = 50$
25. $(x + 16) + (2x - 16) = 180$, $3x = 180$, $x = 60$; $m\angle A = x + 16 = 76$; $m\angle B = 2x - 16 = 104$
26. $(3y + 5) + 2y = 90$, $5y = 85$, $y = 17$; $m\angle C = 3y + 5 = 56$; $m\angle D = 2y = 34$

27. $(y - 8) + (3y + 2) = 90$, $4y = 96$, $y = 24$; $m\angle C = y - 8 = 16$; $m\angle D = 3y + 2 = 74$

28. $x = 2(180 - x)$; $3x = 360$; $x = 120$

29. $x = \frac{1}{2}(90 - x)$; $2x = 90 - x$; $3x = 90$; $x = 30$

30. $180 - x = 2x + 12$, $3x = 168$, $x = 56$; $180 - x = 124$

31. $180 - x = 6(90 - x)$, $5x = 360$, $x = 72$; $180 - x = 108$; $90 - x = 18$

32. $x + (3x - 8) = 180$, $4x = 188$, $x = 47$; $2y - 17 = 3x - 8$, $2y - 17 = 3(47) - 8$, $2y = 150$, $y = 75$

33. $2x - 16 = 50$, $2x = 66$, $x = 33$; $x = 3x - y$, $33 = 3(33) - y$, $y = 66$

C **34.** No. If $90 - x = \frac{1}{2}(180 - x)$, then $\frac{1}{2}x = 0$ and $x = 0$.

35. No conclusion is possible. For any acute angle with measure x, it is true that $x = (180 - x) - 2(90 - x)$, or $x = x$. The angle could be any acute angle.

Page 55 • APPLICATION

1. north **2.** right **3.** Ray **4.** It gives them the greatest margin of safety.

Page 57 • CLASSROOM EXERCISES

1. Given; ⊥ lines; ≅ ∠s

2. $\angle AOD$, $\angle DOB$, $\angle COB$, $\angle COA$; $\angle EOH$, $\angle HOF$, $\angle GOF$, $\angle GOE$

3. $\angle POZ$, $\angle ZOQ$, $\angle ZOX$, $\angle ZOY$; $\angle POY$, $\angle YOQ$, $\angle POX$, $\angle XOQ$ **4.** 50; 40; 50; 130

5. $90 - x$; x; $90 - x$; $90 + x$ **6.** Def. of ⊥ lines **7.** Def. of ⊥ lines

8. If 2 lines are ⊥, then they form ≅ adj. ∠s. **9.** Def. of ⊥ lines

10. If 2 lines form ≅ adj. ∠s, then the lines are ⊥. **11.** Def. of ⊥ lines

Pages 58–60 • WRITTEN EXERCISES

A **1.** **a.** $90 - x$ **b.** $180 - x$

2. 1. Given 2. ∠ Add. Post. 3. Substitution Prop. 4. Div. Prop. of = 5. $l \perp n$

3. Def. of ⊥ lines **4.** Def. of ⊥ lines

5. If the ext. sides of 2 adj. ∠s are ⊥, then the ∠s are comp. **6.** Def. of comp. ∠s

7. Def. of ⊥ lines

8. If 2 lines form ≅ adj. ∠s, then the lines are ⊥.

9. $(2x - 15) + x = 90$; $3x = 105$; $x = 35$

10. $3x + (4x - 1) = 90$; $7x = 91$; $x = 13$

11. $(3x - 12) + (2x + 2) = 90$, $5x = 100$, $x = 20$; or $(2x + 2) + (2x + 8) = 90$, $4x = 80$, $x = 20$

12. Methods of solution may vary; for example, $6x + (3x + 9) = 90$, $9x = 81$, $x = 9$.

13. 1. Given 3. $\angle$ Add. Post. 4. $m\angle AOB + m\angle BOC = 90$ 5. $\angle AOB$ and $\angle BOC$ are comp. $\angle$s.

B **14.** $x + 90$ **15.** $180 - y$ **16.** $x + y$ **17.** $90 - (x + y)$ **18.** Yes **19.** No **20.** No **21.** No **22.** No **23.** Yes **24.** Yes **25.** No

26, 27. Answers may vary. Examples are given.

26. $m\angle 1 = 45$; $m\angle 4 = 45$; $\angle DAC$ is a rt. $\angle$; $\angle ECA$ is a rt. $\angle$; $\overrightarrow{AD} \perp \overleftrightarrow{AC}$; $\overrightarrow{CE} \perp \overleftrightarrow{AC}$

27. $m\angle 2 = m\angle 3$; $m\angle DAC = 90$; $m\angle ECA = 90$

28.

Statements	Reasons
1. $\overleftrightarrow{AO} \perp \overleftrightarrow{CO}$	1. Given
2. $\angle 1$ and $\angle 2$ are comp. $\angle$s.	2. If the ext. sides of 2 adj. acute $\angle$s are $\perp$, then the $\angle$s are comp.
3. $m\angle 1 + m\angle 2 = 90$	3. Def. of comp. $\angle$s
4. $m\angle 2 = m\angle 3$	4. Vert. $\angle$s are $\cong$.
5. $m\angle 1 + m\angle 3 = 90$	5. Substitution Prop.
6. $\angle 1$ and $\angle 3$ are comp. $\angle$s.	6. Def. of comp. $\angle$s

C **29.** Prove: $\overleftrightarrow{XD} \perp \overleftrightarrow{XF}$

Statements	Reasons
1. $\overleftrightarrow{YD} \perp \overleftrightarrow{YF}$	1. Given
2. $\angle 5$ and $\angle 6$ are comp. $\angle$s.	2. If the ext. sides of 2 adj. acute $\angle$s are $\perp$, then the $\angle$s are comp.
3. $m\angle 5 + m\angle 6 = 90$	3. Def. of comp. $\angle$s
4. $m\angle 7 = m\angle 5$; $m\angle 8 = m\angle 6$	4. Given
5. $m\angle 7 + m\angle 8 = 90$	5. Substitution Prop. (Steps 3 and 4)
6. $m\angle 7 + m\angle 8 = m\angle FXD$	6. $\angle$ Add. Post.
7. $m\angle FXD = 90$	7. Substitution Prop.
8. $\angle FXD$ is a rt. $\angle$.	8. Def. of rt. $\angle$
9. $\overleftrightarrow{XD} \perp \overleftrightarrow{XF}$	9. Def. of $\perp$ lines

Page 60 • MIXED REVIEW EXERCISES

1–7. Answers may vary. Examples are given. **1.** $\angle CBF \cong \angle BCG$ **2.** $AC = BD$ **3.** $\angle 3 \cong \angle 4$ **4.** $m\angle 1 = m\angle 2 = 45$ **5.** $CE = BE$ **6.** $\overleftrightarrow{AB} \perp \overline{BF}$ **7.** $m\angle 5 = 90$

Page 62 • CLASSROOM EXERCISES

1. a. $\angle 6 \cong \angle 7$ **b.** If 2 $\angle$s are comps. of the same $\angle$, then the 2 $\angle$s are $\cong$.

2. **a.** $\angle 5 \cong \angle 7$ **b.** Def. of $\cong$ ∠s
3. **a.** $\angle 8 \cong \angle 9$ **b.** If 2 lines are $\perp$, then they form $\cong$ adj. ∠s.
4. **a.** $\angle 6 \cong \angle 7$ **b.** Def. of $\angle$ bis.
5. **a.** $\angle 2 \cong \angle 4$ **b.** If 2 ∠s are supps. of $\cong$ ∠s, then the 2 ∠s are $\cong$.
6. **a.** $\angle 1 \cong \angle 3$ **b.** Vert. ∠s are $\cong$.

7–10. Answers may vary.

7. Show that $\angle 2 \cong \angle 1$ and $\angle 4 \cong \angle 3$. Then $\angle 1 \cong \angle 4$ by Substitution.
8. Show that $\angle 3$ and $\angle 4$ are supps. of $\cong$ ∠s.
9. Show that $\angle 3$ and $\angle 2$ are comps. of the same angle, $\angle 1$.
10. Show that $\angle 2$ and $\angle 3$ are supps. of $\cong$ ∠s.

Pages 63–65 • WRITTEN EXERCISES

A **1.** Seg. Add. Post. **2.** $\angle$ Add. Post. **3.** Vert. ∠s are $\cong$. **4.** Midpt. Thm. **5.** Def. of $\angle$ bis. **6.** $\angle$ Add. Post. **7.** Def. of $\perp$ lines **8.** Def. of $\angle$ bis. **9.** Def. of comp. ∠s **10.** Def. of supp. ∠s

11. If 2 lines are $\perp$, then they form $\cong$ adj. ∠s.

12. If 2 ∠s are comps. of the same $\angle$, then the 2 ∠s are $\cong$. **13.** Def. of $\perp$ lines

14. If 2 lines form $\cong$ adj. ∠s, then the lines are $\perp$.

15. 1. $\angle 5$ are supplementary; Given 2. Def. of supp. ∠s 3. Substitution Prop. 5. Subtr. Prop. of $=$

16. **a.** $\angle 2$; Vert. ∠s are $\cong$. **b.** $\angle 2$, $\angle 6$, $\angle 8$

17. **a.** 1. Given 2. If the ext. sides of 2 adj. acute ∠s are $\perp$, then the ∠s are comp. 3. Given 4. If 2 ∠s are comps. of $\cong$ ∠s, then the 2 ∠s are $\cong$. **b.** Show that $\angle 3$ and $\angle 6$ are supps. of $\cong$ ∠s.

B **18.** Given: $\angle 1$ and $\angle 2$ are comp.;
$\angle 3$ and $\angle 4$ are comp.;
$\angle 2 \cong \angle 4$ (or $m\angle 2 = m\angle 4$)
Prove: $\angle 1 \cong \angle 3$ (or $m\angle 1 = m\angle 3$)

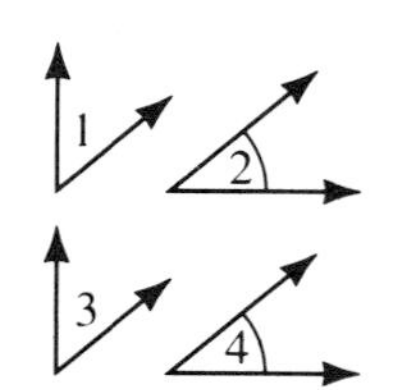

Statements	Reasons
1. $\angle 1$ and $\angle 2$ are comp.; $\angle 3$ and $\angle 4$ are comp.	1. Given
2. $m\angle 1 + m\angle 2 = 90$; $m\angle 3 + m\angle 4 = 90$	2. Def. of comp. ∠s
3. $m\angle 1 + m\angle 2 = m\angle 3 + m\angle 4$	3. Substitution Prop.
4. $m\angle 2 = m\angle 4$	4. Given
5. $m\angle 1 = m\angle 3$, or $\angle 1 \cong \angle 3$	5. Subtr. Prop. of $=$

19.

Statements	Reasons
1. $\angle 2 \cong \angle 3$	1. Given
2. $\angle 1 \cong \angle 2$	2. Vert. $\angle$s are $\cong$.
3. $\angle 1 \cong \angle 3$	3. Substitution Prop.
4. $\angle 3 \cong \angle 4$	4. Vert. $\angle$s are $\cong$.
5. $\angle 1 \cong \angle 4$	5. Trans. Prop.

20.

Statements	Reasons
1. $\angle 3$ is supp. to $\angle 1$; $\angle 4$ is supp. to $\angle 2$.	1. Given
2. $\angle 1 \cong \angle 2$	2. Vert. $\angle$s are $\cong$.
3. $\angle 3 \cong \angle 4$	3. If 2 $\angle$s are supps. of $\cong$ $\angle$s, then the 2 $\angle$s are $\cong$.

21.

Statements	Reasons
1. $\overline{AC} \perp \overline{BC}$	1. Given
2. $\angle 2$ is comp. to $\angle 1$.	2. If the ext. sides of 2 adj. acute $\angle$s are $\perp$, then the 2 $\angle$s are comp.
3. $\angle 3$ is comp. to $\angle 1$.	3. Given
4. $\angle 3 \cong \angle 2$	4. If 2 $\angle$s are comps. of the same $\angle$, then the 2 $\angle$s are $\cong$.

22.

Statements	Reasons
1. $m\angle 1 = m\angle 2$; $m\angle 3 = m\angle 4$	1. Given
2. $m\angle 1 + m\angle 3 = m\angle 2 + m\angle 4$	2. Add. Prop. of =
3. $m\angle 1 + m\angle 3 = m\angle XYS$; $m\angle 2 + m\angle 4 = m\angle SYZ$	3. $\angle$ Add. Post.
4. $m\angle XYS = m\angle SYZ$, or $\angle XYS \cong \angle SYZ$	4. Substitution Prop.
5. $\overrightarrow{YS} \perp \overleftrightarrow{XZ}$	5. If 2 lines form $\cong$ adj. $\angle$s, then the lines are $\perp$.

23. $\overrightarrow{OF}$ bisects $\angle COD$.

Statements	Reasons
1. $\overrightarrow{OE}$ bisects $\angle AOB$.	1. Given
2. $\angle 1 \cong \angle 2$	2. Def. of $\angle$ bis.
3. $\angle 2 \cong \angle 3$	3. Vert. $\angle$s are $\cong$.
4. $\angle 1 \cong \angle 3$	4. Trans. Prop.
5. $\angle 1 \cong \angle 4$	5. Vert. $\angle$s are $\cong$.
6. $\angle 3 \cong \angle 4$	6. Substitution Prop.
7. $\overrightarrow{OF}$ bisects $\angle COD$.	7. Def. of $\angle$ bis.

C 24.

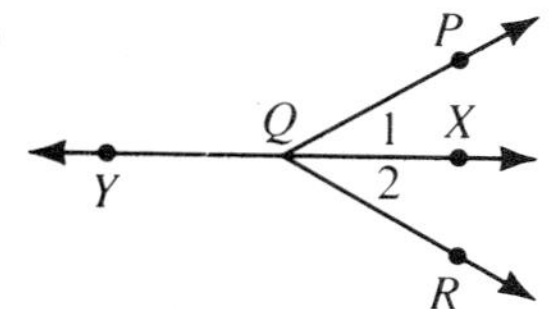

Statements	Reasons
1. $\overrightarrow{QX}$ bisects $\angle PQR$.	1. Given
2. $\angle 1 \cong \angle 2$	2. Def. of $\angle$ bis.
3. $m\angle 1 + m\angle PQY = 180$; $m\angle 2 + m\angle RQY = 180$	3. $\angle$ Add. Post.
4. $\angle PQY$ is supp. to $\angle 1$; $\angle RQY$ is supp. to $\angle 2$.	4. Def. of supp. $\angle$s
5. $\angle PQY \cong \angle RQY$	5. If 2 $\angle$s are supps. of $\cong$ $\angle$s, then the 2 $\angle$s are $\cong$.

25.

Statements	Reasons
1. $m\angle DBA = 45$; $m\angle DEB = 45$	1. Given
2. $\angle DBA \cong \angle DEB$	2. Def. of $\cong$ $\angle$s
3. $m\angle DBA + m\angle DBC = 180$; $m\angle DEB + m\angle FEB = 180$	3. $\angle$ Add. Post.
4. $\angle DBC$ is supp. to $\angle DBA$; $\angle FEB$ is supp. to $\angle DEB$.	4. Def. of supp. $\angle$s
5. $\angle DBC \cong \angle FEB$	5. If 2 $\angle$s are supps. of $\cong$ $\angle$s, then the 2 $\angle$s are $\cong$.

Page 65 • SELF-TEST 2

1. Answers may vary; $90 \leq m\angle HOK < 180$

2. a. $3x - 5 = x + 25$; $2x = 30$; $x = 15$ **b.** $m\angle 1 = 3(15) - 5 = 40$

3. $m\angle 4 = 90 - 37 = 53$; $m\angle 5 = 90 - 53 = 37$; $m\angle 6 = 90 - 37 = 53$

4. $m\angle 4 = 90 - t$; $m\angle 5 = 90 - (90 - t) = t$; $m\angle 6 = 90 - t$ **5.** Def. of $\perp$ lines

6. Vert. $\angle$s are $\cong$. **7.** If the ext. sides of 2 adj. $\angle$s are $\perp$, then the $\angle$s are comp.

8. Show that $\angle 1$ and $\angle 2$, which are adjacent angles, are congruent angles.

9.

Statements	Reasons
1. $\angle 1$ is supp. to $\angle 3$; $\angle 2$ is supp. to $\angle 3$.	1. Given
2. $\angle 1 \cong \angle 2$	2. If 2 $\angle$s are supps. of the same $\angle$, then the 2 $\angle$s are $\cong$.
3. $j \perp k$	3. If 2 lines form $\cong$ adj. $\angle$s, then the lines are $\perp$.

Page 66 • EXTRA

1. None of it; one
2. The result is a non-Möbius (2-sided) band that is twice as long. The band separates into 2 non-Möbius bands that are linked together.
3. 2; The result is 2 non-Möbius (2-sided) bands linked together.
4. The result is a short 1 cm wide Möbius band linked with a longer 1 cm wide non-Möbius band.
5. The result is a rectangular frame.

Pages 67–68 • CHAPTER REVIEW

1. H: $m\angle 1 = 120$, C: $\angle 1$ is obtuse **2.** If $\angle 1$ is obtuse, then $m\angle 1 = 120$.

3. Answers may vary; for example, $m\angle 1 = 100$

4. An angle is a straight angle if and only if the measure of the angle is 180.

5. Substitution Prop. **6.** Subtr. Prop. of = **7.** Div. Prop. of = **8.** Trans. Prop.

9. Def. of midpt. **10.** Def. of $\angle$ bisector **11.** $\angle$ Bis. Thm. **12.** 32; 32; 122

13. $\angle BOA$ or $\angle DOE$

14. $180 - x = 4(90 - x)$; $180 - x = 360 - 4x$; $3x = 180$; $x = 60$

15. Def. of $\perp$ lines **16.** Def. of $\perp$ lines

17. If 2 lines are $\perp$, then they form $\cong$ adj. $\angle$s.

18. If the ext. sides of 2 adj. acute $\angle$s are $\perp$, then the $\angle$s are comp.

19. Show that $\angle 3$ and $\angle 4$ are supps. of $\cong$ $\angle$s.

20.

Statements	Reasons
1. $\angle 3$ is a supplement of $\angle 1$; $\angle 4$ is a supplement of $\angle 2$.	1. Given
2. $\angle 1 \cong \angle 2$	2. Vert. $\angle$s are $\cong$.
3. $\angle 3 \cong \angle 4$	3. If 2 $\angle$s are supps. of $\cong$ $\angle$s, then the 2 $\angle$s are $\cong$.

Pages 68–69 • CHAPTER TEST

1. a. they are vert. $\angle$s **b.** If 2 $\angle$s are $\cong$, then they are vert. $\angle$s.

2. Answers may vary; for example, $x = -3$

3. If $\angle$s are $\cong$, then their measures are =. If $\angle$s have = measures, then they are $\cong$.

4. 3. Substitution Prop. 4. Subtr. Prop. of = 5. Div. Prop. of =

5. 30 **6.** 16; 8; 24 **7. a.** $\angle 3$ and $\angle 4$ **b.** $\angle 1$ and $\angle 2$

8. a. They are $\cong$. **b.** If 2 $\angle$s are supps. of the same $\angle$, then the 2 $\angle$s are $\cong$.

9. $(3x + 5) + (6x + 13) = 180$; $9x + 18 = 180$; $9x = 162$; $x = 18$

10. Vert. $\angle$s are $\cong$. **11.** Trans. Prop.

12.

Statements	Reasons
1. $\overrightarrow{DC} \perp \overleftrightarrow{BD}$	1. Given
2. $m\angle 2 = 90$	2. Def. of $\perp$ lines
3. $\angle 1 \cong \angle 2$, or $m\angle 1 = m\angle 2$	3. Given
4. $m\angle 1 = 90$	4. Substitution Prop.
5. $\overrightarrow{BA} \perp \overleftrightarrow{BD}$	5. Def. of $\perp$ lines

Page 69 • ALGEBRA REVIEW

1. $y = 3x$, $5x + y = 24$; $5x + 3x = 24$, $8x = 24$, $x = 3$; $y = 3x = 9$

2. $y = 2x + 5$, $3x - y = 4$; $3x - (2x + 5) = 4$, $x - 5 = 4$, $x = 9$; $y = 2x + 5 = 23$

3. $x = 8 + 3y$, $2x - 5y = 8$; $2(8 + 3y) - 5y = 8$, $16 + 6y - 5y = 8$, $y = -8$; $x = 8 + 3y = -16$

4. $3x + 2y = 71$, $y = 4 + 2x$; $3x + 2(4 + 2x) = 71$, $3x + 8 + 4x = 71$, $7x = 63$, $x = 9$; $y = 4 + 2x = 22$

5. $4x - 5y = 92$, $x = 7y$; $4(7y) - 5y = 92$, $23y = 92$, $y = 4$; $x = 7y = 28$

6. $y = 3x + 8$, $x = y$; $x = 3x + 8$, $2x = -8$, $x = -4$; $x = y$, $y = -4$

7. $8x + 3y = 26$, $2x = y - 4$; $y = 2x + 4$; $8x + 3(2x + 4) = 26$, $14x = 14$, $x = 1$; $2x = y - 4$, $2 = y - 4$, $y = 6$

8. $x - 7y = 13$, $3x - 5y = 23$; $x = 13 + 7y$; $3(13 + 7y) - 5y = 23$, $16y = -16$, $y = -1$; $x - 7y = 13$, $x + 7 = 13$, $x = 6$

9. $3x + y = 19$, $2x - 5y = -10$; $y = 19 - 3x$; $2x - 5(19 - 3x) = -10$, $2x - 95 + 15x = -10$, $17x = 85$, $x = 5$; $3x + y = 19$, $15 + y = 19$, $y = 4$

10.
$$\begin{array}{r} 5x - y = 20 \\ 3x + y = 12 \\ \hline 8x \qquad = 32 \end{array}$$
$x = 4$; $3x + y = 12$, $12 + y = 12$, $y = 0$

11.
$$\begin{array}{r} x + 3y = 7 \\ x + 2y = 4 \\ \hline y = 3 \end{array}$$
$x + 2y = 4$, $x + 6 = 4$, $x = -2$

12.
$$\begin{array}{r} 3x - 2y = 11 \\ 3x - y = 7 \\ \hline -y = 4 \end{array}$$
$y = -4$; $3x - y = 7$, $3x + 4 = 7$, $3x = 3$, $x = 1$

13.
$$\begin{array}{r} 7x + y = 29 \\ 5x + y = 21 \\ \hline 2x \qquad = 8 \end{array}$$
$x = 4$; $5x + y = 21$, $20 + y = 21$, $y = 1$

14. $\begin{array}{r} 8x - y = 17 \\ 6x + y = 11 \\ \hline 14x \quad = 28, \end{array}$ $x = 2; 6x + y = 11, 12 + y = 11, y = -1$

15. $\begin{array}{r} 9x - 2y = 50 \\ 6x - 2y = 32 \\ \hline 3x \quad = 18, \end{array}$ $x = 6; 6x - 2y = 32, 36 - 2y = 32, 2y = 4, y = 2$

16. $\begin{array}{r} 7y = 2x + 35 \\ 3y = 2x + 15 \\ \hline 4y = \quad 20, \end{array}$ $y = 5; 3y = 2x + 15, 15 = 2x + 15, 2x = 0, x = 0$

17. $\begin{array}{r} 2y = 3x - 1 \\ 2y = x + 21 \\ \hline 0 = 2x - 22, \end{array}$ $2x = 22, x = 11; 2y = 3x - 1, 2y = 33 - 1, 2y = 32, y = 16$

18. $\begin{array}{r} 19 = 5x + 2y \\ 1 = 3x - 4y \\ \hline \end{array}$; $\begin{array}{r} 38 = 10x + 4y \\ 1 = 3x - 4y \\ \hline 39 = 13x, \end{array}$ $x = 3; 1 = 3x - 4y, 1 = 9 - 4y, 4y = 8, y = 2$

Page 70 • PREPARING FOR COLLEGE ENTRANCE EXAMS

1. C **2.** C. $5x + 15 + 10x = 90; 15x = 75; x = 5; m\angle 1 = 5x + 15 = 40$ **3.** D
4. E **5.** D **6.** B. $10x - 20 = 8x + 2; 2x = 22; x = 11; m\angle 1 = 10x - 20 = 90$
7. E **8.** C

Page 71 • CUMULATIVE REVIEW: CHAPTERS 1 AND 2

A **1.** Div. Prop. of = **2.** Trans. Prop. **3.** Def. of ⊥ lines **4.** Substitution Prop.
5. Subtr. Prop. of = **6.** Through any 2 points there is exactly one line.
7. If 2 planes intersect, then their intersection is a line. **8.** Midpt. Thm.
9. Seg. Add. Post. **10.** False **11.** True **12.** True **13.** False **14.** True
15–20. Examples may vary. **15.** False; 3 collinear points **16.** True **17.** True
18. False; 20° 70°

B **19.** True **20.** False; R, U, N, and S need not be coplanar.
21. $2x + 2x + x = 180; 5x = 180; x = 36$ **22.** $2x + (6x + 2) = 90; 8x = 88; x = 11$
23. $6x + 9 = 2x + 49; 4x = 40; x = 10$
24. $3x + 3x + (2x - 4) = 180; 8x = 184; x = 23$
25. Equations may vary. $(x - 8) + (2x + 5) = 90; 3x = 93; x = 31$

CHAPTER 3 • Parallel Lines and Planes

Page 75 • CLASSROOM EXERCISES

1. **a.** $\angle 1, \angle 5; \angle 2, \angle 6; \angle 3, \angle 7; \angle 4, \angle 8$ **b.** $\angle 2, \angle 8; \angle 3, \angle 5$ **c.** $\angle 2, \angle 5; \angle 3, \angle 8$ **d.** $\angle 1, \angle 7; \angle 4, \angle 6$ **e.** $\angle 1, \angle 6; \angle 4, \angle 7$
2. s-s. int. ∠s **3.** corr. ∠s **4.** none **5.** alt. int. ∠s **6.** none
7. corr. ∠s **8.** s-s. int. ∠s **9.** alt. int. ∠s
10. **a.** $\parallel$ **b.** $\parallel$ **c.** skew **d.** int. **e.** skew **f.** skew
11. $\overleftrightarrow{AB}, \overleftrightarrow{EJ}, \overleftrightarrow{FK}, \overleftrightarrow{HM}, \overleftrightarrow{IN}, \overleftrightarrow{DC}$
12. $\overleftrightarrow{HI}, \overleftrightarrow{ID}, \overleftrightarrow{FE}, \overleftrightarrow{EA}, \overleftrightarrow{JK}, \overleftrightarrow{JB}, \overleftrightarrow{MN}, \overleftrightarrow{NC}, \overleftrightarrow{BC}, \overleftrightarrow{AD}$
13. $\overleftrightarrow{EJ}, \overleftrightarrow{FK}, \overleftrightarrow{GL}, \overleftrightarrow{HM}, \overleftrightarrow{IN}$ **14.** Answers may vary; for example, $\overline{EF}, \overline{HI}; \overline{BJ}, \overline{LM}$
15. never **16.** always **17.** sometimes **18.** never
19. **a.** sometimes **b.** sometimes **c.** sometimes

Pages 76–77 • WRITTEN EXERCISES

A **1.** alt. int. ∠s **2.** corr. ∠s **3.** s-s. int. ∠s **4.** alt. int. ∠s **5.** corr. ∠s
6. corr. ∠s **7.** $\overleftrightarrow{PQ}, \overleftrightarrow{SR}; \overleftrightarrow{SQ}$ **8.** $\overleftrightarrow{SP}, \overleftrightarrow{QR}; \overleftrightarrow{SQ}$ **9.** $\overleftrightarrow{PQ}, \overleftrightarrow{SR}; \overleftrightarrow{PS}$
10. $\overleftrightarrow{PS}, \overleftrightarrow{QR}; \overleftrightarrow{SR}$ **11.** $\overleftrightarrow{PQ}, \overleftrightarrow{SR}; \overleftrightarrow{QR}$ **12.** corr. ∠s **13.** corr. ∠s
14. alt. int. ∠s **15.** s-s. int. ∠s **16.** s-s. int. ∠s **17.** corr. ∠s **18.** Corr. ∠s are ≅.
19. Alt. int. ∠s are ≅. **20.** S-s. int. ∠s are supp.

B **21.** **a.** Answers may vary. **b.** Same as $m\angle 1 + m\angle 2$ **c.** Same as $m\angle 1 + m\angle 2$
d. When 2 nonparallel lines are cut by transversals, the sum of the measures of s-s. int. ∠s is a constant.
22. Check students' drawings. **23.** $\overleftrightarrow{BH}, \overleftrightarrow{CI}, \overleftrightarrow{DJ}, \overleftrightarrow{EK}, \overleftrightarrow{FL}$ **24.** $\overleftrightarrow{GH}, \overleftrightarrow{ED}, \overleftrightarrow{KJ}$
25. Answers may vary. $\overleftrightarrow{FL}, \overleftrightarrow{EK}, \overleftrightarrow{DJ}, \overleftrightarrow{CI}, \overleftrightarrow{GL}, \overleftrightarrow{LK}, \overleftrightarrow{JI}, \overleftrightarrow{IH}$ **26.** *CDJI*; *GHIJKL*
27. *ABHG*, *BCIH*, *CDJI*, *DEKJ* **28.** 4
29. If the top and bottom lie in $\parallel$ planes, then $\overline{CD}$ and $\overline{IJ}$ are the lines of intersection of *DCIJ* with 2 $\parallel$ planes, and are therefore $\parallel$.
30. always **31.** sometimes **32.** never **33.** always **34.** sometimes
35. sometimes **36.** sometimes **37.** always **38.** sometimes **39.** sometimes

C **40–42.** Sketches may vary.

40.

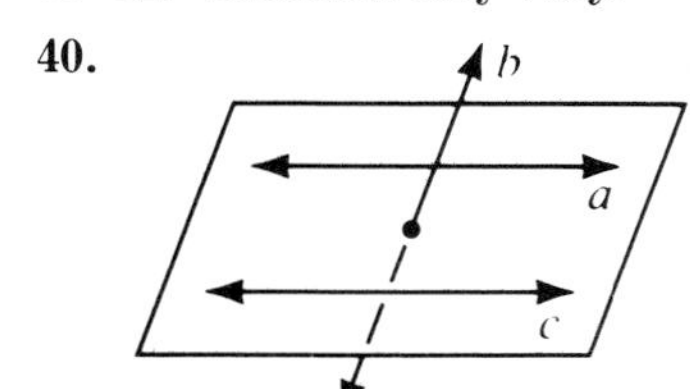

41.

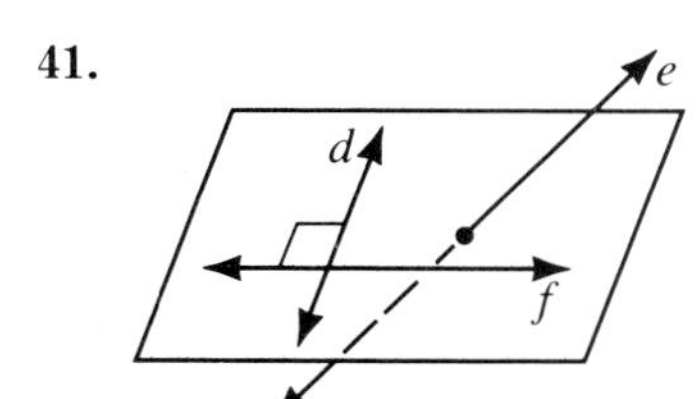

42. 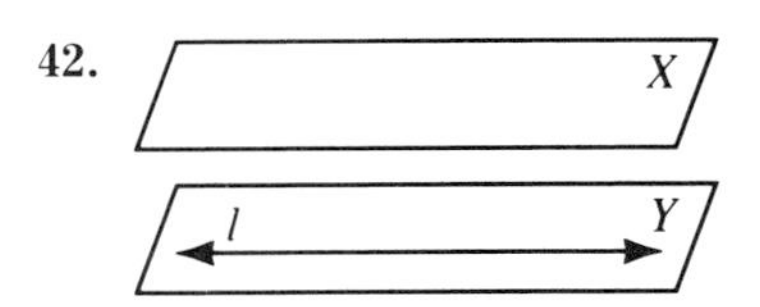

Page 78 • EXPLORATIONS

$\cong$ ∠s: corr. ∠s, alt. int. ∠s, vert. ∠s, alt. ext. ∠s
supp. ∠s: s-s. int. ∠s, s-s. ext. ∠s, adj. ∠s

Page 80 • CLASSROOM EXERCISES

1. $l \parallel p$ 2. If 2 $\parallel$ lines are cut by a trans., then corr. ∠s are $\cong$.
3. If 2 $\parallel$ lines are cut by a trans., then alt. int. ∠s are $\cong$.
4. If 2 $\parallel$ lines are cut by a trans., then s-s. int. ∠s are supp.
5. If 2 $\parallel$ lines are cut by a trans., then corr. ∠s are $\cong$.
6. If 2 $\parallel$ lines are cut by a trans., then alt. int. ∠s are $\cong$. **7.** Vert. ∠s are $\cong$.
8. If a trans. is $\perp$ to one of two $\parallel$ lines, then it is $\perp$ to the other one also.
9. If 2 $\parallel$ lines are cut by a trans., then s-s. int. ∠s are supp.
10. $m\angle 4 = m\angle 5 = m\angle 8 = 130; m\angle 2 = m\angle 3 = m\angle 6 = m\angle 7 = 50$
11. $m\angle 4 = m\angle 5 = m\angle 8 = x; m\angle 2 = m\angle 3 = m\angle 6 = m\angle 7 = 180 - x$
12. $m\angle 3 + m\angle 4 = 180; 3m\angle 3 = 180; m\angle 3 = 60$, so $m\angle 6 = 60$
13. $m\angle 5 + m\angle 6 = 180; 2m\angle 6 + 20 = 180; m\angle 6 = 80 = m\angle 2; m\angle 1 = 100$
14. In Step 2 he used Thm. 3-2, which relies on Post. 10.

Pages 80–82 • WRITTEN EXERCISES

A 1. $\angle 3, \angle 6, \angle 8$ **2.** $\angle 6, \angle 9, \angle 14$ **3.** $\angle 4, \angle 5, \angle 7, \angle 10, \angle 12, \angle 13, \angle 15$
4. $\angle 1, \angle 3, \angle 6, \angle 8, \angle 9, \angle 11, \angle 14, \angle 16$ **5.** 110, 70 **6.** x, $180 - x$
7. $x = 60, y = 61$ **8.** $4x + 14x = 180, x = 10; 2y = 90; y = 45$
9. $120 + x = 180, x = 60; 60 = 3y + 6, y = 18$
10. $x = 70; 50 + 70 + y = 180, y = 60$
11. $3x = 42, x = 14; 3(14) + 6y - 6 = 90, y = 9$
12. $x = 55; y + 55 + 50 = 180, y = 75$
13. 1. Given 2. Def. of $\perp$ lines 3. $l \parallel n$ 4. If 2 $\parallel$ lines are cut by a trans., then corr. ∠s are $\cong$. 5. $m\angle 2 = 90$ 6. Def. of $\perp$ lines

B 14. $x = 56; 56 + 24 + y = 180, y = 100; 56 + 24 + 4z = 180, z = 25$
15. $x = 70; 5y + 10 = 70, y = 12; z + 32 = 5(12) + 10, z = 38$

16. $3x = 90$, $x = 30$; $8y + 4 = 68$, $y = 8$; $2z + 8(8) + 4 = 90$, $z = 11$

17. **a.** $m\angle DAB + 116 = 180$, $m\angle DAB = 64$; $m\angle KAB = 32$; $m\angle DKA = 32$

b. More information is needed.

18. $2x + y = 60$, $2x - y = 40$; $4x = 100$, $x = 25$; $y = 10$

19. $4x - 2y = 110$, $4x + 2y = 130$; $8x = 240$, $x = 30$; $y = 5$

20.

Statements	Reasons
1. $k \parallel l$	1. Given
2. $\angle 2 \cong \angle 4$	2. If 2 ∥ lines are cut by a trans., then corr. ∠s are ≅.
3. $\angle 4 \cong \angle 7$	3. Vert. ∠s are ≅.
4. $\angle 2 \cong \angle 7$	4. Trans. Prop.

21.

Statements	Reasons
1. $k \parallel l$	1. Given
2. $\angle 1 \cong \angle 8$, or $m\angle 1 = m\angle 8$	2. If 2 ∥ lines are cut by a trans., then alt. int. ∠s are ≅.
3. $m\angle 8 + m\angle 7 = 180$	3. ∠ Add. Post.
4. $m\angle 1 + m\angle 7 = 180$	4. Substitution Prop.
5. $\angle 1$ is supp. to $\angle 7$.	5. Def. of supp. ∠s

22.

Statements	Reasons
1. $k \parallel n$	1. Given
2. $\angle 1 \cong \angle 2$, or $m\angle 1 = m\angle 2$	2. If 2 ∥ lines are cut by a trans., then alt. int. ∠s are ≅.
3. $m\angle 2 + m\angle 4 = 180$	3. ∠ Add. Post.
4. $m\angle 1 + m\angle 4 = 180$	4. Substitution Prop.
5. $\angle 1$ is supp. to $\angle 4$.	5. Def. of supp. ∠s

23. **a.**

Statements	Reasons
1. $\overline{AB} \parallel \overline{DC}$; $\overline{AD} \parallel \overline{BC}$	1. Given
2. $\angle A$ is supp. to $\angle B$; $\angle C$ is supp. to $\angle B$.	2. If 2 ∥ lines are cut by a trans., then s-s. int. ∠s are supp.
3. $\angle A \cong \angle C$	3. If 2 ∠s are supps. of the same ∠, then the 2 ∠s are ≅.

b. Yes, by the same reasoning as in part (a).

C **24.**

Statements	Reasons
1. $\overline{AS} \parallel \overline{BT}$	1. Given
2. $m\angle 1 = m\angle 4$	2. If 2 $\parallel$ lines are cut by a trans., then corr. ∠s are ≅.
3. $m\angle 2 = m\angle 5$	3. If 2 $\parallel$ lines are cut by a trans., then alt. int. ∠s are ≅.
4. $m\angle 4 = m\angle 5$	4. Given
5. $m\angle 1 = m\angle 2$	5. Substitution Prop.
6. $\overrightarrow{SA}$ bisects $\angle BSR$.	6. Def. of ∠ bis.

25. Steps 1–5 of the proof in Ex. 24 prove that $m\angle 1 = m\angle 2$. $\overrightarrow{SB}$ bisects $\angle AST$, so $m\angle 2 = m\angle 3$. Since $m\angle 1 + m\angle 2 + m\angle 3 = 180$, $3m\angle 1 = 180$ by Substitution, and $m\angle 1 = 60$.

Page 82 • MIXED REVIEW EXERCISES

1. **a.** True **b.** If 2 lines form ≅ adj. ∠s, then the lines are ⊥. **c.** True
2. **a.** True **b.** If 2 lines are not skew, then they are $\parallel$. **c.** False
3. **a.** True **b.** If 2 ∠s are supp., then the sum of their measures is 180. **c.** True
4. **a.** True **b.** If 2 planes do not intersect, then they are $\parallel$. **c.** True

Page 86 • CLASSROOM EXERCISES

1. $\overline{KC} \parallel \overline{DE}$. If 2 lines are cut by a trans. and s-s. int. ∠s are supp., then the lines are $\parallel$.
2. $\overline{OX} \parallel \overline{IZ}$. If 2 lines are cut by a trans. and corr. ∠s are ≅, then the lines are $\parallel$.
3. $\overline{LA} \parallel \overline{TS}$. If 2 lines are cut by a trans. and s-s. int. ∠s are supp., then the lines are $\parallel$.
4. $\overline{GA} \parallel \overline{EM}$. If 2 lines are cut by a trans. and alt. int. ∠s are ≅, then the lines are $\parallel$.
5. $\overline{PL} \parallel \overline{AR}$ **6.** $\overline{PA} \parallel \overline{LR}$ **7.** no segs. $\parallel$ **8.** $\overline{PL} \parallel \overline{AR}$ **9.** no segs. $\parallel$
10. $\overline{PL} \parallel \overline{AR}$ **11.** $\overline{PA} \parallel \overline{LR}$
12. Through a pt. outside a line, there is a line $\parallel$ to the given line. Through a pt. outside a line, there is no more than one line $\parallel$ to the given line.
13. Through a point outside a line, there is a line ⊥ to the given line. Through a point outside a line, there is no more than one line ⊥ to the given line.
14. one **15.** one **16.** one **17.** one; Protractor Post. **18.** Infinitely many

19. **a.** False **b.** True **c.** True **d.** True

20. If $k \parallel l$, then $\angle 1 \cong \angle 2$. If $k \parallel n$, then $\angle 1 \cong \angle 3$. Therefore, $\angle 2 \cong \angle 3$ and $l \parallel n$. (Substitution Prop.; if 2 lines are cut by a trans. and corr. $\angle$s are $\cong$, then the lines are $\parallel$.)

Pages 87–88 • WRITTEN EXERCISES

A **1.** $\overline{AB} \parallel \overline{FC}$ **2.** $\overline{AE} \parallel \overline{BD}$ **3.** $\overline{AB} \parallel \overline{FC}$ **4.** $\overline{FB} \parallel \overline{EC}$ **5.** none **6.** $\overline{AE} \parallel \overline{BD}$ **7.** none **8.** none **9.** $\overline{AE} \parallel \overline{BD}$ **10.** $\overline{AE} \parallel \overline{BD}$ **11.** $\overline{AE} \parallel \overline{BD}$ **12.** $\overline{FB} \parallel \overline{EC}$ **13.** $\overline{AE} \parallel \overline{BD}$ **14.** none **15.** $\overline{FB} \parallel \overline{EC}$; $\overline{AE} \parallel \overline{BD}$ **16.** $\overline{AB} \parallel \overline{FC}$; $\overline{AE} \parallel \overline{BD}$

17. 1. Given 2. Vert. $\angle$s are $\cong$. 3. Trans. Prop. 4. If 2 lines are cut by a trans. and corr. $\angle$s are $\cong$, then the lines are $\parallel$.

B **18.** $(x - 40) + (x + 40) = 180$, $2x = 180$, $x = 90$; $(x - 40) + y = 180$, $(90 - 40) + y = 180$, $y = 130$

19. $3x = 105$, $x = 35$; $105 = 180 - (2y + x)$, $105 = 180 - (2y + 35)$, $2y = 40$, $y = 20$

20. $\overline{PQ} \parallel \overline{RS}$. $\angle 1 \cong \angle 2$, $\angle 2 \cong \angle 5$ (Vert. $\angle$s are $\cong$.), and $\angle 5 \cong \angle 4$, so $\angle 1 \cong \angle 4$. Since alt. int. $\angle$s are $\cong$, $\overline{PQ} \parallel \overline{RS}$.

21. $\angle 1 \cong \angle 4$; $\angle 2 \cong \angle 5$. If $\angle 3 \cong \angle 6$, then $\overline{PQ} \parallel \overline{RS}$ because alt. int. $\angle$s are $\cong$. If $\overline{PQ} \parallel \overline{RS}$, then $\angle 1 \cong \angle 4$ because they are alt. int. $\angle$s. $\angle 2 \cong \angle 5$ because vert. $\angle$s are $\cong$.

22.

Statements	Reasons
1. $\angle 1$ is supp. to $\angle 2$.	1. Given
2. $m\angle 2 + m\angle 3 = 180$	2. $\angle$ Add. Post.
3. $\angle 3$ is supp. to $\angle 2$.	3. Def. of supp. $\angle$s
4. $\angle 1 \cong \angle 3$	4. If 2 $\angle$s are supps. of the same $\angle$, then the 2 $\angle$s are $\cong$.
5. $k \parallel n$	5. If 2 lines are cut by a trans. and alt. int. $\angle$s are $\cong$, then the lines are $\parallel$.

23.

Statements	Reasons
1. $k \perp t$; $n \perp t$	1. Given
2. $m\angle 1 = 90$; $m\angle 2 = 90$	2. Def. of $\perp$ lines
3. $m\angle 1 = m\angle 2$, or $\angle 1 \cong \angle 2$	3. Substitution Prop.
4. $k \parallel n$	4. If 2 lines are cut by a trans. and corr. $\angle$s are $\cong$, then the lines are $\parallel$.

24.

Statements	Reasons
1. $\overline{BE}$ bisects $\angle DBA$.	1. Given
2. $\angle 2 \cong \angle 3$	2. Def. of $\angle$ bis.
3. $\angle 3 \cong \angle 1$	3. Given
4. $\angle 2 \cong \angle 1$	4. Trans. Prop.
5. $\overline{CD} \parallel \overline{BE}$	5. If 2 lines are cut by a trans. and alt. int. $\angle$s are $\cong$, then the lines are $\parallel$.

25.

Statements	Reasons
1. $\overline{BE} \perp \overline{DA}$; $\overline{CD} \perp \overline{DA}$	1. Given
2. $\overline{CD} \parallel \overline{BE}$	2. In a plane, 2 lines $\perp$ to the same line are $\parallel$.
3. $\angle 1 \cong \angle 2$	3. If 2 $\parallel$ lines are cut by a trans., then alt. int. $\angle$s are $\cong$.

26.

Statements	Reasons
1. $\angle C \cong \angle 3$	1. Given
2. $\overline{CD} \parallel \overline{BE}$	2. If 2 lines are cut by a trans. and corr. $\angle$s are $\cong$, then the lines are $\parallel$.
3. $\overline{BE} \perp \overline{DA}$	3. Given
4. $\overline{CD} \perp \overline{DA}$	4. If a trans. is $\perp$ to one of 2 $\parallel$ lines, then it is $\perp$ to the other one also.

27. $m\angle RST = 40 + 70 = 110$

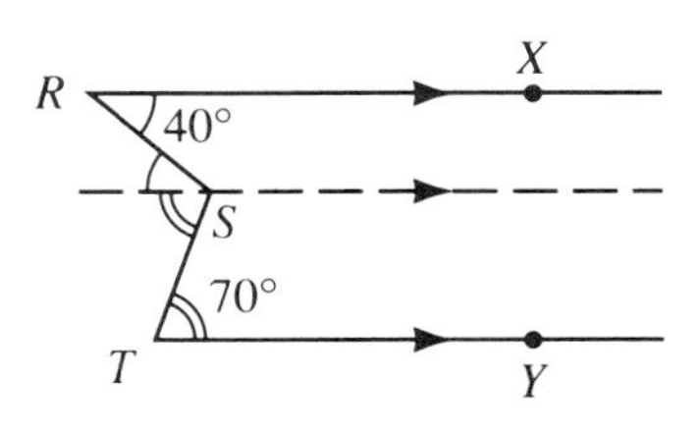

28. $m\angle 1 = 70$, $m\angle 2 = 60$, $m\angle RST = 130$

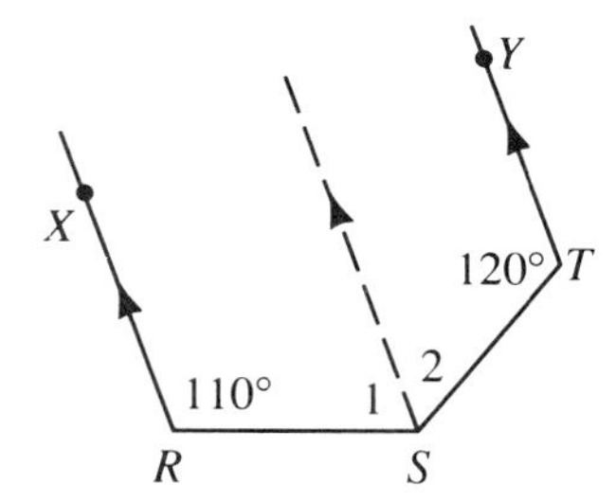

29. $2x = 5y$, $x - y = 30$; $x = 30 + y$, $2(30 + y) = 5y$; $60 + 2y = 5y$, $3y = 60$, $y = 20$; $x - y = 30$, $x - 20 = 30$, $x = 50$

C 30. The bisectors appear to be $\parallel$.
Given: $\overleftrightarrow{BD} \parallel \overleftrightarrow{CF}$; $\overrightarrow{BG}$ bisects $\angle ABD$;
$\overrightarrow{CH}$ bisects $\angle BCF$.
Prove: $\overrightarrow{BG} \parallel \overrightarrow{CH}$

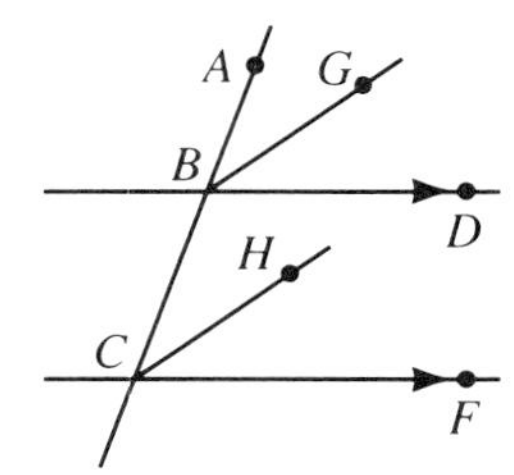

Statements	Reasons
1. $\overleftrightarrow{BD} \parallel \overleftrightarrow{CF}$	1. Given
2. $m\angle ABD = m\angle BCF$	2. If 2 lines are cut by a trans., then corr. ∠s are ≅.
3. $\frac{1}{2}m\angle ABD = \frac{1}{2}m\angle BCF$	3. Mult. Prop. of =
4. $\overrightarrow{BG}$ bisects $\angle ABD$; $\overrightarrow{CH}$ bisects $\angle BCF$.	4. Given
5. $m\angle ABG = \frac{1}{2}m\angle ABD$; $m\angle BCH = \frac{1}{2}m\angle BCF$	5. ∠ Bis. Thm.
6. $m\angle ABG = m\angle BCH$	6. Substitution Prop.
7. $\overrightarrow{BG} \parallel \overrightarrow{CH}$	7. If 2 lines are cut by a trans. and corr. ∠s are ≅, then the lines are ∥.

31. $x^2 + 3x = 180$; $x^2 + 3x - 180 = 0$; $(x + 15)(x - 12) = 0$; $x + 15 = 0$ or $x - 12 = 0$; $x = -15$ (reject) or $x = 12$; $x = 12$

Page 89 • SELF-TEST 1

1. sometimes **2.** never **3.** always **4.** sometimes **5.** always
6. $\angle 3, \angle 6; \angle 4, \angle 5$ **7.** Answers may vary; $\angle 1, \angle 5; \angle 2, \angle 6; \angle 3, \angle 7; \angle 4, \angle 8$
8. $\angle 3, \angle 5$ or $\angle 4, \angle 6$ **9.** $\angle 4; \angle 3$ **10.** $\angle 2, \angle 8; \angle 4, \angle 7$ **11.** $\angle 2, \angle 8$ **12.** 65; 115
13. $\overline{EB} \parallel \overline{DC}$ **14.** none **15.** $\overline{AE} \parallel \overline{BD}$ **16.** one, one

Page 89 • EXPLORATIONS

The sum of the measures of the ∠s inside the △ is 180. The sum of the measures of the ∠s outside the △ is 360.

Page 92 • APPLICATION

1.

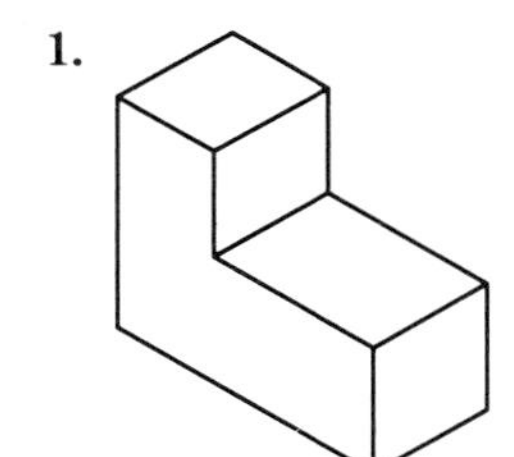

2. d
3. b

4.

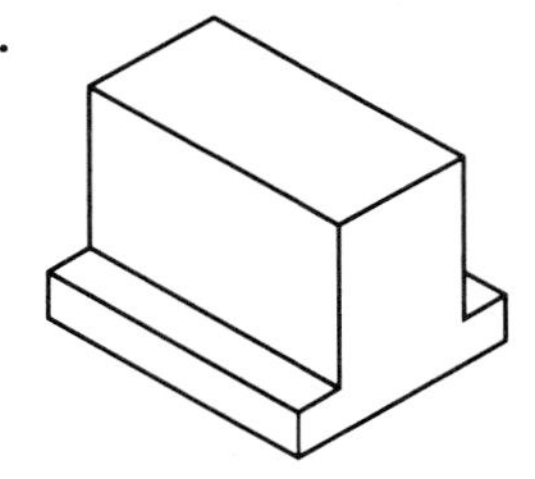

5. c
6. a

7.

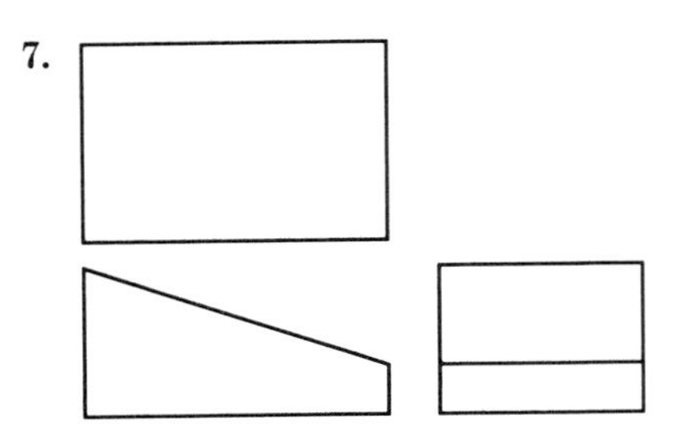

8.

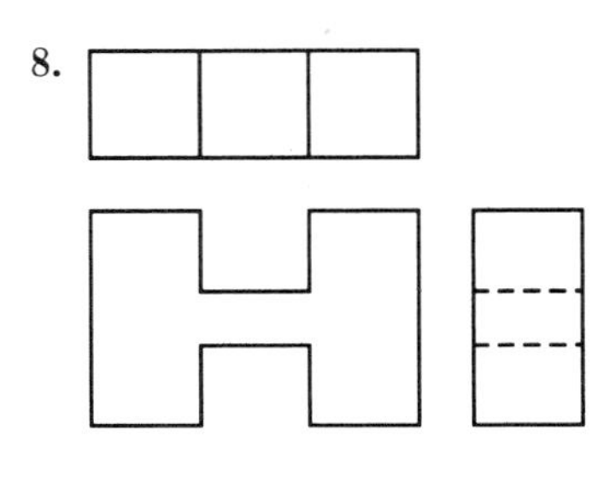

Page 96 • CLASSROOM EXERCISES

1. sometimes 2. always 3. never 4. sometimes
5. The sums of the meas. of the 2 $\angle$s in each $\triangle$ are $=$. The meas. of the third $\angle$ in each $\triangle$ must $= 180 - \text{sum}$.
6. Let meas. of each $\angle = x$; $3x = 180$, $x = 60$
7. In $\triangle ABC$, if $m\angle A \geq 90$ and $m\angle B \geq 90$, then $m\angle A + m\angle B + m\angle C > 180$ since $m\angle C > 0$.
8. In $\triangle ABC$, if $m\angle C = 90$, then $m\angle A + m\angle B = 180 - 90 = 90$.
9. $x = 90$ 10. $x = 105$ 11. $x = 35 + (180 - 140) = 75$
12. The bis. of $\angle J$ may not contain the midpt. of $\overline{PE}$.
13. The line through $P \perp$ to $\overline{JE}$ may not contain the midpt. of $\overline{JE}$.
14. $\overleftrightarrow{PX}$ may not be $\parallel$ to $\overleftrightarrow{JE}$.
15. By the Substitution Prop., $m\angle 1 + m\angle 2 + m\angle 3 = m\angle 3 + m\angle 4$. By the Subtraction Prop. of $=$, $m\angle 1 + m\angle 2 = m\angle 4$, which proves Thm. 3-12.
16. $m\angle 1 + m\angle 2 = 90$, so this illustrates Cor. 4, the acute $\angle$s of a rt. $\triangle$ are comp.

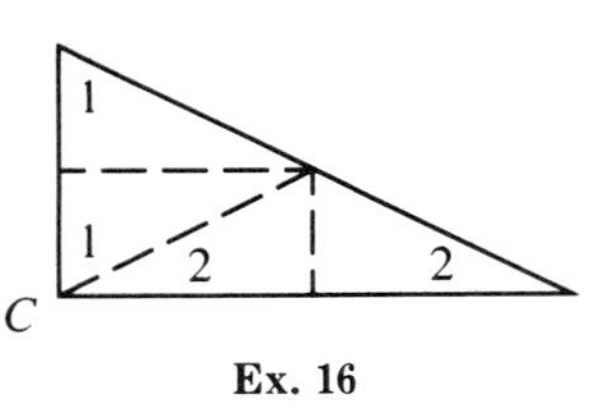

Ex. 16

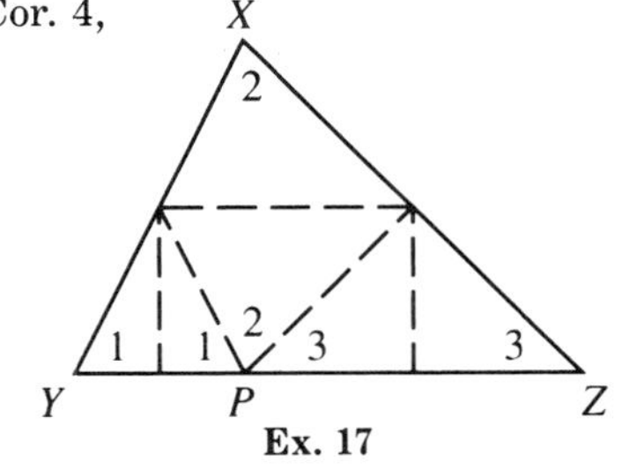

Ex. 17

17. $m\angle 1 + m\angle 2 + m\angle 3 = 180$, so this illustrates Thm. 3-11, the sum of the meas. of the $\angle$s of a $\triangle$ is 180.

Pages 97–99 • WRITTEN EXERCISES

A 1. a.

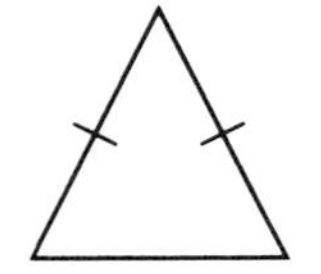

b.

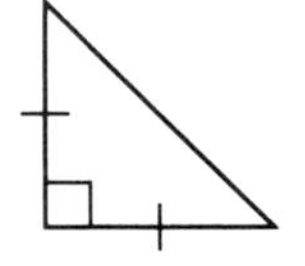

c.

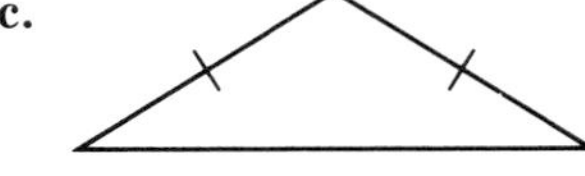

2. a. 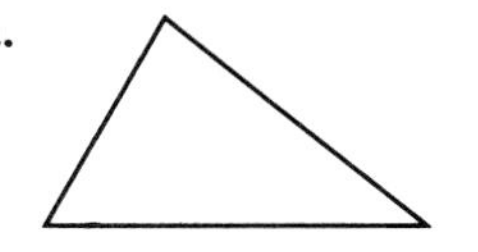 **b.** 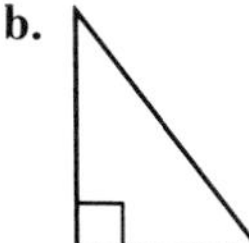 **c.** 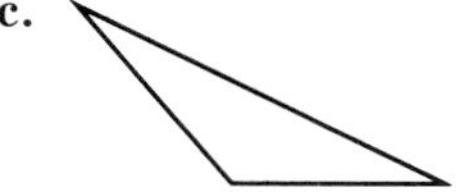

3. not possible **4.** 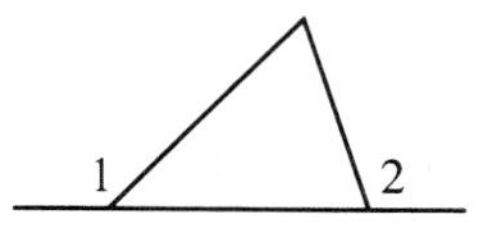

5. 180 **6.** 30 **7.** 95 **8.** $x + (x - 20) = 80; x = 50$

9. $4x + 30 = 6x - 20; x = 25$

10. $m\angle 9 + m\angle 10 + m\angle 11 = (m\angle 7 + m\angle 8) + (m\angle 6 + m\angle 8) + (m\angle 6 + m\angle 7) = 2(m\angle 6 + m\angle 7 + m\angle 8) = 2 \cdot 180 = 360$

11. $x = 30; y = 50 + 30 = 80$ **12.** $x = 110; y = 110 - 40 = 70$

13. $x = 40; y = 90 - 40 = 50$

B 14. $x = 65 + 25 = 90; y = 90 - 65 = 25$

15. $y = 90 - 40 = 50; x = 90 - 50 = 40$

16. $y = 90 - (40 + 20) = 30; x + 30 = 90 - 20, x = 40$

17. Yes; $4n = 2n + 10; n = 5$; the sides are $4(5) = 20, 2(5) + 10 = 20, 7(5) - 15 = 20$.

18. a. $3t = 5t - 12, t = 6; 3t = t + 20, t = 10; 5t - 12 = t + 20, t = 8$

b. No; there is no value of t such that $3t = 5t - 12 = t + 20$.

19. Let x be the measure of the smallest angle; $x + 2x + 3x = 180; 6x = 180; x = 30$; the meas. of the angles are 30, 60, 90.

20. $x + (x + 28) + 2x = 180; 4x + 28 = 180; 4x = 152; x = 38$; 38, 66, 76

21. $m\angle A + m\angle B + m\angle C = 180; m\angle A + m\angle B < 120$, so $m\angle C > 60$.

22. $m\angle R + m\angle S + m\angle T = 180; m\angle R + m\angle S > 110$, so $m\angle T < 70$.

23. a. 22 **b.** 23 **c.** $\angle ABD$ and $\angle C$ are comps. of $\angle CBD$.

24. a. 130 **b.** 130 **c.** If $m\angle E = 80$, then $m\angle FIG$ will always be 130.

25.

Statements	Reasons
1. $\angle ABD \cong \angle AED$	1. Given
2. $\angle A \cong \angle A$	2. Refl. Prop.
3. $\angle C \cong \angle F$	3. If 2 $\angle$s of one $\triangle$ are $\cong$ to 2 $\angle$s of another $\triangle$, then the third $\angle$s are $\cong$.

26. $m\angle MTR = 180 - 85 = 95; m\angle STR = 180 - (30 + 95) = 55$; $m\angle 1 = 90 - 55 = 35; m\angle NRT = 55; m\angle 2 = 180 - 55 = 125$

27. Given: $\triangle ABC$

Prove: $m\angle 1 + m\angle 2 + m\angle 3 = 180$

Statements	Reasons
1. Draw $\overrightarrow{CD}$ through C $\parallel$ to $\overleftrightarrow{AB}$.	1. Through a pt. outside a line, there is exactly 1 line $\parallel$ to the given line.
2. $\angle 2 \cong \angle 5$, or $m\angle 2 = m\angle 5$	2. If 2 $\parallel$ lines are cut by a trans., then alt. int. $\angle$s are $\cong$.
3. $\angle 1 \cong \angle 4$, or $m\angle 1 = m\angle 4$	3. If 2 $\parallel$ lines are cut by a trans., then corr. $\angle$s are $\cong$.
4. $m\angle ACD + m\angle 4 = 180$; $m\angle ACD = m\angle 3 + m\angle 5$	4. $\angle$ Add. Post.
5. $m\angle 3 + m\angle 4 + m\angle 5 = 180$	5. Substitution Prop.
6. $m\angle 1 + m\angle 2 + m\angle 3 = 180$	6. Substitution Prop.

28.

Statements	Reasons
1. $m\angle JGI = m\angle H + m\angle I$	1. The meas. of an ext. $\angle$ of a $\triangle$ = the sum of the meas. of the 2 remote int. $\angle$s.
2. $m\angle H = m\angle I$	2. Given
3. $m\angle JGI = 2m\angle H$	3. Substitution Prop.
4. $\frac{1}{2}m\angle JGI = m\angle H$	4. Div. Prop. of =
5. $\overrightarrow{GK}$ bisects $\angle JGI$.	5. Given
6. $m\angle 1 = \frac{1}{2}m\angle JGI$	6. $\angle$ Bis. Thm.
7. $m\angle 1 = m\angle H$	7. Substitution Prop.
8. $\overline{GK} \parallel \overline{HI}$	8. If 2 lines are cut by a trans. and corr. $\angle$s are $\cong$, then the lines are $\parallel$.

29. $2x + y + 125 = 180$, $2x + y = 55$, $y = 55 - 2x$; $(x + 2y) + (2x + y) = 90$, $(x + 2y) + 55 = 90$, $x + 2y = 35$; $x + 2(55 - 2x) = 35$, $x + 110 - 4x = 35$, $3x = 75$, $x = 25$; $2x + y = 55$, $50 + y = 55$, $y = 5$

30. $(5x + y) + (5x - y) + 100 = 180$, $10x = 80$, $x = 8$; $2x + y = 5x - y$, $2y = 3x$, $2y = 24$, $y = 12$

31. $\angle 1 \cong \angle 2 \cong \angle 5$; $\angle 3 \cong \angle 4 \cong \angle 6$

C 32. $\angle 7 \cong \angle 8$, $\angle 11 \cong \angle 12$

33. a–b. Check students' drawings. See figure at the right.

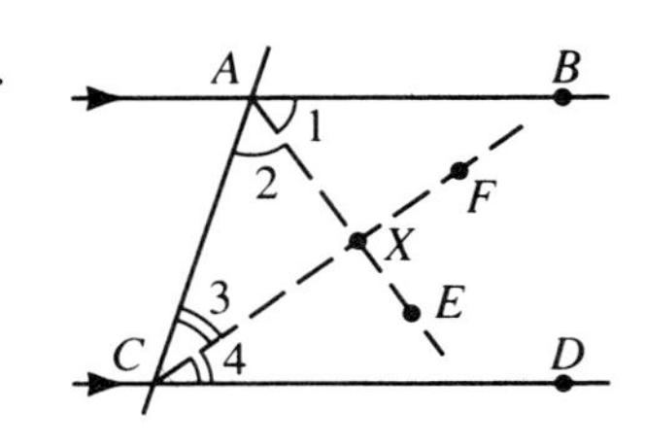

c. The angle measures 90, so the bisectors are $\perp$.

d. Given: $\overleftrightarrow{AB} \parallel \overleftrightarrow{CD}$; $\overrightarrow{AE}$ bisects $\angle BAC$; $\overrightarrow{CF}$ bisects $\angle ACD$.

Prove: $\overleftrightarrow{AE} \perp \overleftrightarrow{CF}$

Statements	Reasons
1. $\overleftrightarrow{AB} \parallel \overleftrightarrow{CD}$	1. Given
2. $m\angle BAC + m\angle ACD = 180$	2. If 2 ∥ lines are cut by a trans., then s-s. int. ∠s are supp.; def. of supp. ∠s
3. $\frac{1}{2}m\angle BAC + \frac{1}{2}m\angle ACD = 90$	3. Div. Prop. of =
4. $\overrightarrow{AE}$ bisects $\angle BAC$; $\overrightarrow{CF}$ bisects $\angle ACD$.	4. Given
5. $m\angle 2 = \frac{1}{2}m\angle BAC$; $m\angle 3 = \frac{1}{2}m\angle ACD$	5. ∠ Bis. Thm.
6. $m\angle 2 + m\angle 3 = 90$	6. Substitution Prop.
7. $m\angle AXF = m\angle 2 + m\angle 3$	7. The meas. of an ext. ∠ of a △ = the sum of the meas. of the 2 remote int. ∠s
8. $m\angle AXF = 90$	8. Substitution Prop.
9. $\overleftrightarrow{AE} \perp \overleftrightarrow{CF}$	9. Def. of ⊥ lines

34. Since $3x$ and $3y$ are meas. of s-s. int. ∠s, $3x + 3y = 180$, and $x + y = 60$. Then $m\angle EDF = m\angle CDA = 180 - (x + y) = 120$. $\angle EBF$ is the third ∠ of a △ with ∠s of meas. $2x$ and $2y$, so $m\angle CBA = 180 - (2x + 2y) = 180 - 120 = 60$. Then, in $ABCD$, $m\angle CDA + m\angle CBA = 120 + 60 = 180$. Also, $\angle BCD$ is an ext. ∠ of $\triangle ECF$ with remote int. ∠s of meas. $2x$ and y, so $m\angle BCD = 2x + y$. Similarly, $m\angle BAD = 2y + x$. So, $m\angle BCD + m\angle BAD = 3x + 3y = 180$. Therefore, in $ABCD$ opp. ∠s are supp.

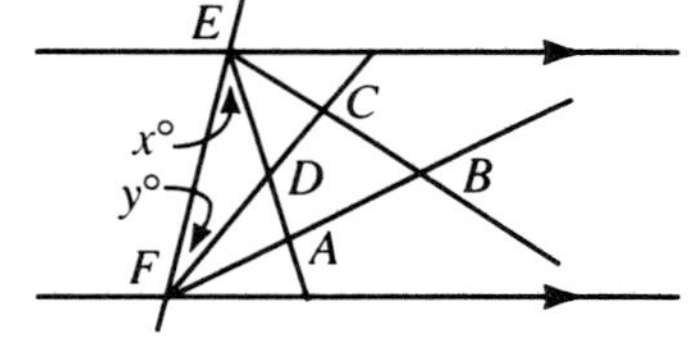

Page 99 • EXPLORATIONS

1–4. Sketches and angle measures will vary. **1.** False; true for acute △s
2. False; true for acute △s **3.** True **4.** False; true for rt. △s

Page 103 • CLASSROOM EXERCISES

1. convex polygon **2.** nonconvex polygon **3.** not a polygon **4.** nonconvex polygon
5. not a polygon **6.** nonconvex polygon **7.** It has the same shape.
8. $(102 - 2)180 = 18{,}000$; 360

9.

No. of sides	6	10	20	36	18	360	4
Meas. of each ext. ∠	60	36	18	10	20	1	90
Meas. of each int. ∠	120	144	162	170	160	179	90

Pages 104–105 • WRITTEN EXERCISES

A **1.** $(4 - 2)180 = 360; 360$ **2.** $(5 - 2)180 = 540; 360$ **3.** $(6 - 2)180 = 720; 360$

4. $(8 - 2)180 = 1080; 360$ **5.** $(10 - 2)180 = 1440; 360$ **6.** $(n - 2)180; 360$

7. 360; yes

8.

No. of sides	9	15	30	60	45	24	180
Meas. of each ext. ∠	40	24	12	6	8	15	2
Meas. of each int. ∠	140	156	168	174	172	165	178

9. Let x = meas. of each of the 2 ≅ ∠s. $2x + 3(90) = (5 - 2)180$, $2x + 270 = 540$, $2x = 270$, $x = 135$

10. Let x = meas. of the fifth ∠. $x + 40 + 80 + 115 + 165 = (5 - 2)180$, $x + 400 = 540$, $x = 140$

11. 120 **12–15.** Sketches may vary. **12.**

13.

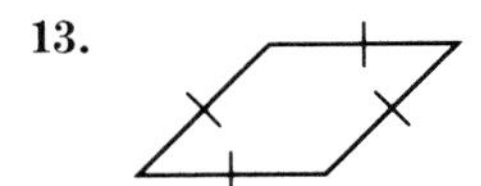

14. not possible

15. not possible (An ext. ∠ would have meas. 50, and 360 is not a multiple of 50.)

B **16.** Let n = number of sides; $(n - 2)180 = 5(360)$, $n = 12$.

17. Let n = number of sides; $\frac{(n - 2)180}{n} = 11\left(\frac{360}{n}\right)$, $n = 24$.

18. **a.** 108 **b.** No (360 is not a multiple of 108.) **19.**

Ex. 19

20. The sum of the meas. of the int. ∠s of 2 hexagons and 1 pentagon at any common vertex is $120 + 120 + 108 = 348$. A sum of 360 is necessary to tile a plane.

21. $x + 2x + 3x + 4x = (4 - 2)180$, $10x = 360$, $x = 36$; $m\angle A = 36$, $m\angle B = 72$, $m\angle C = 108$, $m\angle D = 144$; $m\angle A + m\angle D = 180$ (also, $m\angle B + m\angle C = 180$), so $\overline{AB} \parallel \overline{CD}$.

22. **a.** Let $m\angle R = x$, then $m\angle S = m\angle T = 3x$; $60 + 130 + x + 3x + 3x = (5 - 2)180$, $7x + 190 = 540$, $7x = 350$, $x = 50$; $m\angle R = 50$, $m\angle S = m\angle T = 150$
b. $m\angle Q + m\angle R = 180$, so $\overline{PQ} \parallel \overline{RS}$.

23. $\angle KBC$ and $\angle KCB$ are ext. $\angle$s of a reg. decagon, so $m\angle KBC = m\angle KCB = \frac{360}{10} = 36$. $m\angle K = 180 - (m\angle KBC + m\angle KCB)$, so $m\angle K = 180 - (36 + 36) = 108$.

24. $\angle WBC$ and $\angle WCB$ are each ext. $\angle$s of the n-gon, so $m\angle WBC = m\angle WCB = \frac{360}{n}$.
$m\angle W = 180 - (m\angle WBC + m\angle WCB) = 180 - 2\left(\frac{360}{n}\right) = \frac{180n - 720}{n}$

25. $2100 < (n - 2)180 < 2200$; $13\frac{2}{3} < n < 14\frac{2}{9}$; $n = 14$

C **26.** **a.** $[(n + 1) - 2]180 = [(n - 2) + 1]180 = (n - 2)180 + 180 = S + 180$
b. $(2n - 2)180 = [2(n - 1)180] = 2[(n - 2) + 1]180 = 2[(n - 2)180 + 180] = 2(S + 180)$

27. **a.** Sketches may vary.
b. Yes; $90 + 90 + 50 + 260 + 50 = 540$

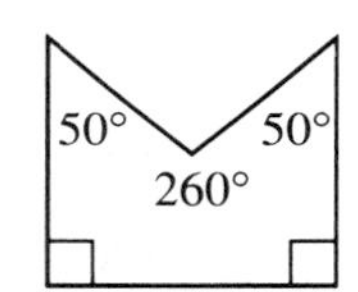

28. **a.** $\frac{(n - 2)180}{n} = x \cdot \frac{360}{n}$; $x = \frac{n - 2}{2}$
b. Even values ≥ 4

Page 107 • CLASSROOM EXERCISES

1. inductive **2.** inductive **3.** deductive **4.** inductive
5. deductive **6.** inductive

Pages 107–109 • WRITTEN EXERCISES

A **1.** 256, 1024 **2.** 6, 3 **3.** $\frac{1}{81}, \frac{1}{243}$ **4.** 25, 36 **5.** 17, 23 **6.** 40, 52 **7.** 15, 4

8. $-\frac{1}{4}, \frac{1}{8}$ **9.** 500, 250 **10.** Chan is older than Sarah. **11.** none

12. Polygon G has 7 sides. **13.** none **14.** No; deductively

15. $1234 \times 9 + 5 = 11111$ **16.** $9876 \times 9 + 4 = 88888$ **17.** $9999^2 = 99980001$

B **18.** True
Given: $\overline{BA} \cong \overline{BC}$
Prove: $\angle A \cong \angle C$

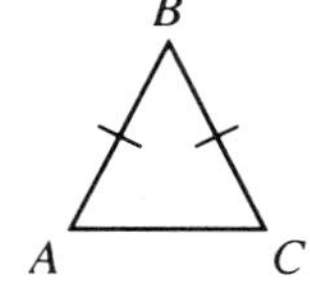

19. True
Given: $\angle A \cong \angle C$
Prove: $\overline{BA} \cong \overline{BC}$

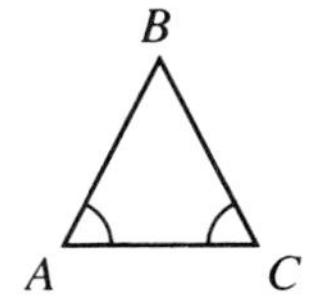

20. False

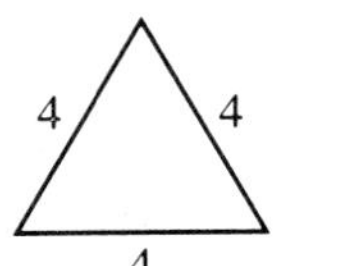

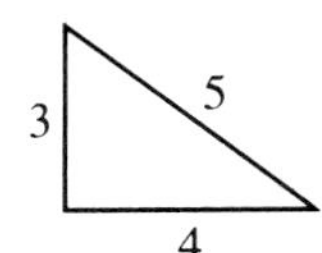

21. True

Given: $ABCDE$ is a reg. pentagon.

Prove: $\overline{AC} \cong \overline{AD} \cong \overline{BE} \cong \overline{BD} \cong \overline{CE}$

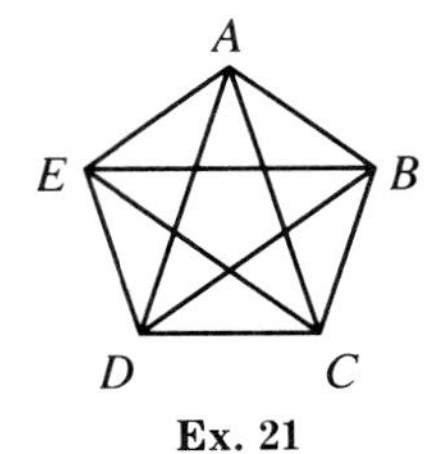

Ex. 21

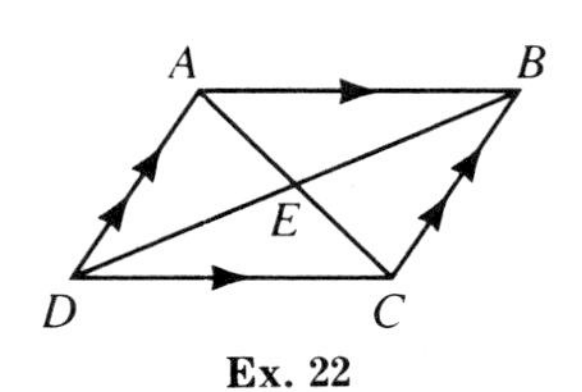

Ex. 22

22. True

Given: $\overline{AB} \parallel \overline{DC}; \overline{AD} \parallel \overline{BC}$

Prove: $\overline{AE} \cong \overline{EC}; \overline{DE} \cong \overline{EB}$

23. False

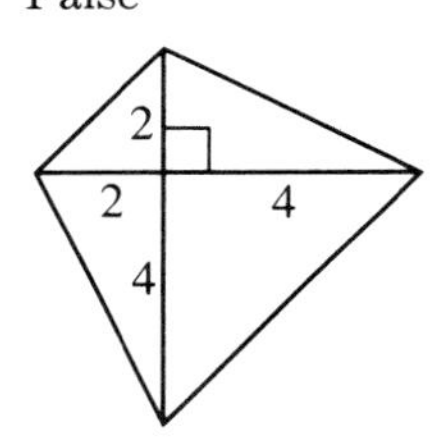

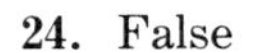

24. False

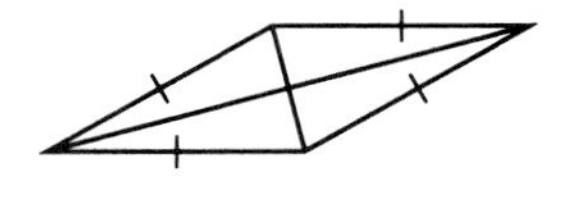

25. True

Given: $ABCD$ is equilateral.

Prove: $\overline{AC} \perp \overline{BD}$

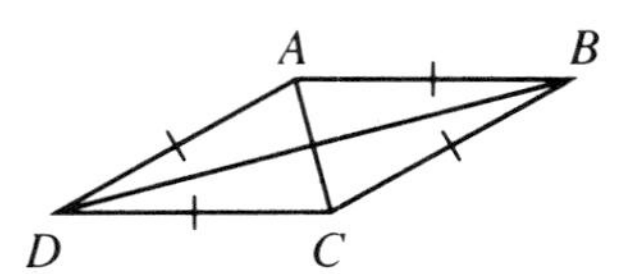

26. **a.** 16 **b.** Guess: 32, Actual Count: 31

27. **a.** If both pairs of opp. sides of a quad. are $\parallel$, then opp. $\angle$s are $\cong$.

b. If both pairs of opp. $\angle$s of a quad. are $\cong$, then opp. sides are $\parallel$.

Given: $ABCD$ is a quad.; $m\angle A = m\angle C; m\angle B = m\angle D$

Prove: $\overline{AD} \parallel \overline{BC}; \overline{AB} \parallel \overline{CD}$

Statements	Reasons
1. $m\angle A + m\angle B + m\angle C + m\angle D = 360$	1. The sum of the meas. of the int. $\angle$s of a quad. is 360.
2. $m\angle A = m\angle C$; $m\angle B = m\angle D$	2. Given
3. $2m\angle A + 2m\angle B = 360$; $2m\angle C + 2m\angle B = 360$	3. Substitution Prop.
4. $m\angle A + m\angle B = 180$; $m\angle C + m\angle B = 180$	4. Div. Prop. of =
5. $\angle A$ and $\angle B$ are supp.; $\angle B$ and $\angle C$ are supp.	5. Def. of supp. $\angle$s
6. $\overline{AD} \parallel \overline{BC}$; $\overline{AB} \parallel \overline{CD}$	6. If 2 lines are cut by a trans. and s-s. int. $\angle$s are supp., then the lines are $\parallel$.

c. Both pairs of opp. $\angle$s of a quad. are $\cong$ if and only if opp. sides are $\parallel$.

C 28. a. 13, 17, 23, 31, 41, 53, 67, 83, 101 **b.** Guess: a prime number **c.** 121, 143, neither of which is prime

29.

No. of sides	3	4	5	6	7	8	n
No. of diagonals	0	2	5	9	14	20	$\frac{n(n-3)}{2}$

30. a. There are 5 small $\triangle$s each with one of the points A, B, C, D, E as one vertex. The other two $\angle$s of each of the $\triangle$s are ext. $\angle$s of a pentagon. There are two complete sets of ext. $\angle$s of the pentagon, with each set having total meas. 360. Then $m\angle A + m\angle B + m\angle C + m\angle D + m\angle E + 360 + 360 = 5(180) = 900$ and $m\angle A + m\angle B + m\angle C + m\angle D + m\angle E = 180$. **b.** Using the same reasoning as in part (a), $m\angle A + m\angle B + m\angle C + m\angle D + m\angle E + m\angle F + 360 + 360 = 6(180) = 1080$ and $m\angle A + m\angle B + m\angle C + m\angle D + m\angle E + m\angle F = 360$. **c.** For each additional point of a star, the sum of the meas. of the $\angle$s increases by 180. The sum of the $\angle$ meas. for an n-pointed star is $180(n - 4)$. **d.** If a star has n points, $m\angle A + m\angle B + m\angle C + \cdots + m\angle N + 360 + 360 = n(180)$ and $m\angle A + m\angle B + m\angle C + \cdots + m\angle N = n(180) - 720 = 180(n - 4)$.

Page 109 • CALCULATOR KEY-IN

1. 1; 121; 12321; $1111 \times 1111 = 1234321$
2. 42; 4422; 444222; $6666 \times 6667 = 44442222$
3. 64; 9604; 996004; $9998 \times 9998 = 99960004$
4. 63; 7623; 776223; $7777 \times 9999 = 77762223$

Page 110 • SELF-TEST 2

1. acute **2.** scalene **3.** 60 **4.** 105, 35 **5.** $(2x + 4) + (3x - 9) = 90; x = 19$
6. $y = 50; 110 = z + 50, z = 60$
7. $2x + 5 = 3x + 10, x = -5$ (reject); $2x + 5 = x + 12, x = 7;$
$3x + 10 = x + 12, x = 1$

8. 8 **9.** equilateral, equiangular **10.** $360; \frac{(10 - 2)180}{10} = 144$

11. $180 - 174 = 6; \frac{360}{6} = 60$ **12.** 32 **13.** 32 **14.** 36 **15.** 16

Pages 111–112 • CHAPTER REVIEW

1. 2 **2.** corr. **3.** alt. int. **4.** No; they can be skew. **5.** 105, 105
6. $70 = 6x - 2; x = 12$ **7.** $(8y - 40) + (2y + 20) = 180; y = 20$
8. $b \perp c$; if a trans. is $\perp$ to one of 2 $\parallel$ lines, then it is $\perp$ to the other one also.
9. $\overleftrightarrow{DE}$; $\angle A$ is supp. to $\angle ADE$, and if 2 lines are cut by a trans. and s-s. int. $\angle$s are supp., then the lines are $\parallel$.
10. $\overleftrightarrow{BE} \parallel \overleftrightarrow{CF}$; both are $\perp$ to $\overleftrightarrow{DF}$.
11. corr. $\angle$s $\cong$; alt. int. $\angle$s $\cong$; s-s. int. $\angle$s supp.; in a plane, both lines are $\perp$ to a third line; both lines are $\parallel$ to a third line.
12. $x + (2x - 15) = 90; x = 35$ **13.** 180 **14.** 100
15. $\angle 3 \cong \angle 6$ (If 2 $\angle$s are supps. of $\cong$ $\angle$s, then the 2 $\angle$s are $\cong$.), $\angle 2 \cong \angle 8$ (If 2 $\angle$s of one $\triangle$ are $\cong$ to 2 $\angle$s of another $\triangle$, then the third $\angle$s are $\cong$.)

16. a.

b. $(6 - 2)180 = 720$
c. 360

17. $\frac{(18 - 2)180}{18} = 160$ **18.** $\frac{360}{24} = 15$ **19.** $\frac{(n - 2)180}{n} = 150; n = 12$

20. 75, 90 **21.** $\frac{1}{100}, -\frac{1}{1000}$

Pages 112–113 • CHAPTER TEST

1. sometimes **2.** sometimes **3.** never **4.** never **5.** never **6.** always
7. $(3x - 20) + x = 180; x = 50$ **8.** $2x + 12 = 4(x - 7); x = 20$
9. $m\angle 1 = m\angle 2 = 60, m\angle 3 = 120$
10. $m\angle 1 = 58, m\angle 2 = 90, m\angle 3 = 32, m\angle 4 = 180 - (32 + 35) = 113, m\angle 5 = 35,$
$m\angle 6 = 55$

11. $m\angle 4 = \frac{(5 - 2)180}{5} = 108$, $m\angle 5 = 108 - 72 = 36$, $m\angle 1 = 180 - 108 = 72$, $m\angle 2 = 180 - 108 = 72$, $m\angle 3 = 180 - (72 + 72) = 36$

12. $\angle EBC \cong \angle 2$ (If 2 lines are cut by a trans. and alt. int. ⦣s are ≅, then the lines are ∥.), or $\angle 5 \cong \angle 3$ (If 2 lines are cut by a trans. and corr. ⦣s are ≅, then the lines are ∥.)

13.

Statements	Reasons
1. $\overrightarrow{BF}$ bisects $\angle ABE$; $\overrightarrow{DG}$ bisects $\angle CDB$.	1. Given
2. $m\angle GDB = \frac{1}{2}m\angle CDB$; $m\angle FBE = \frac{1}{2}m\angle ABE$	2. ∠ Bis. Thm.
3. $\overleftrightarrow{AB} \parallel \overleftrightarrow{CD}$	3. Given
4. $m\angle CDB = m\angle ABE$	4. If 2 ∥ lines are cut by a trans., then corr. ⦣s are ≅.
5. $\frac{1}{2}m\angle CDB = \frac{1}{2}m\angle ABE$	5. Div. Prop. of =
6. $m\angle GDB = m\angle FBE$	6. Substitution Prop.
7. $\overleftrightarrow{BF} \parallel \overleftrightarrow{DG}$	7. If 2 lines are cut by a trans. and corr. ⦣s are ≅, then the lines are ∥.

14. 15, 17

Page 113 • ALGEBRA REVIEW

1. 3 **2.** 2 **3.** (0, 0) **4.** *Z* **5.** (3, 5) **6.** (4, 3) **7.** (4, 0) **8.** (0, 4) **9.** (−5, 0) **10.** (−4, 3) **11.** (−2, 2) **12.** (−4, −2) **13.** (−2, −3) **14.** (3, −2) **15.** *K, O, S* **16.** *O, R, Z* **17.** 3 **18.** c, e **19.** *M, N, P* **20.** *T, U* **21.** *V, W* **22.** *J, Q*

23–34.

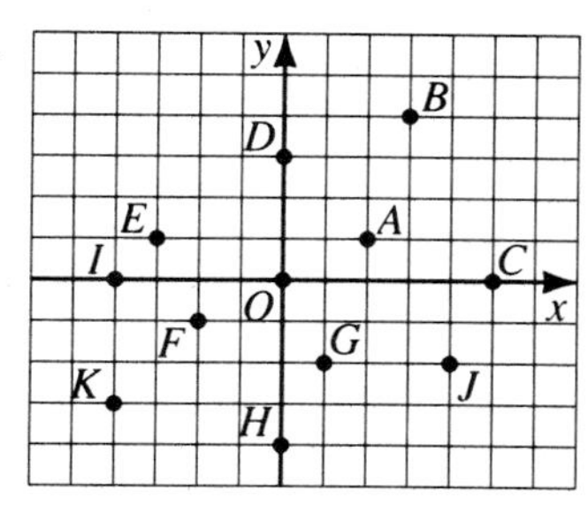

35. (2, 1) **36.** (2, 5) **37.** (0, 3) **38.** (−3, 0) **39.** (−4, −2) **40.** (1, −1)

Pages 114–115 • CUMULATIVE REVIEW: CHAPTERS 1–3

A **1.** sometimes **2.** always **3.** sometimes **4.** always **5.** always

6–8. Sketches may vary.

6.

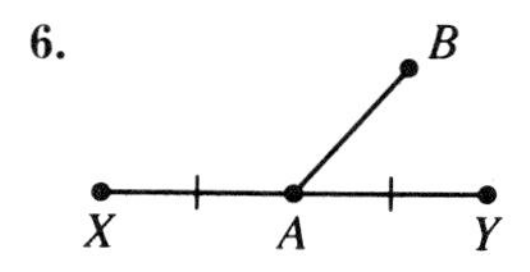

7.

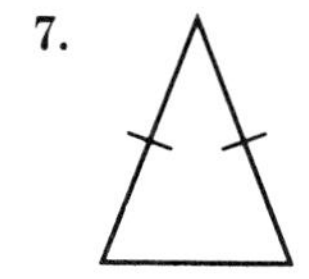

8.

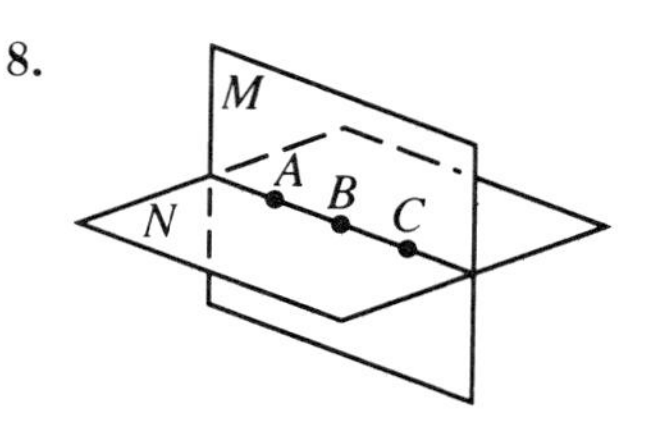

9. not possible **10.** $\frac{-3.5 + 8.5}{2} = \frac{5}{2}$ or 2.5

11. $5x + 13 = 9x - 39$, $x = 13$; $m\angle PQR = 2m\angle PQX = 2[5(13) + 13] = 156$

12. $180 - x = 2(90 - x) + 35$, $x = 35$; $\angle$ measure: 35, supp. measure: 145; comp. measure: 55

13. $x + 5x + 6x = 180$, $x = 15$; $\angle$ measures: 15, 75, 90 **14.** $2(60) = 120$ **15.** 90

16. $90 - 60 = 30$ **17.** 90 **18.** 60 **19.** $180 - 2(60) = 60$

20. $180 - 60 = 120$ **21.** $180 - 120 = 60$ **22.** $180 - 2(60) = 60$

23. False. If 2 lines are $\parallel$, then they do not intersect; true.

24. True. If 2 lines are $\perp$, then they intersect to form rt. $\angle$s; true.

25. True. If an $\angle$ is not obtuse, then it is acute; false.

26. True. If a $\triangle$ is isos., then it is equilateral; false.

27. Vert. $\angle$s are $\cong$. **28.** Seg. Add. Post. **29.** $\angle$ Add. Post.

30. If 2 lines are cut by a trans. and alt. int. $\angle$s are $\cong$, then the lines are $\parallel$.

31. The meas. of an ext. $\angle$ of a $\triangle$ = the sum of the meas. of the 2 remote int. $\angle$s.

32. Def. of $\perp$ lines **33.** The sum of the meas. of the $\angle$s of a $\triangle$ is 180.

34. Def. of $\perp$ lines **35.** X **36.** supp. **37.** $\frac{(n - 2)180}{n} = 108$, $n = 5$; pentagon

38. $2(12) = 24$ **39.** $\cong$ **40.** inductive **41.** biconditional **42.** 360

43. $(8 - 2)180 = 1080$ **44.** acute

B **45.**

Statements	Reasons
1. $\overline{WX} \perp \overline{XY}$	1. Given
2. $\angle 1$ is comp. to $\angle 2$.	2. If the ext. sides of 2 adj. acute $\angle$s are $\perp$, then the $\angle$s are comp.
3. $\angle 1$ is comp. to $\angle 3$.	3. Given
4. $\angle 2 \cong \angle 3$	4. If 2 $\angle$s are comps. of the same $\angle$, then the 2 $\angle$s are $\cong$.

46.

Statements	Reasons
1. $\overline{RU} \parallel \overline{ST}$	1. Given
2. $\angle 1 \cong \angle 2$	2. If 2 $\parallel$ lines are cut by a trans., then alt. int. $\angle$s are $\cong$.
3. $\angle R \cong \angle T$	3. Given
4. $\angle 3 \cong \angle 4$	4. If 2 $\angle$s of one $\triangle$ are $\cong$ to 2 $\angle$s of another $\triangle$, then the third $\angle$s are $\cong$.
5. $\overline{RS} \parallel \overline{UT}$	5. If 2 lines are cut by a trans. and alt. int. $\angle$s are $\cong$, then the lines are $\parallel$.

CHAPTER 4 • Congruent Triangles

Page 119 • CLASSROOM EXERCISES

1. $\overline{FI}, \overline{WE}; \overline{IN}, \overline{EB}; \overline{FN}, \overline{WB}$ **2.** $\angle F, \angle W; \angle I, \angle E; \angle N, \angle B$ **3.** Yes **4.** No

5. $\triangle CDO$ **6.** $\angle C$ **7.** $\overline{CO}$ **8.** DO

9. Yes; O is the midpt. of $\overline{AC}$ and of $\overline{DB}$ because $AO = OC$ and $DO = OB$.

10. If 2 lines are cut by a trans. and alt. int. ∠s are ≅, then the lines are ∥.

11. If $\overline{DB} \perp \overline{DC}$, then $m\angle D = 90$. But $m\angle D = m\angle B$, so $m\angle B = 90$. Therefore, $\overline{DB} \perp \overline{BA}$.

12. R **13.** $ROHES$ **14.** $m\angle C$ **15.** 4 **16.** $\angle A, \angle H$

17. The leaf in the lower left-hand corner is flipped over.

18. **a.** $A(1, 2), B(4, 2), C(2, 4)$ **b.** $D(2, 0)$ **19.** $G(6, 5)$ or $G(6, 1)$

20. $H(7, 5)$ or $H(7, 1)$

Pages 120–121 • WRITTEN EXERCISES

A **1.** $\angle T$ **2.** $m\angle I$ **3.** CA **4.** $\overline{IG}$ **5.** $\triangle ATC$ **6.** $\triangle BGI$

7. $\angle E, \angle F, \angle S, \angle T$ **8.** Def. (of ≅ △s)

9. $\angle L \cong \angle F, \angle X \cong \angle N, \angle R \cong \angle E, \overline{LX} \cong \overline{FN}, \overline{XR} \cong \overline{NE}, \overline{LR} \cong \overline{FE}$

10. **a.** $\triangle KRO$ **b.** $\angle K$, Corr. parts of ≅ △s are ≅. **c.** $\overline{KO}$, Corr. parts of ≅ △s are ≅; $\overline{SK}$ **d.** $\angle R$, Corr. parts of ≅ △s are ≅; If 2 lines are cut by a trans. and alt. int. ∠s are ≅, then the lines are ∥.

11. **a.** $\triangle RLA$ **b.** $\overline{RL}$ **c.** $\angle 3$, Corr. parts of ≅ △s are ≅; $\overline{LR}$, If 2 lines are cut by a trans. and alt. int. ∠s are ≅, then the lines are ∥. **d.** $\angle 4$, Corr. parts of ≅ △s are ≅; $\overline{PL}, \overline{AR}$, If 2 lines are cut by a trans. and alt. int. ∠s are ≅, then the lines are ∥.

12. $C(7, 2)$

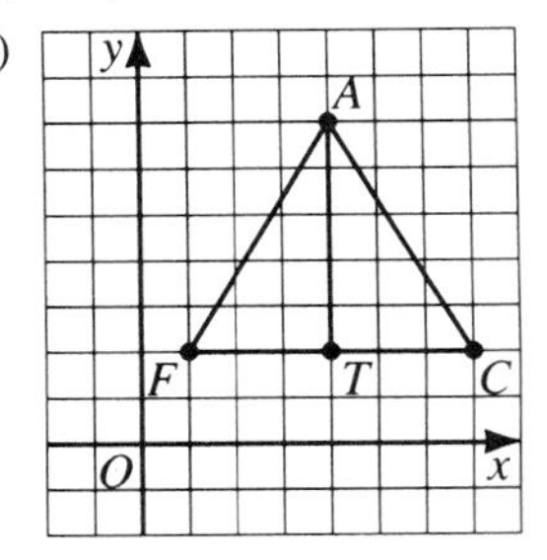

13. $C(7, -1)$

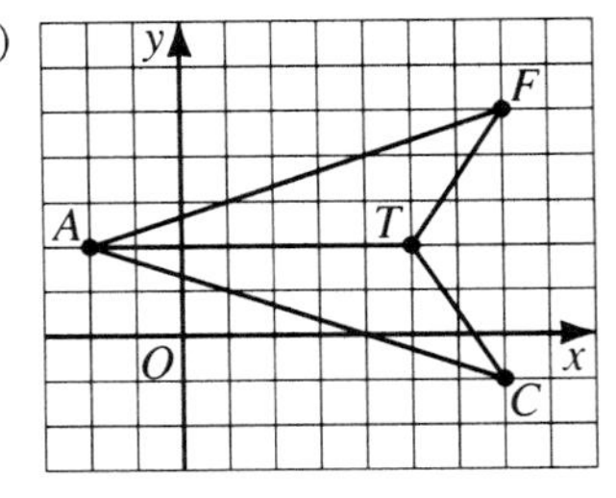

B **14.** $\triangle ABC \cong \triangle FDE$

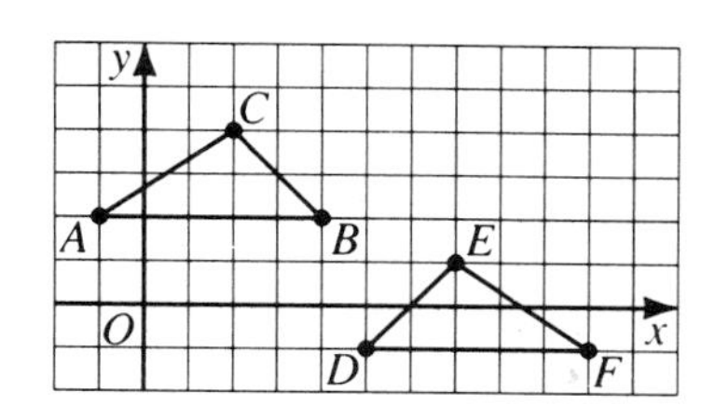

15. $\triangle ABC \cong \triangle EDF$

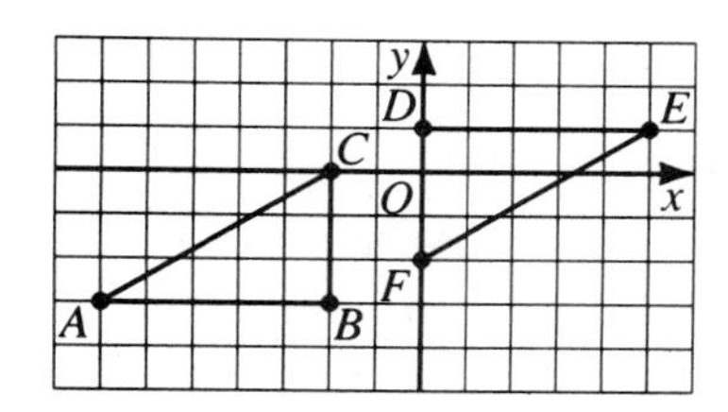

16. $\triangle ABC \cong \triangle FED$

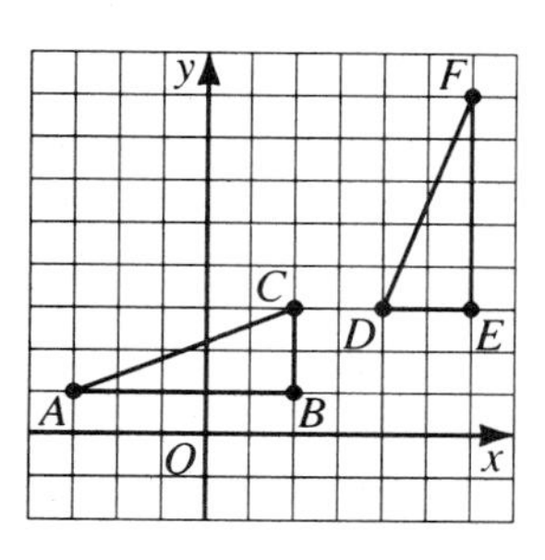

17. $\triangle ABC \cong \triangle FDE$

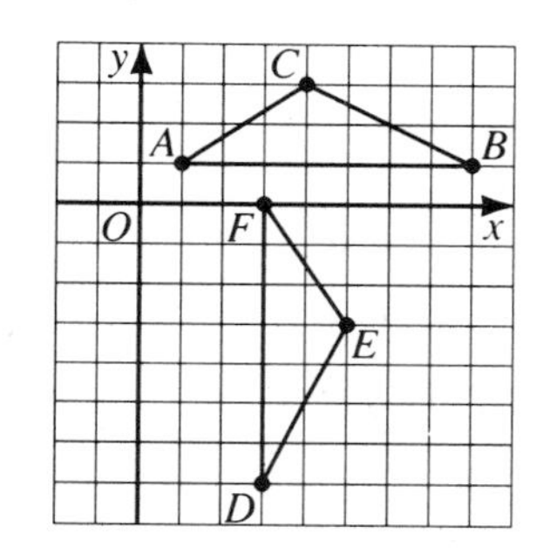

18. $F(4, 5)$ and $F(8, 5)$

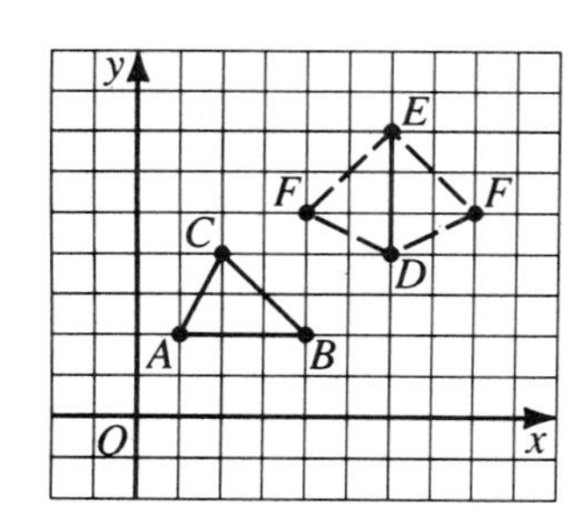

19. $F(2, 5)$ and $F(6, 1)$

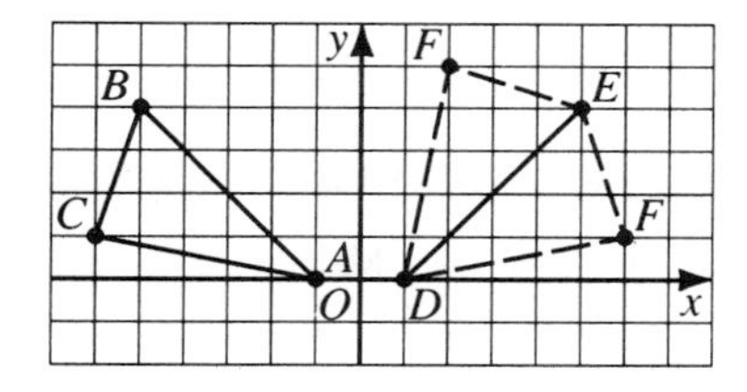

20. $MARO$

21. **a.** Since $NERO \cong MARO$, $\overline{NO} \cong \overline{OM}$. By the def. of midpt., O is the midpt. of $\overline{NM}$. **b.** $\angle NOR$ and $\angle MOR$ are corr. ∠s of ≅ quads. **c.** If 2 lines form ≅ adj. ∠s, then the lines are ⊥.

22. Check students' drawings. **a.** Yes **b.** Yes **c.** No **d.** Yes

23. Yes; yes; yes

C **24.** 24; for each vertex of the given △, there are 8 △s with that vertex at P.

Page 121 • CHALLENGE

a.

b.

Page 121 • MIXED REVIEW EXERCISES

1.

Statements	Reasons
1. $\overline{AD} \perp \overline{BC}$; $\overline{BA} \perp \overline{AC}$	1. Given
2. $\angle BDA$ and $\angle BAC$ are rt. $\angle$s.	2. Def. of $\perp$ lines
3. $\triangle ABC$ and $\triangle DBA$ are rt. $\triangle$s.	3. Def. of rt. $\triangle$
4. $\angle 1$ and $\angle B$ are comp.; $\angle 2$ and $\angle B$ are comp.	4. The acute $\angle$s of a rt. $\triangle$ are comp.
5. $\angle 1 \cong \angle 2$	5. If 2 $\angle$s are comps. of the same $\angle$, then the 2 $\angle$s are $\cong$.

2.

Statements	Reasons
1. $\overline{FC}$ and $\overline{SH}$ bis. each other at A.	1. Given
2. A is the midpt. of $\overline{FC}$ and $\overline{SH}$.	2. Def. of bis.
3. $SA = \frac{1}{2}SH$ and $AC = \frac{1}{2}FC$	3. Midpt. Thm.
4. $FC = SH$	4. Given
5. $\frac{1}{2}FC = \frac{1}{2}SH$	5. Mult. Prop. of =
6. $SA = AC$	6. Substitution Prop.

Pages 123–124 • CLASSROOM EXERCISES

1. Yes **2.** Yes **3.** No **4.** Yes; ASA **5.** Yes; SSS **6.** Yes; SAS **7.** No **8.** No **9.** No **10.** Use vert. $\angle$s and alt. int. $\angle$s to prove the $\triangle$s $\cong$ by ASA. **11. a.** $\overline{TR}$, $\overline{TR}$; $\overline{YT}$, $\overline{XT}$; $\angle R$, $\angle R$ **b.** No

Pages 124–127 • WRITTEN EXERCISES

A **1.** $\triangle ABC \cong \triangle NPY$; ASA **2.** $\triangle ABC \cong \triangle ADC$; SAS **3.** $\triangle ABC \cong \triangle CKA$; SSS **4.** $\triangle ABC \cong \triangle SBC$; SAS **5.** No $\cong$ can be deduced. **6.** No $\cong$ can be deduced. **7.** $\triangle ABC \cong \triangle PQC$; SAS **8.** No $\cong$ can be deduced. **9.** $\triangle ABC \cong \triangle AGC$; ASA **10.** $\triangle ABC \cong CDA$; ASA **11.** $\triangle ABC \cong \triangle BST$; ASA **12.** No $\cong$ can be deduced. **13.** No $\cong$ can be deduced. **14.** $\triangle ABC \cong \triangle CGA$; SAS **15.** $\triangle ABC \cong \triangle MNC$; ASA **16.** 1. Given 2. Refl. Prop. 3. Given 4. If 2 $\parallel$ lines are cut by a trans., then alt. int. $\angle$s are $\cong$. 5. SAS Post.

17. 1. Given 2. T; Def. of $\perp$ lines 3. Def. of $\cong$ ∠s 4. Given 5. $\overline{VT}$; Def. of midpt. 6. UVT; Vert. ∠s are $\cong$. 7. RSV, UTV; ASA Post.

B 18.

Statements	Reasons
1. $\overline{TM} \parallel \overline{RP}$	1. Given
2. $\angle T \cong \angle P$; $\angle M \cong \angle R$	2. If 2 $\parallel$ lines are cut by a trans., then alt. int. ∠s are $\cong$.
3. $\overline{TM} \cong \overline{PR}$	3. Given
4. $\triangle TEM \cong \triangle PER$	4. ASA Post.

19.

Statements	Reasons
1. E is the midpt. of $\overline{TP}$ and $\overline{MR}$.	1. Given
2. $\overline{TE} \cong \overline{PE}$; $\overline{ME} \cong \overline{RE}$	2. Def. of midpt.
3. $\angle TEM \cong \angle PER$	3. Vert. ∠s are $\cong$.
4. $\triangle TEM \cong \triangle PER$	4. SAS Post.

20.

Statements	Reasons
1. Plane M bis. $\overline{AB}$.	1. Given
2. $\overline{AO} \cong \overline{BO}$	2. Def. of bis.
3. $\overline{PA} \cong \overline{PB}$	3. Given
4. $\overline{PO} \cong \overline{PO}$	4. Refl. Prop.
5. $\triangle POA \cong \triangle POB$	5. SSS Post.

21.

Statements	Reasons
1. Plane M bis. $\overline{AB}$.	1. Given
2. $\overline{AO} \cong \overline{BO}$	2. Def. of bis.
3. $\overline{PO} \perp \overline{AB}$	3. Given
4. $\angle POA \cong \angle POB$	4. If 2 lines are $\perp$, then they form $\cong$ adj. ∠s.
5. $\overline{PO} \cong \overline{PO}$	5. Refl. Prop.
6. $\triangle POA \cong \triangle POB$	6. SAS Post.

22. Given: Isos. $\triangle ABC$ with $\overline{AC} \cong \overline{AB}$;
$\overline{AD}$ bisects $\angle CAB$.
Prove: $\triangle ACD \cong \triangle ABD$

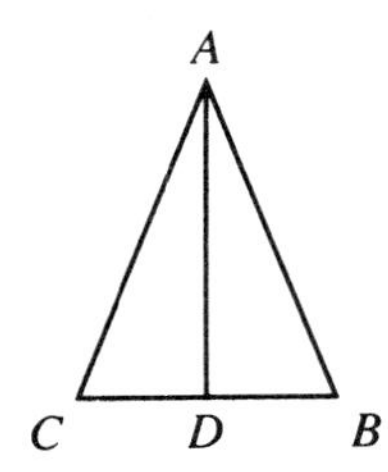

Statements	Reasons
1. $\overline{AC} \cong \overline{AB}$; $\overline{AD}$ bisects $\angle CAB$.	1. Given
2. $\angle CAD \cong \angle BAD$	2. Def. of $\angle$ bis.
3. $\overline{AD} \cong \overline{AD}$	3. Refl. Prop.
4. $\triangle ACD \cong \triangle ABD$	4. SAS Post.

23. Given: Isos. $\triangle ABC$ with $\overline{AC} \cong \overline{AB}$;
D is the midpt. of $\overline{CB}$.
Prove: $\triangle ACD \cong \triangle ABD$

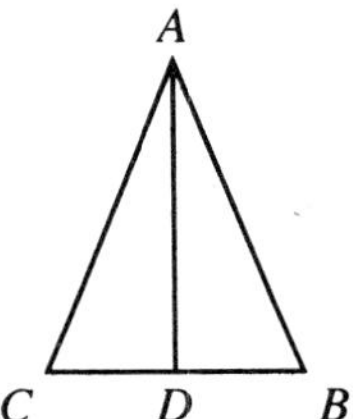

Statements	Reasons
1. $\overline{AC} \cong \overline{AB}$; D is the midpt. of $\overline{CB}$.	1. Given
2. $\overline{CD} \cong \overline{BD}$	2. Def. of midpt.
3. $\overline{AD} \cong \overline{AD}$	3. Refl. Prop.
4. $\triangle ACD \cong \triangle ABD$	4. SSS Post.

24. Given: $l \perp \overline{AB}$ at M;
M is the midpt. of $\overline{AB}$.
Prove: $\triangle PMA \cong \triangle PMB$

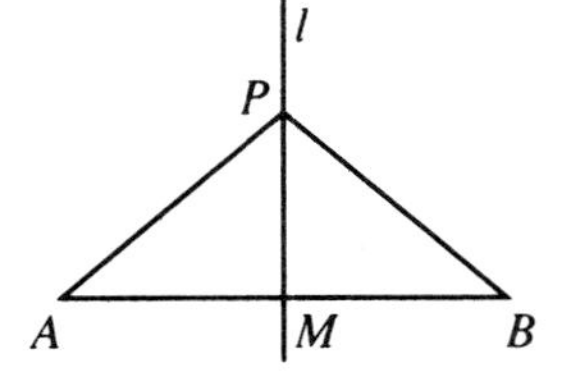

Statements	Reasons
1. $l \perp \overline{AB}$ at M	1. Given
2. $\angle PMA \cong \angle PMB$	2. If 2 lines are $\perp$, then they form $\cong$ adj. $\angle$s.
3. M is the midpt. of $\overline{AB}$.	3. Given
4. $\overline{AM} \cong \overline{BM}$	4. Def. of midpt.
5. $\overline{PM} \cong \overline{PM}$	5. Refl. Prop.
6. $\triangle PMA \cong \triangle PMB$	6. SAS Post.

25. Given: $ABCDE$ is equilateral;
$\angle B$ and $\angle E$ are rt. $\angle$s.
Prove: $\triangle ABC \cong \triangle AED$

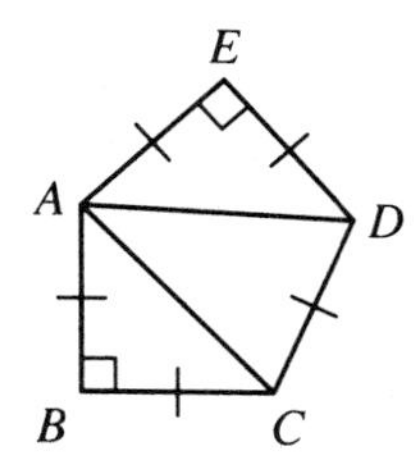

Statements	Reasons
1. $ABCDE$ is equilateral.	1. Given
2. $\overline{AB} \cong \overline{AE} \cong \overline{BC} \cong \overline{ED}$	2. Def. of equilateral
3. $\angle B$ and $\angle E$ are rt. $\angle$s.	3. Given
4. $m\angle B = 90$; $m\angle E = 90$	4. Def. of rt. $\angle$
5. $\angle B \cong \angle E$	5. Def. of $\cong$ $\angle$s
6. $\triangle ABC \cong \triangle AED$	6. SAS Post.

C 26. SAS ($\overline{AB} \cong \overline{BC}$, $\angle ABF \cong \angle BCG$, $\overline{BF} \cong \overline{CG}$)

27. SSS ($\overline{AB} \cong \overline{BC}$, $\overline{VA} \cong \overline{VB}$, $\overline{VB} \cong \overline{VC}$)

Page 129 • CLASSROOM EXERCISES

1. Prove $\triangle PQR \cong \triangle PSR$ by SAS, so corr. parts $\angle Q$ and $\angle S$ are $\cong$.
2. Prove $\triangle PQR \cong \triangle PSR$ by ASA, so corr. parts $\overline{RQ}$ and $\overline{RS}$ are $\cong$.
3. Prove $\triangle ZWX \cong \triangle XYZ$ by SSS, so corr. parts $\angle 1$ and $\angle 2$ are $\cong$.
   $\overline{WX} \parallel \overline{ZY}$ because alt. int. $\angle$s are $\cong$.
4. Prove $\triangle ZWX \cong \triangle XYZ$ by SAS, so corr. parts $\angle 1$ and $\angle 2$ are $\cong$.
   $\overline{ZY} \parallel \overline{WX}$ because alt. int. $\angle$s are $\cong$.
5. Prove $\triangle CAD \cong \triangle CBD$ by SAS, so corr. parts $\overline{CA}$ and $\overline{CB}$ are $\cong$.
6. $\overline{AP} \cong \overline{BP}$; Prove $\triangle APM \cong \triangle BPM$ by SAS, so corr. parts $\overline{AP}$ and $\overline{BP}$ are $\cong$.

Pages 130–132 • WRITTEN EXERCISES

A 1. 1. Given 2. Given 3. Def. of midpt. 4. Vert. $\angle$s are $\cong$. 5. ASA Post.
6. Corr. parts of $\cong$ $\triangle$s are $\cong$. 7. Def. of midpt.

2. Answers may vary; a, e, c, f, b, d

3.

Statements	Reasons
1. $\overline{WO} \cong \overline{ZO}$; $\overline{XO} \cong \overline{YO}$	1. Given
2. $\angle WOX \cong \angle ZOY$	2. Vert. $\angle$s are $\cong$.
3. $\triangle WOX \cong \triangle ZOY$	3. SAS Post.
4. $\angle W \cong \angle Z$	4. Corr. parts of $\cong$ $\triangle$s are $\cong$.

4.

Statements	Reasons
1. M is the midpt. of $\overline{AB}$.	1. Given
2. $\overline{AM} \cong \overline{BM}$	2. Def. of midpt.
3. $\angle 1 \cong \angle 2$; $\angle 3 \cong \angle 4$	3. Given
4. $\triangle AMC \cong \triangle BMD$	4. ASA Post.
5. $\overline{AC} \cong \overline{BD}$	5. Corr. parts of $\cong$ ⚠s are $\cong$.

5.

Statements	Reasons
1. $\overline{SK} \parallel \overline{NR}$; $\overline{SN} \parallel \overline{KR}$	1. Given
2. $\angle 1 \cong \angle 3$; $\angle 2 \cong \angle 4$	2. If 2 $\parallel$ lines are cut by a trans., then alt. int. ∠s are $\cong$.
3. $\overline{SR} \cong \overline{SR}$	3. Refl. Prop.
4. $\triangle SKR \cong \triangle RNS$	4. ASA Post.
5. $\overline{SK} \cong \overline{NR}$; $\overline{SN} \cong \overline{KR}$	5. Corr. parts of $\cong$ ⚠s are $\cong$.

6.

Statements	Reasons
1. $\overline{SK} \cong \overline{NR}$; $\overline{SN} \cong \overline{KR}$	1. Given
2. $\overline{SR} \cong \overline{SR}$	2. Refl. Prop.
3. $\triangle SNR \cong \triangle RKS$	3. SSS Post.
4. $\angle 1 \cong \angle 3$; $\angle 2 \cong \angle 4$	4. Corr. parts of $\cong$ ⚠s are $\cong$.
5. $\overline{SK} \parallel \overline{NR}$; $\overline{SN} \parallel \overline{KR}$	5. If 2 lines are cut by a trans. and alt. int. ∠s are $\cong$, then the lines are $\parallel$.

7.

Statements	Reasons
1. $\overline{AD} \parallel \overline{ME}$; $\overline{MD} \parallel \overline{BE}$	1. Given
2. $\angle A \cong \angle EMB$; $\angle DMA \cong \angle B$	2. If 2 $\parallel$ lines are cut by a trans., then corr. ∠s are $\cong$.
3. M is the midpt. of $\overline{AB}$.	3. Given
4. $\overline{AM} \cong \overline{MB}$	4. Def. of midpt.
5. $\triangle ADM \cong \triangle MEB$	5. ASA Post.
6. $\overline{MD} \cong \overline{BE}$	6. Corr. parts of $\cong$ ⚠s are $\cong$.

B 8.

Statements	Reasons
1. $\overline{AD} \parallel \overline{ME}$	1. Given
2. $\angle A \cong \angle EMB$	2. If 2 $\parallel$ lines are cut by a trans., then corr. △s are $\cong$.
3. M is the midpt. of $\overline{AB}$.	3. Given
4. $\overline{AM} \cong \overline{MB}$	4. Def. of midpt.
5. $\overline{AD} \cong \overline{ME}$	5. Given
6. $\triangle ADM \cong \triangle MEB$	6. SAS Post.
7. $\angle DMA \cong \angle B$	7. Corr. parts of $\cong$ △ are $\cong$.
8. $\overline{MD} \parallel \overline{BE}$	8. If 2 lines are cut by a trans. and corr. △s are $\cong$, then the lines are $\parallel$.

9. Either (a) $\angle 1 \cong \angle 2$ or (b) $\overline{QR} \cong \overline{SR}$ can be omitted.

a.

Statements	Reasons
1. $\overline{PQ} \cong \overline{PS}$; $\overline{QR} \cong \overline{SR}$	1. Given
2. $\overline{PR} \cong \overline{PR}$	2. Refl. Prop.
3. $\triangle PQR \cong \triangle PSR$	3. SSS Post.
4. $\angle 3 \cong \angle 4$	4. Corr. parts of $\cong$ △ are $\cong$.

b.

Statements	Reasons
1. $\overline{PQ} \cong \overline{PS}$; $\angle 1 \cong \angle 2$	1. Given
2. $\overline{PR} \cong \overline{PR}$	2. Refl. Prop.
3. $\triangle PQR \cong \triangle PSR$	3. SAS Post.
4. $\angle 3 \cong \angle 4$	4. Corr. parts of $\cong$ △ are $\cong$.

10. Either (a) $\overrightarrow{KO}$ bisects $\angle MKN$ or (b) $\overline{LM} \cong \overline{LN}$ can be omitted.

a.

Statements	Reasons
1. $\overline{LM} \cong \overline{LN}$; $\overline{KM} \cong \overline{KN}$	1. Given
2. $\overline{LK} \cong \overline{LK}$	2. Refl. Prop.
3. $\triangle LKM \cong \triangle LKN$	3. SSS Post.
4. $\angle MLK \cong \angle NLK$	4. Corr. parts of $\cong$ △ are $\cong$.
5. $\overrightarrow{LO}$ bis. $\angle MLN$.	5. Def. of $\angle$ bis.

b.

Statements	Reasons
1. $\overline{KM} \cong \overline{KN}$; $\overrightarrow{KO}$ bis. $\angle MKN$.	1. Given
2. $\angle MKO \cong \angle NKO$	2. Def. of $\angle$ bis.
3. $m\angle LKM + m\angle MKO = 180$; $m\angle LKN + m\angle NKO = 180$	3. $\angle$ Add. Post.
4. $\angle LKM$ and $\angle MKO$ are supp.; $\angle LKN$ and $\angle NKO$ are supp.	4. Def. of supp. $\angle$s
5. $\angle LKM \cong \angle LKN$	5. If 2 $\angle$s are supps. of $\cong$ $\angle$s, then the 2 $\angle$s are $\cong$.
6. $\overline{LK} \cong \overline{LK}$	6. Refl. Prop.
7. $\triangle LKM \cong \triangle LKN$	7. SAS Post.
8. $\angle MLK \cong \angle NLK$	8. Corr. parts of $\cong$ $\triangle$s are $\cong$.
9. $\overrightarrow{LO}$ bis. $\angle MLN$.	9. Def. of $\angle$ bis.

11. 1, 2

12. (a) $\angle 3 \cong \angle 4$ and (b) $\angle 1 \cong \angle 2$

a.

Statements	Reasons
1. $\overline{WX} \perp \overline{UV}$	1. Given
2. $\angle 5 \cong \angle 6$	2. If 2 lines are $\perp$, then they form $\cong$ adj. $\angle$s.
3. $\overline{WU} \cong \overline{WV}$	3. Given
4. $\overline{WX} \cong \overline{WX}$	4. Refl. Prop.
5. $\triangle XWU \cong \triangle XWV$	5. SAS Post.
6. $\angle 3 \cong \angle 4$	6. Corr. parts of $\cong$ $\triangle$s are $\cong$.

b.

Statements	Reasons
1. $\angle 3 \cong \angle 4$	1. Part (a) above
2. $\overline{WX} \perp \overline{YZ}$	2. Given
3. $\angle 1$ and $\angle 3$ are comp.; $\angle 2$ and $\angle 4$ are comp.	3. If the ext. sides of 2 adj. acute $\angle$s are $\perp$, then the $\angle$s are comp.
4. $\angle 1 \cong \angle 2$	4. If 2 $\angle$s are comps. of $\cong$ $\angle$s, then the 2 $\angle$s are $\cong$.

13.

Statements	Reasons
1. $\overline{RS} \perp$ plane Y	1. Given
2. $\overline{RS} \perp \overline{ST}$; $\overline{RS} \perp \overline{SV}$	2. Def. of a line $\perp$ to a plane
3. $m\angle RST = 90$; $m\angle RSV = 90$	3. Def. of $\perp$ lines
4. $\angle RST \cong \angle RSV$	4. Def. of $\cong$ &angles;
5. $\angle TRS \cong \angle VRS$	5. Given
6. $\overline{RS} \cong \overline{RS}$	6. Refl. Prop.
7. $\triangle RST \cong \triangle RSV$	7. ASA Post.
8. $\overline{RT} \cong \overline{RV}$	8. Corr. parts of $\cong$ △s are $\cong$.
9. $\triangle RTV$ is isos.	9. Def. of isos. $\triangle$

14.

Statements	Reasons
1. $\overline{PA}$ and $\overline{QB}$ are $\perp$ to plane X.	1. Given
2. $\overline{PA} \perp \overline{AB}$; $\overline{QB} \perp \overline{AB}$	2. Def. of a line $\perp$ to a plane
3. $m\angle A = 90$; $m\angle B = 90$	3. Def. of $\perp$ lines
4. $\angle A \cong \angle B$	4. Def. of $\cong$ &angles;
5. O is the midpt. of $\overline{AB}$.	5. Given
6. $\overline{AO} \cong \overline{BO}$	6. Def. of midpt.
7. $\angle POA \cong \angle QOB$	7. Vert. &angles; are $\cong$.
8. $\triangle POA \cong \triangle QOB$	8. ASA Post.
9. $\overline{PO} \cong \overline{QO}$	9. Corr. parts of $\cong$ △s are $\cong$.
10. O is the midpt. of $\overline{PQ}$.	10. Def. of midpt.

15. The wires are of equal length, so $PA = PB = PC$. The stakes are equidistant from the base of the tree, so $TA = TB = TC$. $PT = PT = PT$ by the Refl. Prop. and $\triangle PTA \cong \triangle PTB \cong \triangle PTC$ by SSS. The &angles; that the 3 wires make with the ground are corr. parts of $\cong$ △s.

C 16. ASA. A is the soldier's eye, C his feet, D the point on the opposite bank in line with the tip of his visor, and B is the point on the ground in line with the tip of his visor. $\overline{AC} \cong \overline{AC}$ and $\angle BCA$ and $\angle ACD$ are both rt. &angles;. By keeping his visor in the same position, the soldier makes $\angle BAC \cong \angle DAC$. Then $\triangle BAC \cong \triangle DAC$ and $\overline{CD} \cong \overline{CB}$.

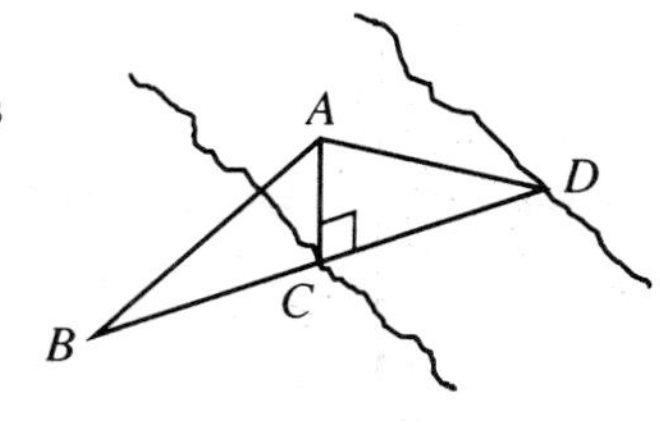

Pages 132–133 • SELF-TEST 1

1. $\angle P \cong \angle T$; Corr. parts of $\cong$ ⚠ are $\cong$. **2.** $\overline{KO}$, $\overline{MA}$; $\overline{OP}$, $\overline{AT}$; $\overline{KP}$, $\overline{MT}$

3. $\triangle JKX \cong \triangle JKY$; SAS **4.** No $\cong$ can be deduced. **5.** $\triangle TRP \cong \triangle TRS$; ASA

6.

Statements	Reasons
1. $\angle 1 \cong \angle 2$; $\angle 3 \cong \angle 4$	1. Given
2. $\overline{DB} \cong \overline{DB}$	2. Refl. Prop.
3. $\triangle ADB \cong \triangle CBD$	3. ASA Post.

7.

Statements	Reasons
1. $\overline{CD} \cong \overline{AB}$; $\overline{CB} \cong \overline{AD}$	1. Given
2. $\overline{DB} \cong \overline{DB}$	2. Refl. Prop.
3. $\triangle ADB \cong \triangle CBD$	3. SSS Post.
4. $\angle 1 \cong \angle 2$	4. Corr. parts of $\cong$ ⚠ are $\cong$.

8.

Statements	Reasons
1. $\overline{AD} \parallel \overline{BC}$	1. Given
2. $\angle 4 \cong \angle 3$	2. If 2 $\parallel$ lines are cut by a trans., then alt. int. ∠s are $\cong$.
3. $\overline{AD} \cong \overline{CB}$	3. Given
4. $\overline{DB} \cong \overline{DB}$	4. Refl. Prop.
5. $\triangle ADB \cong \triangle CBD$	5. SAS Post.
6. $\angle 1 \cong \angle 2$	6. Corr. parts of $\cong$ ⚠ are $\cong$.
7. $\overline{DC} \parallel \overline{AB}$	7. If 2 lines are cut by a trans. and alt. int. ∠s are $\cong$, then the lines are $\parallel$.

Page 134 • EXPLORATIONS

In each isos. $\triangle$ the ∠s opp. the $\cong$ sides are $\cong$. In each $\triangle$ with 2 $\cong$ ∠s, the sides opp. the $\cong$ ∠s are $\cong$.

Page 136 • CLASSROOM EXERCISES

1. A, D **2.** OBC, OCB **3.** 45 **4.** b, c, d, f **5.** $\overline{KX}$, $\overline{KY}$

6. $\overline{KM}$, $\overline{KN}$ **7.** $\overline{KG}$, $\overline{KH}$ **8.** True

9. $\triangle ABC$ is equilateral. $\overline{AB} \cong \overline{AC}$, so $\angle B \cong \angle C$. $\overline{AB} \cong \overline{BC}$, so $\angle A \cong \angle C$. Then $\angle A \cong \angle B \cong \angle C$.

10. $m\angle A = m\angle B = m\angle C$ and $m\angle A + m\angle B + m\angle C = 180$. Then $3m\angle A = 180$ and $m\angle A = 60$. Similarly, $m\angle B = m\angle C = 60$.

11. Use the diagram for Thm. 4-1. Since $\triangle BAD \cong \triangle CAD$, $\angle BDA \cong \angle CDA$. If 2 lines form $\cong$ adj. ∠s, then the lines are $\perp$. $\overline{BD} \cong \overline{CD}$, so D is the midpt. of $\overline{BC}$.

12. $\triangle ABC$ is equiangular, $\angle A \cong \angle B$, so $\overline{BC} \cong \overline{AC}$. $\angle A \cong \angle C$, so $\overline{BC} \cong \overline{AB}$. Then $\overline{AB} \cong \overline{BC} \cong \overline{AC}$.

Pages 137–139 • WRITTEN EXERCISES

A **1.** $x = 180 - 2(50) = 80$ **2.** $2x = 90; x = 45$ **3.** $2x = 180 - 74; x = 53$

4. $180 - (54 + 63) = 63; x = 11$ **5.** $5x - 8 = 2x + 7; x = 5$

6. $4x - 6 = 18; x = 6$

7. The △s are $\cong$ by SSS. $2x + 98 = 180; 2x = 82; x = 41$

8. In the $\triangle$, the base ∠s are $\cong$ and measure $2x$ and $90 - x$. $2x = 90 - x$; $3x = 90; x = 30$

9. Answers may vary; c, d, b, a **10.** Answers may vary; a, c, d, e, b

11.

Statements	Reasons
1. $\overline{AB} \cong \overline{AC}$	1. Given
2. Let the bis. of $\angle A$ int. $\overline{BC}$ at D.	2. By the Protractor Post., an ∠ has exactly one bis.
3. $\angle BAD \cong \angle CAD$	3. Def. of ∠ bis.
4. $\overline{AD} \cong \overline{AD}$	4. Refl. Prop.
5. $\triangle BAD \cong \triangle CAD$	5. SAS Post.
6. $\angle B \cong \angle C$	6. Corr. parts of $\cong$ △s are $\cong$.

12.

Statements	Reasons
1. $\angle B \cong \angle C$	1. Given
2. Let the bis. of $\angle A$ int. $\overline{BC}$ at D.	2. By the Protractor Post., an ∠ has exactly one bis.
3. $\angle BAD \cong \angle CAD$	3. Def. of ∠ bis.
4. $\angle BDA \cong \angle CDA$	4. If 2 ∠s of one $\triangle$ are $\cong$ to 2 ∠s of another $\triangle$, then the third ∠s are $\cong$.
5. $\overline{AD} \cong \overline{AD}$	5. Refl. Prop.
6. $\triangle BAD \cong \triangle CAD$	6. ASA Post.
7. $\overline{AB} \cong \overline{AC}$	7. Corr. parts of $\cong$ △s are $\cong$.

13.

Statements	Reasons
1. $\angle 1 \cong \angle 2$	1. Given
2. $\overline{JG} \cong \overline{JM}$	2. If 2 ∠s of a △ are ≅, then the sides opp. those ∠s are ≅.
3. M is the midpt. of $\overline{JK}$.	3. Given
4. $\overline{JM} \cong \overline{MK}$	4. Def. of midpt.
5. $\overline{JG} \cong \overline{MK}$	5. Trans. Prop.

14.

Statements	Reasons
1. $\overline{XY} \cong \overline{XZ}$	1. Given
2. $\angle 3 \cong \angle 4$	2. Isos. △ Thm.
3. $\angle 4 \cong \angle 5$	3. Vert. ∠s are ≅.
4. $\angle 3 \cong \angle 5$	4. Trans. Prop.

B **15.** 1, 3 **16.** 1, 2

17.

Statements	Reasons
1. $\overline{XY} \cong \overline{XZ}$	1. Given
2. $\angle XYZ \cong \angle XZY$, or $m\angle XYZ = m\angle XZY$	2. Isos. △ Thm.
3. $m\angle XYZ = m\angle 1 + m\angle 2$; $m\angle XZY = m\angle 3 + m\angle 4$	3. ∠ Add. Post.
4. $m\angle 1 + m\angle 2 = m\angle 3 + m\angle 4$	4. Substitution Prop.
5. $\overline{OY} \cong \overline{OZ}$	5. Given
6. $\angle 2 \cong \angle 3$, or $m\angle 2 = m\angle 3$	6. Isos. △ Thm.
7. $m\angle 1 = m\angle 4$	7. Subtr. Prop. of =

18.

Statements	Reasons
1. $\overline{XY} \cong \overline{XZ}$	1. Given
2. $\angle XYZ \cong \angle XZY$, or $m\angle XYZ = m\angle XZY$	2. Isos. △ Thm.
3. $\overrightarrow{YO}$ bis. $\angle XYZ$; $\overrightarrow{ZO}$ bis. $\angle XZY$.	3. Given
4. $m\angle XYZ = 2m\angle 2$; $m\angle XZY = 2m\angle 3$	4. ∠ Bis. Thm.
5. $2m\angle 2 = 2m\angle 3$	5. Substitution Prop.
6. $m\angle 2 = m\angle 3$	6. Div. Prop. of =
7. $\overline{YO} \cong \overline{ZO}$	7. If 2 ∠s of a △ are ≅, then the sides opp. those ∠s are ≅.

19.

Statements	Reasons
1. $\overline{AB} \cong \overline{AC}$	1. Given
2. $\angle B \cong \angle C$	2. Isos. △ Thm.
3. $\overline{AL}$ and $\overline{AM}$ trisect $\angle BAC$, so $\angle 1 \cong \angle 3$.	3. Given
4. $\triangle BLA \cong \triangle CMA$	4. ASA Post.
5. $\overline{AL} \cong \overline{AM}$	5. Corr. parts of ≅ ⚠ are ≅.

20.

Statements	Reasons
1. $\angle 4 \cong \angle 7$; $\angle 1 \cong \angle 3$	1. Given
2. $\angle B \cong \angle C$	2. If 2 ∠s of one △ are ≅ to 2 ∠s of another △, then the third ∠s are ≅.
3. $\overline{AB} \cong \overline{AC}$	3. If 2 ∠s of a △ are ≅, then the sides opp. those ∠s are ≅.
4. $\triangle ABC$ is isosceles.	4. Def. of isos. △

21.

Statements	Reasons
1. $\overline{OP} \cong \overline{OQ}$; $\angle 3 \cong \angle 4$	1. Given
2. $\angle POS \cong \angle QOR$	2. Vert. ∠s are ≅.
3. $\triangle POS \cong \triangle QOR$	3. ASA Post.
4. $\overline{OS} \cong \overline{OR}$	4. Corr. parts of ≅ ⚠ are ≅.
5. $\angle 5 \cong \angle 6$	5. Isos. △ Thm.

22. a. $m\angle 2 = m\angle 1 = 40$; $m\angle 7 = 180 - (40 + 40) = 100$; $m\angle 5 = m\angle 6 = \frac{1}{2}(180 - 100) = 40$; yes, $\overline{PQ} \parallel \overline{SR}$ because alt. int. ∠s are ≅.

b. $m\angle 2 = m\angle 1 = k$; $m\angle 7 = 180 - (k + k) = 180 - 2k$; $m\angle 5 = m\angle 6 = \frac{1}{2}(180 - (180 - 2k)) = \frac{1}{2}(2k) = k$; yes

23. a. 40, 40, 60 **b.** $2x, 2x, 180 - ((180 - 4x) + x) = 3x$

24. a. $m\angle 2 = m\angle 1 = 35$; $m\angle 3 = 35 + 35 = 70$; $m\angle 5 = \frac{1}{2}(180 - 70) = 55$; $m\angle ABC = 35 + 55 = 90$

b. $m\angle 2 = m\angle 1 = x$; $m\angle 3 = m\angle 1 + m\angle 2 = 2x$; $m\angle 5 = \frac{1}{2}(180 - m\angle 3) = \frac{1}{2}(180 - 2x) = 90 - x$; $m\angle ABC = m\angle 2 + m\angle 5 = x + 90 - x = 90$

25. **a.** $m\angle 2 = m\angle 1 = 23$; $m\angle 3 = 180 - (23 + 23) = 134$; $m\angle 4 = 180 - 134 = 46$; $m\angle 5 = m\angle 6 = \frac{1}{2}(180 - 46) = 67$; $m\angle 7 = 180 - (23 + 67) = 90$

b. $m\angle 2 = m\angle 1 = k$; $m\angle 3 = 180 - 2k$; $m\angle 4 = 180 - m\angle 3 = 180 - (180 - 2k) = 2k$; $m\angle 5 = m\angle 6 = \frac{1}{2}(180 - m\angle 4) = \frac{1}{2}(180 - 2k) = 90 - k$; $m\angle 7 = 180 - (m\angle 2 + m\angle 5) = 180 - (k + 90 - k) = 90$

26. **a.** Yes; $m\angle B = m\angle C = \frac{1}{2}(180 - 80) = 50$ (The sum of the meas. of the $\angle$s of a $\triangle$ is 180.) $m\angle XAC = \frac{1}{2}(50 + 50) = 50$ (The meas. of an ext. $\angle$ equals the sum of the meas. of the 2 remote int. $\angle$s.) Thus, $\overrightarrow{AX} \parallel \overline{BC}$ (If 2 lines are cut by a trans. and alt. int. $\angle$s are $\cong$, then the lines are $\parallel$.) **b.** No

27. $4x - y = 7$, so $y = 4x - 7$; $2x + 3y = 7$; substituting, $2x + 3(4x - 7) = 7$, $2x + 12x - 21 = 7$, $14x = 28$, $x = 2$; $y = 4x - 7 = 1$

28. $x + y = 60$, so $y = 60 - x$; $2x - y = 60$; substituting, $2x - (60 - x) = 60$, $3x = 120$, $x = 40$; $y = 60 - x = 20$

29. $2x - y = x + 2y$, so $x = 3y$; $(2x + 2y) + (2x - y) + (x + 2y) = 180$, $5x + 3y = 180$; substituting, $5x + x = 180$, $x = 30$; $3y = 30$, $y = 10$

30. $\triangle DAC \cong \triangle DAB$

Statements	Reasons
1. $\angle ACB \cong \angle ABC$; $\angle DCB \cong \angle DBC$	1. Given
2. $\overline{AB} \cong \overline{AC}$; $\overline{DB} \cong \overline{DC}$	2. If 2 $\angle$s of a $\triangle$ are $\cong$, then the sides opp. those $\angle$s are $\cong$.
3. $\overline{DA} \cong \overline{DA}$	3. Refl. Prop.
4. $\triangle DAC \cong \triangle DAB$	4. SSS Post.

31. **a.**

Statements	Reasons
1. $\overline{JL} \perp$ plane Z	1. Given
2. $\overline{JL} \perp \overline{KM}$; $\overline{JL} \perp \overline{KN}$	2. Def. of a line $\perp$ to a plane
3. $\angle JKM \cong \angle JKN$; $\angle LKM \cong \angle LKN$	3. If 2 lines are $\perp$, then they form $\cong$ adj. $\angle$s.
4. $\overline{KM} \cong \overline{KN}$	4. Given
5. $\overline{JK} \cong \overline{JK}$; $\overline{LK} \cong \overline{LK}$	5. Refl. Prop.
6. $\triangle JKM \cong \triangle JKN$; $\triangle LKM \cong \triangle LKN$	6. SAS Post.
7. $\overline{JM} \cong \overline{JN}$; $\overline{LM} \cong \overline{LN}$	7. Corr. parts of $\cong$ $\triangle$s are $\cong$.
8. $\triangle JMN$ and $\triangle LMN$ are isos.	8. Def. of isos. $\triangle$

b. No. They are $\cong$ if and only if $\overline{KJ} \cong \overline{KL}$.

32. It is isosceles. In isos. $\triangle ABC$, $\overline{AB} \cong \overline{AC}$, so $\angle B \cong \angle C$. D, E, and F are midpts. of $\overline{AB}$, $\overline{BC}$, and $\overline{AC}$, respectively. $DB = \frac{1}{2}AB$ and $FC = \frac{1}{2}AC$, so $\overline{DB} \cong \overline{FC}$; $\overline{BE} \cong \overline{EC}$; $\triangle DBE \cong \triangle FCE$ by SAS, and corr. parts $\overline{ED}$ and $\overline{EF}$ are $\cong$. Therefore, $\triangle DEF$ is an isos. $\triangle$.

A
D F
B E C

C 33. $m\angle AED = 108$; $m\angle DEF = 90$; $m\angle AEF = 360 - (108 + 90) = 162$; $\overline{EA} \cong \overline{EF}$, so $m\angle EAF = m\angle EFA = \frac{1}{2}(180 - m\angle AEF) = \frac{1}{2}(180 - 162) = 9$; $\overline{ED} \cong \overline{EF}$, so $m\angle EFD = m\angle EDF = \frac{1}{2}(180 - m\angle DEF) = \frac{1}{2}(180 - 90) = 45$; $m\angle AFD = m\angle AFE + m\angle EFD = 9 + 45 = 54$; $\overline{AE} \cong \overline{ED}$, so $m\angle DAE = \frac{1}{2}(180 - m\angle AED) = \frac{1}{2}(180 - 108) = 36$; $m\angle DAF = m\angle DAE + m\angle EAF = 36 + 9 = 45$

34. $\triangle DEF$ is equilateral.

Statements	Reasons
1. $\triangle ABC$ is equilateral.	1. Given
2. $\overline{AB} \cong \overline{BC} \cong \overline{AC}$	2. Def. of equilateral
3. $m\angle BAC = m\angle ABC = m\angle ACB$	3. An equilateral $\triangle$ is also equiangular.
4. $m\angle BAC = m\angle BAE + m\angle CAD$; $m\angle ABC = m\angle FBC + m\angle ABE$; $m\angle ACB = m\angle ACD + m\angle BCF$	4. $\angle$ Add. Post.
5. $m\angle BAE + m\angle CAD = m\angle FBC + m\angle ABE = m\angle ACD + m\angle BCF$	5. Substitution Prop.
6. $m\angle CAD = m\angle ABE = m\angle BCF$	6. Given
7. $m\angle BAE = m\angle FBC = m\angle ACD$	7. Subtr. Prop. of =
8. $\triangle BAE \cong \triangle CBF \cong \triangle ACD$	8. ASA Post.
9. $\angle AEB \cong \angle CDA \cong \angle BFC$	9. Corr. parts of $\cong$ $\triangle$s are $\cong$.
10. $m\angle AEB + m\angle FED = 180$; $m\angle CDA + m\angle EDF = 180$; $m\angle BFC + m\angle EFD = 180$	10. $\angle$ Add. Post.
11. $\angle AEB$ and $\angle FED$ are supp.; $\angle CDA$ and $\angle EDF$ are supp.; $\angle BFC$ and $\angle EFD$ are supp.	11. Def. of supp. $\angle$s
12. $\angle FED \cong \angle EDF \cong \angle EFD$	12. If 2 $\angle$s are supps. of $\cong$ $\angle$s, then the 2 $\angle$s are $\cong$.
13. $\triangle DEF$ is equilateral.	13. An equiangular $\triangle$ is also equilateral.

Page 139 • CHALLENGE

a.
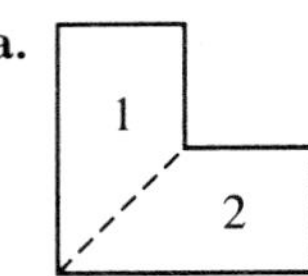

b.
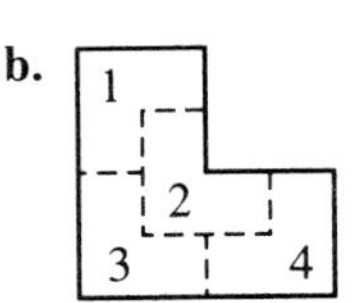

Pages 142–143 • CLASSROOM EXERCISES

1. AAS, ASA **2.** AAS, HL **3.** ASA **4.** none **5.** SAS, SSS **6.** HL
7. none **8.** AAS, ASA, SAS, HL **9.** none **10.** $\triangle ABC \cong \triangle DCB$ by SSS
11. $\triangle LMN \cong \triangle PNM$ by ASA **12.** $\triangle UWY \cong \triangle VZX$ by HL
13. $\triangle ADB \cong \triangle AEC$ by AAS, $\triangle EBC \cong \triangle DCB$ by AAS
14. Check students' drawings. **a.** SAS **b.** AAS **c.** AAS or ASA

Pages 143–145 • WRITTEN EXERCISES

A **1.** 1. Given 2. Def. of rt. $\triangle$ 3. Given 4. $\overline{XZ} \cong \overline{XZ}$ 5. $\triangle XYZ$; HL
6. $\overline{WZ} \cong \overline{YZ}$; Corr. parts of $\cong$ $\triangle$s are $\cong$.
2. Answers may vary; a, d, f, c, b, h, g, e

3.

Statements	Reasons
1. $\overline{EF} \perp \overline{EG}$; $\overline{HG} \perp \overline{EG}$	1. Given
2. $\angle HGE$ and $\angle FEG$ are rt. $\angle$s.	2. Def. of $\perp$ lines
3. $\triangle HGE$ and $\triangle FEG$ are rt. $\triangle$s.	3. Def. of rt. $\triangle$
4. $\overline{EH} \cong \overline{GF}$	4. Given
5. $\overline{EG} \cong \overline{EG}$	5. Refl. Prop.
6. $\triangle HGE \cong \triangle FEG$	6. HL Thm.
7. $\angle H \cong \angle F$	7. Corr. parts of $\cong$ $\triangle$s are $\cong$.

4.

Statements	Reasons
1. $\overline{RT} \cong \overline{AS}$; $\overline{RS} \cong \overline{AT}$	1. Given
2. $\overline{ST} \cong \overline{ST}$	2. Refl. Prop.
3. $\triangle TSA \cong \triangle STR$	3. SSS Post.
4. $\angle TSA \cong \angle STR$	4. Corr. parts of $\cong$ $\triangle$s are $\cong$.

5. SAS **6.** AAS **7.** HL

B **8. a.** Yes; $\triangle AOB \cong \triangle AOC$ by SSS, so corr. parts $\angle AOB$ and $\angle AOC$ are $\cong$. **b.** No

9. a.

Statements	Reasons
1. $\overline{PR} \cong \overline{PQ}$	1. Given
2. $\angle PQR \cong \angle PRQ$	2. Isos. $\triangle$ Thm.
3. $\overline{SR} \cong \overline{TQ}$	3. Given
4. $\overline{RQ} \cong \overline{RQ}$	4. Refl. Prop.
5. $\triangle RQS \cong \triangle QRT$	5. SAS Post.
6. $\overline{QS} \cong \overline{RT}$	6. Corr. parts of $\cong$ $\triangle$s are $\cong$.

b.

Statements	Reasons
1. $\overline{PR} \cong \overline{PQ}$, or $PR = PQ$; $\overline{SR} \cong \overline{TQ}$, or $SR = TQ$	1. Given
2. $PR = PS + SR$; $PQ = PT + TQ$	2. Seg. Add. Post.
3. $PS + SR = PT + TQ$	3. Substitution Prop.
4. $PS = PT$, or $\overline{PS} \cong \overline{PT}$	4. Subtr. Prop. of =
5. $\angle P \cong \angle P$	5. Refl. Prop.
6. $\triangle PQS \cong \triangle PRT$	6. SAS Post.
7. $\overline{QS} \cong \overline{RT}$	7. Corr. parts of $\cong$ $\triangle$s are $\cong$.

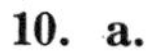

10. a.

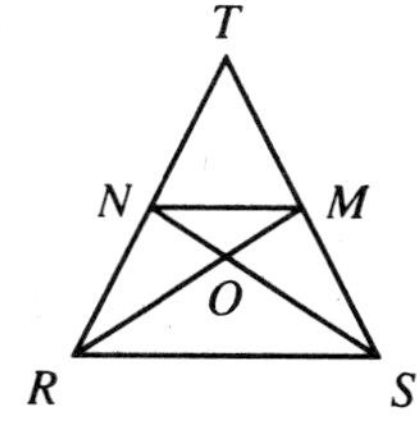

b. $\triangle TRM \cong \triangle TSN$; $\triangle MNR \cong \triangle NMS$; $\triangle NRS \cong \triangle MSR$; $\triangle ONR \cong \triangle OMS$

11. $\overline{PR} \cong \overline{PS}$, $\overline{PQ} \cong \overline{PT}$, $\overline{QR} \cong \overline{TS}$; SSS

12. $\angle 3 \cong \angle 4$, $\angle QXP \cong \angle TYP$, $\overline{PQ} \cong \overline{PT}$; AAS

13. $\angle 3 \cong \angle 4$, $\overline{PQ} \cong \overline{PT}$, $\angle 6 \cong \angle 5$; AAS

14.

Statements	Reasons
1. Draw $\overline{QS}$.	1. Through any 2 pts. there is exactly one line.
2. $\overline{RS} \parallel \overline{QT}$	2. Given
3. $\angle RSQ \cong \angle TQS$	3. If 2 $\parallel$ lines are cut by a trans., then alt. int. $\angle$s are $\cong$.
4. $\angle R \cong \angle T$	4. Given
5. $\overline{QS} \cong \overline{QS}$	5. Refl. Prop.
6. $\triangle RSQ \cong \triangle TQS$	6. AAS Thm.
7. $\overline{RS} \cong \overline{TQ}$	7. Corr. parts of $\cong$ $\triangle$s are $\cong$.

15.

Statements	Reasons
1. $\angle 1 \cong \angle 2 \cong \angle 3$	1. Given
2. $\overline{ME} \cong \overline{MD}$	2. If 2 $\angle$s of a $\triangle$ are $\cong$, then the sides opp. those $\angle$s are $\cong$.
3. $\overline{EN} \cong \overline{DG}$	3. Given
4. $\triangle MEN \cong \triangle MDG$	4. SAS Post.
5. $\angle 4 \cong \angle 5$	5. Corr. parts of $\cong$ $\triangle$s are $\cong$.

16. 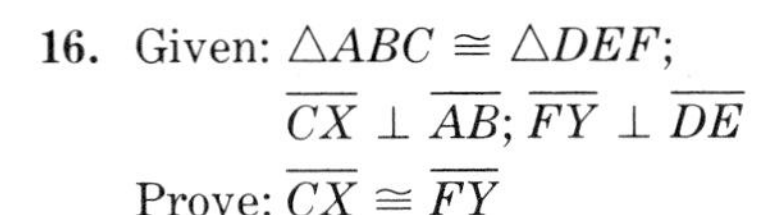
Given: $\triangle ABC \cong \triangle DEF$;
$\overline{CX} \perp \overline{AB}$; $\overline{FY} \perp \overline{DE}$
Prove: $\overline{CX} \cong \overline{FY}$

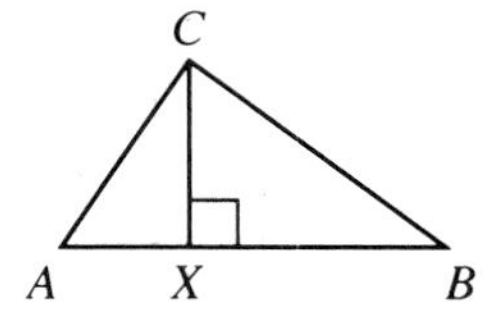

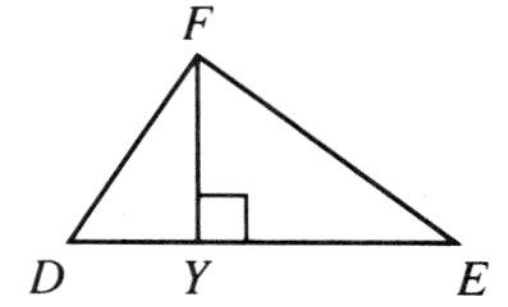

Statements	Reasons
1. $\triangle ABC \cong \triangle DEF$	1. Given
2. $\overline{CB} \cong \overline{FE}$; $\angle B \cong \angle E$	2. Corr. parts of $\cong$ $\triangle$s are $\cong$.
3. $\overline{CX} \perp \overline{AB}$; $\overline{FY} \perp \overline{DE}$	3. Given
4. $m\angle CXB = 90$; $m\angle FYE = 90$	4. Def. of $\perp$ lines
5. $\angle CXB \cong \angle FYE$	5. Def. of $\cong$ $\angle$s
6. $\triangle CXB \cong \triangle FYE$	6. AAS Thm.
7. $\overline{CX} \cong \overline{FY}$	7. Corr. parts of $\cong$ $\triangle$s are $\cong$.

17. Given: Isos. $\triangle XYZ$ with $\overline{XY} \cong \overline{XZ}$;
$\overline{ZA} \perp \overline{XY}$; $\overline{YB} \perp \overline{XZ}$
Prove: $\overline{ZA} \cong \overline{YB}$

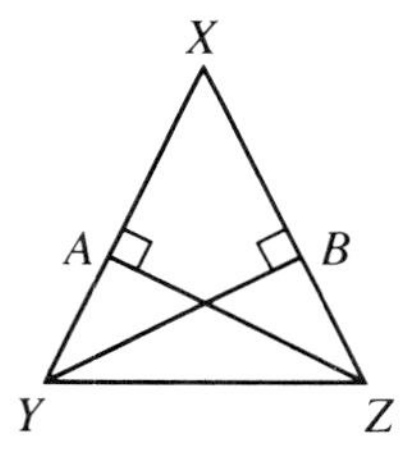

Statements	Reasons
1. $\overline{ZA} \perp \overline{XY}$; $\overline{YB} \perp \overline{XZ}$	1. Given
2. $m\angle XBY = 90$; $m\angle XAZ = 90$	2. Def. of $\perp$ lines
3. $\angle XBY \cong \angle XAZ$	3. Def. of $\cong$ $\angle$s
4. $\angle X \cong \angle X$	4. Refl. Prop.
5. $\overline{XY} \cong \overline{XZ}$	5. Given
6. $\triangle XBY \cong \triangle XAZ$	6. AAS Thm.
7. $\overline{ZA} \cong \overline{YB}$	7. Corr. parts of $\cong$ $\triangle$s are $\cong$.

18. Given: Isos. $\triangle ABC$ with $\overline{CA} \cong \overline{CB}$;
$\overline{AX}$ bis. $\angle CAB$; $\overline{BY}$ bis. $\angle CBA$.
Prove: $\overline{AX} \cong \overline{BY}$

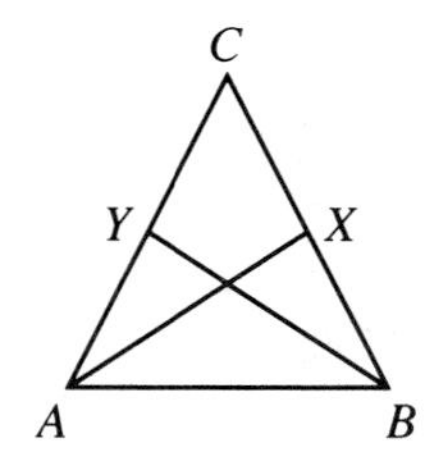

Statements	Reasons
1. $\overline{CA} \cong \overline{CB}$	1. Given
2. $\angle CAB \cong \angle CBA$, or $m\angle CAB = m\angle CBA$	2. Isos. △ Thm.
3. $\frac{1}{2}m\angle CAB = \frac{1}{2}m\angle CBA$	3. Mult. Prop. of =
4. $\overline{AX}$ bis. $\angle CAB$; $\overline{BY}$ bis. $\angle CBA$.	4. Given
5. $m\angle BAX = \frac{1}{2}m\angle CAB$; $m\angle ABY = \frac{1}{2}m\angle CBA$	5. ∠ Bis. Thm.
6. $m\angle BAX = m\angle ABY$, or $\angle BAX \cong \angle ABY$	6. Substitution Prop.
7. $\overline{AB} \cong \overline{AB}$	7. Refl. Prop.
8. $\triangle BAX \cong \triangle ABY$	8. ASA Post.
9. $\overline{AX} \cong \overline{BY}$	9. Corr. parts of ≅ ⚠ are ≅.

19. Given: Isos. $\triangle XYZ$ with $\overline{XY} \cong \overline{XZ}$;
M is the midpt. of $\overline{XY}$;
N is the midpt. of $\overline{XZ}$;
$\overline{MP} \perp \overline{YZ}$; $\overline{NQ} \perp \overline{YZ}$
Prove: $\overline{MP} \cong \overline{NQ}$

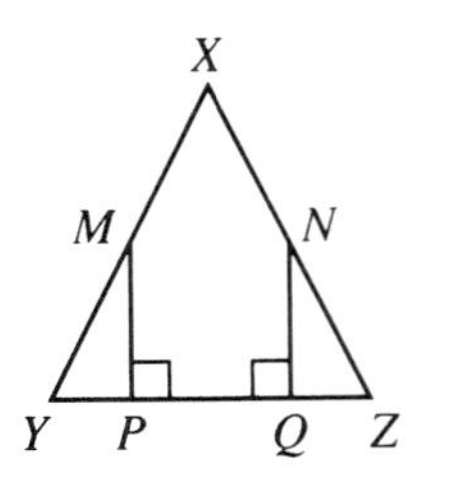

Statements	Reasons
1. $\overline{XY} \cong \overline{XZ}$, or $XY = XZ$	1. Given
2. $\angle Y \cong \angle Z$	2. Isos. $\triangle$ Thm.
3. $\frac{1}{2}XY = \frac{1}{2}XZ$	3. Mult. Prop. of $=$
4. M is the midpt. of $\overline{XY}$; N is the midpt. of $\overline{XZ}$.	4. Given
5. $MY = \frac{1}{2}XY$; $NZ = \frac{1}{2}XZ$	5. Midpt. Thm.
6. $MY = NZ$	6. Substitution Prop.
7. $\overline{MP} \perp \overline{YZ}$; $\overline{NQ} \perp \overline{YZ}$	7. Given
8. $m\angle MPY = 90$; $m\angle NQZ = 90$	8. Def. of $\perp$ lines
9. $\angle MPY \cong \angle NQZ$	9. Def. of $\cong$ $\angle$s
10. $\triangle MPY \cong \triangle NQZ$	10. AAS Thm.
11. $\overline{MP} \cong \overline{NQ}$	11. Corr. parts of $\cong$ $\triangle$s are $\cong$.

20. Plan for Proof: $\triangle FSK$ is isos., so $\angle F \cong \angle K$. Add LA to both FL and KA, so that $FA = KL$ and $\overline{FA} \cong \overline{KL}$. Use the Midpt. Thm. and Substitution Prop. to prove $\overline{FM} \cong \overline{KN}$. Then $\triangle FAM \cong \triangle KLN$ by SAS, and corr. parts $\overline{AM}$ and $\overline{LN}$ are $\cong$.

C 21.

Statements	Reasons
1. $\angle ADC \cong \angle FED \cong \angle NCE$	1. Def. of square
2. $\angle CDE \cong \angle DEC \cong \angle ECD$	2. An equilateral $\triangle$ is also equiangular.
3. $m\angle ADC + m\angle CDE = m\angle FED + m\angle DEC = m\angle NCE + m\angle ECD$	3. Add. Prop. of $=$
4. $m\angle ADE = m\angle ADC + m\angle CDE$; $m\angle FEC = m\angle FED + m\angle DEC$; $m\angle NCD = m\angle NCE + m\angle ECD$	4. $\angle$ Add. Post.
5. $\angle ADE \cong \angle FEC \cong \angle NCD$	5. Substitution Prop.
6. $\overline{AD} \cong \overline{DE} \cong \overline{FE} \cong \overline{EC} \cong \overline{NC} \cong \overline{CD}$	6. Def. of square and equilateral $\triangle$
7. $\triangle ADE \cong \triangle FEC \cong \triangle NCD$	7. SAS Post.
8. $\overline{AE} \cong \overline{FC} \cong \overline{ND}$	8. Corr. parts of $\cong$ $\triangle$s are $\cong$.

22.

Statements	Reasons
1. $\overline{AE} \cong \overline{FC} \cong \overline{ND}$	1. Ex. 21
2. $m\angle AED = m\angle FCE = m\angle NDC$	2. Ex. 21; Corr. parts of $\cong$ △ are $\cong$.
3. $m\angle FED = m\angle NCE = m\angle ADC$	3. Def. of square
4. $m\angle AED + m\angle FED = m\angle FCE + m\angle NCE = m\angle NDC + m\angle ADC$	4. Add. Prop. of =
5. $m\angle AEF = m\angle AED + m\angle FED$; $m\angle FCN = m\angle FCE + m\angle NCE$; $m\angle NDA = m\angle NDC + m\angle ADC$	5. $\angle$ Add. Post.
6. $\angle AEF \cong \angle FCN \cong \angle NDA$	6. Substitution Prop.
7. $\overline{FE} \cong \overline{NC} \cong \overline{AD}$	7. Def. of square
8. $\triangle AEF \cong \triangle FCN \cong \triangle NDA$	8. SAS Post.
9. $\overline{FA} \cong \overline{NF} \cong \overline{AN}$	9. Corr. parts of $\cong$ △ are $\cong$.
10. $\triangle FAN$ is equilateral.	10. Def. of equilateral △

Page 146 • SELF-TEST 2

1. $2x = 180 - 40; x = 70$ **2.** $5x - 11 = 3x + 3; 2x = 14; x = 7$

3. $x = 180 - (60 + 90) = 30$

4. Since $\overline{BN} \perp \overline{AC}$ and $\overline{CM} \perp \overline{AB}$, $\angle ANB$ and $\angle AMC$ are rt. $\angle$s and are $\cong$. $\overline{AB} \cong \overline{AC}$ and $\angle A \cong \angle A$, so $\triangle ABN \cong \triangle ACM$ by AAS Thm.

5.

Statements	Reasons
1. $\overline{BN} \perp \overline{AC}$; $\overline{CM} \perp \overline{AB}$	1. Given
2. $\angle BMC$ and $\angle CNB$ are rt. $\angle$s.	2. Def. of $\perp$ lines
3. $\triangle BMC$ and $\triangle CNB$ are rt. △.	3. Def. of rt. △
4. $\overline{MB} \cong \overline{NC}$	4. Given
5. $\overline{BC} \cong \overline{BC}$	5. Refl. Prop.
6. $\triangle BMC \cong \triangle CNB$	6. HL Thm.
7. $\overline{CM} \cong \overline{BN}$	7. Corr. parts of $\cong$ △ are $\cong$.

Page 148 • CLASSROOM EXERCISES

1. a. SAS **b.** Corr. parts of $\cong$ △ are $\cong$. **c.** SAS **d.** Corr. parts of $\cong$ △ are $\cong$.

2. a. SAS **b.** Corr. parts of $\cong$ △ are $\cong$. **c.** ASA **d.** Corr. parts of $\cong$ △ are $\cong$.

3. a. SSS **b.** Corr. parts of $\cong$ △ are $\cong$. **c.** SAS **d.** Corr. parts of $\cong$ △ are $\cong$.

4. Prove $\triangle CPE \cong \triangle GQE$ by AAS, so corr. parts $\overline{CP}$ and $\overline{GQ}$ are $\cong$. Then prove $\triangle CDP \cong \triangle GFQ$ by HL, so corr. parts $\angle D$ and $\angle F$ are $\cong$.

Pages 148–151 • WRITTEN EXERCISES

A **1. a.** SSS **b.** Corr. parts of $\cong$ △ are $\cong$. **c.** SAS **d.** Corr. parts of $\cong$ △ are $\cong$.

2. a. SSS **b.** Corr. parts of $\cong$ △ are $\cong$. **c.** AAS **d.** Corr. parts of $\cong$ △ are $\cong$.

3. a. AAS **b.** Corr. parts of $\cong$ △ are $\cong$. **c.** SAS **d.** Corr. parts of $\cong$ △ are $\cong$.

4. a. SAS **b.** Corr. parts of $\cong$ △ are $\cong$. **c.** ASA **d.** Corr. parts of $\cong$ △ are $\cong$.

5. a. SAS **b.** Corr. parts of $\cong$ △ are $\cong$. **c.** HL **d.** Corr. parts of $\cong$ △ are $\cong$.

6. a. If 2 ∠ of a $\triangle$ are $\cong$, then the sides opp. those ∠ are $\cong$. **b.** SSS

c. Corr. parts of $\cong$ △ are $\cong$. **d.** SAS **e.** Corr. parts of $\cong$ △ are $\cong$.

B **7. a.** 1. $\triangle FLA \cong \triangle FKA$ (SSS) 2. $\angle 1 \cong \angle 2$ (Corr. parts of $\cong$ △ are $\cong$.)
3. $\triangle FLJ \cong \triangle FKJ$ (SAS) 4. $\overline{LJ} \cong \overline{KJ}$ (Corr. parts of $\cong$ △ are $\cong$.)

b.

Statements	Reasons
1. $\overline{LF} \cong \overline{KF}$; $\overline{LA} \cong \overline{KA}$	1. Given
2. $\overline{FA} \cong \overline{FA}$	2. Refl. Prop.
3. $\triangle FLA \cong \triangle FKA$	3. SSS Post.
4. $\angle 1 \cong \angle 2$	4. Corr. parts of $\cong$ △ are $\cong$.
5. $\overline{FJ} \cong \overline{FJ}$	5. Refl. Prop.
6. $\triangle FLJ \cong \triangle FKJ$	6. SAS Post.
7. $\overline{LJ} \cong \overline{KJ}$	7. Corr. parts of $\cong$ △ are $\cong$.

8. a. 1. $\triangle PSR \cong \triangle PTR$ (ASA) 2. $\overline{SR} \cong \overline{TR}$ (Corr. parts of $\cong$ △ are $\cong$.)
3. $\triangle SRQ \cong \triangle TRQ$ (SAS Post.) 4. $\angle 3 \cong \angle 4$ (Corr. parts of $\cong$ △ are $\cong$.)
5. $\overline{PR}$ bis. $\angle SQT$ (Def. of $\angle$ bis.)

b. $\triangle PSR \cong \triangle PTR$ by ASA, so corr. parts $\overline{SR}$ and $\overline{TR}$ are $\cong$. $\overline{SR}$ and $\overline{TR}$ are also corr. parts of $\triangle SRQ$ and $\triangle TRQ$, which can now be proved $\cong$ by SAS. So corr. parts $\angle 3$ and $\angle 4$ are $\cong$, and $\overline{PR}$ bis. $\angle SQT$.

9.

Statements	Reasons
1. $\triangle RST \cong \triangle XYZ$	1. Given
2. $\overline{ST} \cong \overline{YZ}$; $\angle T \cong \angle Z$; $\angle RST \cong \angle XYZ$, or $m\angle RST = m\angle XYZ$	2. Corr. parts of $\cong$ △ are $\cong$.
3. $\frac{1}{2}m\angle RST = \frac{1}{2}m\angle XYZ$	3. Mult. Prop. of =
4. $\overrightarrow{SK}$ bis. $\angle RST$; $\overrightarrow{YL}$ bis. $\angle XYZ$.	4. Given
5. $m\angle KST = \frac{1}{2}m\angle RST$; $m\angle LYZ = \frac{1}{2}m\angle XYZ$	5. $\angle$ Bis. Thm.
6. $m\angle KST = m\angle LYZ$	6. Substitution Prop.
7. $\triangle KST \cong \triangle LYZ$	7. ASA Post.
8. $\overline{SK} \cong \overline{YL}$	8. Corr. parts of $\cong$ △ are $\cong$.

10.

Statements	Reasons
1. Draw $\overline{AC}$ and $\overline{AE}$.	1. Through any 2 pts. there is exactly one line.
2. $\overline{CD} \cong \overline{ED}$; $\angle CDA \cong \angle EDA$	2. Given
3. $\overline{AD} \cong \overline{AD}$	3. Refl. Prop.
4. $\triangle CDA \cong \triangle EDA$	4. SAS Post.
5. $\overline{AC} \cong \overline{AE}$	5. Corr. parts of $\cong$ △s are $\cong$.
6. $\overline{AB} \cong \overline{AF}$; $\overline{BC} \cong \overline{FE}$	6. Given
7. $\triangle ABC \cong \triangle AFE$	7. SSS Post.
8. $\angle B \cong \angle F$	8. Corr. parts of $\cong$ △s are $\cong$.

11.

Statements	Reasons
1. $\overline{DE} \cong \overline{FG}$; $\overline{GD} \cong \overline{EF}$	1. Given
2. $\overline{GE} \cong \overline{GE}$	2. Refl. Prop.
3. $\triangle GDE \cong \triangle EFG$	3. SSS Post.
4. $\angle DEH \cong \angle FGK$	4. Corr. parts of $\cong$ △s are $\cong$.
5. $\angle HDE$ and $\angle KFG$ are rt. ∠s.	5. Given
6. $m\angle HDE = 90$; $m\angle KFG = 90$	6. Def. of rt. $\angle$
7. $\angle HDE \cong \angle KFG$	7. Def. of $\cong$ ∠s
8. $\triangle HDE \cong \triangle KFG$	8. ASA Post.
9. $\overline{DH} \cong \overline{FK}$	9. Corr. parts of $\cong$ △s are $\cong$.

12.

Statements	Reasons
1. $\overline{PQ} \perp \overline{QR}$; $\overline{PS} \perp \overline{SR}$	1. Given
2. $\angle PQR$ and $\angle PSR$ are rt. ∠s.	2. Def. of $\perp$ lines
3. $\triangle PQR$ and $\triangle PSR$ are rt. △s.	3. Def. of rt. $\triangle$
4. $\overline{PQ} \cong \overline{PS}$	4. Given
5. $\overline{PR} \cong \overline{PR}$	5. Refl. Prop.
6. $\triangle PQR \cong \triangle PSR$	6. HL Thm.
7. $\angle QPO \cong \angle SPO$	7. Corr. parts of $\cong$ △s are $\cong$.
8. $\overline{PO} \cong \overline{PO}$	8. Refl. Prop.
9. $\triangle QPO \cong \triangle SPO$	9. SAS Post.
10. $\overline{QO} \cong \overline{SO}$	10. Corr. parts of $\cong$ △s are $\cong$.
11. O is the midpt. of $\overline{QS}$.	11. Def. of midpt.

13. Given: $\overline{KL}$ and $\overline{MN}$ bis. each other at O.
Prove: O is the midpt. of $\overline{PQ}$.

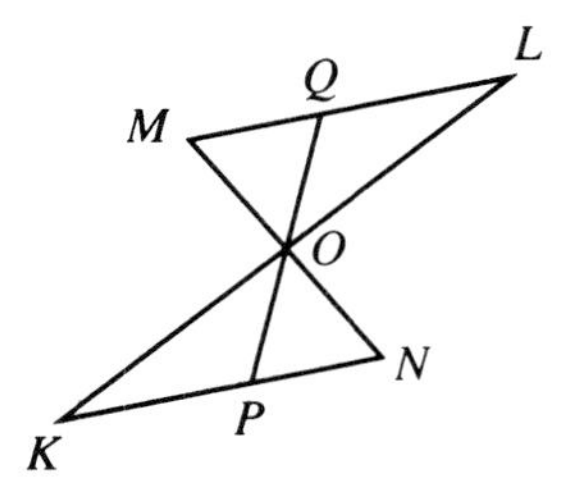

Statements	Reasons
1. $\overline{KL}$ and $\overline{MN}$ bis. each other at O.	1. Given
2. O is the midpt. of $\overline{KL}$ and $\overline{MN}$.	2. Def. of bis.
3. $KO = OL$ or $\overline{KO} \cong \overline{OL}$; $MO = ON$ or $\overline{MO} \cong \overline{ON}$	3. Def. of midpt.
4. $\angle KON \cong \angle LOM$	4. Vert. $\angle$s are $\cong$.
5. $\triangle KON \cong \triangle LOM$	5. SAS Post.
6. $\angle K \cong \angle L$	6. Corr. parts of $\cong$ $\triangle$s are $\cong$.
7. $\angle KOP \cong \angle LOQ$	7. Vert. $\angle$s are $\cong$.
8. $\triangle KOP \cong \triangle LOQ$	8. ASA Post.
9. $\overline{PO} \cong \overline{QO}$	9. Corr. parts of $\cong$ $\triangle$s are $\cong$.
10. O is the midpt. of $\overline{PQ}$.	10. Def. of midpt.

14. Since $DB = EC$ and $BA = CA$, then $DA = EA$. Using $\angle A$ in both $\triangle$s, $\triangle DAC \cong \triangle EAB$ by SAS. So corr. parts $\angle D$ and $\angle E$ are $\cong$, and corr. parts $\overline{DC}$ and $\overline{EB}$ are $\cong$. Then $\triangle DBC \cong \triangle ECB$ by SAS, so corr. parts $\angle DBC$ and $\angle ECB$ are $\cong$. Therefore, $\angle ABC \cong \angle ACB$ because they are supps. of $\cong$ $\angle$s.

C 15.

Statements	Reasons
1. $\angle MDC \cong \angle MCD$	1. Given
2. $\overline{MD} \cong \overline{MC}$	2. If 2 $\angle$s of a $\triangle$ are $\cong$, then the sides opp. those $\angle$s are $\cong$.
3. $\overline{AM} \cong \overline{MB}$; $\overline{AD} \cong \overline{BC}$	3. Given
4. $\triangle AMD \cong \triangle BMC$	4. SSS Post.
5. $\angle MAD \cong \angle MBC$	5. Corr. parts of $\cong$ $\triangle$s are $\cong$.
6. $\overline{AB} \cong \overline{AB}$	6. Refl. Prop.
7. $\triangle DAB \cong \triangle CBA$	7. SAS Post.
8. $\overline{AC} \cong \overline{BD}$	8. Corr. parts of $\cong$ $\triangle$s are $\cong$.

16.

Statements	Reasons
1. $\angle 3 \cong \angle 4$	1. Given
2. $m\angle 3 + m\angle 7 = 180$; $m\angle 4 + m\angle 8 = 180$	2. $\angle$ Add. Post.
3. $\angle 3$ and $\angle 7$ are supp.; $\angle 4$ and $\angle 8$ are supp.	3. Def. of supp. $\angle$s
4. $\angle 7 \cong \angle 8$	4. If 2 $\angle$s are supps. of $\cong$ $\angle$s, then the 2 $\angle$s are $\cong$.
5. $\overline{AG} \cong \overline{AF}$	5. If 2 $\angle$s of a $\triangle$ are $\cong$, then the sides opp. those $\angle$s are $\cong$.
6. $\angle 1 \cong \angle 2$	6. Given
7. $\triangle BAG \cong \triangle EAF$	7. ASA Post.
8. $\overline{AB} \cong \overline{AE}$	8. Corr. parts of $\cong$ $\triangle$s are $\cong$.
9. $\angle 5 \cong \angle 6$	9. Given
10. $\overline{AC} \cong \overline{AD}$	10. If 2 $\angle$s of a $\triangle$ are $\cong$, then the sides opp. those $\angle$s are $\cong$.
11. $\triangle BAC \cong \triangle EAD$	11. SAS Post.
12. $\overline{BC} \cong \overline{ED}$	12. Corr. parts of $\cong$ $\triangle$s are $\cong$.

17. isos.; $\overline{AX} \cong \overline{AY}$, $\overline{AZ} \cong \overline{AZ}$, and $\angle XAZ \cong \angle YAZ$, so $\triangle XAZ \cong \triangle YAZ$ by SAS. Then corr. parts $\overline{XZ}$ and $\overline{YZ}$ are $\cong$, and $\triangle XYZ$ is isosceles.

Page 151 • MIXED REVIEW EXERCISES

1. Two sides of a $\triangle$ are $\cong$ if and only if the $\angle$s opp. those sides are $\cong$. **2.** sometimes

3. sometimes **4.** always

5. a. A M B **b.** A M B

6. a.

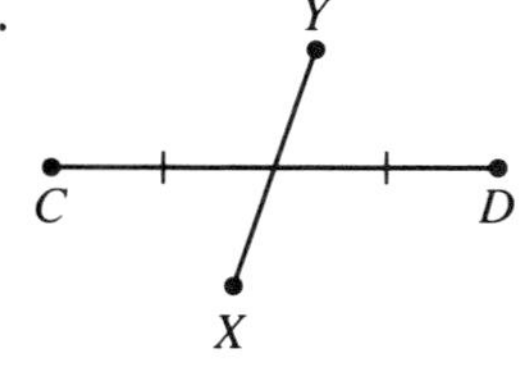

b.

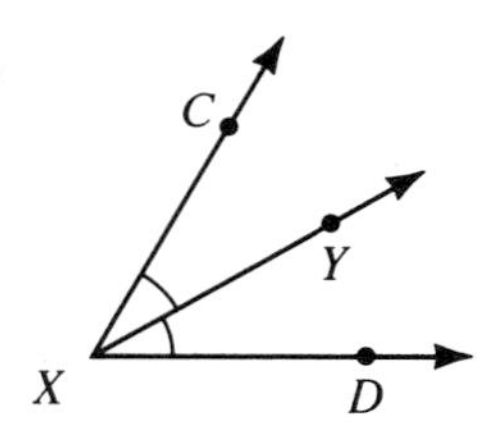

7. a.

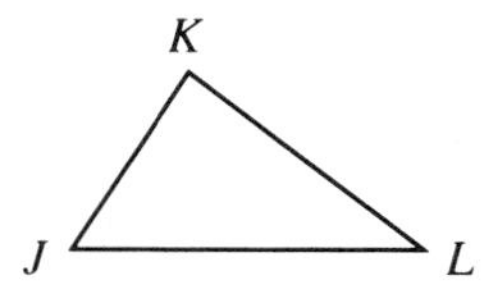

b.

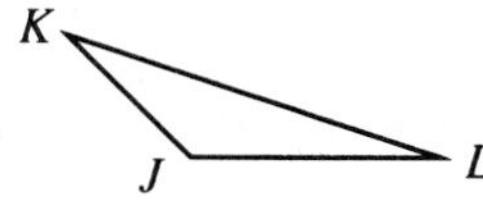

8. a.

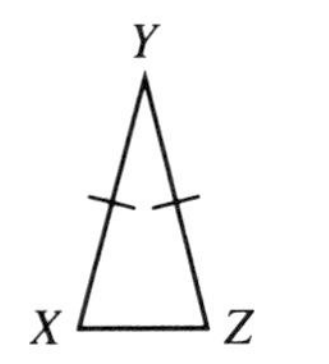

b.

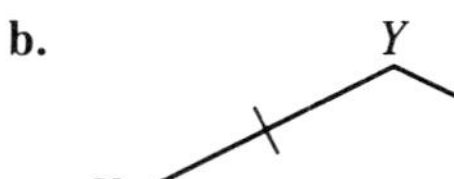

9. a.

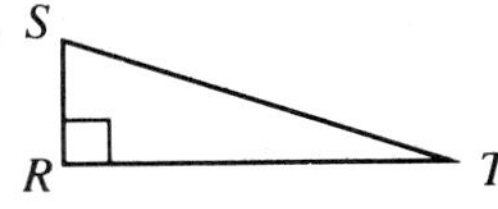

b.

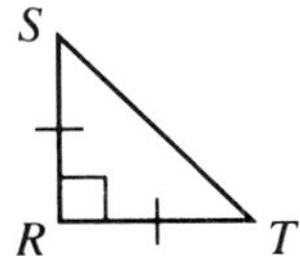

10. a.

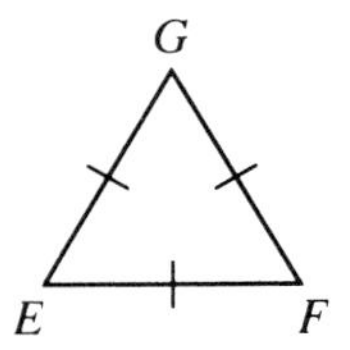

b.

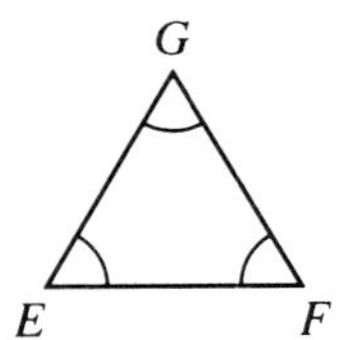

11.

Statements	Reasons
1. $\overline{BE} \cong \overline{CD}$; $\overline{BD} \cong \overline{CE}$	1. Given
2. $\overline{BC} \cong \overline{BC}$	2. Refl. Prop.
3. $\triangle EBC \cong \triangle DCB$	3. SSS Post.
4. $\angle EBC \cong \angle DCB$	4. Corr. parts of $\cong$ ⚠ are $\cong$.
5. $\overline{AB} \cong \overline{AC}$	5. If 2 ∡ of a $\triangle$ are $\cong$, then the sides opp. those ∡ are $\cong$.
6. $\triangle ABC$ is isos.	6. Def. of isos. $\triangle$

Page 155 • CLASSROOM EXERCISES

1. median **2.** altitude **3.** $\perp$ bis.

4. a. SAS **b.** isos. **5.** S, T; RS, RT

6. a. $\overline{BE}$ **b.** $\overline{AD}$ **c.** $\overrightarrow{CF}$

7. a.

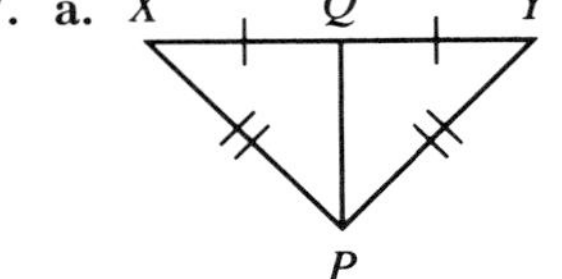

b. SSS Post.

c. Corr. parts of $\cong$ ⚠ are $\cong$. **d.** If 2 lines form $\cong$ adj. ∡, then the lines are $\perp$. **e.** $\perp$ bis.

8. a. Yes **b.** Yes **9.** equilateral

10. a. P is equidistant from the sides of $\angle N$. **b.** Q lies on $\overrightarrow{NO}$.

11. $\triangle AOC \cong \triangle BOC$ by SAS; $\triangle AOD \cong \triangle BOD$ by SAS; $\triangle ACD \cong \triangle BCD$ by SSS.

Pages 156–158 • WRITTEN EXERCISES

A **1–5.** Check students' drawings. **1. b.** No **5.** Yes; at the midpt. of the hyp.
6. The 3 $\angle$ bisectors meet in a pt. **7.** $\overrightarrow{KS}$, $\overrightarrow{KN}$ **8.** $\overrightarrow{NS}$, $\overrightarrow{NK}$ **9.** bis. of $\angle S$
10. L, A **11.** A, F **12.** $\perp$ bis. of $\overline{LF}$

13.

Statements	Reasons
1. P is on the $\perp$ bisectors of $\overline{AB}$ and $\overline{BC}$.	1. Given
2. $PA = PB$; $PB = PC$	2. If a pt. lies on the $\perp$ bis. of a seg., then it is equidistant from the endpts. of the seg.
3. $PA = PC$	3. Trans. Prop.

B **14.**

Statements	Reasons
1. l is the $\perp$ bis. of $\overline{BC}$.	1. Given
2. $l \perp \overline{BC}$; X is the midpt. of $\overline{BC}$.	2. Def. of $\perp$ bis.
3. $\angle AXB \cong \angle AXC$	3. If 2 lines are $\perp$, then they form $\cong$ adj. $\angle$s.
4. $\overline{BX} \cong \overline{CX}$	4. Def. of midpt.
5. $\overline{AX} \cong \overline{AX}$	5. Refl. Prop.
6. $\triangle AXB \cong \triangle AXC$	6. SAS Post.
7. $\overline{AB} \cong \overline{AC}$ or $AB = AC$	7. Corr. parts of $\cong$ $\triangle$s are $\cong$.

15.

Statements	Reasons
1. Let X be the midpt. of $\overline{BC}$.	1. Ruler Post.
2. $\overline{XB} \cong \overline{XC}$	2. Def. of midpt.
3. Draw $\overline{AX}$.	3. Through any 2 pts. there is exactly one line.
4. $\overline{AX} \cong \overline{AX}$	4. Refl. Prop.
5. $AB = AC$ or $\overline{AB} \cong \overline{AC}$	5. Given
6. $\triangle AXB \cong \triangle AXC$	6. SSS Post.
7. $\angle 1 \cong \angle 2$	7. Corr. parts of $\cong$ $\triangle$s are $\cong$.
8. $\overline{AX} \perp \overline{BC}$	8. If 2 lines form $\cong$ adj. $\angle$s, then the lines are $\perp$.
9. $\overleftrightarrow{AX}$ is the $\perp$ bis. of $\overline{BC}$.	9. Def. of $\perp$ bis.
10. A is on the $\perp$ bis. of $\overline{BC}$.	10. A is on $\overleftrightarrow{AX}$.

16.

Statements	Reasons
1. $\overrightarrow{BZ}$ bis. $\angle ABC$.	1. Given
2. $\angle PBX \cong \angle PBY$	2. Def. of $\angle$ bis.
3. $\overline{PX} \perp \overrightarrow{BA}$; $\overline{PY} \perp \overrightarrow{BC}$	3. Given
4. $m\angle PXB = 90$; $m\angle PYB = 90$	4. Def. of $\perp$ lines
5. $\angle PXB \cong \angle PYB$	5. Def. of $\cong$ ∠s
6. $\overline{PB} \cong \overline{PB}$	6. Refl. Prop.
7. $\triangle PXB \cong \triangle PYB$	7. AAS Thm.
8. $\overline{PX} \cong \overline{PY}$ or $PX = PY$	8. Corr. parts of $\cong$ △s are $\cong$.

17.

Statements	Reasons
1. $\overline{PX} \perp \overrightarrow{BA}$; $\overline{PY} \perp \overrightarrow{BC}$	1. Given
2. $\angle PXB$ and $\angle PYB$ are rt. ∠s.	2. Def. of $\perp$ lines
3. $\triangle PXB$ and $\triangle PYB$ are rt. △s.	3. Def. of rt. $\triangle$
4. $\overline{PB} \cong \overline{PB}$	4. Refl. Prop.
5. $PX = PY$ or $\overline{PX} \cong \overline{PY}$	5. Given
6. $\triangle PXB \cong \triangle PYB$	6. HL Thm.
7. $\angle PBX \cong \angle PBY$	7. Corr. parts of $\cong$ △s are $\cong$.
8. $\overrightarrow{BP}$ bis. $\angle ABC$.	8. Def. of $\angle$ bis.

18.

Statements	Reasons
1. S and V are equidistant from E and D.	1. Given
2. S and V are on the $\perp$ bis. of $\overline{ED}$.	2. If a pt. is equidistant from the endpts. of a seg., then it is on the $\perp$ bis. of the seg.
3. $\overleftrightarrow{SV}$ is the $\perp$ bis. of $\overline{ED}$.	3. Through any 2 pts. there is exactly one line.

19. **a.**

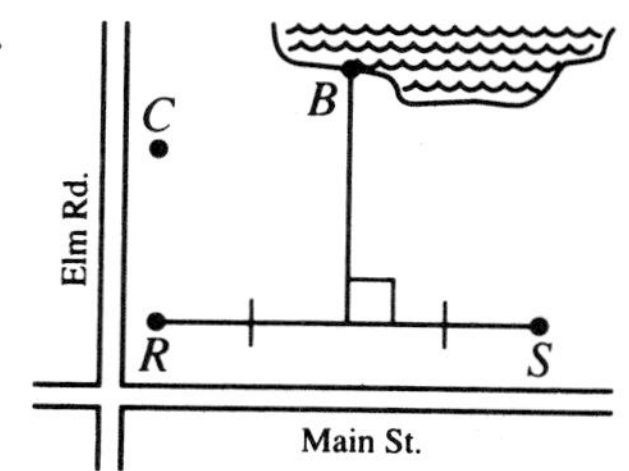

b.

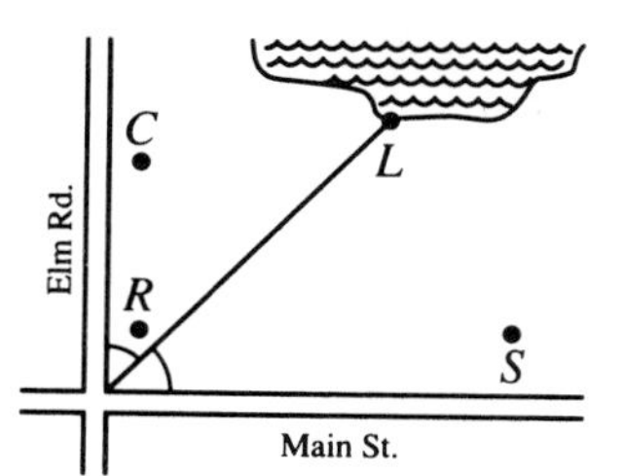

c.

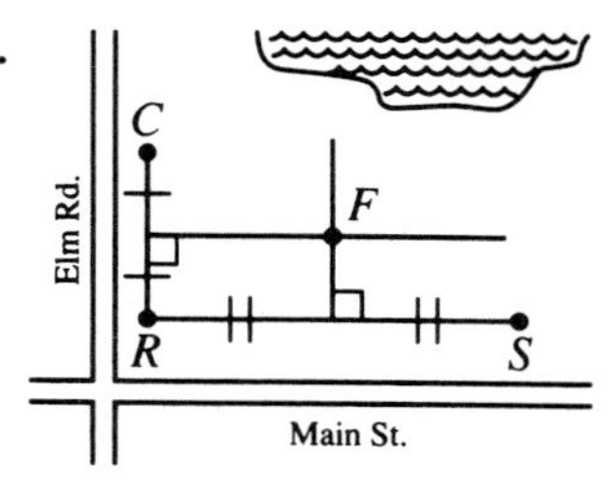

20.

Statements	Reasons
1. $\triangle LMN \cong \triangle RST$	1. Given
2. $\angle N \cong \angle T$; $\overline{LN} \cong \overline{RT}$	2. Corr. parts of $\cong$ △s are $\cong$.
3. $\overline{LX}$ and $\overline{RY}$ are altitudes.	3. Given
4. $\overline{LX} \perp \overline{XN}$; $\overline{RY} \perp \overline{YT}$	4. Def. of altitude
5. $m\angle X = 90$; $m\angle Y = 90$	5. Def. of $\perp$ lines
6. $\angle X = \angle Y$	6. Def. of $\cong$ ∠s
7. $\triangle LXN \cong \triangle RYT$	7. AAS Thm.
8. $\overline{LX} \cong \overline{RY}$	8. Corr. parts of $\cong$ △s are $\cong$.

21. **a.**

Statements	Reasons
1. $\overline{AB} \cong \overline{AC}$; $\overline{BD} \perp \overline{AC}$; $\overline{CE} \perp \overline{AB}$	1. Given
2. $m\angle BDA = 90$; $m\angle CEA = 90$	2. Def. of $\perp$ lines
3. $\angle BDA \cong \angle CEA$	3. Def. of $\cong$ ∠s
4. $\angle A \cong \angle A$	4. Refl. Prop.
5. $\triangle ADB \cong \triangle AEC$	5. AAS Thm.
6. $\overline{BD} \cong \overline{CE}$	6. Corr. parts of $\cong$ △s are $\cong$.

b. The altitudes drawn to the legs of an isos. $\triangle$ are $\cong$.

22. Given: $\overline{AB} \cong \overline{AC}$; $\overline{CM}$ and $\overline{BN}$ are medians.
Prove: $\overline{CM} \cong \overline{BN}$

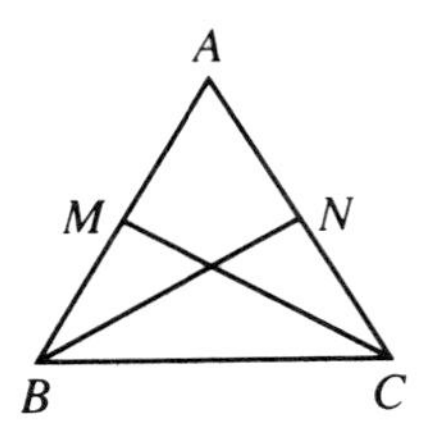

Statements	Reasons
1. $\overline{AB} \cong \overline{AC}$ or $AB = AC$	1. Given
2. $\angle ABC \cong \angle ACB$	2. Isos. △ Thm.
3. $\frac{1}{2}AB = \frac{1}{2}AC$	3. Mult. Prop. of =
4. $\overline{CM}$ and $\overline{BN}$ are medians.	4. Given
5. M is the midpt. of $\overline{AB}$; N is the midpt. of $\overline{AC}$.	5. Def. of median
6. $MB = \frac{1}{2}AB$; $NC = \frac{1}{2}AC$	6. Midpt. Thm.
7. $MB = NC$ or $\overline{MB} \cong \overline{NC}$	7. Substitution Prop.
8. $\overline{BC} \cong \overline{BC}$	8. Refl. Prop.
9. $\triangle MBC \cong \triangle NCB$	9. SAS Post.
10. $\overline{CM} \cong \overline{BN}$	10. Corr. parts of ≅ ▵ are ≅.

23. Q is on the ⊥ bis. of $\overline{PS}$, so $PQ = SQ$. S is on the ⊥ bis. of $\overline{QT}$, so $QS = TS$. Then $PQ = TS$ by the Trans. Prop.

24. P is on the bis. of $\angle ADE$, so the distance from P to $\overline{BA}$ equals the distance from P to $\overline{DE}$. P is on the bis. of $\angle DEC$, so the distance from P to $\overline{DE}$ equals the distance from P to $\overline{BC}$. By the Trans. Prop. the distance from P to $\overline{BA}$ equals the distance from P to $\overline{BC}$. Thus P is on the bis. of $\angle ABC$ and $\overrightarrow{BP}$ is the bis. of $\angle ABC$.

25. **a.** $\overline{OD}$ is a ⊥ bis. of $\overline{AB}$, so $\overline{AD} \cong \overline{BD}$. **b.** $\overline{OC}$ is a ⊥ bis. of $\overline{AB}$, so $\overline{AC} \cong \overline{BC}$.
c. By parts (a) and (b) above, $\overline{AD} \cong \overline{BD}$ and $\overline{AC} \cong \overline{BC}$. Then since $\overline{CD} \cong \overline{CD}$, $\triangle CAD \cong \triangle CBD$ by SSS, and corr. parts $\angle CAD$ and $\angle CBD$ are ≅.

C 26. $\overline{MN}$ is the ⊥ bisector of $\overline{TS}$, so $\overline{MT} \cong \overline{MS}$ and $\overline{TN} \cong \overline{NS}$. Rt. $\triangle MNT \cong$ rt. $\triangle MNS$ by HL, so corr. parts $\angle TMN$ and $\angle SMN$ are ≅. Also, since $\overline{TS} \perp \overline{RT}$ and $\overline{TS} \perp \overline{MN}$, $\overline{RT} \parallel \overline{MN}$. Then $\angle RTM \cong \angle TMN$ (alt. int. ∠s) and $\angle R \cong \angle SMN$ (corr. ∠s). By the Substitution Prop., $\angle RTM \cong \angle R$. Hence, in $\triangle RMT$, $\overline{MR} \cong \overline{MT}$. Since $\overline{MT} \cong \overline{MS}$, then $\overline{MR} \cong \overline{MS}$ by the Trans. Prop. and $\overline{TM}$ is a median.

27. a. Since $\overline{EH}$ and $\overline{FJ}$ are medians, $\overline{HG} \cong \overline{HF}$ and $\overline{JG} \cong \overline{JE}$. Also, $\overline{EH} \cong \overline{HP}$ and $\overline{JQ} \cong \overline{FJ}$. Vert. $\angle$s are $\cong$, so $\triangle EHF \cong \triangle PHG$ and $\triangle FJE \cong \triangle QJG$ by SAS. Then corr. parts $\overline{EF}$ and $\overline{PG}$ are $\cong$ and $\overline{EF}$ and $\overline{GQ}$ are $\cong$. Then $\overline{GQ} \cong \overline{GP}$. **b.** Corr. parts of $\cong$ $\triangle$s are $\cong$, so $\angle HEF \cong \angle HPG$ and $\angle JFE \cong \angle JQG$. Then $\overline{GP} \parallel \overline{EF}$ and $\overline{GQ} \parallel \overline{EF}$ (If 2 lines are cut by a trans. and alt. int. $\angle$s are $\cong$, then the lines are $\parallel$.). **c.** P, G, and Q are collinear since through a pt. outside a line there is exactly one line $\parallel$ to a given line.

28. a. $\overline{AE} \cong \overline{BC}$ and $\overline{AB} \cong \overline{AB}$. Since $\overline{AD} \cong \overline{BD}$, $\angle DAB \cong \angle DBA$. Using alt. int. $\angle$s, $\angle EAD \cong \angle ADB$ and $\angle ADB \cong \angle DBC$, so $\angle EAD \cong \angle DBC$. $m\angle EAB = m\angle EAD + m\angle DAB$ and $m\angle CBA = m\angle CBD + m\angle DBA$, so $\angle EAB \cong \angle CBA$. Then $\triangle EAB \cong \triangle CBA$ by SAS, and corr. parts $\overline{AC}$ and $\overline{BE}$ are $\cong$.

b. Using the $\cong$ $\triangle$s from part (a) above, $\angle EBA \cong \angle CAB$, so $OA = OB$. Since $BE = AC$, $EO = CO$ and thus $\angle CEO \cong \angle ECO$. $\angle EOC \cong \angle AOB$ (vert. $\angle$s), so $m\angle CEO = \frac{1}{2}(180 - m\angle EOC)$ and $m\angle EBA = \frac{1}{2}(180 - m\angle EOC)$. Then $\angle CEO \cong \angle EBA$ and $\overline{EC} \parallel \overline{AB}$.

29. Since $\overleftrightarrow{AM}$ is the $\perp$ bis. of $\overline{BC}$, $\overline{AB} \cong \overline{AC}$. $\angle 1 \cong \angle 2$ so $\overrightarrow{DA}$ bisects $\angle BDF$ and $\overline{AE} \cong \overline{AF}$ (If a pt. lies on the bis. of an $\angle$, then the pt. is equidistant from the sides of the $\angle$.). Since $\overline{AE} \perp \overline{BD}$ and $\overline{AF} \perp \overline{DF}$, $\triangle AEB$ and $\triangle AFC$ are rt. $\triangle$s. $\triangle AEB \cong \triangle AFC$ by HL, so corr. parts $\overline{BE}$ and $\overline{CF}$ are $\cong$.

Page 158 • EXPLORATIONS

Check students' drawings.

1. False; true for vertex $\angle$s of isos. $\triangle$s. **2.** False; true for vertex $\angle$s of isos. $\triangle$s.

3. False; true from rt. $\angle$ vertex in rt. $\triangle$s.

Page 159 • SELF-TEST 3

1. $\overline{EA} \cong \overline{DB}$; $\angle AEB \cong \angle BDA$

2.

Statements	Reasons
1. $\triangle MPQ \cong \triangle PMN$	1. Given
2. $\overline{MN} \cong \overline{QP}$; $\angle MPQ \cong \angle PMN$	2. Corr. parts of $\cong$ $\triangle$s are $\cong$.
3. $\overline{MS} \cong \overline{PR}$	3. Given
4. $\triangle MSN \cong \triangle PRQ$	4. SAS Post.

3. a. $\overline{LJ}$ or $\overline{KJ}$ **b.** $\overline{KZ}$ **4.** No

5. If a pt. lies on the bis. of an angle, then the pt. is equidistant from the sides of the angle.
6. If a pt. is equidistant from the endpts. of a seg., then the pt. lies on the $\perp$ bis. of the seg.

Pages 160–161 • CHAPTER REVIEW

1. $\triangle QPR$ 2. $\triangle TSW$ 3. $\angle W$ 4. WT 5. Yes; SSS 6. No 7. Yes; ASA
8. Yes; SAS

9.

Statements	Reasons
1. $\overline{JM} \cong \overline{LM}$; $\overline{JK} \cong \overline{LK}$	1. Given
2. $\overline{MK} \cong \overline{MK}$	2. Refl. Prop.
3. $\triangle MJK \cong \triangle MLK$	3. SSS Post.
4. $\angle MJK \cong \angle MLK$	4. Corr. parts of $\cong$ ⚠ are $\cong$.

10.

Statements	Reasons
1. $\angle JMK \cong \angle LMK$; $\overline{MK} \perp$ plane P	1. Given
2. $\overline{MK} \perp \overline{JK}$; $\overline{MK} \perp \overline{LK}$	2. Def. of a line $\perp$ to a plane
3. $m\angle MKJ = 90$; $m\angle MKL = 90$	3. Def. of $\perp$ lines
4. $\angle MKJ \cong \angle MKL$	4. Def. of $\cong$ ∡s
5. $\overline{MK} \cong \overline{MK}$	5. Refl. Prop.
6. $\triangle MKJ \cong \triangle MKL$	6. ASA Post.
7. $\overline{JK} \cong \overline{LK}$	7. Corr. parts of $\cong$ ⚠ are $\cong$.

11. $\overline{ER}$, $\overline{EV}$ 12. equilateral 13. $3x = 75; x = 25$
14. $3y + 5 = 25 - y; 4y = 20; y = 5$

15.

Statements	Reasons
1. $\overline{GH} \perp \overline{HJ}$; $\overline{KJ} \perp \overline{HJ}$	1. Given
2. $m\angle GHJ = 90$; $m\angle KJH = 90$	2. Def. of $\perp$ lines
3. $\angle GHJ \cong \angle KJH$	3. Def. of $\cong$ ∡s
4. $\angle G \cong \angle K$	4. Given
5. $\overline{HJ} \cong \overline{HJ}$	5. Refl. Prop.
6. $\triangle GHJ \cong \triangle KJH$	6. AAS Thm.

16.

Statements	Reasons
1. $\overline{GH} \perp \overline{HJ}$; $\overline{KJ} \perp \overline{HJ}$	1. Given
2. $\angle GHJ$ and $\angle KJH$ are rt. $\angle$s.	2. Def. of $\perp$ lines
3. $\triangle GHJ$ and $\triangle KJH$ are rt. $\triangle$s.	3. Def. of rt. $\triangle$
4. $\overline{GJ} \cong \overline{KH}$	4. Given
5. $\overline{HJ} \cong \overline{HJ}$	5. Refl. Prop.
6. $\triangle GHJ \cong \triangle KJH$	6. HL Thm.
7. $\overline{GH} \cong \overline{KJ}$	7. Corr. parts of $\cong$ $\triangle$s are $\cong$.

17. 1. ASA 2. Corr. parts of $\cong$ $\triangle$s are $\cong$. 3. HL 4. Corr. parts of $\cong$ $\triangle$s are $\cong$. 5. If 2 lines are cut by a trans. and alt. int. $\angle$s are $\cong$, then the lines are $\parallel$.

18. **a.** $\overline{DG}$ **b.** $\overline{FH}$ **c.** $\overline{KJ}$

19. If a pt. lies on the $\perp$ bis. of a seg., then the pt. is equidistant from the endpts. of the seg.

20. If a pt. is equidistant from the sides of an $\angle$, then the pt. lies on the bis. of the $\angle$.

Pages 162–163 • CHAPTER TEST

1. $\overline{PT}$, $\triangle DBA$ **2.** $\overline{GE}$, $\overline{GF}$; 43 **3.** $\angle A$, $\angle X$; $\overline{BC}$, $\overline{YZ}$ **4.** HL **5.** 120
6. median **7.** the sides of the angle **8.** $7x + 8 = 38 - 3x$; $10x = 30$; $x = 3$
9. Yes; ASA or AAS **10.** No **11.** Yes; AAS or ASA **12.** Yes; HL **13.** Yes; SSS
14. Yes; SAS **15.** Y, Z **16.** W, X **17.** $\triangle YWX$, $\triangle ZWX$, $\triangle WYZ$, $\triangle XYZ$
18. 8

19.

Statements	Reasons
1. $\angle 1 \cong \angle 2$; $\angle PQR \cong \angle SRQ$	1. Given
2. $\overline{QR} \cong \overline{QR}$	2. Refl. Prop.
3. $\triangle PQR \cong \triangle SRQ$	3. ASA Post.
4. $\overline{PR} \cong \overline{SQ}$	4. Corr. parts of $\cong$ $\triangle$s are $\cong$.

20.

Statements	Reasons
1. $\angle 1 \cong \angle 2$	1. Given
2. $\overline{WX} \cong \overline{WY}$	2. If 2 $\angle$s of a $\triangle$ are $\cong$, then the sides opp. those $\angle$s are $\cong$.
3. $\angle 3 \cong \angle 4$	3. Given
4. $\overline{WZ} \cong \overline{WZ}$	4. Refl. Prop.
5. $\triangle WZX \cong \triangle WZY$	5. SAS Post.
6. $\overline{ZX} \cong \overline{ZY}$	6. Corr. parts of $\cong$ $\triangle$s are $\cong$.
7. $\triangle ZXY$ is isos.	7. Def. of isos. $\triangle$

Page 163 • ALGEBRA REVIEW

1. $x^2 + 5x - 6 = 0$; $(x + 6)(x - 1) = 0$; $x + 6 = 0$ or $x - 1 = 0$; $x = -6, 1$
2. $n^2 - 6n + 8 = 0$; $(n - 2)(n - 4) = 0$; $n - 2 = 0$ or $n - 4 = 0$; $n = 2, 4$
3. $y^2 - 7y - 18 = 0$; $(y + 2)(y - 9) = 0$; $y + 2 = 0$ or $y - 9 = 0$; $y = -2, 9$
4. $x^2 + 8x = 0$; $x(x + 8) = 0$; $x = 0$ or $x = -8$; $x = 0, -8$
5. $y^2 = 13y$; $y^2 - 13y = 0$; $y(y - 13) = 0$; $y = 0$ or $y - 13 = 0$; $y = 0, 13$
6. $2z^2 + 7z = 0$; $z(2z + 7) = 0$; $z = 0$ or $2z + 7 = 0$; $z = 0, -3.5$
7. $n^2 - 144 = 25$; $n^2 - 169 = 0$; $(n + 13)(n - 13) = 0$; $n + 13 = 0$ or $n - 13 = 0$; $n = -13, 13$
8. $50x^2 = 200$; $50x^2 - 200 = 0$; $50(x^2 - 4) = 0$; $(x + 2)(x - 2) = 0$; $x + 2 = 0$ or $x - 2 = 0$; $x = -2, 2$
9. $50x^2 = 2$; $50x^2 - 2 = 0$; $2(25x^2 - 1) = 0$; $(5x + 1)(5x - 1) = 0$; $5x + 1 = 0$ or $5x - 1 = 0$; $x = -0.2, 0.2$
10. $49z^2 = 1$; $49z^2 - 1 = 0$; $(7z + 1)(7z - 1) = 0$; $7z + 1 = 0$ or $7z - 1 = 0$; $z = -\frac{1}{7}, \frac{1}{7}$
11. $y^2 - 6y + 9 = 0$; $(y - 3)(y - 3) = 0$; $y = 3$
12. $x^2 - 7x + 12 = 0$; $(x - 3)(x - 4) = 0$; $x - 3 = 0$ or $x - 4 = 0$; $x = 3, 4$
13. $y^2 + 8y + 12 = 0$; $(y + 6)(y + 2) = 0$; $y + 6 = 0$ or $y + 2 = 0$; $y = -6, -2$
14. $t^2 + 5t - 24 = 0$; $(t + 8)(t - 3) = 0$; $t + 8 = 0$ or $t - 3 = 0$; $t = -8, 3$
15. $v^2 - 10v + 25 = 0$; $(v - 5)(v - 5) = 0$; $v = 5$
16. $x^2 - 3x - 4 = 0$; $(x + 1)(x - 4) = 0$; $x + 1 = 0$ or $x - 4 = 0$; $x = -1, 4$
17. $t^2 - t - 20 = 0$; $(t + 4)(t - 5) = 0$; $t + 4 = 0$ or $t - 5 = 0$; $t = -4, 5$
18. $y^2 - 20y + 36 = 0$; $(y - 2)(y - 18) = 0$; $y - 2 = 0$ or $y - 18 = 0$; $y = 2, 18$
19. $3x^2 + 3x - 4 = 0$; $x = \dfrac{-3 \pm \sqrt{9 + 48}}{6} = \dfrac{-3 \pm \sqrt{57}}{6}$
20. $4y^2 - 17y + 15 = 0$; $(4y - 5)(y - 3) = 0$; $4y - 5 = 0$ or $y - 3 = 0$; $y = 1.25, 3$
21. $x^2 + 5x + 2 = 0$; $x = \dfrac{-5 \pm \sqrt{25 - 8}}{2} = \dfrac{-5 \pm \sqrt{17}}{2}$
22. $x^2 + 2x - 1 = 0$; $x = \dfrac{-2 \pm \sqrt{4 + 4}}{2} = -1 \pm \sqrt{2}$
23. $x^2 - 5x + 3 = 0$; $x = \dfrac{5 \pm \sqrt{25 - 12}}{2} = \dfrac{5 \pm \sqrt{13}}{2}$
24. $x^2 + 3x - 2 = 0$; $x = \dfrac{-3 \pm \sqrt{9 + 8}}{2} = \dfrac{-3 \pm \sqrt{17}}{2}$

25. $y^2 - 10y + 25 = 16; y^2 - 10y + 9 = 0; (y - 1)(y - 9) = 0; y - 1 = 0$ or $y - 9 = 0; y = 1, 9$

26. $z^2 = 8z - 12; z^2 - 8z + 12 = 0; (z - 2)(z - 6) = 0; z - 2 = 0$ or $z - 6 = 0;$ $z = 2, 6$

27. $x^2 + 5x - 14 = 0; (x + 7)(x - 2) = 0; x + 7 = 0$ or $x - 2 = 0; x = -7, 2$

28. $x(x - 50) = 0; x = 0$ (reject) or $x - 50 = 0; x = 50$

29. $x^2 - 400 = 0; (x + 20)(x - 20) = 0; x + 20 = 0$ or $x - 20 = 0;$ $x = -20$ (reject), $x = 20$

30. $x^2 - 17x + 72 = 0; (x - 8)(x - 9) = 0; x - 8 = 0$ or $x - 9 = 0; x = 8, 9$

31. $2x^2 + x - 3 = 0; (2x + 3)(x - 1) = 0; 2x + 3 = 0$ or $x - 1 = 0;$ $x = -\frac{3}{2}$ (reject), $x = 1$

32. $2x^2 - 7x - 4 = 0; (2x + 1)(x - 4) = 0; 2x + 1 = 0$ or $x - 4 = 0;$ $x = -\frac{1}{2}$ (reject), $x = 4$

33. $6x^2 - 5x - 6 = 0; (3x + 2)(2x - 3) = 0; 3x + 2 = 0$ or $2x - 3 = 0;$ $x = -\frac{2}{3}$ (reject), $x = 1.5$

Page 164 • PREPARING FOR COLLEGE ENTRANCE EXAMS

1. A. $(2x + 10) + 3x + (8x - 25) = 180; 13x = 195; x = 15$; substituting, $2x + 10 = 40, 3x = 45, 8x - 25 = 95$, so the triangle is obtuse.

2. C. $(n - 2)180 = 120n; 180n - 360 = 120n; 60n = 360; n = 6$

3. D **4.** C **5.** B **6.** C

7. E. $2(2x - 36) + x + 2 = 180; 4x - 72 + x + 2 = 180; 5x = 250; x = 50;$ substituting, $2x - 36 = 64$

8. D **9.** B

Page 165 • CUMULATIVE REVIEW: CHAPTERS 1–4

A **1.** Seg. Add. Post. **2.** parallel **3.** obtuse **4.** perpendicular

5. 16 **6.** If $3x = 27$, then $x = 9$.

7. $(n - 2)180 = 144n; 180n - 360 = 144n; 36n = 360; n = 10$

8. $(y + 10) + (2y - 31) = 180; 3y = 201; y = 67$; substituting, $y + 10 = 77$, $2y - 31 = 103, 2y - 40 = 94; m\angle G = 360 - (77 + 103 + 94) = 86$

9. SSS

10. $m\angle 1 = 140 - 65 = 75; m\angle 3 = m\angle 2 = 180 - 140 = 40;$ $m\angle 4 = 180 - (75 + 40) = 65$

11. $m\angle 5 = 90; m\angle 7 = 36; m\angle 6 = m\angle 8 = 180 - (36 + 90) = 54$

12. Yes; $c \parallel d$ **13.** No **14.** Yes; $a \parallel b$ **15.** Yes; $a \parallel b$

B 16.

Statements	Reasons
1. $\overline{MN} \cong \overline{MP}$; $\angle NMO \cong \angle PMO$	1. Given
2. $\overline{MO} \cong \overline{MO}$	2. Refl. Prop.
3. $\triangle NMO \cong \triangle PMO$	3. SAS Post.
4. $\overline{NO} \cong \overline{PO}$	4. Corr. parts of $\cong$ $\triangle$s are $\cong$.
5. M and O lie on the $\perp$ bis. of $\overline{NP}$.	5. If a pt. is equidistant from the endpts. of a seg., then the pt. lies on the $\perp$ bis. of the seg.
6. $\overleftrightarrow{MO}$ is the $\perp$ bis. of $\overline{NP}$.	6. Through any 2 pts. there is exactly one line.

17.

Statements	Reasons
1. $\overline{MO} \perp \overline{NP}$; $\overline{NO} \cong \overline{PO}$	1. Given
2. $\angle NQO$ and $\angle PQO$ are rt. $\angle$s.	2. Def. of $\perp$ lines
3. $\triangle NQO$ and $\triangle PQO$ are rt. $\triangle$s.	3. Def. of rt. $\triangle$
4. $\overline{QO} \cong \overline{QO}$	4. Refl. Prop.
5. $\triangle NQO \cong \triangle PQO$	5. HL Thm.
6. $\angle NOQ \cong \angle POQ$	6. Corr. parts of $\cong$ $\triangle$s are $\cong$.
7. $\overline{MO} \cong \overline{MO}$	7. Refl. Prop.
8. $\triangle MNO \cong \triangle MPO$	8. SAS Post.
9. $\overline{MN} \cong \overline{MP}$	9. Corr. parts of $\cong$ $\triangle$s are $\cong$.

18. Since $\overline{AX}$ is a median, $\overline{BX} \cong \overline{CX}$. Since $\overline{AX}$ is an altitude, $\angle AXB$ and $\angle AXC$ are rt. $\angle$s. $\overline{AX} \cong \overline{AX}$ by the Refl. Prop. So $\triangle AXB \cong \triangle AXC$ by SAS, and corr. parts $\overline{AB}$ and $\overline{AC}$ are $\cong$. Therefore, $\triangle ABC$ is isos.

CHAPTER 5 • Quadrilaterals

Page 168 • CLASSROOM EXERCISES

1. **a.** $\overline{GR} \parallel \overline{MA}$, and if 2 $\parallel$ lines are cut by a trans., then s-s. int. $\angle$s are supp.
b. $\overline{GM} \parallel \overline{RA}$, and if 2 $\parallel$ lines are cut by a trans., then s-s. int. $\angle$s are supp.
c. supp., $\cong$

2. They are also rt. $\angle$s. **3.** $x = y = 65$; $z = 180 - 65 = 115$

4. $x = 56$; $y = 180 - (85 + 56)$, $y = 39$; $z = 85$

5. $x = 110$; $y = 180 - 2(70) = 40$; $z = 70$

6. Yes; $\overline{HG} \parallel \overline{EF}$ and $\overline{HE} \parallel \overline{GF}$ since corr. $\angle$s are $\cong$.

7. No; no; $\angle E \not\cong \angle G$. **8.** No; yes; you only know that $\overline{HG} \parallel \overline{EF}$.

9. Def. of ▱. **10.** If 2 $\parallel$ lines are cut by a trans., then alt. int. $\angle$s are $\cong$.

11. Opp. $\angle$s of a ▱ are $\cong$. **12.** Opp. sides of a ▱ are $\cong$.

13, 14. Diags. of a ▱ bis. each other. **15.**

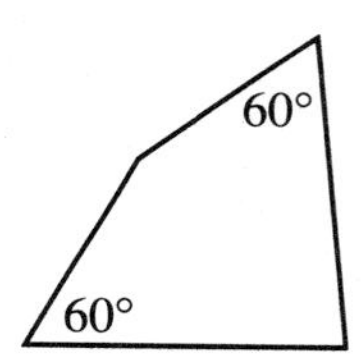

16. **a.** If a quad. is a ▱, then opp. sides are $\cong$.
b. If a quad. is a ▱, then opp. $\angle$s are $\cong$.
c. If a quad. is a ▱, then the diags. bis. each other.

17. **a.** $ABCD$ is a ▱. **b.** Diags. of a ▱ bis. each other. **18.** The quad. is a ▱.

Pages 169–171 • WRITTEN EXERCISES

A **1.** $\overline{CR}, \overline{CE}$ **2.** $\angle REC, \angle WCE, \angle WEC$ **3.** $\overline{ER}, \overline{RC}, \overline{CW}$

4. $\angle CRW, \angle CWR, \angle EWR$ **5.** $a = 8$; $b = 10$; $x = 180 - 62 = 118$; $y = 62$

6. $a = 8$; $b = 15$; $x = 80$; $y = 180 - (80 + 30) = 70$

7. $a = 5$; $b = 3$; $x = 120$; $y = 180 - (120 + 38) = 22$

8. $a = 9$; $b = 11$; $x = 33$; $y = 180 - (50 + 70 + 33) = 27$

9. $a = 8$; $b = 8$; $x = 56$; $y = 180 - 2(56) = 68$

10. $a = 6 + 4 = 10$; $b = 4$; $x = 90$; $y = 45$ **11.** $2(17) + 2(13) = 60$

12. $x + (x + 1) = 27$, $x = 13$; $ST = 14$; $SP = 13$

13.

Statements	Reasons
1. $EFGH$ is a ▱.	1. Given
2. Draw $\overline{EG}$.	2. Through any 2 pts. there is exactly one line.
3. $\overline{HG} \parallel \overline{EF}$; $\overline{HE} \parallel \overline{GF}$	3. Def. of ▱
4. $\angle 1 \cong \angle 2$; $\angle 3 \cong \angle 4$	4. If 2 ∥ lines are cut by a trans., then alt. int. ∠s are ≅.
5. $\overline{EG} \cong \overline{EG}$	5. Refl. Prop.
6. $\triangle GHE \cong \triangle EFG$	6. ASA Post.
7. $\overline{EF} \cong \overline{HG}$; $\overline{FG} \cong \overline{EH}$	7. Corr. parts of ≅ △s are ≅.

14. Given: ▱$EFGH$

Prove: $\angle H \cong \angle F$; $\angle E \cong \angle G$

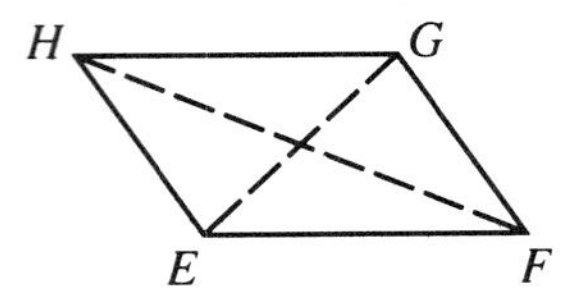

Statements	Reasons
1. $EFGH$ is a ▱.	1. Given
2. $\overline{EF} \cong \overline{GH}$; $\overline{EH} \cong \overline{GF}$	2. Opp. sides of a ▱ are ≅.
3. Draw $\overline{EG}$ and $\overline{FH}$.	3. Through any 2 pts. there is exactly one line.
4. $\overline{EG} \cong \overline{EG}$; $\overline{FH} \cong \overline{FH}$	4. Refl. Prop.
5. $\triangle GHE \cong \triangle EFG$; $\triangle GHF \cong \triangle EFH$	5. SSS Post.
6. $\angle H \cong \angle F$; $\angle E \cong \angle G$	6. Corr. parts of ≅ △s are ≅.

15.

Statements	Reasons
1. $QRST$ is a ▱.	1. Given
2. $\overline{QR} \parallel \overline{TS}$	2. Def. of ▱
3. $\angle 1 \cong \angle 2$; $\angle 3 \cong \angle 4$	3. If 2 ∥ lines are cut by a trans., then alt. int. ∠s are ≅.
4. $\overline{QR} \cong \overline{TS}$	4. Opp. sides of a ▱ are ≅.
5. $\triangle QMR \cong \triangle SMT$	5. ASA Post.
6. $\overline{QM} \cong \overline{MS}$; $\overline{RM} \cong \overline{MT}$	6. Corr. parts of ≅ △s are ≅.
7. $\overline{QS}$ and $\overline{TR}$ bis. each other.	7. Def. of bis.

16.

Statements	Reasons
1. $ABCX$ is a ▱; $DXFE$ is a ▱.	1. Given
2. $\angle E \cong \angle FXD$; $\angle B \cong \angle CXA$	2. Opp. ∠s of a ▱ are ≅.
3. $\angle FXD \cong \angle CXA$	3. Vert. ∠s are ≅.
4. $\angle B \cong \angle E$	4. Substitution Prop.

17. $D(3,2)$

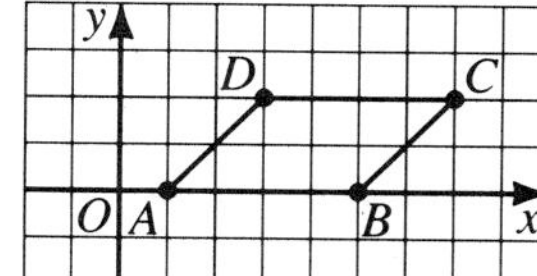

18. $C(5,5)$

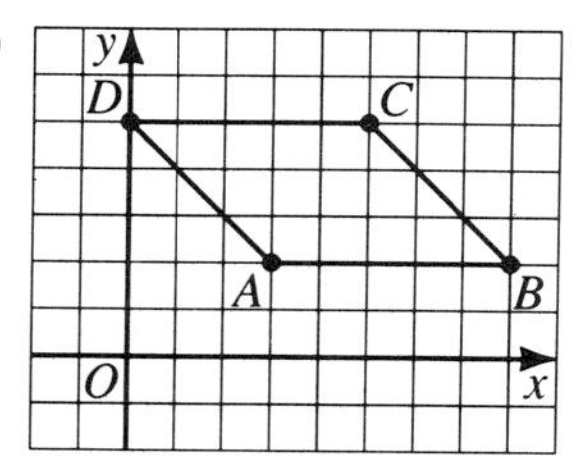

19. $2x + 8 = 14,\ x = 3;\ 3y = 15,\ y = 5$

20. $3x + 4 = 25,\ x = 7;\ 2y - 8 = 28,\ y = 18$

21. $5x = 180 - (80 + 35),\ x = 13;\ 6y + 5 = 35,\ y = 5$

B 22. $x + y = 85;\ 130 = (x - y) + 85,\ x - y = 45,\ x = y + 45;\ (y + 45) + y = 85,$
$2y = 40,\ y = 20;\ x + y = 85,\ x + 20 = 85,\ x = 65$

23.
$$\begin{array}{l} 2x + 5y = 30 \\ \underline{2x + 2y = 18} \\ \quad 3y = 12,\quad y = 4;\ 2x + 5y = 30,\ 2x + 20 = 30,\ 2x = 10,\ x = 5 \end{array}$$

24.
$$\begin{array}{l} 4x - y = 24 \\ \underline{2x + y = 36} \\ 6x \quad = 60,\quad x = 10;\ 2x + y = 36,\ 20 + y = 36,\ y = 16 \end{array}$$

25. $2x + y = 12,\ y = 12 - 2x;\ x + 2y = 9,\ x + 2(12 - 2x) = 9,$
$x + 24 - 4x = 9,\ -3x = -15,\ x = 5;\ 2x + y = 12,\ 10 + y = 12,\ y = 2$

26. $2x - y = x + y,\ x = 2y;\ 3x - 2y = 12,\ 3(2y) - 2y = 12,\ 4y = 12,\ y = 3;$
$2x - y = x + y,\ 2x - 3 = x + 3,\ x = 6;$ perimeter of $\square DECK = 2(3 + 6) +$
$2(12) = 42$

27. $3x = x^2 - 70,\ x^2 - 3x - 70 = 0,\ (x - 10)(x + 7) = 0,\ x = 10$ or
$x = -7$ (reject); $m\angle CED = m\angle CKD = m\angle 1 + m\angle 2 = 3(10) + 4(10) = 70$

28. $m\angle CED = m\angle 1 + m\angle 2$, so $13x = 42 + x^2;\ x^2 - 13x + 42 = 0,$
$(x - 7)(x - 6) = 0,\ x = 7$ or $x = 6;\ m\angle 2 = x^2,\ m\angle 2 = 49$ or 36

29.

Statements	Reasons
1. $PQRS$ is a $\square$; $\overline{PJ} \cong \overline{RK}$	1. Given
2. $\angle P \cong \angle R$	2. Opp. $\angle$s of a $\square$ are $\cong$.
3. $\overline{SP} \cong \overline{QR}$	3. Opp. sides of a $\square$ are $\cong$.
4. $\triangle SPJ \cong \triangle QRK$	4. SAS Post.
5. $\overline{SJ} \cong \overline{QK}$	5. Corr. parts of $\cong$ $\triangle$s are $\cong$.

30.

Statements	Reasons
1. $JQKS$ is a ▱; $\overline{PJ} \cong \overline{RK}$	1. Given
2. $\overline{SJ} \cong \overline{QK}$	2. Opp. sides of a ▱ are $\cong$.
3. $\angle SJQ \cong \angle SKQ$	3. Opp. $\angle$s of a ▱ are $\cong$.
4. $\angle 1 \cong \angle 2$	4. If 2 $\angle$s are supps. of $\cong$ $\angle$s, then the 2 $\angle$s are $\cong$.
5. $\triangle SPJ \cong \triangle QRK$	5. SAS Post.
6. $\angle P \cong \angle R$	6. Corr. parts of $\cong$ $\triangle$s are $\cong$.

31.

Statements	Reasons
1. $ABCD$ is a ▱; $\overline{CD} \cong \overline{CE}$	1. Given
2. $\overline{AB} \parallel \overline{CD}$	2. Def. of ▱
3. $\angle CDE \cong \angle A$	3. If 2 $\parallel$ lines are cut by a trans., then corr. $\angle$s are $\cong$.
4. $\angle CDE \cong \angle E$	4. Isos. $\triangle$ Thm.
5. $\angle A \cong \angle E$	5. Substitution Prop.

32.

Statements	Reasons
1. $ABCD$ is a ▱.	1. Given
2. $\overline{AB} \cong \overline{CD}$	2. Opp. sides of a ▱ are $\cong$.
3. $\overline{AB} \parallel \overline{CD}$	3. Def. of ▱
4. $\angle CDE \cong \angle A$	4. If 2 $\parallel$ lines are cut by a trans., then corr. $\angle$s are $\cong$.
5. $\angle A \cong \angle E$	5. Given
6. $\angle CDE \cong \angle E$	6. Trans. Prop.
7. $\overline{CD} \cong \overline{CE}$	7. If 2 $\angle$s of a $\triangle$ are $\cong$, then the sides opp. those $\angle$s are $\cong$.
8. $\overline{AB} \cong \overline{CE}$	8. Trans. Prop.

33. Answers may vary; for example, $\overline{DX} \parallel \overline{BY}$.

Statements	Reasons
1. *ABCD* is a ▱; $\angle 1 \cong \angle 2$	1. Given
2. $\overline{DA} \cong \overline{CB}$	2. Opp. sides of a ▱ are ≅.
3. $\overline{DA} \parallel \overline{CB}$	3. Def. of ▱
4. $\angle 3 \cong \angle 4$	4. If 2 ∥ lines are cut by a trans., then alt. int. ∠s are ≅.
5. $\triangle DXA \cong \triangle BYC$	5. ASA Post.
6. $\angle DXA \cong \angle BYC$	6. Corr. parts of ≅ △s are ≅.
7. $\angle DXY \cong \angle BYX$	7. If 2 ∠s are supps. of ≅ ∠s, then the 2 ∠s are ≅.
8. $\overline{DX} \parallel \overline{BY}$	8. If 2 lines are cut by a trans. and alt. int. ∠s are ≅, then the lines are ∥.

34. Answers may vary; for example, $\triangle JHI \cong \triangle GEF$.

Statements	Reasons
1. *EFIH* is a ▱; *EGJH* is a ▱.	1. Given
2. $\overline{HJ} \cong \overline{EG}$; $\overline{HI} \cong \overline{EF}$	2. Opp. sides of a ▱ are ≅.
3. $\angle 1 \cong \angle 2$	3. Given
4. $\triangle JHI \cong \triangle GEF$	4. SAS Post.

C 35. (6,0), (0,8), (12,8)

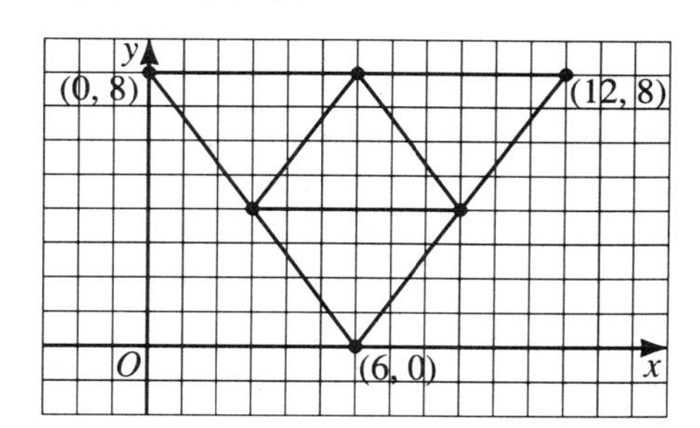

36. (5,0), (−1, −4), (−1,4)

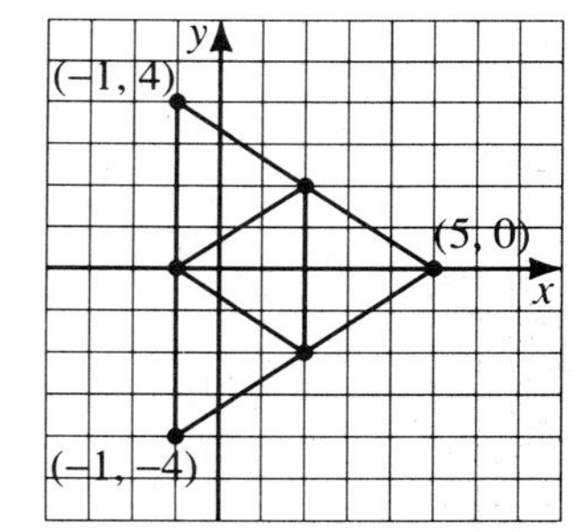

37. a.

Statements	Reasons
1. $j \parallel k$	1. Given
2. $\overleftrightarrow{AB}$ is in P; $\overleftrightarrow{XY}$ is in Q.	2. If 2 pts. are in a plane, then the line joining the pts. is in the plane.
3. A, B, X, and Y are coplanar, say they are in plane M.	3. By def., $\parallel$ lines are coplanar.
4. M int. P in $\overleftrightarrow{AB}$; M int. Q in $\overleftrightarrow{XY}$.	4. If 2 planes int., then their int. is a line.
5. Plane $P \parallel$ plane Q	5. Given
6. $\overleftrightarrow{AB} \parallel \overleftrightarrow{XY}$	6. If 2 $\parallel$ planes are cut by a third plane, then the lines of int. are $\parallel$.
7. $ABYX$ is a ▱.	7. Def. of ▱
8. $AX = BY$	8. Opp. sides of a ▱ are $\cong$.

b. If $\parallel$ planes intersect $\parallel$ lines, then they cut off $\cong$ segments.

38. Given: $RSTQ$ is a ▱.
Prove: $\overline{YO} \cong \overline{XO}$

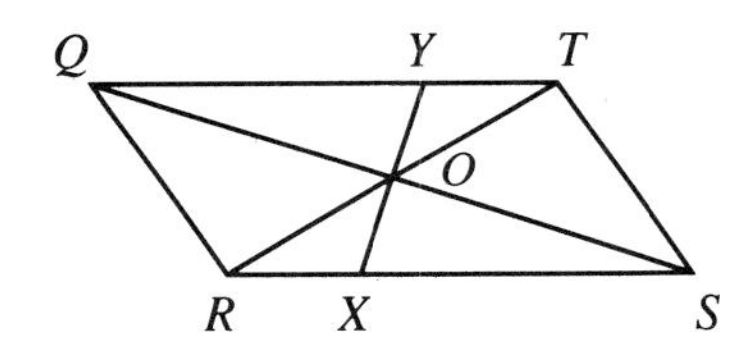

Statements	Reasons
1. $RSTQ$ is a ▱.	1. Given
2. $\overline{QT} \parallel \overline{RS}$	2. Def. of ▱
3. $\angle OYQ \cong \angle OXS$	3. If 2 $\parallel$ lines are cut by a trans., then alt. int. ∠s are $\cong$.
4. $\overline{QO} \cong \overline{OS}$	4. Diags. of a ▱ bis. each other.
5. $\angle YOQ \cong \angle XOS$	5. Vert. ∠s are $\cong$.
6. $\triangle YOQ \cong \triangle XOS$	6. AAS Thm.
7. $\overline{YO} \cong \overline{XO}$	7. Corr. parts of $\cong$ △s are $\cong$.

39. Given: $\overline{AB} \cong \overline{AC}$; X is on $\overline{BC}$; $\overline{XY} \perp \overline{AB}$; $\overline{XZ} \perp \overline{AC}$; $\overline{BD} \perp \overline{AC}$. Prove: $BD = XY + XZ$ Proof: Draw $\overline{XW} \perp \overline{BD}$. Then $XZDW$ is a ▱, and $WD = XZ$. Since $\overline{XW} \parallel \overline{CD}$, $\angle BXW \cong \angle C$. Since $\overline{AB} \cong \overline{AC}$, $\angle C \cong \angle ABC$. Thus, $\angle BXW \cong \angle ABC$. Also, $\overline{BX} \cong \overline{BX}$ and $\angle BYX \cong \angle XWB$. By AAS, $\triangle BYX \cong \triangle XWB$, and thus $XY = BW$. Then $BD = BW + WD = XY + XZ$.

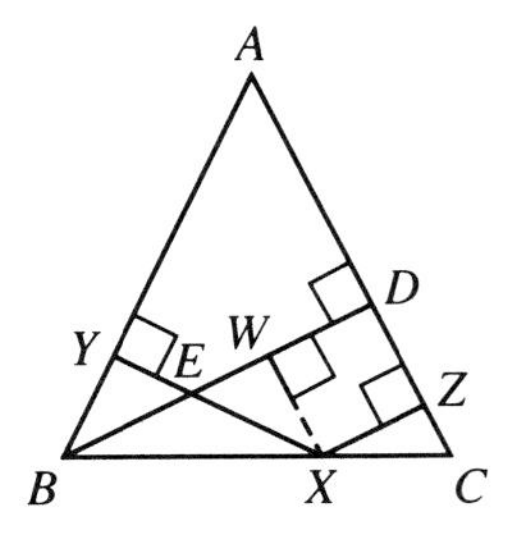

Page 173 • CLASSROOM EXERCISES

1. Yes; if both pairs of opp. sides of a quad. are $\cong$, then the quad. is a ▱.
2. Yes; if one pair of opp. sides of a quad. are both $\cong$ and $\parallel$, then the quad. is a ▱.
3. No 4. Yes; if the diags. of a quad. bis. each other, then the quad. is a ▱.
5. No 6. Yes; if both pairs of opp. $\angle$s of a quad. are $\cong$, then the quad. is a ▱.
7. Yes; def. of ▱ (A ▱ is a quad. with both pairs of opp. sides $\parallel$.)
8. No 9. Yes; def. of ▱ (A ▱ is a quad. with both pairs of opp. sides $\parallel$.)

10. 

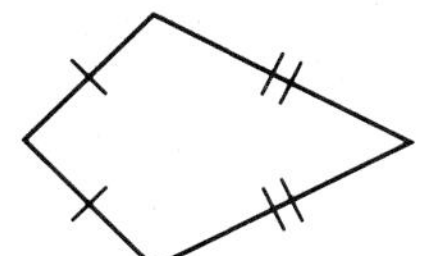

11. 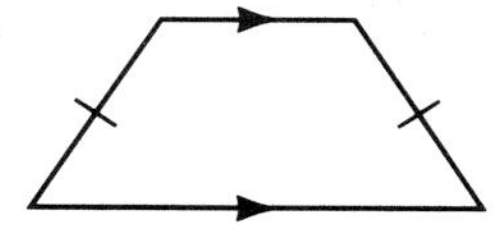

12. Since $EF = HG$ and $HE = GF$, $EFGH$ is a ▱ and $\overleftrightarrow{HG} \parallel \overleftrightarrow{EF}$.
13. The dashed lines shown bis. each other, so the quad. formed by their endpts. is a ▱. The jaws are $\parallel$ to 2 opp. sides of the ▱.

Pages 174–176 • WRITTEN EXERCISES

A 1. Def. of ▱

2. If both pairs of opp. sides of a quad. are $\cong$, then the quad. is a ▱.
3. If one pair of opp. sides of a quad. are both $\cong$ and $\parallel$, then the quad. is a ▱.
4. If the diags. of a quad. bis. each other, then the quad. is a ▱.
5. If both pairs of opp. $\angle$s of a quad. are $\cong$, then the quad. is a ▱.
6. Answers may vary. For example, since $\triangle SOK \cong \triangle COA$, $SO = CO$ and $KO = AO$. Quad. $SACK$ must be a ▱ because the diags. bis. each other.
7. Thm. 5-7; the diags. of quad. $ABDR$ bis. each other, so $ABDR$ is a ▱ and $\overline{AR} \parallel \overline{BD}$.
8. Show that one pair of opp. sides are both $\cong$ and $\parallel$.
9. **a.** Thm. 5-4 **b.** Thm. 5-6 **c.** Thm. 5-7
10. 1. Given 2. The sum of the meas. of the $\angle$s of a quad. is 360.
    3. Div. Prop. of = 4. If 2 lines are cut by a trans. and s-s. int. $\angle$s are supp., then the lines are $\parallel$. 4. Def. of ▱

B **11.** Given: $\overline{TS} \cong \overline{QR}$; $\overline{TQ} \cong \overline{SR}$
Prove: Quad. $QRST$ is a ▱.

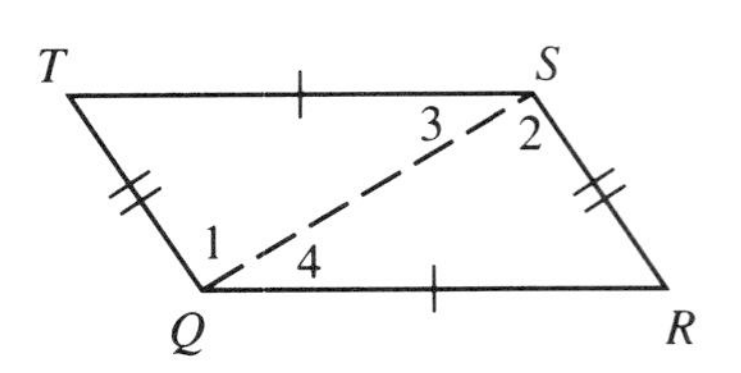

Statements	Reasons
1. $\overline{TS} \cong \overline{QR}$; $\overline{TQ} \cong \overline{SR}$	1. Given
2. $\overline{QS} \cong \overline{QS}$	2. Refl. Prop.
3. $\triangle TSQ \cong \triangle RQS$	3. SSS Post.
4. $\angle 1 \cong \angle 2$; $\angle 3 \cong \angle 4$	4. Corr. parts of $\cong$ △s are $\cong$.
5. $\overline{TQ} \parallel \overline{SR}$; $\overline{TS} \parallel \overline{QR}$	5. If 2 lines are cut by a trans. and alt. int. ∠s are $\cong$, then the lines are $\parallel$.
6. Quad. $QRST$ is a ▱.	6. Def. of ▱

12. Given: $\overline{AB} \cong \overline{CD}$; $\overline{AB} \parallel \overline{CD}$
Prove: $ABCD$ is a ▱.

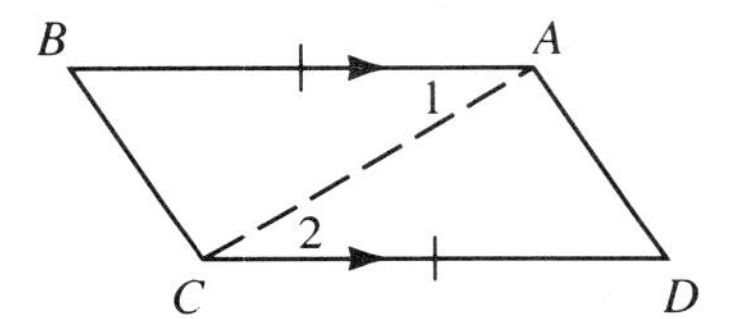

Statements	Reasons
1. $\overline{AB} \cong \overline{CD}$; $\overline{AB} \parallel \overline{CD}$	1. Given
2. $\angle 1 \cong \angle 2$	2. If 2 $\parallel$ lines are cut by a trans., then alt. int. ∠s are $\cong$.
3. $\overline{AC} \cong \overline{AC}$	3. Refl. Prop.
4. $\triangle ABC \cong \triangle CDA$	4. SAS Post.
5. $\overline{BC} \cong \overline{DA}$	5. Corr. parts of $\cong$ △s are $\cong$.
6. $ABCD$ is a ▱.	6. If both pairs of opp. sides of a quad. are $\cong$, then the quad. is a ▱.

13. Given: $\overline{AC}$ and $\overline{BD}$ bis. each other.
Prove: $ABCD$ is a ▱.

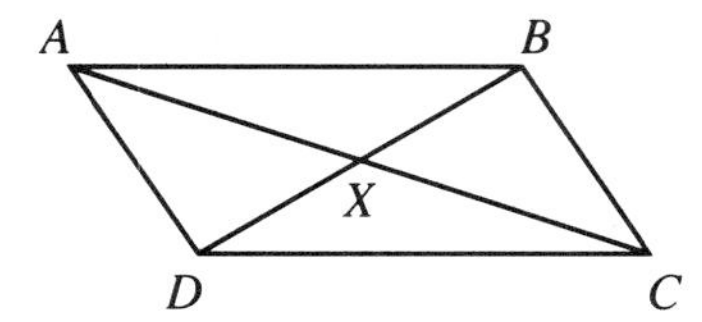

Statements	Reasons
1. $\overline{AC}$ and $\overline{BD}$ bis. each other.	1. Given
2. $\overline{AX} \cong \overline{XC}$; $\overline{DX} \cong \overline{XB}$	2. Def. of bis.
3. $\angle AXD \cong \angle CXB$; $\angle AXB \cong \angle CXD$	3. Vert. $\angle$s are $\cong$.
4. $\triangle AXD \cong \triangle CXB$; $\triangle AXB \cong \triangle CXD$	4. SAS Post.
5. $\overline{AD} \cong \overline{BC}$; $\overline{AB} \cong \overline{DC}$	5. Corr. parts of $\cong$ $\triangle$s are $\cong$.
6. $ABCD$ is a ▱.	6. If both pairs of opp. sides of a quad. are $\cong$, then the quad. is a ▱.

14. $\overline{NC} \parallel \overline{AM}$ and $NC = \frac{1}{2}DC = \frac{1}{2}AB = AM$. If one pair of opp. sides of a quad. are both $\cong$ and $\parallel$, then the quad. is a ▱. So $AMCN$ is a ▱.

15. $m\angle DAB = m\angle BCD$, so $m\angle NAM = \frac{1}{2}m\angle DAB = \frac{1}{2}m\angle BCD = m\angle NCM$. $m\angle DNA = m\angle NAM = m\angle NCM$, so $\overline{AN}$ and $\overline{CM}$ are $\parallel$. $\overline{CN}$ and $\overline{AM}$ are $\parallel$ because $ABCD$ is a ▱. Then $AMCN$ is a ▱, by def. of ▱.

16. $OD = OB$, $OA = OC$ (Diags. of a ▱ bis. each other.); $OZ = \frac{1}{2}OD = \frac{1}{2}OB = OX$ and $OW = \frac{1}{2}OA = \frac{1}{2}OC = OY$. If the diags. of a quad. bis. each other, then the quad. is a ▱. So $WXYZ$ is a ▱.

17. Draw $\overline{AC}$ int. $\overline{DB}$ at Z. Since $DZ = ZB$ and $DE = FB$, $EZ = DZ - DE = ZB - FB = ZF$. Also, $AZ = ZC$. If the diags. of a quad. bis. each other, then the quad. is a ▱. So $AFCE$ is a ▱.

18. $FK = HL$ and $KO = LO$, so $FO = OH$. Also, $JO = GO$ since $KGLJ$ is a ▱. If the diags. of a quad. bis. each other, then the quad. is a ▱. So $FGHJ$ is a ▱.

19. $3y = 42$, $y = 14$; $(8x - 6) + 42 = 180$, $8x + 36 = 180$, $8x = 144$, $x = 18$

20. $3x - 40 = x$, $2x = 40$, $x = 20$; $y^2 = y + 30$, $y^2 - y - 30 = 0$, $(y - 6)(y + 5) = 0$, $y = 6$ or -5

21. $4x + y = 42$, $y = 42 - 4x$; $3x - 2y = 26$, $3x - 2(42 - 4x) = 26$, $3x - 84 + 8x = 26$, $11x = 110$, $x = 10$; $4x + y = 42$, $40 + y = 42$, $y = 2$

22. $\begin{array}{l} 9y = 4x + 1 \\ \underline{3x = 7y - 2} \end{array}$, $\begin{array}{r} -4x + 9y = 1 \\ \underline{3x - 7y = -2} \end{array}$, $\begin{array}{r} -12x + 27y = 3 \\ \underline{12x - 28y = -8} \end{array}$

$-y = -5$, $y = 5$;
$3x = 7y - 2$, $3x = 35 - 2 = 33$, $x = 11$

23.

Statements	Reasons
1. $ABCD$ is a ▱.	1. Given
2. $\overline{AD} \cong \overline{BC}$; $\overline{AD} \parallel \overline{BC}$	2. Opp. sides of a ▱ are ≅ and ∥.
3. $\angle BCF \cong \angle DAE$	3. If 2 ∥ lines are cut by a trans., then alt. int. ∠s are ≅.
4. $\overline{DE} \perp \overline{AC}$; $\overline{BF} \perp \overline{AC}$	4. Given
5. $m\angle DEA = 90$; $m\angle BFC = 90$	5. Def. of ⊥ lines
6. $\angle DEA \cong \angle BFC$	6. Def. of ≅ ∠s
7. $\triangle DAE \cong \triangle BCF$	7. AAS Thm.
8. $\overline{DE} \cong \overline{BF}$	8. Corr. parts of ≅ △s are ≅.
9. $\overline{DE} \parallel \overline{BF}$	9. In a plane, 2 lines ⊥ to the same line are ∥.
10. $DEBF$ is a ▱.	10. If one pair of opp. sides of a quad. are both ≅ and ∥, then the quad. is a ▱.

24.

Statements	Reasons
1. Plane X ∥ plane Y	1. Given
2. $\overline{LM} \parallel \overline{ON}$	2. If 2 ∥ planes are cut by a third plane, then their lines of int. are ∥.
3. $\overline{LM} \cong \overline{ON}$	3. Given
4. $LMNO$ is a ▱.	4. If one pair of opp. sides of a quad. are both ≅ and ∥, then the quad. is a ▱.

C 25. $ABCD$ and $BEDF$ are ▱s. In ▱$ABCD$, draw diags. $\overline{BD}$ and $\overline{AC}$, bisecting each other at O. (The diags. of a ▱ bis. each other.) In ▱$BEDF$, draw diag. $\overline{EF}$. Since $\overline{BD}$ is also a diag. of ▱$BEDF$, $\overline{EF}$ must also be bisected at O. $\overline{EF}$ and $\overline{AC}$ are then bisected at O, and $AECF$ is a ▱. (If the diags. of a quad. bis. each other, the quad. is a ▱.)

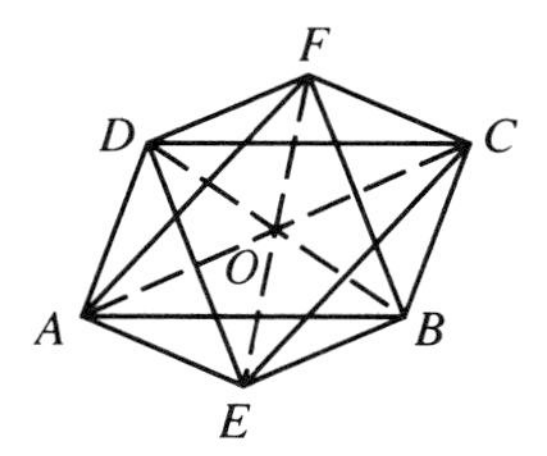

Page 176 • EXPLORATIONS

$AE = EC$; $\angle AED \cong \angle ACB$; $\overline{DE} \parallel \overline{BC}$; $DE = \frac{1}{2}BC$

Page 179 • CLASSROOM EXERCISES

1. The lines on the paper are ∥, and they cut off ≅ segs. on a trans. (vert. paper edge). So they cut off ≅ segs. on trans. $\overline{AB}$. (Thm 5-9)

2. 5 **3.** 14 **4.** 8 **5.** $\frac{1}{2}k$

6. **a.** $TN = 5, NZ = 7, TZ = 4$ **b.** $MY = 5, YN = 7, MN = 4$ **c.** $XM = 5, MT = 7, XT = 4$ **d.** $NT = 5, TM = 7, NM = 4$

7. The segs. joining the midpts. of the sides of a △ divide the △ into 4 ≅ △s. **8.** 3

9. If 2 lines are ∥, then all pts. on one line are equidistant from the other line. (Thm. 5-8)

Pages 180–182 • WRITTEN EXERCISES

A **1.** 12, 12 **2.** $2k, k$ **3.** $5x - 8 = 3x; 4$ **4.** $2(3x + 2) = 8x; 2$

5. **a.** 40 **b.** 20 **c.** 26 **d.** 34

6. **a.** $\triangle NOM, \triangle LMK, \triangle MLN$ **b.** 4(area of $\triangle NLM$) = 69.28 cm^2

7. D, E **8.** D, E **9.** D, F **10.** 15 **11.** 6 **12.** $5x = 2x + 12; 4$

13. $22 - x = 3x - 22; 11$

B **14.** $2x - y = 15, x + y = 15; 10, 5$ **15.** $2x + 3y = 12, 8x = 24; 3, 2$

16. $12x - 8 = 2(3x + 5), 4y + 2 = 7(y - 1); x = 3; y = 3$

17. $3x - 2y = 8, 5x - 3y = 14; x = 4; y = 2$

18.

Statements	Reasons
1. A is the midpt. of $\overline{OX}$; $\overline{AB} \parallel \overline{XY}$	1. Given
2. B is the midpt. of $\overline{OY}$.	2. A line that contains the midpt. of one side of a △ and is ∥ to another side passes through the midpt. of the third side.
3. $\overline{BC} \parallel \overline{YZ}$	3. Given
4. C is the midpt. of $\overline{OZ}$.	4. A line that contains the midpt. of one side of a △ and is ∥ to another side passes through the midpt. of the third side.
5. $\overline{AC} \parallel \overline{XZ}$	5. The seg. that joins the midpts. of 2 sides of a △ is ∥ to the third side.

19.

Statements	Reasons
1. $\overline{BE} \parallel \overline{MD}$; M is the midpt. of $\overline{AB}$.	1. Given
2. D is the midpt. of $\overline{AE}$.	2. A line that contains the midpt. of one side of a △ and is ∥ to another side passes through the midpt. of the third side.
3. $\overline{DE} \cong \overline{AD}$	3. Def. of midpt.
4. $ABCD$ is a ▱.	4. Given
5. $\overline{AD} \cong \overline{BC}$	5. Opp. sides of a ▱ are ≅.
6. $\overline{DE} \cong \overline{BC}$ or $DE = BC$	6. Trans. Prop.

20.

Statements	Reasons
1. $\overline{PQ}$, $\overline{RS}$, and $\overline{TU}$ are each ⊥ to $\overleftrightarrow{UQ}$.	1. Given
2. $\overline{PQ} \parallel \overline{RS} \parallel \overline{TU}$	2. In a plane, lines ⊥ to the same line are ∥ to each other.
3. R is the midpt. of $\overline{PT}$.	3. Given
4. $\overline{PR} \cong \overline{RT}$	4. Def. of midpt.
5. $\overline{US} \cong \overline{SQ}$	5. If 3 ∥ lines cut off ≅ segs. on one trans., then they cut off ≅ segs. on every trans.
6. $\overleftrightarrow{RS}$ is the ⊥ bis. of $\overline{UQ}$.	6. Def. of ⊥ bis.
7. R is equidistant from U and Q.	7. If a pt. lies on the ⊥ bis. of a seg., then the pt. is equidistant from the endpts. of the seg.

21. Given: ▱$EFGH$; M is the midpt. of $\overline{FG}$.

Prove: $MP = \frac{1}{2}EF$

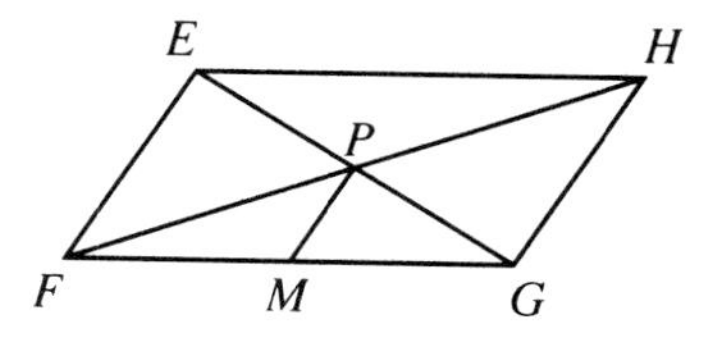

Statements	Reasons
1. $EFGH$ is a ▱.	1. Given
2. $\overline{EP} \cong \overline{PG}$	2. Diags. of a ▱ bis. each other
3. P is the midpt. of $\overline{EG}$.	3. Def. of midpt.
4. M is the midpt. of $\overline{FG}$.	4. Given
5. $MP = \frac{1}{2}EF$	5. The seg. that joins the midpts. of 2 sides of a △ is half as long as the third side.

22. By Thm. 5-11, $\overline{ON} \parallel \overline{WK}$ and $\overline{PM} \parallel \overline{WK}$. So $\overline{ON} \parallel \overline{PM}$ (2 lines $\parallel$ to a third line are $\parallel$ to each other). Also by Thm. 5-11, $ON = \frac{1}{2}WK$ and $PM = \frac{1}{2}WK$. So $ON = PM$ and $\overline{ON} \cong \overline{PM}$. Since $\overline{ON}$ and $\overline{PM}$ are both $\parallel$ and $\cong$, $PMNO$ is a ▱ (Thm. 5-5).

23. Given: X, Y, and Z are the midpts. of $\overline{AB}$, $\overline{AC}$, and $\overline{BC}$, resp.; P and Q are the midpts. of $\overline{BZ}$ and $\overline{CZ}$, resp.

Prove: $PX = QY$

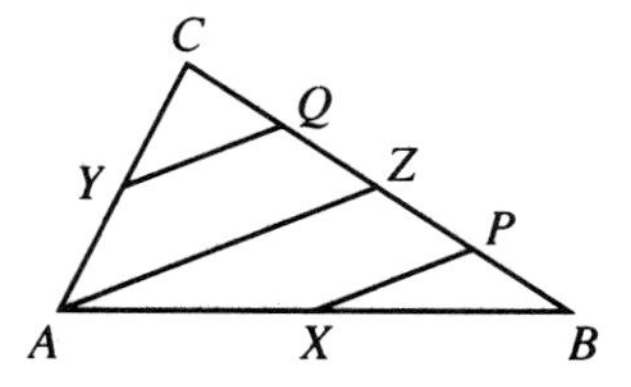

Statements	Reasons
1. X and P are the midpts. of $\overline{AB}$ and $\overline{BZ}$, resp.; Y and Q are the midpts. of $\overline{AC}$ and $\overline{CZ}$, resp.	1. Given
2. $PX = \frac{1}{2}AZ$; $QY = \frac{1}{2}AZ$	2. The seg. that joins the midpts. of 2 sides of a △ is half as long as the third side.
3. $PX = QY$	3. Substitution Prop.

24. Given: D and E are the midpts. of $\overline{AB}$ and $\overline{CD}$, resp.; $\overline{DG} \parallel \overline{EF}$

Prove: $BG = GF = FC$

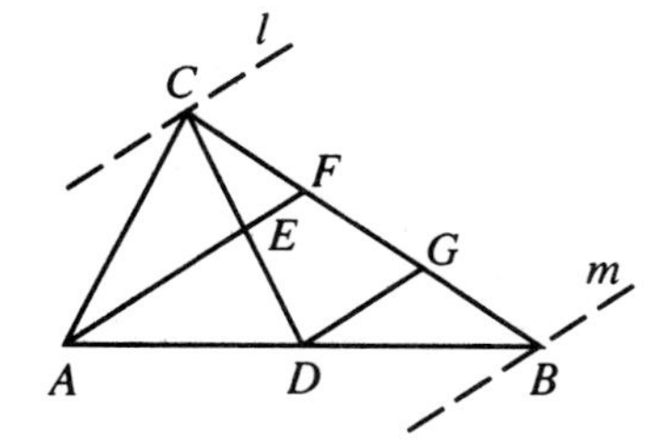

Statements	Reasons
1. D and E are the midpts. of $\overline{AB}$ and $\overline{CD}$, resp.; $\overline{DG} \parallel \overline{EF}$	1. Given
2. $\overline{AD} \cong \overline{DB}$; $\overline{DE} \cong \overline{EC}$	2. Def. of midpt.
3. Draw l through C and $\parallel \overline{AF}$; draw m through B and $\parallel \overline{DG}$.	3. Through a pt. outside a line, there is exactly one line $\parallel$ to the given line.
4. $\overline{GF} \cong \overline{FC}$; $\overline{BG} \cong \overline{GF}$	4. If 3 $\parallel$ lines cut off $\cong$ segs. on one trans., then they cut off $\cong$ segs. on every trans.
5. $BG = GF = FC$	5. Def. of $\cong$ segs.

C 25.

Statements	Reasons
1. Draw $\overline{AF}$, int. plane Q at X.	1. Through any 2 pts. there is exactly one line.
2. Let M be the plane containing $\overleftrightarrow{AC}$ and $\overleftrightarrow{AF}$; let N be the plane containing $\overleftrightarrow{AF}$ and $\overleftrightarrow{DF}$.	2. If 2 lines intersect, then exactly one plane contains them.
3. $\overleftrightarrow{BX}$ is the int. of M and Q; $\overleftrightarrow{CF}$ is the int. of M and R; $\overleftrightarrow{XE}$ is the int. of N and Q; $\overleftrightarrow{AD}$ is the int. of N and P.	3. If 2 planes intersect, then they intersect in a line.
4. P, Q, and R are $\parallel$ planes.	4. Given
5. $\overleftrightarrow{BX} \parallel \overleftrightarrow{CF}$; $\overleftrightarrow{XE} \parallel \overleftrightarrow{AD}$	5. If 2 $\parallel$ planes are cut by a third plane, then the lines of int. are $\parallel$.
6. $AB = BC$	6. Given
7. B is the midpt. of $\overline{AC}$.	7. Def. of midpt.
8. X is the midpt. of $\overline{AF}$.	8. A line that cont. the midpt. of one side of a $\triangle$ and is $\parallel$ to another side passes through the midpt. of the third side.
9. E is the midpt. of $\overline{DF}$.	9. A line that cont. the midpt. of one side of a $\triangle$ and is $\parallel$ to another side passes through the midpt. of the third side.
10. $DE = EF$	10. Def. of midpt.

Page 182 • SELF-TEST 1

1. may be true **2.** must be true **3.** must be true **4.** cannot be true

5. show both pairs of opp. sides $\parallel$; show both pairs of opp. sides $\cong$; show one pair of opp. sides both $\cong$ and $\parallel$; show both pairs of opp. $\angle$s $\cong$; show that the diags. bis. each other.

6. a. If 3 $\parallel$ lines cut off $\cong$ segs. on one trans., then they cut off $\cong$ segs. on every trans. **b.** $3x - 7 = 11$, $5x - y = 11$; $x = 6$; $y = 19$

7.

Statements	Reasons
1. $ABCD$ is a ▱.	1. Given
2. $\overline{AC}$ and $\overline{BD}$ bis. each other.	2. Diags. of a ▱ bis. each other.
3. O is the midpt. of $\overline{BD}$.	3. Def. of bis.
4. M is the midpt. of $\overline{AB}$.	4. Given
5. $MO = \frac{1}{2}AD$	5. The seg. that joins the midpts. of 2 sides of a $\triangle$ is half as long as the third side.

8.

Statements	Reasons
1. $PQRS$ is a $\square$.	1. Given
2. $\overline{SR} \parallel \overline{PQ}$; $\overline{SP} \parallel \overline{RQ}$	2. Def. of $\square$
3. $\angle QPR \cong \angle SRP$ or $m\angle QPR = m\angle SRP$	3. If 2 $\parallel$ lines are cut by a trans., then alt. int. $\angle$s are $\cong$.
4. $\overline{PX}$ bis. $\angle QPR$; $\overline{RY}$ bis. $\angle SRP$.	4. Given
5. $m\angle RPX = \frac{1}{2}m\angle QPR$; $m\angle PRY = \frac{1}{2}m\angle SRP$	5. $\angle$ Bis. Thm.
6. $\frac{1}{2}m\angle QPR = \frac{1}{2}m\angle SRP$	6. Mult. Prop. of =
7. $m\angle RPX = m\angle PRY$	7. Substitution Prop.
8. $\overline{YR} \parallel \overline{PX}$	8. If 2 lines are cut by a trans. and alt. int. $\angle$s are $\cong$, then the lines are $\parallel$.
9. $RYPX$ is a $\square$.	9. Def. of $\square$

Page 183 • COMPUTER KEY-IN

1. a, b. I, II

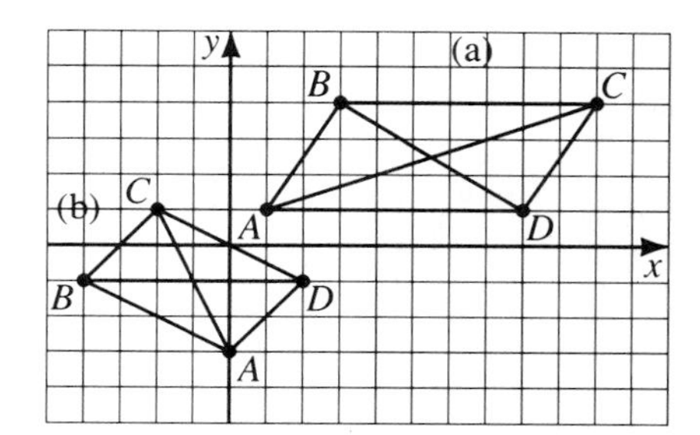

2. a. I, II, III, V

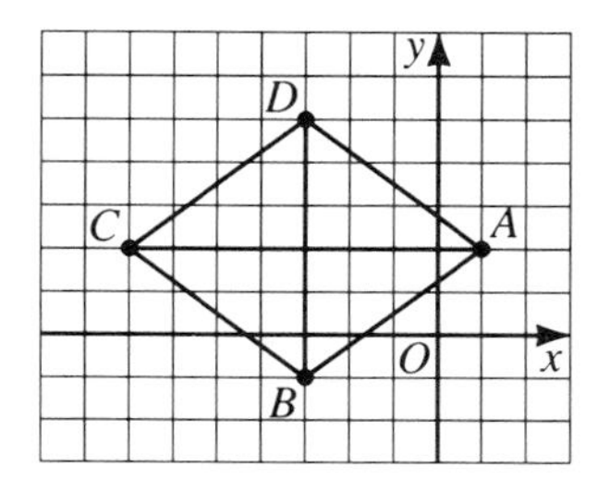

b. I, II, III, V

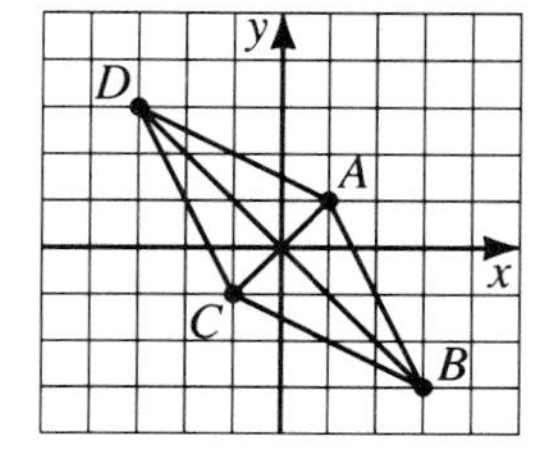

3. **a.** I, II, IV, VI **b.** I, II, IV, VI

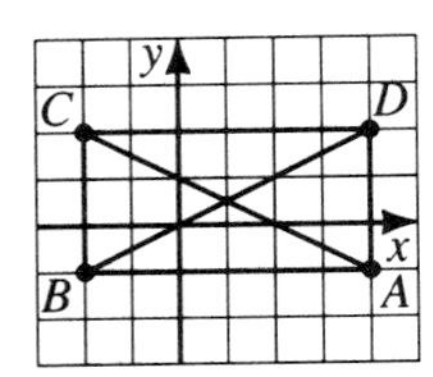

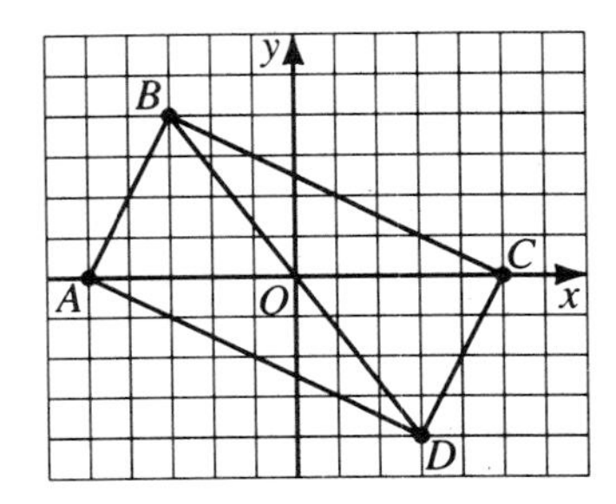

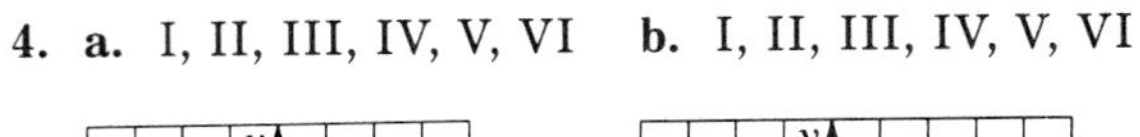
4. **a.** I, II, III, IV, V, VI **b.** I, II, III, IV, V, VI

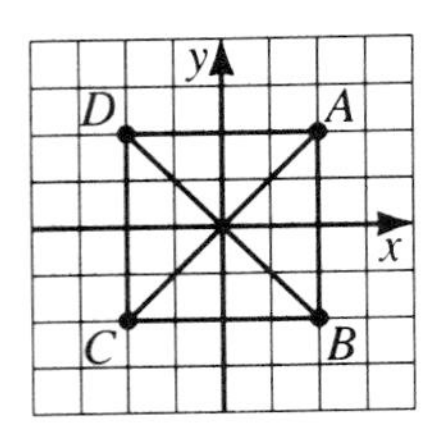

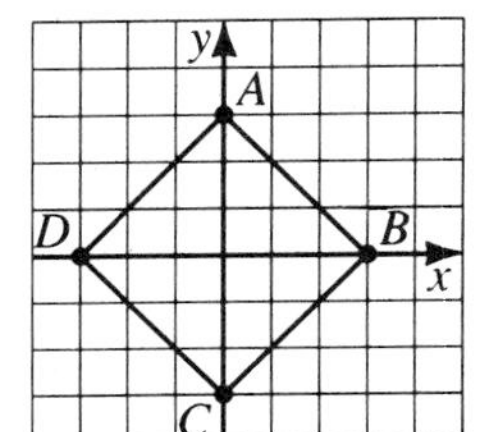

5. **a.** None **b.** IV, V

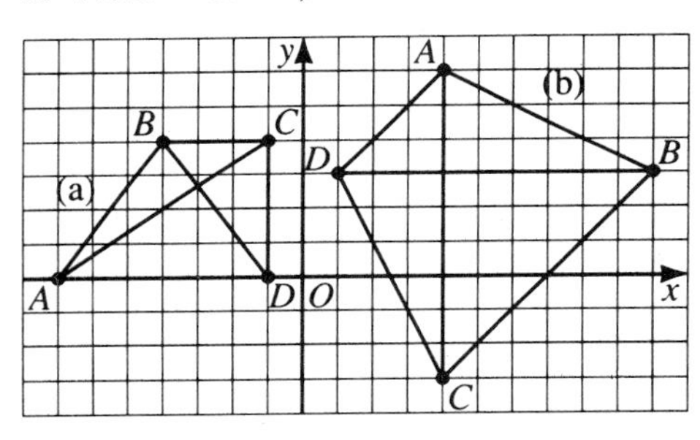

6. **a.** V **b.** IV, V

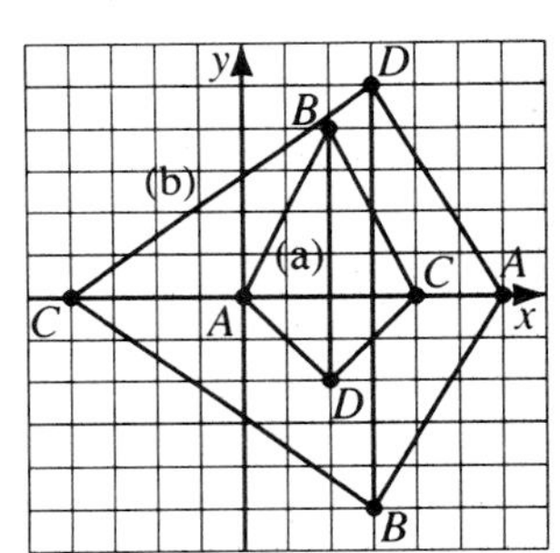

Page 186 • CLASSROOM EXERCISES

1. **a.** 1, 2, 4, 5, 7, 8, 9, 10, 12 **b.** 2, 4, 8, 9, 10 **c.** 2, 5, 9, 12 **d.** 2, 9

2. 2, 9 **3.** 4, 8, 10 **4.** 5, 12

5. Since $AOIE$ is a ▱, $\angle I \cong \angle A$. So $m\angle I = 90$. Since $\overline{OI} \parallel \overline{AE}$, $m\angle O + m\angle A = 180$. So $m\angle O = 90$ and $m\angle E = 90$. Then $AOIE$ is a rect.

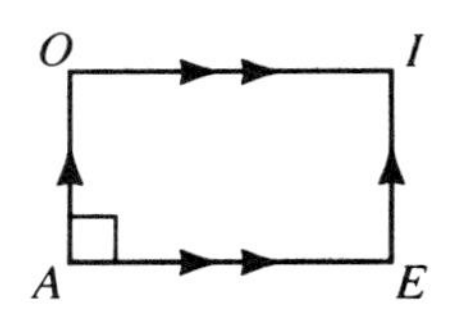

6. Since $CTGN$ is a ▱, $\overline{CT} \cong \overline{NG}$ and $\overline{CN} \cong \overline{TG}$. Then $\overline{CT} \cong \overline{NG} \cong \overline{CN} \cong \overline{TG}$ and $CTGN$ is a rhombus.

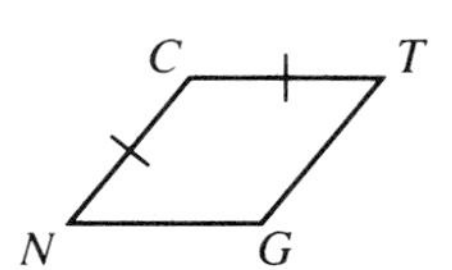

7. **a–c.** $\perp$ bisector **d.** The diags. of a rhombus are $\perp$.

8. $6\frac{1}{2}$ **9.** $2t$ **10.** 40

11. a. They are parallelograms. **b.** Draw the diags. of the quad. and use Thm. 5-11 to prove both pairs of opp. sides $\parallel$ or $\cong$, or one pair of opp. sides $\parallel$ and $\cong$. For example, $\overline{MN}$ joins the midpts. of 2 sides of $\triangle ABC$, so $\overline{MN} \parallel \overline{AC}$ and $MN = \frac{1}{2}AC$. Also, $\overline{PO}$ joins the midpts. of 2 sides of $\triangle ADC$, so $\overline{PO} \parallel \overline{AC}$ and $PO = \frac{1}{2}AC$. Therefore, $\overline{MN} \parallel \overline{PO}$ and $\overline{MN} \cong \overline{PO}$, so $MNOP$ is a ▱.

Pages 187–189 • WRITTEN EXERCISES

	Property	▱	Rect.	Rhombus	Square
A 1.	Opp. sides are $\parallel$.	√	√	√	√
2.	Opp. sides are $\cong$.	√	√	√	√
3.	Opp. ∠s are $\cong$.	√	√	√	√
4.	A diag. forms 2 $\cong$ △s.	√	√	√	√
5.	Diags. bis. each other.	√	√	√	√
6.	Diags. are $\cong$.		√		√
7.	Diags. are $\perp$.			√	√
8.	A diag. bis. 2 ∠s.			√	√
9.	All ∠s are rt. ∠s.		√		√
10.	All sides are $\cong$.			√	√

11. $m\angle 2 = 25$; $m\angle 3 = 65$; $m\angle 4 = 65$; $m\angle 5 = 90$ **12.** $3x + 8 = 11x - 24$; $x = 4$
13. $3x + 1 + 7x - 11 = 90$; $x = 10$ **14.** $m\angle 2 = 18$; $m\angle 3 = 72$; $m\angle 4 = 72$
15. 13.5 **16.** $4y + 7 = 15$; $y = 2$
17. $m\angle 2 = 32$; $m\angle 3 = 58$; $m\angle 4 = 58$ **18.** $7x - 3 = 6(x + 1)$; $x = 9$
19. $12 - 8y = 2(2y + 3)$; $y = \frac{1}{2}$

20. $Q(2, 5)$; no **21.** $D(2, 5)$; no **22.** $G(0, 3)$; no **23.** $J(4, 6)$; yes

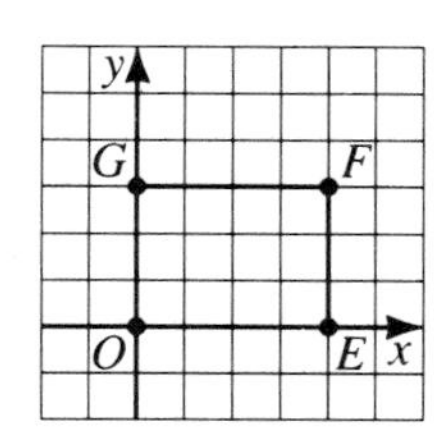

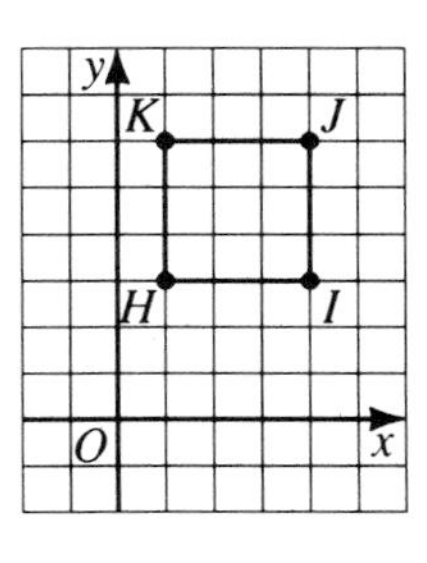

B **24.** 10 **25.** 15 **26.** 30 **27.** 60

28.

Statements	Reasons
1. $\overline{ZY} \cong \overline{BX}$; $\angle 1 \cong \angle 2$	1. Given
2. $\overline{BX} \cong \overline{BZ}$	2. If 2 ∠s of a △ are ≅, then the sides opp. those ∠s are ≅.
3. $\overline{ZY} \cong \overline{BZ}$	3. Trans. Prop.
4. $ABZY$ is a ▱.	4. Given
5. $ABZY$ is a rhombus.	5. If 2 consec. sides of a ▱ are ≅, then the ▱ is a rhombus.

29.

Statements	Reasons
1. $ABZY$ is a ▱.	1. Given
2. $\overline{BZ} \cong \overline{AY}$	2. Opp. sides of a ▱ are ≅.
3. $\overline{AY} \cong \overline{BX}$	3. Given
4. $\overline{BZ} \cong \overline{BX}$	4. Trans. Prop.
5. $\angle 1 \cong \angle 2$	5. Isos. △ Thm.
6. $\overline{BZ} \parallel \overline{AY}$	6. Def. of ▱
7. $\angle 2 \cong \angle 3$	7. If 2 ∥ lines are cut by a trans., then corr. ∠s are ≅.
8. $\angle 1 \cong \angle 3$	8. Trans. Prop.

30.

Statements	Reasons
1. $QRST$ is a rect.	1. Given
2. $\overline{QS} \cong \overline{TR}$	2. The diags. of a rect. are ≅.
3. $RKST$ is a ▱.	3. Given
4. $\overline{TR} \cong \overline{SK}$	4. Opp. sides of a ▱ are ≅.
5. $\overline{QS} \cong \overline{SK}$	5. Trans. Prop.
6. $\triangle QSK$ is isos.	6. Def. of isos. △

31.

Statements	Reasons
1. $QRST$ is a rect.; $RKST$ and $JQST$ are ▱s.	1. Given
2. $\overline{KS} \cong \overline{RT}$; $\overline{JT} \cong \overline{QS}$	2. Opp. sides of a ▱ are ≅.
3. $\overline{RT} \cong \overline{QS}$	3. The diags. of a rect. are ≅.
4. $\overline{KS} \cong \overline{JT}$	4. Substitution Prop.

32. Given: $QRST$ is a rect.
Prove: $\overline{RT} \cong \overline{QS}$

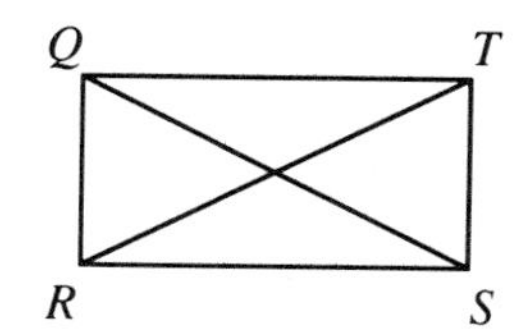

Statements	Reasons
1. $QRST$ is a rect.	1. Given
2. $\angle TQR$ and $\angle SRQ$ are rt. ∠s.	2. Def. of rect.
3. $m\angle TQR = 90$; $m\angle SRQ = 90$	3. Def. of rt. ∠
4. $\angle TQR \cong \angle SRQ$	4. Def. of ≅ ∠s
5. $\overline{TQ} \cong \overline{SR}$	5. Opp. sides of a ▱ are ≅.
6. $\overline{QR} \cong \overline{QR}$	6. Refl. Prop.
7. $\triangle TQR \cong \triangle SRQ$	7. SAS Post.
8. $\overline{RT} \cong \overline{QS}$	8. Corr. parts of ≅ △s are ≅.

33. Given: $EFGH$ is a rhombus.
Prove: $\overline{EG}$ bis. $\angle E$ and $\angle G$.

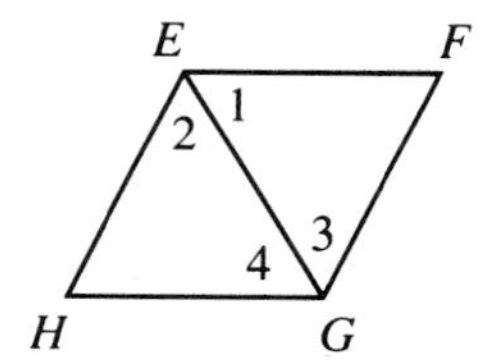

Statements	Reasons
1. $EFGH$ is a rhombus.	1. Given
2. $\overline{FE} \cong \overline{FG}$; $\overline{HE} \cong \overline{HG}$	2. Def. of rhombus
3. $\angle 1 \cong \angle 3$; $\angle 2 \cong \angle 4$	3. If 2 sides of a △ are ≅, then the ∠s opp. those sides are ≅.
4. $\overline{EF} \parallel \overline{HG}$; $\overline{HE} \parallel \overline{GF}$	4. Def. of ▱
5. $\angle 1 \cong \angle 4$; $\angle 2 \cong \angle 3$	5. If 2 ∥ lines are cut by a trans., then alt. int. ∠s are ≅.
6. $\angle 1 \cong \angle 2$; $\angle 3 \cong \angle 4$	6. Substitution Prop.
7. $\overline{EG}$ bis. $\angle E$ and $\angle G$.	7. Def. of ∠ bis.

34. Given: ▱$ABCD$; $\overline{AC} \perp \overline{DB}$
Prove: $ABCD$ is a rhombus.

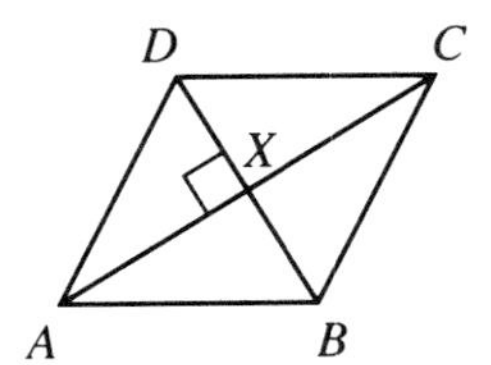

Statements	Reasons
1. $ABCD$ is a ▱.	1. Given
2. $\overline{AC}$ and $\overline{DB}$ bis. each other.	2. Diags. of a ▱ bis. each other.
3. X is the midpt. of $\overline{AC}$ and $\overline{DB}$.	3. Def. of bis.
4. $\overline{DX} \cong \overline{XB}$	4. Def. of midpt.
5. $\overline{AC} \perp \overline{DB}$	5. Given
6. $\angle AXD \cong \angle AXB$	6. If 2 lines are $\perp$, then they form $\cong$ adj. ∠s.
7. $\overline{AX} \cong \overline{AX}$	7. Refl. Prop.
8. $\triangle AXD \cong \triangle AXB$	8. SAS Post.
9. $\overline{AD} \cong \overline{AB}$	9. Corr. parts of $\cong$ △s are $\cong$.
10. $ABCD$ is a rhombus.	10. If 2 consec. sides of a ▱ are $\cong$, then the ▱ is a rhombus.

35. Given: ▱$ABCD$; $\overline{AC} \cong \overline{BD}$
Prove: $ABCD$ is a rect.

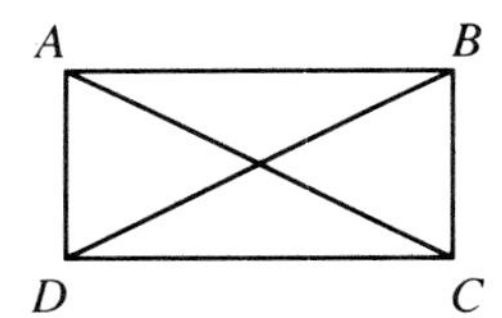

Statements	Reasons
1. $ABCD$ is a ▱; $\overline{AC} \cong \overline{BD}$	1. Given
2. $\overline{AB} \cong \overline{DC}$	2. Opp. sides of a ▱ are $\cong$.
3. $\overline{AD} \cong \overline{AD}$	3. Refl. Prop.
4. $\triangle BAD \cong \triangle CDA$	4. SSS Post.
5. $\angle BAD \cong \angle CDA$ or $m\angle BAD = m\angle CDA$	5. Corr. parts of $\cong$ △s are $\cong$.
6. $\overline{AB} \parallel \overline{DC}$	6. Def. of ▱
7. $m\angle BAD + m\angle CDA = 180$	7. If 2 $\parallel$ lines are cut by a trans., then s-s. int. ∠s are supp.
8. $2m\angle BAD = 180$	8. Substitution Prop.
9. $m\angle BAD = 90$	9. Div. Prop. of =
10. $\angle BAD$ is a rt. $\angle$.	10. Def. of rt. $\angle$
11. $ABCD$ is a rect.	11. If an $\angle$ of a ▱ is a rt. $\angle$, then the ▱ is a rect.

36. **a.** rectangle **b.** Since $\angle D$ and $\angle A$ are bisected and are supp.,

$m\angle ADW + m\angle DAW = \frac{1}{2}(180) = 90$. Then $m\angle DWA = 180 - 90 = 90 = m\angle W$.

Similarly, $m\angle X = m\angle Y = m\angle Z = 90$. Since quad. $WXYZ$ has four rt. ∠s, $WXYZ$ is a rect.

37. square

38. $V(9, 3)$

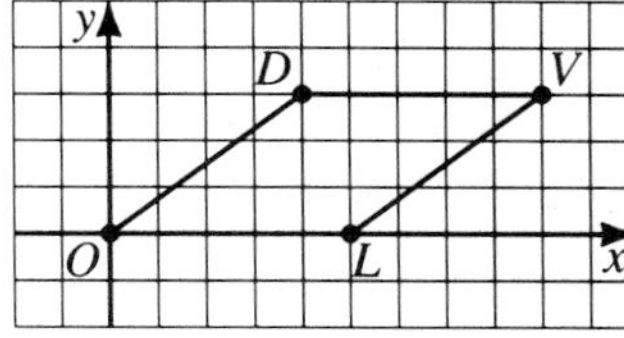

39. $W(6, 8)$

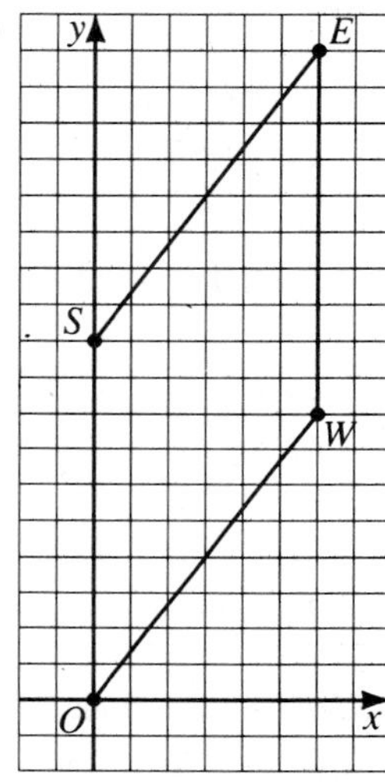

Ex. 39

C 40. **a.** Yes, it must have 2 ≅ consecutive sides.

Given: $\overline{AB} \parallel \overline{DC}$; $\overline{BD}$ bis. $\angle D$.

Prove: $\overline{AB} \cong \overline{AD}$

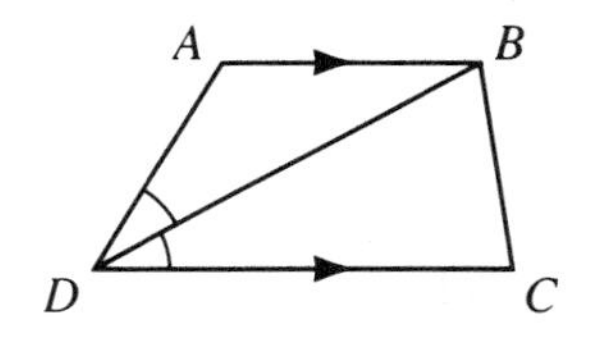

Statements	Reasons
1. $\overline{AB} \parallel \overline{DC}$	1. Given
2. $\angle ABD \cong \angle CDB$	2. If 2 ∥ lines are cut by a trans., then alt. int. ∠s are ≅.
3. $\overline{BD}$ bis. $\angle D$.	3. Given
4. $\angle CDB \cong \angle ADB$	4. Def. of ∠ bis.
5. $\angle ABD \cong \angle ADB$	5. Trans. Prop.
6. $\overline{AB} \cong \overline{AD}$	6. If 2 ∠s of a △ are ≅, then the sides opp. those ∠s are ≅.

b. Yes, the quad. must be a rhombus.

Given: $\overline{AB} \parallel \overline{DC}$; $\overline{BD}$ bis. $\angle D$ and $\angle B$.

Prove: $ABCD$ is a rhombus.

Statements	Reasons
1. $\overline{BD}$ bis. $\angle D$ and $\angle B$.	1. Given
2. $\angle CDB \cong \angle ADB$; $\angle CBD \cong \angle ABD$	2. Def. of $\angle$ bis.
3. $\overline{DB} \cong \overline{DB}$	3. Refl. Prop.
4. $\triangle ADB \cong \triangle CDB$	4. ASA Post.
5. $\overline{AB} \cong \overline{CB}$; $\overline{AD} \cong \overline{CD}$	5. Corr. parts of $\cong$ $\triangle$s are $\cong$.
6. $\overline{AB} \parallel \overline{DC}$	6. Given
7. $\angle ABD \cong \angle CDB$	7. If 2 $\parallel$ lines are cut by a trans., then alt. int. $\angle$s are $\cong$.
8. $\angle ABD \cong \angle ADB$; $\angle CDB \cong \angle CBD$	8. Substitution Prop.
9. $\overline{AB} \cong \overline{AD}$; $\overline{CD} \cong \overline{CB}$	9. If 2 $\angle$s of a $\triangle$ are $\cong$, then the sides opp. those $\angle$s are $\cong$.
10. $\overline{AB} \cong \overline{AD} \cong \overline{CD} \cong \overline{CB}$	11. Substitution Prop.
11. $ABCD$ is a rhombus.	12. Def. of rhombus

41. $AXDE$ is a rhombus. Consider isos. $\triangle$s ABC and BCD. $\angle BAC \cong \angle BCA$ and $\angle CBD \cong \angle CDB$. Each of these 4 $\angle$s has meas. $\frac{1}{2}(180 - 108) = 36$. Then $m\angle AXD = m\angle BXC = 180 - (m\angle DBC + m\angle BCA) = 180 - 72 = 108 = m\angle E$. Also, $m\angle EAX = 108 - m\angle BAC = 72$ and $m\angle EDX = 108 - m\angle CDB = 72$. Since both pairs of opp. $\angle$s of $AXDE$ are $\cong$, $AXDE$ is a ▱. $\overline{AE} \cong \overline{DE}$ ($ABCDE$ is reg.), so $AXDE$ is a rhombus. (If 2 consecutive sides of a ▱ are $\cong$, then the ▱ is a rhombus.)

A E B X D C

42. $\triangle RYZ$ is equilateral. Proof: Since $RSTW$ is a ▱ and $\triangle$s YWT and STZ are equilateral, $\overline{WR} \cong \overline{SZ} \cong \overline{TZ}$ and $\overline{WY} \cong \overline{SR} \cong \overline{TY}$. $m\angle YWR = 60 + 90 = 150$, $m\angle RSZ = 90 + 60 = 150$, and $m\angle YTZ = 360 - (90 + 60 + 60) = 150$. Then $\triangle YWR \cong \triangle RSZ \cong \triangle YTZ$ and $\overline{YR} \cong \overline{RZ} \cong \overline{YZ}$.

Page 189 • EXPLORATIONS

$\overline{CD} \parallel \overline{FE}$; if 2 lines are cut by a trans. and corr. $\angle$s are $\cong$, then the lines are $\parallel$; $\overline{FE} \parallel \overline{BA}$; 2 lines $\parallel$ to a third line are $\parallel$ to each other; $FE = \frac{1}{2}(BA + CD)$

Page 189 • MIXED REVIEW EXERCISES

1. 13 **2.** 20 **3.** 11 **4.** 9 **5.** 8.2 **6.** -1.5 **7.** 1 **8.** 3.45

9. a. 23 **b.** 2 **c.** 4 **d.** -6

Page 192 • CLASSROOM EXERCISES

1. $\overline{XZ} \parallel \overline{AB}$ because the seg. that joins the midpts. of 2 sides of a $\triangle$ is $\parallel$ to the third side, and $\overline{AX} \nparallel \overline{BZ}$.

2. $ACZY, XYBC$ 3. $XYBC$; 22 4. 11 5. 7.5 6. $\frac{1}{2}(11 - x + 11 + x) = 11$

7.

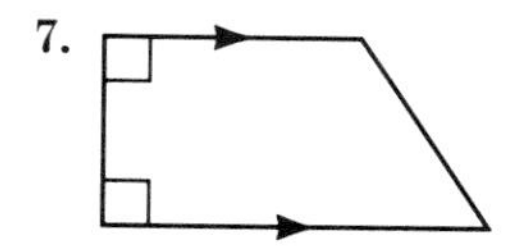

8.

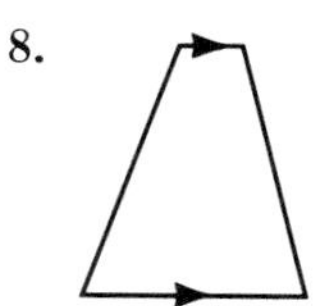

9. Not possible; if both bases were both $\cong$ and $\parallel$, the trap. would be a $\square$.

10. Not possible; 2 of the $\angle$s must be supp., and 2 supp. $\angle$s cannot both be acute.

11.

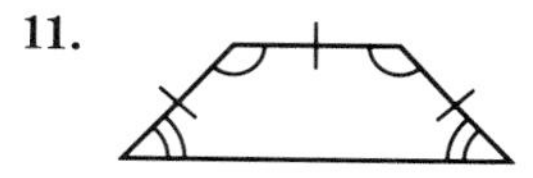

Pages 192–194 • WRITTEN EXERCISES

A 1. 12 2. 19 3. 15 4. 5 5. 9 6. 5 7. $2x - 1 + 7x + 3 = 2(19)$; 4

8. $2(2x + 4) = 3x + 2 + 2x + 1$; 5 9. $10x = 3x + 5x + 12$; 6

10. 57, 123, 123 11. $3x + 10 = 5x - 10$; $x = 10$; 40, 40, 140, 140 12. $AD = \frac{1}{2}BE$

13. $BE = \frac{1}{2}(AD + CF)$ 14. 14; 21 15. 13; 39 16. 6; 18

B 17. $2(x + y) = x + 3 + 36$; $x + y = 2(x + 3)$; 9, 15

18. $2(x + y) = 20$, $40 = x + y + 4x - y$; $x = 8$, $y = 2$; 30

19. $CF = 3 \cdot AD$, but $17 \neq 3 \cdot 5$

20. Since $BE = 2 \cdot AD$ and $CF = 3 \cdot AD$, any $x > -\frac{1}{2}$ is a solution.

21. rectangle

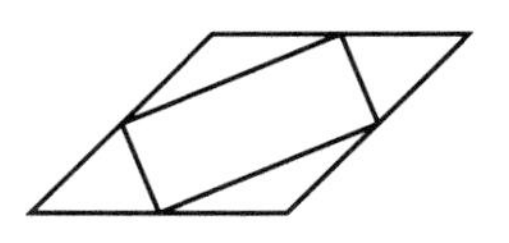

22. rhombus

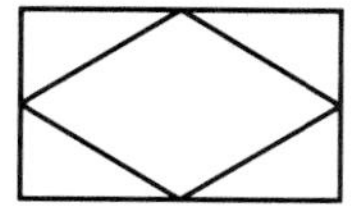

23. rhombus

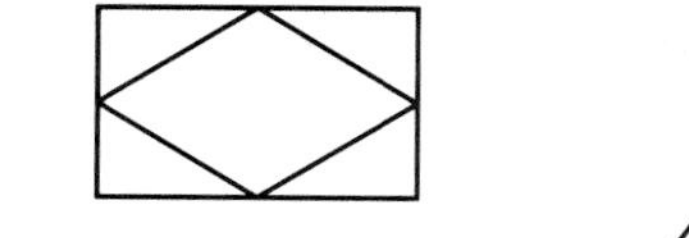

24. parallelogram

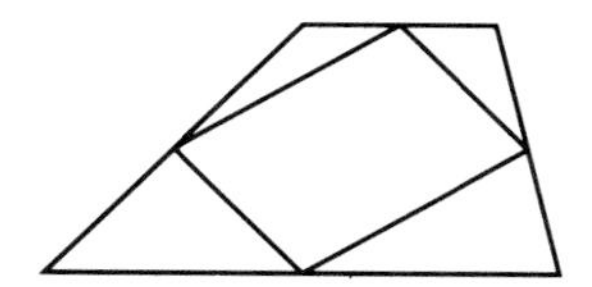

25. parallelogram

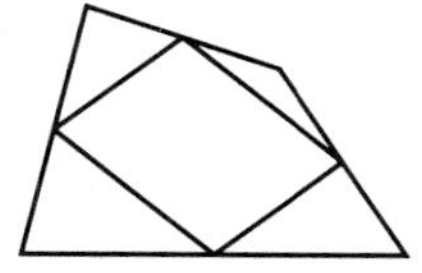

26. The diags. of an isos. trap. are $\cong$.
Given: Trap. $ABCD$ with $\overline{AD} \cong \overline{BC}$
Prove: $\overline{AC} \cong \overline{BD}$

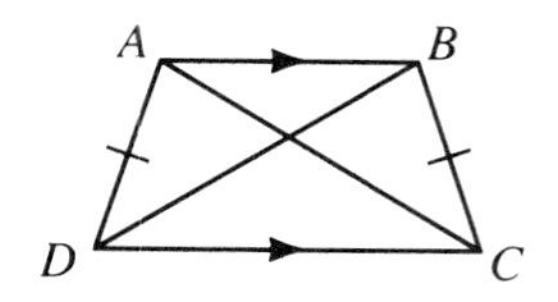

Statements	Reasons
1. $ABCD$ is a trap. with $\overline{AD} \cong \overline{BC}$.	1. Given
2. $\angle ADC \cong \angle BCD$	2. Base $\angle$s of an isos. trap. are $\cong$.
3. $\overline{DC} \cong \overline{DC}$	3. Refl. Prop.
4. $\triangle ADC \cong \triangle BCD$	4. SAS Post.
5. $\overline{AC} \cong \overline{BD}$	5. Corr. parts of $\cong$ $\triangle$s are $\cong$.

27. Two different proofs of Thm. 5-18 are given.
Given: Trap. $ABXY$ with $\overline{BX} \cong \overline{AY}$
Prove: $\angle 1 \cong \angle 3$; $\angle ABX \cong \angle A$

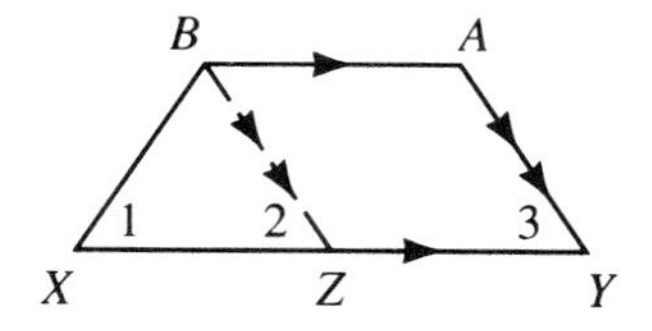

Statements	Reasons
1. Draw $\overline{BZ} \parallel \overline{AY}$ with Z on $\overline{XY}$.	1. Through a pt. outside a given line there is exactly one line $\parallel$ to the given line.
2. $ABXY$ is a trap. with $\overline{BX} \cong \overline{AY}$.	2. Given
3. $\overline{BA} \parallel \overline{XY}$	3. Def. of trap.
4. $ABZY$ is a $\square$.	4. Def. of $\square$
5. $\overline{AY} \cong \overline{BZ}$	5. Opp. sides of a $\square$ are $\cong$.
6. $\overline{BX} \cong \overline{BZ}$	6. Trans. Prop.
7. $\angle 1 \cong \angle 2$	7. Isos. $\triangle$ Thm.
8. $\angle 2 \cong \angle 3$	8. If 2 $\parallel$ lines are cut by a trans., then corr. $\angle$s are $\cong$.
9. $\angle 1 \cong \angle 3$	9. Trans. Prop.
10. $\angle ABX$ is supp. to $\angle 1$; $\angle A$ is supp. to $\angle 3$.	10. If 2 $\parallel$ lines are cut by a trans., then s-s. int. $\angle$s are supp.
11. $\angle ABX \cong \angle A$	11. If 2 $\angle$s are supps. of $\cong$ $\angle$s, then the 2 $\angle$s are $\cong$.

Given: Trap. $ABXY$ with $\overline{BX} \cong \overline{AY}$
Prove: $\angle X \cong \angle Y$; $\angle ABX \cong \angle BAY$

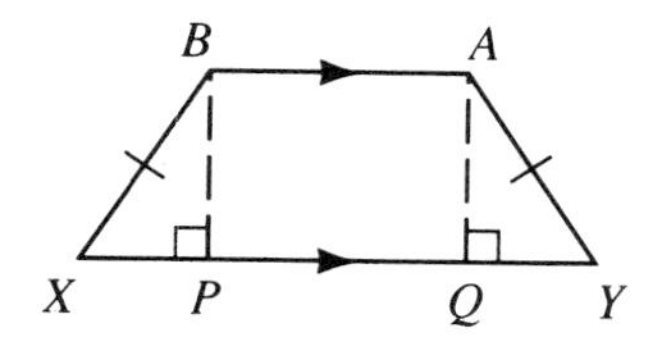

Statements	Reasons
1. Draw $\overline{BP} \perp \overline{XY}$ and $\overline{AQ} \perp \overline{XY}$.	1. Through a pt. outside a line, there is exactly one line $\perp$ to the given line.
2. $\angle BPX$ and $\angle AQY$ are rt. $\angle$s.	2. Def. of $\perp$ lines
3. $\triangle BPX$ and $\triangle AQY$ are rt. $\triangle$s.	3. Def. of rt. $\triangle$
4. $ABXY$ is a trap. with $\overline{BX} \cong \overline{AY}$.	4. Given
5. $\overline{BA} \parallel \overline{XY}$	5. Def. of trap.
6. $BP = AQ$ or $\overline{BP} \cong \overline{AQ}$	6. If 2 lines are $\parallel$, then all pts. on one line are equidistant from the other line.
7. $\triangle BPX \cong \triangle AQY$	7. HL Thm.
8. $\angle X \cong \angle Y$	8. Corr. parts of $\cong$ $\triangle$s are $\cong$.
9. $\angle ABX$ is supp. to $\angle X$; $\angle BAY$ is supp. to $\angle Y$.	9. If 2 $\parallel$ lines are cut by a trans., then s-s. int. $\angle$s are supp.
10. $\angle ABX \cong \angle BAY$	10. If 2 $\angle$s are supps. of $\cong$ $\angle$s, then the 2 $\angle$s are $\cong$.

28. Answers may vary; examples are given. (a) The longer diagonal of a kite is the $\perp$ bis. of the shorter diagonal. Proof: $\overline{BA} \cong \overline{BC}$ or $BA = BC$. So B is equidistant from the endpts. of $\overline{AC}$, and B lies on the $\perp$ bis. of $\overline{AC}$ (Thm. 4-6). Similarly, since $\overline{DA} \cong \overline{DC}$, D also lies on the $\perp$ bis. of $\overline{AC}$. Therefore, $\overline{BD}$ is the $\perp$ bis. of $\overline{AC}$.

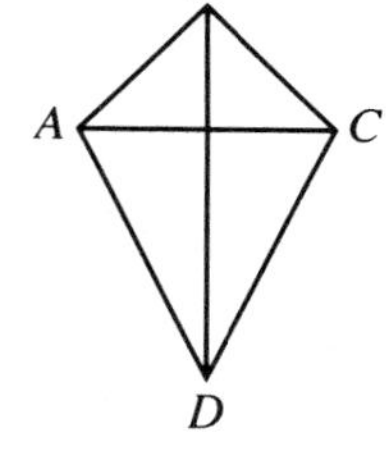

(b) One pair of opposite angles of a kite are $\cong$. Proof: Since $\overline{BA} \cong \overline{BC}$ and $\overline{DA} \cong \overline{DC}$ (Given), and $\overline{BD} \cong \overline{BD}$ by the Refl. Prop., then $\triangle BAD \cong \triangle BCD$ by SSS. Corr. parts $\angle BAD$ and $\angle BCD$ are $\cong$.

29. a.

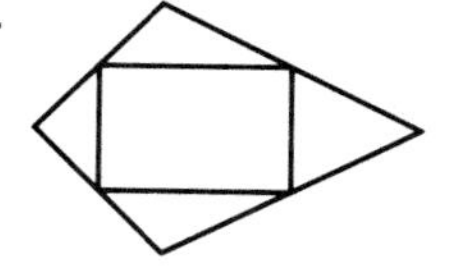

rectangle

b.

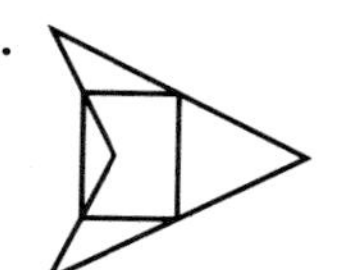

rectangle

30. 3; 3; $EF = MN - 6 = 5$

31. Since $EF = MN - (ME + FN)$ and $ME = \frac{1}{2}DC = FN$, while $MN = \frac{1}{2}(AB + DC)$, then by substituting, $EF = \frac{1}{2}(AB + DC) - \left(\frac{1}{2}DC + \frac{1}{2}DC\right) = \frac{1}{2}AB - \frac{1}{2}DC = \frac{1}{2}(AB - DC)$.

32. Ex. 31 proved that $EF = \frac{1}{2}(AB - DC)$, so $7 = \frac{1}{2}(2x^2 - 3x)$; $2x^2 - 3x - 14 = 0$; $(x + 2)(2x - 7) = 0$; $x + 2 = 0$ or $2x - 7 = 0$; $x = -2$ (reject), $x = 3.5$

C 33. *JKLM* is a rhombus. $KJ = \frac{1}{2}VE = LM$ (Thm. 5-11) and $\overline{KJ} \parallel \overline{VE}$ and $\overline{LM} \parallel \overline{VE}$ so $\overline{KJ} \parallel \overline{LM}$. Since $\overline{KJ}$ and $\overline{LM}$ are both $\cong$ and $\parallel$, *JKLM* is a ▱. Also, $KJ = \frac{1}{2}VE = \frac{1}{2}FG = KL$, so *JKLM* is a rhombus.

34. **a.** Check students' drawings. **b.** The diags. must be $\perp$.

35. **a.** Check students' drawings. **b.** The diags. must be $\cong$.

36. Given: Quad. *ABCD*;
P, Q, R, and S are midpts. of $\overline{AB}$, $\overline{BC}$, $\overline{DC}$, and $\overline{DA}$, resp.
Prove: O is the midpt. of both $\overline{PR}$ and $\overline{SQ}$.

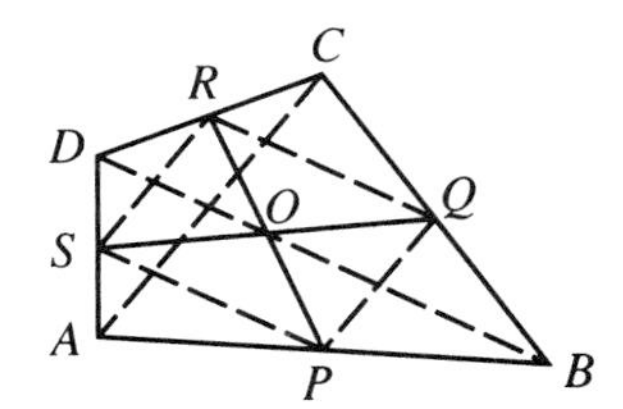

Statements	Reasons
1. Draw $\overline{PQ}$, $\overline{RQ}$, $\overline{RS}$, $\overline{SP}$, $\overline{AC}$, and $\overline{DB}$.	1. Through any 2 pts. there is exactly one line.
2. P, Q, R, and S are midpts. of $\overline{AB}$, $\overline{BC}$, $\overline{DC}$, and $\overline{DA}$, resp.	2. Given
3. $\overline{PS} \parallel \overline{DB}$ and $\overline{RQ} \parallel \overline{DB}$; $\overline{SR} \parallel \overline{AC}$ and $\overline{PQ} \parallel \overline{AC}$	3. The seg. that joins the midpts. of 2 sides of a △ is $\parallel$ to the third side.
4. $\overline{PS} \parallel \overline{RQ}$ and $\overline{SR} \parallel \overline{PQ}$	5. 2 lines $\parallel$ to a third line are $\parallel$ to each other.
6. *PQRS* is a ▱.	6. Def. of ▱
7. $\overline{RO} \cong \overline{OP}$; $\overline{SO} \cong \overline{OQ}$	7. The diags. of a ▱ bis. each other.
8. O is the midpt. of both $\overline{PR}$ and $\overline{SQ}$.	8. Def. of midpt.

Page 194 • CHALLENGE

a.

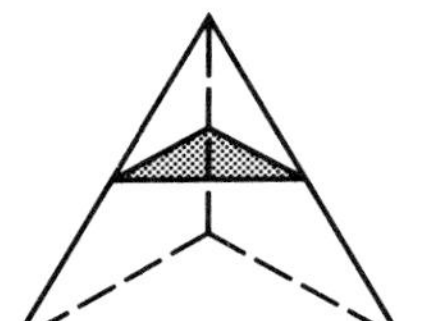

b.

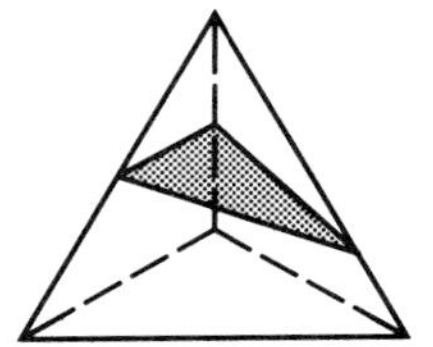

c.

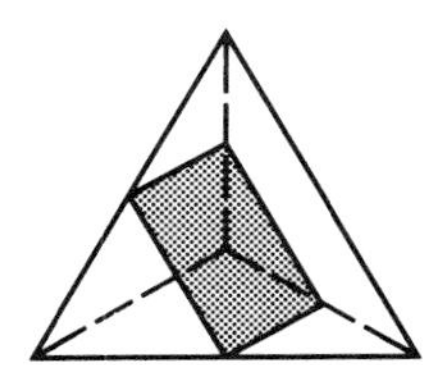

d.

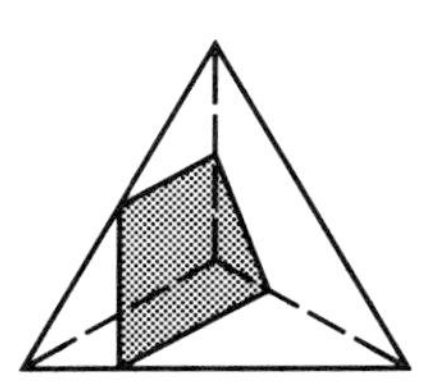

Page 195 • EXPLORATIONS

trapezoid; The seg. that joins the midpts. of 2 sides of a △ is ∥ to the third side; ▱; (same as above)

Page 195 • SELF-TEST 2

1. ▱ **2.** trapezoid **3.** rectangle **4.** square **5.** 11 **6.** 17; 67

7.

Statements	Reasons
1. $\angle 1 \cong \angle 2 \cong \angle 3 \cong \angle 4$	1. Given
2. $\overline{HG} \parallel \overline{EF}$; $\overline{HE} \parallel \overline{GF}$	2. If 2 lines are cut by a trans. and alt. int. ∠s are ≅, then the lines are ∥.
3. *EFGH* is a ▱.	3. Def. of ▱
4. $\overline{HG} \cong \overline{HE}$	4. If 2 ∠s of a △ are ≅, then the sides opp. those ∠s are ≅.
5. *HGFE* is a rhombus.	5. If 2 consec. sides of a ▱ are ≅, then the ▱ is a rhombus.

8. a. ▱

b.

Statements	Reasons
1. *PQRS* is a ▱.	1. Given
2. $\overline{PQ} \parallel \overline{SR}$	2. Def. of ▱
3. *X* is the midpt. of $\overline{PQ}$; *Y* is the midpt. of $\overline{SR}$.	3. Given
4. $XQ = \frac{1}{2}PQ$; $YR = \frac{1}{2}SR$	4. Midpt. Thm.
5. $\overline{PQ} \cong \overline{SR}$ or $PQ = SR$	5. Opp. sides of a ▱ are ≅.
6. $\frac{1}{2}PQ = \frac{1}{2}SR$	6. Mult. Prop. of =
7. $XQ = YR$	7. Substitution Prop.
8. *XQRY* is a ▱.	8. If one pair of opp. sides of a quad. are both ≅ and ∥, then the quad. is a ▱.

c. trapezoid

Pages 197–198 • CHAPTER REVIEW

1. 110 **2.** 38 **3.** 28 **4.** 6 **5.** $GS = 5$ or $\overline{SA} \parallel \overline{GN}$ **6.** $\angle SAN \cong \angle SGN$
7. $\overline{AZ} \cong \overline{GZ}$ **8.** $GN = 17$ or $\overline{GS} \parallel \overline{NA}$

9. A line that contains the midpt. of one side of a △ and is ∥ to another side passes through the midpt. of the third side.

10. The seg. that joins the midpts. of 2 sides of a △ is ∥ to the third side.

11. The seg. that joins the midpts. of 2 sides of a △ is half as long as the third side.

12.

Statements	Reasons
1. $CDEF$ is a ▱.	1. Given
2. $\overline{FE} \parallel \overline{CD}$	2. Def. of ▱
3. $\overline{FE} \parallel \overline{CR}$ and $\overline{FS} \parallel \overline{DR}$	3. $\overline{FE} \parallel \overline{CD}$
4. S and T are midpts. of $\overline{EF}$ and $\overline{ED}$.	4. Given
5. $\overline{SR} \parallel \overline{FD}$	5. The seg. that joins the midpts. of 2 sides of a △ is ∥ to the third side.
6. $FSRD$ is a ▱.	6. Def. of ▱
7. $\overline{SR} \cong \overline{FD}$	7. Opp. sides of a ▱ are ≅.

13. ▱ **14.** rhombus **15.** rectangle **16.** square

17.

Statements	Reasons
1. $ABCD$ is a rhombus.	1. Given
2. $DO = BO$; $AO = CO$	2. Diags. of a ▱ bis. each other.
3. $DO = DE + EO$; $BO = BF + FO$	3. Seg. Add. Post.
4. $DE + EO = BF + FO$	4. Substitution Prop.
5. $DE = BF$	5. Given
6. $EO = FO$	6. Subtr. Prop. of =
7. $AECF$ is a ▱.	7. If the diags. of a quad. bis. each other, then the quad. is a ▱.
8. $\overline{BD} \perp \overline{AC}$	8. Diags. of a rhombus are ⊥.
9. $\angle COE \cong \angle COF$	9. If 2 lines are ⊥, then they form ≅ adj. ∠s.
10. $\overline{CO} \cong \overline{CO}$	10. Refl. Prop.
11. $\triangle COE \cong \triangle COF$	11. SAS Post.
12. $\overline{CE} \cong \overline{CF}$	12. Corr. parts of ≅ △s are ≅.
13. $AECF$ is a rhombus.	13. If 2 consec. sides of a ▱ are ≅, then the ▱ is a rhombus.

18. Given: $\overline{PX}$ and $\overline{QY}$ are altitudes of $\triangle PQR$; Z is the midpt. of $\overline{PQ}$.
Prove: $\triangle XYZ$ is isos.

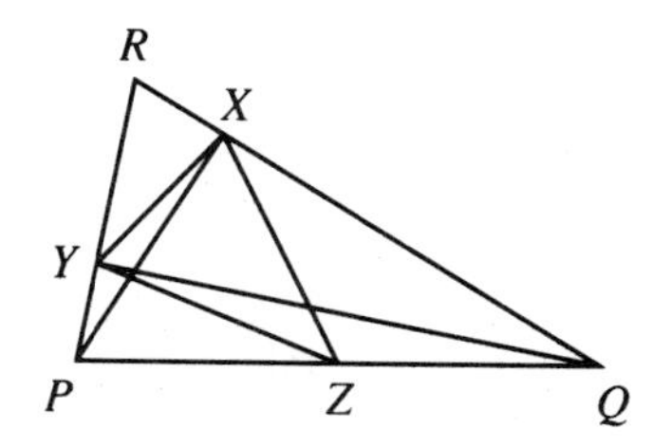

Statements	Reasons
1. $\overline{PX}$ and $\overline{QY}$ are altitudes.	1. Given
2. $\overline{PX} \perp \overline{RQ}$; $\overline{QY} \perp \overline{PR}$	2. Def. of altitude
3. $\angle PXQ$ and $\angle PYQ$ are rt. $\angle$s.	3. Def. of $\perp$ lines
4. $\triangle PXQ$ and $\triangle PYQ$ are rt. $\triangle$s.	4. Def. of rt. $\triangle$
5. Z is the midpt. of $\overline{PQ}$.	5. Given
6. $XZ = PZ$ or $\overline{XZ} \cong \overline{PZ}$; $PZ = YZ$ or $\overline{PZ} \cong \overline{YZ}$	6. The midpt. of the hypotenuse of a rt. $\triangle$ is equidistant from the 3 vertices.
7. $\overline{XZ} \cong \overline{YZ}$	7. Trans. Prop.
8. $\triangle XYZ$ is isos.	8. Def. of isos. $\triangle$

19. $\overline{ZO}$, $\overline{DI}$ **20.** 14 **21.** 4 **22.** 100

Page 199 • CHAPTER TEST

1. always **2.** sometimes **3.** sometimes **4.** always **5.** sometimes **6.** never
7. always **8.** sometimes **9.** 28 **10.** 4.5 **11.** $7j + 2k$

12. Yes; if both pairs of opp. $\angle$s of a quad. are $\cong$, then the quad. is a $\square$.

13. Yes; if one pair of opp. sides of a quad. are both $\cong$ and $\parallel$, then the quad. is a $\square$.

14. Yes; if the diags. of a quad. bisect each other, then the quad. is a $\square$.

15. Yes; if both pairs of opp. sides of a quad. are $\cong$, then the quad. is a $\square$.

16. 14; 15; 26

17. $5x - 4 = 3(x + 4)$, $x = 8$; $6y + 2 = 2y^2 - 6$, $y = 4$; $z = \frac{1}{2}(54 + 30) = 42$

18.

Statements	Reasons
1. $PQRS$ is a $\square$.	1. Given
2. $\overline{PS} \cong \overline{QR}$	2. Opp. sides of a $\square$ are $\cong$.
3. $\angle P \cong \angle R$	3. Opp. $\angle$s of a $\square$ are $\cong$.
4. $PA = RB$ or $\overline{PA} \cong \overline{RB}$	4. Given
5. $\triangle PAS \cong \triangle RBQ$	5. SAS Post.
6. $\overline{AS} \cong \overline{BQ}$ or $AS = BQ$	6. Corr. parts of $\cong$ $\triangle$s are $\cong$.

19.

Statements	Reasons
1. $\overline{PR} \parallel \overline{VO}$; $\overline{RO} \parallel \overline{PV}$	1. Given
2. $PROV$ is a ▱.	2. Def. of ▱
3. $\overline{PR} \cong \overline{RO}$	3. Given
4. $PROV$ is a rhombus.	4. If 2 consec. sides of a ▱ are ≅, then the ▱ is a rhombus.
5. $\overline{RE} \cong \overline{EV}$	5. The diags. of a ▱ bis. each other.
6. $\overline{EO} \cong \overline{EO}$	6. Refl. Prop.
7. $\overline{RO} \cong \overline{VO}$	7. Def. of rhombus
8. $\triangle ROE \cong \triangle VOE$	8. SSS Post.
9. $\overline{OE} \perp \overline{RV}$	9. The diags. of a rhombus are ⊥.
10. $\angle VEO$ is a rt. ∠.	10. Def. of ⊥ lines
11. $\triangle VEO$ is a rt. △.	11. Def. of rt. △
12. $\angle 1$ and $\angle VOE$ are comp.	12. The acute ∠s of a rt. △ are comp.
13. $m\angle 1 + m\angle VOE = 90$	13. Def. of comp. ∠s
14. $\angle 2 \cong \angle VOE$ or $m\angle 2 = m\angle VOE$	14. Corr. parts of ≅ △s are ≅.
15. $m\angle 1 + m\angle 2 = 90$	15. Substitution Prop.
16. $\angle 1$ and $\angle 2$ are comp.	16. Def. of comp. ∠s

Pages 200–201 • CUMULATIVE REVIEW: CHAPTERS 1–5

A **1.** one **2. a.** yes; skew lines **b.** no

3. If you enjoy winter weather, then you are a member of the skiing club.

4. −1 **5.** Trans. Prop.

6. 180; The sum of the meas. of the ∠s of a △ is 180. **7.** 180; ∠ Add. Post.

8. 5; The meas. of an ext. ∠ of a △ equals the sum of the meas. of the 2 remote int. ∠s.

9. ∠1; If 2 ∥ lines are cut by a trans., then corr. ∠s are ≅.

10. $\overline{EB}$; If 2 ∠s of a △ are ≅, then the sides opp. those ∠s are ≅.

11. bisects; ⊥ **12. a.** A and B **b.** $\overrightarrow{SR}$ and $\overrightarrow{ST}$

13. a. $\triangle RTA$ **b.** $\overline{DB}$ **c.** $m\angle E$ **14.** $\frac{38(180)}{40} = 171$ **15.** 150, 150

B **16.** $2x + 7 = 4x - 1$; $2x = 8$; $x = 4$; $SU = 2(4) + 7 = 15$; $UN = 4(4) - 1 = 15$; $SN = 3(4) + 4 = 16$

17. $MN = \frac{1}{2}[(2r + s) + (4r - 3s)] = \frac{1}{2}(6r - 2s) = 3r - s$

18. median **19.** bisector **20.** isos.

21. $m\angle DAC + 2m\angle ADC = 180$, $36 + 2m\angle ADC = 180$, $2m\angle ADC = 144$, $m\angle ADC = 72$; $m\angle ADF = \frac{1}{2}(72) = 36$

22. isos. **23.** ABC, BAC, ACD, CFD

24. In $\triangle VOZ$, $m\angle V = 180 - (90 + 30) = 60$, so $m\angle 1 = 180 - (90 + 60) = 30$; $\overline{OY} \cong \overline{YZ}$, so $m\angle 4 = m\angle Z = 30$; $m\angle 1 + m\angle 2 + m\angle 3 + m\angle 4 = 30 + m\angle 2 + m\angle 3 + 30 = 90$, so $m\angle 2 + m\angle 3 = 30$, and since $\overrightarrow{OX}$ bis. $\angle VOZ$, $m\angle 2 = m\angle 3 = 15$.

25. $m\angle 1 = m\angle 4 = k$; $m\angle 2 = m\angle 3 = 45 - k$

26. ▱; If both pairs of opp. sides of a quad. are ≅, then the quad. is a ▱.

27. $\angle NOM$, $\angle LMO$, $\angle NMO$; Each diag. of a rhombus bisects 2 ∠s of the rhombus.

28. midpt., $\overline{MN}$; A line that contains the midpt. of one side of a △ and is ∥ to another side passes through the midpt. of the third side.

29. PQ, ON; The median of a trap. has a length equal to the average of the base lengths.

30.

Statements	Reasons
1. $WP = ZP$; $PY = PX$	1. Given
2. $WP + PY = ZP + PX$	2. Add. Prop. of =
3. $WY = WP + PY$; $XZ = ZP + PX$	3. Seg. Add. Post.
4. $WY = XZ$ or $\overline{WY} \cong \overline{XZ}$	4. Substitution Prop.
5. $\angle PXY \cong \angle PYX$	5. Isos. △ Thm.
6. $\overline{XY} \cong \overline{XY}$	6. Refl. Prop.
7. $\triangle WXY \cong \triangle ZYX$	7. SAS Post.
8. $\angle WXY \cong \angle ZYX$	8. Corr. parts of ≅ △s are ≅.

31.

Statements	Reasons
1. $\overline{AD} \cong \overline{BC}$; $\overline{AD} \parallel \overline{BC}$	1. Given
2. $ABCD$ is a ▱.	2. If one pair of opp. sides of a quad. are both ≅ and ∥, then the quad. is a ▱.
3. $\overline{DF} \cong \overline{BF}$	3. Diags. of a ▱ bis. each other.
4. $\angle DFG \cong \angle BFE$	4. Vert. ∠s are ≅.
5. $\overline{DC} \parallel \overline{AB}$	5. Def. of ▱
6. $\angle CDB \cong \angle ABD$	6. If 2 ∥ lines are cut by a trans., then alt. int. ∠s are ≅.
7. $\triangle DFG \cong \triangle BFE$	7. ASA Post.
8. $\overline{EF} \cong \overline{FG}$	8. Corr. parts of ≅ △s are ≅.

CHAPTER 6 • Inequalities in Geometry

Page 205 • CLASSROOM EXERCISES

1. False **2.** True **3.** True **4.** True **5.** True **6.** True **7.** False **8.** True
9. True **10.** False **11.** True **12.** True **13.** False **14.** False **15.** True
16. False **17. a.** $=$ **b.** $>$ **c.** $>$ **18. a.** $=$ **b.** $<$ **c.** $>$
19. a. $=$ **b.** $=$ **c.** $>$ **d.** $>$
20. 1. Given 2. The meas. of an ext. $\angle$ of a $\triangle >$ the meas. of either remote int. $\angle$.
3. A Prop. of Ineq. 4. Vert. $\angle$s are $\cong$. 5. Substitution Prop.

Pages 206–207 • WRITTEN EXERCISES

A **1. a.** No **b.** Yes **c.** Yes **d.** No **e.** Yes **f.** No
2. a. No **b.** Yes **c.** Yes **d.** Yes **e.** No **f.** No
3. a. No **b.** No **c.** Yes **d.** Yes
4. a. Yes **b.** Yes **c.** No **d.** No
5. Answers may vary; for example, $j = 2$, $k = 1$, $l = 4$, $m = 3$
6. 1. Given 2. Corr. parts of $\cong$ $\triangle$s are $\cong$. 3. Seg. Add. Post.
4. A Prop. of Ineq. 5. Substitution Prop.
7. 1. Vert. $\angle$s are $\cong$. 2. $\angle$ Add. Post. 3. A Prop. of Ineq. 4. Substitution Prop.

B **8.**

Statements	Reasons
1. $KL > NL$; $LM > LP$	1. Given
2. $KL + LM > NL + LP$	2. A Prop. of Ineq.
3. $KL + LM = KM$; $NL + LP = NP$	3. Seg. Add. Post.
4. $KM > NP$	4. Substitution Prop.

9.

Statements	Reasons
1. $m\angle ROS > m\angle TOV$	1. Given
2. $m\angle SOT = m\angle SOT$	2. Refl. Prop.
3. $m\angle ROS + m\angle SOT > m\angle TOV + m\angle SOT$	3. A Prop. of Ineq.
4. $m\angle ROS + m\angle SOT = m\angle ROT$; $m\angle TOV + m\angle SOT = m\angle SOV$	4. $\angle$ Add. Post.
5. $m\angle ROT > m\angle SOV$	5. Substitution Prop.

10.

Statements	Reasons
1. $\overline{VY} \perp \overline{YZ}$	1. Given
2. $m\angle XYZ = 90$	2. Def. of $\perp$ lines
3. $m\angle VXZ > m\angle XYZ$	3. The meas. of an ext. $\angle$ of a $\triangle$ > the meas. of either remote int. $\angle$.
4. $m\angle VXZ > 90$	4. Substitution Prop.
5. $\angle VXZ$ is an obtuse $\angle$.	5. Def. of obtuse $\angle$

11.

Statements	Reasons
1. $m\angle 1 > m\angle 2$; $m\angle 2 > m\angle 3$	1. The meas. of an ext. $\angle$ of a $\triangle$ > the meas. of either remote int. $\angle$.
2. $m\angle 1 > m\angle 3$	2. A Prop. of Ineq.
3. $\angle 3 \cong \angle 4$, or $m\angle 3 = m\angle 4$	3. Vert. $\angle$s are $\cong$.
4. $m\angle 1 > m\angle 4$	4. Substitution Prop.

12.

Statements	Reasons
1. $\overline{QR}$ and $\overline{ST}$ bis. each other.	1. Given
2. $\overline{QV} \cong \overline{VR}$; $\overline{SV} \cong \overline{VT}$	2. Def. of bis.
3. $\angle QVS \cong \angle RVT$	3. Vert. $\angle$s are $\cong$.
4. $\triangle QVS \cong \triangle RVT$	4. SAS Post.
5. $\angle S \cong \angle T$, or $m\angle S = m\angle T$	5. Corr. parts of $\cong$ $\triangle$s are $\cong$.
6. $m\angle XRT > m\angle T$	6. The meas. of an ext. $\angle$ of a $\triangle$ > the meas. of either remote int. $\angle$.
7. $m\angle XRT > m\angle S$	7. Substitution Prop.

C 13.

Statements	Reasons
1. Extend $\overline{BK}$ to int. $\overline{AC}$ at R.	1. If 2 lines int., then they int. in exactly one pt.
2. $m\angle AKB > m\angle ARK$; $m\angle ARK > m\angle C$	2. The meas. of an ext. $\angle$ of a $\triangle$ > the meas. of either remote int. $\angle$.
3. $m\angle AKB > m\angle C$	3. A Prop. of Ineq.

Page 207 • CHALLENGE

a. 8; 8 **b.** 24; $12(n - 2)$ **c.** 24; $6(n - 2)^2$ **d.** 8; $(n - 2)^3$

Page 210 • CLASSROOM EXERCISES

1. **a.** If you can't dance, then I can't sing. **b.** If I can ride a horse, then you can play baseball. **c.** If $x^2 - 5 \neq 11$, then $x \neq 4$. **d.** If $y = 4$, then $y \geq 3$. **e.** If the sum of the meas. of the $\angle$s of a polygon is not 180, then the polygon is not a $\triangle$.
2. **a.** If I can't sing, then you can't dance. **b.** If you can play baseball, then I can ride a horse. **c.** If $x \neq 4$, then $x^2 - 5 \neq 11$. **d.** If $y \geq 3$, then $y = 4$. **e.** If a polygon is not a $\triangle$, then the sum of the meas. of its $\angle$s is not 180.
3. No; no; yes **4.** No; no; yes
5. True; Inverse: If a $\triangle$ is not equilateral, then it is not equiangular. (true); Contrapositive: If a $\triangle$ is not equiangular, then it is not equilateral. (true)
6. True; Inverse: If $\angle A$ is not acute, then $m\angle A = 100$. (false); Contrapositive: If $m\angle A = 100$, then $\angle A$ is not acute. (true)
7. True; Inverse: If a $\triangle$ is isos., then it is equilateral. (false); Contrapositive: If a $\triangle$ is equilateral, then it is isos. (true)
8. True; Inverse: If 2 planes intersect, then they are not $\parallel$. (true); Contrapositive: If 2 planes are not $\parallel$, then they intersect. (true)
9. If a polygon is a square, then it is a rhombus.
10. If a polygon is a trap., then it is not equiangular.
11. If you are a marathoner, then you have stamina.
12. **a.** Nick has stamina. **b.** No conclusion **c.** Mimi is not a marathoner. **d.** No conclusion

Pages 210–212 • WRITTEN EXERCISES

A

1. **a.** If $4n \neq 68$, then $n \neq 17$. **b.** If $n \neq 17$, then $4n \neq 68$.
2. **a.** If this is not blue, then those are not red and white. **b.** If those are not red and white, then this is not blue.
3. **a.** If $x + 1$ is odd, then x is even. **b.** If x is even, then $x + 1$ is odd.
4. **a.** If Abby is well, then she is here. **b.** If Abby is here, then she is well.
5.

Statement	If I live in L.A., then I live in Calif.	True
Contrapositive	If I don't live in Calif., then I don't live in L.A.	True
Converse	If I live in Calif., then I live in L.A.	False
Inverse	If I don't live in L.A., then I don't live in Calif.	False

6.

Statement	If $\angle 1$ and $\angle 2$ are vert. $\angle$s, then $m\angle 1 = m\angle 2$.	True
Contrapositive	If $m\angle 1 \neq m\angle 2$, then $\angle 1$ and $\angle 2$ are not vert. $\angle$s.	True
Converse	If $m\angle 1 = m\angle 2$, then $\angle 1$ and $\angle 2$ are vert. $\angle$s.	False
Inverse	If $\angle 1$ and $\angle 2$ are not vert. $\angle$s, then $m\angle 1 \neq m\angle 2$.	False

7.

Statement	If $AM = MB$, then M is the midpoint of $\overline{AB}$.	False
Contrapositive	If M is not the midpoint of $\overline{AB}$, then $AM \neq MB$.	False
Converse	If M is the midpoint of $\overline{AB}$, then $AM = MB$.	True
Inverse	If $AM \neq MB$, then M is not the midpoint of $\overline{AB}$.	True

8.

Statement	If a $\triangle$ is scalene, then it has no $\cong$ sides.	True
Contrapositive	If a $\triangle$ has some $\cong$ sides, then it is not scalene.	True
Converse	If a $\triangle$ has no $\cong$ sides, then it is scalene.	True
Inverse	If a $\triangle$ is not scalene, then it has some $\cong$ sides.	True

B 9.

Statement	If $-2n < 6$, then $n > -3$.	True
Contrapositive	If $n \leq -3$, then $-2n \geq 6$.	True
Converse	If $n > -3$, then $-2n < 6$.	True
Inverse	If $-2n \geq 6$, then $n \leq -3$.	True

10.

Statement	If $x^2 > 1$, then $x > 1$.	False
Contrapositive	If $x \leq 1$, then $x^2 \leq 1$.	False
Converse	If $x > 1$, then $x^2 > 1$.	True
Inverse	If $x^2 \leq 1$, then $x \leq 1$.	True

11. If you are a senator, then you are at least 30 years old. **a.** No conclusion
b. She is at least 30 years old. **c.** No conclusion **d.** He is not a senator.

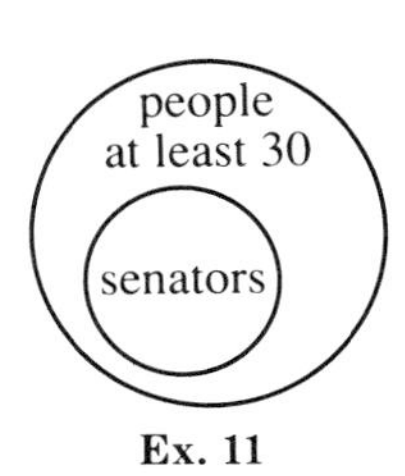

Ex. 11

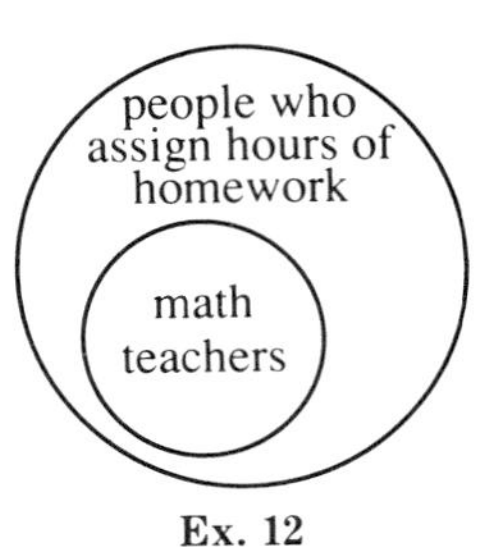

Ex. 12

12. If you are a math teacher, then you assign hours of homework.
a. She assigns hours of homework. **b.** No conclusion
c. He is not a math teacher. **d.** No conclusion

13. **a.** It is raining. **b.** I am happy. **c.** No conclusion **d.** No conclusion

14. **a.** Stu loves geometry. **b.** No conclusion **c.** No conclusion **d.** George is not my student.

15. **a.** No conclusion **b.** $\angle ABC$ and $\angle DBF$ are not vert. $\angle$s. **c.** No conclusion
d. $\angle RVU \cong \angle SVT$, $\angle RVT \cong \angle SVU$

16. **a.** $\overline{JL} \perp \overline{KM}$ **b.** No conclusion **c.** No conclusion **d.** $NOPQ$ is not a rhombus.

17. **a.** Diags. are $\cong$. **b.** No conclusion **c.** No conclusion **d.** $STAR$ is not a rectangle.

18. **a.** No conclusion **b.** $LAST$ is not a rhombus or a square. **c.** $PQRS$ is a rhombus.
d. No conclusion

C **19.** contrapositive **20.** If r, then s.

21. Statement: If $m\angle A + m\angle B \neq 180$, then $m\angle D + m\angle C \neq 180$.
Contrapositive: If $m\angle D + m\angle C = 180$, then $m\angle A + m\angle B = 180$.
Given: $m\angle D + m\angle C = 180$
Prove: $m\angle A + m\angle B = 180$

Statements	Reasons
1. $m\angle D + m\angle C = 180$	1. Given
2. $\overleftrightarrow{AD} \parallel \overleftrightarrow{BC}$	2. If 2 lines are cut by a trans. and s-s. int. $\angle$s are supp., then the lines are $\parallel$.
3. $m\angle A + m\angle B = 180$	3. If 2 $\parallel$ lines are cut by a trans., then s-s. int. $\angle$s are supp.

22. Statement: If n^2 is not a multiple of 3, then n is not a multiple of 3.

Contrapositive: If n is a multiple of 3, then n^2 is a multiple of 3.

Given: n is a multiple of 3, or $n = 3k$ for some integer k.

Prove: n^2 is a multiple of 3, or $n^2 = 3t$ for some integer t.

Statements	Reasons
1. $n = 3k$	1. Given
2. $n^2 = (3k)(3k) = 3(3k^2)$	2. Mult. Prop. of =
3. $n^2 = 3t$, $t = 3k^2$	3. Substitution Prop.

Page 212 • MIXED REVIEW EXERCISES

1. sometimes **2.** sometimes **3.** always **4.** never **5.** always

6. always **7.** sometimes

8. $m\angle 1 = m\angle 4 = 60$; $m\angle 2 = 180 - (60 + 45) = 75$; $m\angle 3 = 180 - (60 + 75) = 45$

9. $x = 60 + 35 = 95$

Pages 215–216 • CLASSROOM EXERCISES

1. b **2.** Assume temp. that $\triangle ABC$ is not equilateral.

3. Assume temp. that Doug isn't a Canadian. **4.** Assume temp. that $a < b$.

5. Assume temp. that Kim is a violinist. **6.** Assume temp. that $m\angle X \leq m\angle Y$.

7. Assume temp. that $\overline{CX}$ is a median of $\triangle ABC$. **8.** $\angle A$ may be a rt. $\angle$.

9. l and m may be $\parallel$. **10.** d, a, e, b, c

Pages 216–217 • WRITTEN EXERCISES

A **1.** Assume temp. that $m\angle B \neq 40$. **2.** Assume temp. that $\overline{DE} \cong \overline{RS}$.

3. Assume temp. that $a - b = 0$. **4.** Assume temp. that $x = y$.

5. Assume temp. that $\overleftrightarrow{EF} \parallel \overleftrightarrow{GH}$.

6. Assume temp. that it's not cold outside. Then when people wearing coats come to the door, they are not cold, and since they are not cold, they are not shivering. But this contradicts the given information. The temp. assumption must be false. It follows that it's cold outside.

7. Assume temp. that $\angle Y$ is a rt. $\angle$. Since $m\angle X = 100$, $\angle X$ is an obtuse $\angle$. This contradicts Thm. 3-11 Cor. 3, which states that in a $\triangle$, there can be at most one rt. $\angle$ or obtuse $\angle$. The temp. assumption must be false. It follows that $\angle Y$ is not a rt. $\angle$.

8. Assume temp. that n is not odd. Then n is even, and $n^2 = n \times n =$ even $\times$ even $=$ even. But this contradicts the given information that n^2 is odd. The temp. assumption must be false. It follows that n is odd.

9. Assume temp. that $a \parallel b$. Then $\angle 1 \cong \angle 3$ (corr. ∠s), and since $\angle 3 \cong \angle 2$ (vert. ∠s), $\angle 1 \cong \angle 2$ (Trans. Prop.). This contradicts the given information that $m\angle 1 \neq m\angle 2$. The temp. assumption must be false. It follows that $a \nparallel b$.

10. Assume temp. that $\overrightarrow{OE}$ bis. $\angle JOK$. Then $\angle 1 \cong \angle 2$, $\overline{OJ} \cong \overline{OK}$, and $\overline{OE} \cong \overline{OE}$, so $\triangle OJE \cong \triangle OKE$ (SAS) and $\overline{JE} \cong \overline{KE}$ (Corr. parts of $\cong$ △s are $\cong$.). This contradicts the given information that $\overline{JE} \ncong \overline{KE}$. The temp. assumption must be false. It follows that $\overrightarrow{OE}$ doesn't bisect $\angle JOK$.

B 11. Assume temp. that planes P and Q do not intersect, that is, they are $\parallel$. The lines in which plane N intersects planes P and Q, $\overleftrightarrow{AB}$ and $\overleftrightarrow{CD}$, must be $\parallel$. This contradicts the given information that $\overleftrightarrow{AB} \nparallel \overleftrightarrow{CD}$. The temp. assumption must be false. It follows that planes P and Q intersect.

12. Assume temp. that $\triangle RST$ is equilateral. Then $RS = ST = TR = RV = VS$. This contradicts the given information that $\triangle RVS$ is not equilateral. The temp. assumption must be false. It follows that $\triangle RST$ is not equilateral.

13. Assume temp. that $EFGH$ is a convex quad. Then the sum of its $\angle$ measures is 360. But this contradicts the given information that the sum of the $\angle$ measures is $93 + 20 + 147 + 34$, or 294. The temp. assumption must be false. It follows that $EFGH$ is not a convex quad.

14. Assume temp. that $\angle ACB$ is a rt. $\angle$. Then $\triangle ACB$ is a rt. $\triangle$ and T is the midpt. of the hyp. The midpt. of the hyp. of a rt. $\triangle$ is equidistant from the 3 vertices. This contradicts the given information that $CT = 4 \neq AT$. The temp. assumption must be false. It follows that $\angle ACB$ is not a rt. $\angle$.

15. Assume temp. that n does not int. k. Since n and k are coplanar, n and k must be $\parallel$. Then P is on n and l, and n and l are both $\parallel$ to k. This contradicts the thm. which states that through a pt. outside a line there is exactly one line $\parallel$ to the given line. The temp. assumption must be false. It follows that n intersects k.

16. Given: $\triangle ABC$ with $\angle A \ncong \angle B$
Prove: $\overline{BC} \ncong \overline{AC}$
Proof: Assume temp. that $\overline{BC} \cong \overline{AC}$.
Then by the Isos. $\triangle$ Thm., $\angle A \cong \angle B$.

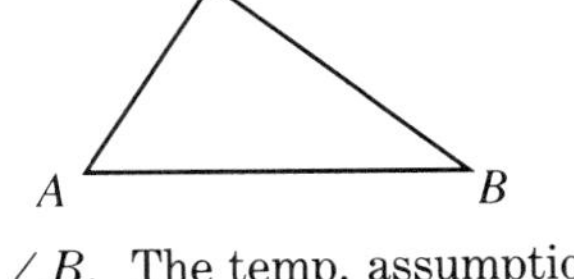

This contradicts the given information that $\angle A \ncong \angle B$. The temp. assumption must be false. It follows that $\overline{BC} \ncong \overline{AC}$.

17. Assume temp. that there is an n-sided reg. polygon with an interior $\angle$ of meas. 155. Then the meas. of each ext. $\angle$ is 25 and $25n = 360$. This contradicts the fact that there is no whole number n such that $25n = 360$. The temp. assumption must be false. It follows that there is no reg. polygon with an interior $\angle$ of meas. 155.

18. Given: Trap. $ABCD$ Prove: The diags. of $ABCD$ do not bis. each other. Proof: Assume temp. that the diags. of $ABCD$ bis. each other. Then $ABCD$ is a ▱. But this contradicts the given information that $ABCD$ is a trap. The temp. assumption must be false. It follows that the diags. of trap. $ABCD$ do not bis. each other.

C 19. Given: $j \perp P$; $k \perp P$ Prove: j does not intersect k. Proof: Assume temp. that j int. k at a point X. Since $j \perp P$, $j \perp \overleftrightarrow{AB}$, and since $k \perp P$, $k \perp \overleftrightarrow{AB}$. This contradicts the fact that through a pt. (such as X) outside a line ($\overleftrightarrow{AB}$) there is exactly one line $\perp$ to the given line. The temp. assumption must be false. It follows that j does not int. k.

20. Assume temp. that $\overleftrightarrow{RT}$ and $\overleftrightarrow{SW}$ are not skew. Case 1: $\overleftrightarrow{RT}$ intersects $\overleftrightarrow{SW}$. Then $\overleftrightarrow{RT}$ and $\overleftrightarrow{SW}$ are coplanar, so R, S, T, and W are coplanar and $\overleftrightarrow{RS}$ and $\overleftrightarrow{TW}$ are coplanar. This contradicts the given information that $\overleftrightarrow{RS}$ and $\overleftrightarrow{TW}$ are skew. Case 2: $\overleftrightarrow{RT} \parallel \overleftrightarrow{SW}$. Then $\overleftrightarrow{RT}$ and $\overleftrightarrow{SW}$ are coplanar, so R, S, T, and W are coplanar and $\overleftrightarrow{RS}$ and $\overleftrightarrow{TW}$ are coplanar. This contradicts the given information that $\overleftrightarrow{RS}$ and $\overleftrightarrow{TW}$ are skew. In each case, the temp. assumption must be false. It follows that $\overleftrightarrow{RT}$ and $\overleftrightarrow{SW}$ are skew.

Page 217 • CHALLENGE

Leo ate the last piece. If only Joan lied, then Ken and Leo both told the truth, which is impossible. If only Ken lied, then Martha told the truth, which is impossible. If only Martha lied, then both Ken and Leo told the truth, which is impossible. Therefore, Leo lied and ate the last piece of lasagna.

Page 218 • SELF-TEST 1

1. True **2.** True **3.** False **4.** False

5. If $\triangle ABC$ is not acute, then $m\angle C = 90$. False

6. If $m\angle C = 90$, then $\triangle ABC$ is not acute. True **7.** C

8. a. $ABCD$ is not a rhombus. **b.** No conclusion **c.** No conclusion **d.** $GHIJ$ is a ▱.

9. Assume temp. that $AC \neq 14$. **10.** d, b, a, c

Pages 221–222 • CLASSROOM EXERCISES

1. largest: $\angle R$; smallest: $\angle S$ **2.** largest: $\angle V$; smallest: $\angle W$

3. largest: $\angle Z$; smallest: $\angle Y$ **4.** longest: $\overline{BC}$; shortest: $\overline{AC}$

5. longest: $\overline{DF}$; shortest: $\overline{DE}$ **6.** longest: $\overline{GI}$; shortest: $\overline{GH}$

7. Yes **8.** No **9.** No **10.** No **11.** Yes **12.** Yes

13. a. **b.**

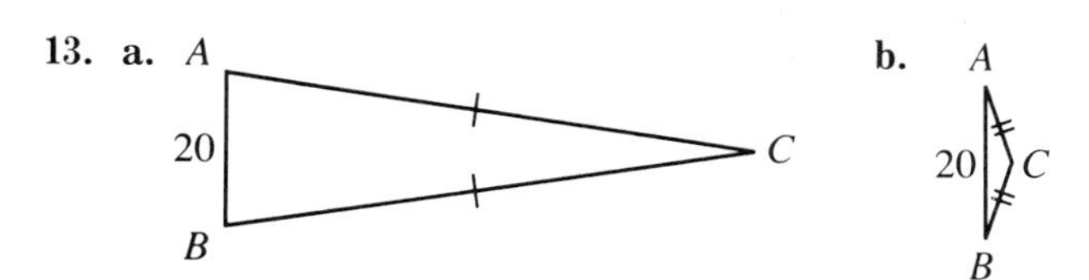

14. The length of each leg must be greater than 6.

15. The length of each diag. must be between 2 and 22. **16.** 5, 35

17. The length of each diag. must be > 100.

18. A △ can have at most one right or obtuse ∠. Then $m\angle A > m\angle B$ and by Thm. 6-3, $PB > PA$.

19. A △ can have at most one right or obtuse ∠. Then $m\angle C > m\angle D$ and by Thm. 6-3, $PD > PC$.

20. The largest ∠ is the right ∠. (The sum of the other 2 ≜ is 90, so they must both be < 90.) The longest side is the hypotenuse. (Thm. 6-3)

Pages 222–223 • WRITTEN EXERCISES

A **1.** 3; 15 **2.** 2; 28 **3.** 0; 200 **4.** $3n$; $17n$ **5.** $a - b$; $a + b$ **6.** 5; $2k + 5$ **7.** $\angle 2$ **8.** $\angle 1$ **9.** $\angle 3$ **10.** $\overline{DF}$ **11.** $\overline{WT}$ **12.** $\overline{OA}$

B **13.** $\overline{WY}$ **14.** $\overline{VB}$ **15.** $c > d > e > b > a$

16. $m\angle 3 > m\angle 1 > m\angle 2$ **17.** $m\angle 2 > m\angle X > m\angle XZY > m\angle Y > m\angle 1$

18.

Statements	Reasons
1. $AB + BC > AC$; $CD + DA > AC$	1. △ Ineq. Thm.
2. $AB + BC + CD + DA > 2(AC)$	2. A Prop. of Ineq.

19.

Statements	Reasons
1. $EFGH$ is a ▱; $EF > FG$	1. Given
2. $\overline{EF} \cong \overline{HG}$ or $EF = HG$; $\overline{FG} \cong \overline{EH}$ or $FG = EH$	2. Opp. sides of a ▱ are ≅.
3. $HG > EH$	3. Substitution Prop.
4. $m\angle 1 > m\angle 2$	4. If one side of a △ is longer than a second side, then the ∠ opp. the first side is larger than the ∠ opp. the second side.

C **20.** The perimeter of a quad. is greater than the sum of the lengths of the diagonals.

Given: Quad. $ABCD$

Prove: $AB + BC + CD + DA > AC + BD$

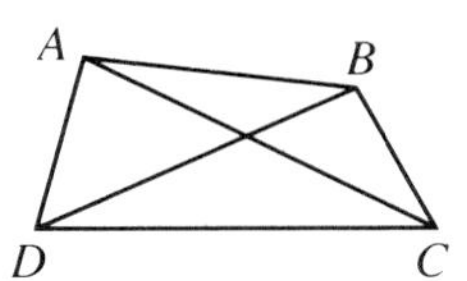

Statements	Reasons
1. $AB + BC > AC$; $CD + DA > AC$	1. $\triangle$ Ineq. Thm.
2. $AB + BC + CD + DA > 2AC$	2. A Prop. of Ineq.
3. $DA + AB > BD$; $BC + CD > BD$	3. $\triangle$ Ineq. Thm.
4. $DA + AB + BC + CD > 2BD$	4. A Prop. of Ineq.
5. $2(AB + BC + CD + DA) > 2AC + 2BD$	5. A Prop. of Ineq.
6. $AB + BC + CD + DA > AC + BD$	6. A Prop. of Ineq.

21. Given: $\overline{AN}$, $\overline{BP}$, and $\overline{CM}$ are medians of $\triangle ABC$.

Prove: $AN + BP + CM > \frac{1}{2}(AB + BC + AC)$

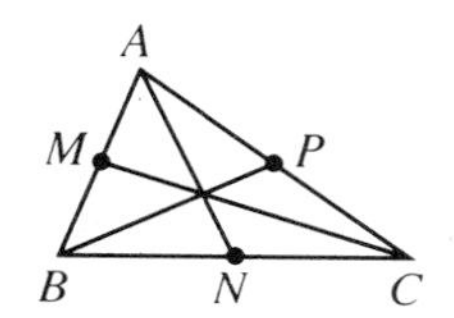

Statements	Reasons
1. $BP + AP > AB$; $CM + MB > BC$; $AN + NC > AC$	1. $\triangle$ Ineq. Thm.
2. $\overline{AN}$, $\overline{BP}$, and $\overline{CM}$ are medians.	2. Given
3. N, M, and P are midpts.	3. Def. of median
4. $AP = \frac{1}{2}AC$; $MB = \frac{1}{2}AB$; $NC = \frac{1}{2}BC$	4. Midpt. Thm.
5. $BP + \frac{1}{2}AC > AB$; $CM + \frac{1}{2}AB > BC$; $AN + \frac{1}{2}BC > AC$	5. Substitution Prop.
6. $BP + \frac{1}{2}AC + CM + \frac{1}{2}AB + AN + \frac{1}{2}BC > AB + BC + AC$	6. A Prop. of Ineq.
7. $BP + CM + AN > \frac{1}{2}(AB + BC + AC)$	7. A Prop. of Ineq.

22. No; consider $\triangle ABC$ with $m\angle A = 170$. Note that the altitudes drawn from $\angle B$ and $\angle C$ would be outside $\triangle ABC$.

23. The length of the longest side of a quad. is less than the sum of the lengths of the other 3 sides.

Given: Quad. $ABCD$ with $\overline{AD}$ the longest side

Prove: $AD < AB + BC + CD$

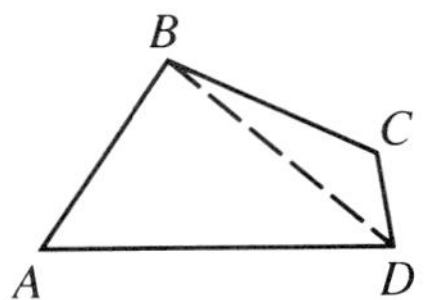

Statements	Reasons
1. Draw $\overline{BD}$.	1. Through any 2 pts. there is exactly one line.
2. $AB + BD > AD$; $BC + CD > BD$	2. △ Ineq. Thm.
3. $AB + BD + BC + CD > AD + BD$	3. A Prop. of Ineq.
4. $AB + BC + CD > AD$, or $AD < AB + BC + CD$	4. A Prop. of Ineq.

24. Given: Point P inside $\triangle XYZ$

Prove: $ZX + ZY > PX + PY$

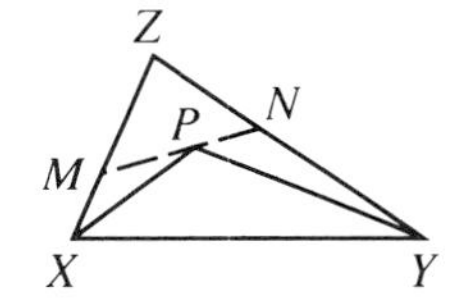

Statements	Reasons
1. Let M be any pt. btwn. Z and X.	1. Ruler Post.
2. Draw $\overleftrightarrow{MP}$ int. $\overline{ZY}$ at N.	2. If 2 lines int., then they int. in exactly one pt.
3. $ZM + ZN > MN$; $MX + MP > PX$; $PN + NY > PY$	3. △ Ineq. Thm.
4. $ZM + ZN + MX + MP + PN + NY > MN + PX + PY$	4. A Prop. of Ineq.
5. $ZM + MX = ZX$; $ZN + NY = ZY$; $MP + PN = MN$	5. Seg. Add. Post.
6. $ZX + ZY + MN > MN + PX + PY$	6. Substitution Prop.
7. $ZX + ZY > PX + PY$	7. A Prop. of Ineq.

Page 225 • APPLICATION

1. 1. Def. of $\perp$ bis. 2. If a pt. lies on the $\perp$ bis. of a seg., then it is equidistant from the endpts. of the seg. 3. Seg. Add. Post. 4. Substitution Prop. 5. If a pt. lies on the $\perp$ bis. of a seg., then it is equidistant from the endpts. of the seg.
   6. The sum of the lengths of any 2 sides of a $\triangle >$ the length of the third side. ($\triangle$ Ineq. Thm.) 7. Substitution Prop.
2. Since $\overline{PC} \cong \overline{PB}$, $\overline{PS} \cong \overline{PS}$, and $\angle CPS \cong \angle BPS$, then $\triangle SPC \cong \triangle SPB$ by SAS. $\angle CSP \cong \angle BSP$ by corr. parts. Also, $\angle QSA \cong \angle CSP$ (vert. $\angle$s). So $\angle QSA \cong \angle PSB$ by substitution.

Page 225 • EXPLORATIONS

Table entries will vary. In the first table, the $\angle$ opp. the longer side got larger. In the second table, the side opp. the larger $\angle$ got longer.

Page 227 • COMPUTER KEY-IN

1–2. Answers will vary. Check students' answers.

3. Computer results will vary. Possible results include, for D = 100, P = 0.22; for D = 400, P = 0.255; for D = 800, P = 0.234. The probability appears to be less than $\frac{1}{2}$.

Page 230 • CLASSROOM EXERCISES

1. $m\angle CAB > m\angle FDE$; SSS Ineq. Thm. **2.** $NO > JK$; SAS Ineq. Thm.
3. $RT > TS$; SAS Ineq. Thm. **4.** $EF > BC$; SAS Ineq. Thm.
5. $\triangle UNG \cong \triangle ONG$; SAS Post. **6.** $BC > AB$; SAS Ineq. Thm.
7. $m\angle 1 > m\angle 2$; SSS Ineq. Thm. **8.** $EF > ED$; SAS Ineq. Thm.
9. **a–b.** Check students' drawings. **a.** about 12.1 cm **b.** The length of the base decreases; equilateral; 7 cm
10. **a–b.** Check students' drawings. **a.** 10 cm **b.** The length of the third side decreases to about 5.6 cm.

Pages 231–232 • WRITTEN EXERCISES

A **1.** $m\angle 1 > m\angle 2$; SSS Ineq. Thm. **2.** $TR > VS$; SAS Ineq. Thm. **3.** $>$; $>$ **4.** $<$
5. $<$; $>$ **6.** $<$ **7.** $<$ **8.** $<$
B **9.** $>$ **10.** $>$

11.

Statements	Reasons
1. $m\angle SUV > m\angle STU$	1. The meas. of an ext. $\angle$ of a $\triangle >$ the meas. of either remote int. $\angle$.
2. $\overline{TU} \cong \overline{US} \cong \overline{SV}$	2. Given
3. $m\angle SVU = m\angle SUV$	3. Isos. $\triangle$ Thm.
4. $m\angle SVU > m\angle STU$	4. Substitution Prop.
5. $ST > SV$	5. If one $\angle$ of a $\triangle$ is larger than a second $\angle$, then the side opp. the first $\angle$ is longer than the side opp. the second $\angle$.

12. Answers may vary; for example, prove that $\overline{XZ}$ is not $\perp$ to plane P by proving $m\angle WYZ > m\angle WYX$.

Statements	Reasons
1. Plane P bis. $\overline{XZ}$ at Y.	1. Given
2. Y is the midpt. of $\overline{XZ}$.	2. Def. of bis.
3. $\overline{XY} \cong \overline{ZY}$	3. Def. of midpt.
4. $\overline{WY} \cong \overline{WY}$	4. Refl. Prop.
5. $WZ > WX$	5. Given
6. $m\angle WYZ > m\angle WYX$	6. SSS Ineq. Thm.

C 13.

Statements	Reasons
1. $\overline{PA} \cong \overline{PC} \cong \overline{QC} \cong \overline{QB}$	1. Given
2. $AC > BC$	2. Diagram
3. $m\angle P > m\angle Q$	3. SSS Ineq. Thm.
4. $m\angle PCA + m\angle A + m\angle P = 180$; $m\angle QCB + m\angle QBC + m\angle Q = 180$	4. The sum of the meas. of the $\angle$s of a $\triangle$ is 180.
5. $\angle PCA \cong \angle A$, or $m\angle PCA = m\angle A$; $\angle QCB \cong \angle QBC$, or $m\angle QCB = m\angle QBC$	5. Isos. $\triangle$ Thm.
6. $2m\angle PCA + m\angle P = 180$; $2m\angle QCB + m\angle Q = 180$	6. Substitution Prop.
7. $m\angle P = 180 - 2m\angle PCA$; $m\angle Q = 180 - 2m\angle QCB$	7. Algebra
8. $180 - 2m\angle PCA > 180 - 2m\angle QCB$	8. Substitution Prop.
9. $m\angle PCA < m\angle QCB$	9. Algebra

14.

Statements	Reasons
1. On $\overline{EK}$, take Z so that $EZ = EJ$, or $\overline{EZ} \cong \overline{EJ}$.	1. Ruler Post.
2. Draw $\overline{DZ}$.	2. Through any 2 pts. there is exactly one line.
3. $\overline{DE} \perp$ plane M	3. Given
4. $\overline{DE} \perp \overline{EZ}$; $\overline{DE} \perp \overline{EJ}$	4. Def. of $\perp$ to a plane
5. $m\angle DEZ = 90$; $m\angle DEJ = 90$	5. Def. of $\perp$ lines
6. $\angle DEZ \cong \angle DEJ$	6. Def. of $\cong$ ∠s
7. $\overline{DE} \cong \overline{DE}$	7. Refl. Prop.
8. $\triangle DEZ \cong \triangle DEJ$	8. SAS Post.
9. $\overline{DZ} \cong \overline{DJ}$, or $DZ = DJ$	9. Corr. parts of $\cong$ △s are $\cong$.
10. $m\angle DZK > m\angle DEZ$	10. The meas. of an ext. $\angle$ of a $\triangle >$ the meas. of either remote int. $\angle$.
11. $m\angle DZK > 90$	11. Substitution Prop.
12. $m\angle K < 90$, or $90 > m\angle K$	12. In a $\triangle$, there can be at most one rt. $\angle$ or obtuse $\angle$.
13. $m\angle DZK > 90 > m\angle K$	13. A Prop. of Ineq.
14. $DK > DZ$	14. If one $\angle$ of a $\triangle$ is larger than a second $\angle$, then the side opp. the first $\angle$ is longer than the side opp. the second $\angle$.
15. $DK > DJ$	15. Substitution Prop.

15. a. $\angle VBC$ and $\angle VAC$ are the largest ∠s, each measuring > 60, but < 120.

b. $\angle BVC$, $\angle BCV$, $\angle AVC$, and $\angle ACV$ are the largest ∠s, each measuring > 60, but < 90.

Page 233 • SELF-TEST 2

1. $\overline{XY}$ **2.** $\overline{OD}$ **3.** $<$ **4.** $=$ **5.** $>$ **6.** 1; 11 **7.** cannot be true
8. must be true **9.** may be true

Pages 235–236 • CHAPTER REVIEW

1. $>$ **2.** $>$ **3.** $=$ **4.** $=$ **5.** $>$ **6.** No conclusion **7.** No conclusion
8. Will is not a registered voter. **9.** Barbara is at least 18 years old. **10.** c, d, a, b
11. $m\angle T$ **12.** RE **13.** $<$ **14.** 2; 14 **15.** $>$ **16.** $<$ **17.** $=$ **18.** $>$

Pages 236–237 • CHAPTER TEST

1. $>$ **2.** $<$ **3.** $>$ **4.** $<$

5. **a.** If point P is not on $\overline{AB}$, then $AB \leq AP$. **b.** If $AB \leq AP$, then point P is not on $\overline{AB}$.

6. **a.** No conclusion **b.** $AB > AP$ **c.** Point P is not on $\overline{AB}$. **d.** No conclusion

7. 6; 36 **8.** $\overline{BC}$ **9.** $\overline{DG}$ **10.** $\overline{JK}$ **11.** VOE; VEO **12.** UO; UE **13.** UE; UO

14. VUE

15. Assume temp. that $\angle C$ and $\angle D$ are both rt. $\angle$s. Then $\overline{AD} \perp \overline{DC}$, $\overline{BC} \perp \overline{DC}$, and $\overline{AD} \parallel \overline{BC}$. (In a plane, 2 lines $\perp$ to the same line are $\parallel$.) But this contradicts the fact that a trap. has only one pair of $\parallel$ sides. The temp. assumption must be false. It follows that $\angle C$ and $\angle D$ are not both rt. $\angle$s.

16.

Statements	Reasons
1. S is the midpt. of $\overline{RT}$.	1. Given
2. $\overline{RS} \cong \overline{ST}$	2. Def. of midpt.
3. $\overline{RX} \cong \overline{TY}$; $XS > YS$	3. Given
4. $m\angle R > m\angle T$	4. SSS Ineq. Thm.

Page 237 • ALGEBRA REVIEW

1. $\frac{1}{5}$ **2.** 5 **3.** $\frac{a}{2}$ **4.** 3 **5.** $\frac{1}{3}$ **6.** $\frac{c}{2b}$ **7.** $-4y^2$ **8.** $-\frac{3r^2}{2}$ **9.** $\frac{ab}{2c}$

10. $\frac{6(a + 2)}{6} = a + 2$ **11.** $\frac{3(3x - 2y)}{3} = 3x - 2y$ **12.** $\frac{11b(3a - 2)}{11b} = 3a - 2$

13. $\frac{x + 2}{3(x + 2)} = \frac{1}{3}$ **14.** $\frac{2(c - d)}{2(c + d)} = \frac{c - d}{c + d}$ **15.** $\frac{(t - 1)(t + 1)}{(t - 1)} = t + 1$

16. $\frac{5(a + b)}{(a - b)(a + b)} = \frac{5}{a - b}$ **17.** $\frac{(b + 5)(b - 5)}{(b - 7)(b - 5)} = \frac{b + 5}{b - 7}$

18. $\frac{(a + 4)(a + 4)}{(a + 4)(a - 4)} = \frac{a + 4}{a - 4}$ **19.** $\frac{3(x^2 - 2x - 8)}{3x^2 + 2x - 8} = \frac{3(x - 4)(x + 2)}{(3x - 4)(x + 2)} = \frac{3(x - 4)}{3x - 4}$

Page 238 • PREPARING FOR COLLEGE ENTRANCE EXAMS

1. A **2.** A **3.** B **4.** B **5.** B **6.** E

7. E. $4(x - 1) = x + 2$; $4x - 4 = x + 2$; $3x = 6$; $x = 2$; $CK = 2[4(x - 1)] = 8$

8. C

Page 239 • CUMULATIVE REVIEW: CHAPTERS 1–6

A **1.** $(x + 38) + (2x - 5) = 90$; $3x + 33 = 90$; $3x = 57$; $x = 19$; $x + 38 = 19 + 38 = 57$

2. $(5 - 2)180 = 540$

3. a. Yes; SAS **b.** Yes; ASA **c.** No **d.** Yes; AAS or ASA

4. a. sometimes **b.** sometimes **c.** never **d.** always **e.** always **f.** always

5. a. $\overline{YZ}$ **b.** $\overline{XZ}$ **6. a.** $\overline{RS}$ **b.** SAS Ineq. Thm.

B **7.** $x - (180 - x) = 38$; $2x - 180 = 38$; $2x = 218$; $x = 109$; $180 - x = 71$ **8.** $z > 3$

9. Assume temp. that $\angle Q$, $\angle R$, and $\angle S$ are all 120° angles. Then $m\angle P > 0$ and $m\angle Q + m\angle R + m\angle S + m\angle P > 360$. This contradicts the thm. that states the sum of the int. $\angle$s of a quad. $= 360$. The temp. assumption must be false. It follows that $\angle Q$, $\angle R$, and $\angle S$ are not all 120° angles.

10.

Statements	Reasons
1. $m\angle B > m\angle A$; $m\angle E > m\angle D$	1. Given
2. $AC > BC$; $CD > CE$	2. If one $\angle$ of a $\triangle$ is larger than a second $\angle$, then the side opp. the first $\angle$ is longer than the side opp. the second $\angle$.
3. $AC + CD > BC + CE$	3. A Prop. of Ineq.
4. $AC + CD = AD$; $BC + CE = BE$	4. Seg. Add. Post.
5. $AD > BE$	5. Substitution Prop.

11.

Statements	Reasons
1. $\overline{DC} \parallel \overline{AB}$; $\overline{CE} \perp \overline{AB}$; $\overline{AF} \perp \overline{AB}$	1. Given
2. $\overline{AF} \parallel \overline{CE}$	2. In a plane, 2 lines $\perp$ to the same line are $\parallel$.
3. $AECF$ is a ▱.	3. Def. of ▱
4. $\angle FAE$ is a rt. $\angle$.	4. Def. of $\perp$ lines
5. $AECF$ is a rect.	5. If an $\angle$ of a ▱ is a rt. $\angle$, then the ▱ is a rect.

CHAPTER 7 • Similar Polygons

Page 243 • CLASSROOM EXERCISES

1. $\frac{3}{5}$ **2.** $\frac{3}{5}$ **3.** $\frac{4}{n}$ **4.** $\frac{n}{4}$ **5.** Sometimes; the ratios are equal when $|a| = |b|$.

6. $100 : 90 = 10 : 9$ **7. a.** $\frac{3}{2}$ **b.** $\frac{3}{2} = \frac{6}{x}$; $x = 4$; 4 cups **8.** $10 : 4 = 5 : 2$

9. $12 : 10 = 6 : 5$ **10.** $16 : 26 = 8 : 13$ **11.** $10 : 16 = 5 : 8$

12. $16 : 14 = 8 : 7$ **13.** $4 : 10 : 16 = 2 : 5 : 8$

14. 1 L = 1000 mL, so 1.5 L = 1500 mL; $750 : 1500 = 1 : 2$

15. No; there is no common unit. **16.** $4 : 3$ **17.** Yes; Yes; No; Yes **18.** b

Pages 243–244 • WRITTEN EXERCISES

A **1.** $15 : 9 = 5 : 3$ **2.** $15 : 15 = 1 : 1$ **3.** $30 : 150 = 1 : 5$ **4.** $150 : 30 = 5 : 1$

5. $9 : 48 = 3 : 16$ **6.** 12 to 10 = 6 to 5 **7.** 24 to 12 = 2 to 1

8. 22 to 24 = 11 to 12 **9.** $\frac{12}{36} = \frac{1}{3}$ **10.** $\frac{22}{34} = \frac{11}{17}$ **11.** $\frac{34}{2} = \frac{17}{1}$

12. $12 : 10 : 24 = 6 : 5 : 12$ **13.** $24 : 12 : 10 = 12 : 6 : 5$

14. $12 : 22 : 34 = 6 : 11 : 17$ **15.** $5 : 45 = 1 : 9$ **16.** $1 : 0.6 = 10 : 6 = 5 : 3$

17. $0.6 : 0.8 = 6 : 8 = 3 : 4$ **18.** 1 m = 100 cm; $100 : 85 = 20 : 17$

19. 1 cm = 10 mm, so 8 cm = 80 mm; $80 : 50 = 8 : 5$

20. 1 m = 1000 mm, so 0.2 m = 200 mm; $40 : 200 = 1 : 5$ **21.** $\frac{3}{4b}$ **22.** $\frac{2d}{5c}$ **23.** $\frac{3}{a}$

B **24.** $4x + 5x = 90$; $9x = 90$; $x = 10$; $4x = 40$; $5x = 50$; 40, 50

25. $11x + 4x = 180$; $15x = 180$; $x = 12$; $11x = 132$; $4x = 48$; 132, 48

26. $3x + 4x + 5x = 180$; $12x = 180$; $x = 15$; $3x = 45$; $4x = 60$; $5x = 75$; 45, 60, 75

27. $5x + 7x = 90$; $12x = 90$; $x = 7.5$; $5x = 37.5$; $7x = 52.5$; 37.5, 52.5

28. $3x + 3x + 2x = 180$; $8x = 180$; $x = 22.5$; $3x = 67.5$; $2x = 45$; 67.5, 67.5, 45

29. $4x + 5x + 5x + 8x + 9x + 9x = (6 - 2)180 = 720$; $40x = 720$; $x = 18$; $4x = 72$; $5x = 90$; $8x = 144$; $9x = 162$; 72, 90, 90, 144, 162, 162

30. $8x + 11x + 14x = 132$; $33x = 132$; $x = 4$; $8x = 32$; $11x = 44$; $14x = 56$; 32 cm, 44 cm, 56 cm

31. $5x + 7x + 11x + 13x = 360$; $36x = 360$; $x = 10$; $5x = 50$; $7x = 70$; $11x = 110$; $13x = 130$; 50, 70, 110, 130; 2 s-s. int. ∠s are supp., so 2 sides of the quad. must be ∥.

32. $\frac{(6 - 2)180}{6} : \frac{360}{6} = 120 : 60 = 2 : 1$; $\frac{(10 - 2)180}{10} : \frac{360}{10} = 144 : 36 = 4 : 1$;

$\frac{(n - 2)180}{n} : \frac{360}{n} = (n - 2) : 2$

33. a. $\frac{320}{1000} = \frac{h}{325}$; $h = 104$ b. $\frac{104}{325 + 10} = \frac{104}{335} \approx 0.310$

C 34. $\frac{24 + x}{30 + x} = 0.85$; $x = 10$ free throws

35. $\frac{AB}{BD} = \frac{3}{4}$, so $AB = \frac{3}{4}BD$; $\frac{AC}{CD} = \frac{5}{6}$, so $CD = \frac{6}{5}AC$; $AB + BD = AD$ and $AC + CD = AD$, so $AB + BD = AC + CD$; $\frac{3}{4}BD + BD = AC + \frac{6}{5}AC$; $\frac{7}{4}BD = \frac{11}{5}AC$; $AC = \frac{5}{11} \cdot \frac{7}{4} \cdot BD = \frac{5}{11} \cdot \frac{7}{4} \cdot 66 = 52.5$

36. $\frac{4}{y} + \frac{3}{x} = \frac{12}{y} - \frac{2}{x}$; $xy\left(\frac{4}{y} + \frac{3}{x}\right) = xy\left(\frac{12}{y} - \frac{2}{x}\right)$; $4x + 3y = 12x - 2y$; $5y = 8x$; $\frac{x}{y} = \frac{5}{8}$ or $5 : 8$

Page 246 • CLASSROOM EXERCISES

1. b 2. a, c, d 3. $2b$ 4. $\frac{7}{4}$ 5. $\frac{f}{9}$ 6. $\frac{8}{h}$ 7. $\frac{2 + 3}{3} = \frac{5}{3}$ 8. $\frac{n}{p}$ or $\frac{q}{r}$ or $\frac{7}{9}$

9. a. fg b. $5e$; fg c. Yes; the product of the means equals the product of the extremes, so each proportion is equivalent to the equation $5e = fg$, and the proportions are equivalent.

10. Use the means-extremes prop.; $3x = 14$ is not equivalent to $7x = 6$. 11. 40; 8

12. $7y$; 36; $\frac{36}{7}$ 13. $t = p$ 14. $y = z$ 15. $ad = bc$; $ad = cb$

16. Apply the means-extremes property; $ad = bc$ is equivalent to $bc = ad$.

Pages 247–248 • WRITTEN EXERCISES

A 1. 6 2. 28 3. 21 4. 36 5. $\frac{4}{7}$ 6. $\frac{8}{3}$ 7. $\frac{y + 3}{3}$ 8. $\frac{5}{x}$ 9. $5x = 12$; $x = 2\frac{2}{5}$

10. $2x = 20$; $x = 10$ 11. $15x = 14$; $x = \frac{14}{15}$ 12. $2x = 40$; $x = 20$

13. $2x + 10 = 4$; $2x = -6$; $x = -3$ 14. $3x + 9 = 8$; $3x = -1$; $x = -\frac{1}{3}$

15. $5x + 10 = 4x + 12$; $x = 2$ 16. $6x + 3 = 8x - 2$; $5 = 2x$; $x = 2\frac{1}{2}$

17. $3x + 9 = 4x - 2$; $x = 11$ 18. $5x + 20 = 6x - 24$; $x = 44$

19. $28x + 42 = 54x - 36$; $78 = 26x$; $x = 3$

20. $21x + 35 = 54x + 15$; $20 = 33x$; $x = \frac{20}{33}$

	KR	RT	KT	KS	SU	KU
21.	12	9	21	16	12	28
22.	8	2	10	12	3	15
23.	16	8	24	20	10	30
24.	6	2	8	9	3	12
B 25.	8	4	12	10	5	15
26.	12	4	16	15	5	20
27.	27	9	36	36	12	48
28.	20	10	30	28	14	42

29. By the means-extremes property, $\frac{a + b}{b} = \frac{c + d}{d}$ is equivalent to $(a + b)d = b(c + d)$; $ad + bd = bc + bd$; $ad = bc$; also, $\frac{a}{b} = \frac{c}{d}$ is equivalent to $ad = bc$. Since both proportions are equivalent to the same equation, they are equivalent to each other.

30. By the properties of proportions, $\frac{x + y}{y} = \frac{r}{s}$ is equivalent to $\frac{x + y}{r} = \frac{y}{s}$ and $\frac{r}{x + y} = \frac{s}{y}$; then $\frac{r}{x + y} = \frac{x - y}{x + y}$ and $r = x - y$.

31. By the means-extremes property, $\frac{a - b}{a + b} = \frac{c - d}{c + d}$ is equivalent to $(a - b)(c + d) = (a + b)(c - d)$; $ac + ad - bc - bd = ac - ad + bc - bd$; $ad - bc = -ad + bc$; $2ad = 2bc$; $ad = bc$; also, $\frac{a}{b} = \frac{c}{d}$ is equivalent to $ad = bc$. Since the two proportions are equivalent to the same equation, they are equivalent to each other.

32. By the means-extremes property, $\frac{a + c}{b + d} = \frac{a - c}{b - d}$ is equivalent to $(a + c)(b - d) = (b + d)(a - c)$; $ab - ad + bc - cd = ab - bc + ad - cd$; $-ad + bc = -bc + ad$; $2bc = 2ad$; $ad = bc$; also, $\frac{a}{b} = \frac{c}{d}$ is equivalent to $ad = bc$. Since the two proportions are equivalent to the same equation, they are equivalent to each other.

33. $x^2 = (x + 5)(x - 4)$; $x^2 = x^2 + x - 20$; $0 = x - 20$; $x = 20$

34. $(x - 2)(x + 3) = x^2$; $x^2 + x - 6 = x^2$; $x - 6 = 0$; $x = 6$

35. $(x + 1)(x - 6) = (x - 2)(x + 5)$; $x^2 - 5x - 6 = x^2 + 3x - 10$; $-8x = -4$; $x = \frac{1}{2}$

C 36. $(x - 1)(x + 2) = (x - 2)(x + 4)$; $x^2 + x - 2 = x^2 + 2x - 8$; $-x = -6$; $x = 6$

37. $5x(x + 5) = 9(4x + 4)$; $5x^2 + 25x - 36x - 36 = 0$; $5x^2 - 11x - 36 = 0$; $(5x + 9)(x - 4) = 0$; $5x + 9 = 0$ or $x - 4 = 0$; $x = -\frac{9}{5}$ or $x = 4$

38. $(x - 1)(3x - 2) = 10(x + 2)$; $3x^2 - 5x + 2 = 10x + 20$; $3x^2 - 15x - 18 = 0$; $3(x^2 - 5x - 6) = 0$; $(x - 6)(x + 1) = 0$; $x - 6 = 0$ or $x + 1 = 0$; $x = 6$ or $x = -1$

39. $7y = 4x - 36$; $3x + 3y = 5x - 5y$, $2x = 8y$, $x = 4y$; $7y = 4(4y) - 36$, $7y = 16y - 36$, $-9y = -36$, $y = 4$; $x = 16$

40. $2x - 6 = 4y + 8$, $2x = 4y + 14$, $x = 2y + 7$; $5x + 5y - 5 = 6x - 6y + 6$, $-x + 11y = 11$, $-2y - 7 + 11y = 11$, $9y = 18$, $y = 2$; $x = 11$

41. Let $\frac{a}{b} = r$. Then $a = br$, $c = dr$, and $e = fr$. Then $\frac{a + c + e}{b + d + f} = \frac{br + dr + fr}{b + d + f} = \frac{r(b + d + f)}{b + d + f} = r = \frac{a}{b}$

42. Suppose there are n terms, the last being $\frac{j}{k}$. Let $\frac{a}{b} = r$. Then $a = br$, $c = dr$, $e = fr$, . . . , and $j = kr$. Then $\frac{a + c + e + \cdots + j}{b + d + f + \cdots + k} = \frac{br + dr + fr + \cdots + kr}{b + d + f + \cdots + k} = \frac{r(b + d + f + \cdots + k)}{b + d + f + \cdots + k} = r = \frac{a}{b}$

43. $b(4a - 9b) = 4a(a - 2b)$; $4ab - 9b^2 = 4a^2 - 8ab$; $4a^2 - 12ab + 9b^2 = 0$; $(2a - 3b)^2 = 0$; $2a - 3b = 0$; $2a = 3b$; $\frac{a}{b} = \frac{3}{2}$; $a : b = 3 : 2$

Page 250 • CLASSROOM EXERCISES

1. No; corr. sides are not in proportion. 2. Yes 3. No; corr. ∠s are not ≅.
4. No; corr. ∠s are not ≅. 5. No 6. No 7. No 8. Yes 9. Yes

10. a. 70; 90 b. $20 : 15 = 4 : 3$ c. $\frac{4}{3} = \frac{DU}{6}$, $DU = 8$; $\frac{4}{3} = \frac{28}{Y'J'}$, $Y'J' = 21$; $\frac{4}{3} = \frac{12t}{J'U'}$, $J'U' = 9t$ d. $20 + 28 + 12t + 8 : 15 + 21 + 9t + 6 = 56 + 12t : 42 + 9t = 4(14 + 3t) : 3(14 + 3t) = 4 : 3$ e. Corr. ∠s are not ≅.

Pages 250–252 • WRITTEN EXERCISES

A 1. always 2. sometimes 3. sometimes 4. sometimes 5. always
6. sometimes 7. sometimes 8. sometimes 9. always 10. sometimes

11. never **12.** never **13.** sometimes **14.** always **15.** $28 : 35 = 4 : 5$

16. Trap.; $\overline{T'U'} \parallel \overline{E'N'}$ **17.** 135 **18.** $360 - (135 + 90 + 90) = 45$

19. $\frac{4}{5} = \frac{UN}{15}$; $UN = 12$ **20.** $\frac{4}{5} = \frac{16}{T'U'}$; $T'U' = 20$ **21.** $\frac{4}{5} = \frac{TE}{5k}$; $TE = 4k$

22. $28 + 4k + 16 + 12 : 35 + 5k + 20 + 15 = 56 + 4k : 70 + 5k =$

$4(14 + k) : 5(14 + k) = 4 : 5$

B **23.** Property 2

24. The scale factor is $\frac{15}{20} = \frac{3}{4}$; $\frac{3}{4} = \frac{21}{x}$, $x = 28$; $\frac{3}{4} = \frac{18}{y}$, $y = 24$; $\frac{3}{4} = \frac{27}{z}$, $z = 36$

25. The scale factor is $\frac{15}{10} = \frac{3}{2}$; $\frac{3}{2} = \frac{12}{x}$, $x = 8$; $\frac{3}{2} = \frac{y}{12}$, $y = 18$; $\frac{3}{2} = \frac{18}{z}$, $z = 12$

26. The scale factor is $\frac{12}{20} = \frac{3}{5}$; $x = 90 - 60 = 30$; $\frac{3}{5} = \frac{y}{40}$, $y = 24$; $\frac{3}{5} = \frac{12\sqrt{3}}{z}$,

$z = 20\sqrt{3}$

27. The scale factor is $\frac{10}{24} = \frac{5}{12}$; $\frac{5}{12} = \frac{x}{15}$, $12x = 75$, $x = 6\frac{1}{4}$; $\frac{5}{12} = \frac{y}{16}$, $12y = 80$, $y = 6\frac{2}{3}$;

$\frac{5}{12} = \frac{z}{12}$, $z = 5$

28–29. Drawings may vary. **28.**

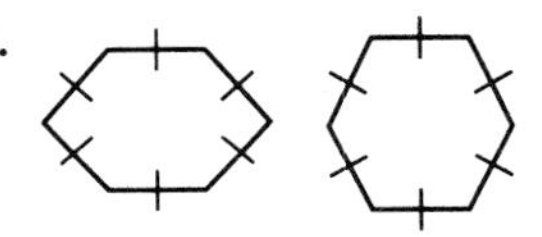

29.

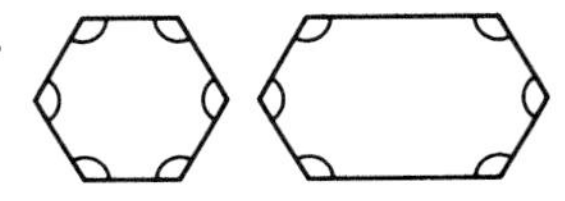

30. $\frac{AB}{DE} = \frac{BC}{EF} = \frac{AC}{DF}$, so $AB = \frac{BC \cdot DE}{EF}$ or $AB = \frac{AC \cdot DE}{DF}$

31. $RS = RS$, but $ZR > XR$, so $\frac{RS}{RS} = 1 \neq \frac{ZR}{XR}$.

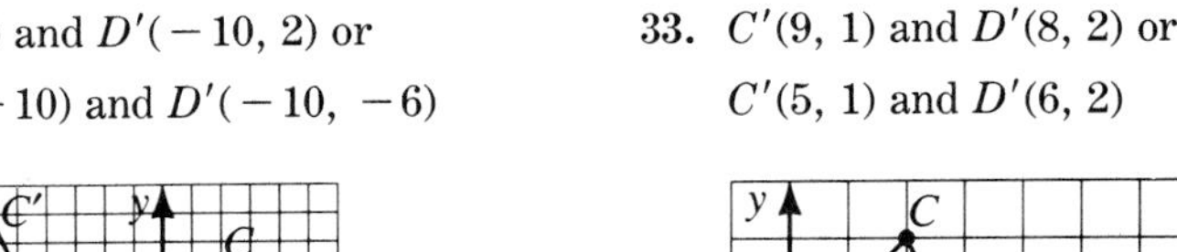

32. $C'(-6, 6)$ and $D'(-10, 2)$ or
$C'(-6, -10)$ and $D'(-10, -6)$

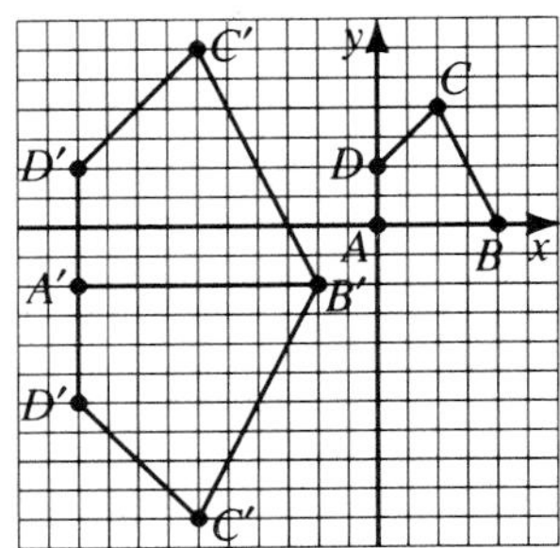

33. $C'(9, 1)$ and $D'(8, 2)$ or
$C'(5, 1)$ and $D'(6, 2)$

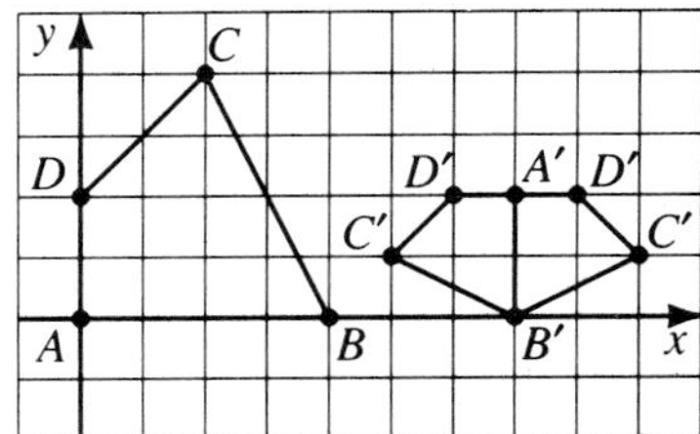

34. $\frac{x}{40} = \frac{10}{x}$; $x^2 = 400$; $x = 20$ **35.** The measure of each $\angle$ is 90; square

C **36.** Answers may vary. For example, $\overline{PS} \parallel \overline{JT}$ and $\overline{RQ} \parallel \overline{NZ}$ since alt. int. ∠s are ≅; $PQRS \sim TNZJ$ since corr. sides are in proportion and corr. ∠s are ≅. $\frac{PQ}{TN} = \frac{QR}{NZ} = \frac{RS}{ZJ} = \frac{SP}{JT} = \frac{2}{3}$; $\angle R \cong \angle Z$, so $\overline{RQ} \parallel \overline{NZ}$ and $\angle Q \cong \angle N$; isos. $\triangle POS \sim$ isos. $\triangle TOJ$, with $\angle OPS \cong \angle OSP \cong \angle OTJ \cong \angle OJT$; then $\angle QPS \cong \angle NTJ \cong \angle ZJT \cong \angle RSP$.

37. a. $\frac{x}{6} = \frac{6 - x}{x}$; $x^2 = 36 - 6x$; $x^2 + 6x - 36 = 0$; by the quadratic formula, $x = \frac{-6 \pm \sqrt{6^2 - 4(-36)}}{2} = -3 \pm 3\sqrt{5}$; since $-3 - 3\sqrt{5}$ is negative, $x = -3 + 3\sqrt{5}$. **b.** $\frac{6}{-3 + 3\sqrt{5}} = \frac{6(-3 - 3\sqrt{5})}{(-3 + 3\sqrt{5})(-3 - 3\sqrt{5})} = \frac{-18 - 18\sqrt{5}}{-36} = \frac{1 + \sqrt{5}}{2} \approx 1.62$

Page 252 • SELF-TEST 1

1. 3 : 5 **2.** 1 m = 100 cm, so 2 m = 200 cm; 60 to 200 = 3 to 10 **3.** $\frac{2a}{3b}$

4. $12x = 72$; $x = 6$ **5.** $4x - 8 = 2x + 12$; $2x = 20$; $x = 10$

6. $8x = 60 - 12x$; $20x = 60$; $x = 3$ **7.** No **8.** Yes **9.** Yes

10. 45; 60; 180 − (45 + 60) = 75 **11.** 6 : 9 = 2 : 3 **12.** $\frac{2}{3} = \frac{8}{x}$; $x = 12$

13. $\frac{2}{3} = \frac{10}{y}$; $y = 15$ **14.** $\frac{2}{3} = \frac{z}{18}$; $z = 12$

15. $5x + 5x + 5x + 6x + 7x + 8x = (6 - 2)180 = 720$; $36x = 720$; $x = 20$; $5x = 100$; $6x = 120$; $7x = 140$; $8x = 160$; 100, 100, 100, 120, 140, 160

Page 253 • CALCULATOR KEY-IN

1. $\frac{AD}{AC} \approx 1.62$, $\frac{AC}{AB} \approx 1.62$, $\frac{AB}{BC} \approx 1.62$ **2.** $\frac{l}{w} = \frac{1 + \sqrt{5}}{2} \approx 1.618$

Page 254 • EXPLORATIONS

Corr. ∠s are ≅. Ratios are equal. The triangles are similar. Ratio of medians = ratio of corr. sides. Ratio of perimeters = ratio of corr. sides

Page 256 • CLASSROOM EXERCISES

1. No **2.** Yes **3.** No **4.** No **5.** Yes **6.** No **7.** Yes **8.** Yes

9. Yes; each triangle has ∠s measuring 37, 53, and 90, so the △s are ~ by the AA ~ Post.

10. $\angle JIK$ and $\angle JHL$, $\angle JKI$ and $\angle JLH$

11. a. $\triangle HLJ$ **b.** $\frac{12}{x} = \frac{9}{24}$; $9x = 288$; $x = 32$ **c.** $\frac{9}{24} = \frac{6}{y + 6}$; $9y + 54 = 144$; $9y = 90$; $y = 10$

12. $\frac{AB}{CD} = \frac{WX}{YZ}; \frac{YZ}{CD} = \frac{WX}{AB}$

13. **a.** $\triangle SCH \sim \triangle STR$ **b.** $\frac{SC}{ST} = \frac{CH}{TR}$ **c.** $\frac{0.8}{5} = \frac{1.6}{t}$; $t = 10$; the tree is about 10 m tall.

Pages 257–260 • WRITTEN EXERCISES

A **1.** Similar **2.** Similar **3.** Similar **4.** Similar **5.** No conclusion **6.** Similar **7.** Similar **8.** Similar **9.** No conclusion

10. **a.** $\triangle JKN \sim \triangle MLN$ **b.** $\frac{JK}{ML} = \frac{JN}{MN} = \frac{KN}{LN}$ **c.** $\frac{15}{20} = \frac{18}{x}$ and $\frac{15}{20} = \frac{12}{y}$

d. $15x = 360$, $x = 24$; $15y = 240$, $y = 16$

11. $\frac{12}{9} = \frac{8}{x}$, $12x = 72$, $x = 6$; $\frac{12}{9} = \frac{12 + y}{12}$, $144 = 108 + 9y$, $9y = 36$, $y = 4$

12. $\frac{4}{6} = \frac{6}{x}$, $4x = 36$, $x = 9$; $\frac{4}{6} = \frac{y}{y + 3}$, $4y + 12 = 6y$, $2y = 12$, $y = 6$

13. $\frac{12}{16} = \frac{x}{12}$, $144 = 16x$, $x = 9$; $\frac{12}{16} = \frac{15}{15 + y}$, $180 + 12y = 240$, $12y = 60$, $y = 5$

B **14.** **a.** $\triangle ACD$, $\triangle CBD$ **b.** $\frac{16}{12} = \frac{20}{x}$, $16x = 240$, $x = 15$; $\frac{16}{12} = \frac{12}{y}$, $16y = 144$, $y = 9$

15. $\frac{36}{20} = \frac{RI}{15}$; $540 = 20 \cdot RI$; $RI = 27$ m

16. $\frac{1.6}{4.4} = \frac{2}{p}$; $1.6p = 8.8$; $p = 5.5$; the pole is about 5.5 m tall.

17. $\frac{6}{2400} = \frac{f}{220}$; $1320 = 2400f$; $f = 0.55$; the film image is about 0.55 cm tall.

18. $\frac{10}{6} = \frac{15}{x}$, $10x = 90$, $x = 9$; $\frac{10}{6} = \frac{9}{y}$, $10y = 54$, $y = 5.4$

19. $\frac{18}{6} = \frac{y}{x}$, $18x = 6y$, $y = 3x$; $x + y = 8$, $x + 3x = 8$, $4x = 8$, $x = 2$, $y = 6$

20. The girl and the flagpole are both $\perp$ to the ground so $\angle A \cong \angle A'$ and the $\triangle$s are $\sim$ by the AA $\sim$ Post. Let h = height of pole in cm; $\frac{120}{450} = \frac{160}{h}$; $120h = 72{,}000$; $h = 600$; the height of the pole is about 600 cm or 6 m.

21. **a.**

Statements	Reasons
1. $\overline{EF} \parallel \overline{RS}$	1. Given
2. $\angle XFE \cong \angle XSR$; $\angle XEF \cong \angle XRS$	2. If 2 $\parallel$ lines are cut by a trans., then corr. $\angle$s are $\cong$.
3. $\triangle FXE \sim \triangle SXR$	3. AA $\sim$ Post.

b.

Statements	Reasons
1. $\triangle FXE \sim \triangle SXR$	1. Part (a)
2. $\frac{FX}{SX} = \frac{EF}{RS}$	2. Corr. sides of ~ ⚠ are in prop.

22. a.

Statements	Reasons
1. $\angle 1 \cong \angle 2$	1. Given
2. $\angle J \cong \angle J$	2. Refl. Prop.
3. $\triangle JIG \sim \triangle JZY$	3. AA ~ Post.

b.

Statements	Reasons
1. $\triangle JIG \sim \triangle JZY$	1. Part (a)
2. $\frac{JG}{JY} = \frac{GI}{YZ}$	2. Corr. sides of ~ ⚠ are in prop.

23.

Statements	Reasons
1. $\angle B \cong \angle C$	1. Given
2. $\angle 1 \cong \angle 2$	2. Vert. ∠s are ≅.
3. $\triangle MLC \sim \triangle MNB$	3. AA ~ Post.
4. $\frac{NM}{LM} = \frac{BM}{CM}$	4. Corr. sides of ~ ⚠ are in prop.
5. $NM \cdot CM = LM \cdot BM$	5. Means-extremes Prop.

24.

Statements	Reasons
1. $\overline{BN} \parallel \overline{LC}$	1. Given
2. $\angle L \cong \angle N$; $\angle B \cong \angle C$	2. If 2 ∥ lines are cut by a trans., then alt. int. ∠s are ≅.
3. $\triangle MLC \sim \triangle MNB$	3. AA ~ Post.
4. $\frac{BN}{CL} = \frac{NM}{LM}$	4. Corr. sides of ~ ⚠ are in prop.
5. $BN \cdot LM = CL \cdot NM$	5. Means-extremes Prop.

25.

Statements	Reasons
1. $\overline{AD}$ and $\overline{XW}$ are altitudes.	1. Given
2. $\overline{AD} \perp \overline{BC}$; $\overline{XW} \perp \overline{YZ}$	2. Def. of alt.
3. $m\angle ADB = 90$; $m\angle XWY = 90$	3. Def. of $\perp$ lines
4. $\angle ADB \cong \angle XWY$	4. Def. of $\cong$ ∠s
5. $\triangle ABC \sim \triangle XYZ$	5. Given
6. $\angle B \cong \angle Y$	6. Corr. ∠s of $\sim$ △s are $\cong$.
7. $\triangle ADB \sim \triangle XWY$	7. AA $\sim$ Post.
8. $\frac{AD}{XW} = \frac{AB}{XY}$	8. Corr. sides of $\sim$ △s are in prop.

26.

Statements	Reasons
1. $\triangle PQR \sim \triangle GHI$	1. Given
2. $\angle QPR \cong \angle HGI$ or $m\angle QPR = m\angle HGI$	2. Corr. ∠s of $\sim$ △s are $\cong$.
3. $\frac{1}{2}m\angle QPR = \frac{1}{2}m\angle HGI$	3. Mult. Prop. of =
4. $\overrightarrow{PS}$ and $\overrightarrow{GJ}$ are ∠ bisectors.	4. Given
5. $m\angle QPS = \frac{1}{2}m\angle QPR$; $m\angle HGJ = \frac{1}{2}m\angle HGI$	5. ∠ Bis. Thm.
6. $m\angle QPS = m\angle HGJ$ or $\angle QPS \cong \angle HGJ$	6. Substitution Prop.
7. $\angle Q \cong \angle H$	7. Corr. ∠s of $\sim$ △s are $\cong$.
8. $\triangle QPS \sim \triangle HGJ$	8. AA $\sim$ Post.
9. $\frac{PS}{GJ} = \frac{PQ}{GH}$	9. Corr. sides of $\sim$ △s are in prop.

27.

Statements	Reasons
1. $\overline{AH} \perp \overline{EH}$; $\overline{AD} \perp \overline{DG}$	1. Given
2. $m\angle AHE = 90$; $m\angle ADG = 90$	2. Def. of $\perp$ lines
3. $\angle AHE \cong \angle ADG$	3. Def. of $\cong$ ∠s
4. $\angle A \cong \angle A$	4. Refl. Prop.
5. $\triangle AHE \sim \triangle ADG$	5. AA $\sim$ Post.
6. $\frac{AE}{AG} = \frac{HE}{DG}$	6. Corr. sides of $\sim$ △s are in prop.
7. $AE \cdot DG = AG \cdot HE$	7. Means-extremes Prop.

28.

Statements	Reasons
1. $\overline{QT} \parallel \overline{RS}$	1. Given
2. $\angle PQU \cong \angle R$; $\angle PUQ \cong \angle PVR$; $\angle PUT \cong \angle PVS$; $\angle PTU \cong \angle S$	2. If 2 ∥ lines are cut by a trans., then corr. ∠s are ≅.
3. $\triangle PQU \sim \triangle PRV$; $\triangle PUT \sim \triangle PVS$	3. AA ~ Post.
4. $\frac{QU}{RV} = \frac{PU}{PV}$; $\frac{UT}{VS} = \frac{PU}{PV}$	4. Corr. sides of ~ △s are in prop.
5. $\frac{QU}{RV} = \frac{UT}{VS}$	5. Substitution Prop.

29.

Statements	Reasons
1. $\angle 1 \cong \angle 2$	1. Given
2. $\angle A \cong \angle A$	2. Refl. Prop.
3. $\triangle ABC \sim \triangle ADB$	3. AA ~ Post.
4. $\frac{AB}{AD} = \frac{AC}{AB}$	4. Corr. sides of ~ △s are in prop.
5. $(AB)^2 = AD \cdot AC$	5. Means-extremes Prop.

30. First note that if 2 ∥ planes are cut by a third plane, then the lines of int. are ∥, so $\overline{A'B'} \parallel \overline{AB}$, $\overline{A'C'} \parallel \overline{AC}$, and $\overline{C'B'} \parallel \overline{CB}$. **a.** $\triangle VA'C' \sim \triangle VAC$; $\frac{15}{35} = \frac{18}{VC}$; $15 \cdot VC = 630$; $VC = 42$ **b.** $\frac{49 - BB'}{49} = \frac{15}{35} = \frac{3}{7}$; $343 - 7BB' = 147$; $7BB' = 196$; $BB' = 28$ **c.** $\frac{15}{35} = \frac{24}{AB}$; $15 \cdot AB = 840$; $AB = 56$

31. $\frac{10}{25} = \frac{A'B'}{20}$, $25 \cdot A'B' = 200$, $A'B' = 8$; $\frac{10}{25} = \frac{A'C'}{16}$, $25 \cdot A'C' = 160$, $A'C' = 6.4$; $\frac{8}{20} = \frac{B'C'}{14}$, $20 \cdot B'C' = 112$, $B'C' = 5.6$. Perimeter is $8 + 6.4 + 5.6 = 20$.

C 32. $\triangle CEF \sim \triangle CAD$ and $\triangle DFE \sim \triangle DCB$, so $\frac{EF}{AD} = \frac{FC}{DC}$ and $\frac{EF}{BC} = \frac{DF}{DC}$; then $\frac{EF}{AD} + \frac{EF}{BC} = \frac{FC}{DC} + \frac{DF}{DC}$; $\frac{EF}{6} + \frac{EF}{12} = \frac{FC + DF}{DC} = 1$;

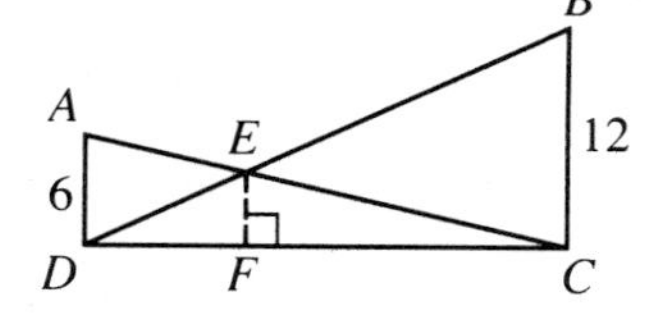

$2 \cdot EF + EF = 12$; $3 \cdot EF = 12$; $EF = 4$; the ropes cross 4 ft above the ground.

33. **a, b.** Check students' drawings; $m\angle EAD = m\angle AEB = m\angle EDA = m\angle ADB = m\angle BDC = m\angle CBD = m\angle DBE = m\angle EBA = 36$; $m\angle DAB = m\angle AKB = m\angle EKD = m\angle DEB = 72$; $m\angle EKA = m\angle DKB = m\angle C = 108$ **c.** By the AA $\sim$ Post., $\triangle DBA \sim \triangle DEK$ and $\triangle KBD \sim \triangle KEA$; then $\frac{DA}{DK} = \frac{DB}{DE}$ and $\frac{DK}{AK} = \frac{DB}{AE}$; $ABCDE$ is reg. so $DE = AE$. So $\frac{DA}{DK} = \frac{DB}{DE} = \frac{DB}{AE} = \frac{DK}{AK}$; $\frac{DA}{DK} = \frac{DK}{AK}$

34. **a.** Draw $\overline{HW}$ and $\overline{HX} \perp$ to $\overline{AB}$ and $\overline{DC}$, resp., and $\overline{HY}$ and $\overline{HZ} \perp$ to $\overline{AD}$ and $\overline{BC}$, resp. $\overline{AD} \parallel \overline{WX} \parallel \overline{BC}$, and $\overline{AB} \parallel \overline{YZ} \parallel \overline{DC}$. If 3 $\parallel$ lines cut off $\cong$ segs. on one trans., then they cut off $\cong$ segs. on every trans. Then $HW = HX$ and $HW + HX = 16$ so $HW = HX = 8$. Also $AY = YD$ so Y is the midpt. of $\overline{AD}$ and H is the midpt. of $\overline{AG}$. (The seg. that joins the midpts. of 2 sides of a $\triangle$ is half as long as the third side.) Then $HY = \frac{1}{2} \cdot 12 = 6$ and $HZ = 16 - 6 = 10$. **b.** $\angle AEH \cong \angle HFZ$ so $\angle YAH \cong \angle ZHF$ and $\triangle YAH \sim \triangle ZHF$. Then $\frac{10}{8} = \frac{ZF}{6}$; $8 \cdot ZF = 60$; $ZF = 7.5$; $BF = 8 - 7.5 = 0.5$; $FC = 16 - 0.5 = 15.5$; $CG = 16 - 12 = 4$; $\triangle AHE \cong \triangle HZF$ by ASA so $EH = ZF = 7.5$. $\triangle YEH \sim \triangle ZFH$ and $\frac{6}{10} = \frac{YE}{7.5}$; $10 \cdot YE = 45$; $YE = 4.5$; $DE = 8 - 4.5 = 3.5$; $EA = 16 - 3.5 = 12.5$. Since $\triangle AYH \sim \triangle HZF$, $\frac{8}{10} = \frac{10}{HF}$; $8 \cdot HF = 100$; $HF = 12.5$

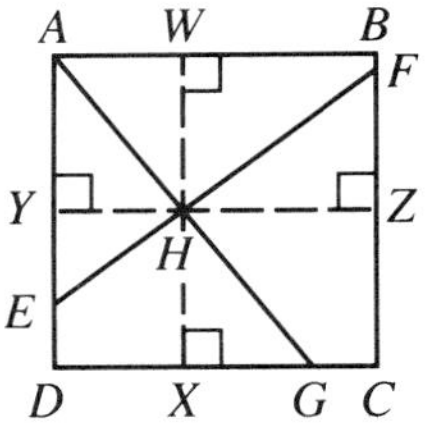

35. **a.** $\overline{TJ} \perp \overline{JO}$ and $\overline{PO} \perp \overline{JO}$ so $\overline{TJ} \parallel \overline{PO}$ and $TJOP$ is a $\square$. $\angle J$ and $\angle I$ are rt. $\angle$s and $\angle TOJ \cong \angle MOI$ (Vert. $\angle$s are $\cong$.). So by the AA $\sim$ Post., $\triangle TJO \sim \triangle MIO$. Since corr. sides of $\sim$ $\triangle$s are in prop., $\frac{OM}{OT} = \frac{OI}{OJ}$ and by the Means-extremes Prop., $OJ \cdot OM = OT \cdot OI$. Since $\overline{TP} \parallel \overline{JO}$, $\angle PTO \cong \angle IOM$ and $\angle TPM \cong \angle OFM$ (If 2 $\parallel$ lines are cut by a trans., then corr. $\angle$s are $\cong$.). So by the AA $\sim$ Post., $\triangle TPM \sim \triangle OFM$. Since corr. sides of $\sim$ $\triangle$s are in prop., $\frac{OF}{TP} = \frac{OM}{TM}$ and by the Means-extremes Prop., $TM \cdot OF = TP \cdot OM$. $TJOP$ is a $\square$ so $TP = OJ$. Then $TM \cdot OF = OJ \cdot OM = OT \cdot OI$. By the Seg. Add. Post., $OM + OT = TM$ and by the Div. Prop. of $=$, $\frac{OM}{OJ \cdot OM} + \frac{OT}{OT \cdot OI} = \frac{TM}{TM \cdot OF}$ or $\frac{1}{OJ} + \frac{1}{OI} = \frac{1}{OF}$.

b. $\frac{1}{OJ} + \frac{1}{OI} = \frac{1}{OF}$; $\frac{OI}{OJ \cdot OI} + \frac{OJ}{OJ \cdot OI} = \frac{1}{OF}$; $\frac{OJ + OI}{OJ \cdot OI} = \frac{1}{OF}$; $OF = \frac{OJ \cdot OI}{OJ + OI}$

Page 260 • CHALLENGES

1.

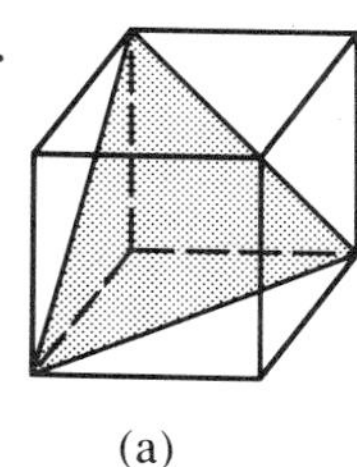
(a)

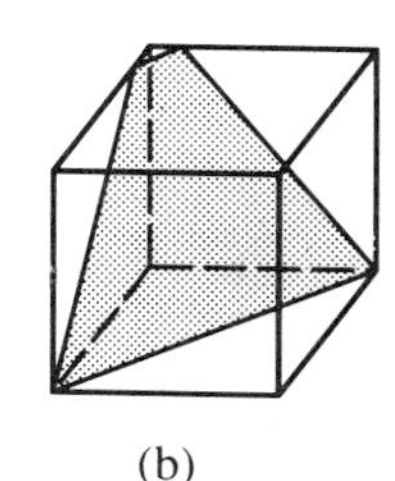
(b)

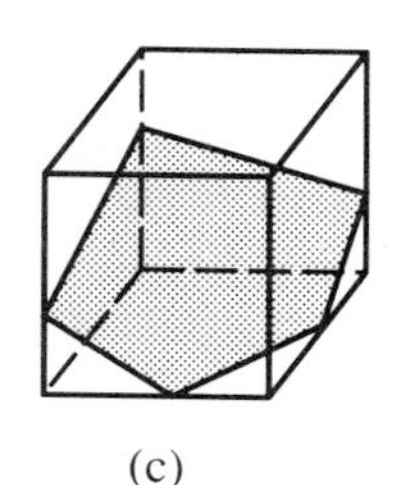
(c)

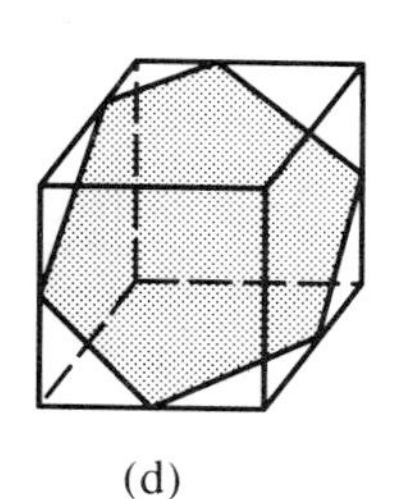
(d)

2. It is a 3-D figure with 8 faces that are ≅ equilateral △ (a reg. octahedron). See the figure at the right.

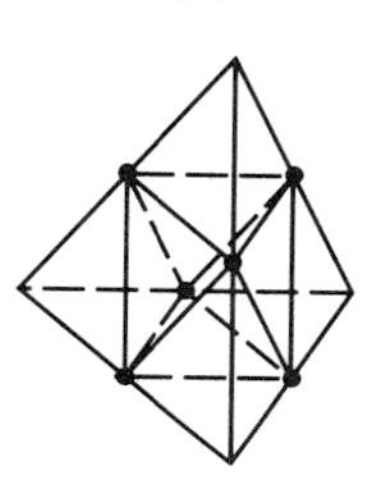

Page 261 • COMPUTER KEY-IN

1, 2. The values of the ratios approach 1.618.

3. The values of the ratios always approach 1.618.

4. The values approach the golden ratio, $\frac{1 + \sqrt{5}}{2} \approx 1.618$.

Page 262 • APPLICATION

1. length $= \frac{3}{4} \cdot 24 = 18$ ft; width $= \frac{1}{2} \cdot 24 = 12$ ft

2. $\frac{9}{24} = \frac{3}{8}$; $\frac{15}{2} \div 24 = \frac{5}{16}$; dim. of rug on floor plan would be $\frac{3}{8}$ in. by $\frac{5}{16}$ in.; no **3.** 2.4

4. actual length of the verandah $= 1 \cdot 24 = 24$ ft; $12x = 24$; $x = 2$; scale : 1 in. = 2 ft

Pages 264–265 • CLASSROOM EXERCISES

1. $\triangle HFG \sim \triangle RXS$; SSS ~ Thm. **2.** No **3.** $\triangle RQS \sim \triangle UTS$; SAS ~ Thm.

4. No **5.** $\triangle LNP \sim \triangle ANL$; SAS ~ Thm. **6.** $\triangle ACB \sim \triangle DCA$; SSS ~ Thm.

7. $\frac{RS}{XY} = \frac{RT}{XZ} = \frac{ST}{YZ}$ **8.** $\frac{RS}{XY} = \frac{RT}{XZ}$

9. a. $\overline{AD} \parallel \overline{BC}$ so $\angle PAD \cong \angle B$ and $\angle ADP \cong \angle E$; $\triangle PBE \sim \triangle PAD$ (AA ~ Post.)
b. 10 : 3 **c.** 10 : 3 **d.** 10 : 3

Pages 266–267 • WRITTEN EXERCISES

A **1.** $\triangle BAC \sim \triangle EDC$; SAS ~ Thm. **2.** $\triangle CAB \sim \triangle JTH$; AA ~ Post.

3. $\triangle LKM \sim \triangle NPO$; SAS ~ Thm. **4.** $\triangle CAB \sim \triangle NXR$; SSS ~ Thm.

5. $\triangle ABC \sim \triangle AEF$; AA ~ Post. **6.** $\triangle SRA \sim \triangle CBA$; SAS ~ Thm.

7. $\triangle ABC \sim \triangle TRI$; 2 : 3 **8.** No **9.** $\triangle ABC \sim \triangle ITR$; 2 : 5

10. $\triangle ABC \sim \triangle IRT$; 4 : 5

11.

Statements	Reasons
1. $\frac{DE}{GH} = \frac{DF}{GI} = \frac{EF}{HI}$	1. Given
2. $\triangle DEF \sim \triangle GHI$	2. SSS ~ Thm.
3. $\angle E \cong \angle H$	3. Corr. ⦞ of ~ ⚠ are ≅.

12.

Statements	Reasons
1. $\frac{DE}{GH} = \frac{EF}{HI}$; $\angle E \cong \angle H$	1. Given
2. $\triangle DEF \sim \triangle GHI$	2. SAS ~ Thm.
3. $\frac{EF}{HI} = \frac{DF}{GI}$	3. Corr. sides of ~ ⚠ are in prop.

B 13.

Statements	Reasons
1. $\frac{VW}{VX} = \frac{VZ}{VY}$	1. Given
2. $\angle V \cong \angle V$	2. Refl. Prop.
3. $\triangle VWZ \sim \triangle VXY$	3. SAS ~ Thm.
4. $\angle 1 \cong \angle 2$	4. Corr. ⦞ of ~ ⚠ are ≅.
5. $\overline{WZ} \parallel \overline{XY}$	5. If 2 lines are cut by a trans. and corr. ⦞ are ≅, then the lines are ∥.

14. (3) $\angle 1 \cong \angle Y$

15.

Statements	Reasons
1. $\frac{JL}{NL} = \frac{KL}{ML}$	1. Given
2. $\angle MLN \cong \angle KLJ$	2. Vert. ⦞ are ≅.
3. $\triangle MLN \sim \triangle KLJ$	3. SAS ~ Thm.
4. $\angle J \cong \angle N$	4. Corr. ⦞ of ~ ⚠ are ≅.

16.

Statements	Reasons
1. $\frac{AB}{SR} = \frac{BC}{RA} = \frac{CA}{AS}$	1. Given
2. $\triangle ABC \sim \triangle SRA$	2. SSS ~ Thm.
3. $\angle BCA \cong \angle RAS$	3. Corr. ⦞ of ~ ⚠ are ≅.
4. $\overline{BC} \parallel \overline{AR}$	4. If 2 lines are cut by a trans. and alt. int. ⦞ are ≅, then the lines are ∥.

17. Given: $\triangle ABC \sim \triangle DEF$;
$\overline{AM}$ and $\overline{DN}$ are medians.

Prove: $\frac{AM}{DN} = \frac{AB}{DE}$

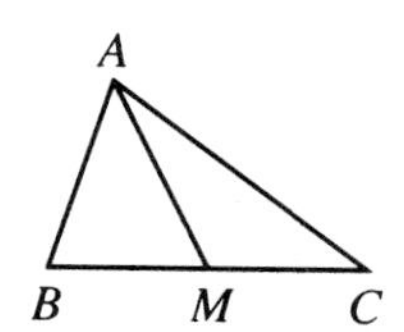

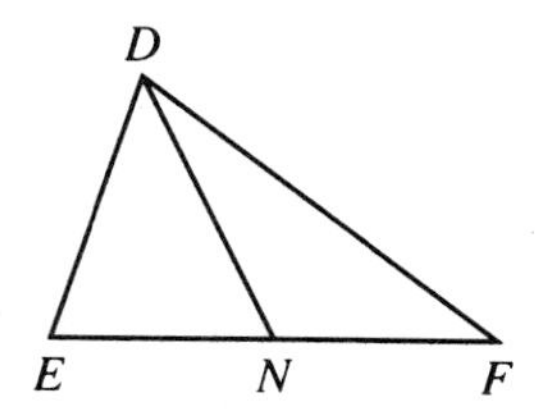

Statements	Reasons
1. $\overline{AM}$ and $\overline{DN}$ are medians.	1. Given
2. M and N are midpts.	2. Def. of median
3. $BM = \frac{1}{2}BC$ or $2BM = BC$; $EN = \frac{1}{2}EF$ or $2EN = EF$	3. Midpt. Thm.
4. $\triangle ABC \sim \triangle DEF$	4. Given
5. $\angle B \cong \angle E$	5. Corr. ∠s of ~ △s are ≅.
6. $\frac{BC}{EF} = \frac{AB}{DE}$	6. Corr. sides of ~ △s are in prop.
7. $\frac{2BM}{2EN} = \frac{AB}{DE}$ or $\frac{BM}{EN} = \frac{AB}{DE}$	7. Substitution Prop.
8. $\triangle ABM \sim \triangle DEN$	8. SAS ~ Thm.
9. $\frac{AM}{DN} = \frac{AB}{DE}$	9. Corr. sides of ~ △s are in prop.

18. Given: $ABCD \sim A'B'C'D'$

Prove: $\frac{BD}{B'D'} = \frac{AB}{A'B'}$

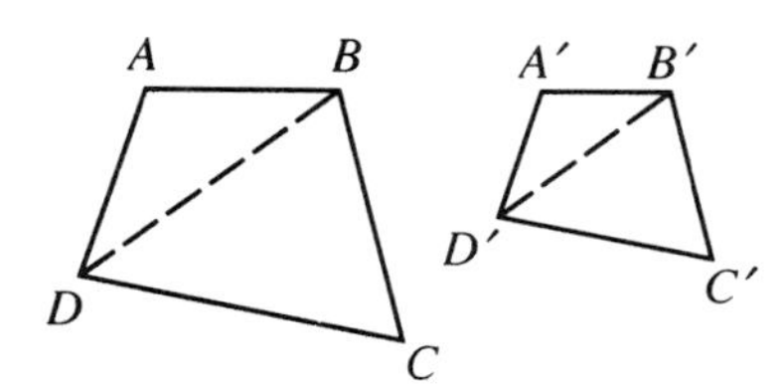

Statements	Reasons
1. $ABCD \sim A'B'C'D'$	1. Given
2. $\angle A \cong \angle A'$	2. Corr. ∠s of ~ polygons are ≅.
3. $\frac{AD}{A'D'} = \frac{AB}{A'B'}$	3. Corr. sides of ~ polygons are in prop.
4. $\triangle ABD \sim \triangle A'B'D'$	4. SAS ~ Thm.
5. $\frac{BD}{B'D'} = \frac{AB}{A'B'}$	5. Corr. sides of ~ △s are in prop.

19. Given: Isos. $\triangle ABC$ with $AB = AC$;
Isos. $\triangle DEF$ with $DE = DF$;
$\angle A \cong \angle D$

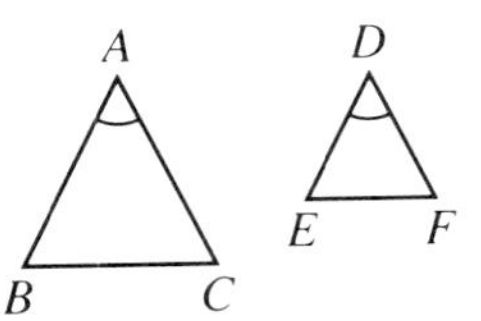

Prove: $\triangle ABC \sim \triangle DEF$

Statements	Reasons
1. $AB = AC; DE = DF$	1. Given
2. $\frac{AB}{DE} = \frac{AC}{DF}$	2. Div. Prop. of =
3. $\angle A \cong \angle D$	3. Given
4. $\triangle ABC \sim \triangle DEF$	4. SAS ~ Thm.

20. $WD = WC = 4 \cdot VA = 4 \cdot VB; \frac{VA}{WD} = \frac{VB}{WC}$ and $m\angle BVA = m\angle CWD = 90$; $\triangle AVB \sim \triangle DWC; \frac{AB}{DC} = \frac{VA}{WD} = \frac{1}{4}; DC = 4 \cdot AB$; length of median of trap. $ABCD = \frac{1}{2}(AB + 4 \cdot AB) = \frac{5}{2} \cdot AB$

21.

Statements	Reasons
1. $OR' = 2 \cdot OR; OS' = 2 \cdot OS;$ $OT' = 2 \cdot OT$	1. Given
2. $OR = \frac{1}{2} \cdot OR'; OS = \frac{1}{2} \cdot OS';$ $OT = \frac{1}{2} \cdot OT'$	2. Div. Prop. of =
3. R, S, and T are midpts.	3. Def. of midpt.
4. $RS = \frac{1}{2} \cdot R'S'; ST = \frac{1}{2} \cdot S'T';$ $RT = \frac{1}{2} \cdot R'T'$	4. The seg. that joins the midpts. of 2 sides of a $\triangle$ is half as long as the third side.
5. $\frac{RS}{R'S'} = \frac{ST}{S'T'} = \frac{RT}{R'T'} = \frac{1}{2}$	5. Div. Prop. of =, Subst. Prop.
6. $\triangle RST \sim \triangle R'S'T'$	6. SSS ~ Thm.

22.

Statements	Reasons
1. M is the midpt. of $\overline{AB}$; N is the midpt. of $\overline{AC}$.	1. Given
2. $AM = \frac{1}{2} \cdot AB$; $AN = \frac{1}{2} \cdot AC$	2. Midpt. Thm.
3. $\frac{AM}{AB} = \frac{1}{2} = \frac{AN}{AC}$	3. Div. Prop. of =, Subst. Prop.
4. $\angle A \cong \angle A$	4. Refl. Prop.
5. $\triangle AMN \sim \triangle ABC$	5. SAS ~ Thm.
6. $\angle AMN \cong \angle B$	6. Corr. ∠s of ~ △s are ≅.
7. $\overline{MN} \parallel \overline{BC}$	7. If 2 lines are cut by a trans. and corr. ∠s are ≅, then the lines are ∥.
8. $\frac{MN}{BC} = \frac{AM}{AB} = \frac{1}{2}$	8. Corr. sides of ~ △s are in prop.
9. $MN = \frac{1}{2} \cdot BC$	9. Mult. Prop. of =

C 23.

Statements	Reasons
1. $WXYZ$ is a ▱.	1. Given
2. $\overline{WX} \parallel \overline{ZY}$; $\overline{ZW} \parallel \overline{YX}$	2. Def. of ▱
3. $\angle WAT \cong \angle YA'T$; $\angle BWT \cong \angle B'YT$	3. If 2 ∥ lines are cut by a trans., then alt. int. ∠s are ≅.
4. $\angle WTA \cong \angle YTA'$; $\angle BTW \cong \angle B'TY$	4. Vert. ∠s are ≅.
5. $\triangle WTA \sim \triangle YTA'$; $\triangle BTW \sim \triangle B'TY$	5. AA ~ Post.
6. $\frac{TW}{TY} = \frac{AT}{A'T}$; $\frac{TW}{TY} = \frac{BT}{B'T}$	6. Corr. sides of ~ △s are in prop.
7. $\frac{AT}{A'T} = \frac{BT}{B'T}$	7. Substitution Prop.
8. $\angle ATB \cong \angle A'TB'$	8. Vert. ∠s are ≅.
9. $\triangle ATB \sim \triangle A'TB'$	9. SAS ~ Thm.

Page 268 • EXPLORATIONS

1. $\triangle ABC \sim \triangle ADE$; $\frac{AE}{AC} = \frac{AD}{AB}$; $\frac{AE}{EC} = \frac{AD}{DB}$; $\frac{EC}{AC} = \frac{DB}{AB}$; the ratios are =.

2. $\frac{BD}{DC} = \frac{AB}{AC}$; $\frac{BD}{DC} = \frac{AB}{AC}$ for any △.

Page 268 • MIXED REVIEW EXERCISES

1. **a.** $\overline{GC}$; $\overline{EF}$ **b.** 18 **c.** $3x + 2 = 7x - 10$; $4x = 12$; $x = 3$ **d.** 90

2. **a.** midpoint; $\overline{RV}$ **b.** $3x + 8 = \frac{1}{2}(12x + 4)$; $3x + 8 = 6x + 2$; $3x = 6$; $x = 2$

   c. $4y + 1 = 9y - 19$; $5y = 20$; $y = 4$

Page 271 • CLASSROOM EXERCISES

1. $\frac{u}{w} = \frac{x}{z}$, $\frac{u}{v} = \frac{x}{y}$; $\frac{u}{x} = \frac{w}{z}$, etc.
2. whole right; upper right, lower right; lower parallel; upper right, whole right
3. $\frac{n}{g} = \frac{a}{b}$ 4. $\frac{10 - x}{x} = \frac{9}{6}$ 5. $\frac{y}{14} = \frac{15}{10}$
6. **a.** $\frac{x}{14} = \frac{6}{16}$; $16x = 84$; $x = 5\frac{1}{4}$ **b.** $\frac{x}{14 - x} = \frac{6}{10}$; $10x = 84 - 6x$; $16x = 84$;

   $x = 5\frac{1}{4}$ **c.** $3y + 5y = 14$; $8y = 14$; $y = 1\frac{3}{4}$; $3y = 5\frac{1}{4}$
7. Since 6 and 10 are in the ratio 3 : 5, so are the lengths of the parts of the unknown side.
8. No
9. Yes; each pair of transversals form 2 ~ △s. Thus corr. ∠s are ≅ and the lines are ∥.

Pages 272–273 • WRITTEN EXERCISES

A 1. **a.** No **b.** Yes **c.** Yes **d.** No **e.** Yes **f.** Yes

2. **a.** No **b.** No **c.** Yes **d.** Yes 3. $\frac{4}{5} = \frac{6}{x}$; $4x = 30$; $x = 7.5$

4. $\frac{12}{21} = \frac{20}{x}$; $12x = 420$; $x = 35$ 5. $\frac{10}{x} = \frac{15}{39}$; $15x = 390$; $x = 26$

6. $\frac{8}{x} = \frac{14}{21}$; $14x = 168$; $x = 12$

7. $\frac{42 - x}{x} = \frac{36}{27} = \frac{4}{3}$; $4x = 126 - 3x$; $7x = 126$; $x = 18$

8. $\frac{18}{x} = \frac{15}{25} = \frac{3}{5}$; $3x = 90$; $x = 30$ 9. $\frac{29}{x} = \frac{22}{11} = \frac{2}{1}$; $2x = 29$; $x = 14.5$

10. $\frac{24 - x}{x} = \frac{18}{9} = \frac{2}{1}$; $2x = 24 - x$; $3x = 24$; $x = 8$

11. $\frac{4x}{3x} = \frac{5x}{15}$; $15x^2 = 60x$; $15x(x - 4) = 0$; $15x = 0$ or $x - 4 = 0$;

    $x = 0$ (reject) or $x = 4$

	AR	RT	AT	AN	NP	AP	RN	TP
B 12.	6	4	10	9	6	15	9	15
13.	?	?	?	10	6	16	?	?
14.	18	6	24	?	?	?	30	40
15.	12	8	20	18	12	30	15	25
16.	9	18	27	13	26	39	12	36
17.	8	8	16	6	6	12	?	?

Note: In Exs. 13, 14, and 17 the question mark indicates that the length cannot be determined.

18.

Statements	Reasons
1. Draw $\overline{TX}$ int. $\overleftrightarrow{SY}$ at N.	1. Through any 2 pts. there is exactly one line.
2. $\overleftrightarrow{RX} \parallel \overleftrightarrow{SY} \parallel \overleftrightarrow{TZ}$	2. Given
3. $\frac{RS}{ST} = \frac{XN}{NT}$; $\frac{XN}{NT} = \frac{XY}{YZ}$	3. $\triangle$ Prop. Thm.
4. $\frac{RS}{ST} = \frac{XY}{YZ}$	4. Trans. Prop.

19.

Statements	Reasons
1. Draw a line through $E \parallel$ to $\overrightarrow{DG}$.	1. Through a pt. outside a line there is exactly one line $\parallel$ to the given line.
2. Extend $\overrightarrow{FD}$ to int. the line at K.	2. If 2 lines int., then they int. in exactly one pt.
3. $\frac{GF}{GE} = \frac{DF}{DK}$	3. $\triangle$ Prop. Thm.
4. $\overrightarrow{DG}$ bis. $\angle FDE$.	4. Given
5. $\angle 1 \cong \angle 2$	5. Def. of $\angle$ bis.
6. $\angle 1 \cong \angle 3$	6. If 2 $\parallel$ lines are cut. by a trans., then alt. int. $\angle$s are $\cong$.
7. $\angle 2 \cong \angle 3$	7. Substitution Prop.
8. $\angle 2 \cong \angle 4$	8. If 2 $\parallel$ lines are cut by a trans., then corr. $\angle$s are $\cong$.
9. $\angle 3 \cong \angle 4$	9. Substitution Prop.
10. $\overline{DK} \cong \overline{DE}$ or $DK = DE$	10. If 2 $\angle$s of a $\triangle$ are $\cong$, then the sides opp. those $\angle$s are $\cong$.
11. $\frac{GF}{GE} = \frac{DF}{DE}$	11. Substitution Prop.

20. $\frac{AB}{25 - AB} = \frac{21}{14} = \frac{3}{2}$; $2 \cdot AB = 75 - 3 \cdot AB$; $5 \cdot AB = 75$; $AB = 15$

21. $\frac{BC}{60 - BC} = \frac{30}{50} = \frac{3}{5}$; $5 \cdot BC = 180 - 3 \cdot BC$; $8 \cdot BC = 180$; $BC = 22.5$

22. $\frac{x}{27} = \frac{\frac{4}{3}x}{x} = \frac{4}{3}$; $3x = 108$; $x = 36$; $BC = 36$; $AC = 27 + 36 = 63$

23. $\frac{2x - 12}{x} = \frac{2x - 4}{x + 5}$; $2x^2 - 2x - 60 = 2x^2 - 4x$; $2x = 60$; $x = 30$; $2x - 12 = 48$; $AC = 48 + 30 = 78$

24. The frontages of the lots on Martin Luther King Ave. are prop. to their resp. frontages on Christa McAuliffe Blvd. Let the frontages on Martin Luther King Ave. be $40x$, $30x$, and $35x$. Then $40x + 30x + 35x = 140$; $x = \frac{4}{3}$. The frontages are $40x = 53.3$ m, $30x = 40$ m, and $35x = 46.7$ m.

25. $\frac{AP}{13 - AP} = \frac{14}{12} = \frac{7}{6}$; $6 \cdot AP = 91 - 7 \cdot AP$; $13 \cdot AP = 91$; $AP = 7$; $AM = \frac{1}{2} \cdot 13 = 6.5$; $MP = 7 - 6.5 = 0.5$

26. Given: $\overleftrightarrow{AX}$ bis. $\angle A$; $\overleftrightarrow{AX}$ bis. $\overline{BC}$.

Prove: $\triangle ABC$ is isos.

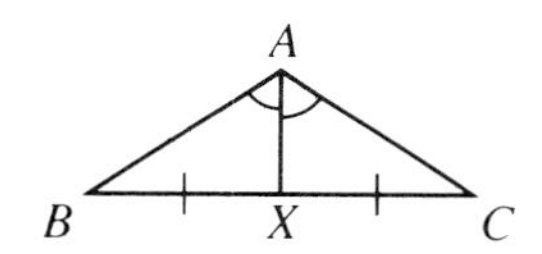

Statements	Reasons
1. $\overleftrightarrow{AX}$ bis. $\angle A$.	1. Given
2. $\frac{XB}{XC} = \frac{AB}{AC}$	2. $\triangle$ $\angle$-Bis. Thm.
3. $\overleftrightarrow{AX}$ bis. $\overline{BC}$.	3. Given
4. X is the midpt. of $\overline{BC}$.	4. Def. of bis.
5. $\overline{XB} \cong \overline{XC}$, or $XB = XC$	5. Def. of midpt.
6. $\frac{XB}{XC} = 1$	6. Div. Prop. of =
7. $\frac{AB}{AC} = 1$	7. Substitution Prop.
8. $AB = AC$, or $\overline{AB} \cong \overline{AC}$	8. Mult. Prop. of =
9. $\triangle ABC$ is isos.	9. Def. of isos. $\triangle$

C 27. If 3 $\parallel$ planes intersect 2 transversals, then they divide the transversals proportionally.

Given: Plane $P \parallel$ plane $Q \parallel$ plane R;
transversals $\overleftrightarrow{AE}$ and $\overleftrightarrow{BF}$

Prove: $\dfrac{AC}{CE} = \dfrac{BD}{DF}$

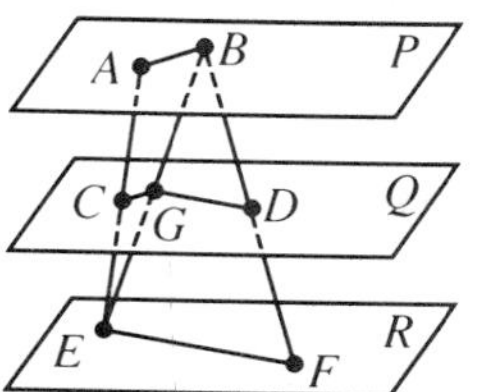

Statements	Reasons
1. Draw $\overline{EB}$, int. Q at G; draw $\overline{AB}$, $\overline{CG}$, $\overline{GD}$, and $\overline{EF}$.	1. Through any 2 pts. there is exactly one line.
2. Exactly one plane contains $\overleftrightarrow{AE}$ and $\overleftrightarrow{BE}$; exactly one plane contains $\overleftrightarrow{BE}$ and $\overleftrightarrow{BF}$.	2. If 2 lines int., then exactly one plane contains them.
3. $\overline{AB} \parallel \overline{CG}$; $\overline{GD} \parallel \overline{EF}$	3. If 2 $\parallel$ planes are cut by a third plane, then the lines of intersection are $\parallel$.
4. $\dfrac{AC}{CE} = \dfrac{BG}{GE}$; $\dfrac{BG}{GE} = \dfrac{BD}{DF}$	4. $\triangle$ Prop. Thm.
5. $\dfrac{AC}{CE} = \dfrac{BD}{DF}$	5. Trans. Prop.

28. Assume temporarily that $\overrightarrow{AD}$ and $\overrightarrow{AE}$ trisect $\angle BAC$ and $BD = DE = EC$. Then $\overrightarrow{AD}$ bisects $\angle BAE$ and $\overrightarrow{AE}$ bisects $\angle DAC$, so $\dfrac{BD}{DE} = \dfrac{AB}{AE}$ and $\dfrac{DE}{EC} = \dfrac{AD}{AC}$. Since $\dfrac{BD}{DE} = \dfrac{DE}{EC} = 1$, then $AB = AE$ and $AD = AC$. Also, $m\angle BAE = 2 \cdot m\angle DAE = m\angle DAC$; since $\triangle BAE$ and $\triangle DAC$ are isos., $m\angle ABE = m\angle AEB = m\angle ADC = m\angle ACD$. $BE = DC$ and $\triangle BAE \cong \triangle DAC$ by ASA. Then $AB = AD = AE = AC$ and $\triangle ADB \cong \triangle ADE$, so $m\angle ADB = m\angle ADE = 90$. Since $\overline{AD} \cong \overline{AE}$, $m\angle AED = m\angle ADE = 90$, so $\overline{AD}$ and $\overline{AE}$ are both $\perp$ to $\overline{BC}$. This contradicts the fact that in a plane, 2 lines $\perp$ to the same line are $\parallel$. Our temp. assumption must be incorrect; it follows that $\overrightarrow{AD}$ and $\overrightarrow{AE}$ cannot trisect both $\angle BAC$ and $\overline{BC}$.

29. No. By $\triangle$ Angle-Bisector Thm., in $\triangle ROE$, $\dfrac{1}{2} = \dfrac{OR}{OE}$; $OE = 2 \cdot OR$. In $\triangle ODS$, $\dfrac{2}{4} = \dfrac{OD}{OS}$; $OS = 2 \cdot OD$. Then $\triangle ROE \sim \triangle DOS$ (SAS $\sim$ Thm.) and $\angle ORD \cong \angle ODE$. This contradicts the fact that the measure of $\angle ODE$, an ext. $\angle$ of $\triangle ORD$, is $=$ to $m\angle ORD + m\angle ROD$.

30. $ZNKJ$ is a trapezoid.

Given: $\angle ZEN$ is obtuse; $\overrightarrow{EX}$ bis. $\angle ZEN$;

$ZJ = ZX$; $NK = NX$

Prove: $ZNKJ$ is a trap.

Statements	Reasons
1. $\overrightarrow{EX}$ bis. $\angle ZEN$.	1. Given
2. $\frac{ZX}{NX} = \frac{ZE}{NE}$	2. $\triangle$ $\angle$-Bis. Thm.
3. $ZE = ZJ + JE$; $NE = NK + KE$	3. Seg. Add. Post.
4. $\frac{ZX}{NX} = \frac{ZJ + JE}{NK + KE}$	4. Substitution Prop.
5. $ZJ = ZX$; $NK = NX$	5. Given
6. $\frac{ZX}{NX} = \frac{ZX + JE}{NX + KE}$	6. Substitution Prop.
7. $\frac{ZX}{NX} = \frac{JE}{KE}$	7. A prop. of proportions
8. $\frac{ZE}{NE} = \frac{JE}{KE}$	8. Substitution Prop.
9. $\frac{JE}{ZE} = \frac{KE}{NE}$	9. A prop. of proportions
10. $\angle ZEN \cong \angle ZEN$	10. Refl. Prop.
11. $\triangle JEK \sim \triangle ZEN$	11. SAS $\sim$ Thm.
12. $\angle EJK \cong \angle EZN$	12. Corr. $\angle$s of $\sim$ $\triangle$s are $\cong$.
13. $\overline{JK} \parallel \overline{ZN}$	13. If 2 lines are cut by a trans. and corr. $\angle$s are $\cong$, then the lines are $\parallel$.
14. $\overleftrightarrow{JZ}$ and $\overleftrightarrow{KN}$ intersect, so $\overline{JK} \nparallel \overline{KN}$.	14. Given
15. $ZNKJ$ is a trap.	15. Def. of trap.

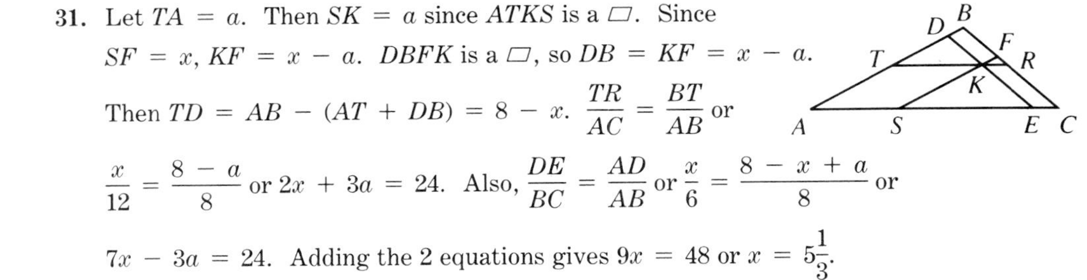

31. Let $TA = a$. Then $SK = a$ since $ATKS$ is a $\square$. Since $SF = x$, $KF = x - a$. $DBFK$ is a $\square$, so $DB = KF = x - a$. Then $TD = AB - (AT + DB) = 8 - x$. $\frac{TR}{AC} = \frac{BT}{AB}$ or $\frac{x}{12} = \frac{8 - a}{8}$ or $2x + 3a = 24$. Also, $\frac{DE}{BC} = \frac{AD}{AB}$ or $\frac{x}{6} = \frac{8 - x + a}{8}$ or $7x - 3a = 24$. Adding the 2 equations gives $9x = 48$ or $x = 5\frac{1}{3}$.

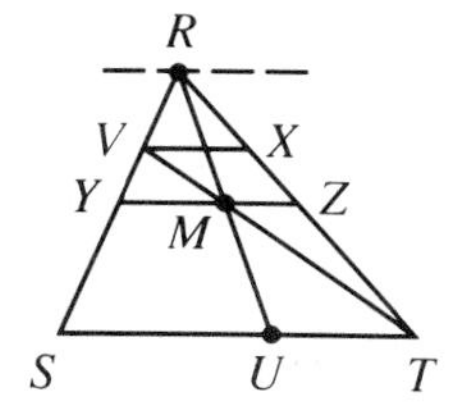

32. Draw lines $\parallel$ to $\overline{ST}$ through R, M, and V. $\frac{RY}{YS} = \frac{RZ}{ZT} = \frac{RM}{MU}$; since M is the midpt. of $\overline{RU}$, Y and Z are the midpts. of $\overline{RS}$ and $\overline{RT}$, and $YM = \frac{1}{2} \cdot SU = \frac{1}{2} \cdot \frac{3}{5} \cdot ST = \frac{3}{10} \cdot ST$; $\frac{YM}{ST} = \frac{3}{10}$; $\triangle VYM \sim \triangle VST$ so $\frac{VY}{VS} = \frac{YM}{ST} = \frac{3}{10}$; then $\frac{VY}{VY + YS} = \frac{3}{10}$ and $\frac{VY}{YS} = \frac{3}{7}$. $RV = RS - (VY + YS) = RS - \left(\frac{10}{3} \cdot VY\right) = RS - \left(\frac{10}{3} \cdot \frac{3}{7} \cdot YS\right) = RS - \left(\frac{10}{7} \cdot YS\right) = RS - \left(\frac{10}{7} \cdot \frac{1}{2} \cdot RS\right) = RS - \frac{5}{7} \cdot RS = \frac{2}{7} \cdot RS$; $RV : RS = 2 : 7$

33.

Statements	Reasons
1. Draw $\overline{AN}$ and $\overline{BM} \parallel$ to $\overline{CX}$.	1. Through a pt. outside a line there is exactly one line $\parallel$ to the given line.
2. $\frac{AX}{XB} = \frac{AP}{PM}$	2. $\triangle$ Prop. Thm.
3. $\overline{AN} \parallel \overline{BM}$	3. 2 lines $\parallel$ to a third line are $\parallel$ to each other.
4. $\angle ANP \cong \angle MBP$; $\angle MBY \cong \angle PCY$; $\angle BMY \cong \angle CPY$; $\angle PCZ \cong \angle NAZ$; $\angle ANZ \cong \angle CPZ$	4. If 2 $\parallel$ lines are cut by a trans., then alt. int. $\angle$s are $\cong$.
5. $\angle NPA \cong \angle BPM$	5. Vert. $\angle$s are $\cong$.
6. $\triangle APN \sim \triangle MPB$; $\triangle BYM \sim \triangle CYP$; $\triangle CZP \sim \triangle AZN$	6. AA $\sim$ Post.
7. $\frac{AP}{PM} = \frac{AN}{MB}$; $\frac{BY}{YC} = \frac{MB}{PC}$; $\frac{CZ}{ZA} = \frac{PC}{AN}$	7. Corr. sides of $\sim$ $\triangle$s are in prop.
8. $\frac{AX}{XB} \cdot \frac{BY}{YC} \cdot \frac{CZ}{ZA} = \frac{AN}{MB} \cdot \frac{MB}{PC} \cdot \frac{PC}{AN} = 1$	8. Substitution Prop., algebra

Page 274 • SELF-TEST 2

1. SSS $\sim$ Thm. **2.** AA $\sim$ Post. **3.** SAS $\sim$ Thm.

4. a. $\triangle EDC$ **b.** $\frac{AB}{ED} = \frac{AC}{EC} = \frac{BC}{DC}$ **c.** $\frac{15}{10} = \frac{21}{x}$; $15x = 210$; $x = 14$

d. $\frac{15}{10} = \frac{y}{12}$; $10y = 180$; $y = 18$

5. r **6.** p **7.** h **8.** a **9.** $\frac{x}{20} = \frac{9}{15}$; $15x = 180$; $x = 12$

10. $\frac{x}{7} = \frac{12}{6} = \frac{2}{1}$; $x = 14$ **11.** $\frac{x}{12} = \frac{5}{9}$; $9x = 60$; $x = 6\frac{2}{3}$

Page 274 • CHALLENGE

$\overline{FB} \parallel \overline{EC}$ so $\frac{FE}{AF} = \frac{BC}{AB}$ (△ Prop. Thm.) and $\overline{BD} \parallel \overline{AE}$ so $\frac{BC}{AB} = \frac{CD}{DE}$. Then $\frac{FE}{AF} = \frac{CD}{DE}$. $\overline{FD} \parallel \overline{AC}$ so $\frac{FE}{AF} = \frac{DE}{CD}$; by subst., $\frac{CD}{DE} = \frac{DE}{CD}$; $(CD)^2 = (DE)^2$; $CD = DE$; D is the midpt. of $\overline{CE}$. Then $\frac{BC}{AB} = 1$, so B is the midpt. of $\overline{AC}$; $\frac{FE}{AF} = 1$, and F is the midpt. of $\overline{AE}$.

Page 276 • EXTRA

1. c **2.** d **3.** b **4.** a **5.** d **6.** {1, 2, 3, 5, 7}, {0, 4, 6, 9}, {8}
7. Check students' drawings.

Pages 277–278 • CHAPTER REVIEW

1. 3 : 5 **2.** 2 : 4 : 3 **3.** $\frac{2y}{3x}$

4. $4x + 4x + 7x = 180$; $15x = 180$; $x = 12$; $4x = 48$; $7x = 84$; 48, 48, 84

5. No **6.** Yes **7.** Yes **8.** Yes **9.** $\angle J$ **10.** SP

11. a. $\frac{5}{3} = \frac{20}{ET}$; $5 \cdot ET = 60$; $ET = 12$ **b.** $\frac{5}{3} = \frac{p}{30}$; $p = 50$

12. $\frac{12}{8} = \frac{3}{2} = \frac{x}{6}$, $2x = 18$, $x = 9$; $\frac{3}{2} = \frac{y}{9}$, $2y = 27$, $y = 13.5$

13. a. $\triangle UVH$ **b.** AA ~ Post. **14.** $\frac{RT}{UV} = \frac{TS}{VH} = \frac{RS}{UH}$ **15.** $\frac{RS}{UH} = \frac{RT}{UV}$

16. $\triangle NCD \sim \triangle NBA$; AA ~ Post. **17.** $\triangle NCD \sim \triangle NAB$; AA ~ Post.

18. $\triangle NCD \sim \triangle NAB$; SAS ~ Thm. **19.** No **20.** $\triangle PXE \sim \triangle HTU$; SSS ~ Thm.

21. (2) **22.** $\frac{10}{OW} = \frac{8}{20} = \frac{2}{5}$; $2 \cdot OW = 50$; $OW = 25$

23. $\frac{VW}{24} = \frac{12}{20} = \frac{3}{5}$; $5 \cdot VW = 72$; $VW = 14.4$

24. $\frac{x}{28 - x} = \frac{18}{24} = \frac{3}{4}$; $4x = 84 - 3x$; $7x = 84$; $x = 12$; $AK = 12$

Page 279 • CHAPTER TEST

1. a. 20 : 32 = 5 : 8 **b.** 104 : 32 = 13 : 4 **2. a.** $\angle C$ **b.** TS

3. $\frac{x}{45} = \frac{4}{9}$, $9x = 180$, $x = 20$; $\frac{y}{45} = \frac{6}{9}$, $9y = 270$, $y = 30$ **4.** $9x = 120$, $x = 13\frac{1}{3}$

5. $\frac{a+b}{b} = \frac{c+10}{10}$ **6.** AA ~ Post. **7.** VN **8.** $\angle NKV$

9. $\frac{5}{8} = \frac{2.5}{VN}$; $5 \cdot VN = 20$; $VN = 4$ **10.** $\frac{10}{6} = \frac{15}{TU}$; $10 \cdot TU = 90$; $TU = 9$

11. $\frac{10}{10 + PR} = \frac{16}{48} = \frac{1}{3}$; $10 + PR = 30$; $PR = 20$

12. $\frac{26}{SU} = \frac{14}{21} = \frac{2}{3}$; $2 \cdot SU = 78$; $SU = 39$ **13.** $\frac{7}{EB} = \frac{5}{8}$; $5 \cdot EB = 56$; $EB = 11.2$

14. $\frac{GK}{30 - GK} = \frac{14}{21} = \frac{2}{3}$; $3 \cdot GK = 60 - 2 \cdot GK$; $5 \cdot GK = 60$; $GK = 12$

15.

Statements	Reasons
1. $\overleftrightarrow{DE} \parallel \overleftrightarrow{FG} \parallel \overleftrightarrow{HJ}$	1. Given
2. $\frac{DF}{FH} = \frac{EG}{GJ}$	2. If 3 ∥ lines int. 2 trans., then they divide them proportionally.
3. $DF \cdot GJ = FH \cdot EG$	3. Means-extremes Prop.

16.

Statements	Reasons
1. $BX = 6$; $AX = 8$; $CX = 9$; $DX = 12$	1. Given
2. $\frac{BX}{CX} = \frac{6}{9} = \frac{2}{3} = \frac{8}{12} = \frac{AX}{DX}$	2. Algebra
3. $\angle BXA \cong \angle CXD$	3. Vert. ∠s are ≅.
4. $\triangle BXA \sim \triangle CXD$	4. SAS ~ Thm.
5. $\angle A \cong \angle D$	5. Corr. ∠s of ~ △s are ≅.
6. $\overline{AB} \parallel \overline{CD}$	6. If 2 lines are cut by a trans. and alt. int. ∠s are ≅, then the lines are ∥.

Page 280 • ALGEBRA REVIEW

1. $\sqrt{36} = 6$ **2.** $\sqrt{81} = 9$ **3.** $\sqrt{24} = \sqrt{4 \cdot 6} = 2\sqrt{6}$ **4.** $\sqrt{98} = \sqrt{49 \cdot 2} = 7\sqrt{2}$

5. $\sqrt{300} = \sqrt{100 \cdot 3} = 10\sqrt{3}$ **6.** $\sqrt{\frac{1}{4}} = \frac{1}{2}$ **7.** $\frac{\sqrt{5}}{\sqrt{3}} = \frac{\sqrt{5}}{\sqrt{3}} \cdot \frac{\sqrt{3}}{\sqrt{3}} = \frac{\sqrt{15}}{3}$

8. $\sqrt{\frac{80}{25}} = \frac{\sqrt{16 \cdot 5}}{5} = \frac{4\sqrt{5}}{5}$ **9.** $\frac{2\sqrt{3}}{\sqrt{12}} = \frac{2\sqrt{3}}{\sqrt{4 \cdot 3}} = \frac{2\sqrt{3}}{2\sqrt{3}} = 1$

10. $\sqrt{\frac{250}{48}} = \frac{\sqrt{25 \cdot 10}}{\sqrt{16 \cdot 3}} = \frac{5\sqrt{10}}{4\sqrt{3}} \cdot \frac{\sqrt{3}}{\sqrt{3}} = \frac{5\sqrt{30}}{4 \cdot 3} = \frac{5\sqrt{30}}{12}$ **11.** $\sqrt{13^2} = 13$

12. $(\sqrt{17})^2 = 17$ **13.** $(2\sqrt{3})^2 = 2\sqrt{3} \cdot 2\sqrt{3} = 4 \cdot 3 = 12$

14. $(3\sqrt{8})^2 = 3\sqrt{8} \cdot 3\sqrt{8} = 9 \cdot 8 = 72$ **15.** $(9\sqrt{2})^2 = 9\sqrt{2} \cdot 9\sqrt{2} = 81 \cdot 2 = 162$

16. $5\sqrt{18} = 5\sqrt{9 \cdot 2} = 5 \cdot 3\sqrt{2} = 15\sqrt{2}$ **17.** $4\sqrt{27} = 4\sqrt{9 \cdot 3} = 4 \cdot 3\sqrt{3} = 12\sqrt{3}$

18. $6\sqrt{24} = 6\sqrt{4 \cdot 6} = 6 \cdot 2\sqrt{6} = 12\sqrt{6}$ **19.** $5\sqrt{8} = 5\sqrt{4 \cdot 2} = 5 \cdot 2\sqrt{2} = 10\sqrt{2}$

20. $9\sqrt{40} = 9\sqrt{4 \cdot 10} = 9 \cdot 2\sqrt{10} = 18\sqrt{10}$

21. $9 + 16 = x^2; 25 = x^2; x = \sqrt{25} = 5$ **22.** $x^2 + 16 = 25; x^2 = 9; x = \sqrt{9} = 3$

23. $25 + x^2 = 169; x^2 = 144; x = \sqrt{144} = 12$ **24.** $x^2 + 9 = 16; x^2 = 7; x = \sqrt{7}$

25. $16 + 49 = x^2; 65 = x^2; x = \sqrt{65}$

26. $x^2 + 25 = 100; x^2 = 75; x = \sqrt{75} = \sqrt{25 \cdot 3} = 5\sqrt{3}$

27. $1 + x^2 = 9; x^2 = 8; x = \sqrt{8} = \sqrt{4 \cdot 2} = 2\sqrt{2}$

28. $x^2 + 25 = 50; x^2 = 25; x = \sqrt{25} = 5$

29. $x^2 + 147 = 4x^2; 147 = 3x^2; 49 = x^2; x = \sqrt{49} = 7$

Page 280 • CHALLENGE

$\overline{HG} \parallel \overline{EB} \parallel \overline{DC}$. $\triangle ODC$ is equilateral, so $OD = 12$. Let $\overline{AD}$ intersect $\overline{HG}$ at T. $\triangle HTD$ is equilateral, so $HT = 6$. Also, $OT = 6$. Then in $\triangle AHT$, $\frac{AO}{AT} = \frac{JO}{HT}$ or $\frac{12}{18} = \frac{JO}{6}$, so $JO = 4$ and $JK = 8$.

Pages 281–283 • CUMULATIVE REVIEW: CHAPTERS 1–7

True-False Exercises

A **1.** F **2.** T **3.** F **4.** T **5.** T **6.** T

B **7.** F **8.** F **9.** F **10.** T **11.** F

Multiple-Choice Exercises

A **1.** d **2.** c **3.** d

B **4.** b **5.** c

Always-Sometimes-Never Exercises

A **1.** S **2.** N **3.** S **4.** A **5.** N **6.** N **7.** A **8.** S **9.** S **10.** N **11.** A

B **12.** S **13.** A **14.** N **15.** S **16.** N

Completion Exercises

A **1.** 120 **2.** comp. **3.** obtuse **4.** $\perp$ **5.** 108 **6.** $\sim$

B **7.** rect. **8.** $\frac{s}{u}$ **9.** 36

Algebraic Exercises

A **1.** $-1 - (-8) = x - (-1); 7 = x + 1; x = 6$

2. $x^2 + 18x = x^2 + 54; 18x = 54; x = 3$

3. $x + (x + 4) + (x + 8) + (x + 12) = 360; 4x + 24 = 360; 4x = 336; x = 84$

4. $7x - 13 = 2x + 17; 5x = 30; x = 6$

5. $6x + (2x + 20) = 180; 8x = 160; x = 20$

6. $\frac{1}{2}(x + x + 8) = 15; x + 4 = 15; x = 11$

7. $3(3x - 1) = 2(4x + 2); 9x - 3 = 8x + 4; x = 7$

8. $30 = 8x - 8; 8x = 38; x = 4\frac{3}{4}$

9. $x(x + 9) = (x + 4)(x + 3); x^2 + 9x = x^2 + 7x + 12; 2x = 12; x = 6$

B 10. $180 - x = 8 + 3(90 - x); 180 - x = 8 + 270 - 3x; 2x = 98; x = 49$

11. $\frac{360}{n} : \frac{(n - 2)180}{n} = 2 : 13; 13 \cdot \frac{360}{n} = 2 \cdot \frac{(n - 2)180}{n}; 4680n = 360(n^2 - 2n);$ $360n^2 - 5400n = 0; 360n(n - 15) = 0; 360n = 0$ or $n - 15 = 0; n = 0$ (reject) or $n = 15$; 15 sides

12. $P = 2(12) + 2(15) = 54; \frac{54}{90} = \frac{3}{5}; \frac{3}{5} = \frac{12}{x}; 3x = 60; x = 20; \frac{3}{5} = \frac{15}{y}; 3y = 75;$ $y = 25$; 20 cm, 25 cm

13. $4x + 5x + 7x = 64; 16x = 64; x = 4; 4x = 16; 5x = 20; 7x = 28$; 16 cm, 20 cm, 28 cm

14. $5x + 5x + 2x = 180; 12x = 180; x = 15; 5x = 75; m\angle Z = 75$

15. $\triangle ADF \sim \triangle BGF, \frac{y}{7} = \frac{4}{8}, 8y = 28, y = 3.5; \triangle EDA \sim \triangle EGC, \frac{x}{9} = \frac{7}{7 + y} = \frac{7}{10.5},$ $10.5x = 63, x = 6$

Proof Exercises

A 1.

Statements	Reasons
1. $\overline{SU} \cong \overline{SV}; \angle 1 \cong \angle 2$	1. Given
2. $\overline{QS} \cong \overline{QS}$	2. Refl. Prop.
3. $\triangle QUS \cong \triangle QVS$	3. SAS Post.
4. $\overline{UQ} \cong \overline{VQ}$	4. Corr. parts of $\cong$ △s are $\cong$.

2.

Statements	Reasons
1. $\overrightarrow{QS}$ bis. $\angle RQT; \angle R \cong \angle T$	1. Given
2. $\angle RQS \cong \angle TQS$	2. Def. of $\angle$ bis.
3. $\overline{QS} \cong \overline{QS}$	3. Refl. Prop.
4. $\triangle RQS \cong \triangle TQS$	4. AAS Thm.
5. $\angle 1 \cong \angle 2$	5. Corr. parts of $\cong$ △s are $\cong$.
6. $\overrightarrow{SQ}$ bis. $\angle RST$.	6. Def. of $\angle$ bis.

B 3.

Statements	Reasons
1. $\triangle QRU \cong \triangle QTV$	1. Given
2. $\overline{QR} \cong \overline{QT}$; $\angle R \cong \angle T$; $\overline{RU} \cong \overline{TV}$, or $RU = TV$	2. Corr. parts of $\cong$ ⚠ are $\cong$.
3. $\overline{US} \cong \overline{VS}$, or $US = VS$	3. Given
4. $RU + US = TV + VS$	4. Add. Prop. of =
5. $RS = RU + US$; $TS = TV + TS$	5. Seg. Add. Prop.
6. $RS = TS$, or $\overline{RS} \cong \overline{TS}$	6. Substitution Prop.
7. $\triangle QRS \cong \triangle QTS$	7. SAS Post.

4.

Statements	Reasons
1. $\overline{QS}$ bis. $\angle UQV$ and $\angle USV$.	1. Given
2. $\angle UQS \cong \angle VQS$; $\angle 1 \cong \angle 2$	2. Def. of $\angle$ bis.
3. $\overline{QS} \cong \overline{QS}$	3. Refl. Prop.
4. $\triangle UQS \cong \triangle VQS$	4. ASA Post.
5. $\overline{QU} \cong \overline{QV}$; $\angle QUS \cong \angle QVS$	5. Corr. parts of $\cong$ ⚠ are $\cong$.
6. $\angle QUR \cong \angle QVT$	6. If 2 ⦣ are supps. of $\cong$ ⦣, then the 2 ⦣ are $\cong$.
7. $\angle R \cong \angle T$	7. Given
8. $\triangle QUR \cong \triangle QVT$	8. AAS Thm.
9. $\overline{RQ} \cong \overline{TQ}$	9. Corr. parts of $\cong$ ⚠ are $\cong$.

5.

Statements	Reasons
1. $\overline{EF} \parallel \overline{JK}$; $\overline{JK} \parallel \overline{HI}$	1. Given
2. $\overline{EF} \parallel \overline{HI}$	2. 2 lines $\parallel$ to a third line are $\parallel$ to each other.
3. $\angle 2 \cong \angle 3$; $\angle F \cong \angle H$	3. If 2 $\parallel$ lines are cut by a trans., then alt. int. ⦣ are $\cong$.
4. $\triangle EFG \sim \triangle IHG$	4. AA $\sim$ Post.

6.

Statements	Reasons
1. $\frac{JG}{HG} = \frac{KG}{IG}$	1. Given
2. $\angle KGJ \cong \angle KGJ$	2. Refl. Prop.
3. $\triangle KGJ \sim \triangle IGH$	3. SAS $\sim$ Thm.
4. $\angle 1 \cong \angle 3$	4. Corr. ⦣ of $\sim$ ⚠ are $\cong$.
5. $\angle 1 \cong \angle 2$	5. Given
6. $\angle 2 \cong \angle 3$	6. Substitution Prop.
7. $\overline{EF} \parallel \overline{HI}$	7. If 2 lines are cut by a trans. and alt. int. ⦣ are $\cong$, then the lines are $\parallel$.

CHAPTER 8 • Right Triangles

Pages 287–288 • CLASSROOM EXERCISES

1. 60, 30, 60 **2.** $90 - k, k, 90 - k$ **3.** *RUS, SUT* **4.** *RTS, STU*

5. $\sqrt{50} = \sqrt{25 \cdot 2} = 5\sqrt{2}$ **6.** $3\sqrt{8} = 3\sqrt{4 \cdot 2} = 6\sqrt{2}$ **7.** $\sqrt{225} = 15$

8. $7\sqrt{63} = 7\sqrt{9 \cdot 7} = 21\sqrt{7}$ **9.** $\sqrt{288} = \sqrt{144 \cdot 2} = 12\sqrt{2}$ **10.** $\sqrt{\frac{3}{4}} = \frac{\sqrt{3}}{2}$

11. $\sqrt{\frac{1}{5}} = \sqrt{\frac{1}{5} \cdot \frac{5}{5}} = \sqrt{\frac{5}{25}} = \frac{\sqrt{5}}{5}$ **12.** $\frac{\sqrt{5}}{\sqrt{2}} = \frac{\sqrt{5}}{\sqrt{2}} \cdot \frac{\sqrt{2}}{\sqrt{2}} = \frac{\sqrt{10}}{\sqrt{4}} = \frac{\sqrt{10}}{2}$

13. $\sqrt{\frac{5}{2}} = \sqrt{\frac{5}{2} \cdot \frac{2}{2}} = \sqrt{\frac{10}{4}} = \frac{\sqrt{10}}{2}$

14. $\frac{3}{4}\sqrt{\frac{28}{3}} = \frac{3}{4}\sqrt{\frac{28}{3} \cdot \frac{3}{3}} = \frac{3}{4}\sqrt{\frac{84}{9}} = \frac{3}{4} \cdot \sqrt{\frac{4 \cdot 21}{9}} = \frac{3}{4} \cdot \frac{2}{3} \cdot \sqrt{21} = \frac{\sqrt{21}}{2}$

15. a. $\frac{2}{x} = \frac{x}{3}$; $x^2 = 6$; $x = \sqrt{6}$ **b.** $\frac{2}{x} = \frac{x}{6}$; $x^2 = 12$; $x = \sqrt{12} = \sqrt{4 \cdot 3} = 2\sqrt{3}$

c. $\frac{4}{x} = \frac{x}{25}$; $x^2 = 100$; $x = \sqrt{100} = 10$

16. a. s, r **b.** k, r **c.** k, s

17. a. 2, 7; $\frac{2}{z} = \frac{z}{7}$; $z^2 = 14$; $z = \sqrt{14}$ **b.** 9, 2; $\frac{9}{x} = \frac{x}{2}$; $x^2 = 18$; $x = \sqrt{18} =$

$\sqrt{9 \cdot 2} = 3\sqrt{2}$ **c.** 9, 7; $\frac{9}{y} = \frac{y}{7}$; $y^2 = 63$; $y = \sqrt{63} = \sqrt{9 \cdot 7} = 3\sqrt{7}$

Pages 288–290 • WRITTEN EXERCISES

A **1.** $\sqrt{12} = \sqrt{4 \cdot 3} = 2\sqrt{3}$ **2.** $\sqrt{72} = \sqrt{36 \cdot 2} = 6\sqrt{2}$ **3.** $\sqrt{45} = \sqrt{9 \cdot 5} = 3\sqrt{5}$

4. $\sqrt{75} = \sqrt{25 \cdot 3} = 5\sqrt{3}$ **5.** $\sqrt{800} = \sqrt{400 \cdot 2} = 20\sqrt{2}$

6. $\sqrt{54} = \sqrt{9 \cdot 6} = 3\sqrt{6}$ **7.** $9\sqrt{40} = 9\sqrt{4 \cdot 10} = 18\sqrt{10}$

8. $4\sqrt{28} = 4\sqrt{4 \cdot 7} = 8\sqrt{7}$

9. $\sqrt{30} \cdot \sqrt{6} = \sqrt{2 \cdot 3 \cdot 5} \cdot \sqrt{2 \cdot 3} = \sqrt{2^2 \cdot 3^2 \cdot 5} = 2 \cdot 3\sqrt{5} = 6\sqrt{5}$

10. $\sqrt{5} \cdot \sqrt{35} = \sqrt{5} \cdot \sqrt{5 \cdot 7} = \sqrt{5^2 \cdot 7} = 5\sqrt{7}$ **11.** $\sqrt{\frac{3}{7}} = \sqrt{\frac{3}{7} \cdot \frac{7}{7}} = \frac{\sqrt{21}}{7}$

12. $\sqrt{\frac{9}{5}} = \sqrt{\frac{9}{5} \cdot \frac{5}{5}} = \frac{\sqrt{9 \cdot 5}}{5} = \frac{3\sqrt{5}}{5}$ **13.** $\frac{18}{\sqrt{3}} = \frac{18}{\sqrt{3}} \cdot \frac{\sqrt{3}}{\sqrt{3}} = \frac{18\sqrt{3}}{3} = 6\sqrt{3}$

14. $\frac{24}{3\sqrt{2}} = \frac{8}{\sqrt{2}} \cdot \frac{\sqrt{2}}{\sqrt{2}} = \frac{8\sqrt{2}}{2} = 4\sqrt{2}$

15. $\frac{\sqrt{15}}{3\sqrt{45}} = \frac{1}{3}\sqrt{\frac{15}{45}} = \frac{1}{3}\sqrt{\frac{1}{3}} = \frac{1}{3}\sqrt{\frac{1}{3} \cdot \frac{3}{3}} = \frac{1}{3} \cdot \frac{\sqrt{3}}{3} = \frac{\sqrt{3}}{9}$

16. $\frac{2}{x} = \frac{x}{18}$; $x^2 = 36$; $x = 6$ **17.** $\frac{3}{x} = \frac{x}{27}$; $x^2 = 81$; $x = 9$

18. $\frac{49}{x} = \frac{x}{25}$; $x^2 = 49 \cdot 25$; $x = \sqrt{49 \cdot 25} = 7 \cdot 5 = 35$

19. $\frac{1}{x} = \frac{x}{1000}$; $x^2 = 1000$; $x = \sqrt{1000} = \sqrt{100 \cdot 10} = 10\sqrt{10}$

20. $\frac{16}{x} = \frac{x}{24}$; $x^2 = 16 \cdot 24$; $x = \sqrt{16 \cdot 24} = \sqrt{16 \cdot 4 \cdot 6} = 4 \cdot 2\sqrt{6} = 8\sqrt{6}$

21. $\frac{22}{x} = \frac{x}{55}$; $x^2 = 22 \cdot 55$; $x = \sqrt{22 \cdot 55} = \sqrt{2 \cdot 11 \cdot 5 \cdot 11} = 11\sqrt{10}$

22. $\frac{x}{4} = \frac{4}{8}$; $8x = 16$; $x = 2$ 23. $\frac{4}{6} = \frac{6}{x}$; $4x = 36$; $x = 9$

24. $\frac{3}{x} = \frac{x}{6}$; $x^2 = 18$; $x = \sqrt{18} = \sqrt{9 \cdot 2} = 3\sqrt{2}$

25. $MK = 9 - 4 = 5$; $\frac{9}{x} = \frac{x}{5}$; $x^2 = 45$; $x = \sqrt{45} = \sqrt{9 \cdot 5} = 3\sqrt{5}$

26. $JK = 3 + 9 = 12$; $\frac{3}{x} = \frac{x}{12}$; $x^2 = 36$; $x = 6$

B 27. $\frac{3}{6} = \frac{6}{3 + x}$; $9 + 3x = 36$; $3x = 27$; $x = 9$

28. $\frac{6}{9} = \frac{9}{x + 6}$; $6x + 36 = 81$; $6x = 45$; $x = \frac{45}{6} = 7.5$

29. $\frac{6}{3\sqrt{6}} = \frac{3\sqrt{6}}{x + 6}$; $6x + 36 = 54$; $6x = 18$; $x = 3$

30. $\frac{6}{7} = \frac{7}{x + 6}$; $6x + 36 = 49$; $6x = 13$; $x = \frac{13}{6} = 2\frac{1}{6}$

31. $\frac{4}{x} = \frac{x}{25}$, $x^2 = 100$, $x = 10$; $\frac{4}{y} = \frac{y}{29}$, $y^2 = 4 \cdot 29$, $y = \sqrt{4 \cdot 29} = 2\sqrt{29}$; $\frac{25}{z} = \frac{z}{29}$, $z^2 = 25 \cdot 29$, $z = \sqrt{25 \cdot 29} = 5\sqrt{29}$

32. $\frac{9}{x} = \frac{x}{7}$, $x^2 = 9 \cdot 7$, $x = \sqrt{9 \cdot 7} = 3\sqrt{7}$; $\frac{9}{y} = \frac{y}{16}$, $y^2 = 9 \cdot 16$, $y = \sqrt{9 \cdot 16} = 3 \cdot 4 = 12$; $\frac{7}{z} = \frac{z}{16}$, $z^2 = 7 \cdot 16$, $z = \sqrt{7 \cdot 16} = 4\sqrt{7}$

33. $\frac{\frac{1}{3}}{x} = \frac{x}{\frac{1}{6}}$, $x^2 = \frac{1}{18}$, $x = \sqrt{\frac{1}{18} \cdot \frac{2}{2}} = \sqrt{\frac{2}{36}} = \frac{\sqrt{2}}{6}$; $\frac{\frac{1}{6}}{y} = \frac{y}{\frac{1}{2}}$, $y^2 = \frac{1}{12}$, $y = \sqrt{\frac{1}{12} \cdot \frac{3}{3}} = \sqrt{\frac{3}{36}} = \frac{\sqrt{3}}{6}$; $\frac{\frac{1}{3}}{z} = \frac{z}{\frac{1}{2}}$, $z^2 = \frac{1}{6}$, $z = \sqrt{\frac{1}{6} \cdot \frac{6}{6}} = \sqrt{\frac{6}{36}} = \frac{\sqrt{6}}{6}$

34. $\frac{8}{x} = \frac{x}{8}$, $x^2 = 64$, $x = 8$; $\frac{8}{y} = \frac{y}{16}$, $y^2 = 8 \cdot 16$, $y = \sqrt{8 \cdot 16} = \sqrt{2 \cdot 4 \cdot 16} = 2 \cdot 4\sqrt{2} = 8\sqrt{2}$; $\frac{8}{z} = \frac{z}{16}$, $z = 8\sqrt{2}$

35. $\frac{x}{9} = \frac{9}{15}$, $15x = 81$, $x = \frac{27}{5} = 5.4$; $y = 15 - 5.4 = 9.6$; $\frac{5.4}{z} = \frac{z}{9.6}$, $z^2 = 51.84$, $z = 7.2$

36. $\frac{2}{6} = \frac{6}{x}$, $2x = 36$, $x = 18$; $\frac{16}{y} = \frac{y}{18}$, $y^2 = 16 \cdot 18$, $y = \sqrt{16 \cdot 18} = \sqrt{16 \cdot 9 \cdot 2} = 4 \cdot 3\sqrt{2} = 12\sqrt{2}$; $\frac{2}{z} = \frac{z}{16}$, $z^2 = 2 \cdot 16$, $z = \sqrt{2 \cdot 16} = 4\sqrt{2}$

37. $\frac{\sqrt{2}}{2} = \frac{2}{x + \sqrt{2}}$, $x\sqrt{2} + 2 = 4$, $x\sqrt{2} = 2$, $x = \frac{2}{\sqrt{2}} = \frac{2}{\sqrt{2}} \cdot \frac{\sqrt{2}}{\sqrt{2}} = \frac{2\sqrt{2}}{2} = \sqrt{2}$; $\frac{\sqrt{2}}{y} = \frac{y}{2\sqrt{2}}$, $y^2 = 2\sqrt{2} \cdot \sqrt{2} = 4$, $y = 2$; $\frac{\sqrt{2}}{z} = \frac{z}{\sqrt{2}}$, $z^2 = 2$, $z = \sqrt{2}$

38. $\frac{x}{12} = \frac{12}{x + 7}$, $x^2 + 7x = 144$, $x^2 + 7x - 144 = 0$, $(x - 9)(x + 16) = 0$, $x = 9$ or $x = -16$ (reject), $x = 9$; $\frac{9}{y} = \frac{y}{25}$, $y^2 = 9 \cdot 25$, $y = \sqrt{9 \cdot 25} = 3 \cdot 5 = 15$; $\frac{16}{z} = \frac{z}{25}$, $z^2 = 16 \cdot 25$, $z = \sqrt{16 \cdot 25} = 4 \cdot 5 = 20$

39. $\frac{x}{x + 2} = \frac{x + 2}{2x + 1}$, $2x^2 + x = x^2 + 4x + 4$, $x^2 - 3x - 4 = 0$, $(x - 4)(x + 1) = 0$, $x = 4$ or $x = -1$ (reject), $x = 4$; $\frac{4}{y} = \frac{y}{5}$, $y^2 = 4 \cdot 5$, $y = \sqrt{4 \cdot 5} = 2\sqrt{5}$; $\frac{5}{z} = \frac{z}{9}$, $z^2 = 5 \cdot 9$, $z = \sqrt{5 \cdot 9} = 3\sqrt{5}$

40.

Statements	Reasons
1. $\angle ACB$ is a rt. $\angle$.	1. Given
2. $m\angle ACB = 90$	2. Def. of rt. $\angle$
3. $\overline{CN}$ is an altitude.	3. Given
4. $\overline{CN} \perp \overline{AB}$	4. Def. of alt.
5. $m\angle ANC = 90$; $m\angle CNB = 90$	5. Def. of $\perp$ lines
6. $\angle ACB \cong \angle ANC \cong \angle CNB$	6. Def. of $\cong$ ∠s
7. $\triangle ACB$ and $\triangle CNB$ are rt. △s.	7. Def. of rt. $\triangle$
8. $\angle A$ is comp. to $\angle B$; $\angle BCN$ is comp. to $\angle B$.	8. The acute ∠s of a rt. $\triangle$ are comp.
9. $\angle A \cong \angle BCN$	9. If 2 ∠s are comps. of the same $\angle$, then the 2 ∠s are $\cong$.
10. $\angle A \cong \angle A$	10. Refl. Prop.
11. $\triangle ACB \sim \triangle ANC \sim \triangle CNB$	11. AA $\sim$ Post.

41. **a.** $a^2 = cd$ and $b^2 = ce$ **b.** $a^2 + b^2 = cd + ce = c(d + e) = c(c) = c^2$

C 42. Given: $\triangle ABC$ with rt. $\angle ACB$; altitude $\overline{CD}$

Prove: $AB \cdot CD = AC \cdot BC$

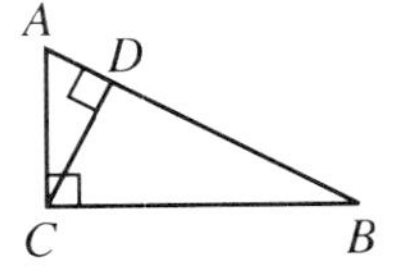

Statements	Reasons
1. $\overline{CD}$ is the alt. to the hyp. of $\triangle ABC$.	1. Given
2. $\triangle ABC \sim \triangle ACD$	2. If the alt. is drawn to the hyp. of a rt. △, then the 2 △s formed are ~ to the orig. △ and to each other.
3. $\frac{AB}{AC} = \frac{BC}{CD}$	3. Corr. sides of ~ △s are in prop.
4. $AB \cdot CD = AC \cdot BC$	4. Means-extremes Prop.

43.

Statements	Reasons
1. $PQRS$ is a rect.	1. Given
2. $m\angle SPQ = m\angle PQR = m\angle R = m\angle S = 90$	2. Def. of rect.
3. $\angle SPQ \cong \angle PQR \cong \angle R \cong \angle S$	3. Def. of ≅ ∠s
4. $PQRS$ is a ▱.	4. If both pairs of opp. ∠s of a quad. are ≅, then the quad. is a ▱.
5. $\overline{PS} \cong \overline{RQ}$, or $PS = RQ$	5. Opp. sides of a ▱ are ≅.
6. PS is the geom. mean btwn. ST and TR.	6. Given
7. $\frac{ST}{PS} = \frac{PS}{TR}$	7. Def. of geom. mean
8. $\frac{ST}{RQ} = \frac{PS}{TR}$	8. Substitution Prop.
9. $\triangle PST \sim \triangle TRQ$	9. SAS ~ Thm.
10. $m\angle PTS = m\angle TQR$	10. Corr. ∠s of ~ △s are ≅.
11. $\triangle TRQ$ is a rt. △.	11. Def. of rt. △
12. $\angle TQR$ and $\angle QTR$ are comp.	12. The acute ∠s of a rt. △ are comp.
13. $m\angle TQR + m\angle QTR = 90$	13. Def. of comp. ∠s
14. $m\angle PTS + m\angle QTR = 90$	14. Substitution Prop.
15. $m\angle PTQ + m\angle PTS + m\angle QTR = 180$	15. ∠ Add. Post.
16. $m\angle PTQ + 90 = 180$, or $m\angle PTQ = 90$	16. Substitution Prop.
17. $\angle PTQ$ is a rt. ∠.	17. Def. of rt. ∠

44.

Statements	Reasons
1. $PQRS$ is a rect.	1. Given
2. $m\angle SPQ = m\angle PQR = m\angle QRS = m\angle PSR = 90$	2. Def. of rect.
3. $\angle SPQ \cong \angle PQR \cong \angle QRS \cong \angle PSR$	3. Def. of $\cong$ $\angle$s
4. $PQRS$ is a $\square$.	4. If both pairs of opp. $\angle$s of a quad. are $\cong$, then the quad. is a $\square$.
5. $\overline{PS} \cong \overline{QR}$, or $PS = QR$	5. Opp. sides of a $\square$ are $\cong$.
6. $\angle A$ is a rt. $\angle$.	6. Given
7. $m\angle A = 90$	7. Def. of rt. $\angle$
8. $m\angle PSB + m\angle PSR = 180$; $m\angle QRS + m\angle QRC = 180$	8. $\angle$ Add. Post.
9. $m\angle PSB + 90 = 180$, or $m\angle PSB = 90$; $90 + m\angle QRC = 180$, or $m\angle QRC = 90$	9. Substitution Prop.
10. $\angle PSB \cong \angle QRC \cong \angle A$	10. Def. of $\cong$ $\angle$s
11. $\angle B \cong \angle B$; $\angle C \cong \angle C$	11. Refl. Prop.
12. $\triangle ABC \sim \triangle SBP \sim \triangle RQC$	12. AA $\sim$ Post.
13. $\frac{BS}{QR} = \frac{PS}{RC}$	13. Corr. sides of $\sim$ $\triangle$s are in prop.
14. $BS \cdot RC = QR \cdot PS$	14. Means-extremes Prop.
15. $BS \cdot RC = QR \cdot PS = (PS)^2$	15. Substitution Prop.

45. a. The midpt. of the hyp. of a rt. $\triangle$ is equidistant from the three vertices; $CM = AM = \frac{AB}{2} = \frac{AH + BH}{2}$; CM is the arith. mean between AH and BH. CH is the geom. mean between AH and BH. (When the alt. is drawn to the hyp. of a rt. $\triangle$, the length of the alt. is the geom. mean between the segments of the hyp.) Since the $\perp$ seg. from a pt. to a line is the shortest seg. from the pt. to the line, $CM > CH$; the arith. mean $>$ the geom. mean. **b.** $(r - s)^2 > 0$; $r^2 - 2rs + s^2 > 0$; $r^2 + 2rs + s^2 > 4rs$; $\frac{r^2 + 2rs + s^2}{4} > rs$; $\frac{(r + s)^2}{4} > rs$; $\frac{r + s}{2} > \sqrt{rs}$

Page 291 • CLASSROOM EXERCISES

1.

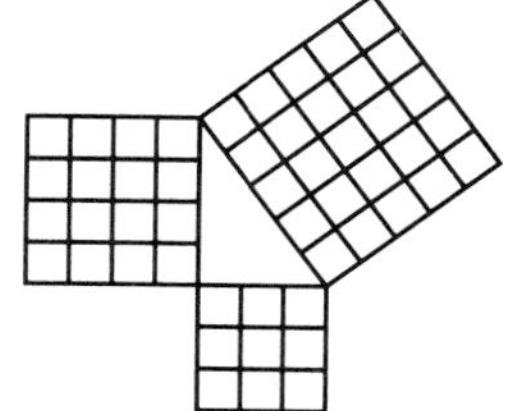

2. c, d

3. $(\sqrt{3})^2 = \sqrt{3} \cdot \sqrt{3} = 3$ **4.** $(3\sqrt{11})^2 = 3^2 \cdot (\sqrt{11})^2 = 9 \cdot 11 = 99$ **5.** 5

6. $2^2 \cdot (\sqrt{7})^2 = 4 \cdot 7 = 28$ **7.** $7^2 \cdot (\sqrt{2})^2 = 49 \cdot 2 = 98$ **8.** $2^2 \cdot n^2 = 4n^2$

9. $\frac{3^2}{(\sqrt{5})^2} = \frac{9}{5}$ **10.** $\frac{(\sqrt{2})^2}{2^2} = \frac{2}{4} = \frac{1}{2}$ **11.** $\frac{n^2}{(\sqrt{3})^2} = \frac{n^2}{3}$ **12.** $\left(\frac{2}{3}\right)^2 \cdot (\sqrt{6})^2 = \frac{4}{9} \cdot 6 = \frac{8}{3}$

13. $x^2 = 10^2 - 7^2; x^2 = 100 - 49 = 51; x = \sqrt{51}$

14. $x^2 = 4^2 + 6^2; x^2 = 16 + 36 = 52; x = \sqrt{52} = \sqrt{4 \cdot 13} = 2\sqrt{13}$

15. $(x + 1)^2 = x^2 + 5^2; x^2 + 2x + 1 = x^2 + 25; 2x = 24; x = 12$

16. $x^2 = 9^2 + 6^2; x^2 = 81 + 36 = 117; x = \sqrt{117} = \sqrt{9 \cdot 13} = 3\sqrt{13}$

17. $x^2 + x^2 = (7\sqrt{2})^2; 2x^2 = 98; x^2 = 49; x = 7$

18. $x^2 + (5\sqrt{3})^2 = (2x)^2; x^2 + 75 = 4x^2; 3x^2 = 75; x^2 = 25; x = 5$

Pages 292–294 • WRITTEN EXERCISES

A **1.** $3^2 + 4^2 = x^2; x^2 = 25; x = 5$ **2.** $12^2 + 5^2 = x^2; x^2 = 169; x = 13$

3. $x^2 + 6^2 = 10^2; x^2 = 64; x = 8$

4. $x^2 + 9^2 = 11^2; x^2 = 40; x = \sqrt{40} = \sqrt{4 \cdot 10} = 2\sqrt{10}$

5. $x^2 + 10^2 = 20^2; x^2 = 300; x = \sqrt{300} = \sqrt{100 \cdot 3} = 10\sqrt{3}$

6. $x^2 + 8^2 = 12^2; x^2 = 80; x = \sqrt{80} = \sqrt{16 \cdot 5} = 4\sqrt{5}$

7. $x^2 = 4^2 + (4\sqrt{3})^2; x^2 = 64; x = 8$

8. $x^2 + 6^2 = (6\sqrt{2})^2; x^2 = 36; x = 6$ **9.** $x^2 = 7^2 + 24^2; x^2 = 625; x = 25$

10. $x^2 + 12^2 = 15^2; x^2 = 81; x = 9$

11. $x^2 = 8^2 + 8^2; x^2 = 2 \cdot 8^2; x = \sqrt{2 \cdot 8^2} = 8\sqrt{2}$

12. $x^2 + x^2 = (6\sqrt{2})^2; 2x^2 = 72; x^2 = 36; x = 6$

13. $x^2 = (2.4)^2 + (1.8)^2 = 9; x = 3$ **14.** $x^2 + (\sqrt{3})^2 = 2^2; x^2 = 1; x = 1$

15. $x^2 = 4^2 + 4^2 = 2 \cdot 4^2; x = \sqrt{2 \cdot 4^2} = 4\sqrt{2}$

16. $x^2 + x^2 = 12^2; 2x^2 = 144; x^2 = 72; x = \sqrt{72} = \sqrt{36 \cdot 2} = 6\sqrt{2}$

17. $x^2 = 8^2 + 15^2 = 289; x = 17; P = 4x = 68$

18. $x^2 + 6^2 = 10^2; x^2 = 64; x = 8$; diag.: 16 cm

B **19.** $x^2 = 5^2 - 4^2 = 9; x = 3$ **20.** $x^2 = 10^2 - 6^2 = 64; x = 8$

21. $x^2 = 3^2 + 6^2 = 45; x = \sqrt{45} = \sqrt{9 \cdot 5} = 3\sqrt{5}$

22. $\left(\frac{x}{2}\right)^2 = 8^2 - 6^2; \frac{x^2}{4} = 28; x^2 = 4 \cdot 28; x = \sqrt{4 \cdot 28} = \sqrt{4 \cdot 4 \cdot 7} = 4\sqrt{7}$

23. $x^2 + 5^2 = 13^2; x^2 = 144; x = 12$

24. $\left(\frac{x - 10}{2}\right)^2 = 15^2 - 12^2; \frac{(x - 10)^2}{4} = 81; (x - 10)^2 = 4 \cdot 81; x - 10 = \sqrt{4 \cdot 81} =$
$2 \cdot 9 = 18; x = 28$

25. $x^2 = 8^2 + 6^2 = 100; x = 10$

26. $x^2 = 12^2 - 4^2 = 128; x = \sqrt{128} = \sqrt{64 \cdot 2} = 8\sqrt{2}$

27. $a^2 = 10^2 - 6^2 = 64; a = 8; x^2 = 8^2 + 15^2 = 289; x = \sqrt{289} = 17$

28. $\frac{3}{x} = \frac{4}{y}$; $y = \frac{4x}{3}$; $3^2 + 4^2 = \left(x + \frac{4}{3}x\right)^2 = \left(\frac{7x}{3}\right)^2$; $25 = \frac{49x^2}{9}$; $x^2 = \frac{9 \cdot 25}{49}$;
$x = \sqrt{\frac{9 \cdot 25}{49}} = \frac{3 \cdot 5}{7} = \frac{15}{7} = 2\frac{1}{7}$

29. $9^2 + 12^2 = y^2$; $y^2 = 225$; $y = 15$; $x^2 = 25^2 - 15^2 = 400$; $x = 20$

30. $a^2 = 17^2 - 8^2 = 225$; $a = 15$; $b^2 = 10^2 - 8^2 = 36$; $b = 6$; $x^2 = 15^2 + 6^2 = 261$;
$x = \sqrt{261} = \sqrt{9 \cdot 29} = 3\sqrt{29}$

31. **a.** $x^2 = 6^2 + 8^2 = 100$; $x = 10$; median: 5 **b.** $\frac{x}{6} = \frac{6}{10}$; $x = 3.6$; $6^2 - (3.6)^2 = h^2$;
$h^2 = 23.04$; $h = 4.8$

32. $w^2 + (w + 2)^2 = 10^2$; $w^2 + w^2 + 4w + 4 = 100$; $2(w^2 + 2w - 48) = 0$;
$2(w + 8)(w - 6) = 0$; $w = 6$; $w + 2 = 8$; $P = 2(6) + 2(8) = 28$ cm

33. $a^2 = 12^2 + 4^2 = 160$; $d^2 = 160 + 3^2 = 169$; $d = 13$

34. $a^2 = 5^2 + 5^2 = 50$; $d^2 = 50 + 2^2 = 54$; $d = \sqrt{54} = \sqrt{9 \cdot 6} = 3\sqrt{6}$

35. $a^2 = e^2 + e^2 = 2e^2$; $d^2 = 2e^2 + e^2 = 3e^2$; $d = e\sqrt{3}$

36. $a^2 = l^2 + w^2$; $d^2 = a^2 + h^2 = l^2 + w^2 + h^2$; $d = \sqrt{l^2 + w^2 + h^2}$

C 37. $h^2 = 20^2 - x^2 = 13^2 - (21 - x)^2$; $400 - x^2 = 169 - 441 + 42x - x^2$; $42x = 672$;
$x = 16$; $h^2 = 20^2 - 16^2 = 144$; $h = 12$

38. $h^2 = 25^2 - x^2 = 30^2 - (x + 11)^2$; $625 = 900 - 22x - 121$; $22x = 154$; $x = 7$;
$h^2 = 25^2 - 7^2 = 576$; $h = 24$

39. Let $VE = t$, $OE = x$. $x^2 = 20^2 - t^2 = 15^2 - (25 - t)^2$; $400 = 225 - 625 + 50t$;
$t = 16$; $x^2 = 20^2 - 16^2 = 144$; $x = 12$

Page 294 • MIXED REVIEW EXERCISES

1. AC **2.** $>$; A **3.** $>$ **4.** $\overline{AB}$ **5.** B; C **6.** AB **7.** BX; CX

Page 294 • CHALLENGE

Yes; answers may vary. Example:

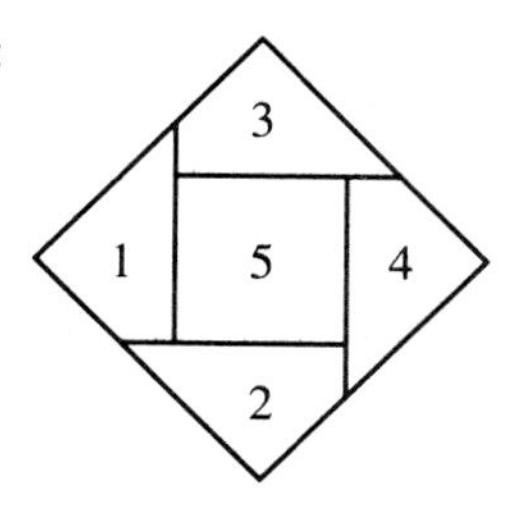

Page 296 • CLASSROOM EXERCISES

1. $10^2 = 100$; $6^2 + 8^2 = 100$; right **2.** $8^2 = 64$; $4^2 + 6^2 = 52$; obtuse

3. $1 + 4 < 6$; not possible **4.** $12^2 = 144$; $8^2 + 10^2 = 164$; acute

5. $(\sqrt{14})^2 = 14$; $(\sqrt{7})^2 + (\sqrt{7})^2 = 14$; right **6.** $8^2 = 64$; $4^2 + (4\sqrt{3})^2 = 64$; right

7. **a.** $x^2 = 6^2 + 8^2 = 100; x = 10$ **b.** $6 + x > 8; x > 2; x^2 < 6^2 + 8^2 = 100;$ $x < 10; 2 < x < 10$ **c.** $x < 8 + 6 = 14; x^2 > 8^2 + 6^2 = 100; x > 10; 10 < x < 14$ **d.** $x = 8$ or $x = 6$ **e.** $x \geq 8 + 6 = 14; x + 6 \leq 8; x \leq 2; x \leq 2$ or $x \geq 14$
8. Since $(5 + 5)^2 = 8^2 + 6^2$, $\triangle ABC$ is a rt. $\triangle$. The midpt. of the hyp. of a rt. $\triangle$ is equidistant from the vertices.
9. $n = \sqrt{3^2 - 2^2} = \sqrt{5}; (\sqrt{5})^2 = (\sqrt{3})^2 + (\sqrt{2})^2$
10. $\frac{9}{6} = \frac{15}{10}$ and $\angle ROS \cong \angle POQ$, so $\triangle ROS \sim \triangle POQ$ (SAS ~ Thm.). Since $15^2 = 9^2 + 12^2$, $\triangle SRO$ is a rt. $\triangle$ and $\angle R$ is a rt. $\angle$. Then $\angle P$ is a rt. $\angle$. (Corr. $\angle$s of ~ $\triangle$s are $\cong$.)

Pages 297–298 • WRITTEN EXERCISES

A

1. $15^2 = 225; 11^2 + 11^2 = 121 + 121 = 242$; acute
2. $13^2 = 169; 9^2 + 9^2 = 81 + 81 = 162$; obtuse
3. $16^2 = 256; 8^2 + (8\sqrt{3})^2 = 64 + 192 = 256$; right
4. $(6\sqrt{2})^2 = 72; 6^2 + 6^2 = 36 + 36 = 72$; right
5. $17^2 = 289; 8^2 + 14^2 = 64 + 196 = 260$; obtuse
6. $1^2 = 1; (0.6)^2 + (0.8)^2 = 0.36 + 0.64 = 1$; right
7. **a.** $(1.3)^2 = 1.69; (0.5)^2 + (1.2)^2 = 0.25 + 1.44 = 1.69$; right **b.** $(13n)^2 = 169n^2;$ $(5n)^2 + (12n)^2 = 25n^2 + 144n^2 = 169n^2$; right
8. **a.** $55^2 = 3025; 33^2 + 44^2 = 1089 + 1936 = 3025$; right **b.** $(5n)^2 = 25n^2;$ $(3n)^2 + (4n)^2 = 9n^2 + 16n^2 = 25n^2$; right
9. $(ST)^2 = 13^2 - 12^2 = 169 - 144 = 25; 25 = 3^2 + 4^2; (ST)^2 = (RS)^2 + (RT)^2$; by the converse of the Pythagorean Thm., $\triangle RST$ is a rt. $\triangle$.
10. $(BC)^2 = 10^2 - 8^2 = 100 - 64 = 36; 11^2 = 121 > 85 = 36 + 7^2;$ $(BD)^2 > (BC)^2 + (CD)^2$; therefore, $\triangle BCD$ is obtuse.

B

11. $(AD)^2 = 13^2 - 12^2 = 25; (DB)^2 = 15^2 - 12^2 = 81; AB = AD + DB =$ $\sqrt{25} + \sqrt{81} = 5 + 9 = 14; 15^2 < 14^2 + 13^2$; acute
12. $(AD)^2 = 10^2 - 8^2 = 36; (DB)^2 = 17^2 - 8^2 = 225; AB = \sqrt{36} + \sqrt{225} =$ $6 + 15 = 21; 21^2 > 10^2 + 17^2$; obtuse
13. $(AD)^2 = 13^2 - 3^2 = 160$, so $AD = \sqrt{160} = 4\sqrt{10}; (DB)^2 = (\sqrt{34})^2 - 3^2 = 25$, so $DB = \sqrt{25} = 5; AB = 4\sqrt{10} + 5; (AB)^2 = (4\sqrt{10} + 5)^2 = 160 + 40\sqrt{10} + 25 =$ $185 + 40\sqrt{10} \approx 311.5; (AC)^2 + (BC)^2 = 13^2 + (\sqrt{34})^2 = 203$; obtuse
14. $(AC)^2 = 2^2 + 4^2 = 20; (BC)^2 = 4^2 + 8^2 = 80; (AB)^2 = (2 + 8)^2 = 10^2 = 100;$ $(AC)^2 + (BC)^2 = 20 + 80 = 100$; right

15. $x^2 + (x + 4)^2 > 20^2$; $x^2 + x^2 + 8x + 16 > 400$; $2x^2 + 8x - 384 > 0$; $x^2 + 4x - 192 > 0$; $(x + 16)(x - 12) > 0$. Either $[(x + 16) > 0$ and $(x - 12) > 0]$ or $[(x + 16) < 0$ and $(x - 12) < 0]$. Then $x > 12$ or $x < -16$ (reject), and since 20 is the longest side of the $\triangle$, $12 < x \leq 16$.

16. $EFGH$ is a ▱. Since the diags. bis. each other, $EX = 12$ and $FX = 5$. $5^2 + 12^2 = 13^2$, so $\angle EXF$ is a rt. $\angle$, $\overline{EG} \perp \overline{FH}$, and $EFGH$ is a rhombus.

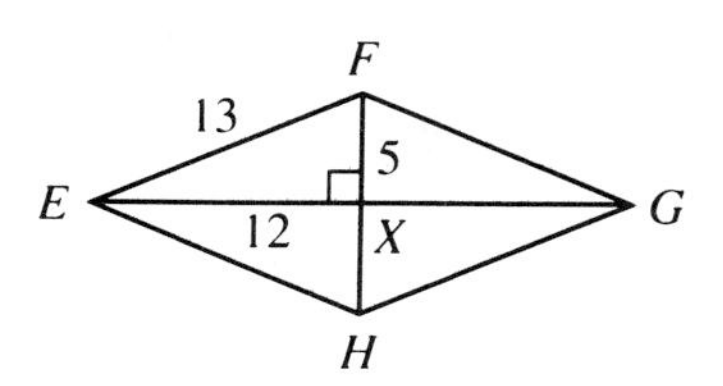

17. $\overline{RM}$. The diags. of a ▱ bis. each other, so $RT = 22$ and $(RT)^2 = 484$. $(RS)^2 + (ST)^2 = 9^2 + 20^2 = 481$. $484 > 481$, so $\angle RST$ is obtuse and its supplement, $\angle SRU$, must be acute. By the SAS Ineq. Thm. applied to $\triangle RST$ and $\triangle SRU$, $RT > SU$ and $RM > SM$.

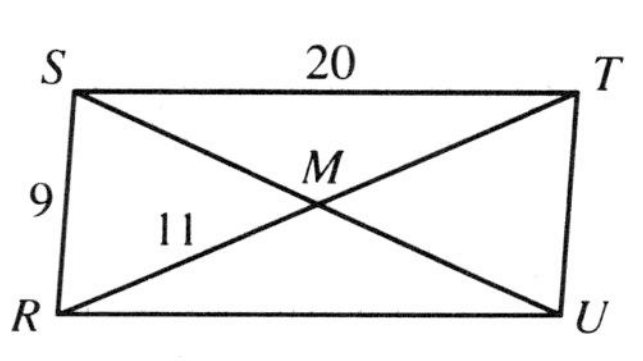

18. $(2xy)^2 + (x^2 - y^2)^2 = 4x^2y^2 + x^4 - 2x^2y^2 + y^4 = x^4 + 2x^2y^2 + y^4$; $(x^2 + y^2)^2 = x^4 + 2x^2y^2 + y^4$; by the converse of the Pythagorean Thm., the $\triangle$ is a rt. $\triangle$.

19. a. If the square of the longest side of a $\triangle$ is greater than the sum of the squares of the other 2 sides, then the $\triangle$ is an obtuse $\triangle$.

b.

Statements	Reasons
1. $n^2 = j^2 + k^2$	1. Pythagorean Thm.
2. $l^2 > j^2 + k^2$	2. Given
3. $l^2 > n^2$ and $l > n$	3. Substitution Prop.
4. $m\angle S > m\angle V = 90$	4. SSS Ineq. Thm.
5. $\triangle RST$ is an obtuse $\triangle$.	5. Def. of obtuse $\triangle$

20. a. If the square of the longest side of a $\triangle$ is less than the sum of the squares of the other 2 sides, then the $\triangle$ is an acute $\triangle$.

b.

Statements	Reasons
1. $n^2 = j^2 + k^2$	1. Pythagorean Thm.
2. $l^2 < j^2 + k^2$	2. Given
3. $l^2 < n^2$ and $l < n$	3. Substitution Prop.
4. $m\angle S < m\angle V = 90$	4. SSS Ineq. Thm.
5. $\triangle RST$ is an acute $\triangle$.	5. Def. of acute $\triangle$

C 21.

Statements	Reasons
1. $\overline{CN} \perp \overline{AB}$	1. Given
2. $\angle CNA$ and $\angle CNB$ are rt. ∠s.	2. Def. of ⊥ lines
3. $\triangle CNA$ and $\triangle CNB$ are rt. △s.	3. Def. of rt. △
4. h is the geom. mean btwn. d and e.	4. Given
5. $\frac{d}{h} = \frac{h}{e}$	5. Def. of geom. mean
6. $de = h^2$	6. Means-extremes Prop.
7. $(AC)^2 = d^2 + h^2$; $(BC)^2 = h^2 + e^2$	7. Pythagorean Thm.
8. $(AC)^2 + (BC)^2 = d^2 + de + de + e^2 = d^2 + 2de + e^2 = (d + e)^2 = (AB)^2$	8. Substitution Prop. and algebra
9. $\triangle ABC$ is a rt. △.	9. If the square of one side of a △ = the sum of the squares of the other 2 sides, then the △ is a rt. △.

22. $s = 9, j = 6$; $9 \geq 2 \cdot 6 - 3$ so the scissors truss is stable. (For the frame, $s = 12, j = 7$; $2j - 3 = 11$, so $s \geq 2j - 3$ and the frame is also stable.) Since $5^2 + 12^2 = 13^2$, $\triangle GBC$ and $\triangle GFE$ are rt. △s. Then $\triangle CFE$ is a rt. △ and $CE = \sqrt{18^2 + 12^2} = \sqrt{468} = 6\sqrt{13}$. $\triangle ABG$ and $\triangle AFG$ are also rt. △s and $\triangle ABG \cong \triangle AFG$ (HL), so $\angle CAD \cong \angle EAD$. Then $\triangle CAD \cong \triangle EAD$ (SAS) and thus $x = \frac{1}{2} \cdot CE = 3\sqrt{13}$. Also, $\triangle AFC$ is a rt. △ and $(y + 12)^2 = y^2 + 18^2$; $y^2 + 24y + 144 = y^2 + 324$; $24y = 180$; $y = 7.5$.

Page 298 • EXPLORATIONS

Check students' drawings.
Parallelogram, rectangle, rhombus, square

Page 299 • COMPUTER KEY-IN

1. (3, 4, 5), (6, 8, 10), (5, 12, 13), (8, 15, 17), (12, 16, 20), (7, 24, 25), (10, 24, 26), (20, 21, 29), (16, 30, 34), (9, 40, 41), (12, 35, 37), (24, 32, 40), (27, 36, 45), (20, 48, 52), (11, 60, 61), (14, 48, 50), (28, 45, 53), (40, 42, 58), (33, 56, 65), (24, 70, 74), (13, 84, 85) Primitive: (3, 4, 5), (5, 12, 13), (8, 15, 17), (7, 24, 25), (20, 21, 29), (9, 40, 41), (12, 35, 37), (11, 60, 61), (28, 45, 53), (33, 56, 65), (13, 84, 85)

2. $(2xy)^2 + (x^2 - y^2)^2 = 4x^2y^2 + x^4 - 2x^2y^2 + y^4 = x^4 + 2x^2y^2 + y^4 = (x^2 + y^2)^2$

3. (3, 4, 5), (5, 12, 13), (7, 24, 25), (9, 40, 41), (11, 60, 61), (33, 56, 65), (13, 84, 85) If the first number is an odd prime, the other 2 numbers differ by 1. **a.** $33 = 2n + 1$; $n = 16$; $2n^2 + 2n = 544$; $(2n^2 + 2n) + 1 = 545$; (33, 544, 545) **b.** $(2n + 1)^2 + (2n^2 + 2n)^2 = 4n^2 + 4n + 1 + 4n^4 + 8n^3 + 4n^2 = 4n^4 + 8n^3 + 8n^2 + 4n + 1$; also $[(2n^2 + 2n) + 1]^2 = 4n^4 + 8n^3 + 4n^2 + 4n^2 + 4n + 1 = 4n^4 + 8n^3 + 8n^2 + 4n + 1$; so $(2n + 1)^2 + (2n^2 + 2n)^2 = (2n^2 + 2n + 1)^2$.

Page 301 • CLASSROOM EXERCISES

1. $6\sqrt{2}$ **2.** 10 **3.** $x\sqrt{2} = 12$; $x = 6\sqrt{2}$ **4.** 10 **5.** 6.5 **6.** $3\sqrt{3}$
7. $8 = x\sqrt{2}$; $x = 4\sqrt{2}$ **8.** $10\sqrt{3}$ **9.** $x\sqrt{2} = 18$; $x = 9\sqrt{2}$
10. $\triangle ABC$ is isos. with $m\angle B = 120$; so $m\angle BAC = m\angle BCA = 30$; then $m\angle ACD = 90$, $m\angle CAD = 30$, $m\angle CDA = 60$. $CD = 8$, so $AD = 16$ and $AC = 8\sqrt{3}$.
11. $PQ = 2a$; $PS = a\sqrt{3}$; $QR = a\sqrt{2}$
12. No. In a 30°–60°–90° △, the sides are in the ratio $1 : \sqrt{3} : 2$, while the ∠s are in the ratio 1 : 2 : 3.

Pages 302–303 • WRITTEN EXERCISES

A

	1.	2.	3.	4.	5.	6.	7.	8.
a	4	$\frac{2}{3}$	$\sqrt{5}$	3	$3\sqrt{2}$	$\sqrt{7}$	$4\sqrt{2}$	$\frac{5}{2}\sqrt{2}$
b	4	$\frac{2}{3}$	$\sqrt{5}$	3	$3\sqrt{2}$	$\sqrt{7}$	$4\sqrt{2}$	$\frac{5}{2}\sqrt{2}$
c	$4\sqrt{2}$	$\frac{2}{3}\sqrt{2}$	$\sqrt{10}$	$3\sqrt{2}$	6	$\sqrt{14}$	8	5

	9.	10.	11.	12.	13.	14.	15.	16.
d	7	$\frac{1}{4}$	5	$2\sqrt{3}$	5	$\frac{13}{2}$	$\sqrt{3}$	$3\sqrt{3}$
e	$7\sqrt{3}$	$\frac{1}{4}\sqrt{3}$	$5\sqrt{3}$	6	$5\sqrt{3}$	$\frac{13}{2}\sqrt{3}$	3	9
f	14	$\frac{1}{2}$	10	$4\sqrt{3}$	10	13	$2\sqrt{3}$	$6\sqrt{3}$

17. $s = 12$; $d = 12\sqrt{2}$ **18.** $s\sqrt{2} = 8$; $s = 4\sqrt{2}$; $p = 16\sqrt{2}$
19. $\frac{s\sqrt{3}}{2} = 6\sqrt{3}$; $\frac{s}{2} = 6$; $s = 12$; $p = 36$ **20.** $\frac{s}{2} = 5$; $h = 5\sqrt{3}$

B **21.** $x = 4$; $y = \frac{4\sqrt{3}}{3}$ **22.** $x = 3$; $x + y = 12$; $y = 9$

23. $x = 6\sqrt{2}$; $y = 6\sqrt{2} \cdot \sqrt{2} = 12$

24. right side of quad. $= 6$; diag. $= 6\sqrt{3}$; $x = 3\sqrt{3}$; $y = 3\sqrt{3} \cdot \sqrt{3} = 9$

25. alt. $= a$; $8 = a\sqrt{2}$; $a = 4\sqrt{2}$; $x = 8\sqrt{2}$; $y = 4\sqrt{2} \cdot \sqrt{3} = 4\sqrt{6}$

26. $x = 6 + 4 = 10$; $y = 4\sqrt{2}$

27. $OB = 1\sqrt{2} = \sqrt{2}$; $OC = \sqrt{2} \cdot \sqrt{2} = 2$; $OD = 2\sqrt{2}$; $OE = 2\sqrt{2} \cdot \sqrt{2} = 4$

28. length $= 4\sqrt{3}$, width $= 4$

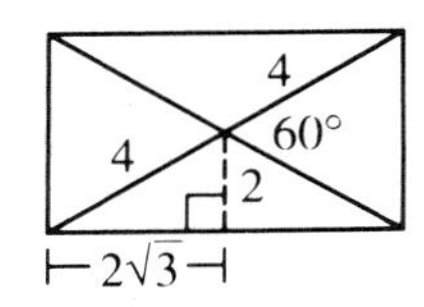

29. 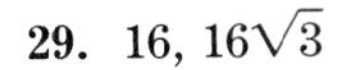 16, $16\sqrt{3}$

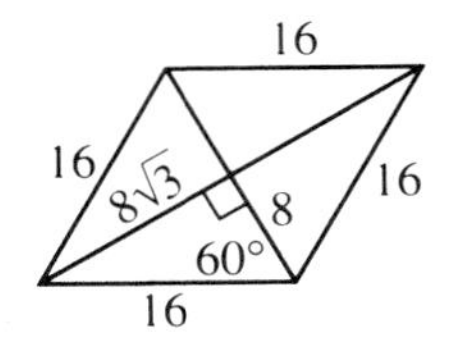

30.

Statements	Reasons
1. The measures of the $\angle$s of the $\triangle$ are 45, 45, and 90.	1. Given
2. The $\triangle$ is a rt. $\triangle$.	2. Def. of rt. $\triangle$
3. The legs have the same length, say a.	3. If 2 $\angle$s of a $\triangle$ are $\cong$, then the sides opp. those $\angle$s are $\cong$.
4. Let hyp. $= c$. Then $c^2 = a^2 + a^2$.	4. Pythagorean Thm.
5. $c = \sqrt{2a^2} = \sqrt{2} \cdot a$	5. Algebra
6. hypotenuse $= \sqrt{2} \cdot$ leg	6. Substitution Prop.

31. Consider a 30°–60°–90° $\triangle ABC$ with $AB = 1$, $BC = \sqrt{3}$, and $AC = 2$. Let $\triangle DEF$ have sides in the ratio $1 : \sqrt{3} : 2$ with $DE = x$, $EF = x\sqrt{3}$, $DF = 2x$. Then $\frac{DE}{AB} = \frac{EF}{BC} = \frac{DF}{AC}$ so $\triangle ABC \sim \triangle DEF$ (SSS $\sim$). Since corr. $\angle$s of $\sim$ $\triangle$s are $\cong$, $\triangle DEF$ is a 30°–60°–90° $\triangle$.

32. $GF = 22$; $DF = 22\sqrt{2}$; $EF = 11\sqrt{2}$; $DE = 11\sqrt{6}$

33. $GI = GH = 6$; $JG = 6\sqrt{3}$; $HI = 6\sqrt{2}$; $JH = 12$

34. $8 = MN \cdot \sqrt{2}$; $MN = 4\sqrt{2} = NL$; $KL = 8\sqrt{2}$; $KN = 4\sqrt{2} \cdot \sqrt{3} = 4\sqrt{6}$

C 35. **a.** $QS = 8\sqrt{2}$; $SV = 4\sqrt{2} = TV$; $RV = 4\sqrt{2} \cdot \sqrt{3} = 4\sqrt{6}$;
$TR = TV + VR = 4\sqrt{2} + 4\sqrt{6}$

b. $RT : QS = 4(\sqrt{2} + \sqrt{6}) : 8\sqrt{2} = (1 + \sqrt{3}) : 2$

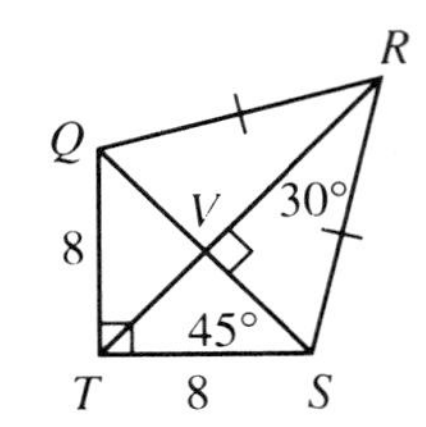

36. $8 = BD\sqrt{2}$; $BD = 4\sqrt{2} = AB$; $AC = 8\sqrt{2}$; $BC = 4\sqrt{2} \cdot \sqrt{3} = 4\sqrt{6}$;
$p = 8 + 4\sqrt{2} + 4\sqrt{6} + 8\sqrt{2} = 8 + 12\sqrt{2} + 4\sqrt{6}$

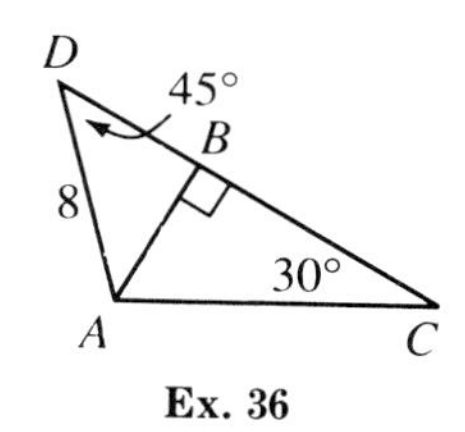

Ex. 36

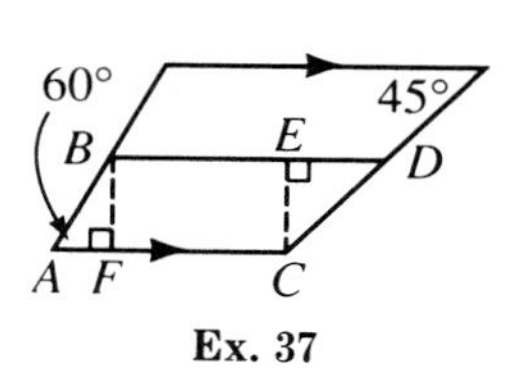

Ex. 37

37. $AB = \frac{1}{2}j$; $AF = \frac{1}{4}j$; $FC = \frac{3}{4}j = BE$; $BF = EC = \frac{1}{4}j \cdot \sqrt{3}$; $m\angle EDC = 45$; $ED = \frac{1}{4}j\sqrt{3}$; median $BD = BE + ED = \frac{3}{4}j + \frac{1}{4}j\sqrt{3} = \frac{3j + j\sqrt{3}}{4}$

38. $BD = \frac{1}{2}$; $AD = \frac{1}{2}\sqrt{3}$; $AC = 2 \cdot AD = \sqrt{3}$; $x = \sqrt{3}$ cm

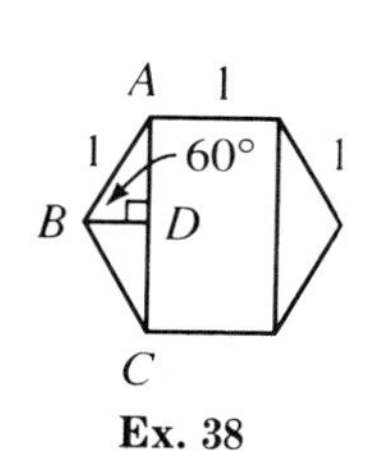

Ex. 38

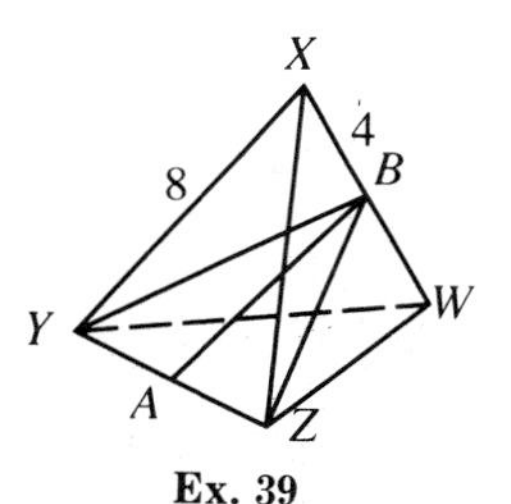

Ex. 39

39. Draw $\overline{BY}$ and $\overline{BZ}$. $\triangle XBY$ and $\triangle XBZ$ are 30°–60°–90° △; $BY = BZ = 4\sqrt{3}$. $\triangle BYZ$ is an isos. △ with $\overline{BA}$ the median to the base. $\triangle BAY$ is a rt. △ and $AB = \sqrt{(4\sqrt{3})^2 - 4^2} = \sqrt{32} = 4\sqrt{2}$.

Page 304 • SELF-TEST 1

1. $\frac{3}{x} = \frac{x}{15}$; $x^2 = 45$; $x = \sqrt{45} = \sqrt{9 \cdot 5} = 3\sqrt{5}$

2. **a.** $\frac{2}{x} = \frac{x}{8}$; $x^2 = 16$; $x = 4$ **b.** $\frac{2}{y} = \frac{y}{10}$; $y^2 = 20$; $y = \sqrt{20} = \sqrt{4 \cdot 5} = 2\sqrt{5}$

c. $\frac{8}{z} = \frac{z}{10}$; $z^2 = 80$; $z = \sqrt{80} = \sqrt{16 \cdot 5} = 4\sqrt{5}$

3. **a.** $61^2 = 3721$; $11^2 + 60^2 = 121 + 3600 = 3721$; right **b.** $11^2 = 121$; $7^2 + 9^2 = 49 + 81 = 130$; acute **c.** $(0.4)^2 = 0.16$; $(0.2)^2 + (0.3)^2 = 0.04 + 0.09 = 0.13$; obtuse

4. $d^2 = 8^2 + 4^2 = 64 + 16 = 80$; $d = \sqrt{80} = \sqrt{16 \cdot 5} = 4\sqrt{5}$

5. $10 = s\sqrt{2}$; $s = 5\sqrt{2}$; $p = 20\sqrt{2}$ cm 6. $h = 6\sqrt{3}$ cm

7. $h^2 = 13^2 - 5^2 = 169 - 25 = 144$; $h = 12$

Pages 306–307 • CLASSROOM EXERCISES

1. $\frac{3}{8}$ 2. $\frac{13}{19}$ 3. $\frac{15}{8}$ 4. $\frac{8}{3}$ 5. $\frac{19}{13}$ 6. $\frac{8}{15}$

7. a. 0.4452 b. 0.8693 c. 28.6363 d. 68° e. 17° f. 39°

8. a. $\frac{4}{4} = \frac{17}{17} = \frac{l}{l} = 1$ b. Yes

9. a. $\frac{3\sqrt{3}}{3} = \frac{5\sqrt{3}}{5} = \frac{t\sqrt{3}}{t} = \sqrt{3}$ b. 1.7321 c. No; yes

10. You can draw a rt. △ with legs 1,000,000 and 1; the tangent of one of its acute ∠s is $\frac{1{,}000{,}000}{1} = 1{,}000{,}000$. There is no upper limit to tangent values.

11. b, c

Pages 308–310 • WRITTEN EXERCISES

A 1. $\tan 32° = \frac{x}{22}$; $x = 22 \tan 32° \approx 13.7$ 2. $\tan 63° = \frac{x}{12}$; $x = 12 \tan 63° \approx 23.6$

3. $\tan 44° = \frac{x}{50}$; $x = 50 \tan 44° \approx 48.3$ 4. $\tan 50° = \frac{x}{1.2}$; $x = 1.2 \tan 50° \approx 1.4$

5. $\tan 61° = \frac{100}{x}$; $x = \frac{100}{\tan 61°} \approx 55.4$ 6. $\tan 25° = \frac{x}{7.1}$; $x = 7.1 \tan 25° \approx 3.3$

7. $\tan y° = \frac{6.1}{4} = 1.525$; $y° \approx 57°$ 8. $\tan y° = \frac{8}{15} = 0.5\bar{3}$; $y° \approx 28°$

9. $\tan y° = \frac{n}{2n} = 0.5$; $y° \approx 27°$ 10. $\tan y° = \frac{3n}{4n} = 0.75$; $y° \approx 37°$

11. $3^2 + x^2 = 34$; $x = \sqrt{34 - 9} = 5$; $\tan y° = \frac{3}{5} = 0.6$; $y° \approx 31°$

12. $x^2 + 2^2 = 13$; $x = \sqrt{13 - 4} = 3$; $\tan y° = \frac{3}{2} = 1.5$; $y° \approx 56°$

13–18. Answers may vary slightly.

B 13. $w = 60$; $\tan 42° = \frac{z}{60}$; $z = 60 \tan 42° \approx 54$

14. $\tan 35° = \frac{200}{w}$; $w = \frac{200}{\tan 35°} \approx 286$; $z = 2w \approx 571$

15. $w = 75$; $\tan 40° = \frac{75}{z}$; $z = \frac{75}{\tan 40°} \approx 89$

16. $w = 82$; $\tan 28° = \frac{82}{z}$; $z = \frac{82}{\tan 28°} \approx 154$

17. $w = 160$; $w + z = 160\sqrt{3}$; $160 + z = 160\sqrt{3}$; $z \approx 117$

18. $900 = w\sqrt{3}$; $w = \frac{900}{\sqrt{3}} = 300\sqrt{3} \approx 520$; $\tan 42° \approx \frac{900}{520 + z}$; $520 \tan 42° +$

$z \tan 42° \approx 900$; $z \approx \frac{900 - 520 \tan 42°}{\tan 42°} \approx 480$

19. $\tan x° = \frac{7}{100} = 0.07$; $x° \approx 4°$ **20.** $\tan 8° \approx 0.14$; grade $\approx 14\%$

21. $\tan x° = \frac{75}{35} \approx 2.1429$; $x° \approx 65°$

22. $\tan\left(\frac{x°}{2}\right) = \frac{2}{5} = 0.4$; $\frac{x°}{2} \approx 22°$; $x° \approx 44°$; $y° = 180° - x° \approx 180° - 44° = 136°$

23. $\tan 35° = \frac{61}{x}$; $x = \frac{61}{\tan 35°} \approx 87$; diag. ≈ 174 cm

24. $\tan\left(\frac{x°}{2}\right) = \frac{10}{40} = 0.25$; $\frac{x°}{2} \approx 14°$; $x° \approx 28°$

25. **a.** 0.7002; 0.4663; 1.1665 **b.** 60; 1.7321 **c.** No **d.** No; $\tan 35° - \tan 25° \approx 0.7002 - 0.4663 = 0.2339$, while $\tan(35° - 25°) = \tan 10° \approx 0.1763$.

26. **a.** 1. $\tan P = \frac{p}{q}$; $\tan Q = \frac{q}{p}$ (Def. of tan) 2. $\tan P \cdot \tan Q = \frac{p}{q} \cdot \frac{q}{p} = 1$

(Subst.; alg.) **b.** $58° = 90° - 32°$; $\tan 32° \cdot \tan 58° = 1$; $\frac{5}{8} \cdot \tan 58° = 1$;

$\tan 58° = \frac{8}{5}$

27. **a.** $BD = \sqrt{3^2 + 4^2} = 5$ **b.** $\tan \angle GBD = \frac{2}{5} = 0.4$; $\angle GBD \approx 22°$

C 28. Let each edge of the cube $= a$. Then $QS = \sqrt{a^2 + a^2} = a\sqrt{2}$;

$\tan \angle TQS = \frac{a}{a\sqrt{2}} = \frac{1}{\sqrt{2}} \approx 0.7071$; $\angle TQS \approx 35°$.

29. $\tan 37° = \frac{40}{WH}$; $WH = \frac{40}{\tan 37°} \approx 53$; $\tan 61° \approx \frac{TH}{53}$; $TH \approx 53 \tan 61° \approx 96$;

$TB = BH + TH \approx 40 + 96 = 136$ ft

30. $\tan 32° = \frac{GF}{355}$; $GF = 355 \tan 32° \approx 222$; $\tan 44° \approx \frac{222}{EF}$; $EF \approx \frac{222}{\tan 44°} \approx 230$

Page 310 • EXPLORATIONS

Check students' drawings and data. As $m\angle A$ increases, sin A increases, cos A decreases, and tan A increases. No.

1. 45 **2.** x **3.** x **4.** 1 **5.** sine and cosine **6.** tangent

Pages 313–314 • CLASSROOM EXERCISES

1. $\sin A = \frac{8}{17}$; $\cos A = \frac{15}{17}$; $\tan A = \frac{8}{15}$ **2.** $\sin A = \frac{7}{25}$; $\cos A = \frac{24}{25}$; $\tan A = \frac{7}{24}$

3. $\sin A = \frac{a}{c}$; $\cos A = \frac{b}{c}$; $\tan A = \frac{a}{b}$ **4.** $\sin B = \frac{15}{17}$; $\cos B = \frac{8}{17}$; $\tan B = \frac{15}{8}$

5. $\sin B = \frac{24}{25}$; $\cos B = \frac{7}{25}$; $\tan B = \frac{24}{7}$ **6.** $\sin B = \frac{b}{c}$; $\cos B = \frac{a}{c}$; $\tan B = \frac{b}{a}$

7. **a.** 0.4067 **b.** 0.5446 **c.** 0.9986 **d.** 15° **e.** 6° **f.** 81°

8. $\sin 41° = \frac{x}{100}$; $\cos 49° = \frac{x}{100}$ **9.** $\sin 35° = \frac{x}{8}$; $\cos 55° = \frac{x}{8}$

10. $\sin 50° = \frac{x}{12}$; $\cos 40° = \frac{x}{12}$ **11.** $\cos A = \frac{b}{c} = \sin B$

12. By the 30°–60°–90° △ Thm., $k = 2j$; $\sin 30° = \frac{j}{k} = \frac{j}{2j} = \frac{1}{2} = 0.5000$.

13. $x = \sqrt{13^2 - 5^2} = \sqrt{144} = 12$; $\cos n° = \frac{12}{13}$; $\tan n° = \frac{5}{12}$

14. Tan 1° is larger. In any rt. △, the hyp. is the longest side, so $\frac{\text{opp.}}{\text{hyp.}} < \frac{\text{opp.}}{\text{adj.}}$.

15. **a.** The hyp. is the longest side, so $\frac{\text{opp.}}{\text{hyp.}} < 1$. **b.** Yes

Pages 314–316 • WRITTEN EXERCISES

A **1.** $\sin 37° = \frac{x}{35}$; $x = 35 \sin 37° \approx 21$; $y \approx \sqrt{35^2 - 21^2} = 28$

2. $\sin 58° = \frac{x}{120}$; $x = 120 \sin 58° \approx 102$; $y \approx \sqrt{120^2 - 102^2} \approx 64$

3. $\tan 50° = \frac{x}{75}$; $x = 75 \tan 50° \approx 89$; $y \approx \sqrt{75^2 + 89^2} \approx 117$

4. $\sin 40° = \frac{x}{84}$; $x = 84 \sin 40° \approx 54$; $y \approx \sqrt{84^2 - 54^2} \approx 64$

5. $\sin 70° = \frac{x}{30}$; $x = 30 \sin 70° \approx 28$; $y \approx \sqrt{30^2 - 28^2} \approx 10$

6. $\cos 65° = \frac{12}{x}$; $x = \frac{12}{\cos 65°} \approx 28$; $y \approx \sqrt{28^2 - 12^2} \approx 26$

7. $\sin v° = \frac{22}{50} = 0.44$; $v° \approx 26°$ **8.** $\cos v° = \frac{11}{28} \approx 0.3929$; $v° \approx 67°$

9. $x = \sqrt{10^2 - (4.5)^2} \approx 9$; $\sin v° \approx \frac{9}{10} = 0.9$; $v° \approx 63°$

10. $x = \sqrt{12^2 - 8^2} = \sqrt{80} \approx 9$; $\cos v° \approx \frac{9}{12} = 0.75$; $v° \approx 42°$

11. $\cos v° = \frac{8}{10} = 0.8$; $v° \approx 37°$; $w° \approx 180 - 2(37) \approx 106°$

12. $x = 20$; $\sin 15° = \frac{\frac{1}{2}y}{20} = \frac{y}{40}$; $y = 40 \sin 15° \approx 10$

13. a. $x = \sqrt{14^2 - 9^2} = \sqrt{115}$ **b.** $\sin y° = \frac{9}{14} \approx 0.6429$; $y° \approx 40°$; $\cos 40° = \frac{x}{14}$; $x = 14 \cos 40° \approx 10.7$ **c.** Yes; $\sqrt{115} \approx 10.7$

B 14. $\sin 65° = \frac{75}{x}$; $x = \frac{75}{\sin 65°} \approx 83$ m **15.** $\cos 65° = \frac{AB}{352}$; $AB = 352 \cos 65° \approx 149$ m

16. $\sin 20° = \frac{x}{1}$; $x = \sin 20° \approx 0.342$ km ≈ 350 m

17. $\sin 70° = \frac{x}{6}$; $x = 6 \sin 70° \approx 5.6$ m; $\sin 60° = \frac{y}{6}$; $y = 6 \sin 60° \approx 5.2$ m; $x - y \approx 5.6 - 5.2 = 0.4$ m

18. a. $h = \sqrt{13^2 - 5^2} = \sqrt{144} = 12$ **b.** $\sin \angle B = \frac{12}{13} \approx 0.9231$; $m\angle B \approx 67$; $m\angle C \approx 67$; $m\angle A \approx 180 - 2(67) = 46$ **c.** $\cos (90° - 67°) = \cos 23° = \frac{h}{10}$; $h = 10 \cos 23° \approx 9$

19. a. $\cos 72° = \frac{5}{AB}$; $AB = \frac{5}{\cos 72°} \approx 16$; $AC \approx 16$ **b.** $\tan 72° = \frac{h}{5}$; $h = 5 \tan 72° \approx 15$

20. $m\angle P = 90 - 24 = 66$; $LM = PM = 6$; $LP = 12$; $\cos 66° = \frac{PA}{12}$; $PA = 12 \cos 66° \approx 5$ cm

21. In rect. $ABCD$, let M be the midpt. of $\overline{BC}$, and let the diags. intersect at X. Then $BX = 9$ and $\sin 17° = \frac{BM}{9}$; $BM = 9 \sin 17° \approx 2.6$, so $BC \approx 5$ cm. $DC \approx \sqrt{18^2 - 5^2} \approx \sqrt{299} \approx 17$ cm

22. $m\angle B = 144$; $\frac{m\angle B}{2} = 72$; $m\angle BAC = 90 - 72 = 18$; $\cos 18° = \frac{AC/2}{16} = \frac{AC}{32}$; $AC = 32 \cos 18° \approx 30$ cm

23. $m\angle XBC = 180 - 144 = 36$; $\cos 36° = \frac{10}{BX}$; $BX = \frac{10}{\cos 36°} \approx 12$ cm

C 24. Draw a $\perp$ from the third vertex to $\overline{AB}$; label it p. Then $\sin A = \frac{p}{b}$ and $\sin B = \frac{p}{a}$, so $\sin A \cdot b = p = \sin B \cdot a$. Therefore, $\frac{a}{\sin A} = \frac{b}{\sin B}$.

25. Choose a pt. S on one side of $\angle R$ and draw $\overline{ST} \perp$ to the other side of $\angle R$, with T on that side. Then $\triangle STR$ is a rt. $\triangle$ and $(ST)^2 + (TR)^2 = (RS)^2$. $\sin R = \frac{ST}{RS}$; $(\sin R)^2 = \frac{(ST)^2}{(RS)^2}$; $\cos R = \frac{TR}{RS}$; $(\cos R)^2 = \frac{(TR)^2}{(RS)^2}$. Thus $(\sin R)^2 + (\cos R)^2 = \frac{(ST)^2 + (TR)^2}{(RS)^2} = \frac{(RS)^2}{(RS)^2} = 1$

26. Because of folding, $\angle DCE \cong \angle ECD'$ and $m\angle DCD' = 2n$. $\cos n^\circ = \frac{DC}{k}$; $k = \frac{DC}{\cos n^\circ}$; since $\overline{AB} \parallel \overline{CD}$, $m\angle CD'B = m\angle DCD' = (2n)^\circ$; $\sin (2n)^\circ = \sin \angle CD'B = \frac{10}{D'C} = \frac{10}{DC}$; $DC = \frac{10}{\sin (2n^\circ)}$; substituting, $k = \frac{\frac{10}{\sin (2n^\circ)}}{\cos n^\circ} = \frac{10}{\sin (2n)^\circ \cos n^\circ}$

Page 316 • CHALLENGE

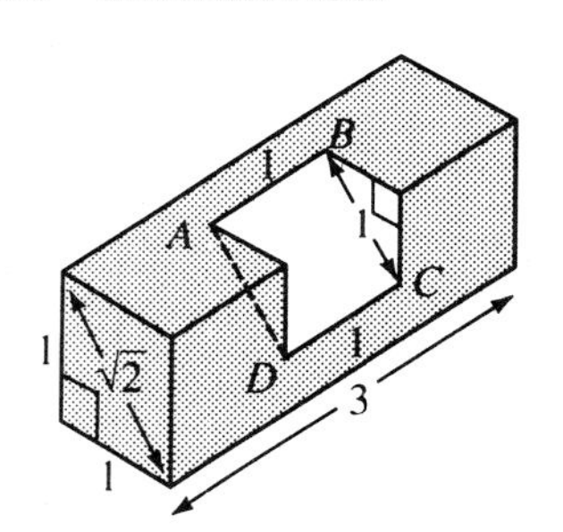

Insert the square end of one block through the other block at A, B, C, and D as shown at the left.

Pages 317–318 • CLASSROOM EXERCISES

1. a. 35° **b.** 35° **c.** 23° **d.** 23° **e.** greater than 35 **2.** $\angle 2$; $\angle$ of elev.
3. $\angle 3$; $\angle$ of depr. **4.** $\angle 5$; $\angle$ of depr. **5.** $\angle 8$; $\angle$ of elev.

6. a. $\tan x^\circ = \frac{15}{100} = 0.15$; $x^\circ \approx 9^\circ$ **b.** $\sin 9^\circ = \frac{y}{12}$; $y = 12 \sin 9^\circ \approx 1.9$ m

7. a. $\sin x^\circ = \frac{50}{130} \approx 0.3846$; $x^\circ \approx 23^\circ$ **b.** 130 m is an estimate of AB.

Pages 318–320 • WRITTEN EXERCISES

A **1.** $\tan 57^\circ = \frac{h}{21}$; $h = 21 \tan 57^\circ \approx 32$ m **2.** $\tan x^\circ = \frac{3}{4} = 0.75$; $x^\circ \approx 37^\circ$

3. $\sin 40^\circ = \frac{h}{80}$; $h = 80 \sin 40^\circ \approx 50$ m **4.** $\tan 8^\circ = \frac{125}{x}$; $x = \frac{125}{\tan 8^\circ} \approx 900$ m

5. $\tan 38° = \frac{h}{3}$; $h = 3 \tan 38° \approx 2.3$ km 6. $\tan 3° = \frac{30{,}000}{x}$; $x = \frac{30{,}000}{\tan 3°} \approx 570{,}000$ ft

B 7. $\tan 70° = \frac{180}{M}$; $M = \frac{180}{\tan 70°} \approx 65.5$; $\tan 35° = \frac{90}{H}$; $H = \frac{90}{\tan 35°} \approx 128.5$; Heidi; $H - M \approx 128.5 - 65.5 \approx 63$ cm

8. $\tan 13° = \frac{x}{40}$; $x = 40 \tan 13° \approx 9.2$; $185 - 9.2 \approx 176$ m

9. $\tan 48° = \frac{x}{400}$; $x = 400 \tan 48° \approx 440$ m

10. **a.** $\tan x° = \frac{10}{100} = 0.1$; $x° \approx 6°$ **b.** $\sin 6° = \frac{x}{2}$; $x = 2 \sin 6° \approx 0.2$ km $= 200$ m

11. $\sin x° = \frac{400}{1600} = 0.25$; $x° \approx 14°$

12. **a.** $f = 3000 \sin 3° \approx 160$ lb **b.** Answers will vary.

13. **a.** $\tan B = \frac{24}{40} = 0.6$; $B \approx 31°$; $\tan\left(\frac{A}{2}\right) = \frac{12}{40} = 0.3$; $\frac{A}{2} \approx 17°$; $A \approx 34°$; $\angle A$ is larger. **b.** A; a player at A has a wider $\angle$ over which to aim at the goal.

C 14. $\tan 19° = \frac{x}{d}$; $\tan 29° = \frac{x + 6.3}{d}$; $d = \frac{x}{\tan 19°} = \frac{x + 6.3}{\tan 29°}$; $x \approx 10.3$ m

Page 320 • SELF-TEST 2

1. $\frac{7}{24}$ 2. $\frac{24}{25}$ 3. $\frac{7}{25}$ 4. $\frac{24}{7}$ 5. 74 6. $\cos 42° = \frac{x}{100}$; $x = 100 \cos 42° \approx 74$

7. $\sin 51° = \frac{h}{70}$; $h = 70 \sin 51° \approx 54.4$; $x = 2h \approx 2(54.4) = 109$

8. The $\triangle$ is a rt. $\triangle$; $\tan 38° = \frac{88}{x}$; $x = \frac{88}{\tan 38°} \approx 113$

9. $\tan 24° = \frac{h}{100}$; $h = 100 \tan 24° \approx 45$ m

Pages 321–322 • APPLICATION

1. $90° - 47\frac{1}{2}° + 23\frac{1}{2}° = 66°$; $90° - 47\frac{1}{2}° - 23\frac{1}{2}° = 19°$

2. $90° - 42° + 23\frac{1}{2}° = 71\frac{1}{2}°$; $90° - 42° - 23\frac{1}{2}° = 24\frac{1}{2}°$

3. $90° - 30° + 23\frac{1}{2}° = 83\frac{1}{2}°$; $90° - 30° - 23\frac{1}{2}° = 36\frac{1}{2}°$

4. $90° - 34° + 23\frac{1}{2}° = 79\frac{1}{2}°$; $90° - 34° - 23\frac{1}{2}° = 32\frac{1}{2}°$

5. $90° - 64\frac{1}{2}° + 23\frac{1}{2}° = 49°$; $90° - 64\frac{1}{2}° - 23\frac{1}{2}° = 2°$

6. $90° - 26° + 23\frac{1}{2}° = 87\frac{1}{2}°$; $90° - 26° - 23\frac{1}{2}° = 40\frac{1}{2}°$

7. The sun is more than 90° from the southern horizon and is therefore less than 90° from the northern horizon.

8. The sun doesn't rise at all; it is never seen above the horizon.

9. $\tan 16° = \frac{x}{7}$; $x = 7 \tan 16° \approx 2$ ft

10. The $\angle$ of elevation of the sun at the winter solstice is 27°. $\tan 27° = \frac{y}{2}$; $y = 2 \tan 27° \approx 2(0.5095) \approx 1$ ft. Since the window is 1 ft below the overhang, all of the window is in the sun.

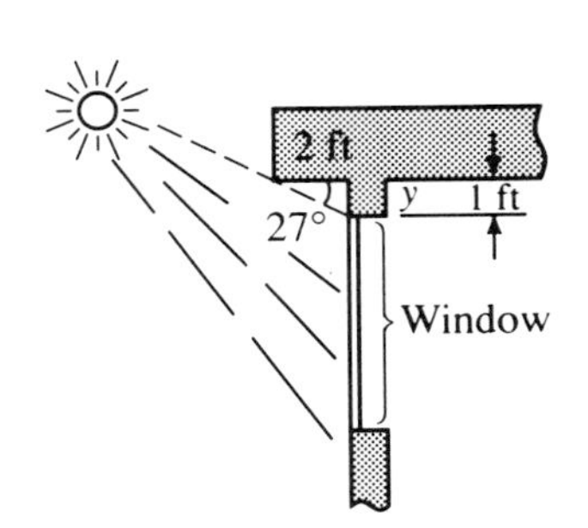

Pages 323–324 • CHAPTER REVIEW

1. $\frac{12}{x} = \frac{x}{3}$; $x^2 = 36$; $x = 6$ **2.** $\frac{10}{x} = \frac{x}{5}$; $x^2 = 50$; $x = 5\sqrt{2}$

3. $\frac{10}{y} = \frac{y}{15}$; $y^2 = 150$; $y = 5\sqrt{6}$ **4.** $\frac{5}{z} = \frac{z}{15}$; $z^2 = 75$; $z = 5\sqrt{3}$

5. $h = \sqrt{3^2 + 6^2} = \sqrt{45} = 3\sqrt{5}$ **6.** $d = \sqrt{10^2 + 8^2} = \sqrt{164} = 2\sqrt{41}$

7. $14 = s\sqrt{2}$; $s = 7\sqrt{2}$ **8.** $\frac{b}{2} = \sqrt{10^2 - 8^2} = \sqrt{36} = 6$; $b = 12$

9. $6^2 = 36$; $4^2 + 5^2 = 41$; acute **10.** $8 + 8 < 17$; not possible

11. $61^2 = 3721$; $11^2 + 60^2 = 3721$; right

12. $6^2 = 36$; $(2\sqrt{3})^2 + (3\sqrt{2})^2 = 12 + 18 = 30$; obtuse

13. $b = 5$; $x = 5\sqrt{3}$ **14.** $x = 7\sqrt{2}$ **15.** $\frac{x\sqrt{3}}{2} = 8\sqrt{3}$; $\frac{x}{2} = 8$; $x = 16$

16. $h = 12\sqrt{2}$; $a = 6\sqrt{2}$ **17. a.** $\frac{9}{6} = 1.5$ **b.** $\frac{6}{9} = \frac{2}{3}$ **c.** 34

18. a. 10 **b.** $\tan 32° = \frac{PN}{10}$; $PN = 10 \tan 32° \approx 6$

19. a. $\frac{24}{26} = \frac{12}{13}$ **b.** $\frac{24}{26} = \frac{12}{13}$ **c.** $\sin K = \frac{12}{13} \approx 0.9231$; $m\angle K \approx 67$

20. a. $\sin 40° = \frac{75}{WX}$; $WX = \frac{75}{\sin 40°} \approx 117$ **b.** $\tan 40° = \frac{75}{VX}$; $VX = \frac{75}{\tan 40°} \approx 89$

21. $\cos y° = \frac{13}{24} \approx 0.5417$; $y° \approx 57°$

22. $\sin 76° = \frac{h}{20}$; $h = 20 \sin 76° \approx 19.4$; $\tan 53° \approx \frac{x}{19.4}$; $x \approx 19.4 \tan 53° \approx 26$

23. $\sin\left(\frac{y°}{2}\right) = \frac{2}{10} = 0.2$; $\frac{y°}{2} \approx 11.5$; $y° \approx 23°$

24. $\sin 35° = \frac{h}{500}$; $h = 500 \sin 35° \approx 290$ ft

Pages 324–325 • CHAPTER TEST

1. $\frac{5}{x} = \frac{x}{20}$; $x^2 = 100$; $x = 10$ **2.** $\frac{6}{x} = \frac{x}{8}$; $x^2 = 48$; $x = 4\sqrt{3}$ **3.** DEN; NEF

4. DE; EF **5.** EF; DF **6.** $\frac{10}{ND} = \frac{ND}{25}$; $ND = \sqrt{250} = 5\sqrt{10}$

7. $x = \sqrt{2^2 + 3^2} = \sqrt{13}$; $y = \sqrt{(\sqrt{13})^2 - 1^2} = \sqrt{12} = 2\sqrt{3}$

8. $x = \sqrt{17^2 - 8^2} = 15$; $y = \sqrt{15^2 - 9^2} = 12$ **9.** $3 + 4 < 8$; not possible

10. $13^2 = 169$; $11^2 + 12^2 = 265$; acute **11.** $10^2 = 100$; $7^2 + 7^2 = 98$; obtuse

12. $1^2 = 1$; $\left(\frac{3}{5}\right)^2 + \left(\frac{4}{5}\right)^2 = \frac{25}{25} = 1$; right

13. $x = 11\sqrt{2}$ **14.** $x = 9$ **15.** $x = 14$ **16.** $\sqrt{6} = x\sqrt{2}$; $x = \sqrt{3}$

17. $x^2 + 7^2 = (x + 1)^2$; $x = 24$ **18.** $\frac{4}{2\sqrt{6}} = \frac{2\sqrt{6}}{x}$; $4x = 24$; $x = 6$

19. $\sin 41° = \frac{x}{32}$; $x = 32 \sin 41° \approx 21$ **20.** $\tan 42° = \frac{28}{x}$; $x = \frac{28}{\tan 42°} \approx 31$

21. $\cos 50° = \frac{x}{44}$; $x = 44 \cos 50° \approx 28$ **22.** $\tan x° = \frac{3}{4} = 0.75$; $x° \approx 37°$

23. $\cos x° = \frac{17}{20}$; $x° \approx 32°$ **24.** $h = 50$; $\tan 58° = \frac{50}{x}$; $x = \frac{50}{\tan 58°} \approx 31$

25. $\frac{d}{2} = \sqrt{4^2 - 2^2} = 2\sqrt{3}$; $d = 4\sqrt{3}$

26. $TS = \sqrt{17^2 - 8^2} = 15$; because $15^2 = 12^2 + 9^2$, $\triangle TVS$ is a rt. $\triangle$. The largest $\angle$ of a $\triangle$ is opp. the longest side, so $m\angle V = 90$.

27. $\tan 4° = \frac{18}{x}$; $x = \frac{18}{\tan 4°} \approx 260$ m

Page 326 • PREPARING FOR COLLEGE ENTRANCE EXAMS

1. A. $2x + 5x + 5x = 180$; $12x = 180$; $x = 15$; $m\angle B = 5x = 75$ **2.** C **3.** B

4. C. $\frac{10}{x + 10} = \frac{x + 3}{2x + 9}$; $20x + 90 = x^2 + 13x + 30$; $x^2 - 7x - 60 = 0$;
$(x + 5)(x - 12) = 0$; $x = 12$

5. E. $\frac{x}{14 - x} = \frac{6}{14}$; $14x = 84 - 6x$; $20x = 84$; $x = 4.2$

6. A. $\frac{2x}{m} = \frac{m}{2y}$; $m^2 = 4xy$; $m = \sqrt{4xy} = 2\sqrt{xy}$

7. C. $41^2 = 1681$; $8^2 + 40^2 = 64 + 1600 = 1664$; obtuse

8. A. Let the shorter diag. $= d$; then the longer diag. is $2\left(\frac{d\sqrt{3}}{2}\right) = d\sqrt{3}$.
Ratio $= d\sqrt{3} : d = \sqrt{3} : 1$

9. B. $\tan A = \frac{k}{j}$; $k = j \tan A$ 10. C. $\cos 65° = \frac{\frac{1}{2}b}{4}$; $b = 8 \cos 65° \approx 3.36 \approx 3.4$

Page 327 • CUMULATIVE REVIEW: CHAPTERS 1–8

A **1.** Seg. Add. Post. **2.** postulate **3.** corollary **4.** conclusion **5.** contrapositive

6. $x + (2x + 2) > 2x + 3$; $x > 1$; $(2x + 3)^2 > x^2 + (2x + 2)^2$; $4x^2 + 12x + 9 > x^2 + 4x^2 + 8x + 4$; $x^2 - 4x - 5 < 0$; $(x - 5)(x + 1) < 0$; $x - 5 < 0$; $x < 5$; $1 < x < 5$

7. $1 : \sqrt{2}$ **8.** $l^2 = 17^2 - 8^2$; $l = 15$; $\cos B = \frac{15}{17}$

9. **a.** If a △ is equiangular, then it is isos.

b. If a △ is isos., then it is equiangular.

10. 92, 141 **11.** $x^2 = 6x$; $x = 6$; $x^2 = 6x = 36$

B **12.** **a.** isos. **b.** $3x + 3x + 4x = 180$; $10x = 180$; $x = 18$; $3x = 54$; $4x = 72$; acute
c. $\overline{XY}$

13. $\frac{x - 5}{x - 2} = \frac{x}{x + 4}$; $(x - 5)(x + 4) = x(x - 2)$; $x^2 - x - 20 = x^2 - 2x$; $x = 20$

14. $s^2 = 9^2 + 12^2 = 81 + 144 = 225$; $s = 15$

15. Since $\overline{AX}$ is a median, $\overline{BX} \cong \overline{CX}$. Since $\overline{AX}$ is an altitude, $\angle AXB \cong \angle AXC$. Thus, $\triangle AXB \cong \triangle AXC$ by SAS, and corr. parts $\overline{AB}$ and $\overline{AC}$ are $\cong$. By def., $\triangle ABC$ is isos.

16. Given: $NPQRST$ is a reg. hexagon.
Prove: $NPRS$ is a rect.

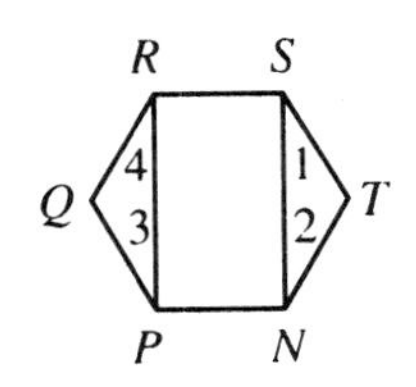

Statements	Reasons
1. $NPQRST$ is a reg. hexagon.	1. Given
2. $\overline{PQ} \cong \overline{QR} \cong \overline{ST} \cong \overline{TN}$	2. Def. of reg. polygon
3. $\angle 1 \cong \angle 2$; $\angle 3 \cong \angle 4$	3. Isos. △ Thm.
4. $m\angle T = m\angle Q = 120$	4. The sum of the meas. of the ∠s of a convex polygon with n sides is $(n - 2)180$; def. of hexagon
5. $m\angle 1 = m\angle 2 = m\angle 3 = m\angle 4 = 30$	5. The sum of the meas. of the ∠s of a △ is 180.
6. $m\angle NPR = m\angle PRS = m\angle RSN = m\angle SNP = 120 - 30 = 90$	6. ∠ Add. Post.; Subtr. Prop. of =
7. $NPRS$ is a rect.	7. Def. of rect.

17.

Statements	Reasons
1. $\angle WXY \cong \angle XZY$	1. Given
2. $\angle Y \cong \angle Y$	2. Refl. Prop.
3. $\triangle XYW \sim \triangle ZYX$	3. AA $\sim$ Post.
4. $\frac{XY}{ZY} = \frac{WY}{XY}$	4. Corr. sides of $\sim$ ⚠ are in prop.
5. $(XY)^2 = WY \cdot ZY$	5. Means-extremes Prop.

Page 330 • CLASSROOM EXERCISES

1. $\overline{OT}$, $\overline{OR}$, $\overline{OL}$ **2.** $\overline{RL}$ **3.** $\overline{RS}$ is a chord; $\overleftrightarrow{RS}$ is a secant. **4.** K is not on $\odot O$.
5. $\overleftrightarrow{LH}$ **6.** Point of tangency **7.** $\overleftrightarrow{EF}$ **8.** Secant: $\overleftrightarrow{AB}$; chord: $\overline{AB}$
9. $\overline{QA}$, $\overline{QB}$, $\overline{QC}$, $\overline{QF}$ **10.** 16; 10.4; $8\sqrt{3}$; $2j$ **11.** 7; 6.5; 2.8; $3n$

Pages 330–331 • WRITTEN EXERCISES

A **1, 2.** Check students' drawings. **1.** The midpts. lie on a diam. $\perp$ to the given chords.
2. $m\angle OTS = 90$
3. a. Check students' drawings. **b.** It is equidistant from the 3 vertices. **c.** At the midpt. of the hypotenuse. **d.** $d = \sqrt{6^2 + 8^2} = \sqrt{100} = 10; r = 5$
4. a. All radii of a sphere are $\cong$. **b.** The pts. in the intersection lie in a plane and are equidistant from Q; by def., the intersection is a $\odot$.
5. $AC = 15 + 7 = 22$, or $AC = 15 - 7 = 8$

6.
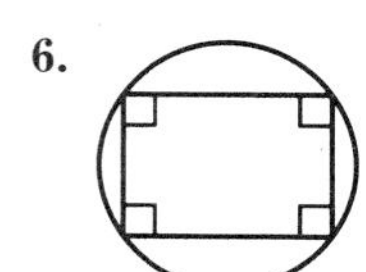

7.

8.
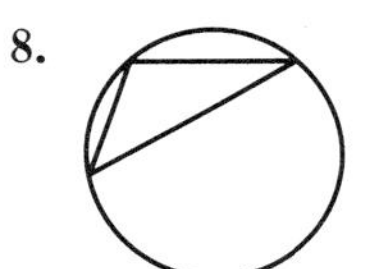

9.

10.

11.
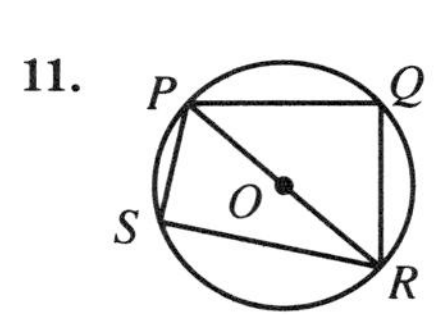

B **12.** $12\sqrt{2}$ **13.** 24 **14.** 12 **15.** $12\sqrt{3}$
16. the $\perp$ bis. of $\overline{AB}$. The centers are the set of all pts. equidistant from A and B, and all pts. equidistant from the endpts. of a seg. lie on its $\perp$ bis.
17. a. rhombus; $\odot Q \cong \odot R$, so $\overline{QC}$, $\overline{QD}$, $\overline{RC}$, and $\overline{RD}$ are $\cong$. **b.** The diags. of a rhombus are $\perp$ bisectors of each other. **c.** $\frac{CD}{2} = \sqrt{17^2 - 15^2} = 8; CD = 16$
18. Check students' drawings; $\frac{x}{2} = 3\sqrt{3}; x = 6\sqrt{3}$

C **19.** $x^2 + y^2 = 5^2; (6 - x)^2 + y^2 = 7^2; (6 - x)^2 + 5^2 - x^2 = 7^2; x = 1; y = 2\sqrt{6};$
$AB = 2y = 4\sqrt{6}$

20. Check students' drawings. **a.** At the center of the ⊙. **b.** The ⊥ bisectors intersect at a pt. equidistant from A, B, and C, by def. the center of a ⊙ on which A, B, and C lie.

Page 332 • EXTRA

1. 4 odd, 1 even; cannot be traced **2.** 0 odd, 6 even; can be traced
3. 2 odd, 6 even; can be traced **4. a.** even **b.** Yes; yes **c.** No
5. There are more than 2 odd vertices.

Page 335 • CLASSROOM EXERCISES

1. a. 2 **b.** 2 **c.** 2 **d.** 2 **e.** 1 **f.** 0
2. a. 2 **b.** 1 **c.** 0 **d.** 0 **e.** 0 **f.** 0 **3. a.** b **b.** e
4. By Thm. 9-1, $\angle A$ and $\angle B$ are rt. $\angle$s. Since $\overline{OA} \cong \overline{OB}$ and $\overline{OP} \cong \overline{OP}$, $\triangle OAP \cong \triangle OBP$ (HL Thm.). Then $\overline{PA} \cong \overline{PB}$.
5. $\cong$: $\angle A$ and $\angle B$, $\angle OPA$ and $\angle OPB$, $\angle AOP$ and $\angle BOP$; comp.: $\angle AOP$ and $\angle APO$, $\angle BOP$ and $\angle BPO$, $\angle AOP$ and $\angle BPO$, $\angle BOP$ and $\angle APO$; supp.: $\angle A$ and $\angle B$, $\angle APB$ and $\angle AOB$

Pages 335–337 • WRITTEN EXERCISES

A **1.** $\sqrt{10^2 - 6^2} = 8$ **2.** $\sqrt{10^2 + 6^2} = 2\sqrt{34}$ **3.** 12 **4.** $\sqrt{17^2 - 8^2} = 15$
5. 8.2 **6.** $RS = 4.7 + 7.3 = 12 = TU$
7. a. $\overline{AB} \cong \overline{CD}$

Statements	Reasons
1. Draw $\overleftrightarrow{AB}$ and $\overleftrightarrow{CD}$ int. at Z.	1. Through any 2 pts. there is exactly one line.
2. $ZA + AB = ZB$; $ZC + CD = ZD$	2. Seg. Add. Post.
3. $ZB = ZD$	3. Tangents to a ⊙ from a pt. are $\cong$.
4. $ZA + AB = ZC + CD$	4. Substitution Prop.
5. $ZA = ZC$	5. Tangents to a ⊙ from a pt. are $\cong$.
6. $AB = CD$ or $\overline{AB} \cong \overline{CD}$	6. Subtr. Prop. of =

b. Yes

8. a. $TR = TS$ so $m\angle TRS = m\angle TSR = \frac{180 - 36}{2} = 72$ **b.** $m\angle ORS = m\angle OSR = 90 - 72 = 18$ **c.** $m\angle ROS = 180 - 2(18) = 144$
d. Yes; $\angle RTS$ and $\angle ROS$ are supp.

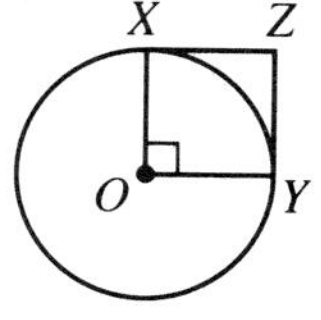
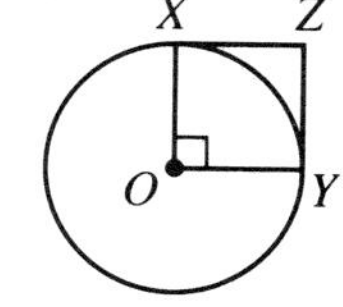

Ex. 9

9. **a.** $\overline{XZ} \perp \overline{OX}$, so $\overline{XZ} \parallel \overline{OY}$. Similarly, $\overline{ZY} \parallel \overline{OX}$, so $OXZY$ is a rect. Since $OX = OY$, 2 consec. sides of $OXZY$ are $\cong$, and $OXZY$ is a square. **b.** $5\sqrt{2}$

10. **a.** OS; SP **b.** OS; OP **c.** $\sqrt{6 \cdot 24} = 12$; $\sqrt{24 \cdot 30} = 12\sqrt{5}$

11. $\overline{AR} \perp \overleftrightarrow{RS}$ and $\overline{BS} \perp \overleftrightarrow{RS}$ (If a line is tan. to a ⊙, then the line is $\perp$ to the radius drawn to the pt. of tangency.), so $\overline{AR} \parallel \overline{BS}$. Then $\angle A \cong \angle B$ and $\triangle ARC \sim \triangle BSC$ (AA ~ Post.), so $\dfrac{AC}{BC} = \dfrac{RC}{SC}$ (Corr. sides of ~ ⊿s are in prop.).

B **12.** Two lines tan. to a ⊙ at the endpts. of a diam. are $\parallel$.

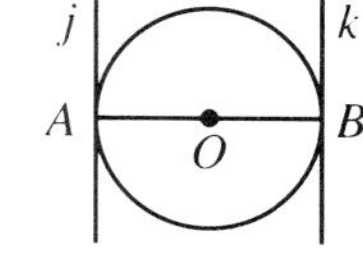

Given: $\overline{AB}$, a diam. of ⊙O; j tan. to ⊙O at A;
k tan. to ⊙O at B.

Prove: $j \parallel k$

Statements	Reasons
1. j is tan. to ⊙O at A; k is tan. to ⊙O at B.	1. Given
2. $j \perp \overline{AB}$; $k \perp \overline{AB}$	2. If a line is tan. to a ⊙, then it is $\perp$ to the radius drawn to the pt. of tangency.
3. $j \parallel k$	3. In a plane, 2 lines $\perp$ to the same line are $\parallel$.

13. Two planes tan. to a sphere at the endpts. of a diam. are $\parallel$.

14. $AB + DC = AD + BC$

Statements	Reasons
1. Let W, X, Y, and Z be the pts. of tangency of $\overline{AB}$, $\overline{BC}$, $\overline{CD}$, and $\overline{AD}$, resp.	1. Def. of a fig. circumscribed about a ⊙
2. $AW = AZ$; $BW = BX$; $CX = CY$; $DY = DZ$	2. Tangents to a ⊙ from a pt. are $\cong$.
3. $AW + BW + CY + DY = AZ + BX + CX + DZ$	3. Add. Prop. of =
4. $AW + BW = AB$; $BX + CX = BC$; $CY + DY = DC$; $AZ + DZ = AD$	4. Seg. Add. Post.
5. $AB + DC = AD + BC$	5. Substitution Prop.

15. $RA = RC$ and $SB = SC$ since tangents to a ⊙ from a pt. are ≅. Also, $RS = RC + CS$; so $PR + RS + SP = PR + RA + SB + SP = PA + PB$ by substitution.

16. $\triangle RTQ \sim \triangle RSP$; $QR = \sqrt{8^2 + 6^2} = 10$; $PQ = 30 - 10 = 20$; $\frac{PS}{6} = \frac{30}{10}$, $10 \cdot PS = 180$, $PS = 18$; $\frac{10}{20} = \frac{8}{ST}$, $10 \cdot ST = 160$, $ST = 16$

17. $JPQK$ is a trap. Let X be on $\overline{JP}$ such that $JX = KQ = 3$; $JP = 11$; $PX = 11 - 3 = 8$; $JK = XQ = \sqrt{17^2 - 8^2} = 15$

18. Let X be on $\overline{AP}$ such that $AX = BQ$; $AB = XQ = \sqrt{8^2 - 4^2} = 4\sqrt{3}$

19. **a.** G is the midpt. of $\overline{EF}$.

Statements	Reasons
1. $\overline{GE}$, $\overline{GH}$, and $\overline{GF}$ are tangents.	1. Given
2. $\overline{GE} \cong \overline{GH}$ or $GE = GH$; $\overline{GH} \cong \overline{GF}$ or $GH = GF$	2. Tangents to a ⊙ from a pt. are ≅.
3. $GE = GF$ or $\overline{GE} \cong \overline{GF}$	3. Trans. Prop.
4. G is the midpt. of $\overline{EF}$.	4. Def. of midpt.

b. $m\angle EHF = 90$

Statements	Reasons
1. $\overline{GE}$, $\overline{GH}$, and $\overline{GF}$ are tangents.	1. Given
2. $\overline{GE} \cong \overline{GH}$; $\overline{GH} \cong \overline{GF}$	2. Tangents to a ⊙ from a pt. are ≅.
3. $\angle E \cong \angle GHE$ or $m\angle E = m\angle GHE$; $\angle F \cong \angle GHF$ or $m\angle F = m\angle GHF$	3. Isos. △ Thm.
4. $m\angle E + m\angle GHE + m\angle F + m\angle GHF = 180$	4. The sum of the meas. of the ∠s of a △ is 180.
5. $2m\angle GHE + 2m\angle GHF = 180$, or $m\angle GHE + m\angle GHF = 90$	5. Substitution Prop.
6. $m\angle GHE + m\angle GHF = m\angle EHF$	6. ∠ Add. Post.
7. $m\angle EHF = 90$	7. Substitution. Prop.

20. 8 2 + 3 + 3 = 8

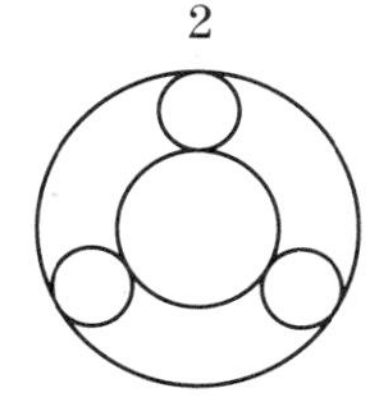

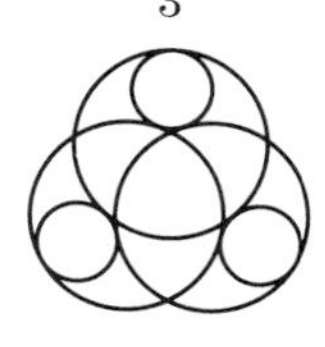

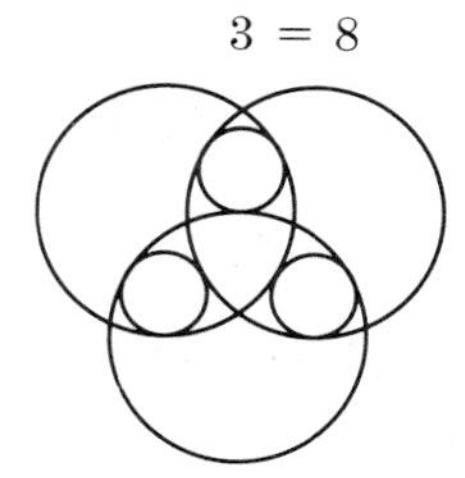

C 21. a. 8. Let the 3 spheres be A, B, and C. There are 2 $\parallel$ tan. planes, X and X', with spheres A, B, and C between them, as shown. There are also 2 tan. planes, Y and Y', between sphere B and spheres A and C, as shown. Similarly, A can be separated from B and C by 2 tan. planes, and C can be separated from A and B by 2 tan. planes. The total is 8 planes tan. to the 3 spheres.

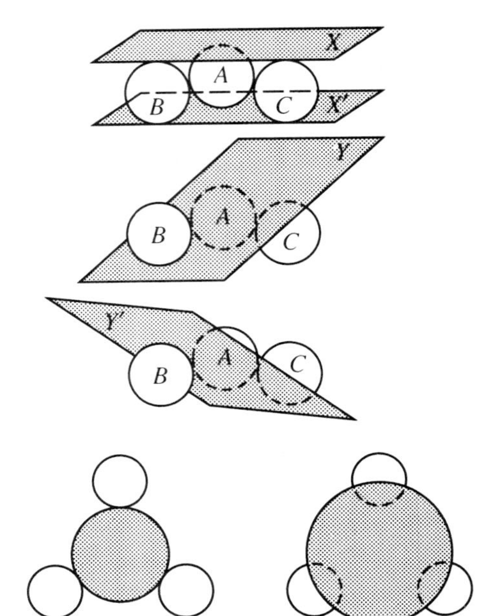

b. Infinitely many. Any sphere with a large enough radius can be placed so that it is tangent to the 3 given spheres. The figures show top views of two such arrangements.

22. Let l be in the plane of $\odot Q$, $l \perp \overline{QR}$ at R, and assume temp. that l is not tan. to $\odot Q$. Since l int. $\odot Q$ but is not tan. to $\odot Q$, it must int. $\odot Q$ in some other pt., say S. $\overline{QR}$ and $\overline{QS}$ are radii of $\odot Q$ so $\overline{QR} \cong \overline{QS}$, and $m\angle QSR = m\angle QRS = 90$. This contradicts the fact that $\triangle QRS$ can have at most one rt. $\angle$. Our temp. assumption must be false; it follows that l is tan. to $\odot Q$.

23. Let O be the ctr. of the $\odot$ and D, E, and F the pts. of tangency of $\overline{BC}$, $\overline{AB}$, and $\overline{AC}$, resp. $AE = AF$ so $BE = CF$; but $BE = BD$ and $CF = CD$ so $BD = DC$ and D is the midpt. of $\overline{BC}$, the base of isos. $\triangle ABC$. Then $\overline{AD} \perp \overline{BC}$ and $\overline{OD} \perp \overline{BC}$ so O is on $\overline{AD}$.

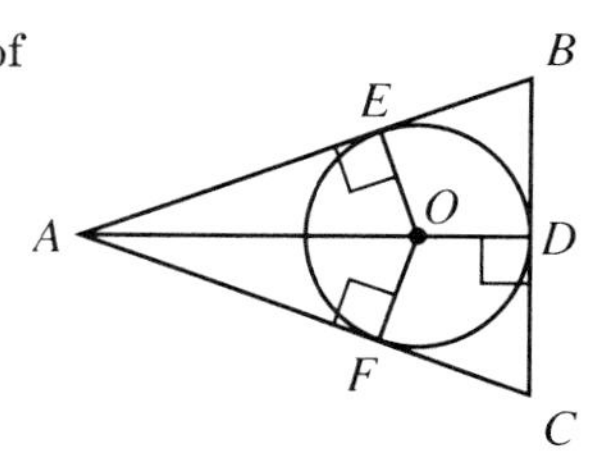

$\triangle AEO \sim \triangle ADB$; $\dfrac{AO}{12} = \dfrac{OE}{4}$; $4 \cdot AO = 12 \cdot OE$; $AO = 3 \cdot OE$; $\dfrac{AE}{AD} = \dfrac{OE}{BD}$;

$\dfrac{8}{AD} = \dfrac{OE}{4}$; $AD = AO + OD = AO + OE = 4 \cdot OE$; $\dfrac{8}{4 \cdot OE} = \dfrac{OE}{4}$;

$4(OE)^2 = 32$; $(OE)^2 = 8$; $OE = \sqrt{8} = 2\sqrt{2}$

Page 337 • MIXED REVIEW EXERCISES

1. $AB = 5\sqrt{3} \cdot \sqrt{3} = 15$ **2.** $AB = 9\sqrt{2}$

3. $AB = \sqrt{8^2 - 6^2} = \sqrt{28} = \sqrt{4 \cdot 7} = 2\sqrt{7}$

Page 338 • EXPLORATIONS

Check students' drawings. $ABCD \sim FGHE$; $ABCD \sim FGHE$ is true for all quads. except the rect. and the isos. trap. for which E, F, G, and H are the same pt.

Page 341 • CLASSROOM EXERCISES

1. a. $\angle DOC$, $\angle COB$, $\angle COA$, $\angle DOB$, $\angle AOB$ **b.** $\overset{\frown}{DCA}$, $\overset{\frown}{DBA}$ **c.** $\overset{\frown}{DC}$, $\overset{\frown}{CB}$, $\overset{\frown}{CA}$, $\overset{\frown}{DB}$, $\overset{\frown}{AB}$ **d.** $\overset{\frown}{ADC}$, $\overset{\frown}{ADB}$, $\overset{\frown}{BDC}$, $\overset{\frown}{BAD}$, $\overset{\frown}{CAD}$

2. 50 **3.** 130 **4.** 180 **5.** 230 **6.** 230 **7.** 280 **8.** 60 **9.** 50 **10.** 110 **11.** 110 **12.** 250 **13.** 310

Pages 341–343 • WRITTEN EXERCISES

A **1.** 85 **2.** $360 - 280 = 80$ **3.** 150 **4.** $180 - 130 = 50$ **5.** $360 - (240 + 68) = 52$ **6.** $90 - 35 = 55$ **7.** 30 **8.** 4, 8

9. a.

A
B
C

b. No

10.

$m\overset{\frown}{CB}$	60	70	56	50	$2x$
$m\angle 1$	60	70	56	50	$2x$
$m\angle 2$	30	35	28	25	x

11.

$m\overset{\frown}{CB}$	70	60	66	60	p
$m\overset{\frown}{BD}$	30	28	34	44	q
$m\angle COD$	100	88	100	104	$p + q$
$m\angle CAD$	50	44	50	52	$\frac{1}{2}(p + q)$

12. a. Check students' drawings. **b.** $m\angle APB = m\angle AQB = m\angle ARB$ **c.** $m\angle AOB = 2m\angle APB$

13. **a.** Check students' drawings. **b, c.** $m\angle A + m\angle C = m\angle B + m\angle D = 180$

d. The opp. $\angle$s of an inscribed quad. are supplementary.

B 14.

Statements	Reasons
1. Draw $\overline{OY}$.	1. Through any 2 pts. there is exactly one line.
2. $\overline{OY} \cong \overline{OZ}$	2. All radii of a $\odot$ are $\cong$.
3. $\angle OZY \cong \angle OYZ$	3. Isos. $\triangle$ Thm.
4. $\overline{OX} \parallel \overline{ZY}$	4. Given
5. $\angle XOY \cong \angle OYZ$	5. If 2 $\parallel$ lines are cut by a trans., then alt. int. $\angle$s are $\cong$.
6. $\angle WOX \cong \angle OZY$	6. If 2 $\parallel$ lines are cut by a trans., then corr. $\angle$s are $\cong$.
7. $\angle WOX \cong \angle XOY$	7. Substitution Prop.
8. $\widehat{WX} \cong \widehat{XY}$	8. In the same $\odot$, 2 minor arcs are $\cong$ if and only if their central $\angle$s are $\cong$.

15.

Statements	Reasons
1. Draw $\overline{OY}$.	1. Through any 2 pts. there is exactly one line.
2. $m\widehat{WX} = m\widehat{XY} = n$	2. Given
3. $m\widehat{WX} + m\widehat{XY} = m\widehat{WY} = 2n$	3. Arc Add. Post.
4. $m\angle WOY = m\widehat{WY} = 2n$	4. Def. of meas. of an arc
5. $m\angle WOY = m\angle Z + m\angle OYZ$	5. The meas. of an ext. $\angle$ of a $\triangle$ = the sum of the meas. of the 2 remote int. $\angle$s.
6. $\overline{OZ} \cong \overline{OY}$	6. All radii of a $\odot$ are $\cong$.
7. $\angle Z \cong \angle OYZ$ or $m\angle Z = m\angle OYZ$	7. Isos. $\triangle$ Thm.
8. $m\angle WOY = 2m\angle Z = 2n$	8. Substitution Prop.
9. $m\angle Z = n$	9. Div. Prop. of =

16. **a.** 35; $35 + 35 = 70$; 70 **b.** $n + n = 2n$ **c.** $3k$

17. $90 - 43 = 47$; $\sin 47^\circ = \frac{r}{6400}$; $r = 6400 \cdot \sin 47^\circ \approx 4700$ km

18. $90 - 40 = 50$; $\sin 50^\circ = \frac{r}{6400}$; $r = 6400 \cdot \sin 50^\circ \approx 4900$ km

19. $90 - 34 = 56$; $\sin 56° = \frac{r}{6400}$; $r = 6400 \cdot \sin 56° \approx 5300$ km

20. $90 - 23 = 67$; $\sin 67° = \frac{r}{6400}$; $r = 6400 \cdot \sin 67° \approx 5900$ km

C 21.

Statements	Reasons
1. Draw $\overline{OQ}$.	1. Through any 2 pts. there is exactly one line.
2. $\overline{OR} \cong \overline{OS}$; $\overline{QR} \cong \overline{QS}$	2. All radii of a ⊙ are ≅.
3. $\overline{OQ} \cong \overline{OQ}$	3. Refl. Prop.
4. $\triangle ORQ \cong \triangle OSQ$	4. SSS Post.
5. $m\angle ROQ = m\angle SOQ = \frac{1}{2}m\angle ROS$; $m\angle RQO = m\angle SQO = \frac{1}{2}m\angle RQS$	5. Corr. parts of ≅ △ are ≅, ∠Bis. Thm.
6. $m\overset{\frown}{RVS} = 60$; $m\overset{\frown}{RUS} = 120$	6. Given
7. $m\angle ROS = 60$; $m\angle RQS = 120$	7. Def. of meas. of an arc
8. $m\angle ROQ = 30$; $m\angle RQO = 60$	8. Substitution Prop.
9. $m\angle ORQ = 180 - (m\angle ROQ + m\angle RQO) = 90$	9. The sum of the meas. of the ∠s of a △ is 180.
10. $\overline{OR} \perp \overline{RQ}$	10. Def. of ⊥ lines
11. $\overline{OR}$ is tan. to ⊙Q; $\overline{QR}$ is tan. to ⊙O.	11. If a line in the plane of a ⊙ is ⊥ to a radius at its outer endpt., then the line is tan. to the ⊙.

22. Either $\overline{JK}$ is a diam. of ⊙Z or $\overline{JK} \parallel \overline{AB}$.
Case I: $\overset{\frown}{AJ}$ and $\overset{\frown}{BK}$ lie on opp. sides of $\overline{AB}$.
Given: $\overline{AB}$ is a diam. of ⊙Z; $m\overset{\frown}{AJ} = m\overset{\frown}{BK}$
Prove: $\overline{JK}$ is a diam. of ⊙Z.

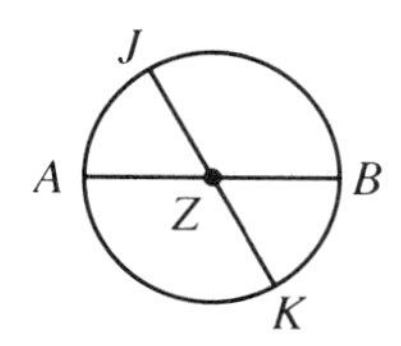

Statements	Reasons
1. Draw $\overline{JK}$.	1. Through any 2 pts. there is exactly one line.
2. $m\overset{\frown}{JK} = m\overset{\frown}{JB} + m\overset{\frown}{BK}$	2. Arc Add. Post.
3. $m\overset{\frown}{AJ} = m\overset{\frown}{BK}$	3. Given
4. $m\overset{\frown}{JK} = m\overset{\frown}{JB} + m\overset{\frown}{AJ} = m\overset{\frown}{AB} = 180$	4. Substitution, Arc Add. Post.
5. $\overset{\frown}{JK}$ is a semicircle; $\overline{JK}$ is a diam.	5. Def. of semicircle

See next page for Case II.

Case II: $\widehat{AJ}$ and $\widehat{BK}$ lie on the same side of $\overline{AB}$.
Given: $\overline{AB}$ is a diam. of $\odot Z$; $m\widehat{AJ} = m\widehat{BK}$.
Prove: $\overline{JK} \parallel \overline{AB}$

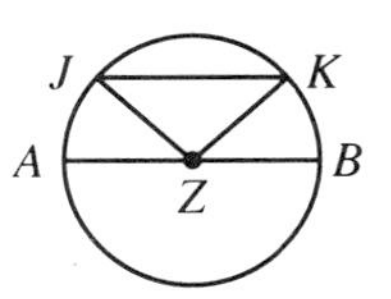

Statements	Reasons
1. Draw $\overline{JK}$, $\overline{ZJ}$, and $\overline{ZK}$.	1. Through any 2 pts. there is exactly one line.
2. $\overline{ZJ} \cong \overline{ZK}$	2. All radii of a $\odot$ are $\cong$.
3. $\angle KJZ \cong \angle JKZ$ or $m\angle KJZ = m\angle JKZ$	3. Isos. $\triangle$ Thm.
4. $m\angle KJZ + m\angle JZK + m\angle ZKJ = 180$; $m\angle KJZ = \frac{1}{2}(180 - m\angle JZK)$	4. The sum of the meas. of the $\angle$s of a $\triangle$ is 180; algebra.
5. $\widehat{AJ} \cong \widehat{BK}$	5. Given
6. $\angle AZJ \cong \angle BZK$ or $m\angle AZJ = m\angle BZK$	6. In the same $\odot$ 2 minor arcs are $\cong$ if and only if their central $\angle$s are $\cong$.
7. $m\angle AZJ + m\angle JZK + m\angle BZK = 180$	7. $\angle$ Add. Post.
8. $m\angle JZK = 180 - 2m\angle AZJ$	8. Substitution Prop., algebra
9. $m\angle KJZ = \frac{1}{2}[180 - (180 - 2m\angle AZJ)] = m\angle AZJ$	9. Substitution Prop., algebra
10. $\overline{JK} \parallel \overline{AB}$	10. If 2 lines are cut by a trans. and alt. int. $\angle$s are $\cong$, then the lines are $\parallel$.

23. $\cos \frac{n^\circ}{2} \approx \frac{6400}{6700} \approx 0.9552$; $\frac{n^\circ}{2} \approx 17^\circ$; $n^\circ \approx 34^\circ$; $m\widehat{XTY} \approx \frac{34}{360} \cdot 40{,}200 \approx 3800$ km

24. Calculator solution: ≈ 5300 km; $\cos \frac{n^\circ}{2} \approx \frac{6400}{7000} \approx 0.9143$; $\frac{n^\circ}{2} \approx 24^\circ$; $n^\circ \approx 48^\circ$;

$m\widehat{XTY} \approx \frac{48}{360} \cdot 40{,}200 \approx 5400$ km

Page 346 • CLASSROOM EXERCISES

1. Yes; in the same $\odot$, $\cong$ chords have $\cong$ arcs. **2.** No; you can't assume $\odot M \cong \odot N$.

3. a. A diam. that is $\perp$ to a chord bis. the chord. **b.** In the same $\odot$, $\cong$ chords are equally distant from the center. **c.** In the same $\odot$, $\cong$ chords have $\cong$ arcs.

4. $r = \sqrt{8^2 + 6^2} = 10$ **5.** $OM = \sqrt{13^2 - 5^2} = 12$

6. 1. Through any 2 pts. there is exactly one line. 2. Given 3. Given
4. Mult. Prop of = 5. A diam. that is ⊥ to a chord bis. the chord.
6. Substitution Prop. 7. All radii of a ⊙ are ≅. 8. HL Thm.
9. Corr. parts of ≅ ⚠ are ≅.

7. Yes

Pages 347–348 • WRITTEN EXERCISES

A **1.** $XY = 2\sqrt{5^2 - 3^2} = 8$ **2.** $OM = \sqrt{13^2 - 12^2} = 5$ **3.** $OR = 9\sqrt{2}$

4. $m\angle 1 = \frac{1}{2}(110) = 55$ **5.** $m\widehat{BC} = \frac{1}{3}(360 - 120) = 80$ **6.** $m\widehat{CD} = \frac{1}{2}(90) = 45$

7. $OA = 24$ **8.** $EF = CD = 2 \cdot 6 = 12$

9. $r = \sqrt{9^2 + 12^2} = 15$; $CD = 2 \cdot \sqrt{15^2 - 10^2} = 10\sqrt{5}$

10. Closer **11.** $c = \sqrt{10^2 - 4^2} = 2\sqrt{21}$ cm

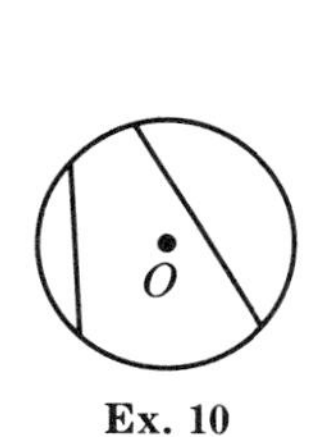

Ex. 10

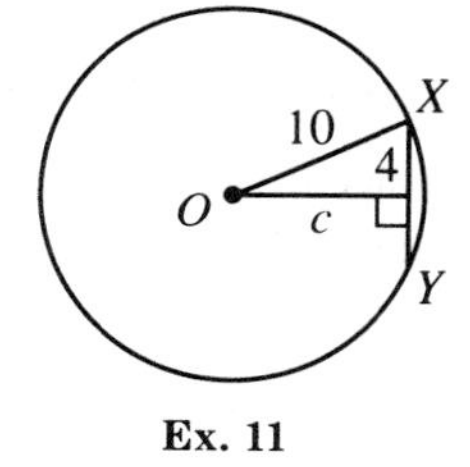

Ex. 11

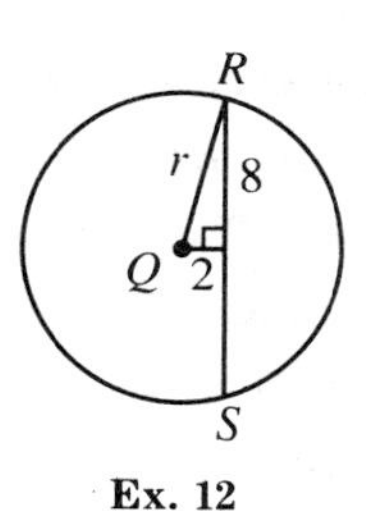

Ex. 12

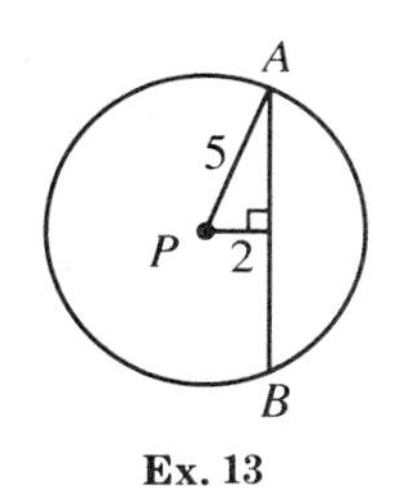

Ex. 13

12. $r = \sqrt{2^2 + 8^2} = 2\sqrt{17}$ cm **13.** $AB = 2\sqrt{5^2 - 2^2} = 2\sqrt{21}$ cm

14.

Statements	Reasons
1. $\widehat{JZ} \cong \widehat{KZ}$	1. Given
2. $\overline{JZ} \cong \overline{KZ}$	2. In the same ⊙, ≅ arcs have ≅ chords.
3. $\angle J \cong \angle K$	3. Isos. △ Thm.

15. Given: $\angle J \cong \angle K$
Prove: $\widehat{JZ} \cong \widehat{KZ}$

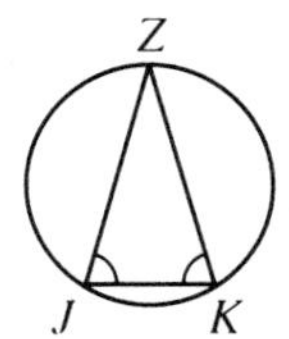

Statements	Reasons
1. $\angle J \cong \angle K$	1. Given
2. $\overline{JZ} \cong \overline{KZ}$	2. If 2 ∡ of a △ are ≅, then the sides opp. those ∡ are ≅.
3. $\widehat{JZ} \cong \widehat{KZ}$	3. In the same ⊙, ≅ chords have ≅ arcs.

B **16. a.** Given: $\odot O$; $\overline{RS} \cong \overline{TU}$

Prove: $\overparen{RS} \cong \overparen{TU}$

Proof: Draw radii $\overline{OR}$, $\overline{OS}$, $\overline{OT}$, and $\overline{OU}$. $\overline{OR} \cong \overline{OT}$ and $\overline{OS} \cong \overline{OU}$ because they are all radii of the same circle. Also, $\overline{RS} \cong \overline{TU}$. So $\triangle ROS \cong \triangle TOU$ by SSS and corr. parts $\angle ROS$ and $\angle TOU$ are $\cong$. Then $\overparen{RS} \cong \overparen{TU}$ because their central $\angle$s are $\cong$.

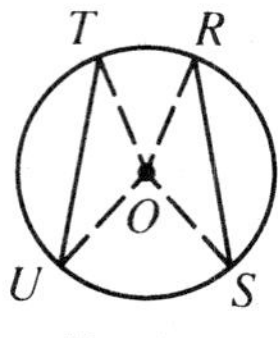

Ex. 16a

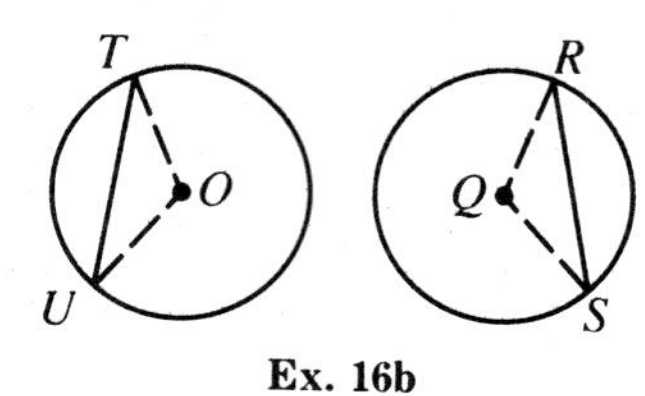

Ex. 16b

b. Given: $\odot O \cong \odot Q$; $\overline{RS} \cong \overline{TU}$

Prove: $\overparen{RS} \cong \overparen{TU}$

Proof: Draw radii $\overline{QR}$, $\overline{QS}$, $\overline{OT}$, and $\overline{OU}$. $\overline{QR} \cong \overline{OT}$ and $\overline{QS} \cong \overline{OU}$ because they are radii of $\cong$ $\odot$s. Also, $\overline{RS} \cong \overline{TU}$. So $\triangle RQS \cong \triangle TOU$ by SSS and corr. parts $\angle RQS$ and $\angle TOU$ are $\cong$. Then $\overparen{RS} \cong \overparen{TU}$ because their central $\angle$s are $\cong$.

17. Draw $\overline{OX} \perp \overline{JK}$ at X. Then $\triangle JOX$ is a 30°–60°–90° $\triangle$, $OX = 5$, $JX = 5\sqrt{3}$, and $JK = 2 \cdot 5\sqrt{3} = 10\sqrt{3}$.

18. Draw $\overline{OF}$. $\triangle OEF$ is a 30°–60°–90° $\triangle$, so $EF = 8$ and radius $OF = 16$. Then diameter $HG = 2 \cdot 16 = 32$.

19. $OT = 5$; $TS = 12$; $r = \sqrt{5^2 + 12^2} = 13$; $d = 2r = 26$ cm

20. $TS = 10$; $OS = 15$; $OT = \sqrt{15^2 - 10^2} = 5\sqrt{5}$

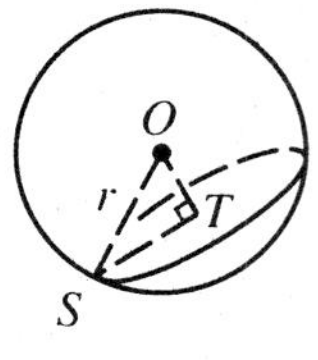

Exs. 19, 20

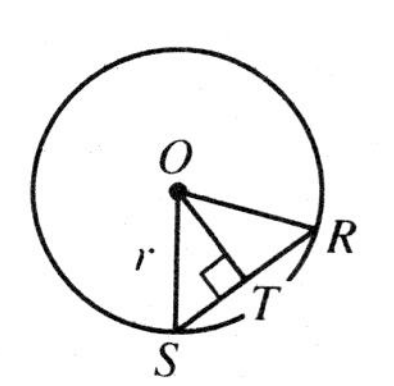

Exs. 21, 22

21. $TS = 6$; $OS = 10$; $\sin \angle SOT = \dfrac{6}{10} = 0.6$;

$m\angle SOT \approx 37$; $m\angle SOR = m\overparen{SR} \approx 2 \cdot 37 \approx 74$

22. $TS = 10$; $m\angle SOT = 35$; $\sin 35° = \dfrac{10}{r}$; $r = \dfrac{10}{\sin 35°} \approx 17.4$

C 23. Answers may vary. Example: If 2 ⊙s are concentric and a chord of the outer ⊙ is tan. to the inner circle, then the pt. of tangency is the midpt. of the chord.
Given: Concentric ⊙s with center O; chord $\overline{AC}$ of outer ⊙ tan. to inner ⊙ at B.
Prove: B is the midpt. of $\overline{AC}$.

Statements	Reasons
1. Draw $\overline{OB}$.	1. Through any 2 pts. there is exactly one line.
2. $\overline{AC}$ is tan. to inner $\odot O$ at B.	2. Given
3. $\overline{AC} \perp \overline{OB}$	3. If a line is tan. to a ⊙, then the line is ⊥ to the radius drawn to the pt. of tangency.
4. $\overline{OB}$ bis. $\overline{AC}$.	4. A diam. that is ⊥ to a chord bis. the chord.
5. B is the midpt. of $\overline{AC}$.	5. Def. of seg. bis.

24. Answers may vary. Example: If a chord of a ⊙ is ∥ to a line tan. to the ⊙, then the pt. of tangency is the midpt. of the arc of the chord.
Given: l tan. to $\odot O$ at F; $\overline{DE}$ a chord of $\odot O$; $\overline{DE} \parallel l$
Prove: F is the midpt. of $\overset{\frown}{DE}$.

Statements	Reasons
1. Draw $\overline{OF}$.	1. Through any 2 pts. there is exactly one line.
2. l is tan. to $\odot O$ at F.	2. Given
3. $l \perp \overline{OF}$	3. If a line is tan. to a ⊙, then the line is ⊥ to the radius drawn to the pt. of tangency.
4. $\overline{DE} \parallel l$	4. Given
5. $\overline{OF} \perp \overline{DE}$	5. If a trans. is ⊥ to one of 2 ∥ lines, then it is ⊥ to the other one also.
6. $\overline{OF}$ bis. $\overline{DE}$ and $\overset{\frown}{DE}$.	6. A diam. that is ⊥ to a chord bis. the chord and its arc.
7. F is the midpt. of $\overset{\frown}{DE}$.	7. Def. of arc bis.

25. $m\angle AOB = \frac{1}{3} \cdot 360 = 120$; $m\angle DOB = \frac{1}{2} \cdot 120 = 60$;
$OB = 6$; $OD = 3$; $DB = 3\sqrt{3}$; $AB = 6\sqrt{3}$; perimeter of $\triangle ABC = 3 \cdot 6\sqrt{3} = 18\sqrt{3}$

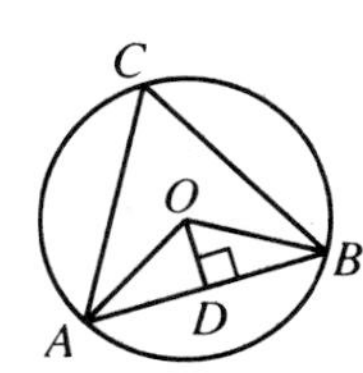

26. Let $\overline{AB}$ and $\overline{CD}$ be chords of $\odot O$ with radius r such that $AB = x$ and $CD = 2x$. Distance from O to $\overline{AB} = \sqrt{r^2 - \left(\frac{x}{2}\right)^2} = \sqrt{r^2 - \frac{x^2}{4}} = \sqrt{\frac{4r^2 - x^2}{4}} =$

$\frac{\sqrt{4r^2 - x^2}}{2}$; Distance from O to $\overline{CD} = \sqrt{r^2 - x^2}$. If the ratio of the distances is $2 : 1$, $\frac{\sqrt{4r^2 - x^2}}{2} = 2\sqrt{r^2 - x^2}$; $\sqrt{4r^2 - x^2} = 4\sqrt{r^2 - x^2}$;

$4r^2 - x^2 = 16(r^2 - x^2)$; $4r^2 - x^2 = 16r^2 - 16x^2$; $15x^2 = 12r^2$; $x^2 = \frac{4}{5}r^2$;

$x = \sqrt{\frac{4}{5}}r = \frac{2\sqrt{5}}{5}r$; $AB = \frac{2\sqrt{5}}{5}r$; $CD = \frac{4\sqrt{5}}{5}r$

27. Refer to semicircle O with diameter $\overline{AB}$. Draw radius $\overline{OY}$. $OY = 10$ and $EY = 8$, so $OE = \sqrt{10^2 - 8^2} = 6$. Draw radius $\overline{OX}$. $OX = 10$ and $FX = 6$, so $OF = \sqrt{10^2 - 6^2} = 8$. Then $EF = GX = 8 - 6 = 2$, and $GY = 8 - 6 = 2$. $XY = 2\sqrt{2} \approx 2.8$ cm.

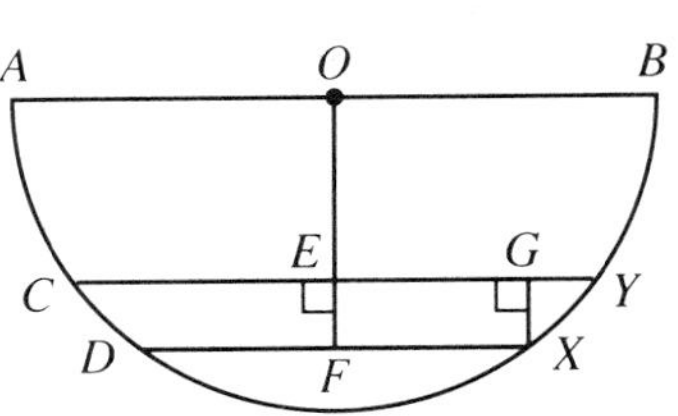

Page 349 • SELF-TEST 1

1. a. $\overline{QB}$, $\overline{QC}$ **b.** $\overline{BC}$ **c.** chord: $\overline{AC}$ or $\overline{BC}$; secant: $\overleftrightarrow{AC}$

2. a.

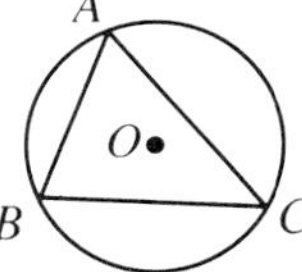

b.

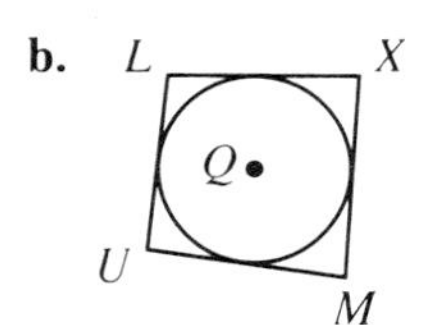

3. $OP = \sqrt{25^2 - 20^2} = 15$ **4.** Two concentric circles

5. $\frac{x}{2} = \sqrt{7^2 - 3^2} = 2\sqrt{10}$; $x = 4\sqrt{10}$ cm

6. a. 50; 310 **b.** In the same $\odot$, $\cong$ chords have $\cong$ arcs.

Pages 352–353 • CLASSROOM EXERCISES

1. Say the measure of the intercepted arc is n; then the measure of each angle is $\frac{1}{2}n$.

2. $\angle X$ intercepts a semicircle, so $m\angle X = \frac{1}{2}(180) = 90$

3. a. 360 **b.** $x + y = \frac{1}{2}$(meas. of red arc) $+ \frac{1}{2}$(meas. of blue arc) $= \frac{1}{2}(360) = 180$

c. If a quad. is inscribed in a $\odot$, then its opp. $\angle$s are supp.

4. $x = 38$; $y = 38$

5. $x = 25; y = \frac{1}{2}(180 - 50) = 65$ **6.** $x = 70; y = 95$ **7.** $x = 120; y = 60$

8. $y = 40; x = \frac{1}{2}[360 - (150 + 80)] = 65$ **9.** $x = y = \frac{1}{2}\left[\frac{1}{2}(360 - 80)\right] = 70$

10. **a.** If the opp. ∡s of a quad. are not supp., then the quad. cannot be inscribed in a ⊙.

b. No; $\angle P$ and $\angle R$ are opp. ∡s and are not supp.

11. $2n$; $2n$. In the same ⊙, ≅ inscribed ∡s int. ≅ arcs.

12. Draw diameter $\overline{BD}$. Let $m\angle ABD = x$ and $m\angle CBD = y$. Then by the ∠ Add. Post., $m\angle ABC = x + y$. By Case I, $m\widehat{AD} = 2x$ and $m\widehat{CD} = 2y$. Then by the Arc Add. Post., $m\widehat{AC} = 2x + 2y$. Using Case I again, $m\angle ABC = x + y = \frac{1}{2}(2x + 2y) = \frac{1}{2}m\widehat{AC}$.

13. Draw diameter $\overline{BD}$. Let $m\angle ABD = x$ and $m\angle CBD = y$. Then by the ∠ Add. Post. and Subtr., $m\angle ABC = x - y$. By Case I, $m\widehat{AD} = 2x$ and $m\widehat{CD} = 2y$. Then by the Arc Add. Post. and Subtr., $m\widehat{AC} = 2x - 2y$. Using Case I again, $m\angle ABC = x - y = \frac{1}{2}(2x - 2y) = \frac{1}{2}m\widehat{AC}$.

14. $ABCD$ is a rhombus. $\overline{AO} \perp \overline{AD}$ and $\overline{AO}$ bis. $\angle BAC$. So $m\angle OAC = 30$, and $m\angle CAD = 60$. Similarly, $m\angle ACD = 60$, so $\triangle ACD$ is equilateral. Then $AD = DC = AC = BC = AB$, and $ABCD$ is a rhombus.

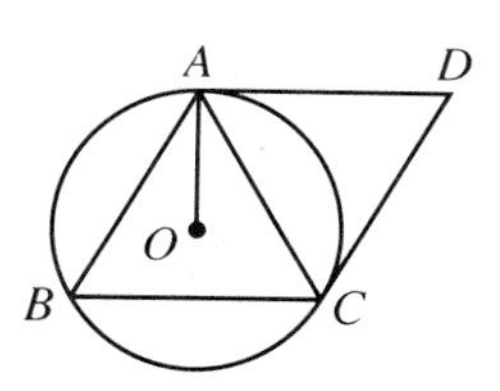

Pages 354–356 • WRITTEN EXERCISES

A **1.** $x = 180 - (100 + 50) = 30; z = \frac{1}{2}x = 15; y = \frac{1}{2} \cdot 50 = 25$

2. $x = 2 \cdot 65 = 130; y = 2 \cdot 60 = 120; z = 2[180 - (60 + 65)] = 110$

3. $x = 110; y = 100; z = 240 - 140 = 100$

4. $x = 70; y = 110; z = 2[180 - (55 + 70)] = 110$

5. $x = 50; y = 2\left(\frac{180 - 50}{2}\right) = 130; z = \frac{180 - 50}{2} = 65$ **6.** $x = 90; y = 90; z = 90$

7. $x = 104; y = 104; z = 52$ **8.** $x = 80; y = 40; z = \frac{1}{2}(360 - 240) = 60$

9. $x = 50; y = 100; z = \frac{1}{2}(360 - 290) = 35$

10.

Statements	Reasons
1. Draw $\overline{BC}$.	1. Through any 2 pts. there is exactly one line.
2. $\overline{AB} \parallel \overline{CD}$	2. Given
3. $\angle ABC \cong \angle BCD$ or $m\angle ABC = m\angle BCD$	3. If 2 $\parallel$ lines are cut by a trans., then alt. int. $\angle$s are $\cong$.
4. $m\angle ABC = \frac{1}{2}m\widehat{AC}$; $m\angle BCD = \frac{1}{2}m\widehat{BD}$	4. The meas. of an inscribed $\angle$ = half the meas. of its intercepted arc.
5. $\frac{1}{2}m\widehat{AC} = \frac{1}{2}m\widehat{BD}$ or $m\widehat{AC} = m\widehat{BD}$ or $\widehat{AC} \cong \widehat{BD}$	5. Substitution Prop., Mult. Prop. of =

11. **a.** If the arcs between two chords of a $\odot$ are $\cong$, then the chords are $\parallel$. **b.** False; the chords may intersect.

12.

Statements	Reasons
1. $\angle U \cong \angle Y$; $\angle X \cong \angle V$	1. If 2 inscribed $\angle$s intercept the same arc, then the $\angle$s are $\cong$.
2. $\triangle UXZ \sim \triangle YVZ$	2. AA ~ Post.

13. 1. If a line is tan. to a $\odot$, then the line is $\perp$ to the radius drawn to the pt. of tangency; def. of $\perp$ lines. 2. Def. of semicircle 3. Substitution Prop.

B **14.**

Statements	Reasons
1. Draw diam. $\overline{TZ}$.	1. Through any 2 pts. there is exactly one line.
2. $m\angle ATZ = \frac{1}{2}m\widehat{AZ}$	2. The meas. of an inscribed $\angle$ = half the meas. of its intercepted arc.
3. $m\angle ZTP = \frac{1}{2}m\widehat{ZNT}$	3. Thm. 9-8, Case I
4. $m\angle ATP = m\angle ATZ + m\angle ZTP$	4. $\angle$ Add. Post.
5. $m\angle ATP = \frac{1}{2}m\widehat{AZ} + \frac{1}{2}m\widehat{ZNT} = \frac{1}{2}(m\widehat{AZ} + m\widehat{ZNT})$	5. Substitution and Distributive Props.
6. $m\widehat{AZ} + m\widehat{ZNT} = m\widehat{ANT}$	6. Arc Add. Post.
7. $m\angle ATP = \frac{1}{2}m\widehat{ANT}$	7. Substitution Prop.

15.

Statements	Reasons
1. Draw diam. $\overline{TZ}$.	1. Through any 2 pts. there is exactly one line.
2. $m\angle ZTA = \frac{1}{2}m\overparen{ZA}$	2. The meas. of an inscribed $\angle$ = half the meas. of its intercepted arc.
3. $m\angle ZTP = \frac{1}{2}m\overparen{ZNT}$	3. Thm. 9-8, Case I
4. $m\angle ZTP = m\angle ZTA + m\angle ATP$ or $m\angle ATP = m\angle ZTP - m\angle ZTA$	4. $\angle$ Add. Post.
5. $m\angle ATP = \frac{1}{2}m\overparen{ZNT} - \frac{1}{2}m\overparen{ZA} = \frac{1}{2}(m\overparen{ZNT} - m\overparen{ZA})$	5. Substitution and Distributive Props.
6. $m\overparen{ZNT} = m\overparen{ZA} + m\overparen{ANT}$ or $m\overparen{ANT} = m\overparen{ZNT} - m\overparen{ZA}$	6. Arc Add. Post.
7. $m\angle ATP = \frac{1}{2}m\overparen{ANT}$	7. Substitution Prop.

16. Given: Inscribed quad. $ABCD$; $\overline{AD} \cong \overline{BC}$
Prove: $\overline{AB} \parallel \overline{DC}$

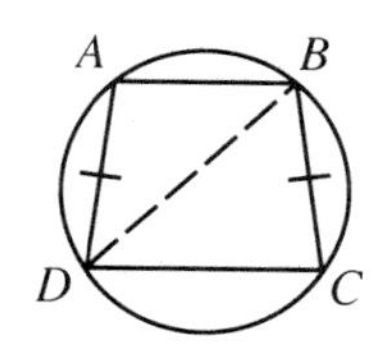

Statements	Reasons
1. Draw $\overline{BD}$.	1. Through any 2 pts. there is exactly one line.
2. $\overline{AD} \cong \overline{BC}$	2. Given
3. $\overparen{AD} \cong \overparen{BC}$ or $m\overparen{AD} = m\overparen{BC}$	3. In the same $\odot$, $\cong$ chords have $\cong$ arcs.
4. $m\angle ABD = \frac{1}{2}m\overparen{AD}$ or $2m\angle ABD = m\overparen{AD}$; $m\angle BDC = \frac{1}{2}m\overparen{BC}$ or $2m\angle BDC = m\overparen{BC}$	4. The meas. of an inscribed $\angle$ = half the meas. of its intercepted arc.
5. $2m\angle ABD = 2m\angle BDC$ or $m\angle ABD = m\angle BDC$ or $\angle ABD \cong \angle BDC$	5. Substitution Prop., Div. Prop. of =
6. $\overline{AB} \parallel \overline{DC}$	6. If 2 lines are cut by a trans. and alt. int. $\angle$s are $\cong$, then the lines are $\parallel$.

17. $\triangle ADE \sim \triangle BCE$ by the AA $\sim$ Post. since $\angle ADE \cong \angle BCE$ (If 2 inscribed $\angle$s intercept the same arc, then the $\angle$s are $\cong$.) and $\angle AED \cong \angle BEC$ (Vert. $\angle$s are $\cong$.). Similarly, $\triangle EDC \sim \triangle EAB$.

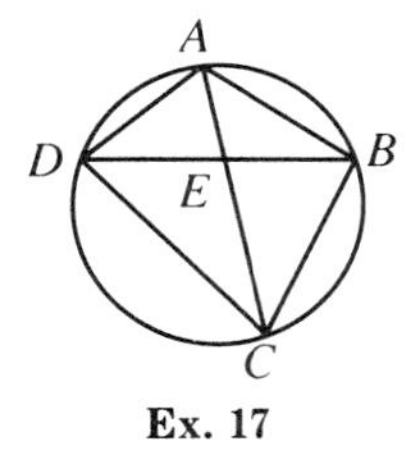

Ex. 17

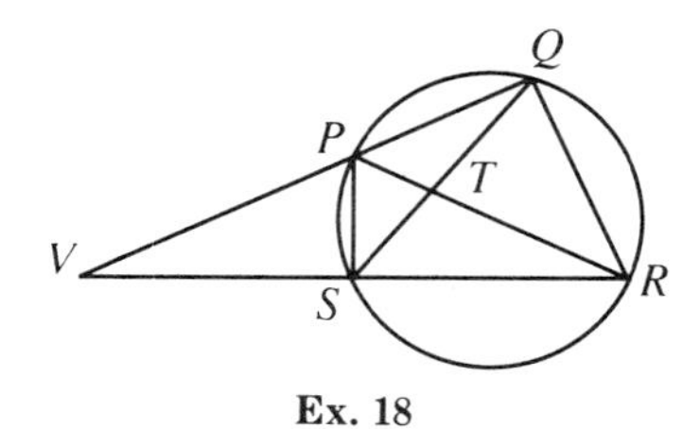

Ex. 18

18. $\triangle VPS \sim \triangle VRQ$ by the AA $\sim$ Post. since $\angle V \cong \angle V$ and $\angle VPS \cong \angle VRQ$ (both are supp. to $\angle SPQ$). Also, $\triangle VSQ \sim \triangle VPR$ by the AA $\sim$ Post. since $\angle V \cong \angle V$ and $\angle VQS \cong \angle VRP$ (both intercept $\overparen{PS}$).

19. $x + (x + 20) = 180; 2x = 160; x = 80; m\angle D = 180 - 2x = 20$

20. $x^2 + 11x = 180; x^2 + 11x - 180 = 0; (x + 20)(x - 9) = 0; x = -20$ (reject) or $x = 9; m\angle D = 180 - (9x - 2) = 101$

21. $m\overparen{ABC} = 2(75) = 150; x^2 + 5x = 150; x^2 + 5x - 150 = 0; (x + 15)(x - 10) = 0;$ $x = -15$ (reject) or $x = 10; m\angle A = \frac{1}{2}(5x + 6x) = 55$

22. $\angle D$ and $\angle A$ are both supps. of $\angle C$, so $\angle D \cong \angle A$. Also, $\angle D$ and $\angle A$ are supp. $\angle$s. So $m\angle D + m\angle A = 2m\angle A = 180$, and $m\angle A = 90$.

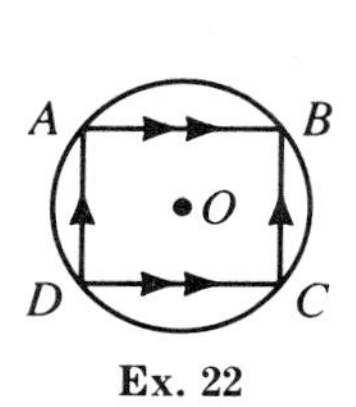

Ex. 22

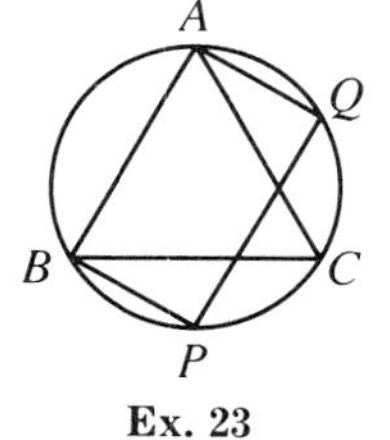

Ex. 23

23. Rectangle. $m\overparen{AB} = 120$ and $m\overparen{AQ} = 60$, so $\overparen{BAQ}$ is a semicircle and $\angle BAQ$ is a rt. $\angle$. Similarly, $\angle AQP$, $\angle QPB$, and $\angle PBA$ are rt. $\angle$s, so by def. $AQPB$ is a rect.

24. **a.** Suppose a ⊙ is circumscribed about a polygon. The sides of the polygon are ≅ chords and so have ≅ arcs. Then the numbered ∠s are all inscribed ∠s intercepting ≅ arcs, so the ∠s are ≅. **b.** yes

C 25. Assume temp. that D is not on $\odot O$. Case I: (D is inside $\odot O$.) Let X be the pt. where $\overrightarrow{AD}$ int. $\odot O$; draw $\overline{CX}$. Quad. $ABCX$ is inscribed in $\odot O$ so $m\angle AXC + m\angle B = 180$. But $m\angle D + m\angle B = 180$ so $m\angle AXC = m\angle D$. Then $\overline{CD} \parallel \overline{CX}$. This contradicts the fact that $\overline{CX}$ and $\overline{CD}$ int. at C. Case II: (D is outside $\odot O$.) Let X be the pt. where $\overrightarrow{AD}$ int. $\odot O$; draw $\overline{CX}$. Again, $m\angle AXC + m\angle B = 180$ and $m\angle D + m\angle B = 180$ so $m\angle AXC = m\angle D$. Then $\overline{CD} \parallel \overline{CX}$. This contradicts the fact that $\overline{CX}$ and $\overline{CD}$ int. at C. In either case, our temp. assumption must be false. It follows that D lies on $\odot O$.

26.

Statements	Reasons
1. Draw $\overline{XY}$.	1. Through any 2 pts. there is exactly one line.
2. $\angle P$ is supp. to $\angle XYS$.	2. If a quad. is inscribed in a ⊙, then its opp. ∠s are supp.
3. $m\angle XYR + m\angle XYS = 180$	3. ∠ Add. Post.
4. $\angle XYR$ is supp. to $\angle XYS$.	4. Def. of supp. ∠s
5. $\angle P \cong \angle XYR$	5. If 2 ∠s are supps. of the same ∠, then the 2 ∠s are ≅.
6. $\angle XYR$ is supp. to $\angle Q$.	6. If a quad. is inscribed in a ⊙, then its opp. ∠s are supp.
7. $m\angle XYR + m\angle Q = 180$	7. Def. of supp. ∠s
8. $m\angle P + m\angle Q = 180$	8. Substitution Prop.
9. $\angle P$ is supp. to $\angle Q$.	9. Def. of supp. ∠s
10. $\overline{PS} \parallel \overline{QR}$	10. If 2 lines are cut by a trans. and s-s. int. ∠s are supp., then the lines are ∥.

27.

Statements	Reasons
1. Let Q be on $\overline{AC}$ such that $\angle ADQ \cong \angle BDC$.	1. Protractor Post.
2. Draw $\overline{DQ}$.	2. Through any 2 pts. there is exactly one line.
3. $\angle DAQ \cong \angle CBD$	3. If 2 inscribed ∠s intercept the same arc, then the ∠s are ≅.
4. $\triangle ADQ \sim \triangle BDC$	4. AA ~ Post.
5. $\frac{AQ}{BC} = \frac{AD}{BD}$	5. Corr. sides of ~ △s are in prop.
6. $AQ = \frac{BC \cdot AD}{BD}$	6. Mult. Prop. of =
7. $m\angle ADQ = m\angle BDC$	7. Step 1
8. $m\angle QDB = m\angle QDB$	8. Refl. Prop.
9. $m\angle ADQ + m\angle QDB = m\angle BDC + m\angle QDB$	9. Add. Prop. of =
10. $m\angle ADB = m\angle ADQ + m\angle QDB$; $m\angle QDC = m\angle BDC + m\angle QDB$	10. ∠ Add. Post.
11. $m\angle ADB = m\angle QDC$ or $\angle ADB \cong \angle QDC$	11. Substitution Prop.
12. $\angle ABD \cong \angle QCD$	12. If 2 inscribed ∠s intercept the same arc, then the ∠s are ≅.
13. $\triangle ADB \sim \triangle QDC$	13. AA ~ Post.
14. $\frac{QC}{AB} = \frac{CD}{BD}$	14. Corr. sides of ~ △s are in prop.
15. $QC = \frac{AB \cdot CD}{BD}$	15. Mult. Prop. of =
16. $AQ + QC = \frac{BC \cdot AD}{BD} + \frac{AB \cdot CD}{BD}$	16. Add. Prop. of =
17. $AC = AQ + QC$	17. Seg. Add. Post.
18. $AC \cdot BD = BC \cdot AD + AB \cdot CD$, or $AB \cdot CD + BC \cdot AD = AC \cdot BD$	18. Substitution Prop., Mult. Prop. of =

28. Given: Equilateral $\triangle ABC$ inscribed in a $\odot$.
Prove: $PA = PB + PC$

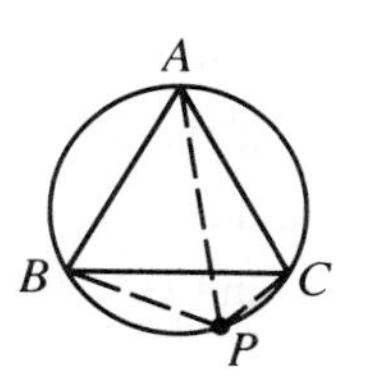

Statements	Reasons
1. Draw $\overline{PA}$, $\overline{PB}$, and $\overline{PC}$.	1. Through any 2 pts. there is exactly one line.
2. $PA \cdot BC = AC \cdot BP + AB \cdot PC$	2. Ptolemy's Thm.
3. $\triangle ABC$ is equilateral.	3. Given
4. $\overline{AB} \cong \overline{BC} \cong \overline{CA}$ or $AB = BC = CA$	4. Def. of equilateral $\triangle$
5. $PA \cdot BC = BC \cdot BP + BC \cdot PC$	5. Substitution Prop.
6. $PA = BP + PC$	6. Div. Prop. of =

29. Let $\overline{CD}$ be the alt. to the hyp. The diam. of the smallest $\odot$ through C that is tan. to $\overline{AB}$ is $\overline{CD}$. The center of the $\odot$ is M, the midpt. of $\overline{CD}$. Since $m\angle JCK = 90$, $\overline{JK}$ is a diam. of $\odot M$ and $JK = CD$. To find CD in terms of a, b, and c, note that $\triangle ABC \sim \triangle ACD$; $\frac{c}{b} = \frac{a}{CD}$; $CD = \frac{ab}{c}$; $JK = CD = \frac{ab}{c}$.

Page 357 • MIXED REVIEW EXERCISES

1. $\overline{LM}$ **2.** $\overleftrightarrow{LM}$ **3.** $\overline{NP}$ **4.** 14 **5.** $360 - x$ **6.** $\frac{4}{x} = \frac{x}{9}$; $x^2 = 36$; $x = 6$

Pages 358–359 • CLASSROOM EXERCISES

1. $\frac{1}{2}(40 + 30) = 35$ **2.** $\frac{1}{2}(120 - 40) = 40$ **3.** $\frac{1}{2}(135 + 140) = 137.5$

4. $\frac{1}{2}(260 - 100) = 80$ **5.** $\frac{1}{2}(170 - 80) = 45$

6. $360 - (160 + 120) = 80$; $\frac{1}{2}(160 - 80) = 40$ **7.** $75 = \frac{1}{2}(100 + x)$; $x = 50$

8. $30 = \frac{1}{2}(x - 70)$; $x = 130$ **9.** $58 = \frac{1}{2}[(360 - x) - x]$; $x = 122$

10. 1. Through any 2 pts. there is exactly one line. 2. The meas. of an ext. ∠ of a △ = the sum of the meas. of the 2 remote int. ∠s. 3. Subtr. Prop. of = 4. The meas. of an inscribed ∠ = half the meas. of its intercepted arc. 5. Substitution and Distributive Props.

Pages 359–361 • WRITTEN EXERCISES

A **1.** $m\angle 1 = 90$ **2.** $m\angle 2 = 90$ **3.** $m\angle 3 = \frac{1}{2}(20 + 30) = 25$

4. $m\angle 4 = \frac{1}{2}(180 - 50) = 65$ **5.** $m\angle 5 = \frac{1}{2}(90 + 20) = 55$

6. $m\angle 6 = 180 - 55 = 125$ **7.** $m\angle 7 = \frac{1}{2}(90 - 20) = 35$

8. $m\angle 8 = \frac{1}{2}(90 + 30) = 60$ **9.** $m\angle 9 = 90$ **10.** $m\angle 10 = \frac{1}{2}(90 + 30) = 60$

11. $\frac{1}{2}(80 + 40) = 60$ **12.** $180 - \frac{1}{2}(130 + 100) = 65$ **13.** $50 = \frac{1}{2}(70 + x); x = 30$

14. $52 = \frac{1}{2}(36 + x); x = 68$ **15.** $\frac{1}{2}(110 - 50) = 30$ **16.** $28 = \frac{1}{2}(x - 46); x = 102$

17. $35 = \frac{1}{2}(110 - x); x = 40$ **18.** $\frac{1}{2}(250 - 110) = 70$ **19.** $\frac{1}{2}(270 - 90) = 90$

20. $360 - t; \frac{1}{2}(360 - 2t) = 180 - t$

21. $65 = \frac{1}{2}[x - (360 - x)]; x = 245; m\widehat{XY} = 360 - x = 115$

B **22.** $42 = \frac{1}{2}(7x - 3x); x = 21$; meas. of third arc $= 360 - 10x = 150$

23. Let $ABCD$ be the quad. with $m\angle A = 80$, $m\angle B = 90$, $m\angle C = 94$, $m\angle D = 96$, and let W, X, Y, and Z be the pts. of tangency of $\overline{AB}$, $\overline{BC}$, $\overline{CD}$, and $\overline{AD}$, resp.

$90 = \frac{1}{2}[(360 - m\widehat{WX}) - m\widehat{WX}]; 180 = 360 - 2m\widehat{WX}; 2m\widehat{WX} = 180; m\widehat{WX} = 90;$

$94 = \frac{1}{2}[(360 - m\widehat{XY}) - m\widehat{XY}]; 188 = 360 - 2m\widehat{XY}; 2m\widehat{XY} = 172; m\widehat{XY} = 86;$

$96 = \frac{1}{2}[(360 - m\widehat{YZ}) - m\widehat{YZ}]; 192 = 360 - 2m\widehat{YZ}; 2m\widehat{YZ} = 168; m\widehat{YZ} = 84;$

$80 = \frac{1}{2}[(360 - m\widehat{WZ}) - m\widehat{WZ}]; 160 = 360 - 2m\widehat{WZ}; 2m\widehat{WZ} = 200; m\widehat{WZ} = 100$

(or $m\widehat{WZ} = 360 - (90 + 86 + 84) = 100$)

24. $m\widehat{AD} = 360 - 3x; 32 = \frac{1}{2}(360 - 4x); x = 74; m\angle A = \frac{1}{2}(148) = 74 = m\angle D;$

$m\angle B = \frac{1}{2}(138 + 74) = 106 = m\angle C$

25. Given: Tangents $\overline{PA}$ and $\overline{PB}$

Prove: $m\angle 1 = \frac{1}{2}(m\widehat{ACB} - m\widehat{AB})$

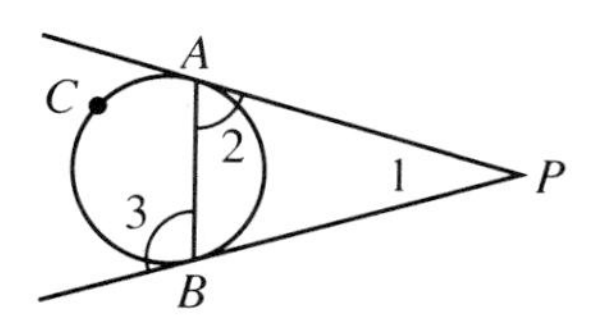

Statements	Reasons
1. Draw chord $\overline{AB}$.	1. Through any 2 pts. there is exactly one line.
2. $m\angle 1 + m\angle 2 = m\angle 3$	2. The meas. of an ext. ∠ of a △ = the sum of the meas. of the 2 remote int. ∠s.
3. $m\angle 1 = m\angle 3 - m\angle 2$	3. Subtr. Prop. of =
4. $m\angle 3 = \frac{1}{2}m\widehat{ACB}$; $m\angle 2 = \frac{1}{2}m\widehat{AB}$	4. The meas. of an ∠ formed by a chord and a tan. = half the meas. of the intercepted arc.
5. $m\angle 1 = \frac{1}{2}m\widehat{ACB} - \frac{1}{2}m\widehat{AB} = \frac{1}{2}(m\widehat{ACB} - m\widehat{AB})$	5. Substitution and Distributive Props.

26. Given: Secant $\overline{PA}$ and tangent $\overline{PC}$

Prove: $m\angle 1 = \frac{1}{2}(m\widehat{AC} - m\widehat{BC})$

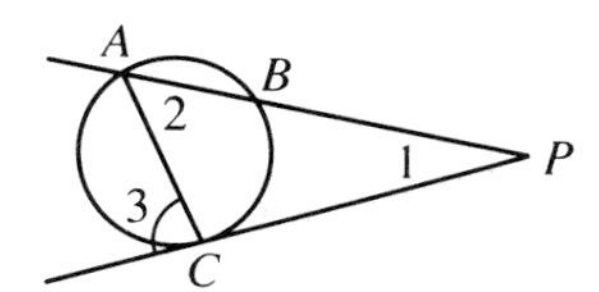

Statements	Reasons
1. Draw chord $\overline{AC}$.	1. Through any 2 pts. there is exactly one line.
2. $m\angle 1 + m\angle 2 = m\angle 3$	2. The meas. of an ext. ∠ of a △ = the sum of the meas. of the 2 remote int. ∠s.
3. $m\angle 1 = m\angle 3 - m\angle 2$	3. Subtr. Prop. of =
4. $m\angle 3 = \frac{1}{2}m\widehat{AC}$	4. The meas. of an ∠ formed by a chord and a tan. = half the meas. of the intercepted arc.
5. $m\angle 2 = \frac{1}{2}m\widehat{BC}$	5. The meas. of an inscribed ∠ = half the meas. of its intercepted arc.
6. $m\angle 1 = \frac{1}{2}m\widehat{AC} - \frac{1}{2}m\widehat{BC} = \frac{1}{2}(m\widehat{AC} - m\widehat{BC})$	6. Substitution and Distributive Props.

27. Let $\angle 1$ be the $\angle$ inscribed in the outer $\odot$; $m\angle 1 = \frac{1}{2}c$; $m\angle 1 = \frac{1}{2}(b - a)$;

$\frac{1}{2}c = \frac{1}{2}(b - a)$; $c = b - a$

28. $n = \frac{1}{2}(x - y)$ so $2n = x - y$; also, $2n = \frac{1}{2}(x + y)$; $x - y = \frac{1}{2}(x + y)$; $2x - 2y =$

$x + y$; $x = 3y$; $\frac{x}{y} = 3$; $x : y = 3 : 1$

29. Given: Isos. $\triangle ABC$ with base $\overline{BC}$.

Prove: $\angle ABP \cong \angle Q$

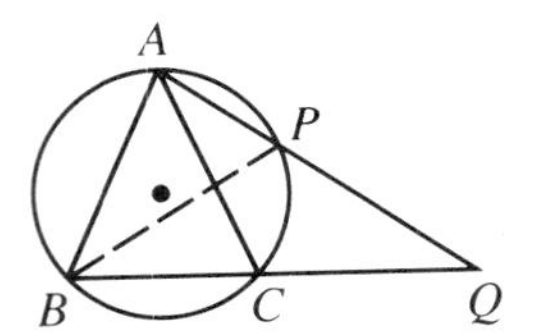

Statements	Reasons
1. Draw chord $\overline{BP}$.	1. Through any 2 pts. there is exactly one line.
2. $\triangle ABC$ is isos. with base $\overline{BC}$.	2. Given
3. $\overline{AB} \cong \overline{AC}$	3. Def. of isos. $\triangle$
4. $\widehat{AB} \cong \widehat{AC}$ or $m\widehat{AB} = m\widehat{AC}$	4. In the same $\odot$, $\cong$ chords have $\cong$ arcs.
5. $m\angle ABP = \frac{1}{2}m\widehat{AP}$	5. The meas. of an inscribed $\angle$ = half the meas. of its intercepted arc.
6. $m\angle Q = \frac{1}{2}(m\widehat{AB} - m\widehat{PC})$	6. The meas. of an $\angle$ formed by 2 secants drawn from a pt. outside a $\odot$ = half the diff. of the intercepted arcs.
7. $m\angle Q = \frac{1}{2}(m\widehat{AC} - m\widehat{PC})$	7. Substitution Prop.
8. $m\widehat{AC} = m\widehat{AP} + m\widehat{PC}$ or $m\widehat{AP} = m\widehat{AC} - m\widehat{PC}$	8. Arc Add. Post.
9. $m\angle Q = \frac{1}{2}m\widehat{AP}$	9. Substitution Prop.
10. $m\angle ABP = m\angle Q$ or $\angle ABP \cong \angle Q$	10. Substitution Prop.

C 30. Let X be a pt. on the $\odot$ not on $\widehat{RST}$. $m\angle P = \frac{1}{2}(m\widehat{RXT} - m\widehat{ST})$; $m\widehat{RXT} =$

$360 - (m\widehat{RS} + m\widehat{ST})$; $160 < m\widehat{RS} + m\widehat{ST} < 180$; $180 < m\widehat{RXT} < 200$;

$90 < m\widehat{RXT} - m\widehat{ST} < 120$; $45 < m\angle P < 60$

31. $m\widehat{CE} = 3m\widehat{BD}$

Statements	Reasons
1. $m\angle CAO = \frac{1}{2}(m\widehat{CE} - m\widehat{BD})$	1. The meas. of an $\angle$ formed by 2 secants drawn from a pt. outside a $\odot$ = half the diff. of the intercepted arcs.
2. $m\angle BOD = m\widehat{BD}$	2. Def. of meas. of an arc
3. $\overline{AB} \cong \overline{OB}$	3. Given
4. $\angle CAO \cong \angle BOD$ or $m\angle CAO = m\angle BOD$	4. Isos. $\triangle$ Thm.
5. $\frac{1}{2}(m\widehat{CE} - m\widehat{BD}) = m\widehat{BD}$	5. Substitution Prop.
6. $m\widehat{CE} - m\widehat{BD} = 2m\widehat{BD}$ or $m\widehat{CE} = 3m\widehat{BD}$	6. Algebra

32. $\widehat{AX} \cong \widehat{XB}$

Given: Tangent $\overline{PT}$; $PK = PT$

Prove: $\widehat{AX} \cong \widehat{XB}$

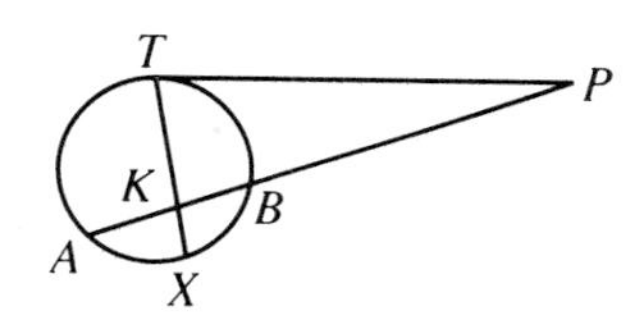

Statements	Reasons
1. $PK = PT$ or $\overline{PK} \cong \overline{PT}$	1. Given
2. $\angle PTK \cong \angle PKT$ or $m\angle PTK = m\angle PKT$	2. Isos. $\triangle$ Thm.
3. $\overline{PT}$ is a tangent.	3. Given
4. $m\angle PTK = \frac{1}{2}m\widehat{TBX}$	4. The meas. of an $\angle$ formed by a chord and a tan. = half the meas. of the intercepted arc.
5. $m\widehat{TBX} = m\widehat{TB} + m\widehat{BX}$	5. Arc Add. Post.
6. $m\angle PTK = \frac{1}{2}(m\widehat{TB} + m\widehat{BX})$	6. Substitution Prop.
7. $m\angle PKT = \frac{1}{2}(m\widehat{TB} + m\widehat{AX})$	7. The meas. of an $\angle$ formed by 2 chords that int. inside a $\odot$ = half the sum of the meas. of the intercepted arcs.
8. $\frac{1}{2}(m\widehat{TB} + m\widehat{BX}) = \frac{1}{2}(m\widehat{TB} + m\widehat{AX})$	8. Substitution Prop.
9. $m\widehat{BX} = m\widehat{AX}$ or $\widehat{AX} \cong \widehat{BX}$	9. Algebra

Page 361 • EXPLORATIONS

Check students' drawings.

1. $w \cdot x = y \cdot z$ **2.** $DE \cdot DB = DF \cdot DC$ **3.** $ED \cdot CE = (BE)^2$

Pages 363–364 • CLASSROOM EXERCISES

1. $3x = 4 \cdot 6$; $3x = 24$; $x = 8$

2. $x(8 - x) = 3 \cdot 4$; $8x - x^2 = 12$; $x^2 - 8x + 12 = 0$; $(x - 6)(x - 2) = 0$; $x = 6$ or $x = 2$

3. $8x = 12 \cdot 4$; $8x = 48$; $x = 6$ **4.** $6(x + 6) = 5 \cdot 12$; $6x + 36 = 60$; $6x = 24$; $x = 4$

5. $9x = 6^2$; $9x = 36$; $x = 4$ **6.** $x^2 = 8 \cdot 2$; $x^2 = 16$; $x = 4$

7. 1. Through any 2 pts. there is exactly one line. 2. If 2 inscribed ∠s intercept the same arc, then the ∠s are ≅. 3. Refl. Prop. 4. AA ~ Post. 5. Corr. sides of ~ △s are in prop. 6. Means-extremes Prop.

Pages 364–366 • WRITTEN EXERCISES

A **1.** $4x = 5 \cdot 8$; $4x = 40$; $x = 10$ **2.** $x^2 = 9 \cdot 16$; $x^2 = 144$; $x = 12$

3. $x^2 = 3 \cdot 7$; $x^2 = 21$; $x = \sqrt{21}$ **4.** $3 \cdot 8 = 2x \cdot 3x$; $24 = 6x^2$; $x^2 = 4$; $x = 2$

5. $8 \cdot 5 = 4(x + 4)$; $40 = 4x + 16$; $4x = 24$; $x = 6$

6. $x^2 = 3 \cdot 9$; $x^2 = 27$; $x = \sqrt{27} = 3\sqrt{3}$

7. $5x = 4 \cdot 10$; $5x = 40$; $x = 8$ **8.** $\frac{9}{2}x = \frac{9}{4} \cdot \frac{16}{3}$; $\frac{9}{2}x = 12$; $x = \frac{8}{3} = 2\frac{2}{3}$

9. $4x \cdot x = 10^2$; $4x^2 = 100$; $x^2 = 25$; $x = 5$

10.

Statements	Reasons
1. Draw chords $\overline{AC}$ and $\overline{BC}$.	1. Through any 2 pts. there is exactly one line.
2. $m\angle A = \frac{1}{2}m\widehat{BC}$	2. The meas. of an inscribed ∠ = half the meas. of its intercepted arc.
3. $m\angle PCB = \frac{1}{2}m\widehat{BC}$	3. The meas. of an ∠ formed by a chord and a tan. = half the meas. of the intercepted arc.
4. $m\angle A = m\angle PCB$ or $\angle A \cong \angle PCB$	4. Substitution Prop.
5. $\angle P \cong \angle P$	5. Refl. Prop.
6. $\triangle PAC \sim \triangle PCB$	6. AA ~ Post.
7. $\frac{r}{t} = \frac{t}{s}$	7. Corr. sides of ~ △s are in prop.
8. $r \cdot s = t^2$	8. Means-extremes Prop.

B **11.**

Statements	Reasons
1. $\overline{UT}$ is tan. to $\odot O$ and $\odot P$.	1. Given
2. $UV \cdot UW = (UT)^2$; $UX \cdot UY = (UT)^2$	2. When a secant seg. and a tan. seg. are drawn to a $\odot$ from an ext. pt., the product of the secant seg. and its ext. seg. = the square of the tan. seg.
3. $UV \cdot UW = UX \cdot UY$	3. Substitution Prop.

12.

Statements	Reasons
1. $\overline{AB}$ is tan. to $\odot Q$; $\overline{AC}$ is tan. to $\odot S$; $\overline{AE}$ and $\overline{AG}$ are secants.	1. Given
2. $AD \cdot AE = (AB)^2$; $AF \cdot AG = (AC)^2$	2. When a secant seg. and a tan. seg. are drawn to a $\odot$ from an ext. pt., the product of the secant seg. and its ext. seg. = the square of the tan. seg.
3. $AD \cdot AE = AF \cdot AG$	3. When 2 secant segs. are drawn to a $\odot$ from an ext. pt., the product of one secant seg. and its ext. seg. = the product of the other secant seg. and its ext. seg.
4. $(AB)^2 = (AC)^2$ or $AB = AC$ or $\overline{AB} \cong \overline{AC}$	4. Substitution Prop., algebra

13. Let $DP = x$. $x(16 - x) = 6 \cdot 8$; $16x - x^2 = 48$; $x^2 - 16x + 48 = 0$; $(x - 4)(x - 12) = 0$; $x = 4$ or $x = 12$

14. Let $AP = x$. $x(11 - x) = 6 \cdot 4$; $11x - x^2 = 24$; $x^2 - 11x + 24 = 0$; $(x - 3)(x - 8) = 0$; $x = 3$ or $x = 8$

15. Let $BP = x$. $x(12 - x) = 9 \cdot 4$; $12x - x^2 = 36$; $x^2 - 12x + 36 = 0$; $(x - 6)^2 = 0$; $x = 6$

16. Let $DP = x$. $x(3x) = 5 \cdot 6$; $3x^2 = 30$; $x^2 = 10$; $x = \sqrt{10}$

17. Let $AB = x$. $3(x + 3) = 6^2$; $3x + 9 = 36$; $3x = 27$; $x = 9$

18. Let $PC = x$. $x(x - 18) = 12^2$; $x^2 - 18x = 144$; $x^2 - 18x - 144 = 0$; $(x + 6)(x - 24) = 0$; $x = -6$ (reject) or $x = 24$

19. Let $PB = x$. $x(x + 11) = 5(5 + 7)$; $x^2 + 11x = 60$; $x^2 + 11x - 60 = 0$; $(x + 15)(x - 4) = 0$; $x = -15$ (reject) or $x = 4$

20. Let $PT = x$ and $PC = y$. $x^2 = 5 \cdot 10$; $x^2 = 50$; $x = \sqrt{50} = 5\sqrt{2}$; $4y = 5 \cdot 10$; $4y = 50$; $y = 12.5$

21. **a.** Pythagorean Thm. **b.** $r^2 + 2rh + h^2 = r^2 + d^2$; $d^2 = 2rh + h^2 = h(2r + h)$ **c.** Thm. 9-13

22. **a.** $TP = \sqrt{6^2 - 3^2} = 3\sqrt{3}$; $(3\sqrt{3})^2 = 3(AB + 3)$, $27 = 3(AB + 3)$, $9 = AB + 3$, $AB = 6$ **b.** Let $\overline{OX}$ be $\perp$ to $\overline{AB}$ at X. $OX = TP = 3\sqrt{3}$ **c.** $AX = 3$; $r^2 = 3^2 + (3\sqrt{3})^2 = 9 + 27 = 36$, $r = 6$

23. $r^2 = 12^2 + (r - 4)^2$; $r^2 = 144 + r^2 - 8r + 16$; $8r = 160$; $r = 20$ m

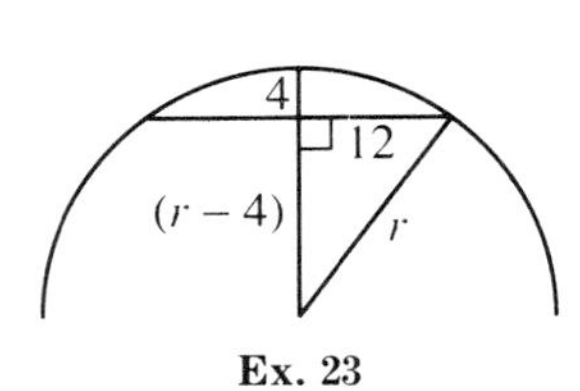

Ex. 23

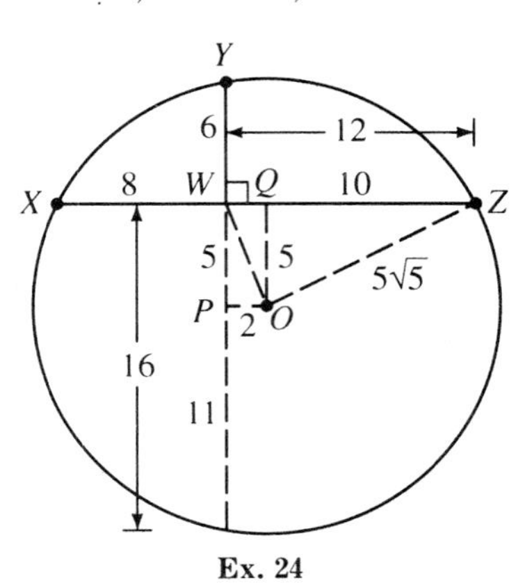

Ex. 24

24. $6x = 8 \cdot 12$, $6x = 96$, $x = 16$; diameters bisect both segments so that $XQ = QZ = 10$, $WQ = 2$, $YP = 11$, and $WP = 5$. **a.** $OZ = \sqrt{10^2 + 5^2} = 5\sqrt{5}$ **b.** $WO = \sqrt{5^2 + 2^2} = \sqrt{29}$

25. Prove: $AX \cdot XB = CX \cdot XD$

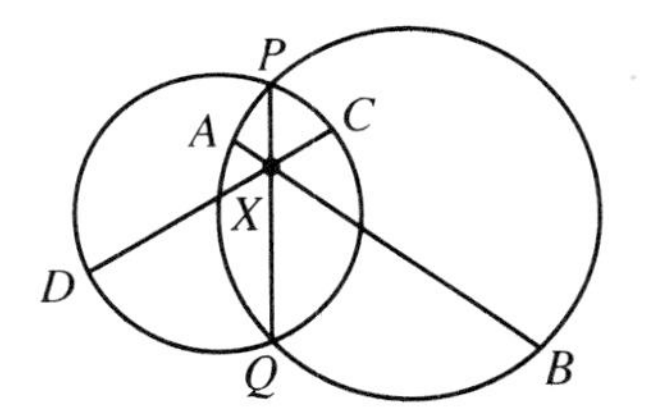

Statements	Reasons
1. $AX \cdot XB = PX \cdot XQ$; $CX \cdot XD = PX \cdot XQ$	1. When 2 chords intersect inside a ⊙, the product of the segs. of one chord = the product of the segs. of the other chord.
2. $AX \cdot XB = CX \cdot XD$	2. Trans. Prop.

26. Prove: T is the midpt. of $\overline{PQ}$.

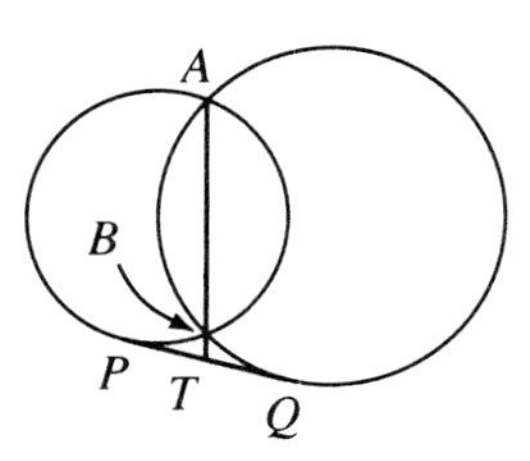

Statements	Reasons
1. $TB \cdot TA = (PT)^2$; $TB \cdot TA = (TQ)^2$	1. When a secant seg. and a tan. seg. are drawn to a ⊙ from an ext. pt., the product of the secant seg. and its ext. seg. = the square of the tan. seg.
2. $(PT)^2 = (TQ)^2$ or $PT = TQ$	2. Substitution Prop.
3. T is the midpt. of $\overline{PQ}$.	3. Def. of midpt.

C 27. Extend $\overrightarrow{PN}$ to intersect $\odot O$ at R; $8(RN + 12) = 12^2$; $8 \cdot RN + 96 = 144$; $8 \cdot RN = 48$; $RN = 6$; let r = radius of $\odot O$; extend $\overrightarrow{ON}$ to intersect $\odot O$ at A and extend $\overrightarrow{NO}$ to intersect $\odot O$ at B; $BN \cdot AN = 6 \cdot 4$; $(BO + ON)NA = 24$; $(r + 4)(r - 4) = 24$; $r^2 - 16 = 24$; $r^2 = 40$; $r = \sqrt{40} = 2\sqrt{10}$.

28. $y^2 = x(x + 16)$; $\angle AFB \cong \angle CFD$ (vert. ∠s) and $\angle BCD \cong \angle ABC$ (each has meas. $\frac{1}{2}\widehat{BC}$), so $\triangle AFB \sim \triangle DFC$. Then $\frac{3}{y} = \frac{6}{x + 10}$, or $y = \frac{1}{2}(x + 10)$.

$\frac{1}{4}(x + 10)^2 = x(x + 16)$; $(3x + 50)(x - 2) = 0$; $x = 2$; $ED = 2$

Page 367 • SELF-TEST 2

1. 40 **2.** 150 **3.** $81 = \frac{1}{2}(80 + x)$; $x = 82$ **4.** $8x = 6 \cdot 12$; $8x = 72$; $x = 9$

5. $x = \frac{1}{2}(250 - 110) = 70$; $40 = \frac{1}{2}(130 - y)$; $y = 50$ **6.** $\frac{1}{2}(100 - 30) = 35$

7. $4 \cdot 15 = 6 \cdot PF$; $60 = 6 \cdot PF$; $PF = 10$ **8.** $(PJ)^2 = 8 \cdot 18 = 144$; $PJ = 12$

Page 368 • APPLICATION

1. $3600\sqrt{16} = 3600(4) = 14{,}400$; about 14 km

2. $3600\sqrt{36} = 3600(6) = 21{,}600$; about 22 km

3. $3600\sqrt{10000} = 3600(100) = 360{,}000$; about 360 km

4. $3600\sqrt{h} = 8000$; $\sqrt{h} = \frac{8000}{3600} \approx 2.2$; $h \approx 5$ m

5. The distance is the same as that from an observer at the top of Mt. Fuji to the horizon; $3600\sqrt{3776} \approx 3600(60) = 216{,}000$; about 220 km

Pages 369–370 • CHAPTER REVIEW

1. chord; secant **2.** radius **3.** diameter **4.** inscribed in **5.** tangent **6.** $\perp$

7. $\sqrt{6^2 + 8^2} = 10$ **8.** $13\sqrt{2}$ **9.** 100 **10.** 260 **11.** $\angle YPW$ **12.** 60 **13.** 120

14. $\sqrt{5^2 + 12^2} = 13$ **15.** In the same ⊙, ≅ chords are equally distant from the center.

16. $180 - 105 = 75$ **17.** 50; 50 **18.** 220; $360 - 220 = 140$ **19.** $\frac{1}{2}(120 + 90) = 105$

20. $\frac{1}{2}(100 - 40) = 30$ **21.** $25 = \frac{1}{2}(90 - x); x = 40$

22. $4x = 8 \cdot 6; 4x = 48; x = 12$

23. $7(x + 7) = 8(8 + 6); 7x + 49 = 112; 7x = 63; x = 9$

24. $x^2 = 5(5 + 6) = 55; x = \sqrt{55}$

Page 371 • CHAPTER TEST

1. False **2.** False **3.** True **4.** False **5.** False **6.** True **7.** True **8.** False

9. $\frac{1}{2}(360 - 100) = 130$ **10.** $\sqrt{17^2 - 15^2} = 8$ **11.** $\frac{1}{2}(360 - 160) = 100$

12. Let $m\widehat{BA} = x$. $110 = \frac{1}{2}(360 - 2x); x = 70; m\angle BCA = \frac{1}{2}(70) = 35$

13.

Statements	Reasons
1. $m\angle BAC = \frac{1}{2}m\widehat{BC}$	1. The meas. of an inscribed $\angle$ = half the meas. of its intercepted arc.
2. $m\angle DBA = \frac{1}{2}m\widehat{AB}$	2. The meas. of an $\angle$ formed by a chord and a tan. = half the meas. of the intercepted arc.
3. $m\widehat{BC} = m\widehat{AB}$	3. Given
4. $\frac{1}{2}m\widehat{BC} = \frac{1}{2}m\widehat{AB}$	4. Mult. Prop. of =
5. $m\angle BAC = m\angle DBA$ or $\angle BAC \cong \angle DBA$	5. Substitution Prop.
6. $\overline{AC} \parallel \overline{DB}$	6. If 2 lines are cut by a trans. and alt. int. $\angle$s are $\cong$, then the lines are $\parallel$.

14. $\frac{1}{2}(40 + 28) = 34$ **15.** $10 \cdot 9 = 15x; x = 6$ **16.** $\frac{1}{2}[160 - (360 - 280)] = 40$

17. $12^2 = x \cdot 18; x = 8$

18. Given: $\square ABCD$ is inscribed in a $\odot$.

Prove: $ABCD$ is a rect.

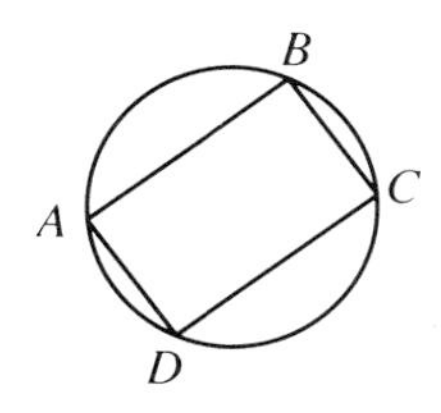

Statements	Reasons
1. $ABCD$ is a $\square$.	1. Given
2. $\angle A \cong \angle C$ or $m\angle A = m\angle C$	2. Opp. $\angle$s of a $\square$ are $\cong$.
3. $\angle A$ and $\angle C$ are supp.	3. If a quad. is inscribed in a $\odot$, then its opp. $\angle$s are supp.
4. $m\angle A + m\angle C = 180$	4. Def. of supp. $\angle$s
5. $2m\angle A = 180$ or $m\angle A = 90$	5. Substitution Prop.
6. $\angle A$ is a rt. $\angle$.	6. Def. of rt. $\angle$
7. $ABCD$ is a rect.	7. If an $\angle$ of a $\square$ is a rt. $\angle$, then the $\square$ is a rect.

Pages 372–373 • CUMULATIVE REVIEW: CHAPTERS 1–9

A **1.** $3 < y < 2x + 3$ **2.** $\frac{180 - x}{90 - x} = \frac{5}{2}$; $360 - 2x = 450 - 5x$; $3x = 90$; $x = 30$

3.

Statements	Reasons
1. $\overline{MN}$ is a median of a trap.	1. Given
2. $\overline{MN} \parallel \overline{ZY} \parallel \overline{WX}$	2. The median of a trap. is $\parallel$ to the bases.
3. V is the midpt. of $\overline{WY}$.	3. A line that contains the midpt. of one side of a $\triangle$ and is $\parallel$ to another side passes through the midpt. of the third side.
4. $\overline{MN}$ bis. $\overline{WY}$.	4. Def. of bis.

4. Given: Rhombus $ABCD$

Prove: $\triangle AEB \cong \triangle CEB \cong \triangle CED \cong \triangle AED$

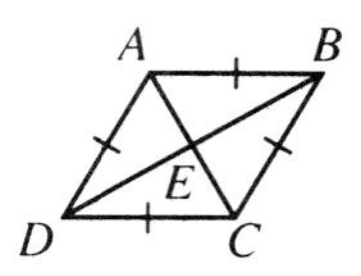

Statements	Reasons
1. $ABCD$ is a rhombus.	1. Given
2. $\overline{AB} \cong \overline{BC} \cong \overline{CD} \cong \overline{DA}$	2. Def. of rhombus
3. $\overline{AC} \perp \overline{BD}$	3. The diags. of a rhombus are $\perp$.
4. $\angle AEB$, $\angle CEB$, $\angle CED$, and $\angle AED$ are rt. $\angle$s.	4. Def. of $\perp$ lines
5. $\triangle AEB$, $\triangle CEB$, $\triangle CED$, and $\triangle AED$ are rt. $\triangle$s.	5. Def. of rt. $\triangle$
6. $\overline{BE} \cong \overline{BE}$; $\overline{CE} \cong \overline{CE}$; $\overline{DE} \cong \overline{DE}$	6. Refl. Prop.
7. $\triangle AEB \cong \triangle CEB \cong \triangle CED \cong \triangle AED$	7. HL Thm.

5. $7, 7\sqrt{3}$ **6.** No

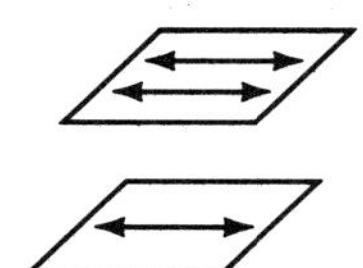

7. $x + 9x + 10x = 180; 20x = 180; x = 9; 9x = 81; 10x = 90; 9, 81, 90$

8. Interior: $\frac{(18 - 2)180}{18} = 160$; Exterior: $180 - 160 = 20$ **9.** $2\sqrt{2}$

10. $x^2 + 6^2 = 10^2, x = 8; 6^2 + 10^2 = x^2, x = 2\sqrt{34}$

11.

Statements	Reasons
1. $\angle 1 \cong \angle 2$; $\angle 2 \cong \angle 3$	1. Given
2. $\overline{AB} \parallel \overline{DC}$	2. If 2 lines are cut by a trans. and alt. int. $\angle$s are $\cong$, then the lines are $\parallel$.
3. $\overline{AD} \parallel \overline{BC}$	3. If 2 lines are cut by a trans. and corr. $\angle$s are $\cong$, then the lines are $\parallel$.
4. $ABCD$ is a $\square$.	4. Def. of $\square$
5. $\overline{AB} \cong \overline{DC}$	5. Opp. sides of a $\square$ are $\cong$.

12. $\frac{8}{x} = \frac{x}{18}$; $x^2 = 144$; $x = 12$

13. **a.** If $\angle A \not\cong \angle C$, then quad. $ABCD$ is not a $\square$. **b.** If quad. $ABCD$ is not a $\square$, then $\angle A \not\cong \angle C$.

14. $2(5t - 7) = 8t + 10; t = 12; m\angle BOC = 5t - 7 = 53$

15. $15 \cdot 12 = 9t; 180 = 9t; t = 20$ **16.** $RS = 3 - (-11) = 14; ST = 7$

17. **a.** inside **b.** on **c.** on **18.** $\frac{18}{x} = \frac{27}{15 - x}$; $x = 6$

19. **a.** Janice likes to dance. **b.** No conclusion **c.** No conclusion **d.** Kim is not Bill's sister.

20. Assume temp. that $ABCD$ is not a rect. **21.** always **22.** never **23.** always

24. sometimes **25.** sometimes **26.** $7 \cdot 10$; $9 \cdot x$; $7\frac{7}{9}$

B **27.** 0; 1

28. **a.** $\triangle ZWS$ **b.** ZW; SZ; WS **c.** $\frac{15}{ZW} = \frac{10}{8}$, $ZW = 12$; $\frac{15}{12} = \frac{12}{WS}$, $WS = 9\frac{3}{5}$; $RS = 15 - 9\frac{3}{5} = 5\frac{2}{5}$

29.

Statements	Reasons
1. $AB > AC$	1. Given
2. $m\angle ACB > m\angle ABC$	2. If one side of a $\triangle$ is longer than a second side, then the $\angle$ opp. the first side $>$ the $\angle$ opp. the second side.
3. $\overline{BD} \cong \overline{EC}$	3. Given
4. $\overline{BC} \cong \overline{BC}$	4. Refl. Prop.
5. $BE > CD$	5. SAS Ineq. Thm.

30.

Statements	Reasons
1. $\frac{PR}{TR} = \frac{SR}{QR}$	1. Given
2. $\angle PRS \cong \angle TRQ$	2. Vert. $\angle$s are $\cong$.
3. $\triangle PRS \sim \triangle TRQ$	3. SAS $\sim$ Thm.
4. $\angle S \cong \angle Q$	4. Corr. $\angle$s of $\sim$ $\triangle$s are $\cong$.

CHAPTER 10 • Constructions and Loci

Page 377 • CLASSROOM EXERCISES

1. Answers may vary. Example: Const. $\overline{AB} \cong \overline{JK}$; const. $\angle CAB \cong \angle J$ and $\angle CBA \cong \angle K$.
2. **a–c.**

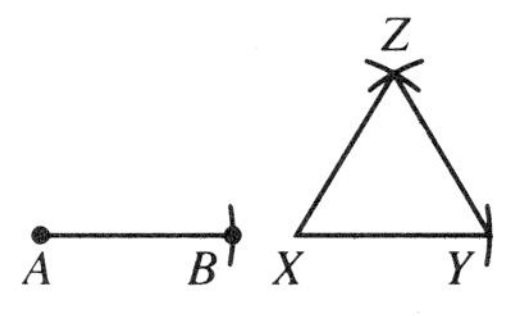

d. equilateral

3. Const. an equilateral △ as in Ex. 2 and bisect one of the ∠s.
4. Answers may vary. Examples: 15°, 45°, 60°, 90°, 75°, 120°
5. No; the sum of the lengths of any 2 sides of a △ is greater than the length of the third side.
6. Yes

Pages 378–379 • WRITTEN EXERCISES

A 1.

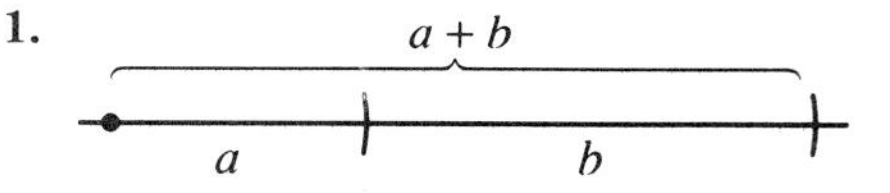

2.

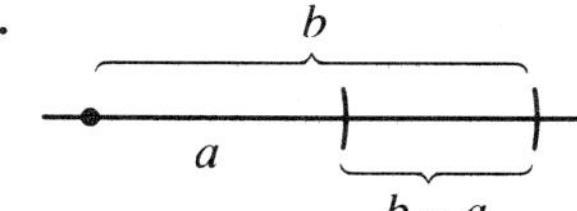

3.

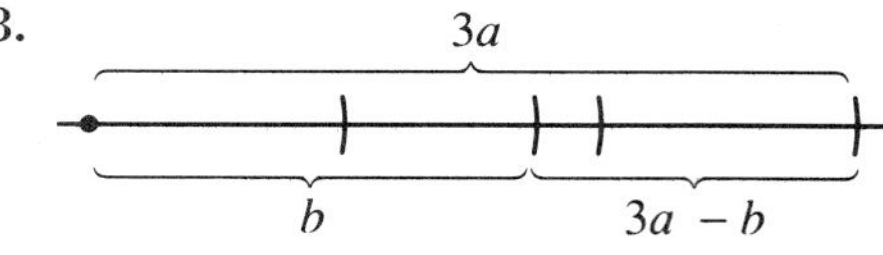

4.

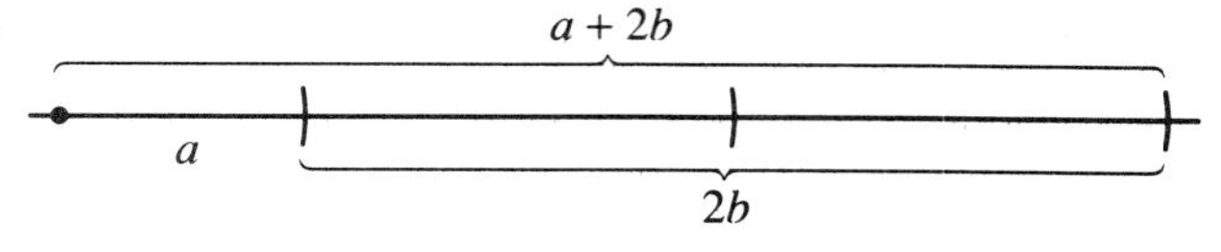

5.

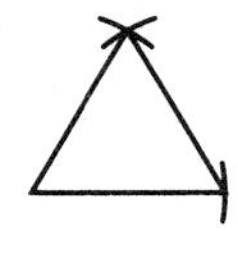

6. **a.**

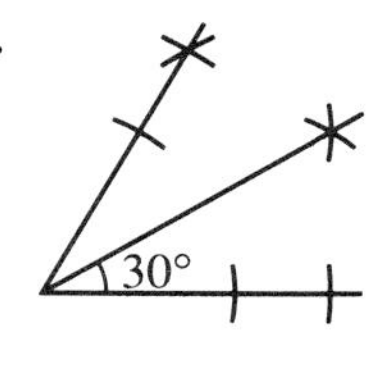

b.

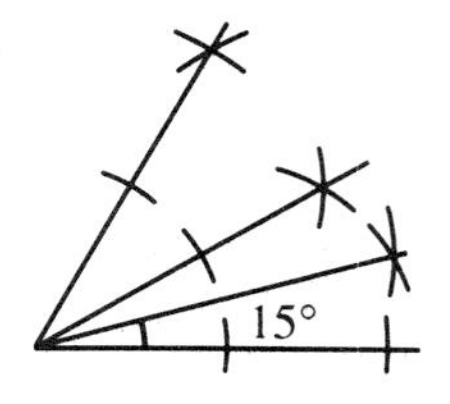

7.

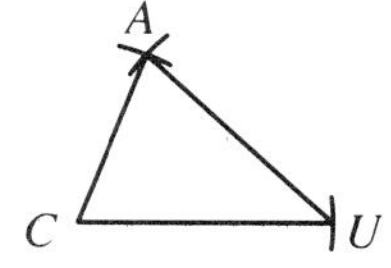

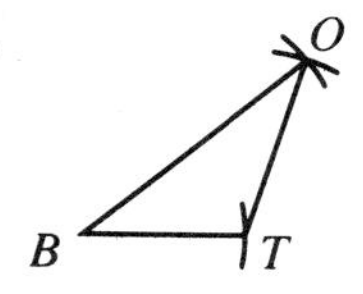

8.

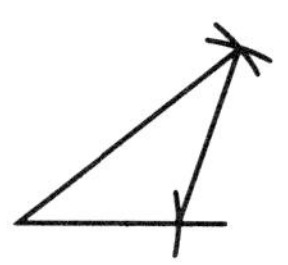

9.

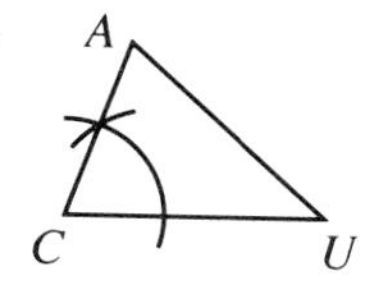

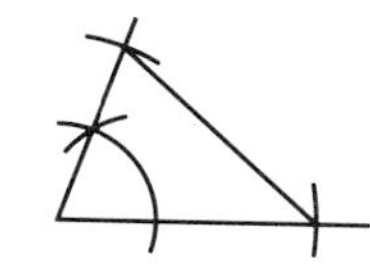

10.

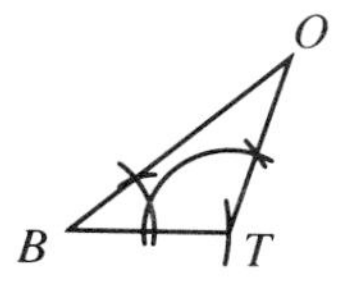

11–14. $\angle ABC$ is the required angle.

11.

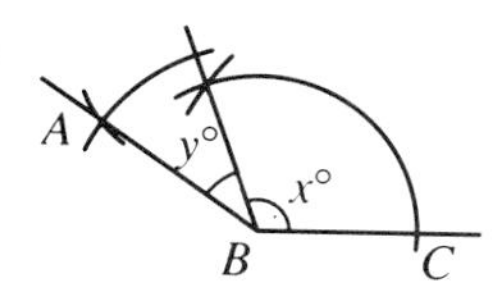

12.

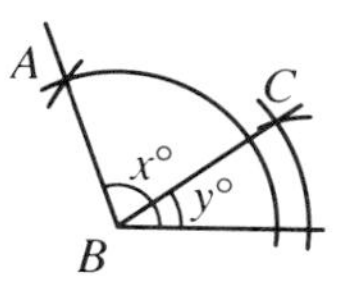

13.

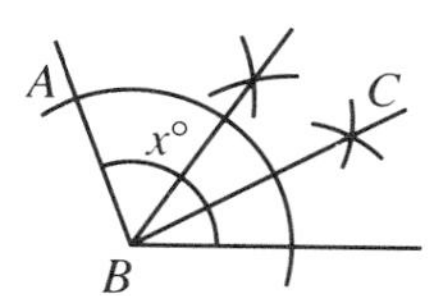

14.

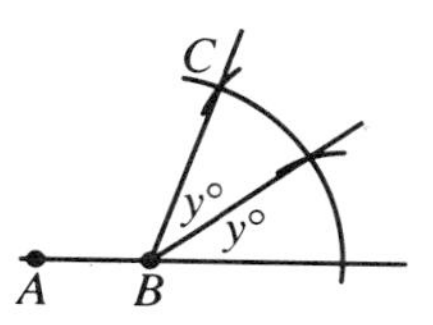

B 15. a.

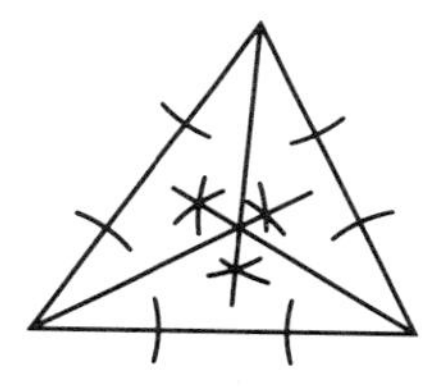

b.

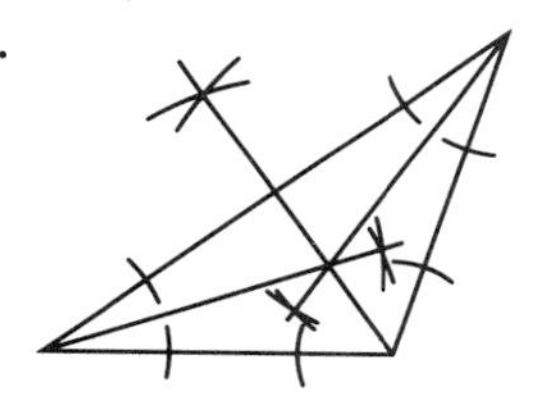

c. They are the same pt., which is equidistant from the sides of the △s.

16.

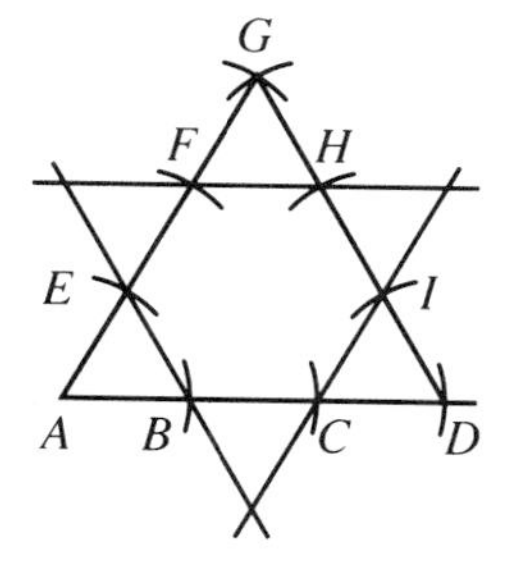

17–26. Methods may vary.

17.

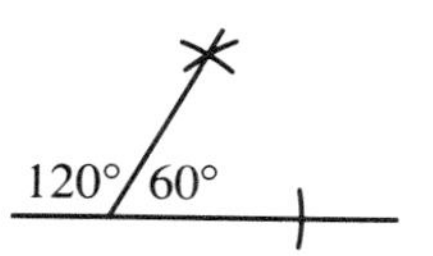

18.

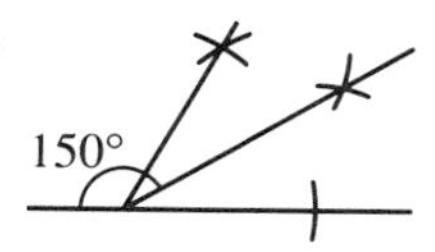

19.

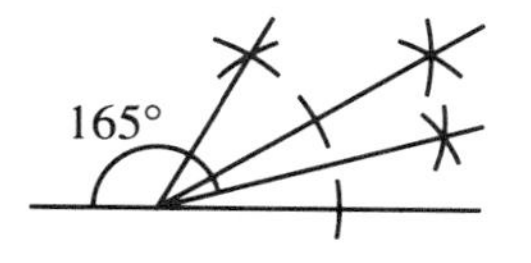

20.

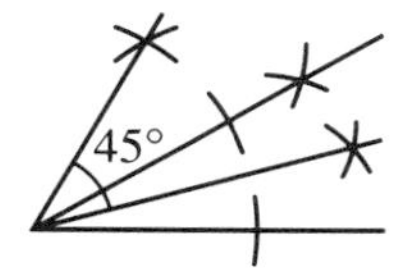

21. Const. $DE = 2AB$. Const. $\angle D \cong \angle A$ and $DF = 2AC$. By SAS $\sim$ Thm., $\triangle DEF \sim \triangle ABC$.

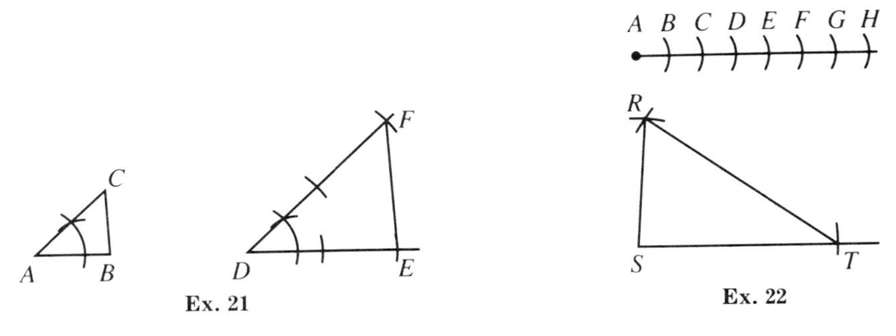

Ex. 21

Ex. 22

22. Draw a line and with any convenient radius r, mark off 7 $\cong$ arcs with length r. Label the pts. A, B, C, D, E, F, G, and H. Const. $\overline{ST}$ such that $\overline{ST} \cong \overline{AG}$; with centers S and T and radii AE and AH, resp., draw arcs int. at R. Draw $\overline{RS}$ and $\overline{RT}$.

23. Const. a leg with length d. At one endpt., const. an $\angle$ with measure n. Mark off another leg with length d, and draw the base.

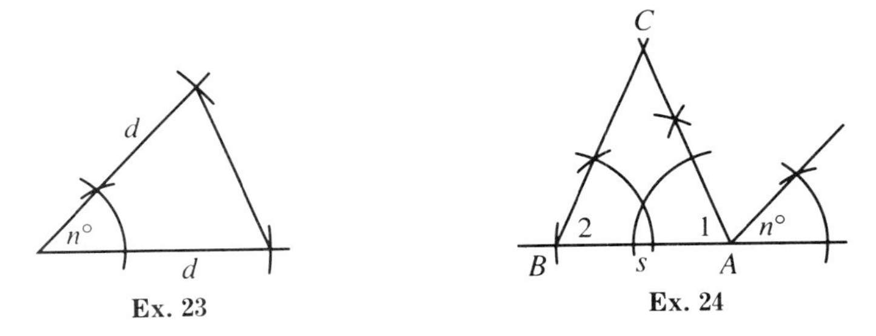

Ex. 23

Ex. 24

24. Const. $\angle A$ with measure n; bisect the supp. of $\angle A$ to const. $\angle 1$ as shown; mark off $AB = s$ with $\overrightarrow{AB}$ a side of $\angle 1$; at B, const. $\angle 2 \cong \angle 1$. Extend the sides of $\angle 1$ and $\angle 2$ to intersect at C.

C 25. Const. $\angle A$ with measure n; const. $\overline{AB}$ so that $AB = s$. Const. $\angle B \cong \angle A$; with center A and radius d, draw an arc int. side of $\angle B$ at C. Const. $\overline{AD} \cong \overline{BC}$; draw $\overline{DC}$. Since $\angle A \cong \angle CBE$, $\overline{AD} \parallel \overline{BC}$; since $\overline{AD} \cong \overline{BC}$ as well, $ABCD$ is a $\square$.

26. Draw a ⊙ with radius r; draw any chord $\overline{AB}$; const. $\angle A \cong \angle 2$, int. ⊙ at C. Draw $\overline{BC}$. Const. an $\angle$ at $B \cong \angle 1$, int. ⊙ at D. Since $\angle D$ and $\angle A$ int. the same arc, $\angle D \cong \angle A \cong \angle 2$. Then $m\angle BCD = 180 - (m\angle DBC + m\angle D) = 180 - (m\angle 1 + m\angle 2) = m\angle 3$.

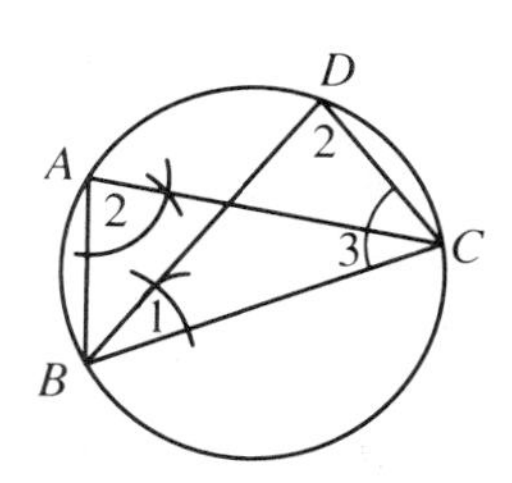

Page 380 • MIXED REVIEW EXERCISES

1. midpt. **2.** ▱ **3.** rect. **4.** rhombus **5.** $5\sqrt{2}$ **6.** 108

Pages 382–383 • CLASSROOM EXERCISES

1. Use Const. 6 to const. $j \perp k$ through P. Then use Const. 5 to const. $l \perp j$ through P; $l \parallel k$.

2. Use Const. 4. **3.** Use Const. 4 to locate the midpt. M of $\overline{AC}$; draw $\overline{MB}$.

4. Use Const. 6 to const. a line $\perp$ to $\overline{AC}$ through B.

5. Extend $\overrightarrow{CB}$. Use Const. 6 to const. a line $\perp$ to $\overleftrightarrow{BC}$ through A.

6. Extend $\overrightarrow{BC}$. Use Const. 5.

7. Const. $\overline{DE} \cong \overline{AC}$; const. $\perp$s to $\overline{DE}$ at D and E; mark off $\overline{DG} \cong \overline{AC}$ and $\overline{EF} \cong \overline{AC}$ on the $\perp$s; draw $\overline{FG}$.

8. Const. $\overline{DX} \cong \overline{AC}$; const. the $\perp$ bis. of $\overline{DX}$ to locate the midpt. Y; const. the $\perp$ bis. of $\overline{DY}$ to locate the midpt. E. Const. $\perp$s to $\overline{DE}$ at D and E; mark off $\overline{DG} \cong \overline{DE}$ and $\overline{EF} \cong \overline{DE}$ on the $\perp$s; draw $\overline{FG}$.

9. Const. $\overline{DE} \cong \overline{BC}$; const. a $\perp$ to $\overline{DE}$ at D; with center E and radius AC, draw an arc int. the $\perp$ at F. Draw $\overline{EF}$.

10. Draw $\overline{DE}$; const. a $\perp$ to $\overline{DE}$ at D; const. $\overline{DF}$ on the $\perp$ with $DF = 2DE$. Draw $\overline{EF}$.

11. on the $\perp$ to l through X **12.** on the $\perp$ bis. of $\overline{XY}$

13. Use Const. 5 to const. a $\perp$ to l at X; use Const. 4 to const. the $\perp$ bis. of $\overline{XY}$; their intersection, O, is the center of the circle; radius $= OX$ (or OY).

Pages 383–385 • WRITTEN EXERCISES

A **1.** Const. 5 **2.** Const. 6 **3.** Const. 4 **4.** Const. 7 **5.** Const. 7 **6.** Const. 5

7. Extend $\overrightarrow{HJ}$; use Const. 6.

8. Methods may vary. Example: Draw $\overrightarrow{NM}$. Const. $\overline{PM} \perp \overline{MN}$; $\angle PMK$ is comp. to $\angle KMN$.

9–12. Methods may vary.

9.

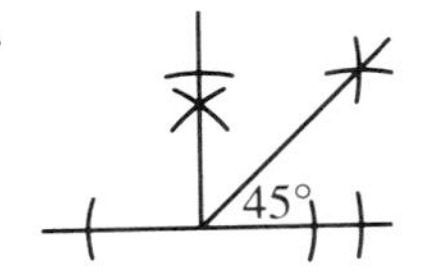

10.

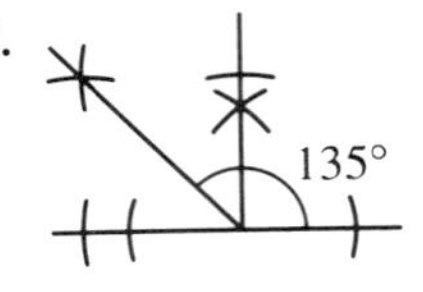

11.

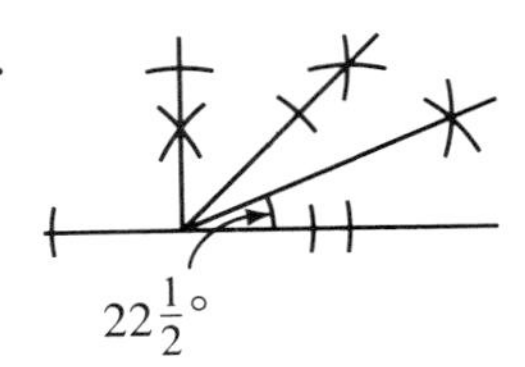

12.

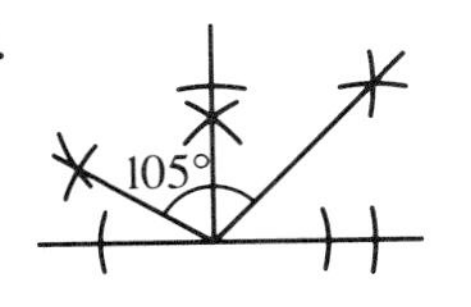

13.

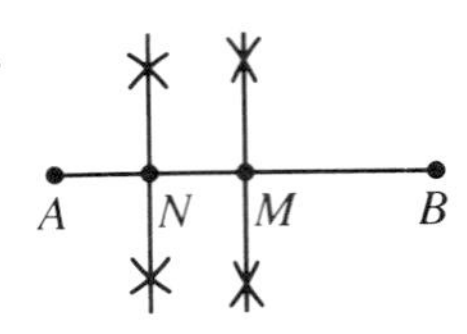

B **14–16.** $\triangle JKL$ is the given $\triangle$.

14. a.

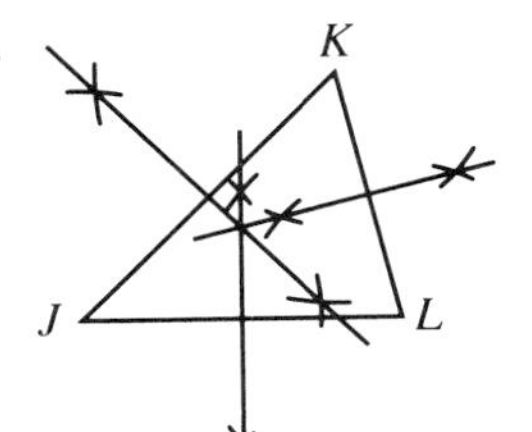

b. Yes

c. Yes

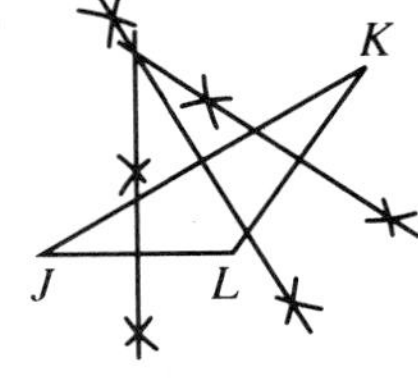

15. a.

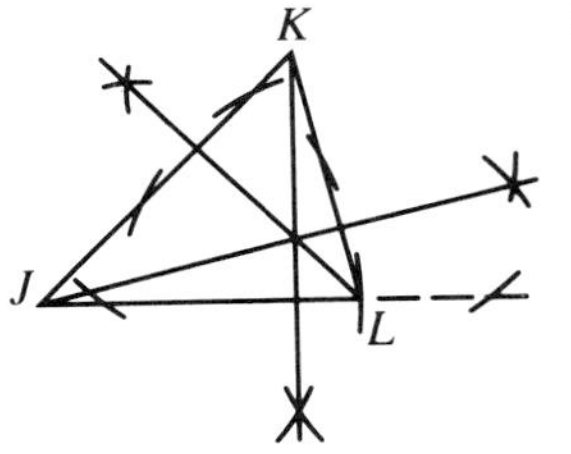

b. Yes

c. Yes

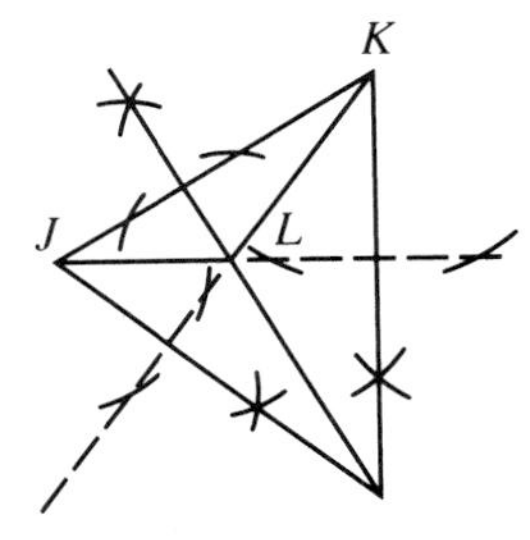

16. a.

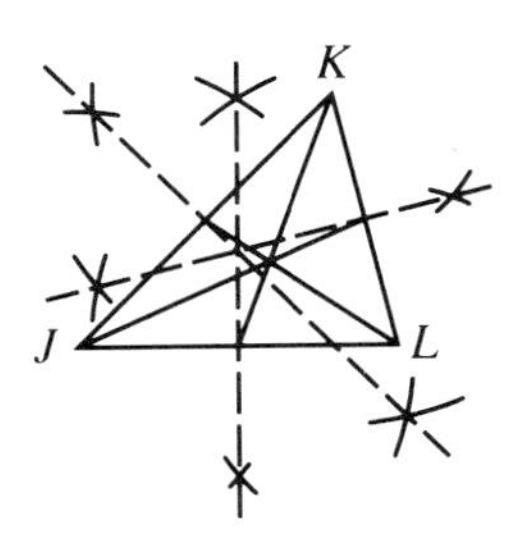

b. Yes

c. Yes

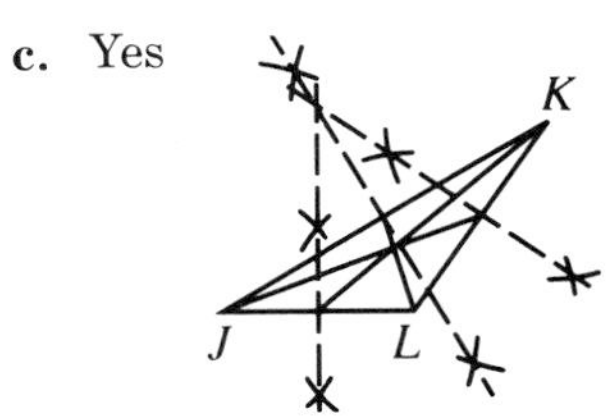

17. Const. $\triangle DAB$ with $m\angle A = n$, $AB = a$, and $AD = b$; with centers D and B and radii a and b, resp., draw arcs int. at C. Draw $\overline{BC}$ and $\overline{DC}$.

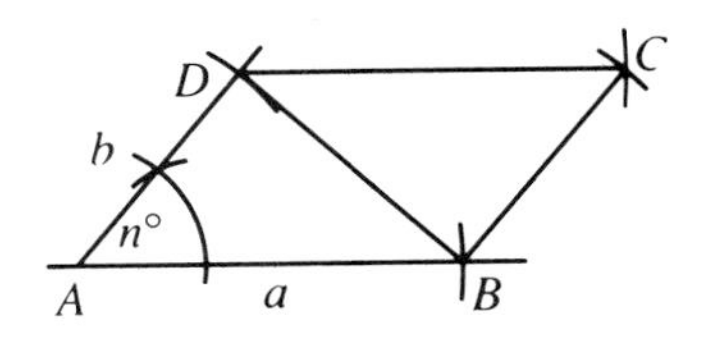

18. Const. 2 $\perp$ lines int. at A; mark off $\overline{AB}$ on one line so that $AB = a$, and $\overline{AD}$ on the other line so that $AD = b$; with centers D and B and radii a and b, resp., draw arcs int. at C; draw $\overline{BC}$ and $\overline{DC}$.

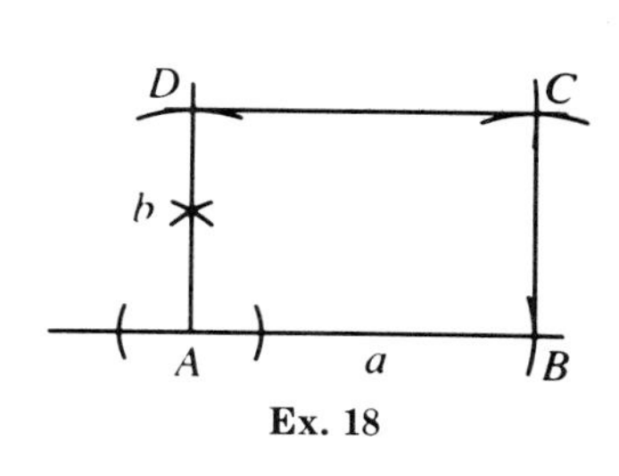

Ex. 18

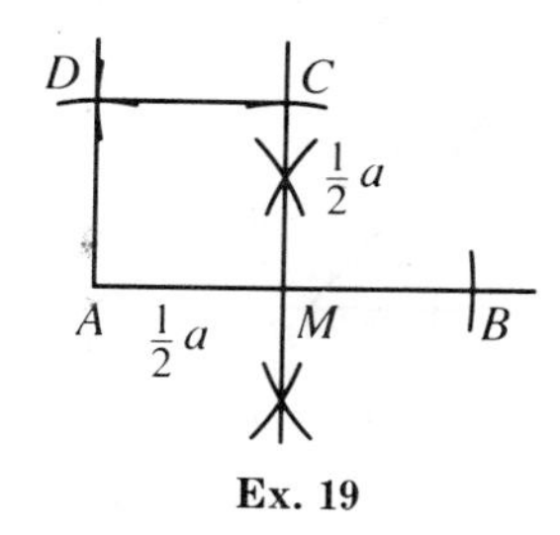

Ex. 19

19. Const. $\overline{AB}$ so that $AB = a$. Const. the $\perp$ bis of $\overline{AB}$ int. $\overline{AB}$ at M, so that $AM = \frac{1}{2}a$. Const. $\overline{MC} \perp \overline{AB}$ so that $MC = \frac{1}{2}a$. With centers A and C and radius AM, draw arcs int. at D. Draw $\overline{AD}$ and $\overline{CD}$.

20. Const. $\overline{AC}$ so that $AC = a$. Const. j, the $\perp$ bis. of $\overline{AC}$ int. $\overline{AC}$ at M. Const. $\overline{MB}$ and $\overline{MD}$, both on j, so that $MB = MD = \frac{1}{2}b$. Draw $ABCD$.

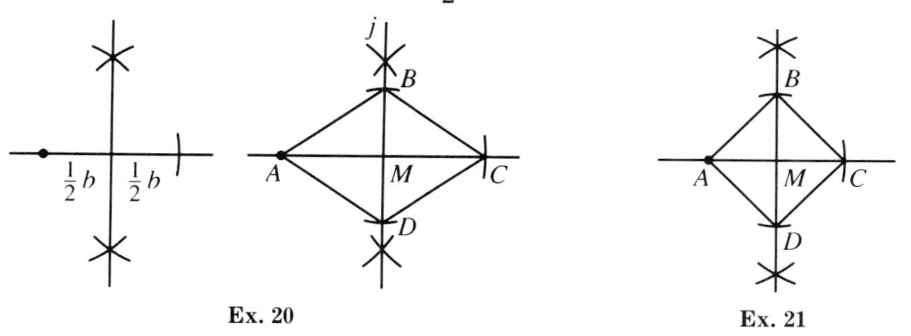

Ex. 20 **Ex. 21**

21. Draw a line and const. $\overline{AC}$ so that $AC = b$. Const. the $\perp$ bis. of $\overline{AC}$ int. $\overline{AC}$ at M. Const. $\overline{MB}$ and $\overline{MD}$ on the $\perp$ bis. of $\overline{AC}$ and both $\cong$ to $\overline{AM}$. Draw $ABCD$.

22. Const. 2 $\perp$ lines at A; mark off $AB = a$ on one line and $AC = b$ on the other line. Draw $\overline{BC}$.

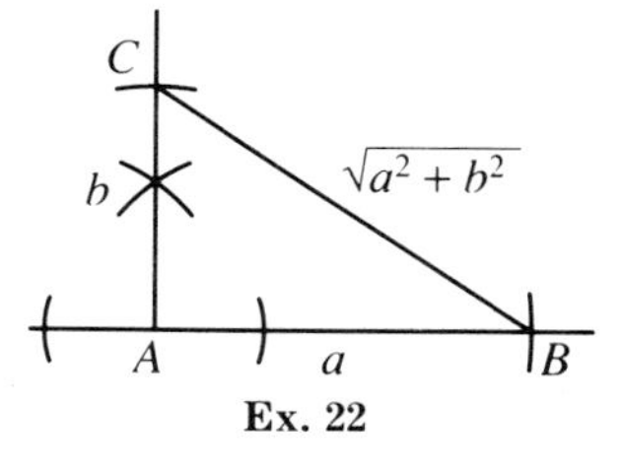

Ex. 22

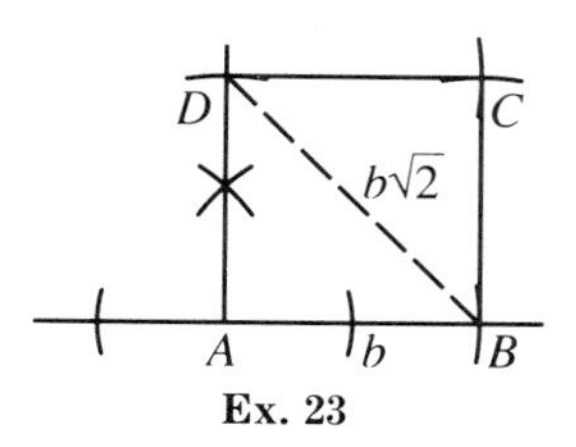

Ex. 23

23. Const. 2 $\perp$ lines at A; mark off $AB = b$ on one line and $AD = b$ on the other line. With centers D and B and radii b, draw arcs int. at C. Draw $\overline{DC}$ and $\overline{BC}$.

24. Const. 2 ⊥ lines at A; mark off $AB = b$ on one line; with center B and radius a, draw an arc int. the other line at C. Draw $\overline{BC}$.

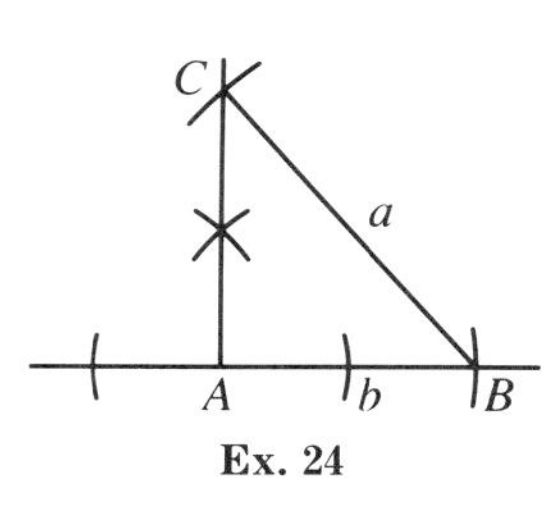

Ex. 24

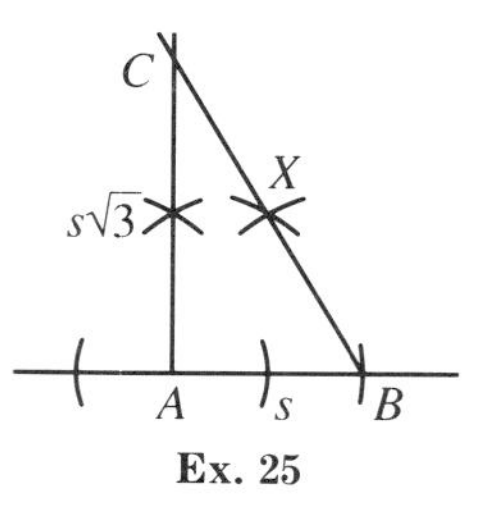

Ex. 25

C 25. Given $AB = s$, const. a ⊥ to $\overline{AB}$ at A. With centers A and B and both radii $= s$, draw arcs int. at X. Draw $\overrightarrow{BX}$ to int. the ⊥ at C. $\triangle ABC$ is a 30°-60°-90° △, and $AC = s\sqrt{3}$.

26. Choose a pt. P on $\overline{RS}$ and const. $\overline{AB} \perp \overline{RS}$ at P as shown. $\overleftrightarrow{RS}$ is the ⊥ bis. of $\overline{AB}$. With centers A and B, draw a pair of arcs int. at T, on the other side of the lake. Using a larger radius, repeat to find a pt. U. Draw $\overleftrightarrow{TU}$. Since T and U are equidistant from A and B, T and U are on the ⊥ bis. of $\overline{AB}$; so $\overleftrightarrow{TU}$ and $\overleftrightarrow{RS}$ are the same.

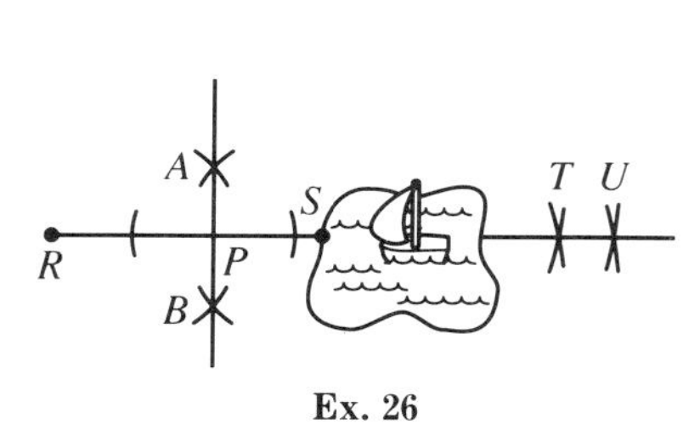

Ex. 26

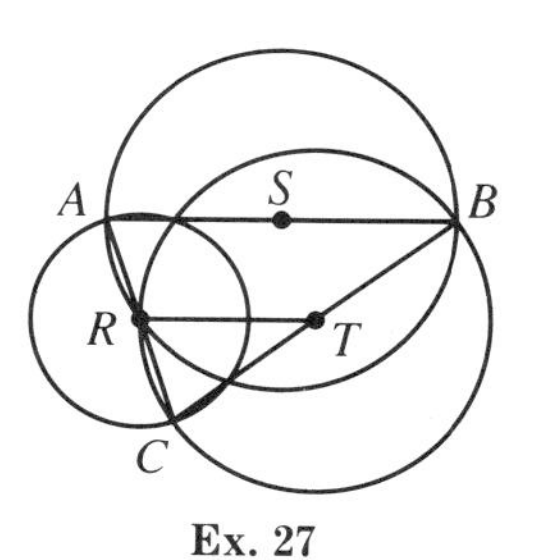

Ex. 27

27. RT = half the length of the side with midpt. S. Const. ⊙S with radius RT, ⊙R with radius ST, and ⊙T with radius RS. R, S, and T are the midpts. of the sides of $\triangle ABC$.

28. **a.** Let $AB = 1$. Const. $\overline{BC} \cong \overline{AB}$ and const. a ⊥ to $\overline{AB}$ at A. Const. $\overline{AD} \cong \overline{AB}$. Draw $\overline{DC}$. $DC = \sqrt{1^2 + 2^2} = \sqrt{5}$. **b.** Draw a line and const. $\overline{EF} \cong \overline{AB}$ and $\overline{FG} \cong \overline{DC}$. Const. the ⊥ bis. of $\overline{EG}$, int. $\overline{EG}$ at H. $EH = \dfrac{1 + \sqrt{5}}{2}$ **c.** Const. $\overline{HI} \cong \overline{AB}$. With centers E and I and radii AB and EH, resp., draw arcs int. at J. Draw $\overline{JI}$ and $\overline{JE}$.

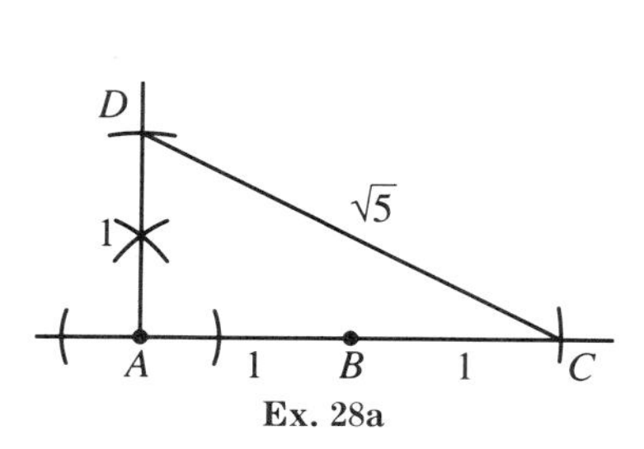

Ex. 28a

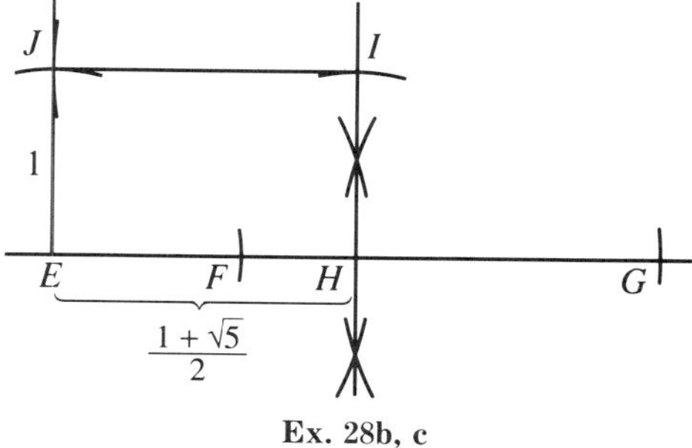

Ex. 28b, c

Page 385 • CHALLENGE

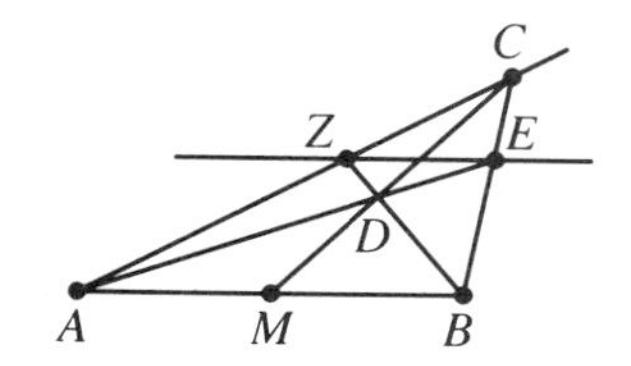

Draw $\overrightarrow{AZ}$ and choose C on $\overrightarrow{AZ}$. Draw $\overline{CB}$, $\overline{CM}$, and $\overline{ZB}$, with $\overline{ZB}$ and $\overline{CM}$ int. at D. Draw $\overrightarrow{AD}$, int. $\overline{CB}$ at E. Draw $\overleftrightarrow{ZE}$; $\overleftrightarrow{ZE} \parallel \overline{AB}$. By Ceva's Thm.,
$\frac{AM}{MB} \cdot \frac{BE}{EC} \cdot \frac{CZ}{ZA} = 1$; $\frac{AM}{MB} = 1$ so $\frac{BE}{EC} \cdot \frac{CZ}{ZA} = 1$;
$\frac{BE}{EC} = \frac{ZA}{CZ}$; $\frac{BE + EC}{EC} = \frac{ZA + CZ}{CZ}$; $\frac{BC}{EC} = \frac{CA}{CZ}$;
then since $\angle ACB \cong \angle ACB$, $\triangle CZE \sim \triangle CAB$ (SAS),
$\angle CZE \cong \angle CAB$ and $\overleftrightarrow{ZE} \parallel \overline{AB}$.

Page 385 • EXPLORATIONS

1. The 3 lengths are equal.
2. **a.** The 3 distances are equal. **b.** Yes **c.** The midpt. of the hypotenuse
3. $\frac{AG}{AD} = \frac{BG}{BE} = \frac{CG}{CF} = \frac{2}{3}$

Page 388 • CLASSROOM EXERCISES

1. Answers may vary. Examples: **a.** An acute $\triangle$ **b.** An obtuse $\triangle$ **c.** A rt. $\triangle$
2. **a.** Any $\triangle$ **b, c.** Not possible **3.** Yes; an equilateral $\triangle$
4. **a.** Yes **b.** $\overline{AM}$ **c.** M **d.** At the midpt. of $\overline{JA}$ **e.** Yes
5. **a.** 10 **b.** 4 **c.** 3; 6
6. **a.** $\overline{XY} \parallel \overline{RS}$ and $XY = \frac{1}{2}RS$; the seg. that joins the midpts. of 2 sides of a $\triangle$ is $\parallel$ to the third side and is half as long as the third side. **b.** $\overline{KJ} \parallel \overline{RS}$ and $KJ = \frac{1}{2}RS$; same as part (a) **c.** $\overline{KJ} \parallel \overline{XY}$ and $KJ = XY$; 2 lines $\parallel$ to a third line are $\parallel$ to each other, and substitution. **d.** $XYJK$ is a ▱; if one pair of opp. sides of a quad. are both $\cong$ and $\parallel$, then the quad. is a ▱. **e.** Diags. of a ▱ bis. each other.
   f. $RX = XG = GJ = \frac{1}{3}RJ$, so $RX + XG = RG = \frac{2}{3}RJ$

Pages 388–389 • WRITTEN EXERCISES

A 1. **a.** Any acute $\triangle$ **b.** Any obtuse $\triangle$ **c.** Any rt. $\triangle$

2. $x = 6$; $y = 2\frac{1}{2}$ **3.** 2; 4 **4.** $\frac{7}{3}$; $\frac{14}{3}$ **5.** 3.8; 5.7 **6.** Const. 4 **7.** Const. 3

8. **a, b.** Check students' drawings. **c.** The 3 segments form an equilateral $\triangle$.

B **9.** The pt. of intersection of the $\perp$ bisectors of $\overline{XY}$, $\overline{XZ}$, and $\overline{YZ}$ is equidistant from all 3 towns. But this pt. is quite far from X, Y, and Z. It would be wiser to build the hall equidistant from X and Z, near Y.

10. Let X, Y, and Z be collinear pts.

11. **a.** $GD = \frac{1}{3} \cdot AD = \frac{1}{3} \cdot BE = GE$ **b.** GB **c.** $\angle GBA$, $\angle GED$, $\angle GDE$

12. $x^2 = 2 \cdot 2x$; $x^2 - 4x = 0$; $x(x - 4) = 0$; $x = 0$ (reject) or $x = 4$

13. $y^2 + 1 = 2(y + 2)$; $y^2 + 1 = 2y + 4$; $y^2 - 2y - 3 = 0$; $(y - 3)(y + 1) = 0$; $y = 3$ or $y = -1$

14. $z^2 - 15 = \frac{2}{3}(2z^2 - 5z - 12)$; $3z^2 - 45 = 4z^2 - 10z - 24$; $z^2 - 10z + 21 = 0$; $(z - 3)(z - 7) = 0$; $z = 3$ or $z = 7$; if $z = 3$, $CP = -6$, so reject $z = 3$; $z = 7$; $CW = 51$, so $PW = 17$

15.

Statements	Reasons
1. Draw $\overline{BD}$ int. $\overline{AC}$ at Y.	1. Through any 2 pts. there is exactly one line.
2. $ABCD$ is a $\square$; M is the midpt. of $\overline{CD}$.	2. Given
3. $\overline{BD}$ and $\overline{AC}$ bis. each other.	3. Diags. of a $\square$ bis. each other.
4. Y is the midpt. of $\overline{AC}$.	4. Def. of bis.
5. $CY = \frac{1}{2}AC$	5. Midpt. Thm.
6. $\overline{BM}$ and $\overline{CY}$ are medians of $\triangle BDC$.	6. Def. of median
7. $CX = \frac{2}{3}CY$	7. The medians of a $\triangle$ int. in a pt. that is $\frac{2}{3}$ of the distance from each vertex to the midpt. of the opp. side.
8. $CX = \frac{2}{3} \cdot \frac{1}{2} \cdot AC = \frac{1}{3}AC$	8. Substitution Prop.

16. Given: $\triangle ABC$ with medians $\overline{CM}$ and $\overline{BN}$; $\overline{CM} \cong \overline{BN}$
Prove: $\triangle ABC$ is isos.

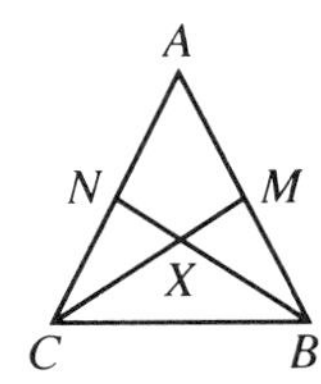

Statements	Reasons
1. $\overline{CM}$ and $\overline{BN}$ are medians.	1. Given
2. $XM = \frac{1}{3}CM$; $XN = \frac{1}{3}BN$	2. The medians of a △ int. in a pt. that is $\frac{2}{3}$ of the distance from each vertex to the midpt. of the opp. side.
3. $\overline{CM} \cong \overline{BN}$ or $CM = BN$	3. Given
4. $\frac{1}{3}CM = \frac{1}{3}BN$	4. Mult. Prop. of =
5. $XM = XN$ or $\overline{XM} \cong \overline{XN}$	5. Substitution Prop.
6. Draw $\overline{MN}$.	6. Through any 2 pts. there is exactly one line.
7. $\angle NMX \cong \angle MNX$	7. Isos. △ Thm.
8. $\overline{MN} \cong \overline{MN}$	8. Refl. Prop.
9. $\triangle MNC \cong \triangle NMB$	9. SAS Post.
10. $\overline{NC} \cong \overline{MB}$ or $NC = MB$	10. Corr. parts of ≅ ▲ are ≅.
11. N and M are midpts. of $\overline{AC}$ and $\overline{AB}$, resp.	11. Def. of median
12. $NC = \frac{1}{2}AC$; $MB = \frac{1}{2}AB$	12. Midpt. Thm.
13. $\frac{1}{2}AC = \frac{1}{2}AB$ or $AC = AB$ or $\overline{AC} \cong \overline{AB}$	13. Substitution Prop., Mult. Prop. of =
14. $\triangle ABC$ is isos.	14. Def. of isos. △

C 17. a. Points in the interior of $\angle XPY$ **b.** Points in the interior of the ∠ vertical to $\angle XPY$

Page 390 • SELF-TEST 1

1. Const. 4

2. Draw $\overline{ST}$. With centers S and T, and radius ST, draw arcs int. at R. Draw $\overrightarrow{SR}$; $m\angle RST = 60$. Use Const. 3.

3. Const. 6 **4.** Const. 7

5. Methods may vary. Example: Const. $\overline{JK}$ such that $JK = 2AB$ (Const. 1). Const. lines ⊥ to $\overline{JK}$ at J and K. Const. $\overline{JM} \cong \overline{KL} \cong \overline{AB}$. Draw $\overline{ML}$.

6. Lines containing the altitudes, medians, ∠ bisectors, and ⊥ bisectors of the sides

7. the midpt. of the hypotenuse

8. $BD = 6$; $AD = 6\sqrt{3}$; $AX = \frac{2}{3} \cdot 6\sqrt{3} = 4\sqrt{3}$; $XD = \frac{1}{3} \cdot 6\sqrt{3} = 2\sqrt{3}$

Page 391 • APPLICATION

1. Check students' work. 2. b. The lines drawn are the medians of the $\triangle$. 3. Yes

Page 391 • MIXED REVIEW EXERCISES

1. $AB = \sqrt{13^2 - 5^2} = \sqrt{144} = 12$ 2. tangent; 10 3. 3

4. $m\angle S = 180 - m\angle Q = 180 - 39 = 141$

Page 392 • EXPLORATIONS

1. $m\angle ABD = \frac{1}{2} m\angle ABC$; $m\angle ACD = \frac{1}{2} m\angle ACB$; $m\angle BAD = \frac{1}{2} m\angle BAC$; $\angle$ bis.

2. $AE = \frac{1}{2}AB$; $BF = \frac{1}{2}BC$; $CG = \frac{1}{2}CA$; the $\perp$ bis. of the sides

Pages 394–395 • CLASSROOM EXERCISES

1. Const. the $\perp$ bis., k, of $\overline{AB}$. The center of the $\odot$ is on k since the center is equidistant from A and B, so k contains a diam. of the $\odot$. k bisects $\overparen{AB}$.
2. Const. the $\perp$ bis. of 2 chords. They intersect at the center of the $\odot$.
3. a. Const. 7 b. Const. a $\perp$ to $\overline{RS}$ through P int. $\odot P$ at Q. Const. a $\parallel$ to $\overline{RS}$ through Q (or a $\perp$ to $\overline{PQ}$ through Q).
4. Let r = radius; choose A on $\odot O$ and mark off an arc with center A and radius r, int. $\odot O$ at B. Continue the process. Six equilateral $\triangle$s have been constructed. $\overline{AB} \cong \overline{BC} \cong \overline{CD}$, and so on. Since $\cong$ chords have $\cong$ arcs, the $\odot$ is divided into 6 $\cong$ arcs.
5. Draw $\triangle AEC$. $\overparen{AE} \cong \overparen{AC} \cong \overparen{CE}$, so $\angle A \cong \angle E \cong \angle C$; hence, $\triangle AEC$ is equiangular and equilateral. Similarly, for $\triangle FBD$.
6. Const. $\perp$s from the center to $\overline{AB}$, $\overline{BC}$, and so forth, int. the chords at U, V, W, X, Y, and Z. The $\odot$ is divided into 12 $\cong$ arcs. Since $\cong$ arcs have $\cong$ chords, $AUBVCWDXEYFZ$ is a reg. 12-sided polygon.
7. Const. a $\perp$ from I to any one of the sides of $\triangle RST$. The length of the $\perp$ is the radius of the circle.

Pages 395–396 • WRITTEN EXERCISES

A 1. Const. 8 2. Const. 9 3. Const. 10 4. Const. 10 5. Const. 10 6. Const. 11 7. Const. 11 8. Const. 11

B **9.** Draw $\odot O$ with radius r. Choose pt. A on $\odot O$, and with center A and radius r, mark off $\overparen{AB}$. With center B and radius r, mark off $\overparen{BC}$. Similarly, mark off $\cong$ arcs $\overparen{CD}$, $\overparen{DE}$, and $\overparen{EF}$. Draw $\overline{AC}$, $\overline{EC}$, and $\overline{AE}$.

10. Draw $\odot O$ and diam. $\overline{AC}$. Const. the $\perp$ bis. of $\overline{AC}$, int. $\odot O$ at B and D. Draw $\overline{AB}$, $\overline{BC}$, $\overline{CD}$, and $\overline{AD}$.

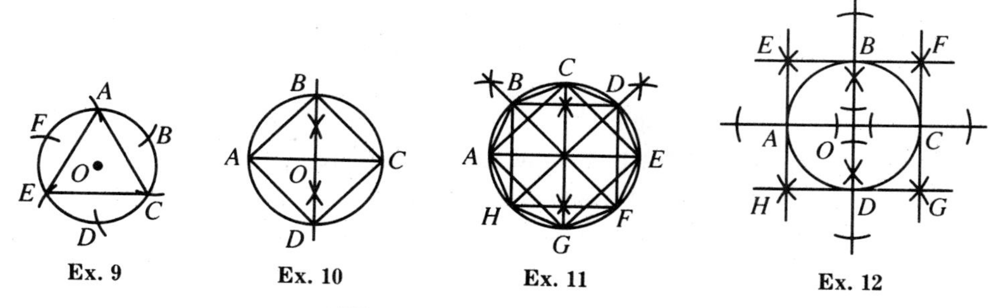

11. **a.** Draw $\odot O$. Draw diam. $\overline{AE}$. Const. $\perp$ diam. $\overline{CG}$. Bis. 2 adj. rt. $\angle$s to form 8 $\cong$ arcs. Connect consec. pts. to form octagon $ABCDEFGH$. **b.** Draw overlapping squares $ACEG$ and $BDFH$.

12. Draw $\odot O$ and diam. $\overline{AC}$. Const. the $\perp$ bis. of $\overline{AC}$ int. $\odot O$ at B and D. Const. tangents to $\odot O$ at A, B, C, and D intersecting at E, F, G, and H.

13. Draw the diags. of the square int. at O. Draw the $\odot$ with center O and radius = half the length of a diag.

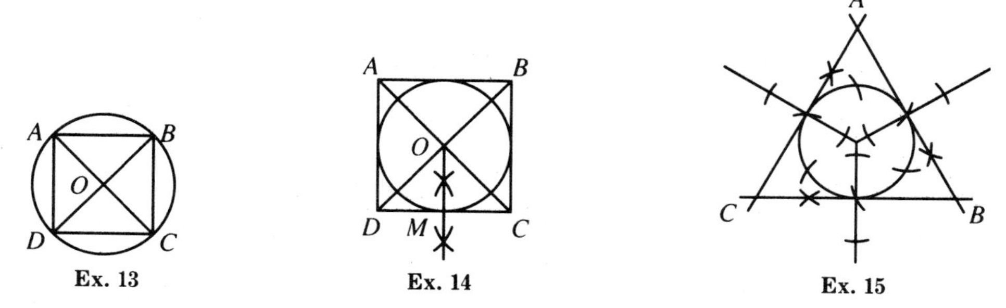

14. Draw the diags. of the square int. at O. Const. the $\perp$ bis. of one side of the square, int. the side at M. Draw the $\odot$ with center O and radius $\overline{OM}$.

15. Divide the $\odot$ into 6 $\cong$ arcs as in Ex. 9. At every other pt., const. a tan. to the $\odot$.

16. Const. a $\perp$ to l through O, int. $\odot O$ at P. Const. a tan. to $\odot O$ at P.

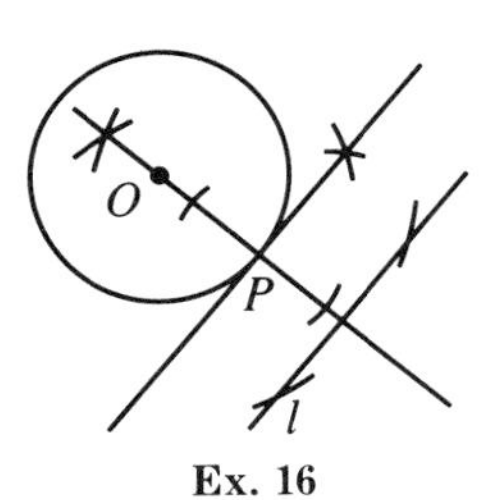

Ex. 16

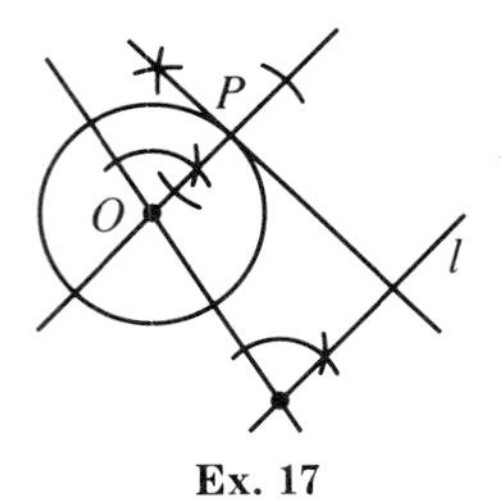

Ex. 17

17. Const. a $\parallel$ to l through O, int. $\odot O$ at P. Const. a tan. to $\odot O$ at P.

C **18.** Const. equilateral $\triangle ABC$. (Draw $\overline{AB}$. With centers A and B and radius AB, draw arcs int. at C. Draw $\overline{CA}$ and $\overline{CB}$.) Const. the $\perp$ bis. of $\overline{AB}$ int. $\overline{AB}$ at D. Draw $\odot$s A, B, and C with radius AD. Const. $\overline{AG} \perp \overline{AC}$, $\overline{CH} \perp \overline{CB}$, and $\overline{BI} \perp \overline{AB}$. Const. a tan. to $\odot A$ at G, a tan. to $\odot C$ at H, and a tan. to $\odot B$ at I. The pts. where the tans. int. are the vertices of an equilateral $\triangle$.

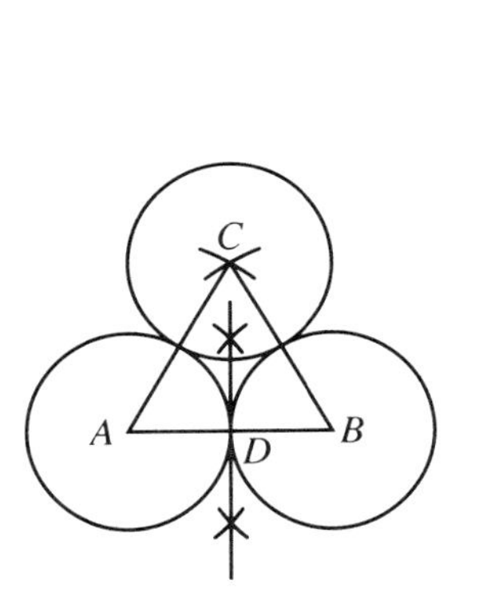

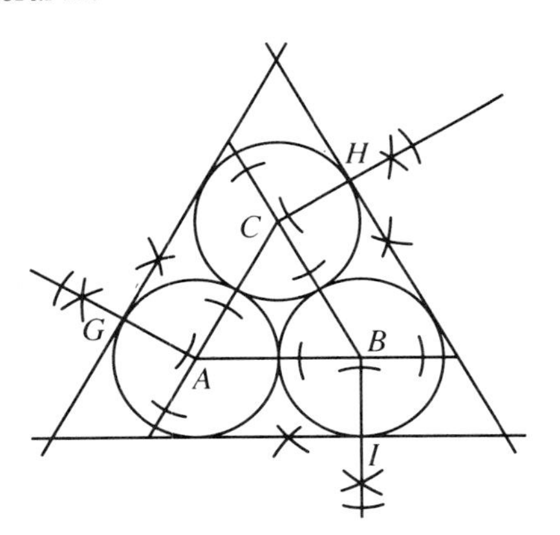

19. Draw $\overleftrightarrow{PQ}$ int. $\odot P$ at X and $\odot Q$ at Y, as shown. Let R be the center of the required $\odot$. $PR = PQ + p = XQ$ and $QR = PQ + q = PY$. With centers P and Q and radii XQ and PY, resp., draw arcs int. at a pt.; label it R. Draw $\odot R$ with radius PQ.

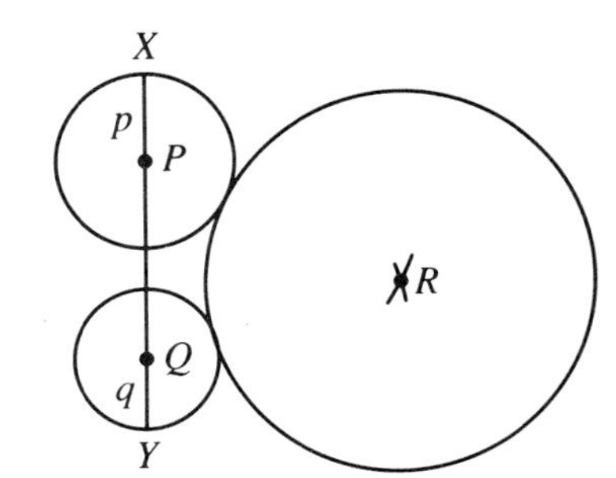

20. $XY = ZQ$; $XZ = p - (p - q) = q = QY$ so $XYQZ$ is a ▱. Since $\angle XZQ$ is a rt. ∠, $XYQZ$ is a rect. Then $\overleftrightarrow{XY}$ is ⊥ to radii $\overline{PX}$ and $\overline{QY}$ and $\overleftrightarrow{XY}$ is a common tan. of ⊙s P and Q.

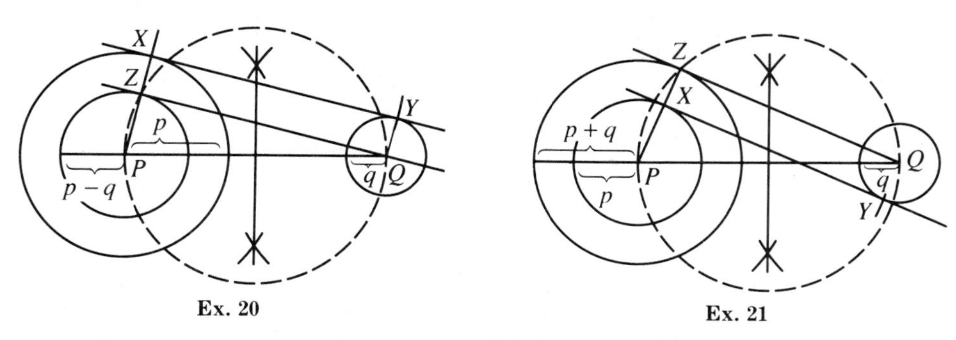

Ex. 20

Ex. 21

21. Draw a ⊙ with center P and radius $p + q$. Const. a tan. to this ⊙ from Q. Let Z be the pt. of tangency. Draw $\overline{PZ}$, int. the inner ⊙P at X. With center X and radius ZQ, draw an arc int. ⊙Q at Y. Draw $\overleftrightarrow{XY}$. $XY = ZQ$ and $XZ = p + q - p = q = YQ$ so $XYQZ$ is a ▱. Since $\angle PZQ$ is a rt. ∠, $XYQZ$ is a rect. Then $\overleftrightarrow{XY}$ is ⊥ to radii $\overline{PX}$ and $\overline{QY}$ and $\overleftrightarrow{XY}$ is a common tan. of ⊙s P and Q.

Page 398 • CLASSROOM EXERCISES

1. Divide the segment $\overline{AD}$ into 3 ≅ parts, $\overline{AB}$, $\overline{BC}$, and $\overline{CD}$. With centers A and B and radius AB, draw arcs int. at E. Draw $\overline{EA}$ and $\overline{EB}$.
2. Const. 13 **3.** Const. a seg. with length $2a$ (or $2b$); then use Const. 14.
4. Const. a seg. with length $5a$ (or $5b$); then use Const. 14.
5. Const. segments with lengths $4a$ and b (or $4b$ and a, or $2a$ and $2b$); then use Const. 14. Or use Const. 14 to find $x = \sqrt{ab}$; then const. a seg. with length $2\sqrt{ab}$.
6. on the ⊥ bis. of $\overline{AB}$ **7.** on the ⊥ to t through K **8.** $(JK)^2 = JB \cdot JA$
9. $\frac{JB}{JK} = \frac{JK}{JA}$ or $\frac{JA}{JK} = \frac{JK}{JB}$ **10.** Const. 14
11. Const. the ⊥ bis. of $\overline{AB}$ and a ⊥ to t at K, int. at O. Draw ⊙O with radius $OA = OB = OK$.

Page 399 • WRITTEN EXERCISES

A **1.** Const. 12 **2. a.** Const. 12 **b.** Const. 4

3. a. Const. 12 **b.** No **c.** Let the 5 ≅ segs. from part (a) be $\overline{AW}$, $\overline{WX}$, $\overline{XY}$, $\overline{YZ}$, and $\overline{ZB}$. $AX : XB = 2 : 3$

4. Use Const. 12 to divide $\overline{AB}$ into 7 $\cong$ parts, $\overline{AU}$, $\overline{UV}$, $\overline{VW}$, $\overline{WX}$, $\overline{XY}$, $\overline{YZ}$, and $\overline{ZB}$. $AW : WB = 3 : 4$

5. Const. 13 **6.** Const. 14 **7.** Const. 14 **8.** Const. 1, 12

B 9. Use $\frac{z}{w} = \frac{y}{x}$ or $\frac{z}{y} = \frac{w}{x}$ with Const. 13. **10.** Const. 13 **11.** Const. 14, 12

12. Const. 1, 14 **13.** Const. 1, 14

14. Draw $\overrightarrow{AX}$ and const. $\overline{AR}$ and $\overline{RS}$ on $\overrightarrow{AX}$ so that $AR = w$ and $RS = y$. Draw $\overline{SB}$. Const. a $\parallel$ to $\overline{SB}$ through R, int. $\overline{AB}$ at T. $AT : TB = w : y$.

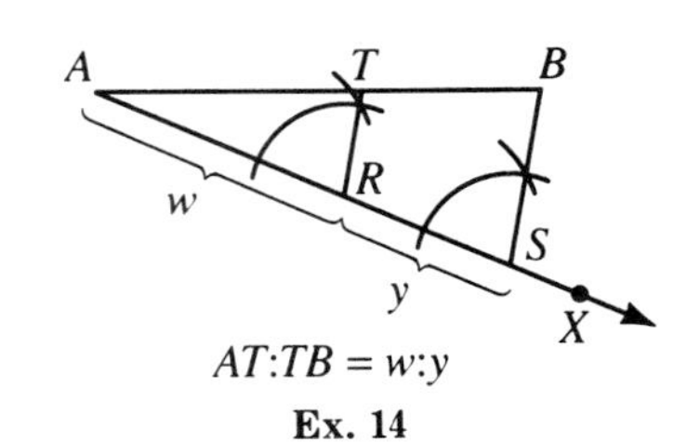

$AT{:}TB = w{:}y$
Ex. 14

15. Draw a line and const. $\overline{AB}$ so that $AB = 3$ and $\overline{BC}$ so that $BC = 5$. Use Const. 14.

16. **a.** an **b.** Use Const. 14 with lengths $2a$ and a; $3a$ and a; $4a$ and a.

C 17. Divide $\overline{CD}$ into 7 $\cong$ parts: $\overline{CU}$, $\overline{UV}$, $\overline{VW}$, $\overline{WX}$, $\overline{XY}$, $\overline{YZ}$, and $\overline{ZD}$. Then $CV : VX : XD = 2 : 2 : 3$. Use SSS to const. $\triangle XED$.

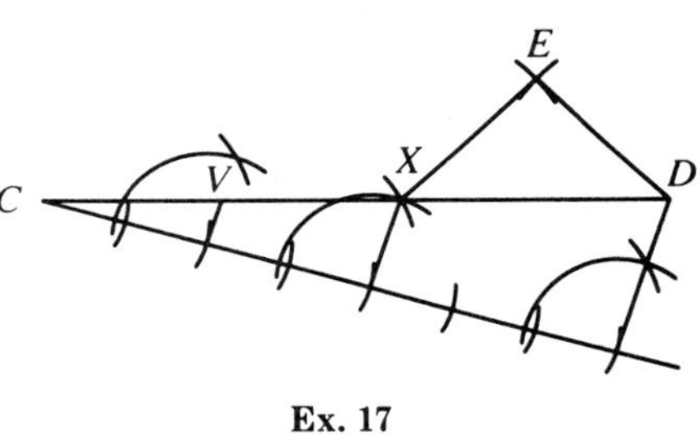

Ex. 17

18. $\overline{GA} \cong \overline{GB}$ so $\angle B \cong \angle A$, and $\overline{BY} \cong \overline{AX}$ so $\triangle GBY \cong \triangle GAX$. Then $\overline{GY} \cong \overline{GX}$ so $m\angle GYA = m\angle GXY$. Consider $\triangle GYA$ and $\triangle GXA$. $m\angle YGA > m\angle 1$ so $m\angle GYA < m\angle GXA$. Then $m\angle GXY < m\angle GXA$. Consider $\triangle GXA$ and $\triangle GXY$. By the SAS Ineq. Thm., $GA > GY$. Assume temp. that $m\angle 1 = m\angle 2$. Then $\overrightarrow{GX}$ bis. $\angle YGA$ and $\frac{YX}{XA} = \frac{GY}{GA}$; $\frac{YX}{XA} = 1 = \frac{GY}{GA}$ and $GY = GA$. This contradicts the fact that $GA > GY$. Our temp. assumption must be false. It follows that $m\angle 1 \neq m\angle 2$ and $\angle G$ is not trisected.

Page 401 • SELF-TEST 2

1. Const. 9 **2.** Const. 11

3. Use Const. 12 to divide a seg. $\overline{AB}$ into 3 $\cong$ parts: $\overline{AX}$, $\overline{XY}$, and $\overline{YB}$. Then $AY : YB = 2 : 1$.

4. Const. 13 **5.** Const. 14 **6.** $\perp$; F; $\perp$; G

7. Const. the $\perp$ bis. of 2 sides of $\triangle TRI$, int. at O. Draw a $\odot$ with center O and radius OT.

Pages 402–403 • CLASSROOM EXERCISES

1. **a, b.** Check students' drawings. **c.** a ⊙ with center A and radius 20 cm **d.** The locus is a sphere with center A and radius 20 cm.
2. **a, b.** Check students' drawings. **c.** The locus is a line $\parallel$ to both lines and halfway between them. **d.** The locus is a plane $\parallel$ to both lines and halfway between them.
3. **a, b.** Check students' drawings. **c.** The locus is the bis. of the $\angle$.
4. The locus is the portion of a plane inside the classroom $\parallel$ to the floor and ceiling, halfway between them.
5. The locus is the portion of a plane inside the classroom $\parallel$ to the floor and 1 m above it.
6. **a.** The locus is a ⊙ with center P and radius 1 m. (Part of the ⊙ may be outside the classroom.) **b.** The locus is a hemisphere with center P and radius 1 m.
7. The locus is a line that is the intersection of 2 planes. One plane is $\parallel$ to and halfway between the floor and ceiling. The second plane is $\parallel$ to and halfway between the 2 side walls. The line joins the centers of the front and back walls.
8. The locus is a line that is the intersection of 2 planes. One plane is $\parallel$ to and halfway between the front and back walls. The second plane is $\parallel$ to and halfway between the 2 side walls. The line joins the centers of the floor and ceiling.
9. the 50-yard line
10. **a.** Check students' drawings. **b.** 2 ⊙s concentric with the given ⊙ and with radii 4 cm and 8 cm **c.** The locus is the surface of a doughnut with a hole 4 cm in radius and an outer radius of 8 cm. (This shape is called a torus in topology.)
11. Answers may vary. Example: The locus of pts. that are equidistant from the vertices of a rect.
12. Answers may vary. Example: The locus of pts. that lie on all 3 sides of a $\triangle$.

Pages 404–405 • WRITTEN EXERCISES

A **1.** the $\perp$ bis. of $\overline{AB}$ **2.** ⊙P with radius 2 cm
3. 2 $\parallel$ lines 4 cm apart with h halfway between them
4. a ⊙ concentric with the given ⊙ with radius half the radius of the given ⊙
5. the seg. joining the midpts. of $\overline{AD}$ and $\overline{BC}$ **6.** diag. $\overline{AC}$ **7.** diag. $\overline{BD}$
8. the int. of $\overline{AC}$ and $\overline{BD}$ **9.** a plane $\parallel$ to both planes and halfway between them
10. 2 $\parallel$ planes 10 cm apart with the given plane halfway between them
11. a sphere with center E and radius 3 cm **12.** a plane $\perp$ to $\overline{CD}$ that bisects $\overline{CD}$

B **13.** **a.** Use Const. 3 to bisect $\angle HEX$. **b.** Use Const. 3 to bisect the $\angle$s formed by j and k; locus is 2 $\perp$ lines.

14. Const. a pair of $\parallel$ lines on either side of n at a distance DE from n.

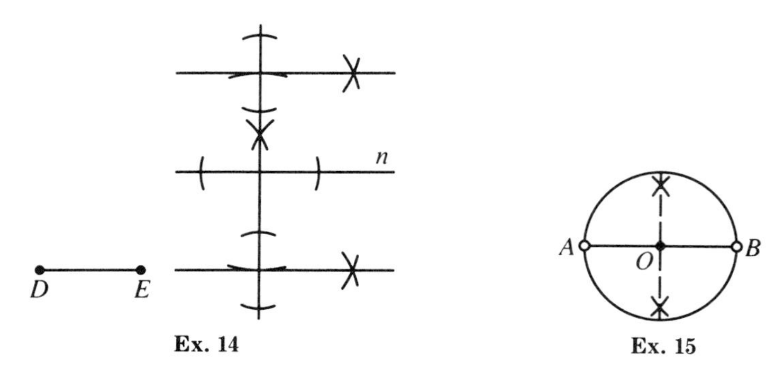

Ex. 14 **Ex. 15**

15. Const. the $\odot$ with diameter $\overline{AB}$, and exclude pts. A and B.

16. Const. the $\perp$ bis. of $\overline{CD}$, and exclude the midpt. of $\overline{CD}$.

Ex. 16 **Ex. 17**

17. Const. 2 $\odot$s with radius EF, one with center E and one with center F, and exclude pts. E and F and the other 2 pts. of the int. of the $\odot$s with $\overleftrightarrow{EF}$.

18. a sphere with the same center as the given sphere and radius half the radius of the given sphere

19. a line $\perp$ to the plane of the square at the int. of the diags.

20. a line $\perp$ to the plane of the $\triangle$ at the int. of the $\perp$ bis. of the sides

C 21. Let P be the pt. on the ground directly below A. The path of M (locus) is a $90°$ arc of $\odot P$ with radius AM.

22. a sphere (excluding pts. C and D) with center the midpt. of $\overline{CD}$ and radius $\frac{1}{2}CD$

23. The region is determined by 5 circles as shown in the diagram: $\odot A$ with radius 5 m, $\odot B$ and $\odot E$ with radius 4 m, and $\odot C$ and $\odot D$ with radius 2 m.

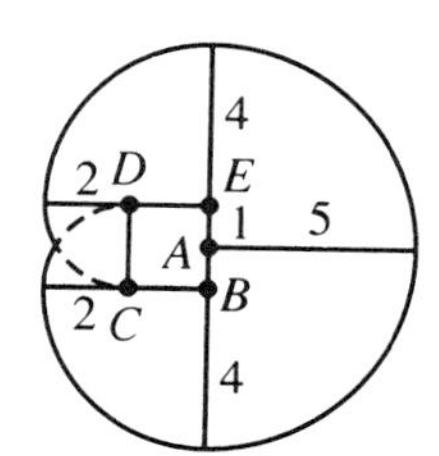

24. $DX = \sqrt{6^2 - 2^2} = \sqrt{32} = 4\sqrt{2}$; the region over which the dog can roam is made up of semicircles A and C, each with radius $4\sqrt{2}$ m, and a rectangle with length $8\sqrt{2}$ m and width $AC = 5$ m.

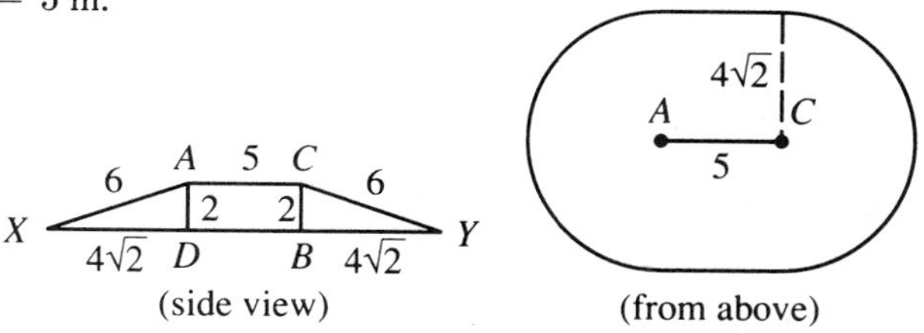

Page 407 • CLASSROOM EXERCISES

1. 0, 1, or 2 pts. **2.** 0, 1, or 2 pts. (assuming the ⊙s do not coincide)

3. 0, 1, 2, 3, or 4 pts., depending on whether the line(s) int. or is(are) tangent to the ⊙

4. 0, 1, 2, 3, or 4 pts.; the 3-pt. case occurs when the ⊙ is tan. to one line and int. the other line in 2 pts.

5. a. the bis. of $\angle A$ **b.** the $\perp$ bis. of $\overline{BC}$

c.

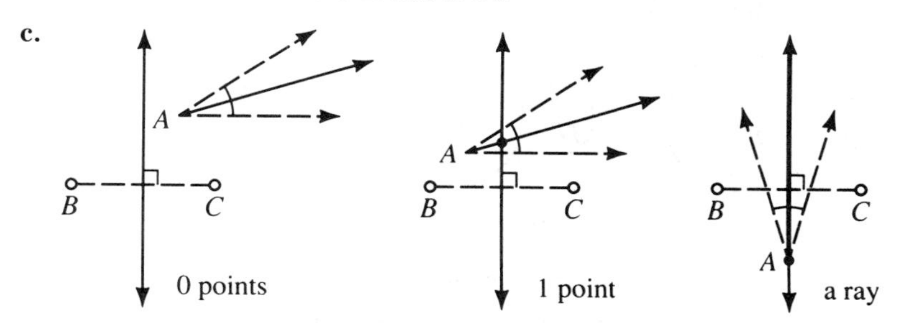

d. 0 pts., 1 pt., or a ray

6. 0 pts., 1 pt., or the line, depending on whether the line is $\parallel$ to the plane, intersects it at one pt., or lies in the plane

7. 0, 1, or 2 pts., depending on whether the line does not int. the sphere, is tangent to it, or int. it at 2 pts.

8. 0 pts., 1 pt., or a ⊙, depending on whether the first sphere does not int. the second, is tangent to it (internally or externally), or int. it in a ⊙ (assuming the 2 spheres do not coincide)

9. 0 pts., 1 pt., or a ⊙, depending on whether the plane does not int. the sphere, is tangent to it, or int. it in a ⊙

10. a. the portion inside the classroom of a sphere that has center C and radius 3 m **b.** the portion inside the classroom of the ⊙ that is the int. of the sphere in part (a) and the plane ∥ to the floor and ceiling halfway between them **c.** the portion inside the classroom of 2 ⊙s, the int. of the sphere in part (a) and 2 planes, each ∥ to the floor, one 1 m from the floor, the other 1 m from the ceiling

Pages 407–410 • WRITTEN EXERCISES

A 1. a.

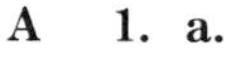

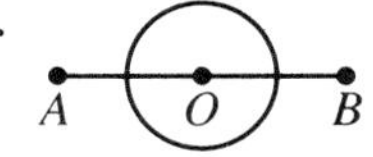

$AO = OB$

b.

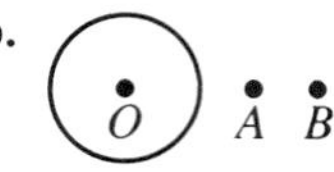

c.

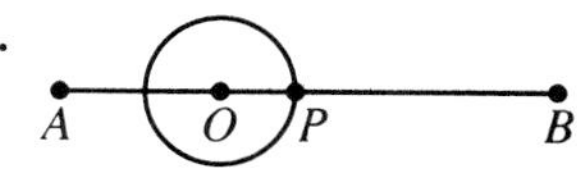

$AP = PB$

2. a.

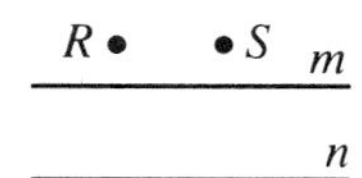

b. R • m n S •

c.

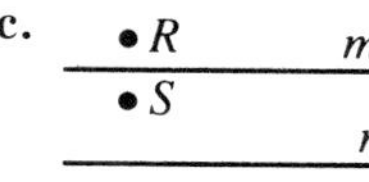

3. a. a ⊙ with center D and radius 1 cm **b.** a ⊙ with center E and radius 2 cm

c.

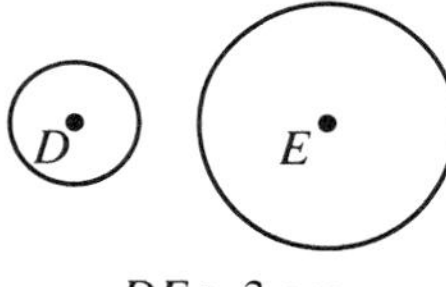

$DE > 3$ cm

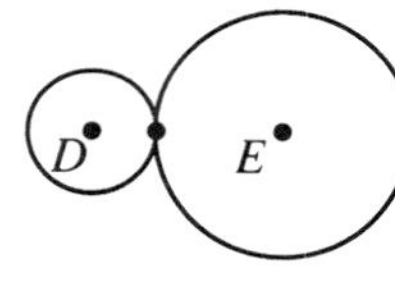

$DE = 3$ cm

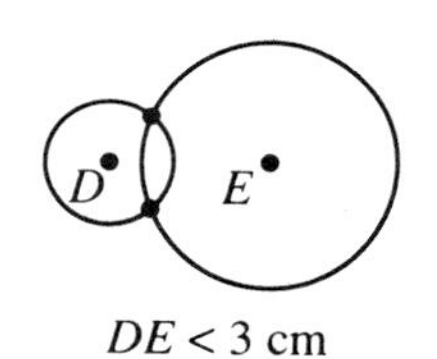

$DE < 3$ cm

d. The locus is 0, 1, or 2 points, depending on the int. of ⊙D and ⊙E.

4. a. a ⊙ with center A and radius 3 cm **b.** 2 ∥ lines 2 cm apart with k halfway between them

c.

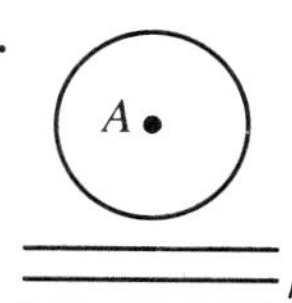

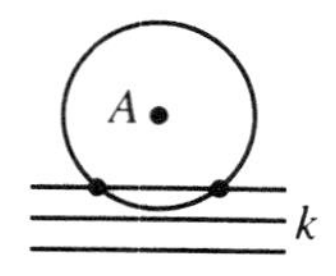

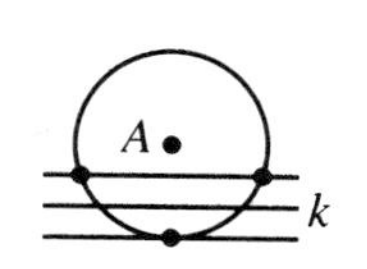

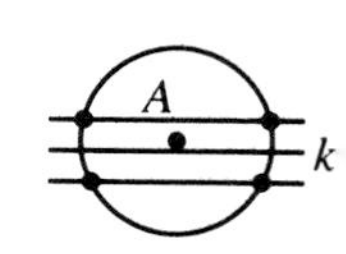

d. The locus is 0, 1, 2, 3, or 4 pts., depending on the int. of ⊙A with radius 3 cm, and 2 lines ∥ to k and 1 cm from k.

5. The locus is the int. of ⊙P, with radius 3 cm, and l. (2 pts.)

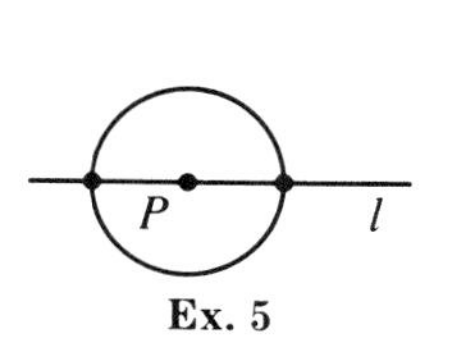

Ex. 5

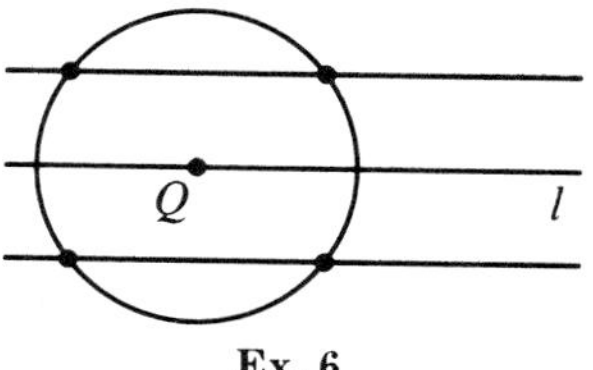

Ex. 6

6. The locus is the int. of ⊙Q, with radius 5 cm, and 2 lines ∥ to l and 3 cm from l. (4 pts.

7. The locus is the int. of $\odot A$, with radius 2 cm, and $\odot B$, with radius 2 cm. (2 pts.)

Ex. 7 **Ex. 8**

8. The locus is the int. of $\odot P$, with radius 2 cm, and the bis. of the $\angle$s formed by j and k. (4 pts.)

9. The locus is the int. of $\odot A$, with radius 2 cm, and the bis. of $\angle A$. (1 pt.)

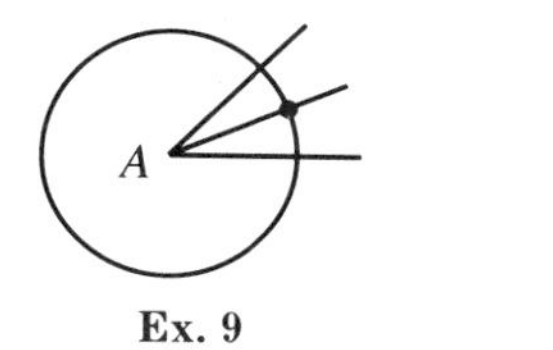

Ex. 9

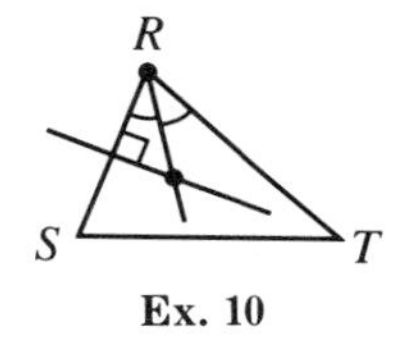

Ex. 10

10. The locus is the int. of the bis. of $\angle R$ and the $\perp$ bis. of $\overline{RS}$. (1 pt.)

B 11. The locus is 0, 1, or 2 pts.

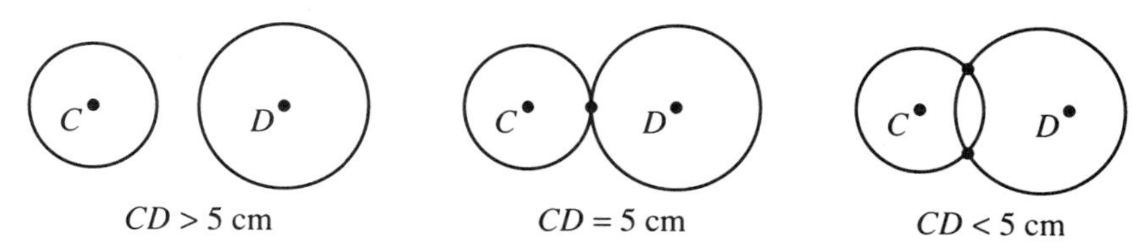

$CD > 5$ cm $CD = 5$ cm $CD < 5$ cm

12. The locus is 0, 1, 2, 3, or 4 pts. Let d be the distance from E to k.

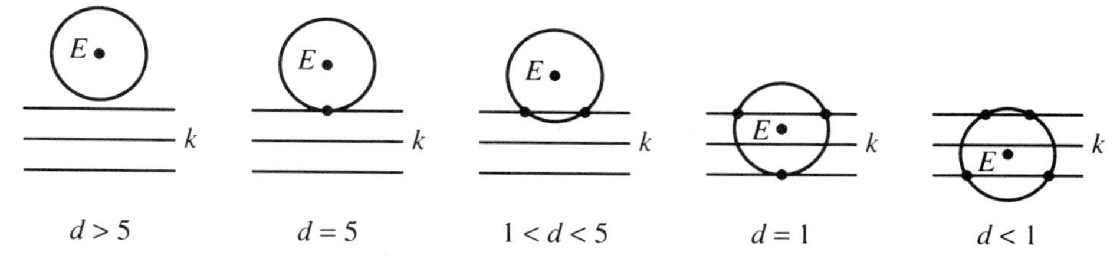

$d > 5$ $d = 5$ $1 < d < 5$ $d = 1$ $d < 1$

13. The locus is 0, 1, or 2 pts.

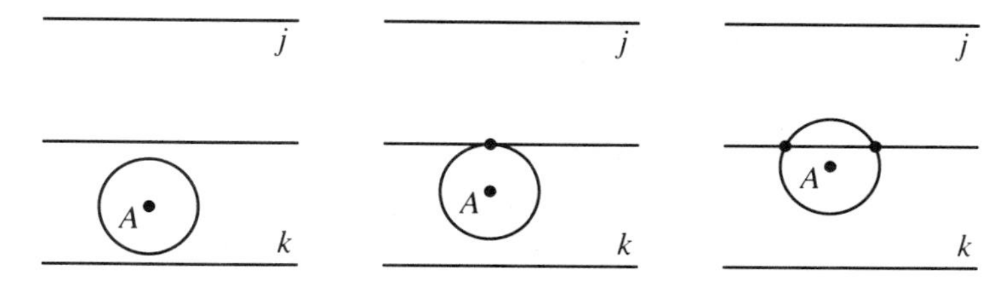

14. The locus is 0 pts., 1 pt., or a line.

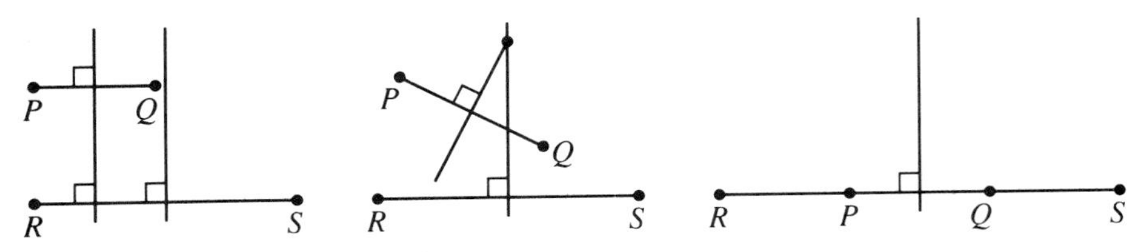

15. The locus is 0 pts., 1 pt., or a ⊙. **16.** The locus is 0 pts., 1 pt., or a ⊙.

17. The locus is 2 circles.

18. The locus is a line ⊥ to the plane of $ABCD$ at the pt. of int. of the diags.

19. The locus is 0 pts. ($d > 5$), 2 pts. ($d = 5$), or 2 circles ($d < 5$).

20. The locus is the set of pts. bounded by $\overset{\frown}{AB}$ (center C and radius 2 cm), $\overset{\frown}{AC}$ (center B and radius 2 cm), and $\overset{\frown}{BC}$ (center A and radius 2 cm).

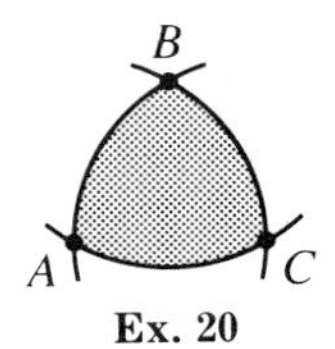

Ex. 20

21. **a.** the ⊥ bis. plane of $\overline{RS}$ **b.** the ⊥ bis. plane of $\overline{RT}$ **c.** line; line **d.** the ⊥ bis. plane of $\overline{RW}$ **e.** point; point

C **22.** **a.** Yes; J, K, L, and M are not coplanar; no 3 collinear **b.** Yes; J, K, L, and M are any 4 pts. on a ⊙. **c.** No **d.** Yes; any 3 collinear

23. **a.** Infinitely many; let H represent Houston and T Toronto. Consider the plane A that determines the great circle containing H and T. Let X be the midpt. of $\overset{\frown}{HT}$. Then consider the plane B containing X that is ⊥ to plane A. The great circle determined by plane B contains infinitely many pts. that are equidistant from H and T. **b.** 2 pts.; let L represent Los Angeles. Consider the plane C that determines the great circle containing L and H. Let Y be the midpt. of $\overset{\frown}{LH}$. Then consider the plane D containing Y that is ⊥ to plane C. The great circle determined by plane D contains the pts. that are equidistant from L and H. Pts. P and Q, which lie on the line of intersection of planes B and D and on the surface of the sphere, are equidistant from H, T, and L. **c.** none

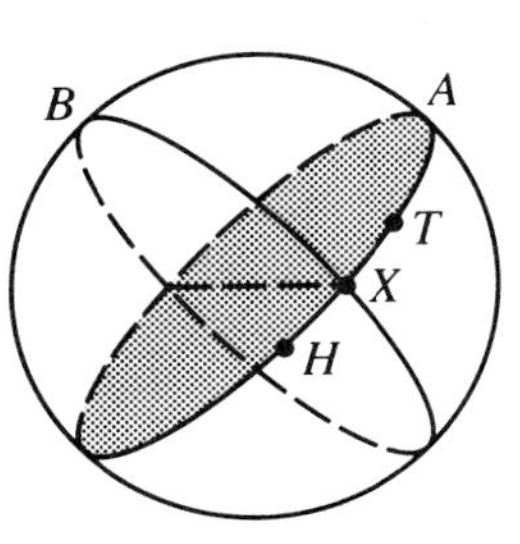

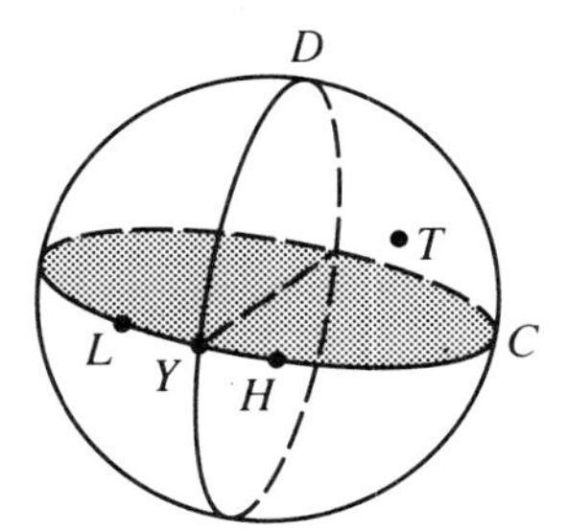

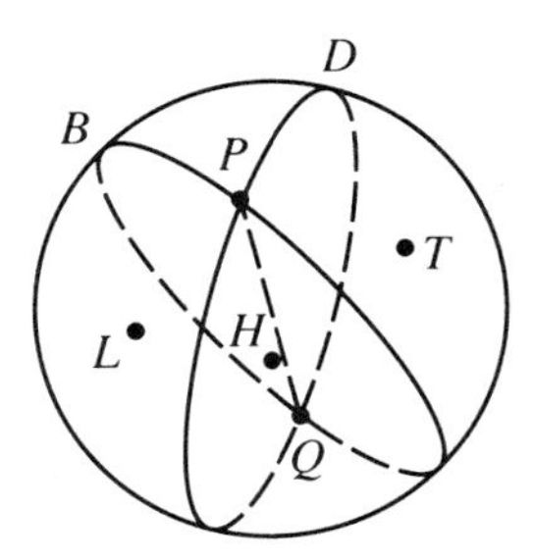

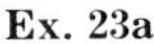
Ex. 23a

Ex. 23b

24. a. Inside the region bounded by $\widehat{AB}$, $\widehat{AC}$, and $\widehat{BC}$ **b.** on $\widehat{AC}$ **c.** at A or G

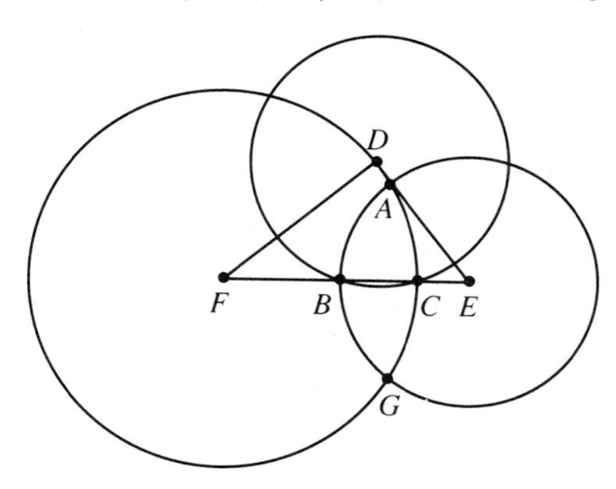

Page 410 • CHALLENGE

Since $\triangle DAM$ and $\triangle CDA$ are isos. △s, $\angle DMA \cong \angle DAM \cong \angle ADC$. Then $\triangle DAM \sim \triangle CDA$ by the AA ~ Post.

$\frac{AM}{DA} = \frac{AD}{CD}$; $\frac{AM}{AB} = \frac{AB}{2 \cdot AB}$;

$AM = \frac{1}{2}AB$ and M is the midpt. of $\overline{AB}$.

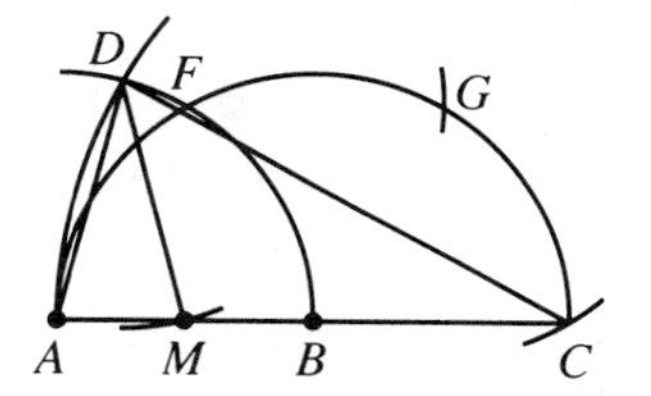

Pages 411–412 • CLASSROOM EXERCISES

1. **a.** on the ⊙ with center R and radius k **b.** on the ⊥ bis. of $\overline{RS}$ **c.** (1) Choose R on the ⊙. With center R and radius k, draw an arc int. the ⊙. (2) Let S be the pt. where the arc int. the ⊙. Draw $\overline{RS}$. (3) Const. the ⊥ bis. of $\overline{RS}$. **d.** (1) Let T be one of the pts. where the ⊥ bis. of $\overline{RS}$ int. the ⊙; draw $\overline{TR}$ and $\overline{TS}$. (2) Let T be the other pt. of intersection.
2. First solution: Const. $\overline{AB}$ so that $AB = s$. Const. the ⊥ bis. of $\overline{AB}$ to locate its midpt., M. Draw a semicircle with center M and radius $\overline{AM}$. Draw an arc with center A and radius r int. the semicircle at C. Draw $\overline{CA}$ and $\overline{CB}$. Since $\angle C$ is inscribed in a semicircle, $\angle C$ is a rt. $\angle$. Second solution: Draw a line and choose pt. C. Const. a ⊥ to the line at C such that $AC = r$. With center A and radius s, draw an arc int. the line at B. Draw $\overline{AB}$.

Pages 412–413 • WRITTEN EXERCISES

A **1.** The locus is the 2 $\parallel$ lines, j and k.

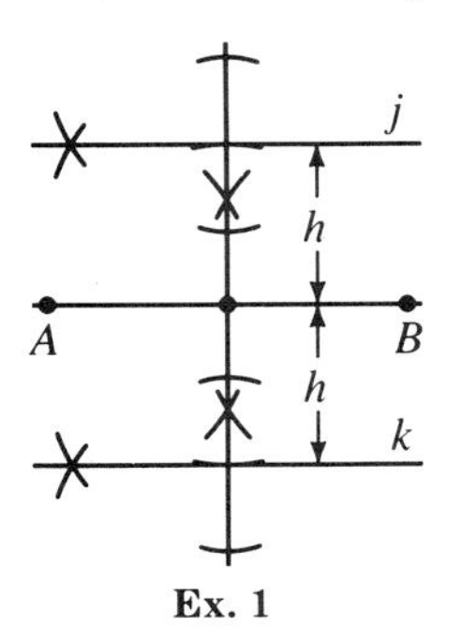

Ex. 1

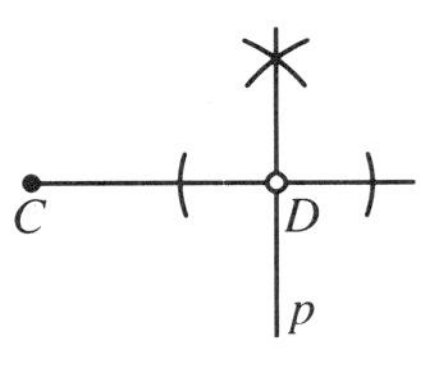

Ex. 2a

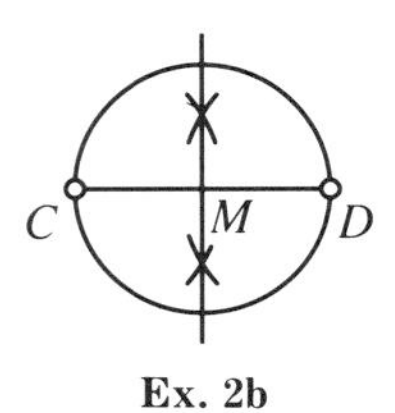

Ex. 2b

2. **a.** The locus is p, the line $\perp$ to $\overline{CD}$ at D, excluding pt. D. **b.** The locus is $\odot M$ with center at the midpt. of $\overline{CD}$ and radius $= \frac{1}{2}CD$, excluding pts. C and D.

3. The $\odot$, of radius a, has center W at the int. pt. of the bis. of $\angle XYZ$ and a line that is $\parallel$ to $\overrightarrow{YZ}$ and a units from $\overrightarrow{YZ}$.

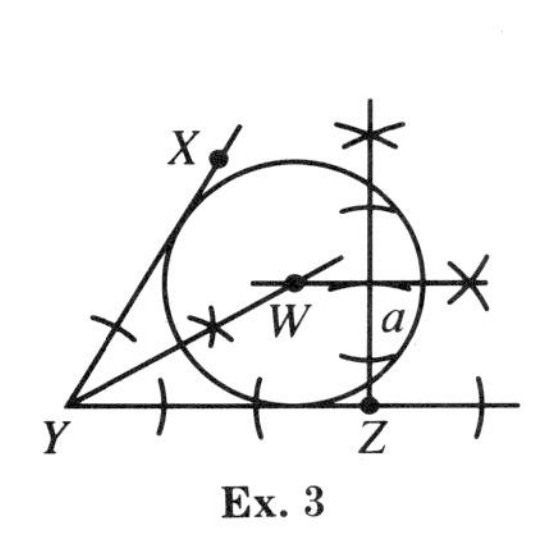

Ex. 3

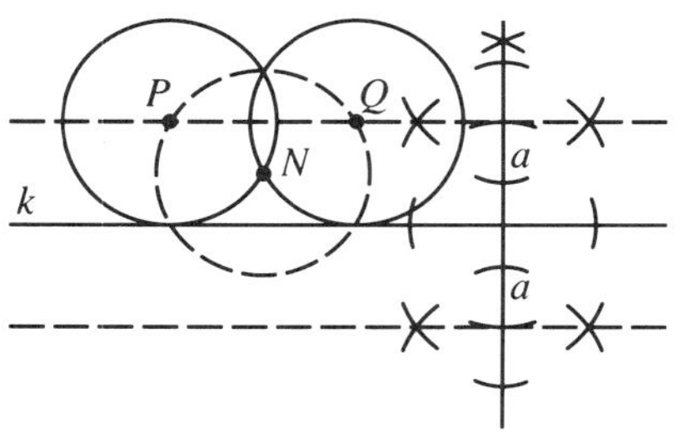

Ex. 4

4. The $\odot$, of radius a, has center at P or Q, the int. of $\odot N$ with radius a and a line $\parallel$ to k, a units from k.

5. The locus is a pair of lines, j and k, both $\parallel$ to $\overline{AB}$ and r units from $\overline{AB}$.

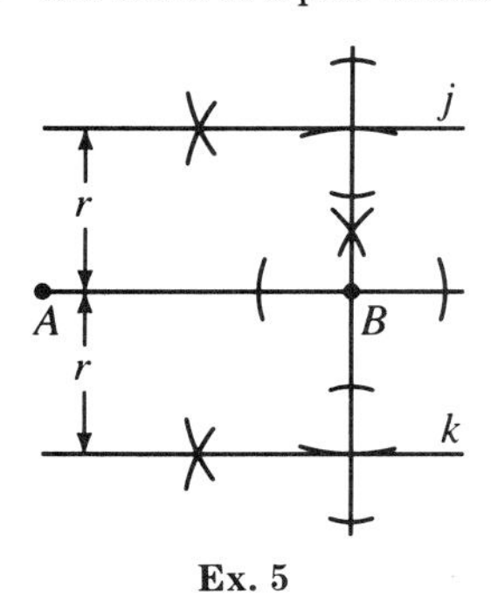

Ex. 5

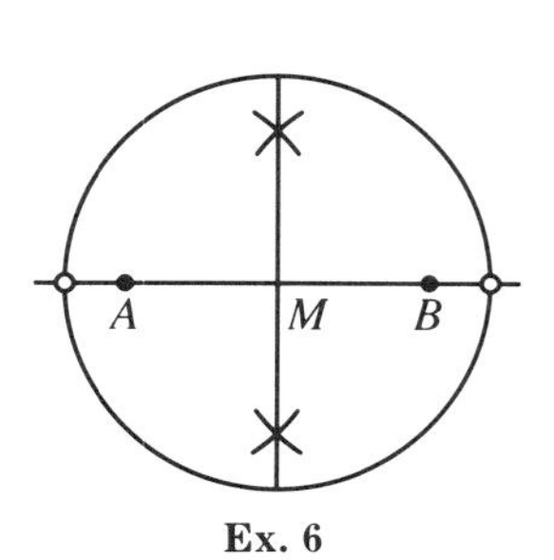

Ex. 6

6. The locus is a $\odot$ with center M, the midpt. of $\overline{AB}$, and radius s, excluding the 2 pts. where $\overleftrightarrow{AB}$ int. $\odot M$.

B **7.** Const. $j \perp k$ at M. Const. $\overline{MA}$ on j so that $MA = s$ and then $\cong$ segs. $\overline{AB}$ and $\overline{AC}$ with B and C on k, so that $AB = AC = t$.

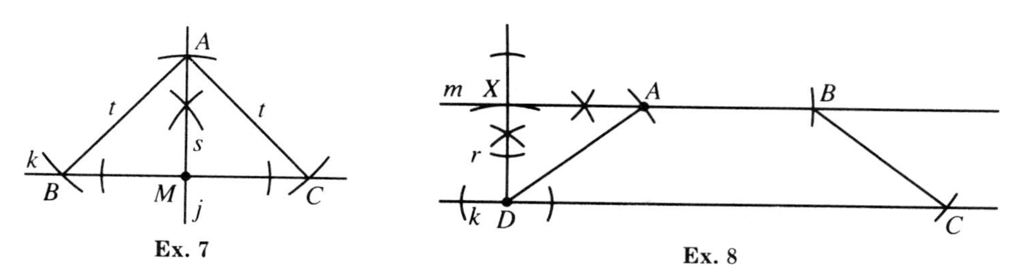

Ex. 7 **Ex. 8**

8. Draw a line k. Choose D on k and const. a $\perp$ to k at D. Const. $\overline{DX}$ on the $\perp$ so that $DX = r$. Const. $m \parallel$ to k through X. With center D and radius t, draw an arc int. the $\parallel$ at A. Draw $\overline{AD}$. On m, const. $\overline{AB} \cong \overline{AD}$. With center B and radius t, draw an arc int. k at C with $\overline{BC} \nparallel \overline{AD}$. Draw $\overline{BC}$.

9. Const. $\overline{AB}$ so that $AB = t$. Const. the $\perp$ bis. of $\overline{AB}$ to locate midpt. M of $\overline{AB}$. Draw an arc with center M and radius r, and an arc with center A and radius s int. at C. Draw $\overline{AC}$ and $\overline{BC}$.

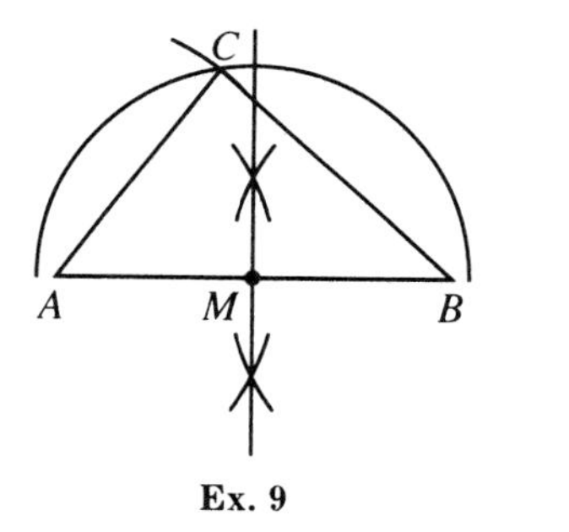

Ex. 9

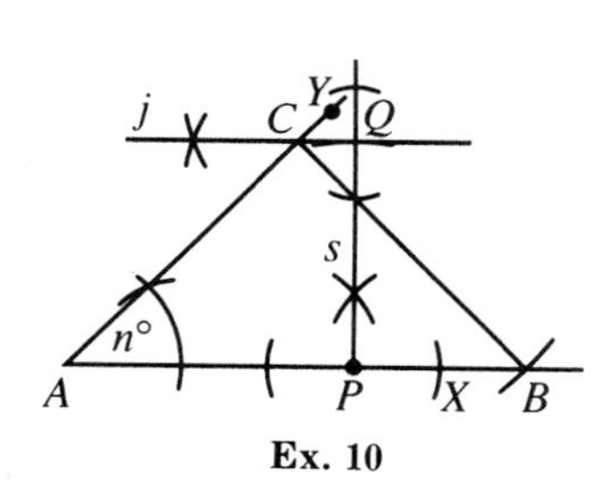

Ex. 10

10. Const. $\angle XAY$ with measure n. Const. line $j \parallel$ to and s units from $\overrightarrow{AX}$; j int. $\overrightarrow{AY}$ at C. With center C and radius AC, draw an arc int. $\overrightarrow{AX}$ at B. Draw $\overline{BC}$.

11. Const. $\angle A$ with measure n. Const. a line $\parallel$ to and s units from $\overrightarrow{AX}$ in order to locate pt. C. Const. $\overline{BC} \perp \overline{AC}$.

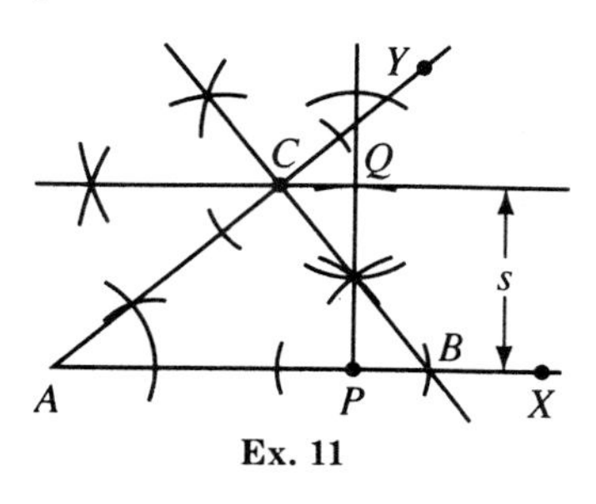

Ex. 11

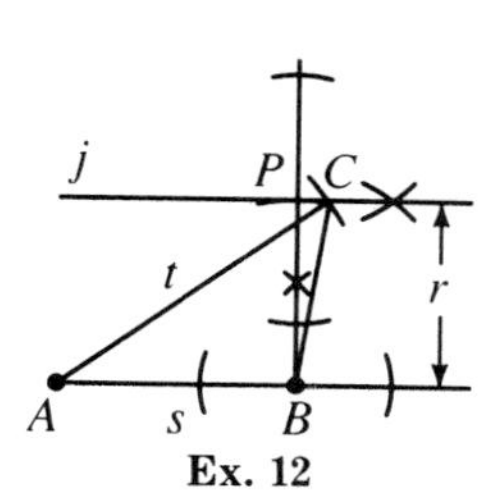

Ex. 12

12. Const. $\overline{AB}$ of length s. Const. line j, $\parallel$ to and r units from $\overleftrightarrow{AB}$. With center A and radius t, draw an arc int. j at C. Draw $\overline{CA}$ and $\overline{CB}$.

13. Const. $\overline{AB}$ such that $AB = t$. Const. the $\perp$ bis. of $\overline{AB}$ to locate midpt. M of $\overline{AB}$. Const. line k, $\parallel$ to and r units from $\overleftrightarrow{AB}$. With center M and radius s, draw an arc int. k at pt. C. Draw $\overline{AC}$ and $\overline{BC}$.

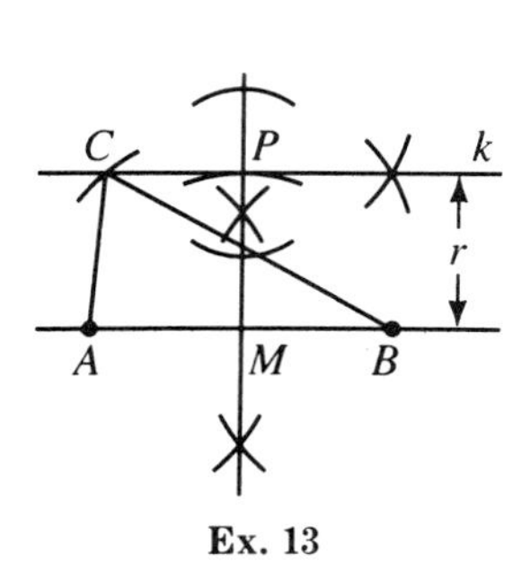

Ex. 13

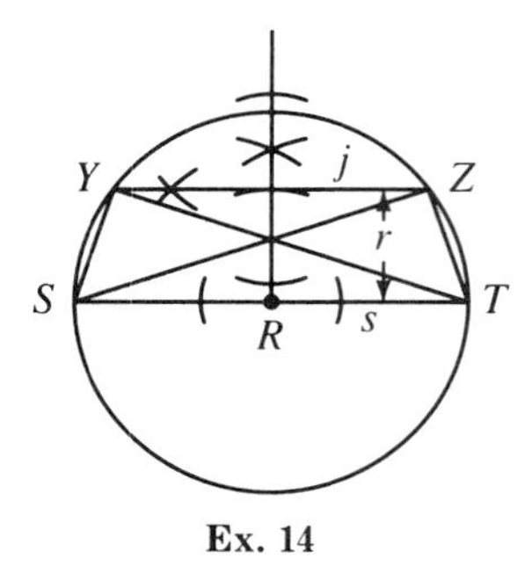

Ex. 14

14. Const. $\odot R$ with radius s. Draw diam. $\overline{ST}$. Const. line j, $\parallel$ to and r units from $\overleftrightarrow{ST}$. Line j int. $\odot R$ at pts. Y and Z. Rt. $\triangle SYT$ and SZT both satisfy the requirements.

15. Const. $\overline{AB}$ of length r. Const. line $j \perp \overline{AB}$ at B. With center A and radius s, draw an arc int. j at C. Draw $\overline{AC}$. Const. $\overline{CE}$ and $\overline{CD}$, both on j, so that $CE = CD = s$. $\triangle ACD$ and $\triangle ACE$ are the two $\triangle$.

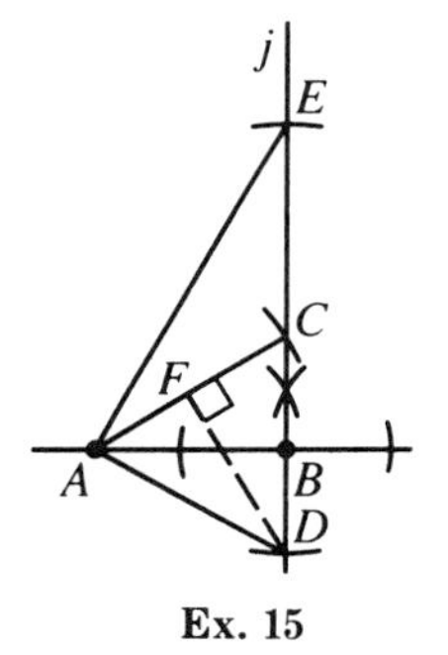

Ex. 15

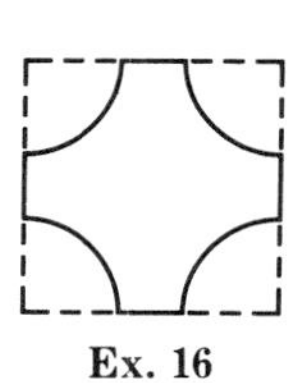

Ex. 16

C 16. To construct the locus, note that when the seg. is not lying on a side of the square, it forms a rt. $\triangle$ with the sides of the square. Then the midpt. of the seg. is equidistant from the vertices of the $\triangle$; it lies on a $\odot$ with center the vertex of the square and radius $\frac{3s}{2}$. The locus is the 4 90° arcs shown and the segments joining them.

17. Const. $\overline{AB}$ and $\overline{BC}$ so that $AB = r$ and $BC = s$. Const. the $\perp$ bis. of $\overline{AC}$, int. $\overline{AC}$ at M. Draw $\odot M$ with radius $\frac{1}{2}AC$, int. the $\perp$ bis. of $\overline{AC}$ at D. Draw $\overrightarrow{DB}$, int. $\odot M$ at E. Draw $\overline{EA}$ and $\overline{EC}$. Since $m\angle AEB = \frac{1}{2}m\widehat{AD} = \frac{1}{2}m\widehat{DC} = m\angle DEC$, $\overrightarrow{ED}$ is the bis. of $\angle AEC$.

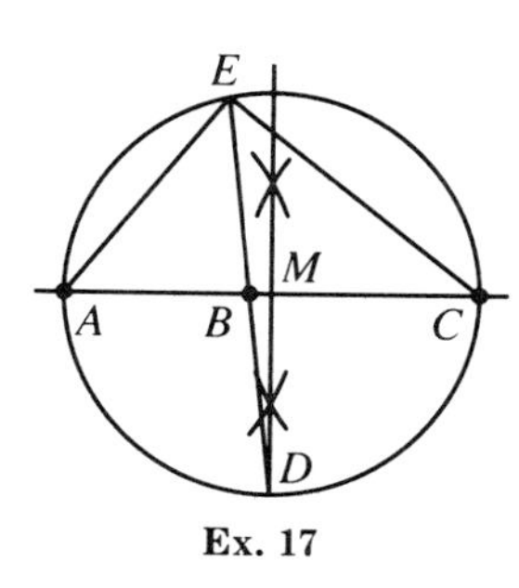

Ex. 17

Ex. 18

18. Const. rt. $\angle A$ and $\overline{AB}$ so that $AB = r$. Const. a $\perp$ to $\overline{AB}$ at B and const. $\overline{BC}$ so that $BC = r$. Draw $\odot C$ with radius r. Draw $\overrightarrow{AC}$, int. $\odot C$ at D. Const. a tan to $\odot C$ at D, int. the sides of $\angle A$ at E and F. $\triangle ABC$ is an isos. rt. $\triangle$ so $m\angle EAD = m\angle FAD = 45$. Then $\triangle EAD \cong \triangle FAD$ and $\overline{AE} \cong \overline{AF}$.

19. Draw a line and const. $\overline{AB}$ so that $AB = t$. Const. $\angle DBC$ on $\overleftrightarrow{AB}$ so that $m\angle DBC = n$. Bisect $\angle ABD$ (the supp. of $\angle DBC$) to const. $\angle ABE$ and $\angle EBD$, each with meas. $x = \frac{1}{2}(180 - n)$. Const. the $\perp$ bisector of $\overline{AB}$ and locate P at the pt. where the $\perp$ bisector of $\overline{AB}$ intersects $\overrightarrow{BE}$. Use Const. 10 to circumscribe a $\odot$ about $\triangle APB$. The locus is $\widehat{APB}$, excluding pts. A and B.

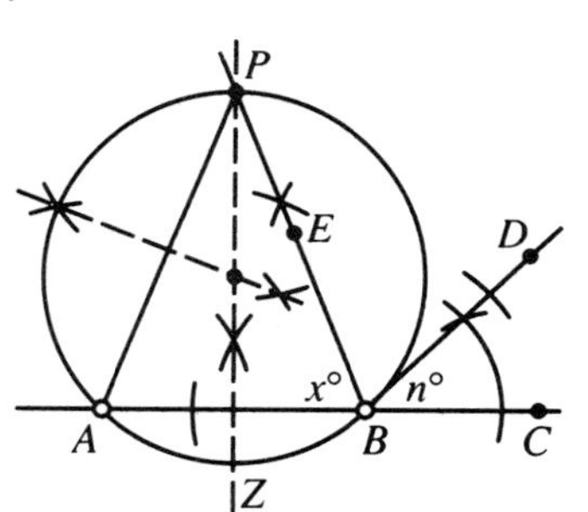

Page 414 • SELF-TEST 3

1. The locus is the bis. of the vert. ∠s formed by j and k. (2 lines)
2. The locus is the sphere with center P and radius t.
3. The locus is the $\perp$ bis. plane of $\overline{WX}$.
4. The locus is the int. of the bis. of $\angle DEF$ and a pair of rays $\parallel$ to $\overrightarrow{EF}$ and each 4 cm from $\overrightarrow{EF}$. (1 pt.)
5. The locus is the int. of $\odot A$ with radius 4 cm and a line $\parallel$ to s and t and halfway between them. (0, 1, or 2 pts.)

6. Use Const. 4 to const. the $\perp$ bis. of 2 sides of $\triangle RST$. The locus is the pt. of intersection of the $\perp$ bisectors.
7. Methods may vary. Example: Const. $\angle X \cong \angle 1$. Const. $\overline{XY} \cong \overline{BC}$ on one side of $\angle X$. Const. a line from $Y \perp$ to the other side of $\angle X$.

Page 415 • EXTRA

1, 2. Check students' constructions.
3. Some of the pts. are the same: L and R, M and S, N and T. The $\odot$ has center H.
4. 3; if $\angle C$ is a rt. $\angle$, the legs are altitudes, and $S = T = Z = C$.
5.

Statements	Reasons
1. $\overline{NM} \parallel \overline{AB}$; $\overline{XY} \parallel \overline{AB}$; $NM = \frac{1}{2}AB$; $XY = \frac{1}{2}AB$	1. The seg. that joins the midpts. of 2 sides of a $\triangle$ is $\parallel$ to the third side and is half as long as the third side.
2. $\overline{NM} \parallel \overline{XY}$	2. 2 lines $\parallel$ to a third line are $\parallel$ to each other.
3. $NM = XY$ or $\overline{NM} \cong \overline{XY}$	3. Substitution Prop.
4. $XYMN$ is a $\square$.	4. If one pair of opp. sides of a quad. are both $\cong$ and $\parallel$, then the quad. is a $\square$.
5. $\overline{CR} \perp \overline{AB}$	5. Def. of alt.
6. $\overline{CR} \perp \overline{NM}$	6. If a trans. is $\perp$ to one of 2 $\parallel$ lines, then it is $\perp$ to the other one also.
7. $\overline{NX} \parallel \overline{CH}$, or $\overline{NX} \parallel \overline{CR}$	7. The seg. that joins the midpts. of 2 sides of a $\triangle$ is $\parallel$ to the third side.
8. $\overline{NM} \perp \overline{NX}$	8. If a trans. is $\perp$ to one of 2 $\parallel$ lines, then it is $\perp$ to the other one also.
9. $\angle MNX$ is a rt. $\angle$.	9. Def. of $\perp$ lines
10. $XYMN$ is a rect.	10. If an $\angle$ of a $\square$ is a rt. $\angle$, then the $\square$ is a rect.

6. If the students' figures are carefully drawn, the ratio of the radius of the nine-point circle to the radius of the circumscribed circle will appear to be 1 : 2. The proof is rather difficult and lengthy.

Pages 416–417 • CHAPTER REVIEW

1.
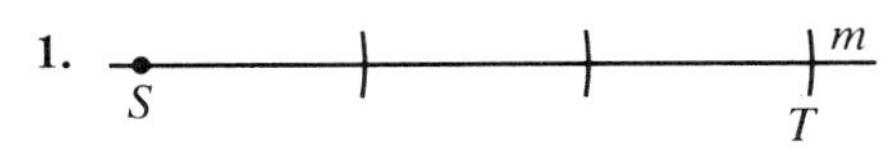

2.
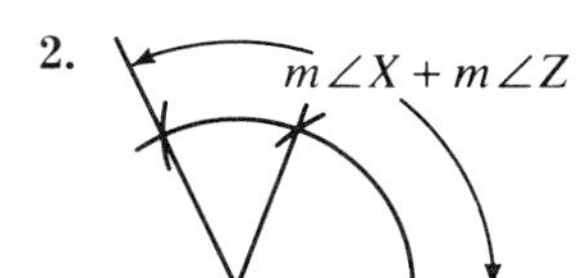

3. Const. 3 **4.** Const. 4 **5.** Const. 5 **6.** Const. 6 **7.** Const. 7

8. $\perp$ bisectors of the sides **9.** $\angle$ bisectors **10.** $12 = \frac{2}{3} \cdot MP; MP = 18$ **11.** $1:2$

12. Const. 8 **13.** Const. 9

14. Const. the bis. of 2 of the $\angle$s. Their int. is the center of the inscribed $\odot$.

15. Const. 10 **16.** Const. 14 **17.** Const. 13 **18.** Const. 1, 12

19. The locus is a line $\parallel$ to l and m and halfway between them.

20. The locus is the $\perp$ bis. plane of $\overline{AB}$.

21. The locus is a plane $\parallel$ to both planes and halfway between them.

22. The locus is a line $\perp$ to the plane of $\triangle HJK$ at the int. of the medians (or altitudes or $\angle$ bisectors).

23. The locus is the int. of the $\perp$ bis. of $\overline{PQ}$ and $\odot P$ with radius 8 cm. (2 pts.)

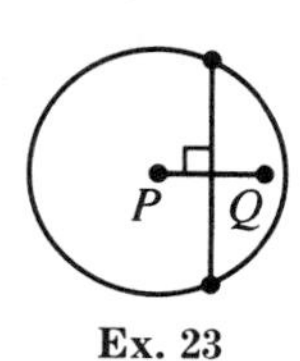

Ex. 23

24. The locus is the int. of the cylindrical surface with axis l and radius 8 cm, and the sphere with center R and radius 8 cm. (a circle)

25. The locus is 0 pts., 1 pt., a $\odot$, a $\odot$ and a pt., or 2 $\odot$s, depending on the int. of 2 planes $\parallel$ to Q and 1 m from Q and a sphere with center Z and radius 2 m.

26. Const. $\overline{AB}$ so that $AB = a$. Const. line m, the $\perp$ bis. of $\overline{AB}$, int. $\overline{AB}$ at C. Draw a semicircle with center C and radius CA, int. m at D. Draw $\overline{AD}$ and $\overline{DB}$. $\triangle ADB$ is an isos. rt. $\triangle$ with hypotenuse of length a.

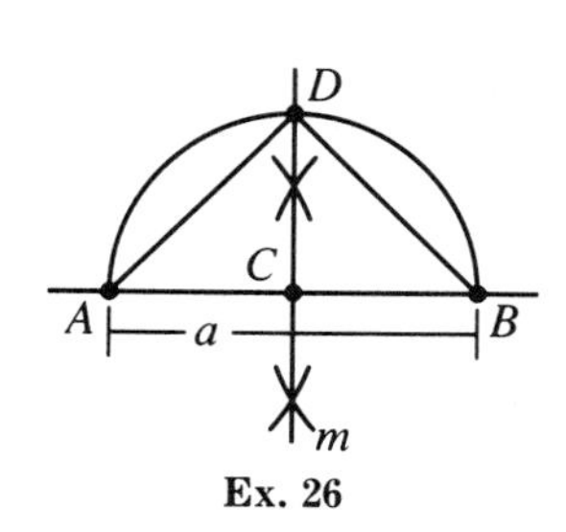

Ex. 26

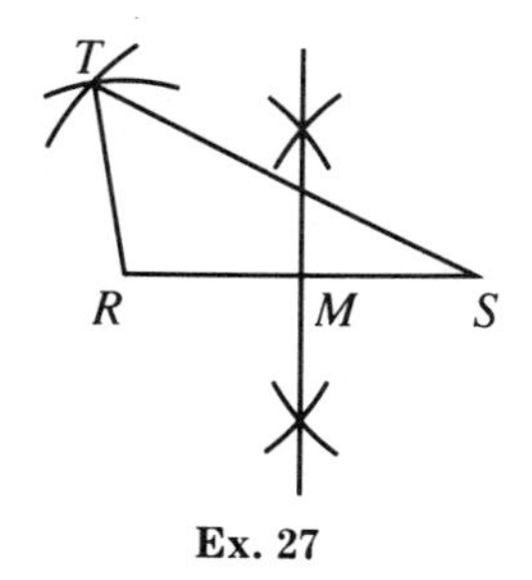

Ex. 27

27. Const. $\overline{RS}$ so that $RS = a$. Const. the $\perp$ bis. of $\overline{RS}$ to locate midpt. M. Draw an arc with center M and radius b and an arc with center R and radius c, int. at T. Draw $\overline{TR}$ and $\overline{TS}$.

Page 418 • CHAPTER TEST

1. Const. 2, 1: Const. $\angle A \cong \angle 1$. With center A and radius z, draw arcs int. sides of $\angle A$ at B and C. Draw $\overline{BC}$.

2. Const. $\overline{AB}$ of length y. Const. line $j \perp$ to $\overline{AB}$ at B. Draw an arc with center A and radius $2y$ that int. j at pt. C. $\triangle ABC$ is a 30°-60°-90° $\triangle$.

3. Const. 14 **4.** Const. 1, 12 **5.** Const. 13

6. Use Const. 9 to const. 2 tangents to the ⊙ from K. Choose a pt. P on the major arc det. by the tangents (other than the endpts. of the arc). Use Const. 8 to const. a tangent to the ⊙ at P. The 3 tangents int. to form the required △.

7. Const. 11 **8. a.** ⊥ bis. of the sides **b.** ∠ bis. or medians **c.** vertex

9. a. 3 **b.** 2, 1 **10. a.** the ⊥ bis. of $\overline{RS}$ **b.** the ⊥ bis. plane of $\overline{RS}$

11. The locus is a ⊙ formed by the int. of a sphere with center T, $r = 6$, and a sphere with center U, $r = 4$.

12. Const. 2 lines, ∥ to and z units from, line l. Const. ⊙A with radius y units. The locus is the int. of the 2 lines and the ⊙. (4 pts.)

Page 419 • ALGEBRA REVIEW

1. $s^2 = (1.3)^2 = 1.69$ **2.** $\sqrt{a^2 + b^2} = \sqrt{15^2 + 20^2} = \sqrt{625} = 25$

3. $2x + 2y = 2\left(\frac{5}{3}\right) + 2\left(\frac{3}{2}\right) = \frac{10}{3} + 3 = \frac{19}{3}$

4. $p = a + b + c = 11.5 + 7.2 + 9.9 = 28.6$

5. $lw = (2\sqrt{6})(3\sqrt{3}) = 6\sqrt{18} = 6 \cdot 3\sqrt{2} = 18\sqrt{2}$

6. $2r + s + t = 2\left(\frac{4}{7}\right) + 1 + \frac{13}{7} = 4$ **7.** $\pi r^2 \approx (3.14)(30)^2 = 2826$

8. $lwh = 8\left(6\frac{1}{4}\right)\left(3\frac{1}{2}\right) = 8\left(\frac{25}{4}\right)\left(\frac{7}{2}\right) = 175$

9. $2(lw + wh + lh) = 2[4.5(3) + 3(1) + 4.5(1)] = 2(13.5 + 3 + 4.5) = 2(21) = 42$

10. $\frac{x - 3}{y + 2} = \frac{3 - 3}{-4 + 2} = \frac{0}{-2} = 0$ **11.** $\frac{x + 5}{y - 2} = \frac{-2 + 5}{-4 - 2} = \frac{3}{-6} = -\frac{1}{2}$

12. $mx + b = \frac{5}{2}(-6) + (-2) = -15 - 2 = -17$ **13.** $6t^2 = 6(3)^2 = 54$

14. $(6t)^2 = (6 \cdot 3)^2 = 18^2 = 324$

15. $\frac{1}{2}h(a + b) = \frac{1}{2}(3)(3\sqrt{2} + 7\sqrt{2}) = \frac{3}{2}(10\sqrt{2}) = 15\sqrt{2}$

16. $\sqrt{(x - 5)^2 + (y - 3)^2} = \sqrt{(1 - 5)^2 + (0 - 3)^2} = \sqrt{16 + 9} = 5$

17. $\frac{1}{3}x^2h = \frac{1}{3}(4\sqrt{3})^2(6) = \frac{1}{3}(48)(6) = 96$

18. $2s^2 + 4sh = 2(\sqrt{6})^2 + 4(\sqrt{6})\left(\frac{5}{2}\sqrt{6}\right) = 2(6) + 10(6) = 72$ **19.** $c(x + y) = cd$

20. $\frac{1}{3}Bh = \frac{1}{3}(\pi r^2)h = \frac{1}{3}\pi r^2 h$ **21.** $\frac{1}{2}pl = \frac{1}{2}(2\pi r)l = \pi rl$

22. $2(l + w) = 2(s + s) = 4s$ **23.** $4\pi r^2 = 4\pi\left(\frac{1}{2}d\right)^2 = 4\pi\left(\frac{1}{4}d^2\right) = \pi d^2$

24. $n\left(\frac{1}{2}sa\right) = ns\left(\frac{1}{2}a\right) = p\left(\frac{1}{2}a\right) = \frac{1}{2}pa$

25. $ax + by = c; ax = c - by; x = \frac{c - by}{a}, a \neq 0$ **26.** $C = \pi d; d = \frac{C}{\pi}$

27. $S = (n - 2)180; S = 180n - 360; 180n = S + 360; n = \frac{S + 360}{180}; n = \frac{S}{180} + 2$

28. $x^2 + y^2 = r^2; y^2 = r^2 - x^2; y = \pm\sqrt{r^2 - x^2}, r^2 - x^2 \geq 0$

29. $\frac{x}{h} = \frac{h}{y}; h^2 = xy; h = \pm\sqrt{xy}$

30. $a^2 + b^2 = (a\sqrt{2})^2; b^2 = 2a^2 - a^2 = a^2; b = \pm a$

31. $A = \frac{1}{2}bh = \frac{bh}{2}; bh = 2A; h = \frac{2A}{b}, b \neq 0$

32. $m = \frac{y + 4}{x - 2}; y + 4 = m(x - 2); y = m(x - 2) - 4$

Page 420 • PREPARING FOR COLLEGE ENTRANCE EXAMS

1. B **2.** C **3.** E. $3x \cdot 10x = 18 \cdot 15; 30x^2 = 270; x^2 = 9; x = 3; 10x = 30$

4. A. $m\widehat{XY} = 360 - (60 + 70 + 70) = 160; m\angle 1 = \frac{1}{2}(160 - 70) = 45$

5. B. $(10 + 6)10 = (8 + x)8; 160 = 64 + 8x; 8x = 96; x = ZY = 12$

6. C **7.** A **8.** C **9.** E

Page 421 • CUMULATIVE REVIEW: CHAPTERS 1–10

A **1.** never **2.** always **3.** sometimes **4.** sometimes **5.** always **6.** sometimes

7. never **8.** sometimes **9.** $m\angle ASD = \frac{1}{2}(m\widehat{AD} + m\widehat{BC}) = \frac{1}{2}(70 + 144) = 107$

10. $12 \cdot 6 = 8 \cdot x; x = 9$ **11.** $x^2 = 9(9 + 16) = 225; x = 15$

12. $m\angle R = \frac{1}{2}(m\widehat{BC} - m\widehat{CD}) = \frac{1}{2}(144 - 66) = 39$

13. Methods may vary, using SSS, SAS, or ASA. Example: Const. $\overline{XY} \cong \overline{MN}$, $\angle X \cong \angle M$, and $\angle Y \cong \angle N$. Extend the sides of $\angle X$ and $\angle Y$ to int. at Z.

B **14.** The locus is 0 pts., 1 pt., a ⊙, a pt. and a ⊙, or 2 ⊙s, depending on the position of sphere J with radius 8 cm relative to 2 $\parallel$ planes, each $\parallel$ to X and 4 cm from X, and with X halfway between them.

15. **a.** $DF = \sqrt{6^2 + 8^2} = 10; \frac{10}{6} = \frac{6}{x}; 10x = 36; x = 3.6$ **b.** $\frac{x}{6} = \frac{10 - x}{8};$

$8x = 60 - 6x; 14x = 60; x = 4\frac{2}{7}$

16. $\frac{(n - 2)180}{n} = 160; n = 18$

17.

Statements	Reasons
1. $m\angle 1 = 45$	1. Given
2. $m\widehat{PQ} = 90$	2. The meas. of an inscribed $\angle$ = half the meas. of its intercepted arc.
3. $m\angle O = 90$	3. Def. of meas. of an arc
4. $\overline{OP} \cong \overline{OQ}$	4. All radii of a $\odot$ are $\cong$.
5. $\angle OQP \cong \angle OPQ$ or $m\angle OQP = m\angle OPQ$	5. Isos. $\triangle$ Thm.
6. $m\angle OQP + m\angle OPQ = 90$; $2m\angle OPQ = 90$; $m\angle OPQ = m\angle OQP = 45$	6. The acute $\angle$s of a rt. $\triangle$ are comp., algebra
7. $\triangle OPQ$ is a 45°-45°-90° $\triangle$.	7. Def. of a 45°-45°-90° $\triangle$

18.

Statements	Reasons
1. $\angle X \cong \angle Y$; $\angle W \cong \angle Z$	1. If 2 inscribed $\angle$s intercept the same arc, then the $\angle$s are $\cong$.
2. $\triangle VXW \sim \triangle VYZ$	2. AA $\sim$ Post.
3. $\frac{WX}{ZY} = \frac{XV}{YV}$	3. Corr. sides of $\sim$ $\triangle$s are in prop.
4. $WX \cdot YV = ZY \cdot XV$	4. Means-extremes Prop.

19. Methods may vary. Two examples are given. (1) Draw line k and pts. P and Q on k so that $PQ < AB$. Const. line $l \perp$ to k at P and line $m \perp$ to k at Q. Draw an arc with center Q and radius AB int. l at S. Draw an arc with center P and radius AB int. m at R. Draw $\overline{RS}$. (2) Const. the $\perp$ bis. of $\overline{AB}$ to locate M, the midpt. of $\overline{AB}$. Const. $\odot M$ with diameter $\overline{AB}$ (radius AM). Choose P on $\odot M$ and draw $\overrightarrow{PM}$, int. $\odot M$ at Q. Draw $APBQ$.

CHAPTER 11 • Areas of Plane Figures

Page 425 • CLASSROOM EXERCISES

1. A: area of a square; s: length of a side
2. A: area of a rectangle; b: length of a base; h: height (length of an altitude to a base)
3. $A = 5^2 = 25\text{ cm}^2$; $p = 4 \cdot 5 = 20$ cm 4. $4s = 28$, $s = 7$; $A = 7^2 = 49\text{ cm}^2$
5. $s^2 = 64$, $s = 8$; $p = 4 \cdot 8 = 32$ cm

	6.	7.	8.	9.	10.	11.	12.	13.
b	8 cm	4 cm	12 m	11 cm	$3\sqrt{2}$	$4\sqrt{2}$	$5\sqrt{3}$	$x + 3$
h	3 cm	1.2 cm	3 m	5 cm	2	$\sqrt{2}$	$2\sqrt{3}$	x
A	24 cm^2	4.8 cm^2	36 m^2	55 cm^2	$6\sqrt{2}$	8	30	$x^2 + 3x$

14. **a.** If 2 figures have the same area, then they are $\cong$. **b.** False. For example, a rect. with length 8 cm and width 5 cm has the same area as a rect. with length 10 cm and width 4 cm, but the 2 rectangles are not $\cong$.
15. **a.** Answers may vary. Examples: 1 cm $\times$ 9 cm (9 cm^2); 2 cm $\times$ 8 cm (16 cm^2); 5 cm $\times$ 5 cm (25 cm^2) **b.** 5 cm $\times$ 5 cm; $l = w = 5$

Pages 426–427 • WRITTEN EXERCISES

A

	1.	2.	3.	4.	5.	6.	7.	8.
b	12 cm	8.2 cm	16 cm	15 m	$3\sqrt{2}$	$\sqrt{6}$	$2x$	$4k - 1$
h	5 cm	4 cm	5 cm	8 m	$4\sqrt{2}$	$\sqrt{2}$	$x - 3$	$k + 2$
A	60 cm^2	32.8 cm^2	80 cm^2	120 m^2	24	$2\sqrt{3}$	$2x^2 - 6x$	$4k^2 + 7k - 2$

	9.	10.	11.	12.	13.
b	9 cm	40 cm	16 cm	$x + 5$	$a + 3$
h	4 cm	10 cm	5 cm	x	$a - 3$
A	36 cm^2	400 cm^2	80 cm^2	$x^2 + 5x$	$a^2 - 9$
p	26 cm	100 cm	42 cm	$4x + 10$	$4a$

	14.	15.	16.
b	$k + 7$	x	$y + 7$
h	$k + 3$	$x - 3$	y
A	$k^2 + 10k + 21$	$x^2 - 3x$	$y^2 + 7y$
p	$4k + 20$	$4x - 6$	$4y + 14$

B **17.** $A = 5(7 + 5 + 2) + 2 \cdot 5 \cdot 6 = 130$ **18.** $A = 10 \cdot 2 \cdot 2.8 = 56$

19. $h = \sqrt{10^2 - 8^2} = 6; A = 8 \cdot 6 = 48$

20. $h = \frac{18}{2} = 9; b = 9\sqrt{3}; A = 9 \cdot 9\sqrt{3} = 81\sqrt{3}$

21. $\sin 26° = \frac{h}{10}; h = 10 \sin 26° \approx 4.384; \cos 26° = \frac{b}{10}; b = 10 \cos 26° \approx 8.988; A \approx 39.4$

22. $s = \frac{7}{\sqrt{2}} = \frac{7\sqrt{2}}{2}; A = \left(\frac{7\sqrt{2}}{2}\right)^2 = \frac{49}{2} = 24.5$

23. $A = 2 \cdot 8y \cdot 2x + 2y \cdot 4x = 32xy + 8xy = 40xy$

24. $A = 2y \cdot y + 8y \cdot 3y + 4y \cdot 2y = 2y^2 + 24y^2 + 8y^2 = 34y^2$

25. $d = s\sqrt{2}; s = \frac{d}{\sqrt{2}}; A = \left(\frac{d}{\sqrt{2}}\right)^2 = \frac{d^2}{2}$

26. $l = w + 12; p = 2(l + w) = 2(w + 12 + w) = 100; 4w + 24 = 100; 4w = 76;$
$w = 19; l = 31; A = 19 \cdot 31 = 589 \text{ cm}^2$

27. $A = 2 \cdot 24 \cdot 2 + 2 \cdot 2 \cdot 12 = 144 \text{ m}^2$ **28.** $11(4 \cdot 12 + 9 \cdot 6) = \$1122$

29. **a.** $A = 2 \cdot 8 \cdot 28 + 2 \cdot 8 \cdot 20 = 768 \text{ ft}^2$ **b.** $\frac{768}{300} = 2.56$; 3 cans

30. **a.** $A = 2 \cdot 6 \cdot 220 = 2640 \text{ ft}^2$ **b.** $\frac{2640}{200} + \frac{2640}{300} = 22$ gal

31. $l = 2w; A = 2w \cdot w = 2w^2 = 392; w = 14; l = 28$; 14 m × 28 m

32. $s^2 + (s + 1)^2 + (s + 2)^2 = 365; s^2 + s^2 + 2s + 1 + s^2 + 4s + 4 = 365;$
$3s^2 + 6s - 360 = 0; 3(s^2 + 2s - 120) = 0; (s + 12)(s - 10) = 0;$
$s = -12$ (reject), $s = 10; p = 4 \cdot 10 + 4 \cdot 11 + 4 \cdot 12 = 132$ cm

33. $2 \cdot A = bh; A = \frac{1}{2}bh$

34. $\frac{d}{2} = 9; \frac{h}{2} = \frac{9}{2}; h = 9; \frac{b}{2} = \frac{9\sqrt{3}}{2}; b = 9\sqrt{3}; A = 9 \cdot 9\sqrt{3} = 81\sqrt{3} \text{ cm}^2$

35. **a.** $l = \frac{1}{2}(40 - 2x) = 20 - x$ **b.** $A = x(20 - x) = 20x - x^2$

c.

x	0	2	4	6	8	10
A	0	36	64	84	96	100

x	12	14	16	18	20
A	96	84	64	36	0

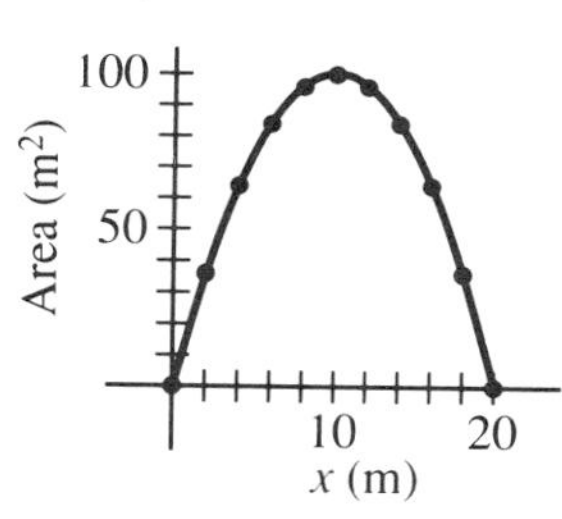

d. 10 m × 10 m

C **36.** **a.** $l = 100 - 2x; A = x(100 - 2x) = 100x - 2x^2$ **b.**

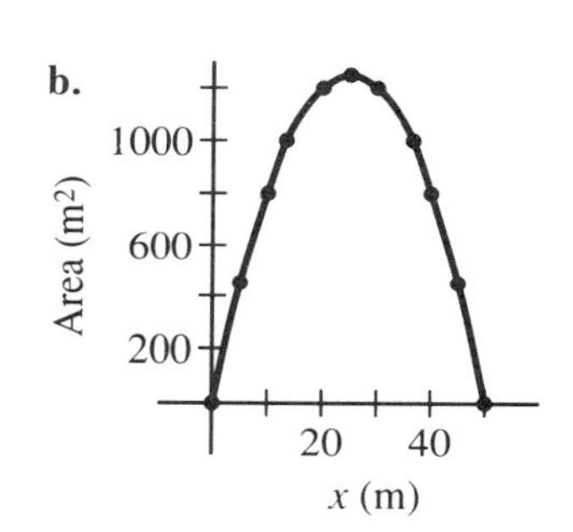

c. 25 m × 50 m

37. Let l and w be the dimensions of the rect. Use Const. 14 to const. $\overline{AB}$ with length $\sqrt{lw}$. Const. $\perp$ lines at A and B, and on the same side of $\overline{AB}$ mark off $\overline{AD}$ and $\overline{BC}$ so that $AD = BC = AB$. Draw $\overline{DC}$. $ABCD$ is a square with area lw.

Page 428 • COMPUTER KEY-IN

1. Area is approximately 0.33835001.
2. Change lines 10, 20, 40, and 50 as follows:

```
10  LET X = 0.001
20  FOR N = 1 TO 1000
40  LET A = A + Y * 0.001
50  LET X = X + 0.001
```

Area is approximately 0.333833494.

3. Less; each rect. has more area than the shaded region below it. But as more rectangles are used, the value given by the computer program gets closer to the actual area of the shaded region.

Pages 430–431 • CLASSROOM EXERCISES

1. **a.** $A = 8 \cdot 3 = 24$ **b.** $A = 4 \cdot 6 = 24$ **2.** For each $\triangle$, $A = \frac{1}{2} \cdot 9 \cdot 4 = 18.$

3. $A = \frac{1}{2}d_1d_2$; since every rhombus is a $\square$, $A = bh$. **4.** $A = \frac{1}{2} \cdot 4 \cdot 2\sqrt{3} = 4\sqrt{3}$

5. $A = 6 \cdot \frac{3\sqrt{3}}{2} = 9\sqrt{3}$ **6.** $A = \frac{1}{2} \cdot 6 \cdot 4 = 12$ **7.** $A = \frac{1}{2} \cdot 4 \cdot 10 = 20$

8. $A = \frac{1}{2} \cdot 8 \cdot 6 = 24$

9. Since $5^2 + 12^2 = 13^2$, the $\triangle$ is a rt. $\triangle$. $A = \frac{1}{2} \cdot 5 \cdot 12 = 30$

Pages 431–433 • WRITTEN EXERCISES

A **1.** $A = \frac{1}{2} \cdot 5.2 \cdot 11.5 = 29.9 \text{ m}^2$ **2.** Rt. $\triangle$; $A = \frac{1}{2} \cdot 4 \cdot 3 = 6$

3. $A = 3\sqrt{2} \cdot 2\sqrt{2} = 12$ **4.** $A = \frac{1}{2} \cdot 4 \cdot 6 = 12$ **5.** $A = \frac{1}{2} \cdot 8 \cdot 4\sqrt{3} = 16\sqrt{3} \text{ ft}^2$

6. $A = \frac{1}{2} \cdot 16 \cdot 6 = 48$ **7.** $A = 5 \cdot 8 = 40$ **8.** $A = 4 \cdot 8 + \frac{1}{2} \cdot 8 \cdot 3 = 44$

9. $h = \sqrt{13^2 - 5^2} = 12$; in upper $\triangle$, $b = \sqrt{15^2 - 12^2} = 9$; $A = \frac{1}{2} \cdot 12 \cdot 5 +$
$\frac{1}{2} \cdot 12 \cdot 9 = 84$

10. $A = \frac{1}{2} \cdot 10 \cdot 12 = 60$ **11.** $A = \frac{1}{2} \cdot \frac{8}{\sqrt{2}} \cdot \frac{8}{\sqrt{2}} = 16$ **12.** $A = \frac{1}{2} \cdot 6 \cdot 3\sqrt{3} = 9\sqrt{3}$

B **13.** $A = 10 \cdot \frac{6}{\sqrt{2}} = 30\sqrt{2}$ **14.** $\frac{d}{2} = 3\sqrt{3}$, $d = 6\sqrt{3}$; $A = \frac{1}{2} \cdot 6\sqrt{3} \cdot 6 = 18\sqrt{3} \text{ cm}^2$

15. $A = \frac{1}{2} \cdot 5 \cdot 5\sqrt{3} = \frac{25\sqrt{3}}{2}$ **16.** $9 = \frac{s}{2}\sqrt{3}$, $s = 6\sqrt{3}$; $A = \frac{1}{2} \cdot 6\sqrt{3} \cdot 9 = 27\sqrt{3}$

17. $\frac{d}{2} = 8$; $A = \frac{1}{2} \cdot 16 \cdot 30 = 240$ **18.** $A = 6 \cdot \frac{1}{2} \cdot 10 \cdot 5\sqrt{3} = 150\sqrt{3}$

19. $A = 4 \cdot \frac{1}{2} \cdot r \cdot r = 2r^2$ **20.** $\frac{h}{2} = 6$, $h = 12$; $A = 16 \cdot 12 = 192$

21. $\tan 20^\circ = \frac{h}{10}$; $h = 10 \tan 20^\circ \approx 3.64$; $A \approx \frac{1}{2} \cdot 10 \cdot 3.64 = 18.2$

22. $\sin 54^\circ = \frac{x}{12}$; $x = 12 \sin 54^\circ \approx 9.708$; $y \approx \sqrt{12^2 - (9.708)^2} \approx 7.054$;
$A \approx \frac{1}{2} \cdot 9.708 \cdot 7.054 \approx 34.2$

23. $\sin 50^\circ = \frac{h}{8}$; $h = 8 \sin 50^\circ \approx 6.128$; $A \approx 12 \cdot 6.128 \approx 73.5$

24. $\tan 16^\circ = \frac{4}{h}$; $h = \frac{4}{\tan 16^\circ} \approx 13.95$; $A = \frac{1}{2} \cdot 8 \cdot 13.95 = 55.8 \text{ cm}^2$

25. $\triangle DFE \sim \triangle DGF \sim \triangle FGE$; $\frac{2}{h} = \frac{h}{8}$; $h = 4$; $\triangle DFE$: $A = \frac{1}{2} \cdot 10 \cdot 4 = 20$;
$\triangle DGF$: $A = \frac{1}{2} \cdot 2 \cdot 4 = 4$; $\triangle FGE$: $A = \frac{1}{2} \cdot 8 \cdot 4 = 16$

26. Area of ▱ $PQRS = bh = 36$; Area of $\triangle RST = \frac{1}{2}bh = 18$

27. $\triangle ABC$: $A = \frac{1}{2} \cdot 16 \cdot 5 = 40$; $\triangle AMB$: $A = \frac{1}{2} \cdot 8 \cdot 5 = 20$

28. The area of $\triangle ABC = \frac{1}{2} \cdot BC \cdot h$; the area of $\triangle AMB = \frac{1}{2} \cdot MB \cdot h$. But since $\overline{AM}$ is a median, $MB = \frac{1}{2} \cdot BC$. Therefore, the area of $\triangle AMB = \frac{1}{2} \cdot \left(\frac{1}{2} \cdot BC\right) \cdot h = \frac{1}{2} \cdot \left(\frac{1}{2} \cdot BC \cdot h\right) = \frac{1}{2} \cdot$ area of $\triangle ABC$.

29. **a.** $\left(\frac{1}{2} \cdot 12 \cdot h\right) : \left(\frac{1}{2} \cdot 18 \cdot h\right) = 2 : 3$ **b.** $240 = \frac{1}{2} \cdot 24 \cdot x$; $x = 20$

30. $h_1 = 3$ cm; $A = \frac{1}{2} \cdot 8 \cdot 3 = 12\text{ cm}^2$; $12 = \frac{1}{2} \cdot 5 \cdot h_2$; $h_2 = h_3 = 4.8$ cm

31. **a.** $A = \frac{1}{2}ab$ **b.** $A = \frac{1}{2}ch$ **c.** $\frac{1}{2}ab = \frac{1}{2}ch$; $h = \frac{ab}{c}$ **d.** $c = \sqrt{6^2 + 8^2} = 10$; $h = \frac{ab}{c} = \frac{6 \cdot 8}{10} = 4.8$; length of median to hyp. $= \frac{1}{2} \cdot 10 = 5$

32. **a.** $A = 30 \cdot 16 = 480$ **b.** $A = \frac{1}{2} \cdot 30 \cdot 16 = 240$ **c.** area of $\triangle OSR = \frac{1}{2} \cdot$ area of $\triangle PSR = 120$ **d.** 120 **e.** 120; $\triangle POQ \cong \triangle ROS$; 120 because $\triangle OQR \cong \triangle OSP$
f. The diags. of a ▱ divide the ▱ into 4 ⚠ with = areas.

33. **a.** $b = s$, $h = \frac{s\sqrt{3}}{2}$; $A = \frac{1}{2} \cdot s \cdot \frac{s\sqrt{3}}{2} = \frac{s^2\sqrt{3}}{4}$ **b.** $A = \frac{7^2}{4}\sqrt{3} = \frac{49\sqrt{3}}{4}$

34. **a.** $h = \frac{10}{2} = 5$; $A = 20 \cdot 5 = 100$

b. $h = \frac{10}{\sqrt{2}} = 5\sqrt{2}$;

$A = 20 \cdot 5\sqrt{2} = 100\sqrt{2}$

c. $h = 5\sqrt{3}$; $A = 20 \cdot 5\sqrt{3} = 100\sqrt{3}$

d. $h = 10$; $A = 20 \cdot 10 = 200$

e. $h = 5\sqrt{3}$; $A = 20 \cdot 5\sqrt{3} = 100\sqrt{3}$ **f.** (b) $A \approx 140$; (c) 170; (e) 170; see graph.

Area: 100, 200; O; 30, 45, 60, 90, 120; Measure of $\angle C$

35. $A = \frac{1}{2} \cdot d \cdot 2d = 100$; $d^2 = 100$; $d = 10$; $2d = 20$; 10, 20

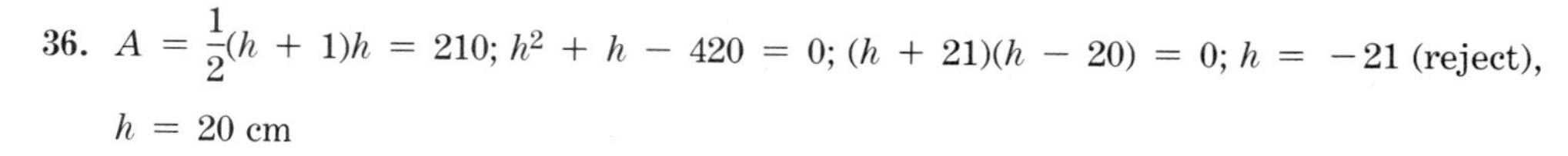

36. $A = \frac{1}{2}(h + 1)h = 210$; $h^2 + h - 420 = 0$; $(h + 21)(h - 20) = 0$; $h = -21$ (reject), $h = 20$ cm

C **37.** Draw rect. $PQRS$ with $\overline{PQ} \parallel x$-axis. Then area of $ABCD$ = area of $PQRS$ − (area of $\triangle ARB$ + area of $\triangle BQC$ + area of $\triangle CPD$ + area of $\triangle DSA$) = $11 \cdot 7 - \left(\frac{1}{2} \cdot 4 \cdot 6 + \frac{1}{2} \cdot 7 \cdot 1 + \frac{1}{2} \cdot 4 \cdot 3 + \frac{1}{2} \cdot 7 \cdot 4\right) = 77 - \left(12 + \frac{7}{2} + 6 + 14\right) = 41.5$

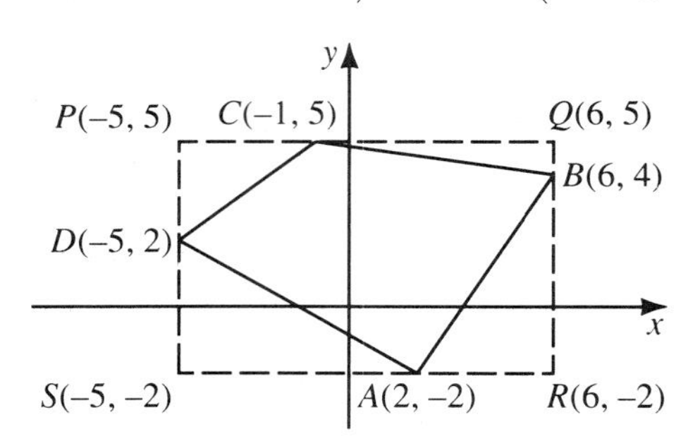

38. Draw $\overline{AX} \perp \overline{BC}$ and $\overline{AY} \perp \overleftrightarrow{DC}$. $\angle BAX$ and $\angle DAY$ are $\cong$ since both are comps. of $\angle XAD$. Then $\triangle BAX \cong \triangle DAY$ by ASA and their areas are =. So area of $ABCD$ = area of $\triangle BAX$ + area of $AXCD$ = area of $\triangle DAY$ + area of $AXCD$ = area of $AXCY$ = $6 \cdot 6 = 36$; 36 cm^2

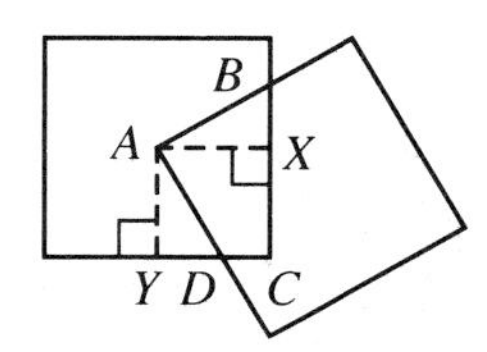

39. Suppose $ABCD$ is a ▱ with $AC = 82$ and $BD = 30$. $\angle A$ and $\angle C$ must be the acute ⦣s. Let $\overline{AX}$ and $\overline{BY}$ be altitudes from $\overline{AB}$ to $\overleftrightarrow{DC}$ and suppose $AX = BY = 18$. Either $\overline{AX}$ intersects $\overleftrightarrow{DC}$ outside of $\overline{DC}$ and Y is on $\overline{DC}$, or $\overline{AX}$ and $\overline{BY}$ both intersect $\overleftrightarrow{DC}$ outside of $\overline{DC}$. (If X and Y are both on $\overline{DC}$, then $\triangle AXD \cong \triangle BYC$, $\angle D \cong \angle C$ and $ABCD$ is a rect. This contradicts the given fact that the diags. are not $\cong$.) If Y is on $\overline{DC}$ and X is not, then $XC = \sqrt{82^2 - 18^2} = 80$; $DY = \sqrt{30^2 - 18^2} = 24$. Since $\triangle AXD \cong \triangle BYC$, $XD = YC$; $YC = \frac{1}{2}(80 - 24) = 28$; $DC = 24 + 28 = 52$;

$A = 18 \cdot 52 = 936\text{ cm}^2$. If $\overline{AX}$ and $\overline{BY}$ both intersect $\overleftrightarrow{DC}$ outside of $\overline{DC}$, again $XC = 80$; also $DY = 24$. But $XY = AB = DC$, so $DC = \frac{1}{2}(XC - YD) = 28$ and $A = 28(18) = 504\text{ cm}^2$. The same results are obtained by considering the altitudes from $\overline{AD}$ to $\overleftrightarrow{BC}$. Then the 2 possible values are 936 cm^2 and 504 cm^2.

40. Given scalene $\triangle ABC$, use Const. 6 to construct $\overline{AX} \perp \overline{BC}$. Area of $\triangle ABC = \frac{1}{2} \cdot BC \cdot AX$. Use Const. 1 to construct $\overline{EF} \cong \overline{BC}$ and use Const. 4 to construct the $\perp$ bis. of $\overline{EF}$, intersecting $\overline{EF}$ at Y. Mark off a seg. $\overline{DY}$ on the $\perp$ bis. such that $DY = AX$. Draw $\overline{DE}$ and $\overline{DF}$. $\triangle DEF$ is isos. and area of $\triangle DEF = \frac{1}{2} \cdot EF \cdot DY = \frac{1}{2} \cdot BC \cdot AX =$ area of $\triangle ABC$.

41. Given $\triangle ABC$, use Const. 6 to construct $\overline{AX} \perp \overline{BC}$. Use Const. 14 to construct $\overline{EF}$ such that EF is the geom. mean between AX and BC. Construct a $\perp$ to $\overline{EF}$ through E using Const. 5 and mark off $\overline{DE}$ such that $DE = EF$. Draw $\overline{DF}$. $\triangle DEF$ is an isos. rt. $\triangle$ and area of $\triangle DEF = \frac{1}{2} \cdot EF \cdot DE = \frac{1}{2}(\sqrt{AX \cdot BC})(\sqrt{AX \cdot BC}) = \frac{1}{2} \cdot AX \cdot BC =$ area of $\triangle ABC$.

42. Given scalene $\triangle ABC$, use Const. 6 to construct $\overline{AX} \perp \overline{BC}$. Construct $\overline{DE} \cong \overline{AX}$ using Const. 1, then extend $\overline{DE}$ through E and on $\overrightarrow{DE}$ mark off EF and $FG = DE$, so that $DG = 3DE$. Use Const. 14 to construct $\overline{HI}$ so that HI is the geom. mean between DE and DG ($HI = \sqrt{3}DE = \sqrt{3}AX$). Use Const. 1 again to construct $\overline{JK} \cong \overline{BC}$ and use Const. 12 to divide $\overline{JK}$ into 3 $\cong$ segs., $\overline{JL}$, $\overline{LM}$, and $\overline{MK}$. Now construct $\overline{NP}$ so that NP is the geom. mean between HI and JM, i.e., $NP = \sqrt{AX\sqrt{3} \cdot \frac{2}{3}BC}$. With centers at N and P and radius NP, draw arcs intersecting at Q. $\triangle QNP$ is an equilateral $\triangle$ with height $= \frac{NP\sqrt{3}}{2}$ and area $= \frac{\sqrt{3}}{4}(NP)^2 = \frac{\sqrt{3}}{4}\left(AX \cdot BC \cdot \frac{2}{3} \cdot \sqrt{3}\right) = \frac{1}{2} \cdot AX \cdot BC =$ area of $\triangle ABC$.

Page 433 • EXPLORATIONS

Area of ▱ $= \frac{1}{2} \cdot$ area of quad., or $\frac{\text{area of ▱}}{\text{area of quad.}} = \frac{1}{2}$

Proof: Draw diags. $\overline{AC}$ and $\overline{BD}$ in quad. $ABCD$. Area of $\triangle AHE = \frac{1}{4} \cdot$ area of $\triangle ADB$, and area of $\triangle FCG = \frac{1}{4} \cdot$ area of $\triangle BCD$.

So area of $\triangle AHE$ + area of $\triangle FCG = \frac{1}{4} \cdot$ (area of $\triangle ADB$ + area of $\triangle BCD$) $= \frac{1}{4} \cdot$ area of $ABCD$. Similarly, area of $\triangle BEF$ + area of $\triangle HDG = \frac{1}{4} \cdot$ area of $ABCD$. Then area of $\triangle AHE$ + area of $\triangle FCG$ + area of $\triangle BEF$ + area of $\triangle HDG$ $= \frac{1}{2} \cdot$ area of $ABCD$. So $\frac{\text{area of ▱ } EFGH}{\text{area of quad. } ABCD} =$

$$\frac{\text{area of } ABCD - \frac{1}{2} \cdot \text{area of } ABCD}{\text{area of } ABCD} = \frac{1}{2}.$$

Page 434 • CALCULATOR KEY-IN

1. $s = 15$; $A = \sqrt{15 \cdot 6 \cdot 5 \cdot 4} = 30\sqrt{2} \approx 42.4$; $h_1 \approx \frac{2(42.4)}{9} \approx 9.42$; $h_2 \approx \frac{2(42.4)}{10} = 8.48$; $h_3 \approx \frac{2(42.4)}{11} \approx 7.71$

2. $s = 10$; $A = \sqrt{10 \cdot 5 \cdot 3 \cdot 2} = 10\sqrt{3} \approx 17.3$; $h_1 \approx \frac{34.6}{5} = 6.92$; $h_2 \approx \frac{34.6}{7} \approx 4.94$; $h_3 \approx \frac{34.6}{8} \approx 4.33$

3. $s = 15$; $A = \sqrt{15 \cdot 9 \cdot 4 \cdot 2} = 6\sqrt{30} \approx 32.9$; $h_1 \approx \frac{65.8}{6} \approx 11.0$; $h_2 \approx \frac{65.8}{11} \approx 5.98$; $h_3 \approx \frac{65.8}{13} \approx 5.06$

4. $s = 24$; $A = \sqrt{24 \cdot 9 \cdot 8 \cdot 7} = 24\sqrt{21} \approx 110.0$; $h_1 \approx \frac{220}{15} \approx 14.7$; $h_2 \approx \frac{220}{16} \approx 13.8$; $h_3 \approx \frac{220}{17} \approx 12.9$

5. $s = 11.8$; $A = \sqrt{11.8 \cdot 5.5 \cdot 4.6 \cdot 1.7} \approx 22.5$; $h_1 \approx \frac{45}{6.3} \approx 7.14$; $h_2 \approx \frac{45}{7.2} = 6.25$; $h_3 \approx \frac{45}{10.1} \approx 4.46$

6. $s = 125; A = \sqrt{125 \cdot 57 \cdot 48 \cdot 20} \approx 2620; h_1 \approx \frac{5240}{68} \approx 77.1; h_2 \approx \frac{5240}{77} \approx 68.1;$

$h_3 \approx \frac{5240}{105} \approx 49.9$

7. $s = 11; A = \sqrt{11 \cdot 5.5 \cdot 4.5 \cdot 1} \approx 16.5; h_1 \approx \frac{33}{5.5} = 6.00; h_2 \approx \frac{33}{6.5} \approx 5.08;$

$h_3 \approx \frac{33}{10} = 3.30$

8. $s = 28.5; A = \sqrt{28.5 \cdot 16.5 \cdot 10.5 \cdot 1.5} \approx 86.1; h_1 \approx \frac{172.2}{12} \approx 14.4;$

$h_2 \approx \frac{172.2}{18} \approx 9.57; h_3 \approx \frac{172.2}{27} \approx 6.38$

9. $A = \sqrt{6 \cdot 3 \cdot 2 \cdot 1} = 6; A = \frac{1}{2} \cdot 3 \cdot 4 = 6$

10. $A = \sqrt{9 \cdot 3 \cdot 3 \cdot 3} = 9\sqrt{3}; A = \frac{1}{2} \cdot 6 \cdot 3\sqrt{3} = 9\sqrt{3}$

11. $A = \sqrt{18 \cdot 5 \cdot 5 \cdot 8} = 60; A = \frac{1}{2} \cdot 10 \cdot 12 = 60$

12. $A = \sqrt{50 \cdot 21 \cdot 21 \cdot 8} = 420; A = \frac{1}{2} \cdot 42 \cdot 20 = 420$

13. $a = 47$, $b = 38$, and $c = 85$ cannot be the sides of a $\triangle$ since $a + b = c$.

14. $s = \frac{1}{2}(10 + 10 + 10 + 20) = 25; A = \sqrt{15 \cdot 15 \cdot 15 \cdot 5} = 75\sqrt{3} \approx 130$

Pages 435–436 • CLASSROOM EXERCISES

1. $A = \frac{1}{2}(5)(7 + 13) = 50$; length of median $= \frac{1}{2}(7 + 13) = 10$

2. $A = \frac{1}{2}(6)(5 + 13) = 54$; length of median $= \frac{1}{2}(5 + 13) = 9$

3. $A = \frac{1}{2}(12)(9 + 14) = 138$; length of median $= \frac{1}{2}(9 + 14) = 11.5$

4. Area of trap. $= \frac{1}{2}h(b_1 + b_2) = h\left[\frac{1}{2}(b_1 + b_2)\right]$ = height × median **5.** No

6. Yes **7. a.** a ▱ **b.** $A = \frac{1}{2} \cdot$ Area of ▱ $= \frac{1}{2} \cdot h(b_1 + b_2)$ = area of trap.

8. $h = 4; A = \frac{1}{2}(4)(6 + 12) = 36$ **9.** $h = 4\sqrt{3}; A = \frac{1}{2}(4\sqrt{3})(12 + 20) = 64\sqrt{3}$

10. $h = 5; A = \frac{1}{2}(5)(6 + 14) = 50$

Pages 436–438 • WRITTEN EXERCISES

A

	1.	2.	3.	4.	5.	6.	7.	8.
b_1	12	6.8	$3\frac{1}{6}$	45	27	3	7	$15k$
b_2	8	3.2	$4\frac{1}{3}$	15	9	5	1	$3k$
h	7	6.1	$1\frac{3}{5}$	10	5	3	$9\sqrt{2}$	$5k$
A	70	30.5	6	300	90	12	$36\sqrt{2}$	$45k^2$
m	10	5	$3\frac{3}{4}$	30	18	4	4	$9k$

9. $m = \frac{54}{6} = 9$ **10.** $b = 5 + 6 = 11; A = \frac{1}{2}(8)(11 + 5) = 64$

11. $h = 9; A = \frac{1}{2}(9)(6 + 18) = 108$

12. $h = 3\sqrt{3}; b = 4; A = \frac{1}{2}(3\sqrt{3})(4 + 10) = 21\sqrt{3}$

13. $h = \frac{3\sqrt{3}}{2}; A = \frac{1}{2}\left(\frac{3\sqrt{3}}{2}\right)(3 + 6) = \frac{27\sqrt{3}}{4}$ **14.** $h = 9; A = \frac{1}{2}(9)(6 + 15) = 94\frac{1}{2}$

15. $h = 4; A = \frac{1}{2}(4)(2 + 10) = 24$ **16.** $h = 12; A = \frac{1}{2}(12)(10 + 20) = 180$

17. $h = 8; A = \frac{1}{2}(8)(10 + 22) = 128$ **18.** $h = 5; A = \frac{1}{2}(5)(8 + 18) = 65$

B **19–21.** Answers may vary. **19.** $h = 6 \tan 35° \approx 4.2012; A \approx \frac{1}{2}(4.2012)(7 + 13) \approx 42.0$

20. $x = \frac{6}{\tan 38°} \approx 7.6796; A = \frac{1}{2}(6)(8 + 15.6796) \approx 71.0$

21. $h = 10 \sin 40° \approx 6.4279; x = 10 \cos 40° \approx 7.6604; b \approx 6 + 2(7.6604) = 21.3208;$

$A \approx \frac{1}{2}(6.4279)(6 + 21.3208) \approx 87.8$

22. $h = \sqrt{10^2 - \left(\frac{1}{2}(21 - 9)\right)^2} = 8; A = \frac{1}{2}(8)(9 + 21) = 120 \text{ cm}^2$; length of each diag. $= \sqrt{15^2 + 8^2} = 17$ cm

23. $300 = \frac{1}{2}h(12 + 28); h = 15$; length of each leg $= \sqrt{15^2 + \left(\frac{1}{2}(28 - 12)\right)^2} = 17$; perimeter $= 2 \cdot 17 + 12 + 28 = 74$

24. **a.** 1 : 1 (Area of $\triangle ABD = 6h =$ area of $\triangle ABC$) **b.** 1 : 1 (Area of $\triangle ACD = 2h =$ area of $\triangle BDC$; area of $\triangle AOD =$ area of $\triangle ACD -$ area of $\triangle DOC =$ area of $\triangle BDC -$ area of $\triangle DOC =$ area of $\triangle BOC$.) **c.** 3 : 1 (Area of $\triangle ABD = 6h$; area of $\triangle ADC = 2h$)

25. $\triangle ABC$ is an isos. $\triangle$ with base ∠s of 30°. Height = 6 and base = $12\sqrt{3}$; area of $\triangle ABC = \frac{1}{2} \cdot 6 \cdot 12\sqrt{3} = 36\sqrt{3}$. $\triangle ACD$ is a rt. $\triangle$ with legs 12 and $12\sqrt{3}$ so area of $\triangle ACD = \frac{1}{2} \cdot 12 \cdot 12\sqrt{3} = 72\sqrt{3}$. Area of $ADEF$ = area of $\triangle ABC$ + area of $\triangle ACD = 108\sqrt{3}$. (Alternatively, find $AD = 24$; $ADEF$ is an isos. trap. with bases 12 and 24 and base ∠s of 60°, so $h = \frac{12}{2}\sqrt{3} = 6\sqrt{3}$ and area $= \frac{1}{2}(6\sqrt{3})(12 + 24) = 108\sqrt{3}$.)

26. Let $AB = 12$ and $DC = 16$ with $\overline{AB}$ and $\overline{DC}$ the bases of trap. $ABCD$ and let O be the ctr. of the circle. Draw $\overline{OA}$, $\overline{OB}$, $\overline{OD}$, and $\overline{OC}$. $\triangle ODC$ is an isos. $\triangle$ with legs 10 and base 16 so its height $= \sqrt{10^2 - 8^2} = 6$. $\triangle OAB$ is an isos. $\triangle$ with legs 10 and base 12 so its height $= \sqrt{10^2 - 6^2} = 8$. The height of $ABCD$ is the sum of the heights of $\triangle ODC$ and $\triangle OAB$. Area of $ABCD = \frac{1}{2}(6 + 8)(12 + 16) = 196$.

27. $AB = 5$, $DC = 15$; $A = \frac{1}{2}h(5 + 15) = 100$; $h = 10$;

$\triangle TAB \sim \triangle TDC$, so $\frac{x}{x + 10} = \frac{5}{15} = \frac{1}{3}$; $x = 5$;

area of $\triangle TAB = \frac{1}{2} \cdot 5 \cdot 5 = 12.5 \text{ cm}^2$;

area of $\triangle TDC = \frac{1}{2} \cdot 15 \cdot 15 = 112.5 \text{ cm}^2$

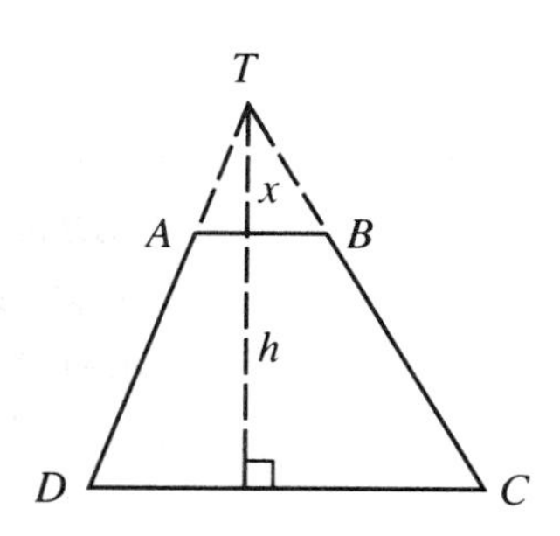

C **28.** Given a non-isos. trap. $ABCD$ with bases $\overline{AB}$ and $\overline{DC}$, use Const. 6 to construct a ⊥ from A to $\overleftrightarrow{DC}$ intersecting $\overleftrightarrow{DC}$ at E. Construct $\overline{FG} \cong \overline{DC}$, using Const. 1. Use Const. 4 to construct the ⊥ bis. of $\overline{FG}$, intersecting $\overline{FG}$ at H. Mark off a seg. $\overline{HI}$ on the ⊥ bis. so that $HI = AE$. Use Const. 7 to construct a ∥ to $\overline{FG}$ through I. Locate the midpt., M, of $\overline{AB}$ by using Const. 4 to construct the ⊥ bis. of $\overline{AB}$. Mark off $\overline{JI}$ and $\overline{IK}$ on the ∥ through I so that $JI = IK = AM$. Draw $\overline{JF}$ and $\overline{KG}$. $JKGF$ is an isos. trap. with the same area as trap. $ABCD$.

29. Let $\overline{ZA}$ and $\overline{YB}$ be ⊥ to $\overline{WX}$. $ZA = \frac{10}{2} = 5$ and $WA = 5\sqrt{3}$; $YB = ZA = 5$ so $BX = 5$. Then $ZY = AB = 20 - (5\sqrt{3} + 5) = 15 - 5\sqrt{3}$ and $A = \frac{1}{2}(5)(20 + 15 - 5\sqrt{3}) = \frac{175 - 25\sqrt{3}}{2}$.

30. Let $\overline{SA}$ and $\overline{RB}$ be $\perp$ to $\overline{PQ}$. $AB = 10$ so $PA + BQ = 14$. Let $PA = x$ and $BQ = 14 - x$. $SA = RB$ so $\sqrt{13^2 - x^2} = \sqrt{15^2 - (14 - x)^2}$; $13^2 - x^2 = 15^2 - (14 - x)^2$; $169 - x^2 = 225 - 196 + 28x - x^2$; $28x = 140$; $x = 5$. $SA = \sqrt{13^2 - 5^2} = 12$; $A = \frac{1}{2}(12)(10 + 24) = 204$

31. Draw $\overline{DE}$ and $\overline{CF} \perp \overline{AB}$ at E and F, resp. $DC = EF = 8$; $DE = CF$; $\triangle ADE \cong \triangle BCF$ (AAS); $AE = FB = 5$. Draw $\overline{GH}$ through O and $\perp$ $\overline{AB}$ and $\overline{DC}$ at tangency pts. G and H, resp. Draw $\overline{AO}$ and $\overline{BO}$. Draw $\overline{OI} \perp \overline{BC}$ at tangency pt. I and $\overline{OJ} \perp \overline{AD}$ at tangency pt. J. $\triangle AOJ \cong \triangle AOG$ (SSS) and $\triangle BOI \cong \triangle BOG$ (SSS), so $\angle OAJ \cong \angle OAG \cong \angle OBI \cong \angle OBG$. Then $\triangle AOG \cong \triangle BOG$ (AAS) and $\overline{AG} \cong \overline{BG}$. Therefore, $AG = AJ = BG = BI = 9$. Similarly, $DH = DJ = CH = CI = 4$. Then $AD = 9 + 4 = 13$. So $DE = \sqrt{13^2 - 5^2} = 12$ and the area of $ABCD = \frac{1}{2}(12)(8 + 18) = 156$.

32. Area of $MNOP = \frac{1}{2}(MN)(PM + ON) = \frac{1}{2}(a + b)(a + b) = \frac{1}{2}(a + b)^2$. Since rt. $\triangle PMR \cong$ rt. $\triangle RNO$, $\angle PRM$ and $\angle ORN$ are comp. Thus $\angle PRO$ is a rt. $\angle$. So area of $MNOP$ = area of $\triangle PMR$ + area of $\triangle RNO$ + area of $\triangle PRO = \frac{1}{2}ab + \frac{1}{2}ab + \frac{1}{2}c^2$. Then $\frac{1}{2}(a + b)^2 = \frac{1}{2}(2ab + c^2)$ or $a^2 + 2ab + b^2 = 2ab + c^2$ and $a^2 + b^2 = c^2$.

33. Let s = length of a side of $ABCD$. $\triangle CHB \cong \triangle DEA$ (AAS) so $HB = EA$. But $HB = HE + EB$ and $EA = EB + BA$ so $HE = BA = s$. $\angle GCD \cong \angle H$ so $\triangle GCD \cong \triangle FHE$. Area of rect. $EFGD$ = area of $\triangle GCD$ + area of $CFED$ = area of $\triangle FHE$ + area of $CFED$ = area of $\square CHED = HE \cdot CB = s \cdot s = s^2$ = area of square $ABCD$.

34. First note that area of $MNOP$ = area of $MNSR$ + area of $NOVS$ − area of $MPTR$ − area of $POVT$. Draw a $\parallel$ to $\overline{RV}$ through M intersecting $\overline{NS}$ at A and a $\parallel$ to $\overline{RV}$ through P intersecting $\overline{OV}$ at B. By extending $\overline{PO}$ in both directions to intersect $\overleftrightarrow{MA}$ and $\overleftrightarrow{RV}$, you can show that $\angle NMA \cong \angle OPB$ and so $\triangle NMA \cong \triangle OPB$ (AAS). $MA = RS = 8$ so $PB = TV = 8$. $NA = 16 - 10 = 6$ so $OB = 6$ and $BV = PT = 3$. Then area of $MNOP = \frac{1}{2} \cdot 8 \cdot 26 + \frac{1}{2} \cdot 11 \cdot 25 - \frac{1}{2} \cdot 11 \cdot 13 - \frac{1}{2} \cdot 8 \cdot 12 = 122$.

Page 439 • COMPUTER KEY-IN

1. Area is approximately 2.33334996. **2.** Area is approximately 2.33333378.
3. Area is approximately 0.335. This answer is much closer to the exact area than the answer obtained on page 428 when rectangles were used instead of trapezoids.

Page 440 • MIXED REVIEW EXERCISES

1. 52 **2.** $2 \cdot 73 = 146$ **3.** $\frac{(8 - 2)180}{8} = 135$ **4.** $n = \frac{360}{20} = 18$

5. $\frac{1}{2} \cdot 20\sqrt{2} = 10\sqrt{2}$ cm **6.** 15 cm, $15\sqrt{3}$ cm **7.** 10 m

8. $BC = \sqrt{17^2 - 8^2} = 15$; $\cos B = \frac{15}{17}$

Pages 442–443 • CLASSROOM EXERCISES

1. a. $\frac{360}{10} = 36$ **b.** $\frac{360}{15} = 24$ **c.** $\frac{360}{360} = 1$ **d.** $\frac{360}{n}$

2. $p = 8 \cdot 4 = 32$; $A = \frac{1}{2}ap = \frac{1}{2} \cdot a \cdot 32 = 16a$

3. $p = 5s$; $A = \frac{1}{2}ap = \frac{1}{2} \cdot 3 \cdot 5s = \frac{15s}{2}$ **4.** $p = 10s$; $A = \frac{1}{2}ap = \frac{1}{2} \cdot a \cdot 10s = 5sa$

5. The apothem is a $\perp$ seg., so it is the shortest distance between the center and a side. (Thm. 6-3, Cor. 1)

6. a. $p = 4 \cdot 4 = 16$ **b.** $\frac{360}{4} = 90$ **c.** $a = \frac{1}{2} \cdot 4 = 2$ **d.** $r = \sqrt{2^2 + 2^2} = 2\sqrt{2}$

e. $A = 4 \cdot 4 = 16$

7. a. $p = 3 \cdot 4 = 12$ **b.** $\frac{360}{3} = 120$ **c.** $a\sqrt{3} = 2$; $a = \frac{2\sqrt{3}}{3}$ **d.** $r = 2a = 2 \cdot \frac{2\sqrt{3}}{3} = \frac{4\sqrt{3}}{3}$ **e.** $A = \frac{1}{2}ap = \frac{1}{2} \cdot \frac{2\sqrt{3}}{3} \cdot 12 = 4\sqrt{3}$

8. a. $p = 6 \cdot 4 = 24$ **b.** $\frac{360}{6} = 60$ **c.** $a = 2\sqrt{3}$ **d.** $r = 2 \cdot 2 = 4$

e. $A = \frac{1}{2} \cdot 2\sqrt{3} \cdot 24 = 24\sqrt{3}$

9. a. $\frac{360}{5} = 72$ **b.** $\triangle AOX \cong \triangle BOX$ so $m\angle AOX = m\angle BOX = \frac{1}{2}m\angle AOB = 36$

c. 10; $a \approx 8.1$ **d.** 10; $s \approx 11.8$ **e.** $p = 5s \approx 5 \cdot 11.8 = 59$;

$A = \frac{1}{2}ap \approx \frac{1}{2} \cdot 8.1 \cdot 59 = 238.95$

Pages 443–444 • WRITTEN EXERCISES

	r	a	A
A 1.	$8\sqrt{2}$	8	256
2.	$5\sqrt{2}$	5	100
3.	$\frac{7\sqrt{2}}{2}$	$\frac{7}{2}$	49
4.	$2\sqrt{3}$	$\sqrt{6}$	24

	r	a	p	A
5.	6	3	$18\sqrt{3}$	$27\sqrt{3}$
6.	8	4	$24\sqrt{3}$	$48\sqrt{3}$
7.	$\frac{4\sqrt{3}}{3}$	$\frac{2\sqrt{3}}{3}$	12	$4\sqrt{3}$
8.	3	$\frac{3}{2}$	$9\sqrt{3}$	$\frac{27\sqrt{3}}{4}$

	r	a	p	A
9.	4	$2\sqrt{3}$	24	$24\sqrt{3}$
10.	10	$5\sqrt{3}$	60	$150\sqrt{3}$
11.	$4\sqrt{3}$	6	$24\sqrt{3}$	$72\sqrt{3}$
12.	$2\sqrt{3}$	3	$12\sqrt{3}$	$18\sqrt{3}$

B 13. $a = 2\sqrt{3}$; $s = 12$; $p = 36$; $A = 36\sqrt{3}$

14. $a = \frac{8}{\sqrt{2}}k = 4k\sqrt{2}$; $s = 8k\sqrt{2}$; $p = 32k\sqrt{2}$; $A = 128k^2$

15. $s = 12$; $a = 6\sqrt{3}$; $A = 216\sqrt{3}$ **16.** $s = \frac{8\sqrt{3}}{3}$; $p = 16\sqrt{3}$; $A = 32\sqrt{3}$

17. a. $\triangle AOX \cong \triangle BOX$ so $m\angle AOX = \frac{1}{2}m\angle AOB = \frac{1}{2}\left(\frac{360}{10}\right) = \frac{1}{2}(36) = 18$

b. 0.3090; $\frac{OX}{1}$; 0.9511 **c.** $p = 10 \cdot 2AX \approx 6.18$ **d.** $A = \frac{1}{2} \cdot OX \cdot 2AX \approx 0.2939$

e. $A \approx 10 \cdot 0.2939 = 2.939$

18. $a = 0.5$; $p = 3\sqrt{3} \approx 5.196$; $A = \frac{1}{2} \cdot \frac{1}{2} \cdot 3\sqrt{3} \approx 1.299$

19. $a = \frac{\sqrt{2}}{2} \approx 0.707$; $p = 4\sqrt{2} \approx 5.656$; $A = \frac{1}{2} \cdot \frac{\sqrt{2}}{2} \cdot 4\sqrt{2} = 2$

20. $a = \frac{\sqrt{3}}{2} \approx 0.866$; $p = 6 \cdot 1 = 6$; $A \approx \frac{1}{2} \cdot 0.866 \cdot 6 = 2.598$

21. $m\angle AOX = \frac{1}{2}m\angle AOB = \frac{1}{2}\left(\frac{360}{12}\right) = 15$; $AX = \sin 15° \approx 0.2588$; $OX =$ $\cos 15° \approx 0.9659$; $p = 12 \cdot 2 \cdot AX \approx 24 \cdot 0.2588 = 6.2112$; $A = \frac{1}{2} \cdot OX \cdot p \approx \frac{1}{2} \cdot 0.9659 \cdot 6.2112 \approx 3$

C 22. a. $\triangle AOX \cong \triangle BOX$ so $m\angle AOX = \frac{1}{2}m\angle AOB = \frac{1}{2}\left(\frac{360}{n}\right) = \frac{180}{n}$ **b.** $\sin\left(\frac{180}{n}\right)^\circ = \frac{AX}{1} = AX$ **c.** $\cos\left(\frac{180}{n}\right)^\circ = \frac{OX}{1} = OX$ **d.** $p = n \cdot AB = n \cdot 2AX = 2n \cdot \sin\left(\frac{180}{n}\right)^\circ$ **e.** $A = \frac{1}{2}ap = \frac{1}{2} \cdot \cos\left(\frac{180}{n}\right)^\circ \cdot 2n \cdot \sin\left(\frac{180}{n}\right)^\circ = n \cdot \sin\left(\frac{180}{n}\right)^\circ \cdot \cos\left(\frac{180}{n}\right)^\circ$

Page 444 • SELF-TEST 1

1. $s = 9$; $A = 9^2 = 81$ **2.** $h = \sqrt{13^2 - 12^2} = 5$; $A = 12 \cdot 5 = 60$

3. $h = 4\sqrt{3}$; $A = 10 \cdot 4\sqrt{3} = 40\sqrt{3}$ **4.** $A = \frac{1}{2} \cdot 4 \cdot 2\sqrt{3} = 4\sqrt{3}$ cm^2

5. $h = \sqrt{7^2 - 6^2} = \sqrt{13}$; $A = \frac{1}{2} \cdot 12 \cdot \sqrt{13} = 6\sqrt{13}$ cm^2 **6.** $A = \frac{1}{2} \cdot 8 \cdot 10 = 40$

7. $h = 3$; $A = \frac{1}{2}(3)(9 + 17) = 39$

8. $a = 5\sqrt{3}$; $p = 6 \cdot 10 = 60$; $A = \frac{1}{2} \cdot 5\sqrt{3} \cdot 60 = 150\sqrt{3}$

9. $A = \frac{1}{2} \cdot y \cdot 10x = 5xy$

10. upper $\triangle$: area $= \frac{1}{2} \cdot 5\sqrt{2} \cdot 5\sqrt{2} = 25$; hyp. $= 5\sqrt{2} \cdot \sqrt{2} = 10$; lower $\triangle$: $x = \sqrt{10^2 - 6^2} = 8$; area $= \frac{1}{2} \cdot 6 \cdot 8 = 24$; $A = 25 + 24 = 49$

Page 445 • CALCULATOR KEY-IN

1. Results may vary, due to rounding.

Number of sides	Perimeter	Area
18	6.2513344	3.0781813
180	6.2828663	3.1409547
1800	6.2831821	3.1415863
18000	6.2831853	3.1415926

2. $p \approx 6.28318 \approx 2\pi$; $A \approx 3.14159 \approx \pi$

Pages 447–448 • CLASSROOM EXERCISES

	1.	2.	3.	4.	5.	6.	7.	8.
r	3	4	0.8	5	9	6	7	12
C	6π	8π	1.6π	10π	18π	12π	14π	24π
A	9π	16π	0.64π	25π	81π	36π	49π	144π

9. $C \approx 2 \cdot 3.14 \cdot 2 \approx 12.6$; $A = \pi \cdot 4 \approx 12.6$

10. $C \approx 2 \cdot 3.14 \cdot 6 \approx 37.7$; $A \approx 3.14 \cdot 36 \approx 113.0$

11. $C \approx 2 \cdot 3.14 \cdot 1.5 \approx 9.4$; $A \approx 3.14 \cdot 1.5^2 \approx 7.1$

12. $C \approx 2 \cdot 3.14 \cdot 1.2 \approx 7.5$; $A \approx 3.14 \cdot 1.2^2 \approx 4.5$

13. 1. A reg. polygon can be inscribed in any ⊙. 2. Def. of perimeter 3. By algebra 4. AA ~ Post. 5. Corr. sides of ~ ⚠ are in prop., and $d = 2r$.

14. $C = \pi d = \pi(2r) = 2\pi r$

Pages 448–450 • WRITTEN EXERCISES

A

	1.	2.	3.	4.	5.	6.	7.	8.
r	7	120	$\frac{5}{2}$	$6\sqrt{2}$	10	6	5	$5\sqrt{2}$
C	14π	240π	5π	$12\pi\sqrt{2}$	20π	12π	10π	$10\pi\sqrt{2}$
A	49π	$14{,}400\pi$	$\frac{25}{4}\pi$	72π	100π	36π	25π	50π

9. **a.** $C \approx \frac{22}{7} \cdot 42 = 132$; $A = \frac{22}{7} \cdot 21^2 = 1386$ **b.** $C \approx \frac{22}{7} \cdot 14k = 44k$; $A = \frac{22}{7} \cdot (7k)^2 = 154k^2$

10. **a.** $C \approx 3.14 \cdot 8 \approx 25.1$; $A \approx 3.14 \cdot 4^2 \approx 50.2$ **b.** $C \approx 3.14 \cdot 4t = 12.56t$; $A \approx 3.14 \cdot (2t)^2 = 12.56t^2$

11. $C \approx 3.14 \cdot 18 = 56.52 \approx 57$ in.; $A = 3.14 \cdot 9^2 = 254.34 \approx 254$ in.^{2}

12. $A \approx \frac{1}{2} \cdot 3.14 \cdot 6^2 + 12 \cdot 15 = 56.52 + 180 \approx 237$ ft^2

13. $A_1 \approx 3.14 \cdot 4^2$; $A_2 \approx 3.14 \cdot 8^2$; $\frac{A_2}{A_1} = \frac{8^2}{4^2} = \frac{64}{16} = \frac{4}{1}$; $4 \cdot 6 = 24$ oz

14. $A_1 \approx 3.14 \cdot 4^2$; $A_2 = 2A_1 \approx 2 \cdot 3.14 \cdot 4^2 = 3.14(2 \cdot 4^2)$; $r_2 = \sqrt{2 \cdot 4^2} = 4\sqrt{2} \approx 5.656$; $d_2 \approx 11.3$ in.

15. $40^2 - \frac{22}{7} \cdot 14^2 = 1600 - 616 = 984 \text{ ft}^2$

16. $3.14 \cdot \left(\frac{d}{2}\right)^2 \approx 1000$; $d^2 \approx \frac{1000 \cdot 4}{3.14}$; $d \approx 36$ ft

17. $3.14 \cdot 5^2 = 78.5 \text{ in.}^2$; $\frac{78.5}{5} = 15.7 \text{ in.}^2/\$1$; $3.14 \cdot 7.5^2 \approx 176.6 \text{ in.}^2$; $\frac{176.6}{9} \approx 19.6 \text{ in.}^2/\$1$; the 15-in. pizza is the better buy.

18. hyp. $= \sqrt{6^2 + 8^2} = \sqrt{100} = 10$; Area I $= \frac{1}{2} \cdot \pi \cdot 3^2 = \frac{9\pi}{2}$; Area II $= \frac{1}{2} \cdot \pi \cdot 4^2 = \frac{16\pi}{2}$; Area III $= \frac{1}{2} \cdot \pi \cdot 5^2 = \frac{25\pi}{2}$; Area I + Area II $= \frac{9\pi}{2} + \frac{16\pi}{2} = \frac{25\pi}{2} =$ Area III.

B **19.** Area I $= \frac{1}{2} \cdot \pi \cdot \left(\frac{a}{2}\right)^2 = \frac{a^2\pi}{8}$; Area II $= \frac{1}{2} \cdot \pi \cdot \left(\frac{b}{2}\right)^2 = \frac{b^2\pi}{8}$; Area III $= \frac{1}{2} \cdot \pi \cdot \left(\frac{c}{2}\right)^2 = \frac{c^2\pi}{8}$. Since $a^2 + b^2 = c^2$, Area I + Area II $= \frac{a^2\pi}{8} + \frac{b^2\pi}{8} = (a^2 + b^2)\frac{\pi}{8} = \frac{c^2\pi}{8} =$ Area III.

20. 4 min $= 4 \cdot 60 = 240$ s; $\frac{240}{20} = 12$ rotations; distance $= 12(2\pi r) \approx 24\left(\frac{22}{7}\right)(21) = 1584$ ft

21. **a.** 5 min $= 5 \cdot 60 = 300$ s; $3 \cdot 300 = 900$ rev.; distance $= 900(2\pi r) \approx 1800\left(\frac{22}{7}\right)(35) = 198{,}000$ cm $= 1980$ m $= 1.98$ km **b.** 1 rev. $= 2\pi r \approx 2\left(\frac{22}{7}\right)(35) = 220$ cm; 22 km $= 22{,}000$ m $= 2{,}200{,}000$ cm; $\frac{2{,}200{,}000}{220} = 10{,}000$ rev.

22. $\triangle ABD \sim \triangle ACE$ with scale factor 1 : 2. So $CE = 4$. $\odot D$ (screen): $C = 2\pi r = 4\pi$ ft; $A = \pi r^2 = 4\pi \text{ ft}^2$; $\odot E$ (wall): $C = 2\pi r = 8\pi$ ft; $A = \pi r^2 = 16\pi \text{ ft}^2$

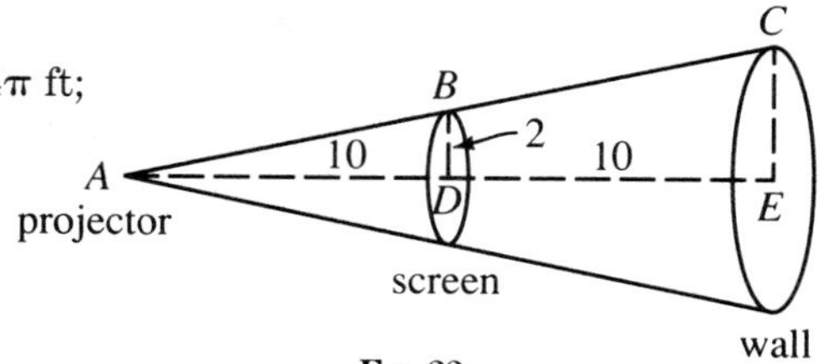

Ex. 22

23. **a.** Area of bull's eye $= \pi r^2 = \pi \cdot 1^2 = \pi$; area of first ring = area of $\odot$ with radius 2 − area of bull's eye $= \pi \cdot 2^2 - \pi = 4\pi - \pi = 3\pi$; area of second ring = area of $\odot$ with radius 3 − area of $\odot$ with radius 2 $= \pi \cdot 3^2 - \pi \cdot 2^2 = 9\pi - 4\pi = 5\pi$; area of third ring = area of $\odot$ with radius 4 − area of $\odot$ with radius 3 $= \pi \cdot 4^2 - \pi \cdot 3^2 = 16\pi - 9\pi = 7\pi$; π, 3π, 5π, 7π **b.** $\pi(n + 1)^2 - \pi \cdot n^2 = \pi(n^2 + 2n + 1 - n^2) = (2n + 1)\pi$

24. The shaded regions combine to form a rectangle with area $= 8 \cdot 4 = 32$.

25. Area of green section $=$ area of blue section $= \frac{1}{2} \cdot \pi \cdot 3^2 - \frac{1}{2} \cdot \pi \cdot 2^2 + \frac{1}{2} \cdot \pi \cdot 1^2 = \frac{1}{2}(9\pi - 4\pi + \pi) = 3\pi$; area of yellow section $= \pi \cdot 3^2 - 2 \cdot 3\pi = 9\pi - 6\pi = 3\pi$; $3\pi, 3\pi, 3\pi$

26. $24 \cdot 12 - 2 \cdot \pi \cdot 6^2 = 288 - 72\pi = 72(4 - \pi)$ **27.** $\frac{1}{2} \cdot \pi \cdot 8^2 = 32\pi$

28. $2 \cdot r \cdot 2r + \frac{1}{2} \cdot \pi \cdot (2r)^2 - \frac{1}{2} \cdot \pi \cdot r^2 = 4r^2 + 2\pi r^2 - \frac{1}{2}\pi r^2 = r^2\left(4 + \frac{3}{2}\pi\right)$

29. If a side of the square $= s$, $OT = TV = \frac{s}{2}$ and $OV = \frac{s}{2}\sqrt{2}$. Then the ratio of the area of the inscribed ⊙ to the area of the circumscribed ⊙ is $\pi \cdot \left(\frac{s}{2}\right)^2 : \pi \cdot \left(\frac{s}{2}\sqrt{2}\right)^2 = 1 : 2$.

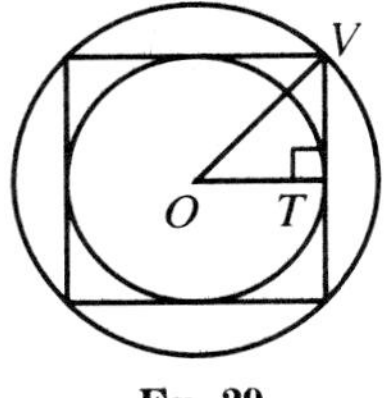

Ex. 29

Ex. 30

30. $\triangle OAT$ is a 30°–60°–90° △. If $OA = 2$, then $OT = 1$. So the ratio of the area of the inscribed ⊙ to the area of the circumscribed ⊙ is $\pi \cdot 1^2 : \pi \cdot 2^2 = 1 : 4$.

C 31. $m\angle O = \frac{360}{12} = 30$; Area $= \pi r^2 - 12\left(\frac{1}{2} \cdot r \cdot \frac{r}{2}\right) = (\pi - 3)r^2 \approx 0.14r^2$

32. $m\angle O = 45$; $\pi r^2 - 8\left(\frac{1}{2} \cdot r \cdot \frac{r\sqrt{2}}{2}\right) = (\pi - 2\sqrt{2})r^2 \approx 0.31r^2$

33. **a.** r **b.** $A = \frac{1}{2} \cdot r \cdot 2\pi r = \pi r^2$; the formula for the area of a ⊙.

34. Let $ABCD$ be the rhombus and O the center of the inscribed circle and draw $\overline{XY}$, a diam. of $\odot O$ that is $\perp$ to $\overline{AB}$ and $\overline{CD}$. $\triangle AOB$ is a rt. △ with legs 6 and 8, so $AB = \sqrt{6^2 + 8^2} = 10$. Area of $ABCD = 4 \cdot$ area of $\triangle AOB = 4\left(\frac{1}{2} \cdot 6 \cdot 8\right) = 96$. Area of $ABCD = AB \cdot XY$ so $XY = \frac{96}{10} = 9.6$. Then $C = 9.6\pi$ cm.

35. Let r be the radius of $\odot O$ and s the radius of $\odot P$. Construct a segment $\overline{BC} \cong$ to a radius of $\odot O$. At C, construct a $\perp$ to $\overline{BC}$ and mark off $CA = s$. Draw $\overline{AB}$. $\triangle ABC$ is a rt. $\triangle$ so $AB = \sqrt{(BC)^2 + (CA)^2} = \sqrt{r^2 + s^2}$. Draw a circle with radius AB. Area of circle $= \pi(\sqrt{r^2 + s^2})^2 = \pi(r^2 + s^2) = \pi r^2 + \pi s^2 =$ area of $\odot O$ + area of $\odot P$.

Page 451 • CALCULATOR KEY-IN

1. $3\frac{1}{7} \approx 3.1429$; $3\frac{10}{71} \approx 3.1408$; yes **2–4.** Answers will vary.

2. Using 6 terms: $\pi \approx 3.1413$ **3.** Using 9 factors: $\pi \approx 2.9722$

4. Using 3 factors: $\pi \approx 3.1214$

Page 451 • ALGEBRA REVIEW

1. $\pi \cdot \left(\frac{2}{3}\sqrt{3}\right)^2 = \pi \cdot \frac{4}{9} \cdot 3 = \frac{4}{3}\pi$ **2.** $\pi \cdot \frac{21}{5} \cdot 15 = 63\pi$

3. $\frac{1}{3}\pi(2\sqrt{6})^2(4) = \frac{\pi}{3} \cdot 4 \cdot 6 \cdot 4 = 32\pi$ **4.** $\frac{4}{3}\pi \cdot 6^3 = \frac{4}{3}\pi \cdot 6 \cdot 36 = 288\pi$

5. $\pi \cdot \sqrt{5} \cdot \sqrt{(\sqrt{5})^2 + (\sqrt{5})^2} = \pi \cdot \sqrt{5} \cdot \sqrt{5 + 5} = \pi\sqrt{50} = 5\pi\sqrt{2}$

6. $2\pi \cdot 10^2 + 2\pi \cdot 10 \cdot 6 = 200\pi + 120\pi = 320\pi$

7. $\pi \cdot 2^2 + \pi(2)\sqrt{2^2 + (2\sqrt{3})^2} = 4\pi + 2\pi\sqrt{4 + 12} = 4\pi + 8\pi = 12\pi$

8. $\pi[6^2 - (3\sqrt{2})^2] = \pi(36 - 18) = 18\pi$

Page 453 • CLASSROOM EXERCISES

1. arc length: $\frac{1}{4}(2\pi) = \frac{\pi}{2}$; area: $\frac{1}{4} \cdot \pi = \frac{\pi}{4}$

2. arc length: $\frac{120}{360} \cdot 2\pi(6) = 4\pi$; area $= \frac{1}{3}\pi(6)^2 = 12\pi$

3. arc length: $\frac{45}{360} \cdot 2\pi(4) = \pi$; area $= \frac{1}{8} \cdot \pi(4)^2 = 2\pi$

4. arc length: $\frac{270}{360} \cdot 2\pi(4) = 6\pi$; area $= \frac{3}{4} \cdot \pi(4)^2 = 12\pi$

5. Check students' drawings. Area of sector $AOB = \frac{1}{6} \cdot \pi \cdot 6^2 = 6\pi$; area of $\triangle AOB = \frac{1}{2} \cdot 6 \cdot 3\sqrt{3} = 9\sqrt{3}$; area of the region bounded by $\overline{AB}$ and $\widehat{AB}$ = $6\pi - 9\sqrt{3}$

6. $\frac{40\pi}{160\pi} = \frac{x}{360}$; $x = 90$

7. **a.** Let the measure of the central $\angle$ be x. Then, if A_S is the area of the smaller sector and A_L the area of the larger one, then $\frac{A_L}{\pi(2r)^2} = \frac{x}{360}$ and $\frac{A_S}{\pi r^2} = \frac{x}{360}$; $A_L = \frac{\pi(2r)^2 x}{360} = \frac{\pi r^2 x}{90}$; $A_S = \frac{\pi r^2 x}{360}$; $\frac{A_L}{A_S} = \frac{\pi r^2 x}{90} \cdot \frac{360}{\pi r^2 x} = \frac{4}{1}$. The area of the larger sector is 4 times the area of the smaller sector. **b.** $\frac{A_L}{\pi r^2} = \frac{2x}{360}$; $A_L = \frac{\pi r^2 x}{180}$; $\frac{A_S}{\pi r^2} = \frac{x}{360}$; $A_S = \frac{\pi r^2 x}{360}$; $A_L = 2 \cdot A_S$.

Pages 453–455 • WRITTEN EXERCISES

A **1–10.** Check students' drawings.

1. length of $\widehat{AB} = \frac{30}{360} \cdot 2\pi \cdot 12 = 2\pi$; area of sector $AOB = \frac{30}{360} \cdot \pi \cdot 12^2 = 12\pi$

2. length of $\widehat{AB} = \frac{45}{360} \cdot 2\pi \cdot 4 = \pi$; area of sector $AOB = \frac{45}{360} \cdot \pi \cdot 4^2 = 2\pi$

3. length of $\widehat{AB} = \frac{120}{360} \cdot 2\pi \cdot 3 = 2\pi$; area of sector $AOB = \frac{120}{360} \cdot \pi \cdot 3^2 = 3\pi$

4. length of $\widehat{AB} = \frac{240}{360} \cdot 2\pi \cdot 3 = 4\pi$; area of sector $AOB = \frac{240}{360} \cdot \pi \cdot 3^2 = 6\pi$

5. length of $\widehat{AB} = \frac{180}{360} \cdot 2\pi \cdot 1.5 = 1.5\pi$; area of sector $AOB = \frac{180}{360} \cdot \pi \cdot 1.5^2 = 1.125\pi$

6. length of $\widehat{AB} = \frac{270}{360} \cdot 2\pi \cdot 0.8 = 1.2\pi$; area of sector $AOB = \frac{270}{360} \cdot \pi \cdot 0.8^2 = 0.48\pi$

7. length of $\widehat{AB} = \frac{40}{360} \cdot 2\pi \cdot \frac{9}{2} = \pi$; area of sector $AOB = \frac{40}{360} \cdot \pi \cdot \left(\frac{9}{2}\right)^2 = \frac{9\pi}{4}$

8. length of $\widehat{AB} = \frac{320}{360} \cdot 2\pi \cdot \frac{6}{5} = \frac{32\pi}{15}$; area of sector $AOB = \frac{320}{360} \cdot \pi \cdot \left(\frac{6}{5}\right)^2 = \frac{32\pi}{25}$

9. length of $\widehat{AB} = \frac{108}{360} \cdot 2\pi \cdot 5\sqrt{2} = 3\pi\sqrt{2}$; area of sector $AOB = \frac{108}{360} \cdot \pi \cdot (5\sqrt{2})^2 = 15\pi$

10. length of $\widehat{AB} = \frac{192}{360} \cdot 2\pi \cdot 3\sqrt{3} = \frac{16\pi\sqrt{3}}{5}$; area of sector $AOB = \frac{192}{360} \cdot \pi \cdot (3\sqrt{3})^2 = \frac{72\pi}{5}$

11. $\frac{10\pi}{\pi r^2} = \frac{100}{360}$; $r = 6$ **12.** $\frac{7\pi \div 2}{\pi r^2} = \frac{315}{360}$; $r = 2$

B 13. area of sector: $\frac{1}{4} \cdot \pi \cdot 4^2 = 4\pi$; shaded area: $4\pi - \frac{1}{2} \cdot 4 \cdot 4 = 4\pi - 8$

14. area of sector: $\frac{1}{6} \cdot \pi \cdot 3^2 = \frac{3\pi}{2}$; shaded area: $\frac{3\pi}{2} - \frac{1}{2} \cdot 3 \cdot \frac{3}{2}\sqrt{3} = \frac{6\pi - 9\sqrt{3}}{4}$

15. Draw radii to endpts. of the chord; area of sector $= \frac{1}{4} \cdot \pi \cdot 4^2 = 4\pi$; shaded area = area of ⊙ − area of sector + area of △ $= \pi \cdot 4^2 - 4\pi + \frac{1}{2} \cdot 4 \cdot 4 = 12\pi + 8$

16. Shaded area = area of semicircle − area of △ $= \frac{1}{2} \cdot \pi \cdot 6^2 - \frac{1}{2} \cdot 6 \cdot 6\sqrt{3} = 18\pi - 18\sqrt{3}$

17. Draw diag. segs. from endpts. of chords. 2 isos. △s and 2 shaded sectors are formed. Radius of circle = length of legs of isos. △s = 2; area of sector $= \frac{60}{360} \cdot \pi \cdot 2^2 = \frac{2\pi}{3}$; area of isos. △ $= \frac{1}{2} \cdot 2\sqrt{3} \cdot 1 = \sqrt{3}$; area of shaded region $= 2\left(\frac{2\pi}{3}\right) + 2(\sqrt{3}) = \frac{4\pi + 6\sqrt{3}}{3}$

18. Vertically connect the endpts. of the chords. Area of the square formed is 36. Shaded area = area of square + $\frac{1}{2}$(area of ⊙ − area of square) = $36 + \frac{1}{2}(\pi \cdot (3\sqrt{2})^2 - 36) = 36 + 9\pi - 18 = 9\pi + 18$.

19. Radius of ⊙ $= \frac{1}{2}$ length of a diag. $= \frac{1}{2}\sqrt{12^2 + 16^2} = \frac{1}{2} \cdot 20 = 10$; area of region inside of ⊙ but outside of rect. $= \pi \cdot 10^2 - 12 \cdot 16 = (100\pi - 192)\text{ cm}^2$

20. $\overline{PA} \cong \overline{PB}$ (Tans. to a ⊙ from a pt. are ≅.) and $\angle A$ and $\angle B$ are rt. ∠s (If a line is tan. to a ⊙, then it is ⊥ to the radius drawn to the pt. of tangency.) so $\triangle OAP$ and $\triangle OBP$ are ≅ 30°–60°–90° △s with short leg 6. Area of $AOBP = 2\left(\frac{1}{2} \cdot 6 \cdot 6\sqrt{3}\right) = 36\sqrt{3}$; area of each sector $= \frac{60}{360} \cdot \pi \cdot 6^2 = 6\pi$; area of region outside ⊙ but inside quad. $= 36\sqrt{3} - 12\pi$.

21. **a.** $\sin \angle AOX = \frac{9}{12} = 0.75$; $m\angle AOX \approx 49$; $m\angle AOB \approx 98$ **b.** Area of sector $= \frac{98}{360} \cdot \pi \cdot 12^2 = \frac{196}{5}\pi$; shaded area $= \frac{196}{5}\pi - \frac{1}{2} \cdot 18 \cdot \sqrt{12^2 - 9^2} = \frac{196}{5}\pi - 27\sqrt{7} \approx 52 \text{ cm}^2$

22. Fair area $= \frac{1}{4} \cdot \pi \cdot 325^2$; foul area $= 2 \cdot 60 \cdot 325 + \frac{1}{4}\pi \cdot 60^2$;

fair : foul $= \frac{105{,}625\pi}{4} : (39{,}000 + 900\pi) = 105{,}625\pi : 4(39{,}000 + 900\pi) \approx 2 : 1$

23. Grazing area $= \frac{180 + 45}{360} \cdot \pi \cdot 25^2 + \frac{1}{2} \cdot 10 \cdot 10 + \frac{45}{360} \cdot \pi \cdot (25 - 10\sqrt{2})^2 + \frac{1}{4} \cdot \pi \cdot 5^2 \approx 1226.56 + 50 + 46.27 + 19.63 \approx 1343 \text{ m}^2$

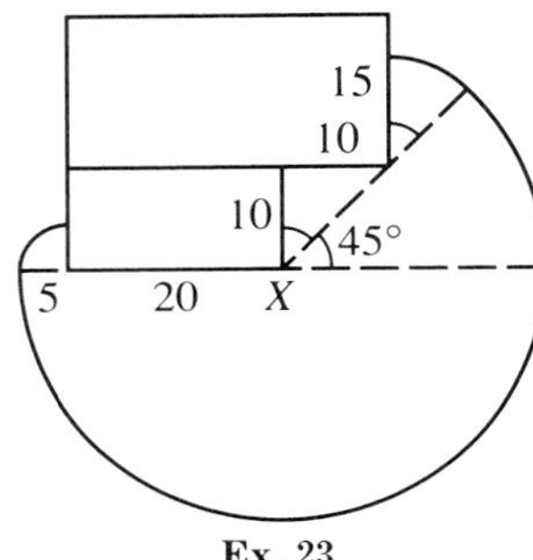

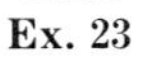
Ex. 23

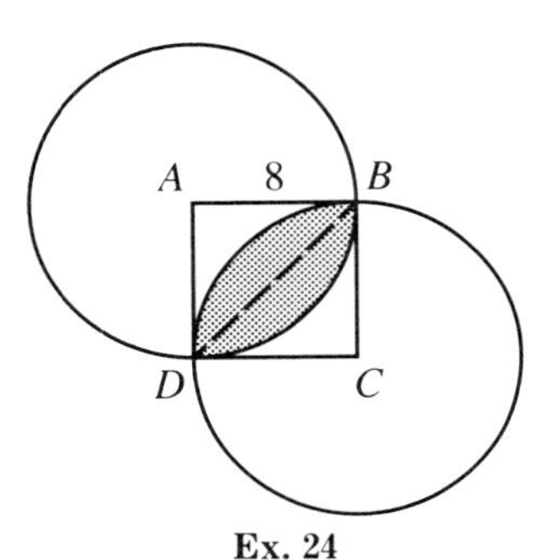

Ex. 24

24. Area of sector $BAD = \frac{90}{360} \cdot \pi \cdot 8^2 = 16\pi$; area of $\triangle BAD = \frac{1}{2} \cdot 8 \cdot 8 = 32$; area of region in $\odot A$ bounded by $\overset{\frown}{BD}$ and $\overline{BD} = 16\pi - 32$. Similarly, area of region in $\odot C$ bounded by $\overset{\frown}{BD}$ and $\overline{BD} = 16\pi - 32$ and area of region inside both circles $= (32\pi - 64) \text{ cm}^2$.

25. Area of region in $\odot O$ bounded by $\overset{\frown}{AB}$ and $\overline{AB} = \frac{120}{360} \cdot \pi \cdot 6^2 - \frac{1}{2} \cdot 6\sqrt{3} \cdot 3 = 12\pi - 9\sqrt{3}$.
Similarly, area inside $\odot P$ bounded by $\overset{\frown}{AB}$ and $\overline{AB} = 12\pi - 9\sqrt{3}$; area of region common to both circles $= (24\pi - 18\sqrt{3}) \text{ cm}^2$.

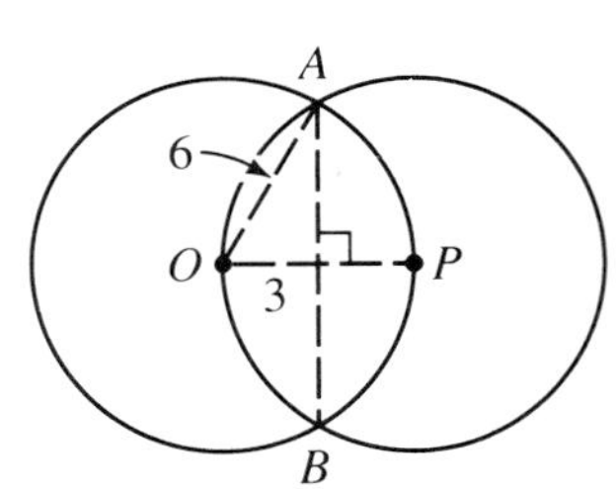

26. **a.** Use Construction 4 to locate the midpts. of the sides of the square. At each of these pts. draw a semicircle inside the square with diam. = s, the length of a side of the square. **b.** The region consists of 4 ≅ figures. We shall find the area of one. Radius of $\odot X$ = radius of $\odot Y = \frac{2}{\sqrt{2}} = \sqrt{2}$; area of sector AYO = $\frac{90}{360} \cdot \pi \cdot (\sqrt{2})^2 = \frac{\pi}{2}$; area of $\triangle AYO = \frac{1}{2}\sqrt{2} \cdot \sqrt{2} = 1$; area of region in $\odot Y$ bounded by $\overset{\frown}{AO}$ and $\overline{AO} = \frac{\pi}{2} - 1$; area of shaded region inside $AYOX = \pi - 2$; area of shaded region $= 4\pi - 8$.

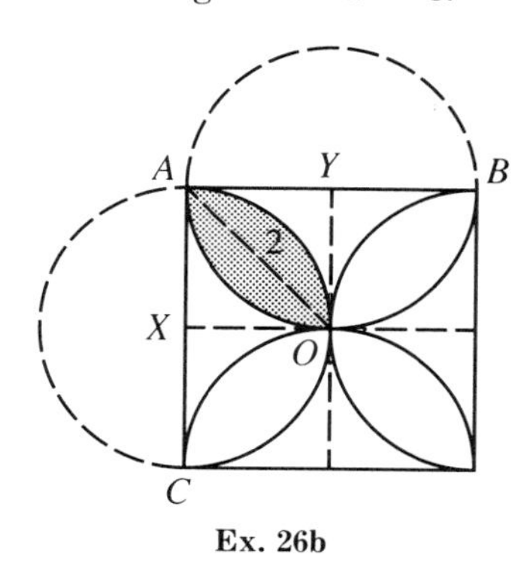

Ex. 26b

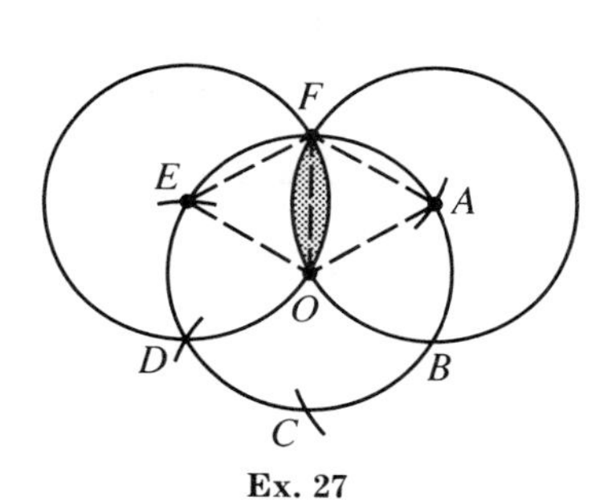

Ex. 27

C 27. **a.** Draw $\odot O$ with radius r. Choose a pt. A on the $\odot$; with radius r draw an arc intersecting the $\odot$ at B and F. With B as ctr. and radius r, draw an arc intersecting the $\odot$ at A and C. Continue drawing ≅ arcs $\overset{\frown}{BD}$, $\overset{\frown}{CE}$, $\overset{\frown}{DF}$, and $\overset{\frown}{EA}$, completing the figure. **b.** The shaded region consists of 6 ≅ figures. We shall find the area of one. $\triangle FAO$ and $\triangle FEO$ are equilateral ▵ with area $\frac{1}{2} \cdot 6 \cdot \frac{6\sqrt{3}}{2} = 9\sqrt{3}$. Area of sector $FAO = \frac{60}{360} \cdot \pi \cdot 6^2 = 6\pi$; area in $\odot A$ bounded by $\overset{\frown}{FO}$ and $\overline{FO} = 6\pi - 9\sqrt{3}$; area of shaded region inside $\odot A$ and $\odot E = 12\pi - 18\sqrt{3}$; area of shaded region = $72\pi - 108\sqrt{3}$.

28. Area of shaded region = area of equilateral $\triangle ABC$ − 3(area of sector XAY) $= \frac{1}{2} \cdot 12 \cdot 6\sqrt{3} -$

$\left(3 \cdot \frac{60}{360} \cdot \pi \cdot 6^2\right) = 36\sqrt{3} - 18\pi$

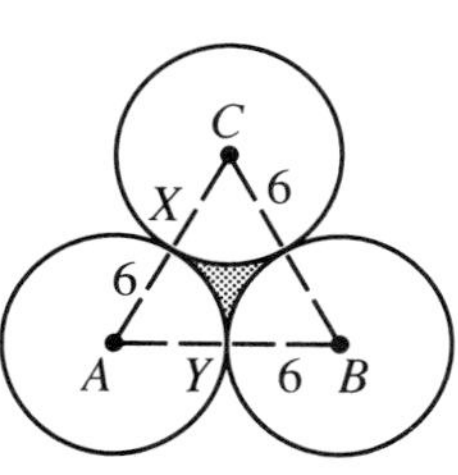

29. Since $\overline{AB}$ is tan. to $\odot X$ and $\odot Y$, $AXYB$ is a trap. (If a line is tan. to a $\odot$, the line is $\perp$ to the radius drawn to the pt. of tangency; in a plane, 2 lines $\perp$ to the same line are $\parallel$.) Draw a $\parallel$ to $\overline{AB}$ through Y, int. $\overline{AX}$ at P. $PX = 4$, $XY = 8$, and $m\angle XPY = 90$; $\triangle XPY$ is a 30°–60°–90° $\triangle$; $PY = AB = 4\sqrt{3}$. Area of $AXYB = \frac{1}{2}(4\sqrt{3})(6 + 2) = 16\sqrt{3}$. Let Q be the pt. of tangency of $\odot X$ and $\odot Y$. Area of sector $AXQ = \frac{60}{360} \cdot \pi \cdot 6^2 = 6\pi$; area of sector $BYQ = \frac{120}{360} \cdot \pi \cdot 2^2 = \frac{4\pi}{3}$; area of shaded region $= 16\sqrt{3} - \left(6\pi + \frac{4\pi}{3}\right) = \frac{48\sqrt{3} - 18\pi - 4\pi}{3} = \frac{48\sqrt{3} - 22\pi}{3}$

30. $\angle PTS$ and $\angle RST$ are rt. $\angle$s; $\triangle PRQ$ is a 30°–60°–90° $\triangle$ so $TS = PR = 20\sqrt{3}$. TS and UV are not $\parallel$ so they intersect at some pt. W. Then $WT = WU$ and $WS = WV$ so $TS = UV = 20\sqrt{3}$. $m\angle WPT = m\angle WPU = 180 - m\angle TPQ = 60$; $m\angle TPU = 120$; length of $\widehat{TU} = \frac{120}{360} \cdot 2\pi \cdot 5 = \frac{10\pi}{3}$. Let X be a pt. on the major arc with endpts. S and V; length of $\widehat{SXV} = \frac{240}{360} \cdot 2\pi \cdot 25 = \frac{100\pi}{3}$; length of belt $= \frac{10\pi}{3} + 2(20\sqrt{3}) + \frac{100\pi}{3} = \left(\frac{120\sqrt{3} + 110\pi}{3}\right)$ cm

Page 455 • CHALLENGE

No matter how many segments are used, the sum of the arc lengths is $\frac{\pi}{2}(XY)$.

Proof: Let n = number of segs.; length of each arc $= \pi r = \pi\left(\frac{XY}{2n}\right)$; sum of arc lengths $= n(\pi)\left(\frac{XY}{2n}\right) = \frac{\pi}{2}(XY)$.

Page 458 • CLASSROOM EXERCISES

1. $\frac{9}{5}$ **2.** Each $\triangle$ has height 12; $\frac{9}{16}$ **3.** $\frac{10}{12}$ or $\frac{5}{6}$

	4.	**5.**	**6.**	**7.**
Scale factor	1 : 3	1 : 5	3 : 4	2 : 3
Ratio of perimeters	1 : 3	1 : 5	3 : 4	2 : 3
Ratio of areas	1 : 9	1 : 25	9 : 16	4 : 9

	8.	9.	10.	11.
Scale factor	4 : 5	3 : 5	4 : 7	6 : 5
Ratio of perimeters	4 : 5	3 : 5	4 : 7	6 : 5
Ratio of areas	16 : 25	9 : 25	16 : 49	36 : 25

12. **a.** Yes **b.** 3 : 4; 9 : 16

13. **a.** No **b.** $\triangle ADE \sim \triangle ABC$ **c.** $2^2 : (2 + 3)^2 = 4 : 25$ **d.** $4x : 25x - 4x = 4 : 21$

14. **a.** 8 : 5 **b.** 8 : 15

15. **a.** 7 : 4 **b.** $7^2 : 4^2 = 49 : 16$

Pages 458–460 • WRITTEN EXERCISES

A

	1.	2.	3.	4.
Scale factor	1 : 4	3 : 2	$r : 2s$	9 : 5
Ratio of perimeters	1 : 4	3 : 2	$r : 2s$	9 : 5
Ratio of areas	1 : 16	9 : 4	$r^2 : 4s^2$	81 : 25

	5.	6.	7.	8.
Scale factor	3 : 13	5 : 1	3 : 8	$\sqrt{2} : 1$
Ratio of perimeters	3 : 13	5 : 1	3 : 8	$\sqrt{2} : 1$
Ratio of areas	9 : 169	25 : 1	9 : 64	2 : 1

9. 1 km = 1000 m = 100,000 cm, so 50 km = 5,000,000 cm; ratio is $1^2 : 5{,}000{,}000^2 = 1 : 25{,}000{,}000{,}000{,}000$

10. $\sqrt{36\pi} : \sqrt{64\pi} = 6 : 8 = 3 : 4$; 3 : 4

11. 1 : 2; 1 : 4 **12.** $\left(\frac{x^2}{xy}\right)^2 = \left(\frac{x}{y}\right)^2 = \frac{x^2}{y^2}$ or $x^2 : y^2$

13. $\triangle ABE \sim \triangle DCE$; areas: $\left(\frac{6}{5}\right)^2 = \frac{36}{25}$; $8 \cdot 5 = 6 \cdot DE$; $DE = \frac{20}{3} = 6\frac{2}{3}$

14. $\triangle ACD \sim \triangle AEB$; areas: $\left(\frac{10}{9}\right)^2 = \frac{100}{81}$; $9(9 + 11) = 10(10 + DE)$; $DE = 8$

15. $\left(\frac{9}{15}\right)^2 = \frac{45}{A}$; $A = 125\ \text{cm}^2$

16. $\sqrt{\frac{48}{27}} = \frac{3 + 4 + 5 + 6 + 7}{p}$; $\frac{4}{3} = \frac{25}{p}$; $p = \frac{75}{4} = 18.75$ m

B **17. a.** $\frac{9}{7}$ **b.** $\frac{10}{8}$ or $\frac{5}{4}$ **18. a.** $\frac{3}{6}$ or $\frac{1}{2}$ **b.** $\left(\frac{3}{6}\right)^2 = \frac{1}{4}$

19. a. $\frac{6}{8}$ or $\frac{3}{4}$ **b.** $\frac{3x}{4x + 3x} = \frac{3}{7}$ **20. a.** $\left(\frac{6}{9}\right)^2 = \frac{4}{9}$ **b.** $\left(\frac{6}{9 + 6}\right)^2 = \frac{4}{25}$

21. Answers may vary. $\triangle ABC \sim \triangle CDA$, 1 : 1; $\triangle ABG \sim \triangle CEG$, $\left(\frac{6}{10}\right)^2 = \frac{9}{25}$; $\triangle ABF \sim \triangle DEF$, $\left(\frac{6}{4}\right)^2 = \frac{9}{4}$; $\triangle AGF \sim \triangle CGB$, $\left(\frac{3}{5}\right)^2 = \frac{9}{25}$ (Since $\frac{ED}{EC} = \frac{FD}{BC}$, $BC = 5$ and $AF = 3$.); $\triangle EFD \sim \triangle EBC$, $\left(\frac{4}{10}\right)^2 = \frac{4}{25}$; $\triangle ABF \sim \triangle CEB$, $\left(\frac{3}{5}\right)^2 = \frac{9}{25}$

22. a. area of $\triangle ABE = \frac{1}{2} \cdot AB \cdot$ height of $\triangle ABE = \frac{1}{2} \cdot AB \cdot$ height of $ABCD = \frac{1}{2} \cdot$ area of $ABCD = 24$ cm^2 **b.** area of $\triangle BEC = \frac{1}{2} \cdot EC \cdot$ height of $\triangle BEC = \frac{1}{2} \cdot \frac{1}{3} \cdot DC \cdot$ height of $ABCD = \frac{1}{6} \cdot$ area of $ABCD = 8$ cm^2 **c.** area of $\triangle ADE = \frac{1}{2} \cdot DE \cdot$ height of $\triangle ADE = \frac{1}{2} \cdot \frac{2}{3} \cdot DC \cdot$ height of $ABCD = \frac{1}{3} \cdot$ area of $ABCD = 16$ cm^2 **d.** $\triangle CEF \sim \triangle DEA$; scale factor = 1 : 2; ratio of areas = 1 : 4; area of $\triangle CEF = 4$ cm^2 **e.** $\triangle DEF$ and $\triangle ECF$ have the same alt. from F to $\overline{DC}$; ratio of areas = 2 : 1; area of $\triangle DEF = 8$ cm^2 **f.** area of $\triangle BEF$ = area of $\triangle BEC$ + area of $\triangle CEF = 8 + 4 = 12$ cm^2

23. a. $\left(\frac{4}{12}\right)^2 = \frac{1}{9}$ **b.** $\left(\frac{4}{8}\right)^2 = \frac{1}{4}$ **c.** $\frac{x}{9x - x} = \frac{1}{8}$

24. a. scale factor = 1 : 2; ratio of areas = 1 : 4 **b.** In $\triangle$I, draw an alt. from lower left corner and note that it is also an alt. for $\triangle$II; ratio of bases is 1 : 2, so ratio of areas is also 1 : 2. **c.** $\triangle$ formed by I and IV has same area as $\triangle$ formed by I and II (same base, same alt.), so $\triangle$II has same area as $\triangle$IV. 1 : 2 **d.** 1 : 1 **e.** $\frac{x}{x + 2x + 4x + 2x} = \frac{1}{9}$

25. a. scale factor = 8 : 18 or 4 : 9; ratio of areas = 16 : 81 **b.** Draw an alt. of $\triangle$I (and $\triangle$II) from the upper left vert. of the trap. From part (a), if base of $\triangle$I $= 4x$, base of $\triangle$II $= 9x$; ratio of areas $= \frac{4xh}{2} : \frac{9xh}{2} = 4 : 9$ **c.** Draw an alt. of $\triangle$I (and $\triangle$IV) from upper rt. vert. of the trap. As in part (b), ratio of areas $= \frac{4xh}{2} : \frac{9xh}{2} =$

4 : 9 **d.** From parts (b) and (c), ratio of areas = 1 : 1

e. $\dfrac{16x}{16x + 2(36x) + 81x} = \dfrac{16}{169}$

26. Area I : (area I + area II) = $8^2 : 18^2 = 4^2 : 9^2 = 16 : 81$; 81 · area I = 16 · area I + 16 · area II; 65 · area I = 16 · area II; ratio of areas = 16 : 65

27. Area I : (area I + area II) = $3^2 : 7^2 = 9 : 49$; 49 · area I = 9 · area I + 9 · area II; 40 · area I = 9 · area II; ratio of areas = 9 : 40

28. Let x = shorter leg of area II; shorter leg of I = $\dfrac{x\sqrt{3}}{3}$; ratio of sides = $\dfrac{x\sqrt{3}}{3} : x = 1 : \sqrt{3}$; ratio of areas = 1 : 3

C **29.** Draw the alt. from A, intersecting $\overline{BC}$ at D. Let $\overrightarrow{AG}$ intersect $\overline{BC}$ at E, and let $\overline{AD}$ intersect the line l, through G and $\parallel$ to $\overline{BC}$, at F. Then $\dfrac{AF}{AD} = \dfrac{AG}{AE} = \dfrac{2}{3}$. Let l intersect $\overline{AB}$ and $\overline{AC}$ at M and N, resp. Area of $\triangle AMN$: area of $\triangle ABC$ = $AF^2 : AD^2 = 4 : 9$ and area of $\triangle AMN$: area of $BMNC = 4x : (9x - 4x) = 4 : 5$.

30. Let the line through J and $\parallel$ to $\overline{MN}$ intersect $\overline{LM}$ and $\overline{LN}$ at P and Q, resp. $\dfrac{\text{Area of } \triangle LPQ}{\text{Area of } \triangle LMN} = \dfrac{1}{2} = \dfrac{LJ^2}{LK^2} = \dfrac{LJ^2}{12^2}$; $LJ = 6\sqrt{2}$ cm

31. Each of the small $\triangle$s has area = $\dfrac{1}{6}$ · area of the orig. $\triangle$. We shall show that area of $\triangle MBE$ = area of $\triangle MEC$ = $\dfrac{1}{6}$ · area of $\triangle ABC$. The other relationships can be demonstrated similarly. Draw lines $\parallel$ to $\overline{BC}$ through A and M and a $\perp$ to $\overline{BC}$ through M. Since the medians of a $\triangle$ int. in a pt. $\dfrac{2}{3}$ of the dist. from each vert. to the opp. side, $ME = \dfrac{1}{3} \cdot AE$. If 3 $\parallel$ lines int. 2 trans., then they divide the trans. prop. so $MY = \dfrac{1}{3} \cdot XY$. Then area of $\triangle MBE$ = $\dfrac{1}{2} \cdot BE \cdot MY = \dfrac{1}{4} \cdot BC \cdot \dfrac{1}{3} \cdot XY = \dfrac{1}{6} \cdot \dfrac{1}{2} BC \cdot XY = \dfrac{1}{6}$ · area of $\triangle ABC$, and area of $\triangle MEC = \dfrac{1}{2} \cdot EC \cdot MY = \dfrac{1}{4} \cdot BC \cdot \dfrac{1}{3} XY = \dfrac{1}{6}$ · area of $\triangle ABC$.

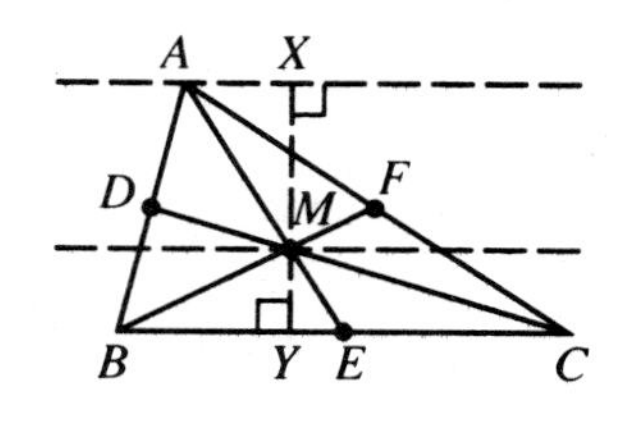

32. Let F represent the area of each quad. $CRYX$, $APZY$, and $BQXZ$. Let G represent the area of each $\triangle$ CXQ, AYR, and BZP. (1) area of $\triangle ABC = \frac{9}{4}\sqrt{3} = 3F + 3G +$ area of $\triangle XYZ$. Since $\triangle CPB$ and $\triangle ABC$ have the same alt. from C, area of $\triangle CPB = \frac{1}{3} \cdot$ area of $\triangle ABC$ or $F + 2G = \frac{3}{4}\sqrt{3}$. Then (2) $F = \frac{3}{4}\sqrt{3} - 2G$ and subst. in (1) yields area of $\triangle XYZ = 3G$. Now, let $CX = r$ and $PZ = t$. Let $\overline{CD}$ be the alt. of $\triangle ABC$ from C. Then $DP = \frac{1}{2}$ and $CD = \frac{3}{2}\sqrt{3}$, so $CP = \sqrt{7}$ and $XZ = \sqrt{7} - r - t$. Since $\triangle CBZ$ and $\triangle BZP$ have the same alt. from B, $\frac{F + G}{G} = \frac{\sqrt{7} - t}{t}$. Subst. in this eq. from (2) above and solving for t yields (3) $t = \frac{4G\sqrt{21}}{9}$. Since $\triangle AXP$ and $\triangle ACX$ have the same alt. from A, and using area of $\triangle XYZ = 3G$, we have $\frac{F + 3G}{F + G} = \frac{\sqrt{7} - r}{r}$. Subst. from (2) and solving for r yields (4) $r = \frac{\sqrt{7}}{2} - \frac{2G\sqrt{21}}{9}$. By the symmetry of the fig., $\triangle XYZ$ is equilateral, so area of $\triangle XYZ = 3G = \frac{(XZ)^2}{4}\sqrt{3} = \frac{(\sqrt{7} - r - t)^2}{4}\sqrt{3}$. Subst. from (3) and (4) into this eq. and simplifying gives $(112\sqrt{3})G^2 - 1800G + 189\sqrt{3} = 0$. Using the quadratic formula, $G = \frac{1800 \pm 1728}{224\sqrt{3}}$, which yields $3G \approx 27$ (reject) or $3G = \frac{9\sqrt{3}}{28}$. The latter gives $7(3G) = \frac{9\sqrt{3}}{4}$ or 7(area of $\triangle XYZ$) = area of $\triangle ABC$, or area of $\triangle XYZ = \frac{1}{7}$(area of $\triangle ABC$).

Page 462 • CLASSROOM EXERCISES

1. a. $\frac{1}{5}$ **b.** $\frac{3}{5}$ **c.** $\frac{4}{5}$ **d.** $\frac{5}{5} = 1$ **e.** 0 **f.** 1

2. $\frac{10}{30} = \frac{1}{3}$ **3.** $\frac{16\pi - \pi}{100} \approx 0.471$; $\frac{\pi}{100} \approx 0.03$ **4.** $\frac{\pi}{10{,}000} \approx 0.0003$

Pages 463–464 • WRITTEN EXERCISES

A **1.** $\frac{1}{4}$ **2. a.** $\frac{1}{2}$ **b.** $\frac{1}{4}$ **c.** $\frac{3}{8}$ **3.** $\frac{1}{4}$ **4. a.** $\frac{1}{6}$ **b.** $\frac{2}{5}$

5. a. $\frac{16\pi}{400\pi} = \frac{1}{25}$ **b.** $\frac{1}{25} = \frac{x}{75}$; $x = 3$

6. a. $\dfrac{\left(\frac{1}{2}s\sqrt{2}\right)^2}{s^2} = \dfrac{1}{2} = 50\%$ **b.** $\dfrac{\pi\left(\frac{s}{2}\right)^2 - \left(\frac{s}{2}\sqrt{2}\right)^2}{s^2} = \dfrac{\pi - 2}{4} \approx 0.285 \approx 29\%$

7. $\dfrac{30(100\pi)}{1200 \cdot 500} = \dfrac{\pi}{200} \approx 0.016$ 8. $\dfrac{100 \cdot 80 - (\pi \cdot 5^2 + \pi \cdot 8^2)}{100 \cdot 80} = \dfrac{8000 - 89\pi}{8000} \approx 0.965$

9. $\dfrac{6^2}{30^2} = \dfrac{1}{25} = 0.04$ **10.** $\dfrac{8^2}{30^2} \approx 0.071$

B 11. $\dfrac{4}{6} = \dfrac{2}{3}$

12. a. $\dfrac{4}{8} = \dfrac{1}{2}$ **b.** $\dfrac{2}{8} = \dfrac{1}{4}$ (The distance of the cut from one end must lie between 3 and 5.)

c. $\dfrac{8}{8} = 1$

13. $\dfrac{10}{80} = \dfrac{x}{1}$; $x = 0.125$ m^2 **14.** $\dfrac{3}{50} = \dfrac{x}{5 \cdot 2}$; $x = 0.6$ m^2; $\dfrac{x}{100} = 0.006$ m^2 or 60 cm^2

15. a. Prob. $= \dfrac{(30 - 2R)^2}{30^2} = \left(\dfrac{30 - 2R}{30}\right)^2$

b. $0.25 = \left(\dfrac{30 - 2R}{30}\right)^2$ or $0.5 = \dfrac{30 - 2R}{30}$; $R = 7.5$ mm

16. a. 1 **b.**

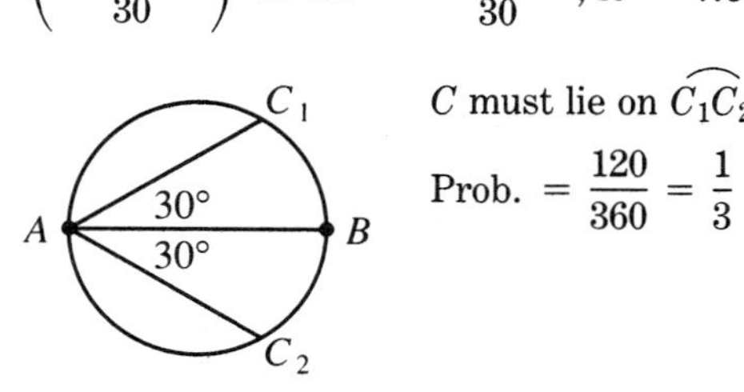

C must lie on $\widehat{C_1C_2}$.

Prob. $= \dfrac{120}{360} = \dfrac{1}{3}$

C 17. a. $\dfrac{13}{50}$ **b.** $\dfrac{5}{50} = \dfrac{1}{10}$

Page 464 • CHALLENGE

$\dfrac{9}{25} = \left(\dfrac{3}{5}\right)^2$; $\dfrac{16}{25} = \left(\dfrac{4}{5}\right)^2$; $\dfrac{16}{A} = \dfrac{(4a)^2}{(12a)^2} = \dfrac{1}{9}$; $A = 144$

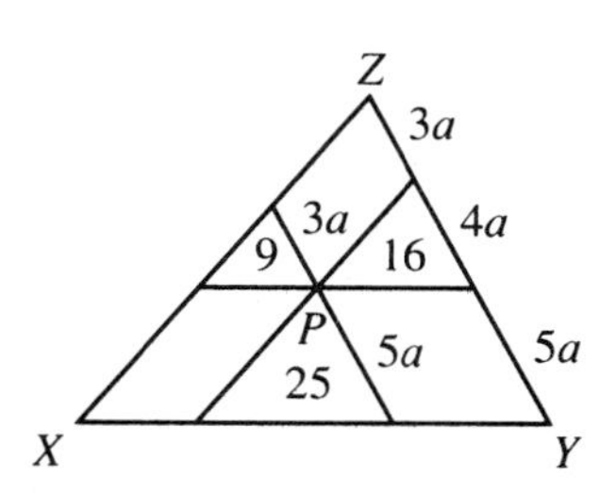

Page 465 • SELF-TEST 2

1. $C = 2\pi r \approx 2 \cdot \frac{22}{7} \cdot 14 = 88$; $A = \pi r^2 \approx \frac{22}{7} \cdot 14^2 = 616$

2. $2\pi r = 18\pi$; $r = 9$; $A = \pi \cdot 9^2 = 81\pi$

3. **a.** $\frac{90}{360} \cdot 2\pi \cdot 12 = 6\pi$ **b.** $\frac{90}{360} \cdot \pi \cdot 12^2 = 36\pi$ **c.** $36\pi - \frac{1}{2} \cdot 12 \cdot 12 = 36\pi - 72$

4. $\frac{4^2}{7^2} = \frac{16}{49}$ **5.** $\frac{\sqrt{36}}{\sqrt{81}} = \frac{6}{9} = \frac{2}{3}$

6. **a.** $\frac{8^2}{12^2} = \frac{4}{9}$ **b.** $\frac{\text{area of } \triangle SOR}{\text{area of } \triangle PTO} = \frac{9}{4}$; $\frac{SO}{TO} = \frac{3}{2}$; $\triangle PSO$ and $\triangle PTO$ have the same alt. from P; $\frac{\text{area of } \triangle PSO}{\text{area of } \triangle PTO} = \frac{3}{2} = \frac{6}{4}$; $\frac{\text{area of } \triangle PSR}{\text{area of } \triangle SPT} = \frac{\text{area of } \triangle PSO + \text{area of } \triangle SOR}{\text{area of } \triangle PSO + \text{area of } \triangle PTO} = \frac{6x + 9x}{6x + 4x} = \frac{3}{2}$

7. $8^2 - 16\pi = 64 - 16\pi$ **8.** $3\left(\frac{120}{360} \cdot 36\pi - \frac{1}{2} \cdot 6\sqrt{3} \cdot 3\right) = 36\pi - 27\sqrt{3}$

9. From Ex. 6, $\frac{OR}{PO} = \frac{3}{2}$ so $\frac{OR}{PR} = \frac{3}{5}$; prob. $= \frac{3}{5}$ **10.** Prob. $= \frac{16\pi}{64} = \frac{\pi}{4}$

Page 466 • EXTRA

1. Draw the altitude, h, from C. Then $\frac{h}{15} = \sin 67° \approx 0.9205$; $h \approx 13.81$; area $= \frac{1}{2}bh \approx \frac{1}{2}(8)(13.81) \approx 55.2$

2. $AC = \sqrt{30^2 - 20^2} = \sqrt{500} = 10\sqrt{5} \approx 22.36$; area $= \frac{1}{2}bh = \frac{1}{2}(20)(22.36) \approx 223.6$

3. semiperimeter $s = \frac{1}{2}(12 + 8 + 10) = \frac{1}{2}(30) = 15$; Area $= \sqrt{s(s - a)(s - b)(s - c)} = \sqrt{15(15 - 12)(15 - 8)(15 - 10)} = \sqrt{15 \cdot 3 \cdot 7 \cdot 5} = \sqrt{1575} \approx 39.7$

4. Draw the altitude from C. Then $\tan 42° = \frac{h}{x}$ and $\tan 28° = \frac{h}{10 - x}$. So $x \tan 42° = h = (10 - x) \tan 28°$; $x(0.9004) \approx 5.3171 - 0.5317x$; $1.4321x \approx 5.3171$; $x \approx 3.713$; $h \approx 3.713 \tan 42° \approx 3.713(0.9004) \approx 3.343$; area $= \frac{1}{2}bh = \frac{1}{2}(10)(3.343) \approx 16.7$.

5. $m\angle C = 180 - (36 + 80) = 64$; draw the altitude from A. Then $\tan 64° = \frac{h}{x}$ and $\tan 80° = \frac{h}{10 - x}$. So $x \tan 64° = h = (10 - x) \tan 80°$; $x(2.0503) \approx 56.713 - 5.6713x$; $7.7216x \approx 56.713$; $x \approx 7.345$; $h \approx 7.345 \tan 64° \approx 15.059$; area $= \frac{1}{2}bh \approx \frac{1}{2}(10)(15.059) \approx 75.3$

6. Draw the altitude from B. Then $\frac{h}{12} = \sin 62° \approx 0.8829$; $h \approx 10.5948$; area $= \frac{1}{2}bh = \frac{1}{2}(20)(10.5948) \approx 105.9$

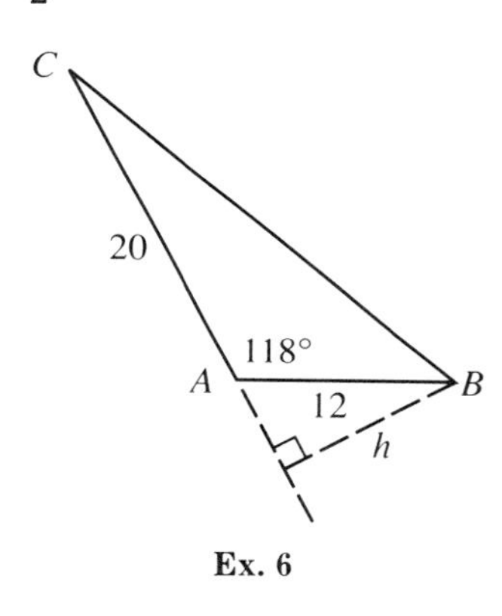

Ex. 6

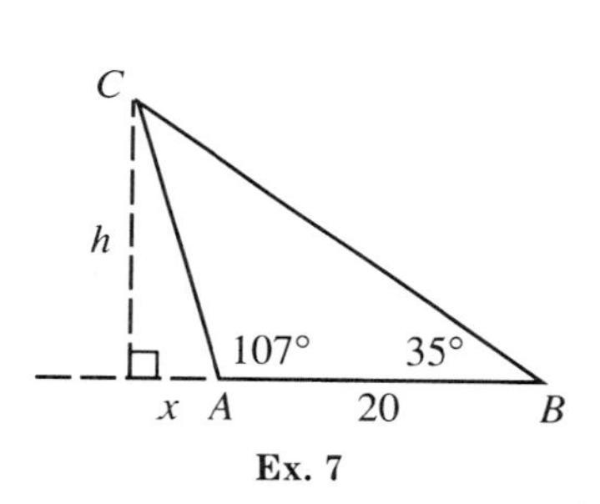

Ex. 7

7. Draw the altitude from C. Then $\tan 73° = \frac{h}{x}$ and $\tan 35° = \frac{h}{20 + x}$; $x \tan 73° = h = (20 + x) \tan 35°$; $3.2709x \approx (20 + x)(0.7002)$; $2.5707x \approx 14.004$; $x \approx 5.448$; $h \approx 5.448 \tan 73° \approx 17.820$; area $= \frac{1}{2}bh \approx \frac{1}{2}(20)(17.820) \approx 178.2$

8. $\tan 35° = \frac{h}{10 - x}$; $h = (10 - x) \tan 35°$;
$h^2 + x^2 = 6^2$; $h = \sqrt{36 - x^2}$;
$(10 - x) \tan 35° = \sqrt{36 - x^2}$;
$(10 - x)^2(\tan 35°)^2 = 36 - x^2$;
$(100 - 20x + x^2)(0.4903) = 36 - x^2$; $1.4903x^2 - 9.806x + 13.03 = 0$. By the quadratic formula and using a calculator, $x = x_1 = 4.732$ and $x = x_2 = 1.848$.

Using $x = x_1$, $h = \sqrt{36 - x^2} \approx 3.689$ and area $= \frac{1}{2}bh \approx \frac{1}{2}(10)(3.689) \approx 18.4$.

Using $x = x_2$, $h = \sqrt{36 - x^2} \approx 5.708$ and area $= \frac{1}{2}bh \approx \frac{1}{2}(10)(5.708) \approx 28.5$.

Page 469 • APPLICATION

1. If $\odot O$ is tan. to $\overleftrightarrow{MA}$, say at X, locate Y on $\overleftrightarrow{MB}$ such that $MY = MX$. $\triangle OMX \cong \triangle OMY$ (SAS) so $\angle X \cong \angle Y$. But $\angle X$ is a rt. $\angle$ (If a line is tan. to a $\odot$, then the line is $\perp$ to the radius drawn to the pt. of tangency.) so $\angle Y$ is a rt. $\angle$ and $\odot O$ is tan. to $\overleftrightarrow{MB}$ at Y. (If a line in the plane of a $\odot$ is $\perp$ to a radius at its outer endpt., then the line is tan. to the $\odot$.)

2. $m\angle AOB = 60$ and length of $\widehat{AB} = \frac{60}{360} \cdot 2\pi \cdot 20{,}000 \approx 20{,}944$ ft 3. near P

4. **a.** Let the vertex be h units below the surface of the Earth. Since $\triangle AEF \sim \triangle ACH$, $\frac{AE}{AC} = \frac{EF}{CH}$;

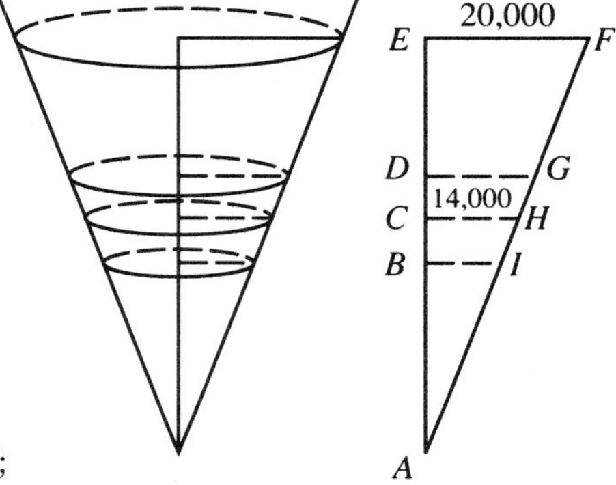

$\frac{30{,}000 + h}{12{,}000 + h} = \frac{20{,}000}{14{,}000} = \frac{10}{7}$;

$120{,}000 + 10h = 210{,}000 + 7h$;

$3h = 90{,}000$; $h = 30{,}000$ ft. **b.** Let DG be the radius of the cone at 15,000 ft above the surface of the Earth. $\triangle ADG \sim \triangle ACH$;

$\frac{15{,}000 + 30{,}000}{12{,}000 + 30{,}000} = \frac{DG}{14{,}000}$; $DG = \frac{15}{14} \cdot (14{,}000) = 15{,}000$ ft.

c. Let x be the height when the radius is 12,000 ft. $\frac{AC}{AB} = \frac{CH}{BI}$;

$\frac{42{,}000}{AB} = \frac{7}{6}$; $AB = \frac{6 \cdot 42{,}000}{7} = 36{,}000$. $x = AB - 30{,}000 = 6{,}000$ ft.

Page 470 • CHAPTER REVIEW

1. $A = 8^2 = 64$ 2. $w = \sqrt{6^2 - 4^2} = 2\sqrt{5}$; $A = 4 \cdot 2\sqrt{5} = 8\sqrt{5}$

3. $A = (3\sqrt{2})^2 = 18\text{ cm}^2$ 4. $\frac{d}{2} = \sqrt{17^2 - 15^2} = 8$; $A = \frac{1}{2}d_1d_2 = \frac{1}{2} \cdot 30 \cdot 16 = 240$

5. $12 \cdot 6 = 8x$; $x = 9$

6. Perimeter $= 8 + 8\sqrt{3} + 16 = 24 + 8\sqrt{3}$; area $= \frac{1}{2} \cdot 8 \cdot 8\sqrt{3} = 32\sqrt{3}$

7. $84 = 12h$; $h = 7$ 8. $h = 3$; $A = \frac{1}{2} \cdot 3 \cdot (4 + 12) = 24$

9. $p = 11 + 4\sqrt{2} + 15 + 4 = 30 + 4\sqrt{2}$; $A = \frac{1}{2} \cdot 4(11 + 15) = 52$

10. $A = (2 \cdot 3)^2 = 36\text{ m}^2$ 11. $s = 6$; $A = \frac{s^2}{4}\sqrt{3} = 9\sqrt{3}$

12. $A = 6\left(\frac{1}{2} \cdot 2 \cdot \sqrt{3}\right) = 6\sqrt{3}$

13. $C = 2\pi r \approx 2 \cdot 3.14 \cdot 30 = 188.4$; $A = \pi r^2 \approx 3.14 \cdot 30^2 = 2826$

14. $121\pi = \pi\left(\frac{d}{2}\right)^2$; $\frac{d}{2} = 11$; $d = 22$ cm

15. $r = \frac{8}{\sqrt{2}} = 4\sqrt{2}$; $C = 2\pi r = 8\pi\sqrt{2}$; $A = \pi r^2 = \pi(4\sqrt{2})^2 = 32\pi$

16. $\frac{135}{360} \cdot 2\pi(24) = 18\pi$ **17.** $\frac{240}{360} \cdot \pi \cdot 6^2 + \frac{1}{2} \cdot 6\sqrt{3} \cdot 3 = 24\pi + 9\sqrt{3}$

18. $\pi\left(\frac{13}{2}\right)^2 - \frac{1}{2} \cdot 5 \cdot 12 = \frac{169}{4}\pi - 30$ **19.** $25\pi - 9\pi = 16\pi$

20. a. $\left(\frac{9}{12}\right)^2 = \frac{9}{16}$ **b.** $\triangle AED$ and $\triangle DEC$ have the same alt. from D. Since, from part (a), $\frac{AE}{EC} = \frac{3}{4}$, then $\frac{\text{area of } \triangle AED}{\text{area at } \triangle DEC} = \frac{3}{4}$.

21. $\left(\frac{16}{32}\right)^2 = \frac{1}{4}$ **22.** $\frac{7^2}{5^2} = \frac{147}{A}$; $A = 75$ **23.** Prob. $= \frac{9\pi}{25\pi} = \frac{9}{25}$

Page 471 • CHAPTER TEST

1. $A = \pi \cdot 5^2 = 25\pi$ **2.** $s = \frac{4}{\sqrt{2}} = 2\sqrt{2}$; $A = (2\sqrt{2})^2 = 8 \text{ cm}^2$

3. $A = \frac{1}{2} \cdot 6 \cdot 6 = 18$ **4.** $2\pi r = 30\pi$; $r = 15$; $A = \pi \cdot 15^2 = 225\pi \text{ m}^2$

5. $A = \frac{1}{2} \cdot 5 \cdot 4 = 10$ **6.** $h = 6$; $A = \frac{1}{2}(6)(6 + 22) = 84$ **7.** $A = bh = 10 \cdot 3 = 30$

8. $A = 6\left(\frac{1}{2} \cdot 4 \cdot 2\sqrt{3}\right) = 24\sqrt{3} \text{ cm}^2$ **9.** $A = \frac{45}{360} \cdot \pi \cdot 4^2 = 2\pi$

10. $\frac{w}{2} = \sqrt{7.5^2 - 6^2} = 4.5$; $w = 9$; $A = 9 \cdot 12 = 108$

11. $A = \frac{10\pi}{2\pi \cdot 12} \cdot \pi \cdot 12^2 = 60\pi$

12. $\frac{s}{2} = \frac{9}{\sqrt{2}}$; $s = 9\sqrt{2}$; $A = (9\sqrt{2})^2 = 162$ **13.** $\frac{8^2}{4}\sqrt{3} + \frac{1}{2} \cdot 8(10 + 14) = 96 + 16\sqrt{3}$

14. Area of smaller $\triangle = \frac{1}{2}(8)(\sqrt{36 - 16}) = 8\sqrt{5}$; area of larger $\triangle = \left(\frac{9}{6}\right)^2 \cdot$ area of smaller $\triangle = \frac{9}{4} \cdot 8\sqrt{5} = 18\sqrt{5}$; shaded area $= 18\sqrt{5} - 8\sqrt{5} = 10\sqrt{5}$

15. $A = \frac{90}{360} \cdot \pi \cdot 8^2 - \frac{1}{2} \cdot 8 \cdot 8 = 16\pi - 32$

16. $\frac{r_1}{r_2} = \frac{\sqrt{100\pi}}{\sqrt{36\pi}} = \frac{5}{3}; \frac{C_1}{C_2} = \frac{2\pi \cdot 5k}{2\pi \cdot 3k} = \frac{5}{3}$ **17.** scale factor $= \frac{14}{3.5} = \frac{4}{1}; \frac{A_1}{A_2} = \left(\frac{4}{1}\right)^2 = \frac{16}{1}$

18. a. $C = 2\pi r = 20\pi$ **b.** $m\widehat{AC} = 360 - 288 = 72$; length of $\widehat{AC} =$

$\frac{72}{360} \cdot 2\pi \cdot 10 = 4\pi$ **c.** $\frac{72}{360} \cdot \pi \cdot 10^2 = 20\pi$

19. $\frac{3}{7}$

20. a. $\frac{1}{2}$ **b.** $\frac{\text{area of } \triangle ABE}{\text{area of } \triangle BDE} = \frac{3}{4}$; prob. $= \frac{\text{area of } \triangle BDE}{\text{area of } \triangle ABC} =$

$\frac{\text{area of } \triangle BDE}{2(\text{area of } \triangle ABE + \text{area of } \triangle BDE)} = \frac{4x}{2(3x + 4x)} = \frac{2}{7}$

Pages 472–473 • CUMULATIVE REVIEW: CHAPTERS 1–11

A **1.** False **2.** False **3.** False **4.** True **5.** False **6.** False **7.** True **8.** False **9.** True **10.** False **11.** False **12.** False **13.** $\|$; skew

14. a. $2x + 10 + 3x - 10 + 4x = 180; 9x = 180; x = 20; m\angle R = 50; m\angle S = 50;$ $m\angle T = 80$ **b.** Isos.; if 2 ∠s of a △ are ≅, then the sides opp. those ∠s are ≅.

15. -45 **16.** SSS **17.** $11 = \frac{4x + 2}{2}; x = 5$ **18.** 4, 5, 6, 7, 8, 9, 10

19. The locus is a sphere with center P and radius 4 cm, along with its interior.

20. Ratio of perimeters is $6\sqrt{3} : 9 = 2\sqrt{3} : 3$; ratio of areas is $(6\sqrt{3})^2 : 9^2 = 108 : 81 = 4 : 3$.

B **21.**

Statements	Reasons
1. $\overline{AB} \perp \overline{BC}; \overline{DC} \perp \overline{BC}; \overline{AC} \cong \overline{BD}$	1. Given
2. $\angle ABC$ and $\angle DCB$ are rt. ∠s.	2. Def. of ⊥ lines
3. $\triangle ABC$ and $\triangle DCB$ are rt. △s.	3. Def. of rt. △
4. $\overline{BC} \cong \overline{BC}$	4. Refl. Prop.
5. $\triangle ABC \cong \triangle DCB$	5. HL Thm.
6. $\angle 1 \cong \angle 2$	6. Corr. parts of ≅ △s are ≅.
7. $\overline{CE} \cong \overline{BE}$	7. If 2 ∠s of a △ are ≅, then the sides opp. those ∠s are ≅.
8. $\triangle BCE$ is isos.	8. Def. of isos. △

22.

Statements	Reasons
1. $\overline{EF} \cong \overline{HG}$; $\overline{EF} \parallel \overline{HG}$	1. Given
2. $EFGH$ is a ▱.	2. If one pair of opp. sides of a quad. are both $\cong$ and $\parallel$, then the quad. is a ▱.
3. $\overline{EH} \parallel \overline{FG}$	3. Def. of ▱
4. Draw $\overline{HF}$.	4. Through any 2 pts. there is exactly one line.
5. $\angle EHF \cong \angle GFH$	5. If 2 $\parallel$ lines are cut by a trans., then alt. int. $\angle$s are $\cong$.

23. Assume temp. that there is a $\triangle$ whose sides have lengths x, y, and $x + y$. Then the length of the longest side equals the sum of the lengths of the other two sides. But this contradicts the $\triangle$ Ineq. Thm., if 2 sides of a $\triangle$ have lengths x and y, then the third side must be greater than $x + y$. Therefore, the temp. assumption must be false. It follows that no $\triangle$ has sides of length x, y, and $x + y$.

24. $\frac{1}{2}\sqrt{4^2 + 8^2} = 2\sqrt{5}$ cm **25.** $\frac{1}{2}(5\sqrt{2} \cdot \sqrt{2}) = 5$ **26.** $1 : 3$

27. $\frac{5}{7 - x} = \frac{9}{x}$; $x = 4.5$ **28.** $\tan 42^\circ = \frac{12}{EF}$; $EF = \frac{12}{\tan 42^\circ} \approx 13$

29. $\frac{XY}{24} = \frac{7}{10}$; $XY \approx 17$ **30.** 45°

31. $58 = \frac{1}{2}(360 - 2m\widehat{QR})$; $m\widehat{QR} = 122$; $m\angle 2 = \frac{1}{2}(122) = 61$

32. $AB = 2$; $BD = 1$; $AD = \sqrt{2^2 + 1^2} = \sqrt{5}$; $AM = \frac{2}{3}\sqrt{5}$

33. Const. a seg. of length $2x$. Use Const. 13 with $a = y$, $b = 2x$, and $c = x$ to find a seg. with length t; $\frac{y}{2x} = \frac{x}{t}$; $ty = 2x^2$; $t = \frac{2x^2}{y}$.

34. $s = 4$; $A = \frac{4^2}{4}\sqrt{3} = 4\sqrt{3}$ cm^2 **35.** $A = \frac{1}{2}\sqrt{7^2 - 5^2} \cdot (11 + 21) = 32\sqrt{6}$

36. **a.** $\frac{200}{360} \cdot 2\pi(12) = \frac{40\pi}{3}$ **b.** $\frac{200}{360} \cdot \pi(12)^2 = 80\pi$

37. **a.** $p = 9 + 9 + 8 + 8 + 12 = 46$ **b.** $\left(\frac{1}{2}\right)^2 = \frac{1}{4}$

CHAPTER 12 • Areas and Volumes of Solids

Page 477 • CLASSROOM EXERCISES

1. hexagonal **2.** 6 **3.** Rectangle

4. Answers may vary. $\overline{PA}$, $\overline{QB}$, $\overline{RC}$, and $\overline{SD}$ are lateral edges and altitudes.

5. height

6. a. L.A. $= ph = 24 \cdot 5 = 120$ cm^2 **b.** $B = \frac{1}{2}ap = \frac{1}{2} \cdot 2\sqrt{3} \cdot 24 = 24\sqrt{3}$ cm^2

c. T.A. = L.A. + $2B = (120 + 48\sqrt{3})$ cm^2 **d.** $V = Bh = 24\sqrt{3} \cdot 5 = 120\sqrt{3}$ cm^3

7. No **8.** 5 **9.** No **10. a.** 9; 27 **b.** 12; 144; 1728 **c.** 100; 10,000; 1,000,000

Pages 478–480 • WRITTEN EXERCISES

A **1.** L.A. $= (2 \cdot 6 + 2 \cdot 4)2 = 40$; T.A. $= 40 + 2 \cdot 6 \cdot 4 = 88$; $V = 6 \cdot 4 \cdot 2 = 48$

2. L.A. $= (2 \cdot 50 + 2 \cdot 30)15 = 2400$; T.A. $= 2400 + 2 \cdot 50 \cdot 30 = 5400$; $V = 50 \cdot 30 \cdot 15 = 22{,}500$

3. $h = \frac{54}{6 \cdot 3} = 3$; L.A. $= (2 \cdot 6 + 2 \cdot 3)3 = 54$; T.A. $= 54 + 2 \cdot 6 \cdot 3 = 90$

4. $l = \frac{360}{8 \cdot 5} = 9$; L.A. $= (2 \cdot 9 + 2 \cdot 8)5 = 170$; T.A. $= 170 + 2 \cdot 9 \cdot 8 = 314$

5. $(2 \cdot 9 + 2w)2 = 60$; $w = 6$; T.A. $= 60 + 2 \cdot 9 \cdot 6 = 168$; $V = 9 \cdot 6 \cdot 2 = 108$

6. L.A. $= (2 \cdot 5x + 2 \cdot 4x)3x = 54x^2$; T.A. $= 54x^2 + 2 \cdot 5x \cdot 4x = 94x^2$; $V = 5x \cdot 4x \cdot 3x = 60x^3$

7. T.A. $= 3^2 \cdot 6 = 54$; $V = 3^3 = 27$ **8.** T.A. $= 6e^2$; $V = e^3$

9. $e^3 = 1000 = 10^3$, so $e = 10$; T.A. $= 10^2 \cdot 6 = 600$

10. $e^3 = 64 = 4^3$, so $e = 4$; T.A. $= 4^2 \cdot 6 = 96$

11. $6e^2 = 150$; $e^2 = 25$; $e = 5$; $V = 5^3 = 125$

12. T.A. $= (2x)^2 \cdot 6 = 24x^2$; $V = (2x)^3 = 8x^3$ **13.** L.A. $= ph = 30 \cdot 13 = 390$

14. $120 = ph = 15h$; $h = 8$ cm **15.** 4; 8 **16.** 9; 9; 27

17–22. Check students' drawings.

17. L.A. $= 3 \cdot 8 \cdot 10 = 240$; T.A. $= 240 + 2 \cdot 16\sqrt{3} = 240 + 32\sqrt{3}$; $V = 16\sqrt{3} \cdot 10 = 160\sqrt{3}$

18. L.A. $= (9 + 12 + 15)10 = 360$; T.A. $= 360 + 2 \cdot \frac{9 \cdot 12}{2} = 468$;

$V = \frac{9 \cdot 12}{2} \cdot 10 = 540$

B **19.** L.A. $= (13 + 13 + 10)7 = 252$; height of base $= \sqrt{13^2 - 5^2} = 12$;

T.A. $= 252 + 2\left(\frac{1}{2} \cdot 10 \cdot 12\right) = 372$; $V = \frac{1}{2} \cdot 10 \cdot 12 \cdot 7 = 420$

20. L.A. $= (10 + 5 + 4 + 5)20 = 480$; height of base $= \sqrt{5^2 - 3^2} = 4$;

T.A. $= 480 + 2\left(\frac{1}{2} \cdot 4 \cdot 14\right) = 536$; $V = \left(\frac{1}{2} \cdot 4 \cdot 14\right)20 = 560$

21. $s = \sqrt{3^2 + 4^2} = 5$; L.A. $= 4 \cdot 5 \cdot 9 = 180$; T.A. $= 180 + 2 \cdot \frac{1}{2} \cdot 6 \cdot 8 = 228$;

$V = 24 \cdot 9 = 216$

22. L.A. $= 6 \cdot 8 \cdot 12 = 576$; apothem of base $= 4\sqrt{3}$; T.A. $= 576 + 2 \cdot \frac{1}{2} \cdot 4\sqrt{3} \cdot 48 =$

$576 + 192\sqrt{3}$; $V = 96\sqrt{3} \cdot 12 = 1152\sqrt{3}$

23. Vol. of rock = vol. of displaced water $= 0.5 \cdot 45 \cdot 30 = 675 \text{ cm}^3$

24. $30 \cdot 5 \cdot 0.03 \cdot \$175 = \$787.50$

25. $V_1 = 20 \cdot 10 \cdot 5 = 1000$; $V_2 = 25 \cdot 15 \cdot 4 = 1500$; $\frac{1.2}{1000} = \frac{x}{1500}$; $1000x = 1800$;

$x = 1.8$; 1.8 kg

26. height of base $= \sqrt{0.5^2 - 0.3^2} = 0.4$; $B = \frac{1}{2}(0.4)(1.4) = 0.28$; $V = 0.28(2) = 0.56$;

about 0.56 metric ton

27. $V = 40 \cdot 20 \cdot 20 - 2 \cdot 12 \cdot 10 \cdot 20 = 11{,}200$; $11{,}200 \text{ cm}^3 = 0.0112 \text{ m}^3$;

weight $= (0.0112) \cdot 1700 \approx 19$ kg

28. $V = 2(0.3)(10)(0.05) + (0.05)(10)(0.2) = 0.4$; $0.4(7860) = 3144$; 3140 kg

29. $V = 4x \cdot 3x \cdot 5x - x \cdot 2x \cdot 5x = 50x^3$; T.A. $= 14x \cdot 5x + 2(4x \cdot 3x - x \cdot 2x) +$

$6x \cdot 5x = 70x^2 + 20x^2 + 30x^2 = 120x^2$

30. $V = 6x \cdot 3y \cdot 4x + 2 \cdot 4x \cdot 3y \cdot x = 72x^2y + 24x^2y = 96x^2y$; T.A. $=$

$4 \cdot 4x \cdot 3y + 8 \cdot x \cdot 3y + 2(4x)^2 + 8 \cdot 4x^2 = 48xy + 24xy + 32x^2 + 32x^2 =$

$64x^2 + 72xy$

31. $l = 2w$; $h = 3w$; $lwh = 2w \cdot w \cdot 3w = 6w^3 = 162$; $w = 3$ cm; T.A. $=$

$2(lh + wh + lw) = 2(6 \cdot 9 + 3 \cdot 9 + 6 \cdot 3) = 198 \text{ cm}^2$

32. $l = w = 3h$; T.A. $= 2(lh + wh + lw) = 2(3h^2 + 3h^2 + 9h^2) = 30h^2 = 750$; $h = 5$;

$V = lwh = 15 \cdot 15 \cdot 5 = 1125 \text{ m}^3$

33. diag. of base $= \sqrt{8^2 + 6^2} = 10$; $\tan 35° = \frac{h}{10}$; $h \approx 10(0.7002) \approx 7$;

$V \approx 7 \cdot 8 \cdot 6 = 336$

34. $\frac{V_1}{V_2} = \frac{Bh_1}{Bh_2} = \frac{48 \cdot 10 \cdot \tan 35°}{48 \cdot 10 \cdot \tan 70°} = \frac{\tan 35°}{\tan 70°}$

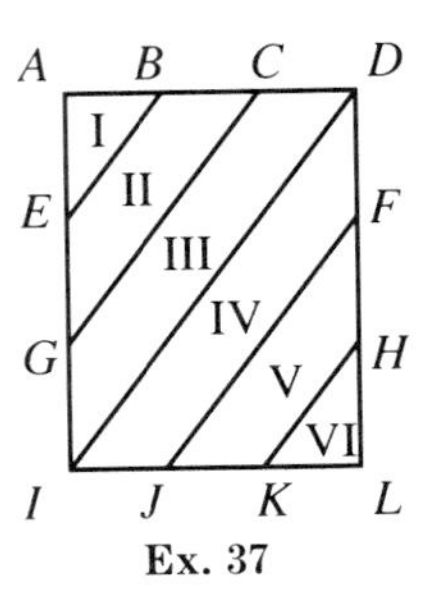

Ex. 37

C **35.** $V = Bh = \frac{1}{2}aph = \frac{1}{2} \cdot \frac{x\sqrt{3}}{6} \cdot 3x \cdot x = \frac{1}{4}x^3\sqrt{3}$

36. $V = Bh = \frac{1}{2}aph = \frac{1}{2} \cdot \frac{s\sqrt{3}}{2} \cdot 6s \cdot h = \frac{3}{2}s^2h\sqrt{3}$

37. The fig. shows the base of the beam and the bases of the resulting prisms. vol. I = vol. VI = $\frac{1}{2} \cdot 6 \cdot 8 \cdot 300 = 7200$ cm^3; base area of II = $\frac{1}{2} \cdot 12 \cdot 16 - \frac{1}{2} \cdot 6 \cdot 8 = 72$; vol. II = vol. V = $72 \cdot 300 = 21{,}600$ cm^3; base area of III $= \frac{1}{2} \cdot 18 \cdot 24 - \frac{1}{2} \cdot 12 \cdot 16 = 120$; vol. III = vol. IV = $120 \cdot 300 = 36{,}000$ cm^3

38. The diag. is the hyp. of a $\triangle$ with one leg an edge of the cube and one leg a diag. of a face. If s = length of an edge, $(4\sqrt{3})^2 = s^2 + (s\sqrt{2})^2$; $48 = s^2 + 2s^2$; $s^2 = 16$; $s = 4$; $V = 4^3 = 64$ cm^3.

39. Let s = length of an edge; bases are equilateral $\triangle$s with height $= \frac{s\sqrt{3}}{2}$;

$B = \frac{1}{2} \cdot s \cdot \frac{s\sqrt{3}}{2} = \frac{s^2\sqrt{3}}{4}$ and $V = \frac{s^2\sqrt{3}}{4} \cdot s = \frac{s^3\sqrt{3}}{4} = 54\sqrt{3}$; $\frac{s^3}{4} = 54$; $s^3 = 216$;

$s = 6$; 6 cm

40. $V_1 = lwh$; $V_2 = (0.8l)(0.8w)(xh) = lwh$; $0.64x = 1$; $x = 1.5625$; h must be increased by about 56%.

Page 480 • CHALLENGE

1. Draw the diags. of each rect. to locate their centers, A and B. Draw $\overleftrightarrow{AB}$.
2. Draw the diags. of each solid to locate their centers, A, B, and C. Draw a plane containing A, B, and C. (If A, B, and C are not collinear, there is only one such plane.)

Page 481 • COMPUTER KEY-IN

1. 1.6 and 1.8
2. Modify line 20: FOR X = 1.6 TO 1.8 STEP 0.01. Maximum volume is 81.9 in.^{3} when $x = 1.7$; length = 8.6 in., width = 5.6 in., height = 1.7 in.
3. **a.** $V = (15 - 2x) \cdot (8 - 2x) \cdot x$ **b.** 90.7 in.^{3} **c.** length = 11.6 in., width = 4.6 in., height = 1.7 in.

Page 484 • CLASSROOM EXERCISES

1. a. $V = \frac{1}{6}(\text{Volume of cube}) = \frac{1}{6}e^3$ **b.** $V = \frac{1}{6}e^3 = \frac{1}{3} \cdot \frac{1}{2} \cdot e^2 \cdot e = \frac{1}{3}\left(e^2\right)\left(\frac{1}{2}e\right) = \frac{1}{3}Bh$

2. 6 **3.** $\sqrt{8^2 + 6^2} = 10$ **4.** $\frac{1}{2} \cdot 12 \cdot 10 = 60$ **5.** $4 \cdot 60 = 240$

6. $\frac{1}{3}Bh = \frac{1}{3} \cdot 12^2 \cdot 8 = 384$ **7.** $\sqrt{10^2 + 6^2} = 2\sqrt{34}$ **8.** $3\sqrt{3}$ cm

9. $\frac{2}{3} \cdot 3\sqrt{3} = 2\sqrt{3}$ cm **10.** $\sqrt{6^2 - (2\sqrt{3})^2} = 2\sqrt{6}$ cm **11.** $\frac{1}{2} \cdot 6 \cdot 3\sqrt{3} = 9\sqrt{3}$ cm^2

12. $\frac{1}{3}Bh = \frac{1}{3} \cdot 9\sqrt{3} \cdot 2\sqrt{6} = 18\sqrt{2}$ cm^3 **13.** $3\sqrt{3}$ cm

14. $3 \cdot B = 3 \cdot 9\sqrt{3} = 27\sqrt{3}$ cm^2 **15.** $4 \cdot B = 4 \cdot 9\sqrt{3} = 36\sqrt{3}$ cm^2

16. No. The height is the length of a leg of a rt. △, and the slant height is the length of the hypotenuse. The length of the hypotenuse is always greater than the length of a leg.

17. No. The slant height is the length of a leg of a rt. △, and the lateral edge is the length of the hypotenuse. The length of the hypotenuse is always greater than the length of a leg.

18. No. The apothem of the base is the leg of a rt. △ with hypotenuse l. Then $a < l$, $\frac{1}{2}ap < \frac{1}{2}pl$, and $B <$ L.A.

Pages 485–487 • WRITTEN EXERCISES

A

	1.	2.	3.	4.	5.	6.
height, h	4	12	24	$3\sqrt{7}$	3	6
slant height, l	5	13	25	12	5	$2\sqrt{17}$
base edge	6	10	14	18	8	$8\sqrt{2}$
lateral edge	$\sqrt{34}$	$\sqrt{194}$	$\sqrt{674}$	15	$\sqrt{41}$	10

7–14. Check students' drawings.

7. $\frac{1}{2} \cdot 3 \cdot 4 \cdot 6 = 36$ **8.** $\frac{1}{2} \cdot 5 \cdot 1.5 \cdot 9 = 33.75$ **9.** $\frac{1}{2} \cdot 4 \cdot 12 \cdot 8 = 192$

10. $\frac{1}{2} \cdot 6 \cdot 10 \cdot 12 = 360$

11. $l = \sqrt{3^2 + 4^2} = 5$; L.A. $= \frac{1}{2} \cdot 4 \cdot 6 \cdot 5 = 60$; T.A. $= 60 + 36 = 96$;

$V = \frac{1}{3} \cdot 36 \cdot 4 = 48$

12. L.A. $= \frac{1}{2} \cdot 4 \cdot 16 \cdot 10 = 320$; T.A. $= 320 + 256 = 576$; $h = \sqrt{10^2 - 8^2} = 6$;

$V = \frac{1}{3} \cdot 256 \cdot 6 = 512$

13. base edge $= 2\sqrt{13^2 - 12^2} = 10$; L.A. $= \frac{1}{2} \cdot 4 \cdot 10 \cdot 13 = 260$; T.A. $= 260 + 100 =$

360; $V = \frac{1}{3} \cdot 100 \cdot 12 = 400$

14. $l = \sqrt{17^2 - 8^2} = 15$; L.A. $= \frac{1}{2} \cdot 4 \cdot 16 \cdot 15 = 480$; T.A. $= 480 + 256 = 736$;

$h = \sqrt{15^2 - 8^2} = \sqrt{161}$; $V = \frac{1}{3} \cdot 256 \cdot \sqrt{161} = \frac{256\sqrt{161}}{3}$

15. $V = \frac{1}{3}Bh$; $32 = \frac{1}{3}(16)h$; $h = 6$ cm **16.** $\frac{60}{8} = \frac{1}{2} \cdot 3 \cdot l$; $l = 5$ m

B 17. a. $VX = \sqrt{12^2 + 9^2} = 15$ cm; $VY = \sqrt{12^2 + 5^2} = 13$ cm

b. $2 \cdot \frac{1}{2} \cdot 18 \cdot 13 + 2 \cdot \frac{1}{2} \cdot 10 \cdot 15 = 234 + 150 = 384$ cm^2. The pyramid is not regular, so the faces are not $\cong$ △s.

18. $h = \sqrt{10^2 - 6^2} = 8$ ft; apothem of base $= 3\sqrt{3}$; $B = \frac{1}{2} \cdot 3\sqrt{3} \cdot 36 = 54\sqrt{3}$;

$V = \frac{1}{3} \cdot 54\sqrt{3} \cdot 8 = 144\sqrt{3}$ ft^3

19. base area of pyramid $= \frac{1}{2}$(base area of the rect. solid); if $Bh =$ volume of rect. solid,

volume of pyramid $= \frac{1}{3}\left(\frac{1}{2}B\right)h = \frac{1}{6}Bh = \frac{1}{6}$(volume of rect. solid).

20. area of pyramid : area of prism $= \frac{1}{3}Bh : Bh = \frac{1}{3} : 1$ or $1 : 3$

21. $AO = \frac{2}{3} \cdot AM = 6$; $h = \sqrt{10^2 - 6^2} = 8$; $OM = \frac{1}{3}AM = 3$; $l = \sqrt{8^2 + 3^2} = \sqrt{73}$

22. a. $AM = \frac{6}{2}\sqrt{3} = 3\sqrt{3}$; $AO = \frac{2}{3} \cdot AM = 2\sqrt{3}$ **b.** $h = \sqrt{4^2 - (2\sqrt{3})^2} = 2$;

$l = \sqrt{2^2 + (\sqrt{3})^2} = \sqrt{7}$

23. a. $OM = \sqrt{5^2 - 4^2} = 3$; $OA = 2 \cdot OM = 6$; $AM = 9$; $BC = 2\left(\frac{9}{\sqrt{3}}\right) = 6\sqrt{3}$

b. L.A. $= \frac{1}{2} \cdot 3 \cdot 6\sqrt{3} \cdot 5 = 45\sqrt{3}$; $V = \frac{1}{3} \cdot \frac{1}{2} \cdot 6\sqrt{3} \cdot 9 \cdot 4 = 36\sqrt{3}$

24. $AO = \sqrt{5^2 - 3^2} = 4$; $OM = 2$; $l = \sqrt{3^2 + 2^2} = \sqrt{13}$; $BC = 2\sqrt{5^2 - (\sqrt{13})^2} = 4\sqrt{3}$; L.A. $= \frac{1}{2} \cdot 3 \cdot 4\sqrt{3} \cdot \sqrt{13} = 6\sqrt{39}$; $V = \frac{1}{3} \cdot \frac{1}{2} \cdot 4\sqrt{3} \cdot 6 \cdot 3 = 12\sqrt{3}$

25. $l = \sqrt{10^2 - 6^2} = 8$; L.A. $= \frac{1}{2} \cdot 3 \cdot 12 \cdot 8 = 144$; $AM = 6\sqrt{3}$; $AO = \frac{2}{3} \cdot 6\sqrt{3} = 4\sqrt{3}$; $h = \sqrt{10^2 - (4\sqrt{3})^2} = 2\sqrt{13}$; $V = \frac{1}{3} \cdot \frac{1}{2} \cdot 12 \cdot 6\sqrt{3} \cdot 2\sqrt{13} = 24\sqrt{39}$

26. $V = \frac{1}{3}Bh = \frac{1}{3}\left(6 \cdot \frac{6^2}{4}\sqrt{3}\right) \cdot 8 = 144\sqrt{3}$ cm^3

27. $\tan 50° = \frac{h}{4}$; $h = 4 \tan 50° \approx 4.7670$; $V \approx \frac{1}{3}\left(6 \cdot \frac{4^2}{4}\sqrt{3}\right)(4.7670) \approx 66$ cubic units

28. $$\frac{V_1}{V_2} = \frac{\frac{1}{3}10^2 \cdot h_1}{\frac{1}{3}10^2 \cdot h_2} = \frac{h_1}{h_2} = \frac{5\sqrt{2}\tan 40°}{5\sqrt{2}\tan 80°} = \frac{\tan 40°}{\tan 80°}$$

C **29.** $$V = \frac{1}{3}\left(\frac{x^2}{4}\sqrt{3}\right)\left(\sqrt{x^2 - \left(\frac{2}{3} \cdot \frac{x}{2} \cdot \sqrt{3}\right)^2}\right) = \frac{x^3\sqrt{2}}{12}$$

30. $V = \frac{1}{3}\left(6 \cdot \frac{y^2}{4}\sqrt{3}\right) \cdot y\sqrt{3} = \frac{3y^3}{2}$ cm^3

31. $\sin 42° = \frac{h}{10}$; $h = 10 \sin 42° \approx 6.6913$; $\frac{d}{2} = \sqrt{10^2 - h^2} \approx \sqrt{100 - (6.6913)^2} \approx 7.4315$; $s = \frac{d}{2}\sqrt{2} \approx 7.4315\sqrt{2} \approx 10.5097$; $V = \frac{1}{3}(10.5097)^2(6.6913) \approx 246$

32. **a.** Since the pyramids have $\cong$ bases and $\cong$ altitudes, they have the same volume.
b. F-$ABCD$; T.A. of F-$ABCD$ = area of $\triangle FAB$ + area of $\triangle FBC$ + area of $\triangle FCD$ + area of $\triangle FDA$ + area of $ABCD$ $= \frac{1}{2} + \frac{1}{2} + \frac{\sqrt{2}}{2} + \frac{\sqrt{2}}{2} + 1 = 2 + \sqrt{2} \approx 3.414$; $MF = \frac{\sqrt{2}}{2}$; $FB = 1$; $MB = \frac{\sqrt{6}}{2}$ and $l = \sqrt{\left(\frac{\sqrt{6}}{2}\right)^2 - \left(\frac{1}{2}\right)^2} = \frac{\sqrt{5}}{2}$; T.A. of M-$ABCD$ $= \frac{1}{2} \cdot 4 \cdot \frac{\sqrt{5}}{2} + 1 = \sqrt{5} + 1 \approx 3.236$

Page 487 • CHALLENGE

1. $x + y + z = h$

2. Again, $x + y + z = h$. Let s = the length of each side of the $\triangle$. Then $A = \frac{1}{2}sh = \frac{1}{2}sx + \frac{1}{2}sy + \frac{1}{2}sz = \frac{1}{2}s(x + y + z)$, so $h = x + y + z$.

3. From any pt. inside an equilateral $\triangle$, the sum of the lengths of the $\perp$ segs. drawn from the pt. to the sides = the height of the $\triangle$.

4. From any pt. inside a regular triangular pyramid with all edges $\cong$, the sum of the lengths of the $\perp$ segs. drawn to the faces = the height of the pyramid.

Page 487 • MIXED REVIEW EXERCISES

	1.	2.	3.	4.	5.	6.	7.	8.
r	6	11	$\frac{1}{2}$	$3\sqrt{3}$	5	9	7	$\sqrt{15}$
C	12π	22π	π	$6\pi\sqrt{3}$	10π	18π	14π	$2\pi\sqrt{15}$
A	36π	121π	$\frac{\pi}{4}$	27π	25π	81π	49π	15π

9–10. Check students' drawings.

9. a. $A = \pi \cdot 12^2 = 144\pi \text{ mm}^2$ **b.** $A = 24^2 = 576 \text{ mm}^2$

10. a. $p = 4 \cdot 4\sqrt{2} \cdot \sqrt{2} = 32$ **b.** $C = 2\pi r = 2 \cdot \pi \cdot 4\sqrt{2} = 8\pi\sqrt{2}$

Page 488 • CALCULATOR KEY-IN

Ratio of top to original pyramid $= \frac{31}{481} \approx \frac{0.0644491}{1}$; ratio of volumes $\approx \frac{0.0002677}{1}$; top which was destroyed was $\approx$ 0.02677% of original volume; $\approx$ 99.97% of original volume remains.

Page 489 • COMPUTER KEY-IN

1. 338.35

2. a. 333.8335

b.

No. of steps	Volume
10	385
100	338.35
500	334.334
750	334.000296
900	333.889095
1000	333.8335

c. Answers may vary; $333.\overline{33}$ **d.** $333.\overline{33}$

e. As the number of steps increases, the volume of the step pyramid approaches the volume of the regular square pyramid.

Pages 491–492 • CLASSROOM EXERCISES

1. **a.** length $= 2\pi r$; width $= h$ **b.** Area $= 2\pi rh$
2. **a.** I: L.A. $= 2\pi \cdot 3 \cdot 4 = 24\pi$; II: L.A. $= 2\pi \cdot 3 \cdot 8 = 48\pi$; III: L.A. $= 2\pi \cdot 6 \cdot 4 = 48\pi$ **b.** Yes **c.** Yes
3. **a.** I: T.A. $= 24\pi + 2\pi \cdot 3^2 = 42\pi$; II: T.A. $= 48\pi + 2\pi \cdot 3^2 = 66\pi$; III: T.A. $= 48\pi + 2\pi \cdot 6^2 = 120\pi$ **b.** No
4. **a.** I: $V = \pi \cdot 3^2 \cdot 4 = 36\pi$; II: $V = \pi \cdot 3^2 \cdot 8 = 72\pi$; III: $V = \pi \cdot 6^2 \cdot 4 = 144\pi$ **b.** Yes **c.** No
5. $l = 5$; L.A. $= \pi \cdot 3 \cdot 5 = 15\pi$; T.A. $= 15\pi + 9\pi = 24\pi$; $V = \frac{1}{3} \cdot \pi \cdot 9 \cdot 4 = 12\pi$
6. $r = 5$; L.A. $= \pi \cdot 5 \cdot 13 = 65\pi$; T.A. $= 65\pi + \pi \cdot 25 = 90\pi$; $V = \frac{1}{3}\pi \cdot 25 \cdot 12 = 100\pi$
7. $h = 8$ cm; L.A. $= \pi \cdot 6 \cdot 10 = 60\pi$ cm^2; T.A. $= 60\pi + \pi \cdot 36 = 96\pi$ cm^2; $V = \frac{1}{3}\pi \cdot 36 \cdot 8 = 96\pi$ cm^3
8. A ⊙ that is parallel to the base and has radius $= \frac{1}{2}$(radius of the base).

Pages 492–495 • WRITTEN EXERCISES

A **1–16.** Check students' drawings.

1. L.A. $= 2\pi \cdot 4 \cdot 5 = 40\pi$; T.A. $= 40\pi + 2\pi \cdot 4^2 = 72\pi$; $V = \pi \cdot 4^2 \cdot 5 = 80\pi$
2. L.A. $= 2\pi \cdot 8 \cdot 10 = 160\pi$; T.A. $= 160\pi + 2 \cdot \pi \cdot 8^2 = 288\pi$; $V = \pi \cdot 8^2 \cdot 10 = 640\pi$
3. L.A. $= 2\pi \cdot 4 \cdot 3 = 24\pi$; T.A. $= 24\pi + 2 \cdot \pi \cdot 4^2 = 56\pi$; $V = \pi \cdot 4^2 \cdot 3 = 48\pi$
4. L.A. $= 2\pi \cdot 8 \cdot 15 = 240\pi$; T.A. $= 240\pi + 2\pi \cdot 8^2 = 368\pi$; $V = \pi \cdot 8^2 \cdot 15 = 960\pi$
5. $64\pi = \pi \cdot r^2 \cdot r$; $r = 4$ **6.** $2\pi \cdot r \cdot 6 = 18\pi$; $r = 1.5$
7. $\pi r^2 \cdot 8 = 72\pi$; $r = 3$; L.A. $= 2\pi \cdot 3 \cdot 8 = 48\pi$
8. $2\pi \cdot r \cdot r + 2\pi \cdot r^2 = 100\pi$; $r = 5$

	r	h	l	L.A.	T.A.	V
9.	4	3	5	20π	36π	16π
10.	8	6	10	80π	144π	128π
11.	12	5	13	156π	300π	240π
12.	$4\sqrt{2}$	2	6	$24\pi\sqrt{2}$	$24\pi\sqrt{2} + 32\pi$	$\frac{64\pi}{3}$
13.	12	9	15	180π	324π	432π
14.	21	20	29	609π	1050π	2940π
15.	15	8	17	255π	480π	600π
16.	9	12	15	135π	216π	324π

17. **a.** $\frac{20\pi}{80\pi} = \frac{1}{4}$ **b.** $\frac{36\pi}{144\pi} = \frac{1}{4}$ **c.** $\frac{16\pi}{128\pi} = \frac{1}{8}$ **18.** $\frac{\text{Vol. of wider}}{\text{Vol. of taller}} = \frac{4\pi r^2 h}{2\pi r^2 h} = \frac{2}{1}$

19. $\frac{1}{3}Bh : Bh = \frac{1}{3} : 1$ or $1 : 3$

B **20.** **a.** can **b.** vol. of can $= \pi(2.5)^2 10 = 62.5\pi\ \text{cm}^3$; vol. of bottle $=$ $\pi \cdot 2^2 \cdot 10 + \frac{1}{3}\pi \cdot 2^2 \cdot 6 = 40\pi + 8\pi = 48\pi\ \text{cm}^3$

21. vol. of cyl. $= \pi \cdot 6^2 \cdot 18 = 648\pi$; $\frac{1}{3}\pi \cdot 9^2 \cdot h = 648\pi$; $27h = 648$; $h = 24$; 24 cm

22. $\pi \cdot 6^2 \cdot 200 - \pi \cdot 5^2 \cdot 200 = 2200\pi \approx 6908\ \text{cm}^3$

23. $V = \frac{1}{3}\pi(2.6)^2\sqrt{(6.8)^2 - (2.6)^2} \approx 44.46$; $\frac{44.46}{1.8} \approx 25$ minutes

24. $\pi\left(\frac{d}{2}\right)^2 \cdot h = \pi \cdot 3^2 \cdot h + \pi \cdot 4^2 \cdot h$; $\left(\frac{d}{2}\right)^2 = 9 + 16 = 25$; $\frac{d^2}{4} = 25$; $d^2 = 100$; $d = 10$; 10 cm

25. $2\pi r \cdot 8 + 2\pi r^2 = 40\pi$; $r^2 + 8r - 20 = 0$; $(r + 10)(r - 2) = 0$; $r = -10$ (reject) or $r = 2$; $r = 2$

26. $2\pi r \cdot 12 + 2\pi r^2 = 90\pi$; $r^2 + 12r - 45 = 0$; $(r + 15)(r - 3) = 0$; $r = -15$ (reject) or $r = 3$; $r = 3$

27. **a.** A cylinder with $r = 6$, $h = 10$; $V = \pi \cdot 6^2 \cdot 10 = 360\pi$ **b.** A cylinder with $r = 10$, $h = 6$; $V = \pi \cdot 10^2 \cdot 6 = 600\pi$

28. **a.** $r = 3; h = 4; l = \sqrt{4^2 + 3^2} = 5$; L.A. $= \pi \cdot 3 \cdot 5 = 15\pi$; $V = \frac{1}{3}\pi \cdot 3^2 \cdot 4 = 12\pi$ **b.** $r = 4; h = 3; l = \sqrt{4^2 + 3^2} = 5$; L.A. $= \pi \cdot 4 \cdot 5 = 20\pi$; $V = \frac{1}{3} \cdot \pi \cdot 4^2 \cdot 3 = 16\pi$ **c.** No

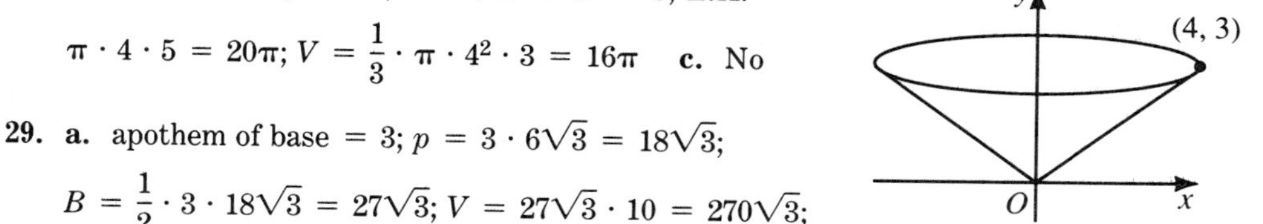

Ex. 28b

29. **a.** apothem of base $= 3$; $p = 3 \cdot 6\sqrt{3} = 18\sqrt{3}$; $B = \frac{1}{2} \cdot 3 \cdot 18\sqrt{3} = 27\sqrt{3}$; $V = 27\sqrt{3} \cdot 10 = 270\sqrt{3}$; L.A. $= 18\sqrt{3} \cdot 10 = 180\sqrt{3}$ **b.** $s = \frac{12}{\sqrt{2}} = 6\sqrt{2}$; $V = (6\sqrt{2})^2 \cdot 10 = 720$; L.A. $= 4 \cdot 6\sqrt{2} \cdot 10 = 240\sqrt{2}$ **c.** apothem of base $= 3\sqrt{3}$; $s = 6$; $p = 36$; $V = \frac{1}{2} \cdot 3\sqrt{3} \cdot 36 \cdot 10 = 540\sqrt{3}$; L.A. $= 36 \cdot 10 = 360$

30. The solid formed is a cone with radius 3 and height $3\sqrt{3}$. $V = \frac{1}{3}Bh = \frac{1}{3} \cdot 9\pi \cdot 3\sqrt{3} = 9\pi\sqrt{3}$ cm^3

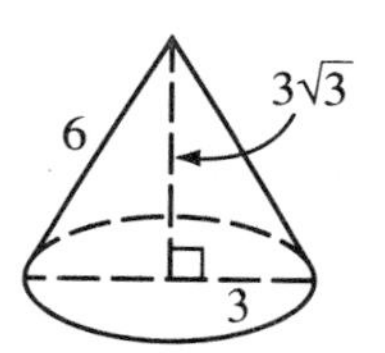

31. The solid formed is a cylinder with radius s and height s; $V = \pi \cdot s^2 \cdot s = \pi s^3$.

32. $V = 12^2 \cdot 10 = 1440$

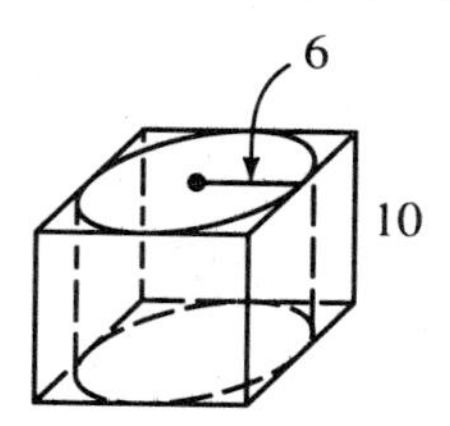

33. $r = \frac{4\sqrt{2}}{2} = 2\sqrt{2}$; $V = \frac{1}{3} \cdot \pi(2\sqrt{2})^2 \cdot 6 = 16\pi$ cm^3

34. **a.** $s = \frac{8}{\sqrt{2}} = 4\sqrt{2}$; $V = \frac{1}{3}(4\sqrt{2})^2 \cdot 4 = \frac{128}{3} = 42\frac{2}{3}$ cm^3 **b.** cone: $\sqrt{4^2 + 4^2} = 4\sqrt{2}$ cm; pyramid: $\sqrt{(2\sqrt{2})^2 + 4^2} = 2\sqrt{6}$ cm

35. $h = \sqrt{9^2 - 3^2} = 6\sqrt{2}$; $V = \frac{1}{3} \cdot \pi \cdot 3^2 \cdot 6\sqrt{2} = 18\pi\sqrt{2}$ cm^3

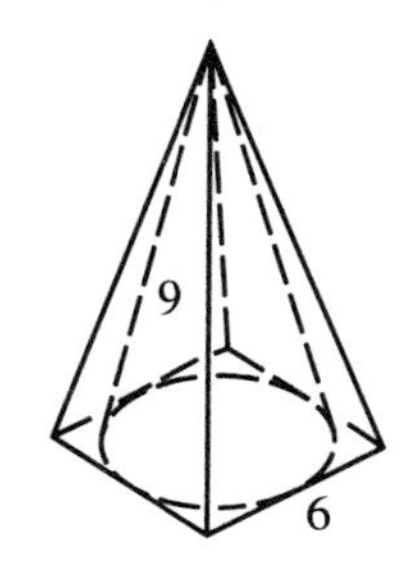

36. $\pi rl = \frac{3}{5}(\pi rl + \pi r^2)$; $5l = 3l + 3r$; $2l = 3r$; $\frac{r}{l} = \frac{2}{3}$

C **37.** $r = 6$; slant height of cone $= \sqrt{6^2 + 8^2} = 10$; L.A. of cone $= \pi \cdot 6 \cdot 10 = 60\pi$; slant height of pyramid $= \sqrt{10^2 - 3^2} = \sqrt{91}$; L.A. of pyramid $=$ $\frac{1}{2} \cdot 36 \cdot \sqrt{91} = 18\sqrt{91}$

38. Circumference of base $= \frac{120}{360} \cdot 2\pi \cdot 6 = 4\pi$, so $r = 2$; $l = 6$ so $h = \sqrt{6^2 - 2^2} =$ $4\sqrt{2}$; $V = \frac{1}{3}\pi \cdot 2^2 \cdot 4\sqrt{2} = \frac{16\sqrt{2}\pi}{3}$ in.^{3}

39. $\triangle ABC$ is a rt. $\triangle$. Let $\overline{AD}$ be the alt. to $\overline{BC}$. The space resulting from the rotation is 2 cones with $r = AD$. One has height BD and one has height DC. Sin $\angle C = \frac{AD}{20} =$ $\frac{15}{25}$ so $AD = 12$; then $BD = \sqrt{15^2 - 12^2} = 9$ and $DC = 16$;

$V = \frac{1}{3}\pi \cdot 12^2 \cdot 9 + \frac{1}{3}\pi \cdot 12^2 \cdot 16 = \frac{144\pi}{3}(25) = 1200\pi$

40. The solid is made up of 2 $\cong$ cones, each with radius $\frac{s}{2}\sqrt{3}$ and height $\frac{s}{2}$.

$V = 2\left[\frac{1}{3}\pi\left(\frac{s}{2}\sqrt{3}\right)^2 \cdot \frac{s}{2}\right] = \frac{1}{4}\pi s^3$

Page 495 • CHALLENGE

Make a right circular cylinder with radius 1 cm and altitude 2 cm. Choose points A, B, C, and D so that quad. $ABCD$ is a square. Then determine points E and F so that $\overline{EF} \perp \overline{AB}$, as shown on the left, below. Cut along two planes: one determined by C, D, and E; the other determined by C, D, and F. The resulting solid looks like the diagram on the right, below. Square $ABCD$ passes through the square hole; circle O passes through the circular hole; triangle EFG passes through the triangular hole.

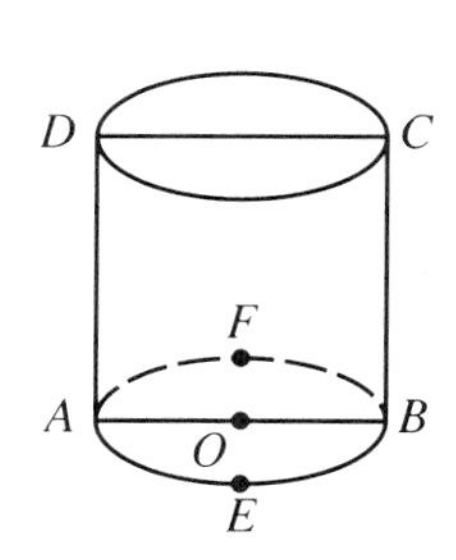

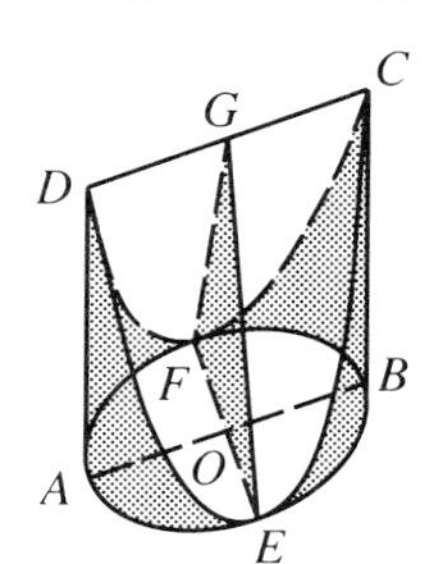

Page 496 • SELF-TEST 1

1. L.A. $= 2 \cdot 10 \cdot 4.5 + 2 \cdot 8 \cdot 4.5 = 162$; T.A. $= 162 + 2 \cdot 8 \cdot 10 = 322$; $V = 8 \cdot 10 \cdot 4.5 = 360$

2. L.A. $= \frac{1}{2} \cdot 4 \cdot 24 \cdot 13 = 624$; T.A. $= 624 + 24^2 = 1200$;

$V = \frac{1}{3} \cdot 24^2 \cdot \sqrt{13^2 - 12^2} = 960$

3. L.A. $= 2\pi \cdot 10 \cdot 7 = 140\pi$ in.^{2}; T.A. $= 140\pi + 2\pi \cdot 10^2 = 340\pi$ in.^{2};
$V = \pi \cdot 10^2 \cdot 7 = 700\pi$ in.^{3}

4. L.A. $= 6 \cdot 6 \cdot 5 = 180$ cm^2; T.A. $= 180 + 2 \cdot \frac{1}{2} \cdot 3\sqrt{3}(6 \cdot 6) = (180 + 108\sqrt{3})$ cm^2;

$V = \frac{108\sqrt{3}}{2} \cdot 5 = 270\sqrt{3}$ cm^3

5. $l = \sqrt{12^2 + 9^2} = \sqrt{225} = 15$; L.A. $= \pi \cdot 9 \cdot 15 = 135\pi$;

T.A. $= 135\pi + \pi \cdot 9^2 = 216\pi$; $V = \frac{1}{3} \cdot \pi \cdot 9^2 \cdot 12 = 324\pi$

6. $6e^2 = 2400$; $e = 20$; $V = 20^3 = 8000$ m^3

7. $V = \pi \cdot 2^2 \cdot 2 = 8\pi$; $\frac{1}{3}\pi(2)^2 \cdot h = 8\pi$; $h = 6$ **8.** $\frac{V_1}{V_2} = \frac{Bh_1}{\frac{1}{3}Bh_2} = \frac{2}{\frac{1}{3} \cdot 5} = \frac{6}{5}$

Page 496 • CALCULATOR KEY-IN

1. a. If the cylindrical layer of paint is cut and unrolled, its shape approximates that of a rect. solid with length = circumference of cylinder = $2\pi r$, width = height of cylinder = h, and height = thickness of paint layer = t. Thus, $V = 2\pi rht$.

b. Since the circumference of the cylinder is less than the circumference of the layer of paint, the figure is not a right prism. Therefore, $V \approx 2\pi rht$.

2. $V_1 \approx 2\pi \cdot 10 \cdot 12 \cdot (0.1) = 24\pi \approx 75.398$; $V_2 = 2\pi \cdot 10 \cdot 12 \cdot (0.01) = 2.4\pi \approx 7.5398$; $V_3 = 2\pi \cdot 10 \cdot 12 \cdot (0.001) = 0.24\pi \approx 0.75398$

3. $V_1 = \pi(10.1)^2 \cdot 12 - \pi(10)^2 \cdot 12 = 12\pi(10.1^2 - 10^2) = 24.12\pi \approx 75.775$; $V_2 = 12\pi(10.01^2 - 10^2) = 2.4012\pi \approx 7.5436$; $V_3 = 12\pi(10.001^2 - 10^2) = 0.240012\pi \approx 0.75402$

4. The exact volume is greater than the approximate volume.

Page 500 • CLASSROOM EXERCISES

	1.	2.	3.	4.	5.	6.
r	1	8	$3t$	3	5	10
A	4π	256π	$36t^2\pi$	36π	100π	400π
V	$\frac{4\pi}{3}$	$\frac{2048\pi}{3}$	$36t^3\pi$	36π	$\frac{500\pi}{3}$	$\frac{4000\pi}{3}$

7. $x = \sqrt{5^2 - 3^2} = 4; A = 16\pi\text{ cm}^2$ **8.** $x = \sqrt{17^2 - 8^2} = 15; A = 225\pi\text{ cm}^2$

9. $x = \sqrt{7^2 - 6^2} = \sqrt{13}; A = 13\pi\text{ cm}^2$

Pages 500–502 • WRITTEN EXERCISES

A

	1.	2.	3.	4.	5.	6.	7.	8.
r	7	5	$\frac{1}{2}$	$\frac{3}{4}k$	4	9	$\sqrt{2}$	6
A	196π	100π	π	$\frac{9k^2\pi}{4}$	64π	324π	8π	144π
V	$\frac{1372\pi}{3}$	$\frac{500\pi}{3}$	$\frac{\pi}{6}$	$\frac{9k^3\pi}{16}$	$\frac{256\pi}{3}$	972π	$\frac{8\pi\sqrt{2}}{3}$	288π

9. 4; 8 **10.** 9; 27 **11.** $4\pi\left(\frac{d}{2}\right)^2 = \pi; d = 1\text{ cm}$

12. $\frac{4}{3}\pi r^3 = 36\pi; r = 3; A = 4\pi \cdot 9 = 36\pi\text{ m}^2$

13. Let x = radius of circle formed; $r = 5; h = 2; x^2 = 5^2 - 2^2; A = \pi x^2 = 21\pi\text{ cm}^2$

14. Let x = radius of circle formed; $r = 8; h = 7; x^2 = 8^2 - 7^2 = 15; A = 15\pi\text{ cm}^2$

15. vol. of sphere $= \frac{4}{3}\pi \cdot 2^3 = \frac{32\pi}{3}$; vol. of hemisphere $= \frac{1}{2} \cdot \frac{4}{3}\pi \cdot 4^3 = \frac{128\pi}{3} =$ 4 · vol. of sphere

16. No; vol. of ice cream $= \frac{4}{3}\pi \cdot 3^3 = 36\pi\text{ cm}^3$; vol. of cone $= \frac{1}{3}\pi(2.5)^2 \cdot 12 = 25\pi\text{ cm}^3$

17. $0.7(4\pi \cdot 6380^2) = 113{,}972{,}320\pi \approx 358$ million km^2

18. a. $V = \frac{4}{3}\pi \cdot 3^3 \approx 113\text{ cm}^3$ **b.** $6^3 - 113 = 216 - 113 = 103\text{ cm}^3$

B **19.** vol. $= \pi \cdot 5^2 \cdot 20 + \frac{1}{2} \cdot \frac{4}{3}\pi \cdot 5^3 = 500\pi + \frac{250\pi}{3} = \frac{1750\pi}{3}; \frac{1750\pi}{3}\text{ m}^3$

20. area of hemisphere $= \frac{1}{2} \cdot 4\pi \cdot 5^2 = 50\pi$; area of cylinder $= 2\pi \cdot 5 \cdot 20 = 200\pi$; 2 cans cover $50\pi\text{ m}^2$; 8 cans are needed to cover $200\pi\text{ m}^2$.

21. Floor area $= \pi r^2$; ceiling area $= \frac{1}{2} \cdot 4\pi r^2 = 2\pi r^2$; 6 cans are needed for the ceiling.

22. Let x = radius of water's surface. $x^2 = 25^2 - 15^2 = 400; A = 400\pi$

23. a. $\frac{4}{3}\pi \cdot 8^3 = \frac{1}{3}\pi \cdot 8^2 \cdot h$; $4 \cdot 8 = h$; $h = 32$ cm **b.** Area of cone $=$ $\pi \cdot 8 \cdot \sqrt{1088} \approx 829$; area of sphere $= 4\pi \cdot 8^2 \approx 804$; $1.03(804) \approx 828$

24. Yes; vol. of 6 balls $= 6\left(\frac{4}{3}\pi r^3\right) = 8\pi r^3 =$ vol. of can

25. Radius $= r$; $h = 2r$; $V = \pi r^2 \cdot 2r = 2\pi r^3$

26. Let $r =$ radius of sphere $=$ radius of cylinder; height of cylinder $= 2r$; area of sphere $= 4\pi r^2$; L.A. of cylinder $= 2\pi rh = 2\pi r(2r) = 4\pi r^2$

27. $V = \pi r^2 \cdot 2r - 2 \cdot \frac{1}{3}\pi r^2 \cdot r = 2\pi r^3 - \frac{2}{3}\pi r^3 = \frac{4}{3}\pi r^3$

28. a. $\frac{4}{3}\pi \cdot 11^3 - \frac{4}{3}\pi \cdot 10^3 = \frac{4}{3}\pi(1331 - 1000) = \frac{1324\pi}{3}$; $\frac{1324\pi}{3}$ cm$^3 \approx 441.3\pi \approx$ 1386 cm^3 **b.** $4\pi \cdot 10^2 \cdot 1 = 400\pi \approx 1257$ cm^3; approx. is less **c.** thin

29. The solid formed is a sphere with $r = 4.5$; $A = 4\pi \cdot 4.5^2 = 81\pi$ in.^{2}; $V = \frac{4}{3}\pi(4.5)^3 = 121.5\pi$ in.^{3}

30. Rad. of cyl. $= \sqrt{10^2 - 6^2} = 8$; $V = \pi \cdot 8^2 \cdot 12 = 768\pi$

C 31. a. radius of base of cylinder $= \sqrt{10^2 - x^2} = \sqrt{100 - x^2}$; $V = \pi(\sqrt{100 - x^2})^2(2x) = 2\pi x(100 - x^2)$ **b.** max. $V =$ $2\pi\left(\frac{10\sqrt{3}}{3}\right)\left(100 - \left(\frac{10\sqrt{3}}{3}\right)^2\right) = 2\pi\left(\frac{10\sqrt{3}}{3}\right)\left(100 - \frac{100}{3}\right) = 2\pi\left(\frac{10\sqrt{3}}{3}\right)\left(\frac{200}{3}\right) =$ $\frac{4000\pi\sqrt{3}}{9}$

c. Answers may vary.

32. a. $h = 10 + x$; $r = \sqrt{10^2 - x^2} = \sqrt{100 - x^2}$; $V = \frac{1}{3}\pi(100 - x^2)(10 + x)$

b. max. $V = \frac{1}{3}\pi\left(100 - \frac{100}{9}\right)\left(10 + \frac{10}{3}\right) = \frac{1}{3}\pi\left(\frac{800}{9}\right)\left(\frac{40}{3}\right) = \frac{32{,}000\pi}{81}$.

c. Answers may vary.

33. $\triangle OXP$ is a rt. $\triangle$ with hyp. $\overline{OP}$.
$OP \cdot XZ = OX \cdot PX$ (Ex. 42, page 289),
so $25 \cdot XZ = 15 \cdot 20 = 300$; $XZ = 12$;
radius of $\odot$ of int. of 2 spheres $= 12$ cm;
$A = 144\pi$ cm^2

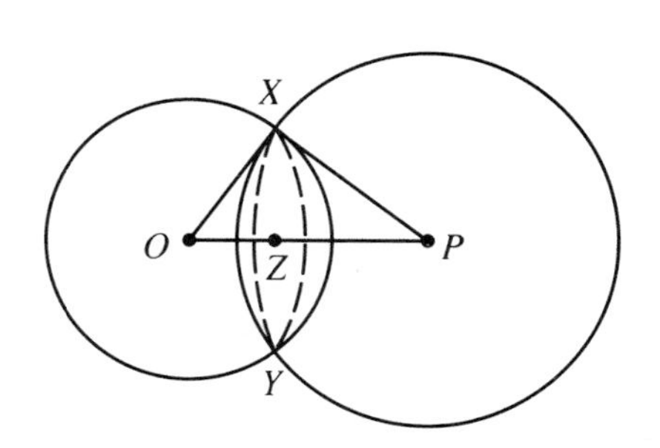

34. Draw $\overline{XY}$, a $\perp$ from ctr. of sphere to $\overline{VZ}$ where $VZ =$ slant height $= \sqrt{6^2 + 8^2} = 10$. Let $r =$ radius of sphere; $XY = XP = r$ and $VX = 8 - r$; $\triangle VPZ \sim \triangle VYX$;

$\frac{10}{8 - r} = \frac{6}{r}$; $10r = 48 - 6r$; $16r = 48$; $r = 3$;

$V = \frac{4}{3}\pi \cdot 3^3 = 36\pi \text{ cm}^3$

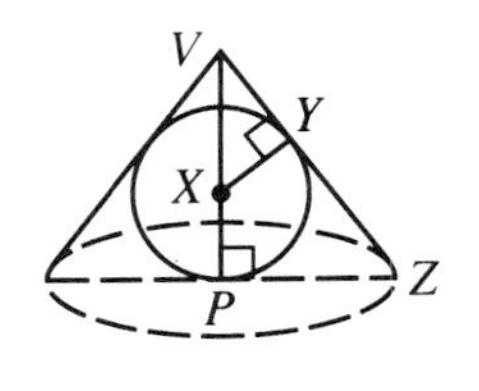

Page 502 • CHALLENGE

The second sub. would have to be in a solid hemisphere of rad. 400 m whose ctr. is the location of the first sub. and which lies in the rect. solid. For this hemisphere, $V_1 = \frac{1}{2} \cdot \frac{4}{3}\pi(400)^3 = \frac{128{,}000{,}000}{3}\pi \text{ m}^3$. For the rect. solid, $V_2 = 6000 \cdot 6000 \cdot 400 =$ $14{,}400{,}000{,}000 \text{ m}^3$. Prob. $= \frac{V_1}{V_2} = \frac{8\pi}{2700} \approx 0.009$

Page 503 • CALCULATOR KEY-IN

1. **a.** $s = 20$ mm; $A = \frac{s^2}{4}\sqrt{3} = 100\sqrt{3} \approx 173 \text{ mm}^2$ **b.** $x = 15$ mm; $A = x^2 = 225 \text{ mm}^2$ **c.** $y = 10$ mm; $A = \frac{1}{2}(60) \cdot 5\sqrt{3} = 150\sqrt{3} \approx 260 \text{ mm}^2$

d. $r = \frac{60}{2\pi} = \frac{30}{\pi} \approx 9.549$ mm; $A = \pi\left(\frac{30}{\pi}\right)^2 = \frac{900}{\pi} \approx 286 \text{ mm}^2$

2. **a.** $s \approx 18.61$ mm; $V \approx 760 \text{ mm}^3$ **b.** $x = 10$ mm; $V = 1000 \text{ mm}^3$ **c.** $r \approx 6.910$ mm; $V \approx 1382 \text{ mm}^3$ **d.** total surface area; sphere; greatest possible volume

3. **a.** $\frac{s^2}{4}\sqrt{3} = 900$; $s \approx 45.59$; $p \approx 137$ cm **b.** $x^2 = 900$; $x = 30$; $p = 120$ cm

c. $6 \cdot \frac{1}{2} \cdot y \cdot \frac{y}{2}\sqrt{3} = 900$; $y \approx 18.61$; $p \approx 112$ cm **d.** $\pi r^2 = 900$; $r \approx 16.93$; $c \approx 106$ cm **e.** Of all plane figures with a fixed area, the $\odot$ has the least possible perimeter (circumference).

4. **a.** $\frac{s^3\sqrt{2}}{12} = 1000$; $s \approx 20.40$; T.A. $\approx s^2\sqrt{3} \approx 721 \text{ cm}^2$ **b.** $x^3 = 1000$; $x = 10$; T.A. $= 6x^2 = 600 \text{ cm}^2$ **c.** $\frac{4}{3}\pi r^3 = 1000$; $r \approx 6.20$; T.A. $= 4\pi r^2 \approx 484 \text{ cm}^2$

d. Of all solid figures with a fixed volume, the sphere has the least possible total surface area.

Pages 504–505 • COMPUTER KEY-IN

1–3. Answers may vary, due to rounding.

1. Vol. of discs ≈ 4492.43; vol. of sphere ≈ 4188.79; $\approx 7\%$

2. Yes **a.** 4343.24818 **b.** 4251.19959 **c.** 4220.09785 **d.** 4191.9272

3. **a.** $V \approx 4157.26605$ **b.** Yes; average: 4188.68195; actual: 4188.7867

Page 506 • APPLICATION

1. **a.** $F = 6, V = 8, E = 12; 6 + 8 - 12 = 2$ **b.** $F = 8, V = 6, E = 12;$ $8 + 6 - 12 = 2$ **c.** $F = 20, V = 12, E = 30; 20 + 12 - 30 = 2$

2. Answers may vary.

3. **a.** No; a framework of hexagons only is not possible. Suppose $F + V - E = 2$. Then $n + \frac{6n}{3} - \frac{6n}{2} = 2$, and $n + 2n - 3n = 2$. But $n + 2n - 3n = 0$, so $F + V - E \neq 2$. **b.** 12 hexagon vertices are lost, but each is a vertex of 3 hexagons, so 4 vertices are lost and $V = \frac{6n}{3} - 4 = \frac{6n - 12}{3}$. 12 edges are lost, but each is an edge of 2 hexagons, so 6 edges are lost and $E = \frac{6n}{2} - 6 = \frac{6n - 12}{2}$. Then $F + V - E = n + \frac{6n - 12}{3} - \frac{6n - 12}{2} = n + 2n - 4 - 3n + 6 = 2$.

Page 507 • MIXED REVIEW EXERCISES

1. $\frac{12}{18} = \frac{2}{3}$ 2. $w = \sqrt{8^2 + 6^2} = 10$ 3. $\frac{10}{x} = \frac{2}{3}; x = 15; \frac{4}{y} = \frac{2}{3}; y = 6; \frac{6}{z} = \frac{2}{3}; z = 9$

4. **a.** perimeter of $ABCD = 10 + 4 + 6 + 12 = 32$; perimeter of $PQRS =$ $15 + 6 + 9 + 18 = 48$ **b.** $\frac{32}{48} = \frac{2}{3}$ **c.** They are both $\frac{2}{3}$.

5. **a.** Area of $ABCD = \frac{1}{2} \cdot 6(4 + 12) = 48$; area of $PQRS = \frac{1}{2} \cdot 9(6 + 18) = 108$ **b.** $\frac{48}{108} = \frac{4}{9}$ **c.** $\frac{4}{9} = \left(\frac{2}{3}\right)^2$; the ratio of the areas is the square of the scale factor.

6. **a.** True **b.** False **c.** True **d.** False **e.** False **f.** True **g.** True **h.** False

Page 510 • CLASSROOM EXERCISES

1. No; $\frac{6}{3} \neq \frac{12}{8}$ 2. Yes; bases are $\sim$ △s; $\frac{12}{8} = \frac{9}{6} = \frac{15}{10} = \frac{3}{2}$

3. **a.** $\frac{3^2}{2^2} = \frac{9}{4}$ **b.** $\frac{3^2}{2^2} = \frac{9}{4}$ **c.** $\frac{3^3}{2^3} = \frac{27}{8}$ 4. **a.** $\frac{12^2}{18^2} = \frac{2^2}{3^2} = \frac{4}{9}$ **b.** $\frac{2^3}{3^3} = \frac{8}{27}$

5. **a.** $\frac{2\pi}{16\pi} = \frac{1}{8}$ **b.** $\frac{a^3}{b^3} = \frac{1}{8} = \frac{1^3}{2^3}$, so $\frac{d_1}{d_2} = \frac{1}{2}$ **c.** $\frac{1^2}{2^2} = \frac{1}{4}$

	6.	7.	8.	9.	10.	11.
Scale factor	3 : 4	5 : 7	2 : 1	1 : 6	2 : 3	2 : 5
Ratio of base circum.	3 : 4	5 : 7	2 : 1	1 : 6	2 : 3	2 : 5
Ratio of slant hts.	3 : 4	5 : 7	2 : 1	1 : 6	2 : 3	2 : 5
Ratio of lat. areas	9 : 16	25 : 49	4 : 1	1 : 36	4 : 9	4 : 25
Ratio of total areas	9 : 16	25 : 49	4 : 1	1 : 36	4 : 9	4 : 25
Ratio of volumes	27 : 64	125 : 343	8 : 1	1 : 216	8 : 27	8 : 125

12. **a.** 1 : 2 **b.** 1 : 4 **c.** 1 : 3 **d.** 1 : 7

13. **a.** 3 : 5 **b.** 9 : 25 **c.** 9 : 16 **d.** 27 : 98

Pages 511–513 • WRITTEN EXERCISES

A 1. Yes 2. No 3. **a.** 3 : 4 **b.** 3 : 4 **c.** 9 : 16 **d.** 27 : 64

4. **a.** 4 : 9 **b.** 4 : 9 **c.** 4 : 9 **d.** 8 : 27 5. **a.** 4 : 1 **b.** 16 : 1 **c.** 64 : 1

6. **a.** 3 : 4 **b.** 9 : 16 **c.** 27 : 64 7. **a.** 2 : 3 **b.** 2 : 3 **c.** 4 : 9

8. **a.** 1 : 5 **b.** 1 : 25 **c.** 1 : 25

9. $1^2 : 200^2$ or 1 : 40,000, so the paint for actual airplane $=$ 40,000 $\times$ paint for model.

10. $\frac{1^3}{48^3} = \frac{90}{x}$; $x = 9{,}953{,}280 \text{ in.}^3 = \frac{9{,}953{,}280}{1728} = 5760 \text{ ft}^3$

11. $\frac{4}{6} = \frac{2}{3}$; $\frac{36\pi}{x} = \frac{4}{9}$; $x = 81\pi \text{ cm}^2$

B 12. $\frac{2}{4\sqrt{3}} = \frac{\sqrt{3}}{6}$; $\frac{(\sqrt{3})^3}{6^3} = \frac{x}{64}$; $x = \frac{8\sqrt{3}}{9} \text{ cm}^3$ 13. $\frac{6}{10} = \frac{3}{5}$; $\frac{3^3}{5^3} = \frac{4}{x}$; $x = 18.5$ kg

14. $\frac{30}{40} = \frac{3}{4}$; $\frac{6}{x} = \frac{27}{64}$; $x = 14\frac{2}{9}$ kg; $\frac{30}{50} = \frac{3}{5}$; $\frac{6}{y} = \frac{27}{125}$; $y = 27\frac{7}{9}$ kg; $6 + 14\frac{2}{9} + 27\frac{7}{9} = 48$ kg

15. $\frac{6}{12} = \frac{1}{2}$; vol. ratio: $\frac{1}{8}$; the larger ball has 8 times as much string; it is the better buy.

16. **a.** $\frac{h^2}{(3h)^2} = \frac{1}{9}$; 9 **b.** $\frac{h^3}{(3h)^3} = \frac{1}{27}$; 27 **c.** the smaller column; strength per unit of vol.: smaller: $\frac{s}{v}$, larger: $\frac{9s}{27v} = \frac{1}{3}\left(\frac{s}{v}\right)$

17. $\frac{8}{18} = \frac{4}{9}$; scale factor: $\frac{2}{3}$; $\frac{2^3}{3^3} = \frac{32}{x}$; $x = 108 \text{ ft}^3$

18. $\frac{12\pi}{96\pi} = \frac{1}{8} = \frac{1^3}{2^3}$, so scale factor $=$ 1 : 2; ratio of lateral areas $= 1^2 : 2^2 =$ 1 : 4; let $a =$ lateral area of larger cone; $\frac{15\pi}{a} = \frac{1}{4}$; $a = 60\pi$

19–20. The subscripts "t," "b," and "w" refer to "top," "bottom," and "whole."

19. a. scale factor $= 9 : 12 = 3 : 4$; $A_t : A_b = 3^2 : 4^2 = 9 : 16$
b. $\text{L.A.}_t : \text{L.A.}_w = 3^2 : 4^2 = 9 : 16$ **c.** $\text{L.A.}_t : (\text{L.A.}_t + \text{L.A.}_b) = 9 : 16$; $\text{L.A.}_t : \text{L.A.}_b = 9 : 7$ **d.** $V_t : V_w = 3^3 : 4^3 = 27 : 64$ **e.** $V_t : (V_t + V_b) = 27 : 64$; $V_t : V_b = 27 : 37$

20. a. scale factor $= 10 : 14 = 5 : 7$; $A_t : A_b = 5^2 : 7^2 = 25 : 49$
b. $\text{L.A.}_t : \text{L.A.}_w = 5^2 : 7^2 = 25 : 49$ **c.** $\text{L.A.}_t : (\text{L.A.}_t + \text{L.A.}_b) = 25 : 49$; $\text{L.A.}_t : \text{L.A.}_b = 25 : 24$ **d.** $V_t : V_w = 5^3 : 7^3 = 125 : 343$ **e.** $V_t : (V_t + V_b) = 125 : 343$; $V_t : V_b = 125 : 218$

21. scale factor of top to whole $= 9 : 15 = 3 : 5$; ratio of vol. of top to vol. of whole $= 3^3 : 5^3 = 27 : 125$. Let v = vol. of top; $\frac{v}{250} = \frac{27}{125}$; $125v = 6750$; $v = 54 \text{ cm}^3$; vol. of bottom $= 250 - 54 = 196 \text{ cm}^3$

22. ratio of areas $= 4\pi a^2 : 4\pi b^2 = a^2 : b^2$

23. ratio of volumes $= \frac{4}{3}\pi a^3 : \frac{4}{3}\pi b^3 = a^3 : b^3$

24. scale factor $= h_1 : h_2$ so $r_1 : r_2 = h_1 : h_2$ and $r_1{}^2 : r_2{}^2 = h_1{}^2 : h_2{}^2$; ratio of volumes $= \frac{1}{3}\pi r_1{}^2 h_1 : \frac{1}{3}\pi r_2{}^2 h_2 = r_1{}^2 h_1 : r_2{}^2 h_2 = h_1{}^3 : h_2{}^3$.

25. scale factor $= r_1 : r_2$ so $l_1 : l_2 = r_1 : r_2$; ratio of lateral areas $= \pi r_1 l_1 : \pi r_2 l_2 = r_1 l_1 : r_2 l_2 = r_1{}^2 : r_2{}^2$.

26. scale factor $= e_1 : e_2$ so $h_1 : h_2 = e_1 : e_2$; ratio of perimeters $= p_1 : p_2 = 5e_1 : 5e_2 = e_1 : e_2$; ratio of lateral areas $= p_1 h_1 : p_2 h_2 = e_1 \cdot e_1 : e_2 \cdot e_2 = e_1{}^2 : e_2{}^2$.

27. ratio of area of bases = ratio of lateral areas $= e_1{}^2 : e_2{}^2$; ratio of volumes $= B_1 h_1 : B_2 h_2 = e_1{}^2 \cdot e_1 : e_2{}^2 \cdot e_2 = e_1{}^3 : e_2{}^3$

C 28. Plane $XYZ \parallel$ plane ABC and plane VAC intersects the $\parallel$ planes in $\overleftrightarrow{XZ}$ and $\overleftrightarrow{AC}$. If 2 $\parallel$ planes are cut by a third plane, the lines of intersection are $\parallel$. Then $\overline{XZ} \parallel \overline{AC}$ and $\triangle VXZ \sim \triangle VAC$ so $\frac{VZ}{VC} = \frac{VX}{VA} = \frac{1}{k}$ or $VC = k \cdot VZ$ and $\frac{XZ}{AC} = \frac{VX}{VA} = \frac{1}{k}$ or $AC = k \cdot XZ$. Similarly, it can be shown that $VB = k \cdot VY$, $CB = k \cdot ZY$, and $AB = k \cdot XY$. By the SSS $\sim$ Thm., the bases of V-XYZ and V-ABC are $\sim$ and corr. lengths are prop., so V-$XYZ \sim V$-ABC.

29. Let h = height of top; scale factor of top to whole $= \frac{h}{12}$; ratio of volumes $= \frac{h^3}{12^3} = \frac{h^3}{1728}$; ratio of volumes of top to whole = ratio of volumes of top to top plus bottom; ratio of top to bottom $= \frac{h^3}{1728 - h^3}$. But volumes are $=$, so $\frac{h^3}{1728 - h^3} = 1$; $h^3 = 1728 - h^3$; $2h^3 = 1728$; $h^3 = 864$; $h = \sqrt[3]{864} = \sqrt[3]{4 \cdot 8 \cdot 27} = 6\sqrt[3]{4}$.

Page 513 • SELF-TEST 2

1. $A = 4\pi r^2 = 4\pi \cdot 9 = 36\pi \text{ cm}^2$; $V = \frac{4}{3}\pi r^3 = \frac{4}{3}\pi \cdot 27 = 36\pi \text{ cm}^3$

2. $\frac{4}{3}\pi r^3 = \frac{32}{3}\pi$; $r = 2$; $A = 4\pi \cdot 4 = 16\pi \text{ m}^2$

3. $V = \frac{4}{3}\pi(10)^3 + \pi \cdot (10)^2 \cdot 60 = \frac{22{,}000\pi}{3} \text{ cm}^3$

4. $A = \pi\left(\sqrt{13^2 - 12^2}\right)^2 = 25\pi \text{ cm}^2$

5. **a.** $\frac{6}{4} = \frac{3}{2} = \frac{8}{x}$; $x = \frac{16}{3}$ **b.** $\frac{3^2}{2^2} = \frac{9}{4}$ 6. **a.** $\frac{32}{200} = \frac{4}{25}$, $\frac{\sqrt{4}}{\sqrt{25}} = \frac{2}{5}$ **b.** $\frac{2^3}{5^3} = \frac{8}{125}$

Page 513 • CHALLENGE

The model is a regular square pyramid.

Page 515 • CALCULATOR KEY-IN

1. Check students' drawings. $VO = 3$; $DB = 4 - x$; $y = \frac{3}{4}(4 - x)$;

   $A = 2xy = 2x\left(\frac{3}{4}(4 - x)\right) = \frac{3x(4 - x)}{2}$

2. The greatest area, 6, occurs when $x = 2$.

Page 515 • COMPUTER KEY-IN

1. $x = 2$ $(V \approx 16.76)$

2. **a.** Check students' drawings. $\frac{y}{3} = \frac{4 - x}{4}$;

   $y = \frac{3(4 - x)}{4}$; $V = \pi x^2 y = \pi x^2 \cdot \frac{3(4 - x)}{4} =$

   $\frac{3}{4}\pi x^2(4 - x)$ **b.** $x = 2.75$ $(V \approx 22.27)$

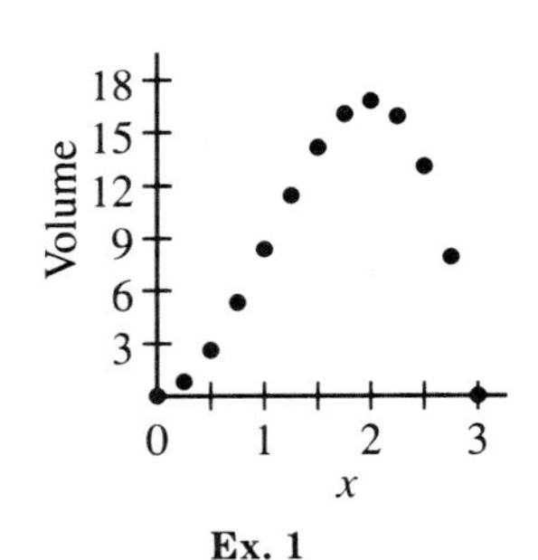

Ex. 1

Page 517 • EXTRA

1. $V = \frac{1}{3}Bh = \frac{1}{3}\left(\frac{1}{2} \cdot 3 \cdot 4 \cdot 11\right) = 22$ 2. $V = \pi \cdot 5^2 \cdot 8 = 200\pi$

3. $V = \frac{1}{3}\pi \cdot 4^2(3.5) = \frac{56\pi}{3}$ 4. $h = \frac{15\sqrt{3}}{2}$; $V = 3^2 \cdot \frac{15\sqrt{3}}{2} = \frac{135\sqrt{3}}{2}$

5. $h = \frac{24}{2} = 12$; $12 \cdot B = 96$; $B = 8$; 8 cm^2

6. The area of each cross section of the sphere equals the difference of the areas of the corresponding cross sections of the cylinder and the double cone. All three solids have equal heights, namely, $2r$. Then by Cavalieri's Principle, the volume of the sphere is equal to the difference between the volumes of the cylinder and the double cone.

Pages 518–519 • CHAPTER REVIEW

1. lateral edge 2. $p = 8 \cdot 7 = 56$; L.A. $= 56 \cdot 12 = 672$
3. T.A. $= 28 \cdot 5 + 2 \cdot 8 \cdot 6 = 236$; $V = 8 \cdot 6 \cdot 5 = 240$
4. $81h = 891$; $h = 11$; T.A. $= 2 \cdot 81 + 36 \cdot 11 = 558$
5. $B = \frac{1}{2} \cdot 8 \cdot 4\sqrt{3} = 16\sqrt{3}$; $V = \frac{1}{3} \cdot 16\sqrt{3} \cdot 10 = \frac{160\sqrt{3}}{3}$
6. $l = \sqrt{5^2 - 3^2} = 4$; L.A. $= \frac{1}{2} \cdot 5 \cdot 6 \cdot 4 = 60$
7. $B = 30^2 = 900$; L.A. $= 1920 - 900 = 1020$; $l = \frac{2 \cdot 1020}{4 \cdot 30} = 17$
8. $h = \sqrt{17^2 - 15^2} = 8$; $V = \frac{1}{3} \cdot 900 \cdot 8 = 2400$
9. L.A. $= 2\pi \cdot 4 \cdot 3 = 24\pi$; T.A. $= 24\pi + 2\pi \cdot 4^2 = 56\pi$
10. L.A. $= \pi \cdot 6 \cdot 10 = 60\pi$ cm^2; T.A. $= 60\pi + \pi \cdot 6^2 = 96\pi$ cm^2;

    $h = \sqrt{10^2 - 6^2} = 8$; $V = \frac{1}{3}\pi \cdot 6^2 \cdot 8 = 96\pi$ cm^3
11. $\frac{1}{3}\pi r^2 \cdot 6 = 8\pi$; $r^2 = 4$; $r = 2$; $l = \sqrt{6^2 + 2^2} = \sqrt{40} = 2\sqrt{10}$ cm
12. $V_1 = \pi r^2 h$; $V_2 = \pi(2r)^2\left(\frac{h}{2}\right) = 2\pi r^2 h$; the volume is doubled.
13. $A = 4\pi \cdot 7^2 = 196\pi \approx 616$ m^2 14. $V = \frac{4}{3}\pi\left(\frac{12}{2}\right)^3 = 288\pi$; 288π ft^3
15. $4\pi r^2 = 484\pi$; $r^2 = 121$; $r = 11$; $V = \frac{4}{3}\pi \cdot 11^3 = \frac{5324\pi}{3}$; $\frac{5324\pi}{3}$ cm^3
16. $1 : 3$ 17. $1^2 : 3^2 = 1 : 9$
18. $\frac{\text{vol. of top}}{\text{vol. of whole}} = \frac{\text{vol. of top}}{\text{vol. of top} + \text{vol. of bottom}} = \frac{1}{27}$; 27(vol. of top) $=$ vol. of top $+$

    vol. of bottom; 26(vol. of top) $=$ vol. of bottom; $\frac{\text{vol. of small pyramid}}{\text{vol. of bottom part}} = 1 : 26$
19. scale factor $= \sqrt{48} : \sqrt{27} = 4\sqrt{3} : 3\sqrt{3} = 4 : 3$; ratio of volumes $= 4^3 : 3^3 =$ $64 : 27$

Page 519 • CHAPTER TEST

1. $V = (2k)^3 = 8k^3$; T.A. $= 6(2k)^2 = 24k^2$ **2.** $135 = \frac{1}{3}Bh = \frac{1}{3} \cdot 9 \cdot h$; $h = 45$ cm

3. $V = \frac{1}{3}\pi \cdot 8^2 \cdot 6 = 128\pi$

4. $l = \sqrt{6^2 + 8^2} = 10$; L.A. $= \pi \cdot 8 \cdot 10 = 80\pi$; T.A. $= 80\pi + \pi \cdot 8^2 = 144\pi$

5. $p = 5 + 12 + 13 = 30$; T.A. $= 30 \cdot 20 + 2\left(\frac{1}{2} \cdot 5 \cdot 12\right) = 660$

6. $\frac{1}{2} \cdot 5 \cdot 12 \cdot 20 = 600$ **7.** L.A. $= 2\pi \cdot 6 \cdot 4 = 48\pi$; 48π cm^2

8. $V = \pi \cdot 6^2 \cdot 4 = 144\pi$; 144π cm^3

9. $\frac{1}{2} \cdot 24l = 60$; $l = 5$; $h = \sqrt{5^2 - 3^2} = 4$; $V = \frac{1}{3} \cdot 6^2 \cdot 4 = 48$; 48 m^3

10. $A = 4\pi \cdot 6^2 = 144\pi$; 144π cm^2; $V = \frac{4}{3}\pi \cdot 6^3 = 288\pi$; 288π cm^3 **11.** No; $\frac{12}{18} \neq \frac{18}{24}$

12. scale factor $= 6 : 18 = 1 : 3$; ratio of areas $= 1^2 : 3^2 = 1 : 9$; let $a =$ total area of smaller pyramid; $\frac{a}{648} = \frac{1}{9}$; $9a = 648$; $a = 72$

13. $\frac{1000}{64} = \frac{125}{8} = \frac{5^3}{2^3}$, so scale factor $= 5 : 2$; ratio of lateral areas $= 5^2 : 2^2 = 25 : 4$

14. ratio of volumes $= \frac{1}{3}\pi \cdot 3^2 \cdot 4 : \pi \cdot 3^2 \cdot 4 = \frac{1}{3} : 1$ or $1 : 3$; $l = \sqrt{3^2 + 4^2} = 5$; ratio of lateral areas $= \pi \cdot 3 \cdot 5 : 2\pi \cdot 3 \cdot 4 = 15 : 24$ or $5 : 8$

15. $4\pi r^2 = 9\pi$; $r^2 = \frac{9}{4}$; $r = \frac{3}{2}$; $V = \frac{4}{3}\pi\left(\frac{3}{2}\right)^3 = \frac{9\pi}{2}$

16. $2\pi \cdot 7^2 + 2\pi \cdot 7h = 168\pi$; $98\pi + 14\pi \cdot h = 168\pi$; $14\pi \cdot h = 70\pi$; $h = 5$; 5 cm

Page 520 • PREPARING FOR COLLEGE ENTRANCE EXAMS

1. A. $V = \frac{1}{3}\pi r^2 \cdot 15 = 320\pi$; $r^2 = 64$; $r = 8$; T.A. $= \pi \cdot 8 \cdot \sqrt{8^2 + 15^2} + \pi \cdot 8^2 = 136\pi + 64\pi = 200\pi$

2. C. $6^2 : (9\sqrt{3})^2 = 36 : 243 = 4 : 27$

3. E. $V = \frac{4}{3}\pi r^3 = 288\pi$; $r^3 = 216$; $r = 6$; $d = 2r = 12$

4. D. $A = \frac{1}{2} \cdot 2\sqrt{3} \cdot (2 + 4) = 6\sqrt{3}$

5. B. $A = (y\sqrt{2})^2 + \frac{1}{2} \cdot y \cdot y = 2y^2 + \frac{1}{2}y^2 = \frac{5}{2}y^2$

6. C. $V_A = \frac{1}{3}Bh = \frac{1}{3} \cdot y^2 \cdot 3x = y^2x$; $V_B = Bh = y^2x$

7. B. $A_A = \frac{1}{2}bh = \frac{1}{2} \cdot 3\pi \cdot 6 = 9\pi$; $A_B = \frac{x}{360} \cdot \pi r^2 = \frac{120}{360} \cdot \pi \cdot 6^2 = 12\pi$

Page 521 • CUMULATIVE REVIEW: CHAPTERS 1–12

A **1.** True **2.** True **3.** False **4.** True **5.** False **6.** True **7.** False **8.** True **9.** True

B **10.** $\frac{3}{2}x = x + 17$; $3x = 2x + 34$; $x = 34$; $x + 17 = 51$; $m\angle K = 180 - 51 = 129$

11.

Statements	Reasons
1. $\overline{WZ} \perp \overline{ZY}$; $\overline{WX} \perp \overline{XY}$; $\overline{WX} \cong \overline{YZ}$	1. Given
2. $\angle X$ and $\angle Z$ are rt. ∠s.	2. Def. of ⊥ lines
3. $\triangle WXY$ and $\triangle YZW$ are rt. △s.	3. Def. of rt. △
4. $\overline{WY} \cong \overline{WY}$	4. Refl. Prop.
5. $\triangle WXY \cong \triangle YZW$	5. HL Thm.
6. $\angle XYW \cong \angle ZWY$	6. Corr. parts of ≅ △s are ≅.
7. $\overline{WZ} \parallel \overline{XY}$	7. If 2 lines are cut by a trans. and alt. int. ∠s are ≅, then the lines are ∥.

12. Given: ▱$ABCD$; $\overline{AC} \perp \overline{BD}$

Prove: $ABCD$ is a rhombus.

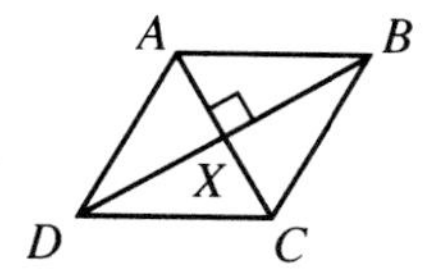

Statements	Reasons
1. $ABCD$ is a ▱.	1. Given
2. $\overline{AX} \cong \overline{CX}$	2. Diags. of a ▱ bis. each other.
3. $\overline{AC} \perp \overline{BD}$	3. Given
4. $\angle AXB \cong \angle CXB$	4. If 2 lines are ⊥, then they form ≅ adj. ∠s.
5. $\overline{BX} \cong \overline{BX}$	5. Refl. Prop.
6. $\triangle AXB \cong \triangle CXB$	6. SAS Post.
7. $\overline{AB} \cong \overline{CB}$	7. Corr. parts of ≅ △s are ≅.
8. $ABCD$ is a rhombus.	8. If 2 consec. sides of a ▱ are ≅, then the ▱ is a rhombus.

13. **a.** AA ~ Post. **b.** If $\triangle JKL \sim \triangle XYZ$, then $\angle J \cong \angle X$ and $\angle K \cong \angle Y$. True

14. $\frac{12}{x} = \frac{9}{14 - x}$; $x = 8$ **15.** $(7.5)(3.2) = x(11 - x)$; $x = 8$

16. The locus is 0 pts., 1 pt. or 2 pts., depending on the int. of a line through the int. of the $\angle$ bis. of $\triangle ABC$ and $\perp$ to the plane of $\triangle ABC$, and a sphere with ctr. A and rad. 4 cm.

17. $AB = 8$

18. $\text{T.A.} = 10^2 + \frac{1}{2}(4 \cdot 10) \cdot \sqrt{12^2 + 5^2} = 360$; $V = \frac{1}{3} \cdot 10^2 \cdot 12 = 400$

19. $100\pi = 2\pi r \cdot r + 2\pi r^2$; $r = h = 5$; $V = \pi \cdot 5^2 \cdot 5 = 125\pi \text{ cm}^3$

20. $r = 0.9$; $A = 4\pi r^2 \approx 4(3.14) \cdot (0.9)^2 \approx 10 \text{ cm}^2$

CHAPTER 13 • Coordinate Geometry

Pages 525–526 • CLASSROOM EXERCISES

1. C is pt. $(-4, 4)$; -4.
2. C is pt. $(-4, 4)$; every pt. on a horiz. line through C has y-coordinate 4; $(2, 4)$, $(0, 4)$, and $(15, 4)$.
3. $OD = 2$; $BF = 5$ **4. a.** $(5, 1)$ **b.** 5, 3 **c.** $\sqrt{5^2 + 3^2} = \sqrt{34}$
5. **a.** $(7, 2)$ **b.** 8, 3 **c.** $\sqrt{8^2 + 3^2} = \sqrt{73}$
6. **a.** $(5, 2)$ **b.** 4, 3 **c.** $\sqrt{4^2 + 3^2} = 5$
7. **a.** $(2, -1)$ **b.** 6, 2 **c.** $\sqrt{6^2 + 2^2} = \sqrt{40} = 2\sqrt{10}$
8. **a.** $(-3, -3)$ **b.** 6, 6 **c.** $\sqrt{6^2 + 6^2} = \sqrt{72} = 6\sqrt{2}$
9. **a.** $(-3, -2)$ **b.** 2, 4 **c.** $\sqrt{4^2 + 2^2} = \sqrt{20} = 2\sqrt{5}$
10. **a.** $\sqrt{(5 - 0)^2 + (-3 - 0)^2} = \sqrt{5^2 + (-3)^2} = \sqrt{25 + 9} = \sqrt{34}$
    b. $|-5 - 3| = 8$ **c.** $\sqrt{(-3 - 4)^2 + (-3 - 4)^2} = \sqrt{(-7)^2 + (-7)^2} = \sqrt{49 + 49} = 7\sqrt{2}$
11. **a.** $(2, 0)$; $\sqrt{1} = 1$ **b.** $(-2, 8)$; $\sqrt{16} = 4$ **c.** $(0, -5)$; $\sqrt{112} = 4\sqrt{7}$
    d. $(-3, -7)$; $\sqrt{14}$
12. **a.** $(x - 2)^2 + (y - 5)^2 = 9$ **b.** $(x + 2)^2 + y^2 = 25$
    c. $(x + 2)^2 + (y - 3)^2 = 100$ **d.** $(x - j)^2 + (y - k)^2 = n^2$

Pages 526–528 • WRITTEN EXERCISES

A 1. $|4 - (-3)| = 7$ **2.** $|-2 - 3| = 5$ **3.** $|-1 - 3| = 4$

4. $\sqrt{3^2 + 4^2} = \sqrt{25} = 5$
5. $\sqrt{(-7 - (-6))^2 + (-5 - (-2))^2} = \sqrt{(-1)^2 + (-3)^2} = \sqrt{1 + 9} = \sqrt{10}$
6. $\sqrt{(5 - 3)^2 + (-2 - 2)^2} = \sqrt{2^2 + (-4)^2} = \sqrt{4 + 16} = \sqrt{20} = 2\sqrt{5}$
7. $\sqrt{(0 - (-8))^2 + (0 - 6)^2} = \sqrt{8^2 + (-6)^2} = \sqrt{64 + 36} = \sqrt{100} = 10$
8. $\sqrt{(0 - 12)^2 + (-6 - (-1))^2} = \sqrt{(-12)^2 + (-5)^2} = \sqrt{144 + 25} = \sqrt{169} = 13$
9. $\sqrt{(1 - 5)^2 + (-2 - 4)^2} = \sqrt{(-4)^2 + (-6)^2} = \sqrt{16 + 36} = \sqrt{52} = 2\sqrt{13}$
10. $\sqrt{(5 - (-2))^2 + (7 - (-2))^2} = \sqrt{7^2 + 9^2} = \sqrt{49 + 81} = \sqrt{130}$
11. $\sqrt{(3 - (-2))^2 + (-2 - 3)^2} = \sqrt{5^2 + (-5)^2} = \sqrt{25 + 25} = 5\sqrt{2}$
12. $|3 - (-1)| = 4$
13. $AB = \sqrt{(-2 - 0)^2 + (1 - 3)^2} = \sqrt{(-2)^2 + (-2)^2} = \sqrt{4 + 4} = 2\sqrt{2}$;
    $BC = \sqrt{(3 - (-2))^2 + (6 - 1)^2} = \sqrt{5^2 + 5^2} = \sqrt{25 + 25} = 5\sqrt{2}$;
    $AC = \sqrt{(3 - 0)^2 + (6 - 3)^2} = \sqrt{3^2 + 3^2} = \sqrt{9 + 9} = 3\sqrt{2}$;
    $AB + AC = 2\sqrt{2} + 3\sqrt{2} = 5\sqrt{2} = BC$, so the pts. are collinear and A lies between B and C.

14. $AB = \sqrt{(0 - 5)^2 + (5 - (-5))^2} = \sqrt{(-5)^2 + 10^2} = \sqrt{25 + 100} = 5\sqrt{5}$;
$BC = \sqrt{(2 - 0)^2 + (1 - 5)^2} = \sqrt{2^2 + (-4)^2} = \sqrt{4 + 16} = 2\sqrt{5}$;
$AC = \sqrt{(2 - 5)^2 + (1 - (-5))^2} = \sqrt{(-3)^2 + 6^2} = \sqrt{9 + 36} = 3\sqrt{5}$;
$BC + AC = 2\sqrt{5} + 3\sqrt{5} = 5\sqrt{5} = AB$, so the pts. are collinear and C lies between A and B.

15. $AB = \sqrt{(0 - (-5))^2 + (2 - 6)^2} = \sqrt{5^2 + (-4)^2} = \sqrt{25 + 16} = \sqrt{41}$;
$BC = \sqrt{(3 - 0)^2 + (0 - 2)^2} = \sqrt{3^2 + (-2)^2} = \sqrt{9 + 4} = \sqrt{13}$;
$AC = \sqrt{(3 - (-5))^2 + (0 - 6)^2} = \sqrt{8^2 + (-6)^2} = \sqrt{64 + 36} = 10$;
$AB + BC \neq AC$, $AB + AC \neq BC$, and $BC + AC \neq AB$, so the pts. are not collinear.

16. $AB = \sqrt{(-3 - 3)^2 + (0 - 4)^2} = \sqrt{(-6)^2 + (-4)^2} = \sqrt{36 + 16} = 2\sqrt{13}$;
$BC = \sqrt{(-1 - (-3))^2 + (1 - 0)^2} = \sqrt{2^2 + 1^2} = \sqrt{4 + 1} = \sqrt{5}$;
$AC = \sqrt{(-1 - 3)^2 + (1 - 4)^2} = \sqrt{(-4)^2 + (-3)^2} = \sqrt{16 + 9} = 5$;
$AB + BC \neq AC$, $AB + AC \neq BC$, and $BC + AC \neq AB$, so the pts. are not collinear.

17. $(x - (-3))^2 + y^2 = 7^2$; center, $(-3, 0)$; radius, 7

18. $(x - (-7))^2 + (y - 8)^2 = \left(\frac{6}{5}\right)^2$; center, $(-7, 8)$; radius, $\frac{6}{5}$

19. $(x - j)^2 + (y - (-14))^2 = (\sqrt{17})^2$; center, $(j, -14)$; radius, $\sqrt{17}$

20. $(x - (-a))^2 + (y - b)^2 = c^2$; center, $(-a, b)$; radius, c

21. $(x - 3)^2 + (y - 0)^2 = 8^2$; $(x - 3)^2 + y^2 = 64$

22. $(x - 0)^2 + (y - 0)^2 = 6^2$; $x^2 + y^2 = 36$

23. $(x - (-4))^2 + (y - (-7))^2 = 5^2$; $(x + 4)^2 + (y + 7)^2 = 25$

24. $(x - (-2))^2 + (y - 5)^2 = \left(\frac{1}{3}\right)^2$; $(x + 2)^2 + (y - 5)^2 = \frac{1}{9}$

25.

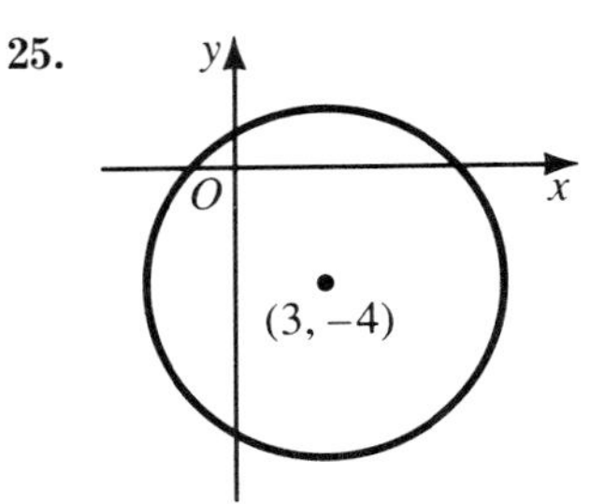

26.

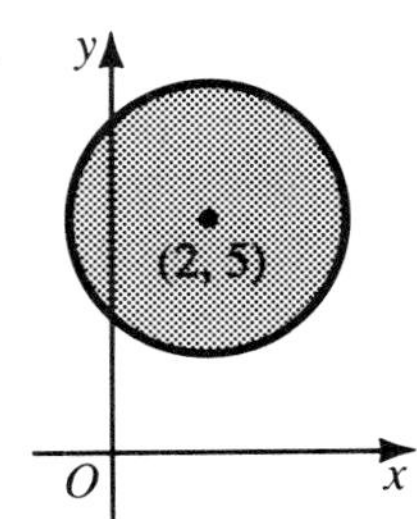

B 27. $AM = \sqrt{(3 - (-3))^2 + (1 - 4)^2} = \sqrt{6^2 + (-3)^2} = \sqrt{36 + 9} = \sqrt{45} = 3\sqrt{5}$;
$AY = \sqrt{(0 - (-3))^2 + (-2 - 4)^2} = \sqrt{3^2 + (-6)^2} = \sqrt{9 + 36} = \sqrt{45} = 3\sqrt{5}$;
$\overline{AM} \cong \overline{AY}$; $\triangle AMY$ is isos.

28. $LA = \sqrt{(-2 - 6)^2 + (4 - (-4))^2} = \sqrt{(-8)^2 + 8^2} = \sqrt{64 + 64} = 8\sqrt{2}$;
$UT = \sqrt{(4 - (-4))^2 + (6 - (-2))^2} = \sqrt{8^2 + 8^2} = \sqrt{64 + 64} = 8\sqrt{2}$; $\overline{LA} \cong \overline{UT}$

29. $JA = |4 - (-2)| = 6$; $AN = \sqrt{(2 - 4)^2 + (2 - (-2))^2} = \sqrt{(-2)^2 + 4^2} = \sqrt{4 + 16} = \sqrt{20} = 2\sqrt{5}$; $JN = \sqrt{(2 - (-2))^2 + (2 - (-2))^2} = \sqrt{4^2 + 4^2} = \sqrt{16 + 16} = 4\sqrt{2}$; $RF = |4 - 1| = 3$; $FK = \sqrt{(6 - 8)^2 + (3 - 4)^2} = \sqrt{(-2)^2 + (-1)^2} = \sqrt{5}$; $RK = \sqrt{(6 - 8)^2 + (3 - 1)^2} = \sqrt{(-2)^2 + 2^2} = \sqrt{4 + 4} = 2\sqrt{2}$; $\frac{JA}{RF} = \frac{AN}{FK} = \frac{JN}{RK} = \frac{2}{1}$; $\triangle JAN \sim \triangle RFK$ (SSS ~ Thm.)

30. congruent; $KA = \sqrt{(2 - 3)^2 + (6 - (-1))^2} = \sqrt{(-1)^2 + 7^2} = \sqrt{50} = 5\sqrt{2}$;
$AT = \sqrt{(5 - 2)^2 + (1 - 6)^2} = \sqrt{3^2 + (-5)^2} = \sqrt{9 + 25} = \sqrt{34}$;
$KT = \sqrt{(5 - 3)^2 + (1 - (-1))^2} = \sqrt{2^2 + 2^2} = \sqrt{4 + 4} = 2\sqrt{2}$;
$IE = \sqrt{(-3 - (-4))^2 + (-6 - 1)^2} = \sqrt{1^2 + (-7)^2} = \sqrt{50} = 5\sqrt{2}$;
$ES = \sqrt{(-6 - (-3))^2 + (-1 - (-6))^2} = \sqrt{(-3)^2 + 5^2} = \sqrt{9 + 25} = \sqrt{34}$;
$IS = \sqrt{(-6 - (-4))^2 + (-1 - 1)^2} = \sqrt{(-2)^2 + (-2)^2} = \sqrt{4 + 4} = 2\sqrt{2}$;
$\overline{KA} \cong \overline{IE}$; $\overline{AT} \cong \overline{ES}$; $\overline{KT} \cong \overline{IS}$; $\triangle KAT \cong \triangle IES$ (SSS Post.)

31. $CB = \sqrt{(8 - 5)^2 + (0 - 2)^2} = \sqrt{3^2 + (-2)^2} = \sqrt{9 + 4} = \sqrt{13}$;
$CR = \sqrt{(-1 - 5)^2 + (-7 - 2)^2} = \sqrt{(-6)^2 + (-9)^2} = \sqrt{36 + 81} = \sqrt{117} = 3\sqrt{13}$; area $= CB \cdot CR = \sqrt{13}(3\sqrt{13}) = 3 \cdot 13 = 39$

32. $DE = \sqrt{(3 - 0)^2 + (1 - 0)^2} = \sqrt{3^2 + 1^2} = \sqrt{9 + 1} = \sqrt{10}$;
$EF = \sqrt{(-2 - 3)^2 + (6 - 1)^2} = \sqrt{(-5)^2 + 5^2} = \sqrt{25 + 25} = \sqrt{50} = 5\sqrt{2}$;
$DF = \sqrt{(-2 - 0)^2 + (6 - 0)^2} = \sqrt{(-2)^2 + 6^2} = \sqrt{4 + 36} = \sqrt{40} = 2\sqrt{10}$;
$(DE)^2 + (DF)^2 = 10 + 40 = 50 = (EF)^2$, so by the converse of the Pythagorean Thm., $\triangle DEF$ is a rt. $\triangle$; area $= \frac{1}{2} \cdot DE \cdot DF = \frac{1}{2} \cdot \sqrt{10} \cdot 2\sqrt{10} = 1 \cdot 10 = 10$

33. All the 6–8–10 ⚠ with one vertex at (0, 0) and one leg on an axis determine 8 of the points: (8, 6), (6, 8), (−6, 8), (−8, 6), (−8, −6), (−6, −8), (6, −8), and (8, −6). The other points are (10, 0), (0, 10), (−10, 0), and (0, −10).

34. **a.** All the 3–4–5 ⚠ with a vertex at (−8, 1) and legs ∥ to the axes determine 8 of the points: (−4, 4), (−5, 5), (−11, 5), (−12, 4), (−12, −2), (−11, −3), (−5, −3), (−4, −2). The other points are (−3, 1), (−8, 6), (−13, 1), and (−8, −4). **b.** $(x - (-8))^2 + (y - 1)^2 = 5^2$; $(x + 8)^2 + (y - 1)^2 = 25$

35. $r = \sqrt{(6 - 0)^2 + (14 - 6)^2} = \sqrt{6^2 + 8^2} = \sqrt{36 + 64} = \sqrt{100} = 10$;
$x^2 + (y - 6)^2 = 100$

36. $r = \sqrt{(3 - (-2))^2 + (8 - (-4))^2} = \sqrt{5^2 + 12^2} = \sqrt{25 + 144} = \sqrt{169} = 13$;
$(x + 2)^2 + (y + 4)^2 = 169$

37. $r = \frac{1}{2} \cdot RS = \frac{1}{2}|3 - (-3)| = \frac{1}{2}(6) = 3$; ctr. = midpt. of $\overline{RS} = (0, 2)$;
$x^2 + (y - 2)^2 = 9$

38. $r =$ length of ⊥ from the ctr. of the ⊙ to the x-axis $= |q|$; $r^2 = q^2$;
$(x - p)^2 + (y - q)^2 = q^2$

39. **a.** $r_1 = \sqrt{25} = 5$; $r_2 = \sqrt{100} = 10$
b. Given $O(0, 0)$ and $P(9, 12)$, $OP = \sqrt{(9 - 0)^2 + (12 - 0)^2} = \sqrt{9^2 + 12^2} = \sqrt{81 + 144} = \sqrt{225} = 15$
c. $OP = 15 = 5 + 10 = r_1 + r_2$, so the distance between the centers = the sum of the radii, and the circles must be externally tangent.

d.

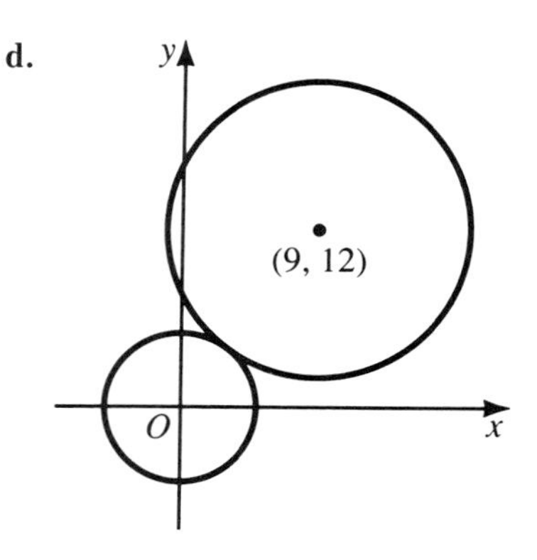

40. **a.** $r_1 = \sqrt{2}$; $r_2 = \sqrt{32} = 4\sqrt{2}$
b. Given $O(0, 0)$ and $P(3, 3)$, $OP = \sqrt{(3 - 0)^2 + (3 - 0)^2} = \sqrt{3^2 + 3^2} = \sqrt{18} = 3\sqrt{2}$ **c.** $OP = 3\sqrt{2} = 4\sqrt{2} - \sqrt{2} = r_2 - r_1$, so the distance between the centers = the difference of the radii, and the circles must be internally tangent.

d.

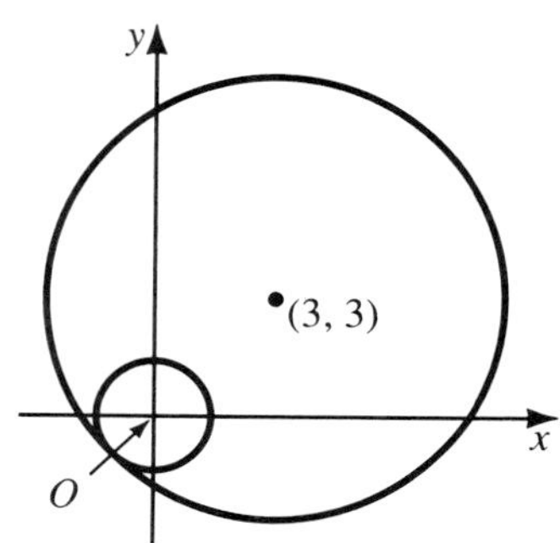

41. $AY = \sqrt{(11 - 1)^2 + (1 - (-3))^2} = \sqrt{10^2 + 4^2} = \sqrt{100 + 16} = \sqrt{116} = 2\sqrt{29}$; $RJ = \sqrt{(9 - (-1))^2 + (-2 - (-6))^2} = \sqrt{10^2 + 4^2} = \sqrt{100 + 16} = \sqrt{116} = 2\sqrt{29}$; $AR = \sqrt{(-1 - 1)^2 + (-6 - (-3))^2} = \sqrt{(-2)^2 + (-3)^2} = \sqrt{4 + 9} = \sqrt{13}$; $YJ = \sqrt{(9 - 11)^2 + (-2 - 1)^2} = \sqrt{(-2)^2 + (-3)^2} = \sqrt{4 + 9} = \sqrt{13}$; $\overline{AY} \cong \overline{RJ}$ and $\overline{AR} \cong \overline{YJ}$; $RAYJ$ is a ▱. (If both pairs of opp. sides of a quad. are ≅, then the quad. is a ▱.)

42. $JK = \sqrt{(-6 - (-3))^2 + (-2 - 4)^2} = \sqrt{(-3)^2 + (-6)^2} = \sqrt{9 + 36} = \sqrt{45} = 3\sqrt{5}$; $KM = \sqrt{(3 - (-3))^2 + (1 - 4)^2} = \sqrt{6^2 + (-3)^2} = \sqrt{36 + 9} = \sqrt{45} = 3\sqrt{5}$; $JM = \sqrt{(3 - (-6))^2 + (1 - (-2))^2} = \sqrt{9^2 + 3^2} = \sqrt{81 + 9} = \sqrt{90} = 3\sqrt{10}$; $\overline{JK} \cong \overline{KM}$; therefore $\triangle JKM$ is isosceles; $(JK)^2 + (KM)^2 = 45 + 45 = 90 = (JM)^2$, so $\triangle JKM$ is a rt. △.

C 43. Let $(4, y)$ be the coordinates of M. If $\overline{GH}$ is the base of $\triangle GHM$, then $MH = MG$; $\sqrt{(-2 - 4)^2 + (7 - y)^2} = \sqrt{(-2 - 4)^2 + (-3 - y)^2}$; $\sqrt{(-6)^2 + 49 - 14y + y^2} = \sqrt{(-6)^2 + 9 + 6y + y^2}$; $36 + 49 - 14y + y^2 = 36 + 9 + 6y + y^2$; $49 - 14y = 9 + 6y$; $20y = 40$; $y = 2$; if $HG = HM$, $|7 - (-3)| = 10 = \sqrt{(4 - (-2))^2 + (y - 7)^2}$; $\sqrt{6^2 + y^2 - 14y + 49} = 10$; $y^2 - 14y + 85 = 100$; $y^2 - 14y - 15 = 0$; $(y - 15)(y + 1) = 0$; $y = 15$ or $y = -1$; if $HG = MG$, $10 = \sqrt{(-2 - 4)^2 + (-3 - y)^2}$;

$\sqrt{(-6)^2 + 9 + 6y + y^2} = 10$; $y^2 + 6y + 45 = 100$; $y^2 + 6y - 55 = 0$; $(y + 11)(y - 5) = 0$; $y = -11$ or $y = 5$; the 5 possible values are 15, 5, 2, -1, and -11.

44. Let the required point be (a, b). $\sqrt{(a - (-2))^2 + (b - 5)^2} = \sqrt{(a - 8)^2 + (b - 5)^2}$; $(a + 2)^2 + (b - 5)^2 = (a - 8)^2 + (b - 5)^2$; $(a + 2)^2 = (a - 8)^2$; $a^2 + 4a + 4 = a^2 - 16a + 64$; $20a = 60$; $a = 3$; also, $\sqrt{(a - 8)^2 + (b - 5)^2} = \sqrt{(a - 6)^2 + (b - 7)^2}$; $(-5)^2 + (b - 5)^2 = (-3)^2 + (b - 7)^2$; $25 + b^2 - 10b + 25 = 9 + b^2 - 14b + 49$; $-10b + 50 = -14b + 58$; $4b = 8$; $b = 2$; the pt. is (3, 2).

45. $(x^2 + 4x + 4) + (y^2 - 8y + 16) = 16 + 4 + 16$; $(x + 2)^2 + (y - 4)^2 = 36 = 6^2$; center, $(-2, 4)$; radius, 6

Page 528 • COMPUTER KEY-IN

1. Programs may vary.

```
10    INPUT "HOW MANY POINTS: ";N
20    FOR I = 1 TO N
30    LET X = RND(1)
40    LET Y = RND(1)
50    LET D = SQR(X ↑ 2 + Y ↑ 2)
60    IF D < 1 THEN LET Q = Q + 1
70    NEXT I
80    PRINT Q; " POINTS INSIDE CIRCLE"
90    PRINT "4 * Q/N = "; 4 * Q/N
100   END
```

2. Results will vary. Sample outputs are given. For $n = 100$, 77 pts. inside ⊙, $A \approx 3.08$; for $n = 500$, 387 pts. inside ⊙, $A \approx 3.096$; for $n = 1000$, 792 pts. inside ⊙, $A \approx 3.168$.

3. $A = \pi r^2 = \pi(1) = \pi \approx 3.14159$; answers may vary.

Pages 531–532 • CLASSROOM EXERCISES

1. $\frac{2 - 1}{3 - (-2)} = \frac{1}{5}$ **2.** $\frac{-2 - (-2)}{4 - (-3)} = 0$ **3.** $\frac{2 - 0}{0 - 2} = \frac{2}{-2} = -1$

4. a. $y_2 > y_1$; positive **b.** $x_2 > x_1$; positive **c.** $\frac{\text{positive}}{\text{positive}} = \text{positive}$

5. a. $y_2 < y_1$; negative **b.** $x_2 > x_1$; positive **c.** $\frac{\text{negative}}{\text{positive}} = \text{negative}$

6. a. positive **b.** not defined **c.** negative **d.** zero

7. a. $\frac{3 - 0}{3 - 1} = \frac{3}{2}$ **b.** $\tan n° = \frac{\text{leg opp. } \angle A}{\text{leg adj. to } \angle A} = \frac{3}{2}$ **c.** True; $\tan n° = \frac{\text{opp.}}{\text{adj.}} = \frac{y_2 - y_1}{x_2 - x_1} = \text{slope of } \overleftrightarrow{AB}$

8. 1. Def. of ⊥ lines; def. of ≅ ∠s 2. If 2 ∥ lines are cut by a trans., then corr. ∠s are ≅. 3. AA ~ Post. 4. Corr. sides of ~ △s are in proportion. 5. Def. of slope 6. Substitution Prop.

Pages 532–534 • WRITTEN EXERCISES

A **1. a.** k **b.** n, r **c.** l, x-axis **d.** s, y-axis

2. a. Slope equals zero. **b.** Slope is not defined. **3.** $\frac{4 - 2}{3 - 1} = \frac{2}{2} = 1$

4. $\frac{-5 - 2}{-2 - 1} = \frac{-7}{-3} = \frac{7}{3}$ **5.** $\frac{5 - 2}{-2 - 1} = \frac{3}{-3} = -1$ **6.** $\frac{1 - 0}{5 - 0} = \frac{1}{5}$

7. $\frac{7 - 2}{2 - 7} = \frac{5}{-5} = -1$ **8.** Slope is not defined. **9.** $\frac{-6 - (-6)}{-6 - 6} = \frac{0}{-12} = 0$

10. $\frac{3 - (-6)}{4 - 6} = \frac{9}{-2} = -\frac{9}{2}$ **11.** $\frac{-6 - (-3)}{-6 - (-4)} = \frac{-3}{-2} = \frac{3}{2}$

12. slope $= \frac{-7 - (-1)}{5 - 3} = \frac{-6}{2} = -3$; $AB = \sqrt{(5 - 3)^2 + (-7 - (-1))^2} = \sqrt{2^2 + (-6)^2} = \sqrt{4 + 36} = \sqrt{40} = 2\sqrt{10}$

13. slope $= \frac{-6 - (-2)}{7 - (-3)} = \frac{-4}{10} = -\frac{2}{5}$; $AB = \sqrt{(7 - (-3))^2 + (-6 - (-2))^2} = \sqrt{10^2 + (-4)^2} = \sqrt{100 + 16} = \sqrt{116} = 2\sqrt{29}$

14. slope $= \frac{-5 - (-7)}{-3 - 8} = \frac{2}{-11} = -\frac{2}{11}$; $AB = \sqrt{(-3 - 8)^2 + (-5 - (-7))^2} = \sqrt{(-11)^2 + 2^2} = \sqrt{121 + 4} = \sqrt{125} = 5\sqrt{5}$

15. slope $= \frac{-3 - (-9)}{8 - 0} = \frac{6}{8} = \frac{3}{4}$; $AB = \sqrt{(8 - 0)^2 + (-3 - (-9))^2} = \sqrt{8^2 + 6^2} = \sqrt{64 + 36} = \sqrt{100} = 10$

16–19. Students' choices of two points may vary.

16.

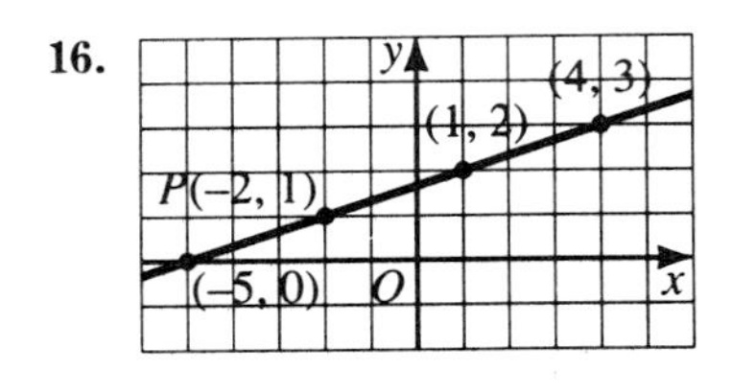

17.

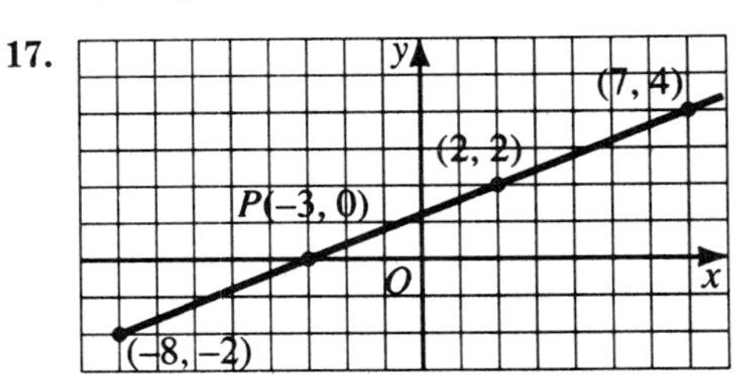

18.

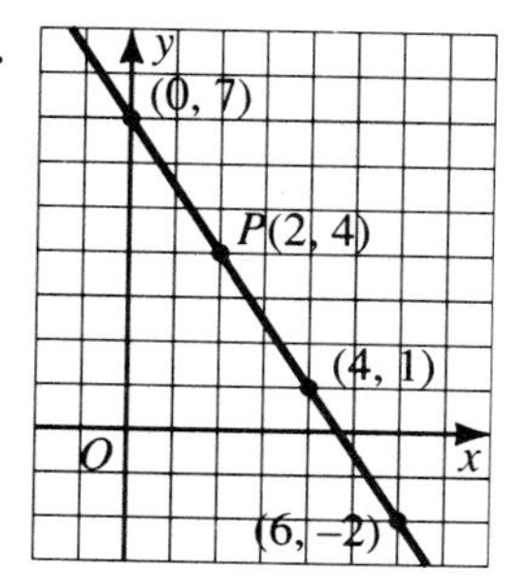

19.

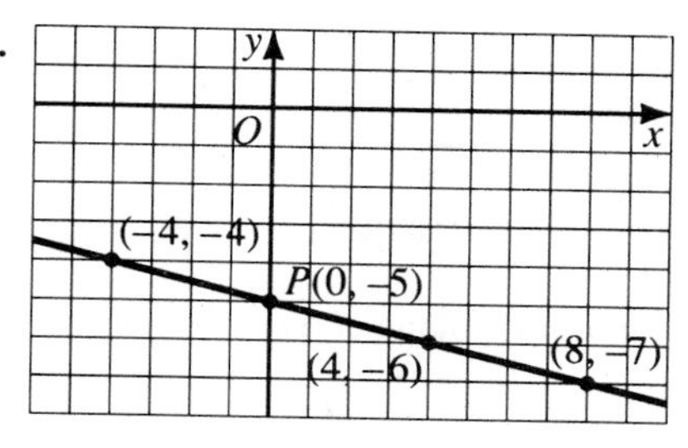

20. slope of $\overline{PQ} = \frac{7 - 3}{2 - (-1)} = \frac{4}{3}$; slope of $\overline{QR} = \frac{15 - 7}{8 - 2} = \frac{8}{6} = \frac{4}{3}$; since the slopes are equal, the points are collinear.

21. slope of $\overline{PQ} = \frac{5 - 6}{-5 - (-8)} = \frac{-1}{3} = -\frac{1}{3}$; slope of $\overline{QR} = \frac{2 - 5}{4 - (-5)} = \frac{-3}{9} = -\frac{1}{3}$; since the slopes are equal, the points are collinear.

B **22.** slope $= \frac{\text{vertical change}}{\text{horizontal change}}$; $\frac{1}{15} = \frac{18}{d}$; $d = 15 \cdot 18 = 270$ in., or $22\frac{1}{2}$ ft

23. $\frac{3}{4} = \frac{y - 3}{10 - 2}$; $4(y - 3) = 24$; $4y - 12 = 24$; $4y = 36$; $y = 9$

24. $-\frac{5}{2} = \frac{6 - (-4)}{x - 7}$; $-5(x - 7) = 20$; $x - 7 = -4$; $x = 3$

25. $m = \frac{y - q}{r - p}$; $m(r - p) = y - q$; $y = m(r - p) + q$

26. **a.** slope of $\overline{OD} = \frac{2 - 0}{3 - 0} = \frac{2}{3}$; slope of $\overline{NF} = \frac{2 - 0}{8 - 5} = \frac{2}{3}$ **b.** SAS Post. ($\overline{OC} \cong \overline{NE}$, $\overline{CD} \cong \overline{EF}$, $\angle OCD \cong \angle NEF$) **c.** Corr. parts of $\cong$ ▵ are $\cong$.
d. If 2 lines are cut by a trans. and corr. ∠s are $\cong$, then the lines are $\parallel$.
e. Slopes of $\parallel$ lines are $=$.

27. **a.** $OR = OA = 5$; $AB = RS = 3$; $m\angle OAB = 90 = m\angle ORS$; $\triangle OAB \cong \triangle ORS$ by the SAS Post. **b.** $m\angle BOS = m\angle BOR + m\angle ROS = m\angle BOR + m\angle AOB = 90$
c. slope of $\overline{OB}$ · slope of $\overline{OS} = \frac{3 - 0}{5 - 0} \cdot \frac{5 - 0}{-3 - 0} = \frac{3}{5}\left(-\frac{5}{3}\right) = -1$

28. **a.** $RS = \sqrt{(-1 - 4)^2 + (7 - 2)^2} = \sqrt{(-5)^2 + 5^2} = \sqrt{25 + 25} = \sqrt{50} = 5\sqrt{2}$; $ST = \sqrt{(1 - (-1))^2 + (1 - 7)^2} = \sqrt{2^2 + (-6)^2} = \sqrt{4 + 36} = \sqrt{40} = 2\sqrt{10}$; $RT = \sqrt{(1 - 4)^2 + (1 - 2)^2} = \sqrt{(-3)^2 + (-1)^2} = \sqrt{9 + 1} = \sqrt{10}$
b. $(RT)^2 + (ST)^2 = 10 + 40 = 50 = (RS)^2$, so $\triangle RST$ is a rt. $\triangle$ with hyp. $\overline{RS}$.
c. slope of $\overline{RT}$ · slope of $\overline{ST} = \frac{2 - 1}{4 - 1} \cdot \frac{7 - 1}{-1 - 1} = \frac{1}{3} \cdot \frac{6}{-2} = -1$

29. **a.** $RS = \sqrt{(-3 - 4)^2 + (6 - 3)^2} = \sqrt{(-7)^2 + 3^2} = \sqrt{49 + 9} = \sqrt{58}$; $RT = \sqrt{(2 - 4)^2 + (1 - 3)^2} = \sqrt{(-2)^2 + (-2)^2} = \sqrt{4 + 4} = \sqrt{8} = 2\sqrt{2}$; $ST = \sqrt{(2 - (-3))^2 + (1 - 6)^2} = \sqrt{5^2 + (-5)^2} = \sqrt{25 + 25} = \sqrt{50} = 5\sqrt{2}$

b. $(RT)^2 + (ST)^2 = 8 + 50 = 58 = (RS)^2$, so $\triangle RST$ is a rt. $\triangle$ with hyp. $\overline{RS}$.

c. slope of $\overline{RT}$ · slope of $\overline{ST} = \frac{3 - 1}{4 - 2} \cdot \frac{6 - 1}{-3 - 2} = \frac{2}{2} \cdot \frac{5}{-5} = -1$

30. **a.** Since $\overline{AB} \perp \overline{BC}$, $\tan A = \frac{BC}{AB} = \frac{4}{5}$; slope of $\overline{AC} = \frac{4 - 0}{5 - 0} = \frac{4}{5} = \tan A$.

b. From the table on p. 311, if $\tan A = \frac{4}{5} = 0.8$, then $m\angle A \approx 39$.

31. Let A be a point on the line in Quadrant I, B be the point where the line intersects the x-axis, and $\overline{AC}$ be the $\perp$ segment from A to the x-axis. Then $\triangle ABC$ is a 45°–45°–90° $\triangle$, and so $AC = BC$. The slope of the line $= \frac{\text{vertical change}}{\text{horizontal change}} = \frac{AC}{BC} = 1$.

C **32.** slope $= \frac{3 - (-1)}{4 - (-2)} = \frac{4}{6} = \frac{2}{3}$. Let the line intersect the x- and y-axes at $(x, 0)$ and $(0, y)$, resp. $\frac{-1 - 0}{-2 - x} = \frac{2}{3}$; $\frac{-1}{-2 - x} = \frac{2}{3}$; $-4 - 2x = -3$; $2x = -1$; $x = -\frac{1}{2}$; $\frac{-1 - y}{-2 - 0} = \frac{2}{3}$; $\frac{-1 - y}{-2} = \frac{2}{3}$; $-4 = -3 - 3y$; $3y = 1$; $y = \frac{1}{3}$; $\left(-\frac{1}{2}, 0\right)$ and $\left(0, \frac{1}{3}\right)$

33. Draw the line $y = 1$ through H; draw the line $x = 5$ through $K(5, 1)$. Draw the line through H that makes a 60° angle with $\overleftrightarrow{HK}$, intersecting $x = 5$ at J. $\triangle HJK$ is a 30°–60°–90° $\triangle$, so $JK = HK\sqrt{3}$; $a - 1 = 2\sqrt{3}$; $a = 2\sqrt{3} + 1$.

34. Let A be $(-3, 4)$, B be $(0, k)$, C be $(k, 10)$; if A, B, and C are collinear, then slope of $\overline{AB}$ = slope of $\overline{BC}$; $\frac{k - 4}{0 - (-3)} = \frac{10 - k}{k - 0}$; $\frac{k - 4}{3} = \frac{10 - k}{k}$; $k^2 - 4k = 30 - 3k$; $k^2 - k - 30 = 0$; $(k - 6)(k + 5) = 0$; $k = 6$ or $k = -5$.

Page 534 • ALGEBRA REVIEW

1. $(-6)^3 = (-6)(-6)(-6) = -216$ **2.** $(-5)^4 = (-5)(-5)(-5)(-5) = 625$

3. $3^{-2} = \frac{1}{3^2} = \frac{1}{9}$ **4.** $2^{-3} = \frac{1}{2^3} = \frac{1}{8}$

5. $(-4)^{-3} = \frac{1}{(-4)^3} = \frac{1}{(-4)(-4)(-4)} = -\frac{1}{64}$ **6.** $\left(\frac{2}{3}\right)^{-2} = \frac{1}{\left(\frac{2}{3}\right)^2} = \frac{1}{\frac{4}{9}} = \frac{9}{4}$

7. $\left(\frac{5}{3}\right)^{-3} = \frac{1}{\left(\frac{5}{3}\right)^3} = \frac{1}{\frac{125}{27}} = \frac{27}{125}$ 8. $15^0 = 1$ 9. $(-1)^{20} = [(-1)^2]^{10} = 1^{10} = 1$

10. $(-1)^{99} = -1 \cdot (-1)^{98} = -1 \cdot [(-1)^2]^{49} = -1 \cdot 1 = -1$

11. $2^3 \cdot 2^2 \cdot 2^{-4} = 2^{3+2+(-4)} = 2^1 = 2$

12. $4^2 \cdot 3^3 \cdot 2^{-3} = 4 \cdot 4 \cdot 3 \cdot 3 \cdot 3 \cdot \frac{1}{2^3} = \frac{16}{8} \cdot 27 = 54$ 13. $r^5 \cdot r^8 = r^{5+8} = r^{13}$

14. $x^{-1} \cdot x^{-2} = x^{-1-2} = x^{-3} = \frac{1}{x^3}$ 15. $\frac{r^9}{r^4} = r^{9-4} = r^5$

16. $\frac{t^3}{t^5} = t^{3-5} = t^{-2} = \frac{1}{t^2}$ 17. $a^1 \cdot a^{-1} = a^{1+(-1)} = a^0 = 1$

18. $(x^2)^{-2} = x^{2(-2)} = x^{-4} = \frac{1}{x^4}$ 19. $(b^4)^2 = b^{4 \cdot 2} = b^8$ 20. $(s^5)^3 = s^{5 \cdot 3} = s^{15}$

21. $(3y^2)(2y^4) = (3 \cdot 2)(y^{2+4}) = 6y^6$

22. $(4x^3y^2)(-3xy) = (4(-3))(x^3 \cdot x)(y^2 \cdot y) = -12x^{3+1}y^{2+1} = -12x^4y^3$

23. $(5a^2b^3)(a^{-2}b) = 5(a^2 \cdot a^{-2})(b^3 \cdot b^1) = 5 \cdot a^{2+(-2)} \cdot b^{3+1} = 5 \cdot a^0 \cdot b^4 =$
$5 \cdot 1 \cdot b^4 = 5b^4$

24. $(-2st^5)(-4st^{-3}) = (-2)(-4) \cdot (s^1 \cdot s^1) \cdot (t^5 \cdot t^{-3}) = 8 \cdot s^{1+1} \cdot t^{5+(-3)} = 8s^2t^2$

Page 536 • CLASSROOM EXERCISES

1. **a.** $-\frac{1}{2}$ **b.** $-\frac{5}{4}$ **c.** $\frac{1}{4}$ **d.** 0 **e.** not defined 2. $\frac{3}{4} = \frac{12}{16}$, so the lines are $\parallel$.

3. $1(-1) = -1$, so the lines are $\perp$. 4. $3 \neq (-3)$; $3(-3) \neq -1$; neither

5. $-\frac{3}{4}\left(\frac{4}{3}\right) = -1$, so the lines are $\perp$. 6. $3\left(\frac{-1}{3}\right) = -1$, so the lines are $\perp$.

7. $\frac{-2}{3} = -\frac{2}{3} = \frac{2}{-3}$, so the lines are $\parallel$. 8. $0 \neq -1$; $0(-1) \neq -1$; neither

9. $\frac{5}{6} \neq \frac{6}{5}$; $\frac{5}{6}\left(\frac{6}{5}\right) \neq -1$; neither

10. If 2 nonvertical lines are $\parallel$, then their slopes are $=$. If 2 nonvertical lines have equal slopes, then the lines are $\parallel$.

11. 1. Through a pt. outside a line, there is exactly 1 line $\perp$ to the given line.
2. When the alt. is drawn to the hyp. of a rt. $\triangle$, the length of the alt. is the geom. mean between the segs. of the hyp. 3. Def. of slope 4. Def. of slope
5. Mult. Prop. of $=$ 6. Substitution Prop.

Pages 537–538 • WRITTEN EXERCISES

A **1. a.** $\frac{4-0}{4-(-2)} = \frac{4}{6} = \frac{2}{3}$ **b.** $\frac{2}{3}$ **c.** $-\frac{3}{2}$

2. a. $\frac{-1-1}{2-(-3)} = \frac{-2}{5} = -\frac{2}{5}$ **b.** $-\frac{2}{5}$ **c.** $\frac{5}{2}$

3. $\overline{OE}$: $\frac{7-0}{2-0} = \frac{7}{2}$; $\overline{GF}$: $\frac{7}{2}$; $\overline{OG}$: 0; $\overline{EF}$: 0

4. $\overline{HI}$: $\frac{-1-(-5)}{6-2} = \frac{4}{4} = 1$; $\overline{JK}$: 1; $\overline{IJ}$: -1; $\overline{KH}$: -1

5. a. $\overline{LM}$: $\frac{2-(-1)}{-4-(-3)} = \frac{3}{-1} = -3$; $\overline{PN}$: $\frac{4-(-2)}{2-4} = \frac{6}{-2} = -3$ **b.** slope of $\overline{LM}$ = slope of $\overline{PN}$ **c.** $\overline{MN}$: $\frac{4-2}{2-(-4)} = \frac{2}{6} = \frac{1}{3}$; $\overline{LP}$: $\frac{-2-(-1)}{4-(-3)} = \frac{-1}{7} = -\frac{1}{7}$

d. slope of $\overline{MN} \neq$ slope of $\overline{LP}$ **e.** trapezoid

6. a. $\overline{RV}$: $\frac{-8-(-6)}{0-(-2)} = \frac{-2}{2} = -1$; $\overline{TV}$: $\frac{-8-(-3)}{0-5} = \frac{-5}{-5} = 1$ **b.** The prod. of their slopes is -1. **c.** If one ∠ of a ▱ is a rt. ∠, then the ▱ is a rect.

d. Let $S = (x, y)$; $\frac{y-(-6)}{x-(-2)} = 1$; $\frac{y+6}{x+2} = 1$; $y + 6 = x + 2$; $y = x - 4$;

$\frac{y-(-3)}{x-5} = -1$; $\frac{y+3}{x-5} = -1$; $y + 3 = -x + 5$; $y = -x + 2$;

$x - 4 = -x + 2$; $2x = 6$; $x = 3$; $y = 3 - 4 = -1$; $S(3, -1)$

7. $\overline{AB}$: $\frac{3-0}{7-0} = \frac{3}{7}$; alt. to $\overline{AB}$: $-\frac{7}{3}$; $\overline{AC}$: $\frac{-5-0}{2-0} = \frac{-5}{2} = -\frac{5}{2}$; alt. to $\overline{AC}$: $\frac{2}{5}$;

$\overline{BC}$: $\frac{-5-3}{2-7} = \frac{-8}{-5} = \frac{8}{5}$; alt. to $\overline{BC}$: $-\frac{5}{8}$

8. $\overline{AB}$: $\frac{-3-4}{-1-1} = \frac{-7}{-2} = \frac{7}{2}$; alt. to $\overline{AB}$: $-\frac{2}{7}$; $\overline{AC}$: $\frac{-5-4}{4-1} = \frac{-9}{3} = -3$; alt. to $\overline{AC}$: $\frac{1}{3}$;

$\overline{BC}$: $\frac{-5-(-3)}{4-(-1)} = \frac{-2}{5} = -\frac{2}{5}$; alt. to $\overline{BC}$: $\frac{5}{2}$

9. slope of $\overline{RS} = \frac{2-(-4)}{2-(-3)} = \frac{6}{5}$; slope of $\overline{ST} = \frac{-8-2}{14-2} = \frac{-10}{12} = -\frac{5}{6}$;

$\frac{6}{5}\left(-\frac{5}{6}\right) = -1$, so $\overline{RS} \perp \overline{ST}$ and $\triangle RST$ is a rt. △.

10. slope of $\overline{RS} = \frac{4-1}{2-(-1)} = \frac{3}{3} = 1$; slope of $\overline{ST} = \frac{1-4}{5-2} = \frac{-3}{3} = -1$;

$1(-1) = -1$, so $\overline{RS} \perp \overline{ST}$ and $\triangle RST$ is a rt. △.

B **11. a.** slope of $\overline{AB} = \frac{2-(-4)}{4-(-6)} = \frac{6}{10} = \frac{3}{5}$; slope of $\overline{DC} = \frac{8-2}{6-(-4)} = \frac{6}{10} = \frac{3}{5}$;

$\overline{AB} \parallel \overline{DC}$ since slopes are =. Slope of $\overline{AD} = \frac{2-(-4)}{-4-(-6)} = \frac{6}{2} = 3$; slope of

$\overline{BC} = \frac{8 - 2}{6 - 4} = \frac{6}{2} = 3$; $\overline{AD} \parallel \overline{BC}$ since slopes are $=$. Thus, by the def. of $\square$, $ABCD$ is a $\square$. **b.** $AB = \sqrt{(4 - (-6))^2 + (2 - (-4))^2} = \sqrt{10^2 + 6^2} = \sqrt{100 + 36} = 2\sqrt{34}$; $DC = \sqrt{(-4 - 6)^2 + (2 - 8)^2} = \sqrt{(-10)^2 + (-6)^2} = \sqrt{100 + 36} = 2\sqrt{34}$; $\overline{AB} \cong \overline{DC}$. $AD = \sqrt{(-4 - (-6))^2 + (2 - (-4))^2} = \sqrt{2^2 + 6^2} = \sqrt{4 + 36} = 2\sqrt{10}$; $BC = \sqrt{(6 - 4)^2 + (8 - 2)^2} = \sqrt{2^2 + 6^2} = \sqrt{4 + 36} = 2\sqrt{10}$; $\overline{AD} \cong \overline{BC}$. Since both pairs of opp. sides of quad. $ABCD$ are $\cong$, $ABCD$ is a $\square$.

12. a. $EF = \sqrt{(2 - (-4))^2 + (3 - 1)^2} = \sqrt{6^2 + 2^2} = \sqrt{36 + 4} = 2\sqrt{10}$;
$FG = \sqrt{(4 - 2)^2 + (9 - 3)^2} = \sqrt{2^2 + 6^2} = \sqrt{4 + 36} = 2\sqrt{10}$;
$HG = \sqrt{(-2 - 4)^2 + (7 - 9)^2} = \sqrt{(-6)^2 + 2^2} = \sqrt{36 + 4} = 2\sqrt{10}$;
$EH = \sqrt{(-2 - (-4))^2 + (7 - 1)^2} = \sqrt{2^2 + 6^2} = \sqrt{4 + 36} = 2\sqrt{10}$;
$\overline{EF} \cong \overline{FG} \cong \overline{HG} \cong \overline{EH}$, so by the def. of rhombus, $EFGH$ is a rhombus.

b. slope of $\overline{EG}$ · slope of $\overline{HF} = \frac{9 - 1}{4 - (-4)} \cdot \frac{7 - 3}{-2 - 2} = \frac{8}{8} \cdot \frac{4}{-4} = 1(-1) = -1$, so $\overline{EG} \perp \overline{HF}$.

13. a. slope of $\overline{RS} = \frac{9 - 5}{-1 - (-4)} = \frac{4}{3}$; slope of $\overline{UT} = \frac{-1 - 3}{4 - 7} = \frac{-4}{-3} = \frac{4}{3}$;

slopes are $=$, so $\overline{RS} \parallel \overline{UT}$. Slope of $\overline{RU} = \frac{-1 - 5}{4 - (-4)} = \frac{-6}{8} = -\frac{3}{4}$;

slope of $\overline{ST} = \frac{3 - 9}{7 - (-1)} = \frac{-6}{8} = -\frac{3}{4}$; slopes are $=$, so $\overline{RU} \parallel \overline{ST}$. Then,

by the def. of a $\square$, $RSTU$ is a $\square$. Slope of $\overline{RS}$ · slope of $\overline{RU} = \frac{4}{3}\left(-\frac{3}{4}\right) = -1$,

so $\overline{RS} \perp \overline{RU}$ and $\angle RST$ is a rt. $\angle$. A $\square$ with a rt. $\angle$ is a rect., so $RSTU$ is a rect.
b. $RT = \sqrt{(7 - (-4))^2 + (3 - 5)^2} = \sqrt{11^2 + (-2)^2} = \sqrt{121 + 4} = 5\sqrt{5}$;
$US = \sqrt{(4 - (-1))^2 + (-1 - 9)^2} = \sqrt{5^2 + (-10)^2} = \sqrt{25 + 100} = 5\sqrt{5}$;
$\overline{RT} \cong \overline{US}$

14. a. slope of $\overline{NO} = \frac{-5 - 0}{-1 - 0} = \frac{-5}{-1} = 5$; slope of $\overline{OP} = \frac{2 - 0}{3 - 0} = \frac{2}{3}$; slope of $\overline{PQ} = \frac{1 - 2}{8 - 3} = \frac{-1}{5} = -\frac{1}{5}$; slope of $\overline{NQ} = \frac{-5 - 1}{-1 - 8} = \frac{-6}{-9} = \frac{2}{3}$; slope of $\overline{OP}$ = slope of $\overline{NQ}$, so $\overline{OP} \parallel \overline{NQ}$; slope of $\overline{NO} \neq$ slope of $\overline{PQ}$, so $\overline{NO} \nparallel \overline{PQ}$; $NOPQ$ is a trap. by def. $NO = \sqrt{(-1 - 0)^2 + (-5 - 0)^2} = \sqrt{(-1)^2 + (-5)^2} = \sqrt{1 + 25} = \sqrt{26}$; $PQ = \sqrt{(8 - 3)^2 + (1 - 2)^2} = \sqrt{5^2 + (-1)^2} = \sqrt{25 + 1} = \sqrt{26}$; $\overline{NO} \cong \overline{PQ}$, so by def., $NOPQ$ is an isos. trap. **b.** $NP = \sqrt{(3 - (-1))^2 + (2 - (-5))^2} = \sqrt{4^2 + 7^2} = \sqrt{16 + 49} = \sqrt{65}$; $OQ = \sqrt{(8 - 0)^2 + (1 - 0)^2} = \sqrt{8^2 + 1^2} = \sqrt{64 + 1} = \sqrt{65}$; $\overline{NP} \cong \overline{OQ}$

15. $HIJK$ is a trap. Proof: Since slope of $\overline{HI} = \frac{0 - 0}{5 - 0} = \frac{0}{5} = 0$, and slope of $\overline{KJ} = \frac{9 - 9}{7 - 1} = \frac{0}{6} = 0$, then $\overline{HI} \parallel \overline{KJ}$. Since slope of $\overline{KH} = \frac{9 - 0}{1 - 0} = \frac{9}{1} = 9$, and slope of $\overline{JI} = \frac{9 - 0}{7 - 5} = \frac{9}{2}$, then $\overline{KH} \nparallel \overline{JI}$. So $HIJK$ is a trap. by def.

16. $HIJK$ is a rhombus. Proof: By the distance formula, $KH = \sqrt{(0 - (-4))^2 + (1 - 3)^2} = \sqrt{4^2 + (-2)^2} = \sqrt{16 + 4} = \sqrt{20} = 2\sqrt{5}$, $HI = \sqrt{(2 - 0)^2 + (-3 - 1)^2} = \sqrt{2^2 + (-4)^2} = \sqrt{4 + 16} = \sqrt{20} = 2\sqrt{5}$, $IJ = \sqrt{(-2 - 2)^2 + (-1 - (-3))^2} = \sqrt{(-4)^2 + 2^2} = \sqrt{16 + 4} = \sqrt{20} = 2\sqrt{5}$, and $JK = \sqrt{(-4 - (-2))^2 + (3 - (-1))^2} = \sqrt{(-2)^2 + 4^2} = \sqrt{4 + 16} = \sqrt{20} = 2\sqrt{5}$. Therefore, $HIJK$ is a rhombus by def.

17. $HIJK$ is a rect. Proof: Slope of $\overline{HI} = \frac{5 - 3}{7 - 8} = \frac{2}{-1} = -2$, slope of $\overline{KJ} = \frac{-1 - 1}{0 - (-1)} = \frac{-2}{1} = -2$, slope of $\overline{KH} = \frac{5 - 1}{7 - (-1)} = \frac{4}{8} = \frac{1}{2}$, and slope of $\overline{JI} = \frac{3 - (-1)}{8 - 0} = \frac{4}{8} = \frac{1}{2}$. Since $-2 \cdot \frac{1}{2} = -1$, then $\overline{KH} \perp \overline{HI}$, $\overline{HI} \perp \overline{JI}$, $\overline{JI} \perp \overline{KJ}$, and $\overline{KJ} \perp \overline{KH}$. So $\angle H$, $\angle I$, $\angle J$, and $\angle K$ are rt. $\angle$s, and $HIJK$ is a rect. by def.

18. $HIJK$ is a ▱. Proof: Since slope of $\overline{HI} = \frac{-6 - (-3)}{-5 - (-3)} = \frac{-3}{-2} = \frac{3}{2}$, and slope of $\overline{KJ} = \frac{-5 - (-2)}{4 - 6} = \frac{-3}{-2} = \frac{3}{2}$, then $\overline{HI} \parallel \overline{KJ}$. Since slope of $\overline{HK} = \frac{-2 - (-3)}{6 - (-3)} = \frac{1}{9}$, and slope of $\overline{IJ} = \frac{-5 - (-6)}{4 - (-5)} = \frac{1}{9}$, then $\overline{HK} \parallel \overline{IJ}$. Therefore, $HIJK$ is a ▱ by def.

19. Let $\odot O$ be the $\odot$ with ctr. (0, 0) and radius 5; slope of $\overline{ON} = \frac{-4 - 0}{3 - 0} = \frac{-4}{3} = -\frac{4}{3}$; the tan. to $\odot O$ at N is $\perp$ to $\overline{ON}$; slope of tan. $= -\frac{1}{\text{slope of } \overline{ON}} = \frac{3}{4}$.

20. Let $\odot O$ be the $\odot$ with ctr. $(-2, 1)$ and radius 10; slope of $\overline{OP} = \frac{7 - 1}{6 - (-2)} = \frac{6}{8} = \frac{3}{4}$; the tan. to $\odot O$ at P is $\perp$ to $\overline{OP}$; slope of tan. $= -\frac{1}{\text{slope of } \overline{OP}} = -\frac{4}{3}$.

21. **a.** True **b.** True **c.** True

22. Yes; Exs. 21(a), 21(b), and 21(c), resp., demonstrate that the relation "is $\parallel$ to" is reflexive, symmetric, and transitive.

C **23.** **a.** Since $\overleftrightarrow{TU} \perp \overleftrightarrow{US}$, $\angle U$ is a rt. $\angle$, $\triangle SUT$ is a rt. $\triangle$, and $(ST)^2 = (TU)^2 + (US)^2$ by the Pythagorean Thm. Using the distance formula, $(ST)^2 = (a - b)^2$, $(TU)^2 = a^2 + c^2$, and $(US)^2 = b^2 + c^2$. Substituting, $(a - b)^2 = a^2 + c^2 + b^2 + c^2$; $a^2 - 2ab + b^2 = a^2 + b^2 + 2c^2$; $-2ab = 2c^2$; $-ab = c \cdot c$; $-1 = \left(-\frac{c}{a}\right) \cdot \left(-\frac{c}{b}\right) =$ slope of $\overleftrightarrow{TU}$ · slope of $\overleftrightarrow{US}$. **b.** Since $\left(-\frac{c}{a}\right) \cdot \left(-\frac{c}{b}\right) = -1$, $\frac{c^2}{ab} = -1$; $c^2 = -ab$; $2c^2 = -2ab$; $a^2 + b^2 + 2c^2 = a^2 - 2ab + b^2$; $a^2 + c^2 + b^2 + c^2 = (a - b)^2$. By the distance formula, $(ST)^2 = (a - b)^2$, $(TU)^2 = a^2 + c^2$, and $(US)^2 = b^2 + c^2$. Substituting, $(ST)^2 = (TU)^2 + (US)^2$. By the converse of the Pythagorean Thm., $\triangle SUT$ is a rt. $\triangle$, and $\overleftrightarrow{TU} \perp \overleftrightarrow{US}$.

Page 541 • CLASSROOM EXERCISES

1. **a.** $(4, -3)$ **b.** $(1, 4)$ **c.** $(-4 - 1, 4 - 4) = (-5, 0)$ **d.** $(-4 - (-4), 2 - 4) = (0, -2)$ **e.** $(5 - 4, 1 - (-3)) = (1, 4)$ **f.** $(-3 - 1, 0 - (-3)) = (-4, 3)$

2. **a.** $\sqrt{4^2 + (-3)^2} = 5$ **b.** $\sqrt{1^2 + 4^2} = \sqrt{17}$ **c.** $\sqrt{(-5)^2 + 0^2} = 5$ **d.** $\sqrt{0^2 + (-2)^2} = 2$ **e.** $\sqrt{1^2 + 4^2} = \sqrt{17}$ **f.** $\sqrt{(-4)^2 + 3^2} = 5$

3. **a.** Yes; slope of $\overrightarrow{BC} = \frac{4}{1} = 4$; slope of $\overrightarrow{OD} = \frac{4}{1} = 4$; since the slopes are equal, the vectors are parallel. **b.** Yes; they have the same magnitude and direction. **c.** parallelogram; $\overline{OD}$ and $\overline{BC}$ are $\parallel$ and $\cong$.

4. **a.** Yes; slope of $\overrightarrow{AG} = \frac{3}{-4} = -\frac{3}{4}$; slope of $\overrightarrow{OB} = \frac{-3}{4} = -\frac{3}{4}$; since the slopes are equal, the vectors are parallel. **b.** No; they do not have the same direction.

5. $|\overrightarrow{ST}| = \sqrt{4^2 + 9^2} = \sqrt{97}$; $\tan \angle S = \frac{\text{change in } y}{\text{change in } x} = \frac{4}{9}$

6. **a.** $(3 + 5, 1 + 6) = (8, 7)$ **b.** $(0 + 7, -6 + 4) = (7, -2)$ **c.** $(-3 + (-5), 10 + (-12)) = (-8, -2)$

7. **a.** $(2 \cdot 3, 2 \cdot 1) = (6, 2)$ **b.** $(3(-5), 3(1)) = (-15, 3)$ **c.** $\left(-\frac{1}{2}(-6), -\frac{1}{2}(0)\right) = (3, 0)$

8. $\overrightarrow{QP}$ or $-\overrightarrow{PQ}$

Pages 541–543 • WRITTEN EXERCISES

A

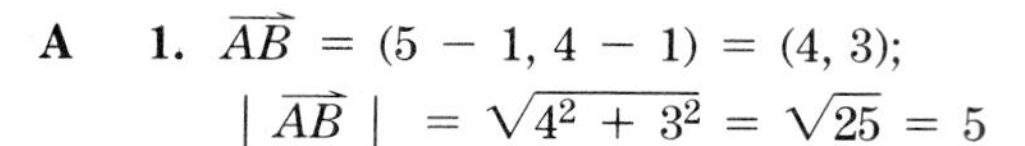

1. $\overrightarrow{AB} = (5 - 1, 4 - 1) = (4, 3)$;
$|\overrightarrow{AB}| = \sqrt{4^2 + 3^2} = \sqrt{25} = 5$

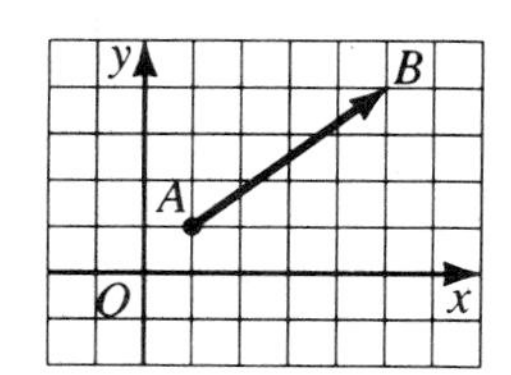

2. $\overrightarrow{AB} = (8 - 2, 8 - 0) = (6, 8)$;
$|\overrightarrow{AB}| = \sqrt{6^2 + 8^2} = \sqrt{100} = 10$

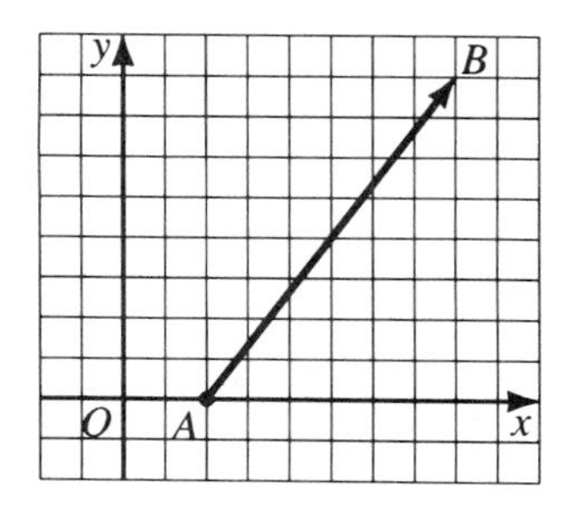

3. $\overrightarrow{AB} = (4 - 6, 3 - 1) = (-2, 2)$;
$|\overrightarrow{AB}| = \sqrt{(-2)^2 + 2^2} = \sqrt{8} = 2\sqrt{2}$

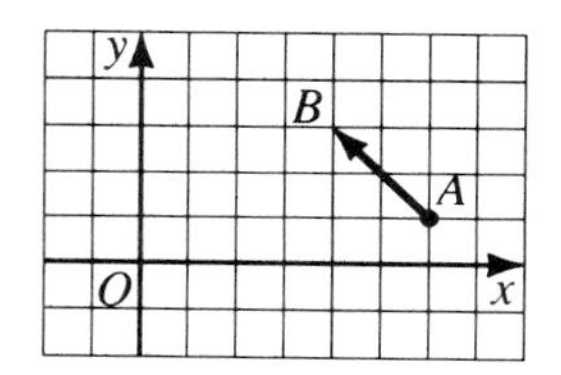

4. $\overrightarrow{AB} = (-3 - 0, 2 - 5) = (-3, -3)$;
$|\overrightarrow{AB}| = \sqrt{(-3)^2 + (-3)^2} = \sqrt{18} = 3\sqrt{2}$

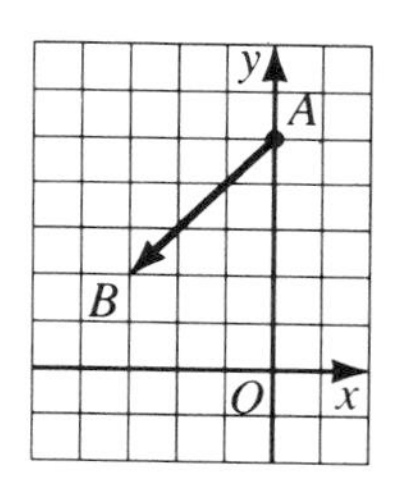

5. $\overrightarrow{AB} = (-1 - 3, 7 - 5) = (-4, 2)$;
$|\overrightarrow{AB}| = \sqrt{(-4)^2 + 2^2} = \sqrt{20} = 2\sqrt{5}$

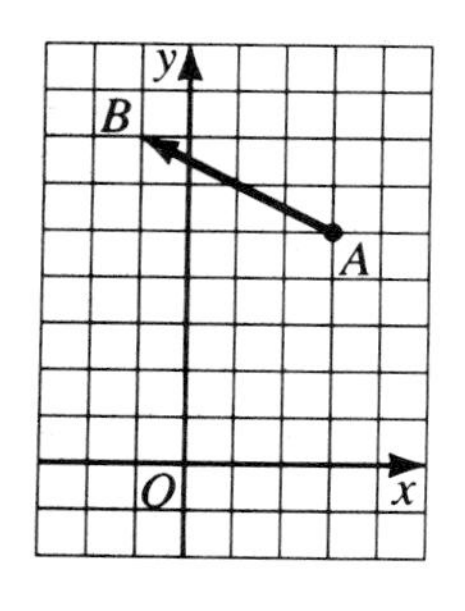

6. $\overrightarrow{AB} = (0 - 4, 0 - (-2)) = (-4, 2)$;
$|\overrightarrow{AB}| = \sqrt{(-4)^2 + 2^2} = \sqrt{20} = 2\sqrt{5}$

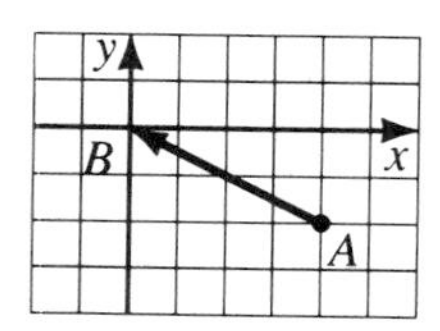

7. $\overrightarrow{AB} = (5 - 0, -9 - 0) = (5, -9)$; $|\overrightarrow{AB}| = \sqrt{5^2 + (-9)^2} = \sqrt{106}$

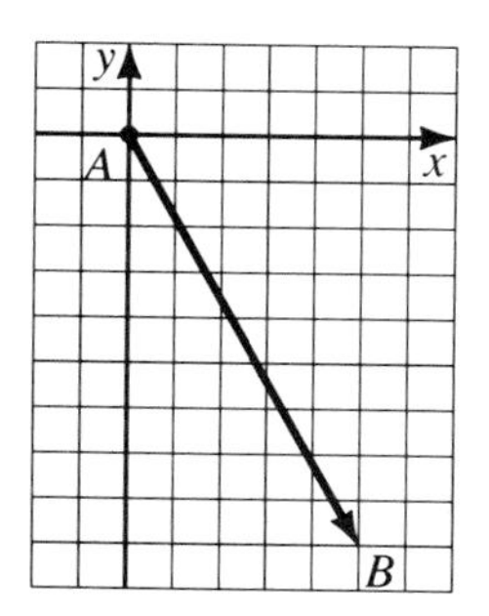

8. $\overrightarrow{AB} = (3 - (-3), 0 - 5) = (6, -5)$; $|\overrightarrow{AB}| = \sqrt{6^2 + (-5)^2} = \sqrt{61}$

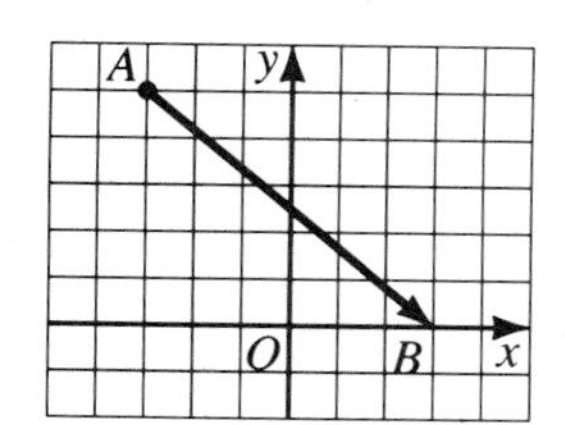

9. $\overrightarrow{AB} = (-4 - (-1), -7 - (-1)) = (-3, -6)$; $|\overrightarrow{AB}| = \sqrt{(-3)^2 + (-6)^2} = \sqrt{45} = 3\sqrt{5}$

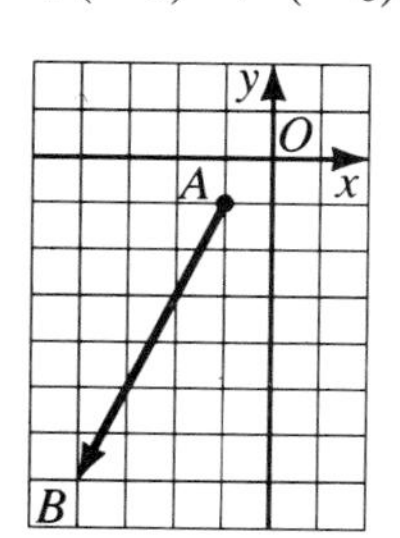

10.

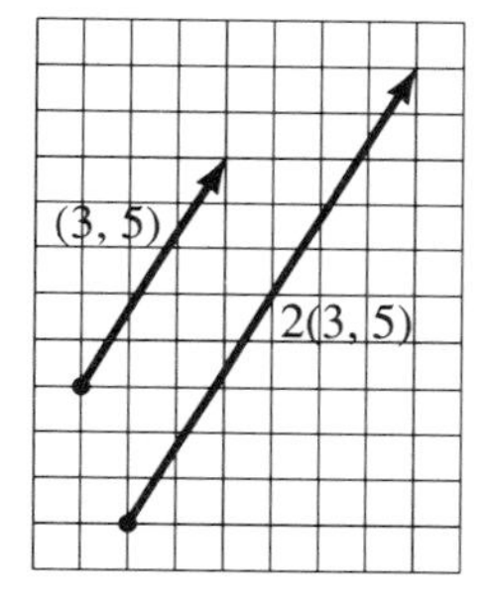

11.

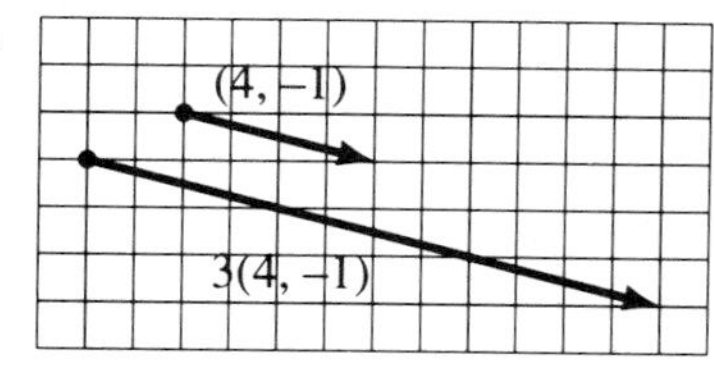

12.

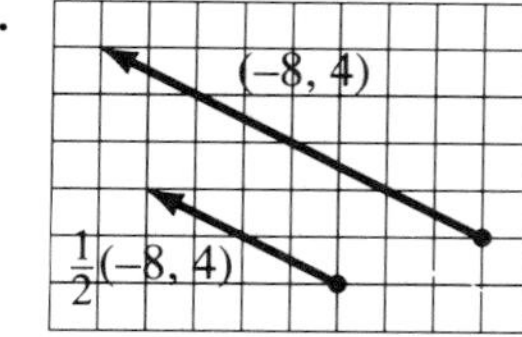

13.

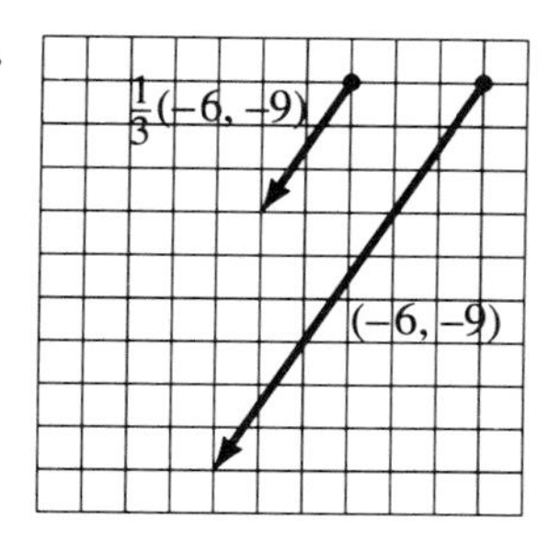

14.

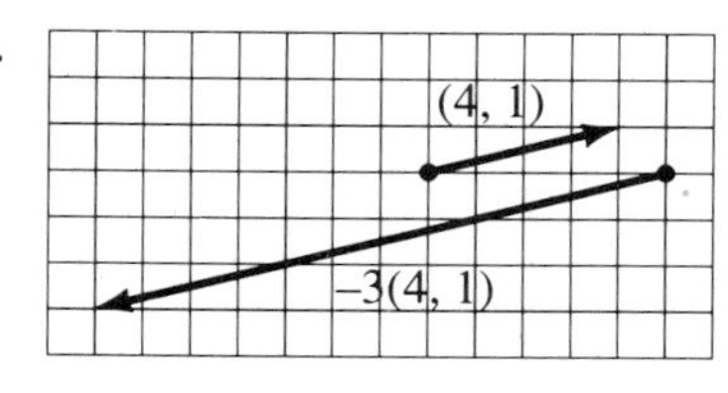

15.

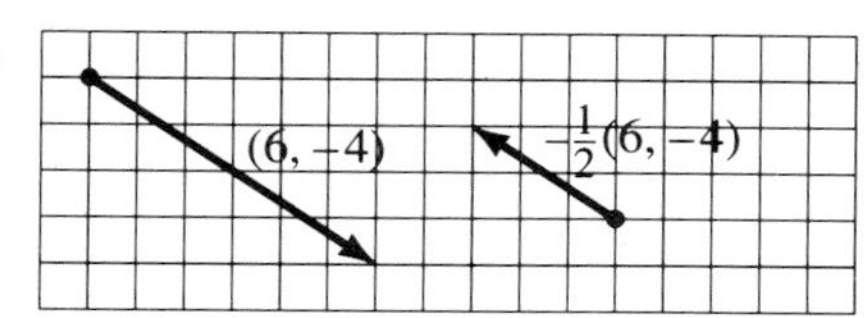

16. Any vector of the form $(3k, -8k)$, $k \neq 0$, is parallel to $(3, -8)$. For example, $(6, -16)$ and $(-3, 8)$ are parallel to $(3, -8)$.

17. slope of $(8, 6) = \frac{6}{8} = \frac{3}{4}; \frac{k}{12} = \frac{3}{4}; k = \frac{3}{4} \cdot 12 = 9$

18. slope of $(4, -5) = \frac{-5}{4} = -\frac{5}{4}$; slope of $(15, 12) = \frac{12}{15} = \frac{4}{5}$; since $-\frac{5}{4} \cdot \frac{4}{5} = -1$, the vectors are $\perp$.

19. slope of $(8, k) \cdot$ slope of $(9, 6) = \frac{k}{8} \cdot \frac{6}{9} = -1; 6k = -72; k = -12$

20. $(2, 1) + (3, 6) = (5, 7)$

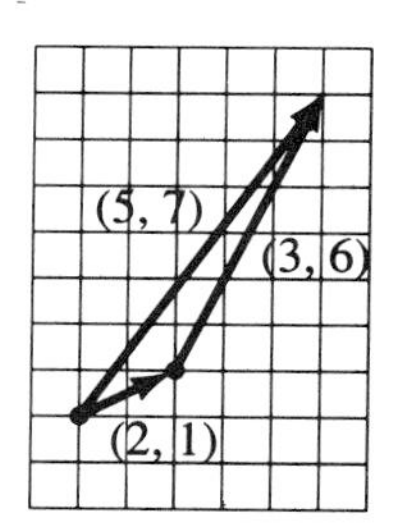

21. $(3, -5) + (4, 5) = (7, 0)$

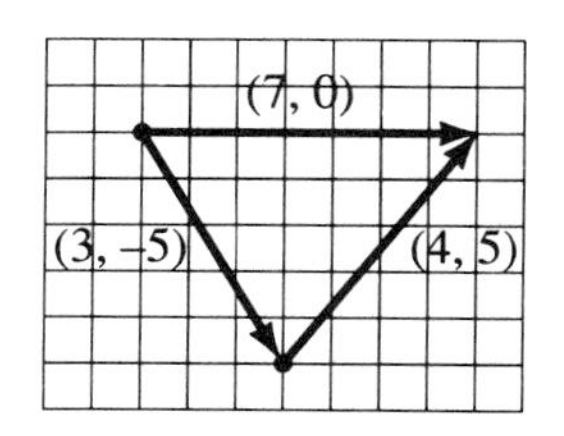

22. $(-8, 2) + (4, 6) = (-4, 8)$

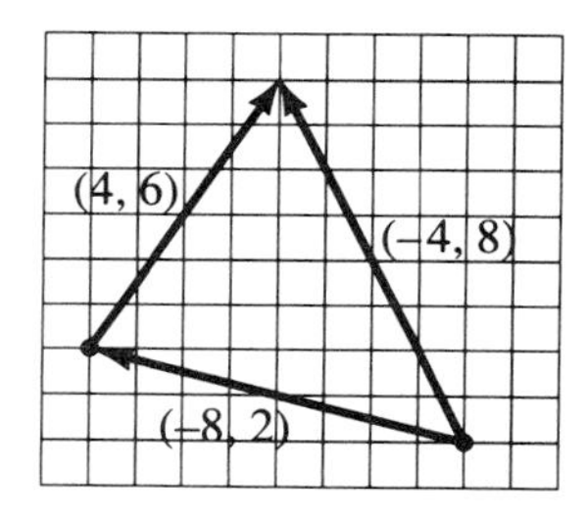

23. $(-3, -3) + (7, 7) = (4, 4)$

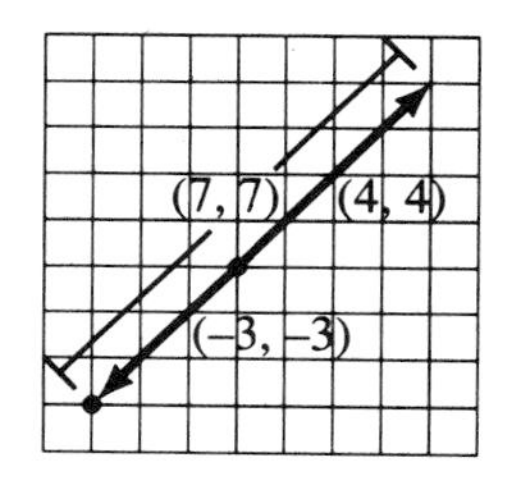

24. $(1, 4) + 2(3, 1) = (7, 6)$

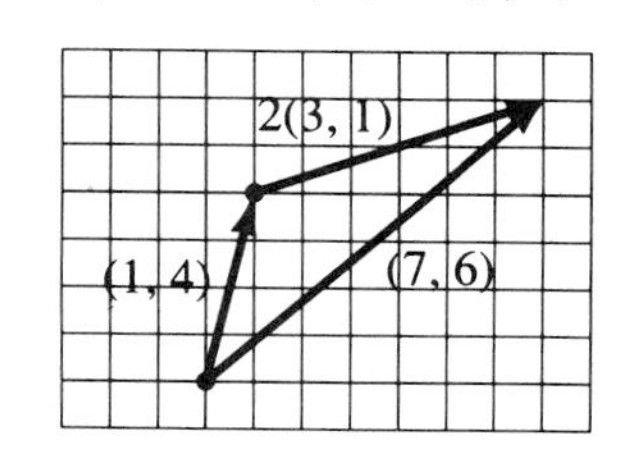

25. $(7, 2) + 3(-1, 0) = (4, 2)$

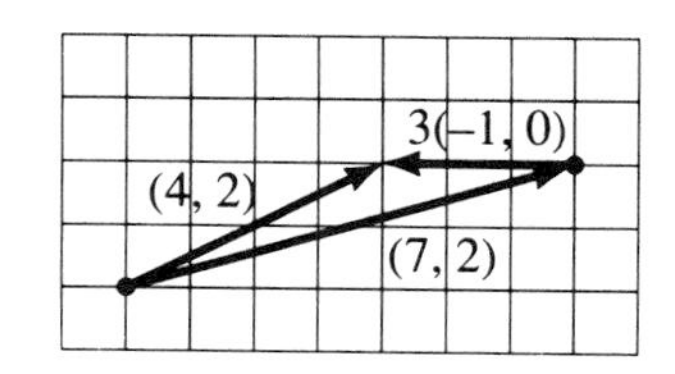

B 26. Since $|\overrightarrow{CD}| = |\overrightarrow{AB}|$ and $\overrightarrow{CD} \parallel \overrightarrow{AB}$, $\overrightarrow{CD}$ and $\overrightarrow{AB}$ are equal vectors. Thus, $\overrightarrow{AD} = \overrightarrow{AC} + \overrightarrow{CD} = \overrightarrow{AC} + \overrightarrow{AB}$.

27. $\overrightarrow{KZ} = \overrightarrow{KX} + \overrightarrow{KY} = (6, 8)$; $|\overrightarrow{KZ}| = \sqrt{6^2 + 8^2} = \sqrt{100} = 10$

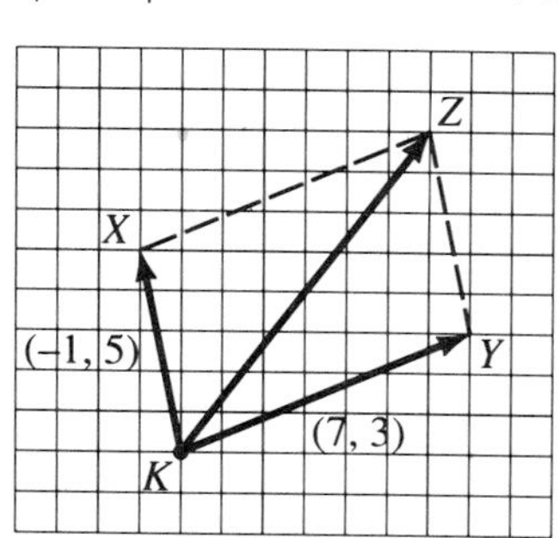

28. Since $\overrightarrow{KY} = -\overrightarrow{KX}$, $\overrightarrow{KZ} = \overrightarrow{KX} + \overrightarrow{KY} = (0, 0)$; $|\overrightarrow{KZ}| = \sqrt{0^2 + 0^2} = 0$

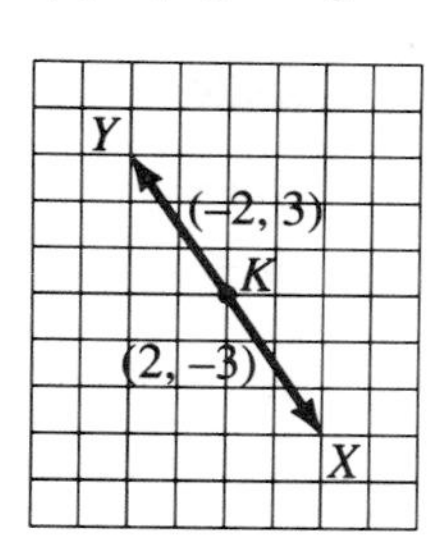

29. a. $\overrightarrow{AB} = (20 - 2, 21 - 3) = (18, 18)$; $\overrightarrow{AM} = \frac{1}{2}\overrightarrow{AB} = \frac{1}{2}(18, 18) = (9, 9)$; $\overrightarrow{AT} = \frac{1}{3}\overrightarrow{AB} = \frac{1}{3}(18, 18) = (6, 6)$ **b.** If $M = (r, s)$, then $(r - 2, s - 3) = \overrightarrow{AM} = (9, 9)$; $r - 2 = 9$ and $s - 3 = 9$; $r = 11$ and $s = 12$, then $M = (11, 12)$. If $T = (u, v)$, then $(u - 2, v - 3) = \overrightarrow{AT} = (6, 6)$; $u - 2 = 6$ and $v - 3 = 6$; $u = 8$ and $v = 9$, then $T = (8, 9)$.

30. a. $\overrightarrow{AB} = (20 - (-10), -15 - 9) = (30, -24)$; $\overrightarrow{AM} = \frac{1}{2}\overrightarrow{AB} = \frac{1}{2}(30, -24) = (15, -12)$; $\overrightarrow{AT} = \frac{1}{3}\overrightarrow{AB} = \frac{1}{3}(30, -24) = (10, -8)$ **b.** If $M = (r, s)$, then $(r + 10, s - 9) = \overrightarrow{AM} = (15, -12)$; $r + 10 = 15$ and $s - 9 = -12$; $r = 5$ and $s = -3$, then $M = (5, -3)$. If $T = (u, v)$, then $(u + 10, v - 9) = \overrightarrow{AT} = (10, -8)$; $u + 10 = 10$ and $v - 9 = -8$; $u = 0$ and $v = 1$, then $T = (0, 1)$.

31. $|(ka, kb)| = \sqrt{(ka)^2 + (kb)^2} = \sqrt{k^2a^2 + k^2b^2} = \sqrt{k^2(a^2 + b^2)} = |k|\sqrt{a^2 + b^2} = |k| \cdot |(a, b)|$

C 32. a. $k[(a, b) + (c, d)] = k(a + c, b + d) = [k(a + c), k(b + d)] = (ka + kc, kb + kd) = (ka, kb) + (kc, kd) = k(a, b) + k(c, d)$

b. Diagrams may vary. An example is given.

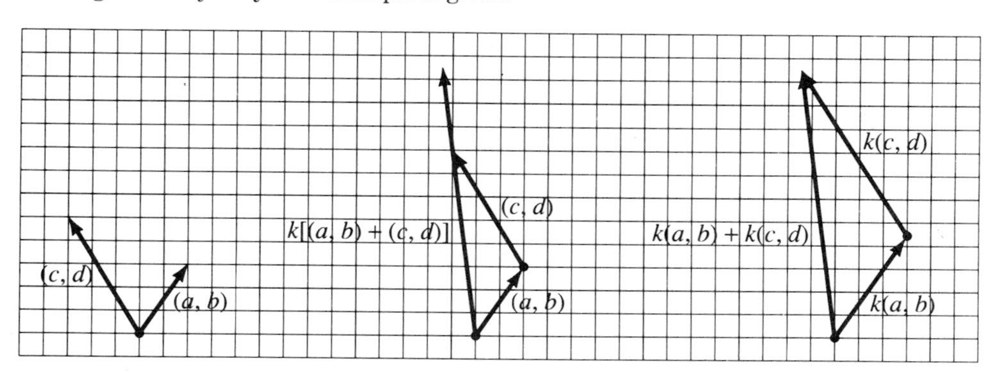

33. a. 1. Def. of vector sum 2. Substitution Prop. 3. $k[(a, b) + (c, d)] = k(a, b) + k(c, d)$ 4. Def. of vector sum **b.** The seg. that joins the midpts. of 2 sides of a $\triangle$ is half as long as the third side.

34. a. slope of (a, b) · slope of $(c, d) = \frac{b}{a} \cdot \frac{d}{c} = -1$; $\frac{bd}{ac} = -1$ **b.** $bd = -ac$; $ac + bd = 0$ **c.** is zero **d.** Ex. 2(b), p. 540: $9 \cdot 2 + (-6)(3) = 18 - 18 = 0$; Ex. 18, p. 542: $4 \cdot 15 + (-5)12 = 60 - 60 = 0$

Page 543 • MIXED REVIEW EXERCISES

1. $\frac{-11 + 7}{2} = \frac{-4}{2} = -2$, or $-11 + \frac{1}{2}(7 - (-11)) = -11 + \frac{1}{2}(18) = -11 + 9 = -2$

2. Since the midpt. of the hyp. of a rt. $\triangle$ is equidistant from the 3 vertices, $MB = MC = AM = 6$.

3. $\frac{1}{2}(12 + 20) = \frac{1}{2} \cdot 32 = 16$

4. The altitude divides the $\triangle$ into 2 $\cong$ 30°–60°–90° $\triangle$s; hyp. $= 2a$; shorter leg $= a$; altitude $=$ longer leg $= a\sqrt{3}$

5. measure of each ext. $\angle = \frac{360}{6} = 60$; measure of each int. $\angle = 180 - 60 = 120$

6. From Ex. 5, $m\angle B = 120$; $AB = BC$, so $m\angle BAC = m\angle BCA = 30$; $m\angle ACD = 120 - 30 = 90$; $m\angle CDA = \frac{1}{2} \cdot 120 = 60$, so $m\angle CAD = 30$. In 30°–60°–90° $\triangle CDA$, $CD = x$, so $AD = 2x$ and $AC = x\sqrt{3}$.

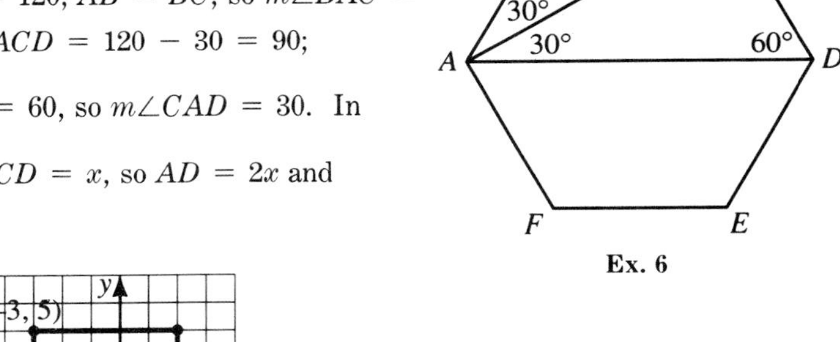

Ex. 6

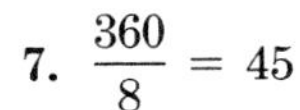

7. $\frac{360}{8} = 45$

8.

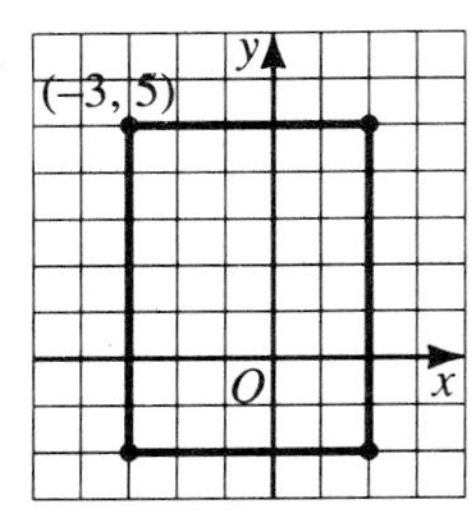

9. Quad. $ABCD$ is a rect.; $AB = |7 - 2| = 5$; $AD = |5 - 0| = 5$; Area $= bh = AB \cdot AD = 5 \cdot 5 = 25$

10. $\triangle PQR$ is a rt. $\triangle$; $b = PQ = |-6 - 0| = 6$; $h = QR = |6 - 0| = 6$; Area $= \frac{1}{2}bh = \frac{1}{2} \cdot 6 \cdot 6 = 18$.

11. a. $(DE)^2 = (-2 - (-5))^2 + (-3 - 1)^2 = 3^2 + (-4)^2 = 9 + 16 = 25$; $(EF)^2 = (6 - (-2))^2 + (3 - (-3))^2 = 8^2 + 6^2 = 64 + 36 = 100$; $(DF)^2 = (6 - (-5))^2 + (3 - 1)^2 = 11^2 + 2^2 = 121 + 4 = 125$; $(DE)^2 + (EF)^2 = 25 + 100 = 125 = (DF)^2$, so $\triangle DEF$ is a rt. $\triangle$ by the converse of the Pythagorean Thm. **b.** slope of $\overline{DE}$ · slope of $\overline{EF} = \frac{-3 - 1}{-2 - (-5)} \cdot \frac{3 - (-3)}{6 - (-2)} = \frac{-4}{3} \cdot \frac{3}{4} = -1$, so $\overline{DE} \perp \overline{EF}$, $\angle DEF$ is a rt. $\angle$, and $\triangle DEF$ is a rt. $\triangle$.

12. a. $\overline{AC}$ has slope $\frac{6 - 0}{2 - 6} = \frac{6}{-4} = -\frac{3}{2}$, so the altitude to $\overline{AC}$ has slope $\frac{-1}{-\frac{3}{2}} = \frac{2}{3}$.

b. $\overline{AB}$ has slope $\frac{8 - 0}{4 - 6} = \frac{8}{-2} = -4$, so the $\perp$ bisector of $\overline{AB}$ has slope $\frac{-1}{-4} = \frac{1}{4}$.

Page 545 • CLASSROOM EXERCISES

1. $\left(\frac{3 + 7}{2}, \frac{5 + 5}{2}\right) = (5, 5)$ **2.** $\left(\frac{0 + 4}{2}, \frac{4 + 3}{2}\right) = \left(2, \frac{7}{2}\right)$

3. $\left(\frac{-2 + 6}{2}, \frac{2 + 4}{2}\right) = (2, 3)$ **4.** $\left(\frac{-3 - 7}{2}, \frac{7 - 5}{2}\right) = (-5, 1)$

5. $\left(\frac{-1 - 3}{2}, \frac{-3 + 6}{2}\right) = \left(-2, \frac{3}{2}\right)$ **6.** $\left(\frac{2b + 4}{2}, \frac{3 - 5}{2}\right) = (b + 2, -1)$

7. $\left(\frac{t + (t + 4)}{2}, \frac{2 - 4}{2}\right) = \left(\frac{2t + 4}{2}, -1\right) = (t + 2, -1)$ **8.** $\left(\frac{a + d}{2}, \frac{n + p}{2}\right)$

9. Let $P_2 = (x_2, y_2)$; $\left(\frac{0 + x_2}{2}, \frac{1 + y_2}{2}\right) = (3, 5)$; $\frac{x_2}{2} = 3$ and $\frac{1 + y_2}{2} = 5$; $x_2 = 6$ and $y_2 = 10 - 1 = 9$; $P_2 = (6, 9)$

10. Let $B = (s, t)$; $\left(\frac{-1 + s}{2}, \frac{3 + t}{2}\right) = (1, -1)$; $\frac{-1 + s}{2} = 1$ and $\frac{3 + t}{2} = -1$; $-1 + s = 2$ and $3 + t = -2$; $s = 3$ and $t = -5$; $B = (3, -5)$

Pages 545–547 • WRITTEN EXERCISES

A **1.** $\left(\frac{0 + 6}{2}, \frac{2 + 4}{2}\right) = (3, 3)$ **2.** $\left(\frac{-2 + 4}{2}, \frac{6 + 3}{2}\right) = \left(1, \frac{9}{2}\right)$

3. $\left(\frac{6 + (-6)}{2}, \frac{-7 + 3}{2}\right) = (0, -2)$ **4.** $\left(\frac{a + a + 2}{2}, \frac{4 + 0}{2}\right) = (a + 1, 2)$

5. $\left(\frac{2.3 + 1.5}{2}, \frac{3.7 - 2.9}{2}\right) = (1.9, 0.4)$ **6.** $\left(\frac{a + c}{2}, \frac{b + d}{2}\right)$

7. $PQ = \sqrt{(-5 - 3)^2 + (2 - (-8))^2} = \sqrt{(-8)^2 + 10^2} = \sqrt{64 + 100} = 2\sqrt{41}$;
slope $= \frac{2 - (-8)}{-5 - 3} = \frac{10}{-8} = -\frac{5}{4}$; midpt. $= \left(\frac{3 - 5}{2}, \frac{-8 + 2}{2}\right) = (-1, -3)$

8. $PQ = \sqrt{(7 - (-3))^2 + (8 - 4)^2} = \sqrt{10^2 + 4^2} = \sqrt{100 + 16} = 2\sqrt{29}$;
slope $= \frac{8 - 4}{7 - (-3)} = \frac{4}{10} = \frac{2}{5}$; midpt. $= \left(\frac{-3 + 7}{2}, \frac{4 + 8}{2}\right) = (2, 6)$

9. $PQ = \sqrt{(1 - (-7))^2 + (-4 - 11)^2} = \sqrt{8^2 + (-15)^2} = \sqrt{64 + 225} =$
$\sqrt{289} = 17$; slope $= \frac{-4 - 11}{1 - (-7)} = \frac{-15}{8} = -\frac{15}{8}$; midpt. $= \left(\frac{-7 + 1}{2}, \frac{11 - 4}{2}\right) =$
$\left(-3, \frac{7}{2}\right)$

10. $\frac{x + 4}{2} = 4$; $x + 4 = 8$; $x = 4$; $\frac{y - 2}{2} = 4$; $y - 2 = 8$; $y = 10$; $B(4, 10)$

11. $\frac{x + 1}{2} = 5$; $x + 1 = 10$; $x = 9$; $\frac{y - 3}{2} = 1$; $y - 3 = 2$; $y = 5$; $B(9, 5)$

12. $\frac{x + r}{2} = 0$; $x + r = 0$; $x = -r$; $\frac{y + s}{2} = 2$; $y + s = 4$; $y = 4 - s$; $B(-r, 4 - s)$

B **13.** *Sol. 1*: midpt. $M = \left(\frac{0 + 8}{2}, \frac{0 + 4}{2}\right) = (4, 2)$; slope of $\overline{AB} = \frac{4 - 0}{8 - 0} = \frac{1}{2}$; slope of
$\overline{PM} = \frac{2 - 6}{4 - 2} = \frac{-4}{2} = -2$; $\frac{1}{2}(-2) = -1$, so $\overline{PM} \perp \overline{AB}$.
Sol. 2: $PA = \sqrt{(2 - 0)^2 + (6 - 0)^2} = \sqrt{2^2 + 6^2} = \sqrt{4 + 36} = 2\sqrt{10}$;
$PB = \sqrt{(2 - 8)^2 + (6 - 4)^2} = \sqrt{(-6)^2 + 2^2} = \sqrt{36 + 4} = 2\sqrt{10}$; $PA = PB$, so
P is on the $\perp$ bisector of $\overline{AB}$.

14. **a.** $RS = \sqrt{(7 - 1)^2 + (4 - 0)^2} = \sqrt{6^2 + 4^2} = \sqrt{36 + 16} = 2\sqrt{13}$;
$ST = \sqrt{(11 - 7)^2 + (-2 - 4)^2} = \sqrt{4^2 + (-6)^2} = \sqrt{16 + 36} = 2\sqrt{13}$;
$\overline{RS} \cong \overline{ST}$, so $\triangle RST$ is isos. **b.** $K =$ midpt. of $\overline{RT} = \left(\frac{1 + 11}{2}, \frac{0 - 2}{2}\right) =$
$(6, -1)$.

15. $\overline{DE}$ and $\overline{CF}$ have slope 0, so $\overline{DE} \parallel \overline{CF}$; let $M =$ midpt. of $\overline{CD}$;
M is $\left(\frac{-1 - 4}{2}, \frac{4 - 3}{2}\right) = \left(-\frac{5}{2}, \frac{1}{2}\right)$; let $N =$ midpt. of $\overline{EF}$;
N is $\left(\frac{4 + 7}{2}, \frac{4 - 3}{2}\right) = \left(\frac{11}{2}, \frac{1}{2}\right)$; length of median $=$
$MN = \sqrt{\left(\frac{11}{2} - \left(-\frac{5}{2}\right)\right)^2 + \left(\frac{1}{2} - \frac{1}{2}\right)^2} = \sqrt{8^2 + 0^2} = 8$

16. Midpt. M of $\overline{XY} = \left(\frac{-2 + 6}{2}, \frac{3 - 3}{2}\right) = (2, 0)$; $MZ = \sqrt{(4 - 2)^2 + (7 - 0)^2} = \sqrt{2^2 + 7^2} = \sqrt{4 + 49} = \sqrt{53}$; midpt. N of $\overline{YZ} = \left(\frac{6 + 4}{2}, \frac{-3 + 7}{2}\right) = (5, 2)$; $XN = \sqrt{(-2 - 5)^2 + (3 - 2)^2} = \sqrt{(-7)^2 + 1^2} = \sqrt{49 + 1} = 5\sqrt{2}$; midpt. P of $\overline{XZ} = \left(\frac{-2 + 4}{2}, \frac{3 + 7}{2}\right) = (1, 5)$; $YP = \sqrt{(6 - 1)^2 + (-3 - 5)^2} = \sqrt{5^2 + (-8)^2} = \sqrt{25 + 64} = \sqrt{89}$; the longest median, $\overline{YP}$, has length $\sqrt{89}$.

17. a. Midpt. of $\overline{OQ} = \left(\frac{0 + 9}{2}, \frac{0 + 9}{2}\right) = \left(\frac{9}{2}, \frac{9}{2}\right)$; midpt. of $\overline{PR} = \left(\frac{2 + 7}{2}, \frac{6 + 3}{2}\right) = \left(\frac{9}{2}, \frac{9}{2}\right)$ **b.** parallelogram **c.** slope of $\overline{PQ} = \frac{9 - 6}{9 - 2} = \frac{3}{7}$; slope of $\overline{OR} = \frac{3 - 0}{7 - 0} = \frac{3}{7}$; $\overline{PQ} \parallel \overline{OR}$; slope of $\overline{PO} = \frac{6 - 0}{2 - 0} = 3$; slope of $\overline{QR} = \frac{9 - 3}{9 - 7} = \frac{6}{2} = 3$; $\overline{PO} \parallel \overline{QR}$
d. $PQ = \sqrt{(9 - 2)^2 + (9 - 6)^2} = \sqrt{7^2 + 3^2} = \sqrt{49 + 9} = \sqrt{58}$;
$OR = \sqrt{(7 - 0)^2 + (3 - 0)^2} = \sqrt{7^2 + 3^2} = \sqrt{49 + 9} = \sqrt{58}$; $\overline{PQ} \cong \overline{OR}$;
$PO = \sqrt{(2 - 0)^2 + (6 - 0)^2} = \sqrt{2^2 + 6^2} = \sqrt{4 + 36} = 2\sqrt{10}$;
$QR = \sqrt{(9 - 7)^2 + (9 - 3)^2} = \sqrt{2^2 + 6^2} = \sqrt{4 + 36} = 2\sqrt{10}$;
$\overline{PO} \cong \overline{QR}$

18. a. Slope of $\overline{AD} = \frac{4 - 0}{-3 - (-5)} = \frac{4}{2} = 2$; slope of $\overline{BC} = \frac{6 - 2}{5 - 3} = \frac{4}{2} = 2$; $\overline{AD} \parallel \overline{BC}$; $AD = \sqrt{(-3 - (-5))^2 + (4 - 0)^2} = \sqrt{2^2 + 4^2} = \sqrt{4 + 16} = \sqrt{20} = 2\sqrt{5}$; $BC = \sqrt{(5 - 3)^2 + (6 - 2)^2} = \sqrt{2^2 + 4^2} = \sqrt{4 + 16} = \sqrt{20} = 2\sqrt{5}$; $\overline{AD} \cong \overline{BC}$ **b.** midpt. of $\overline{AC} = \left(\frac{-5 + 5}{2}, \frac{0 + 6}{2}\right) = (0, 3)$; midpt. of $\overline{BD} = \left(\frac{3 - 3}{2}, \frac{2 + 4}{2}\right) = (0, 3)$

y
C(5, 6)
D(–3, 4)
B(3, 2)
A(–5, 0)
O
x

19. a. $\left(\frac{0 - 6}{2}, \frac{8 + 0}{2}\right) = (-3, 4)$ **b.** $MA = \sqrt{(0 - (-3))^2 + (8 - 4)^2} = \sqrt{3^2 + 4^2} = \sqrt{9 + 16} = \sqrt{25} = 5$; $MT = \sqrt{(-6 - (-3))^2 + (0 - 4)^2} = \sqrt{(-3)^2 + (-4)^2} = \sqrt{9 + 16} = \sqrt{25} = 5$; $MO = \sqrt{(0 - (-3))^2 + (0 - 4)^2} = \sqrt{3^2 + (-4)^2} = \sqrt{9 + 16} = \sqrt{25} = 5$; $MA = MT = MO$ **c.** The midpt. of the hyp. of a rt. $\triangle$ is equidistant from the vertices. **d.** The circle has ctr. $M(-3, 4)$ and radius 5; $(x + 3)^2 + (y - 4)^2 = 25$.

20. **a.** $D = \left(\frac{1 + 13}{2}, \frac{1 + 9}{2}\right) = (7, 5)$; $E = \left(\frac{1 + 3}{2}, \frac{1 + 7}{2}\right) = (2, 4)$ **b.** slope of $\overline{DE} = \frac{5 - 4}{7 - 2} = \frac{1}{5}$; slope of $\overline{BC} = \frac{9 - 7}{13 - 3} = \frac{2}{10} = \frac{1}{5}$; $\overline{DE} \parallel \overline{BC}$

c. $DE = \sqrt{(7 - 2)^2 + (5 - 4)^2} = \sqrt{5^2 + 1^2} = \sqrt{25 + 1} = \sqrt{26}$;

$BC = \sqrt{(13 - 3)^2 + (9 - 7)^2} = \sqrt{10^2 + 2^2} = \sqrt{100 + 4} = 2\sqrt{26}$; $DE = \frac{1}{2}BC$

21. **a.** $J = \left(\frac{-1 + 0}{2}, \frac{3 + 0}{2}\right) = \left(-\frac{1}{2}, \frac{3}{2}\right)$; $K = \left(\frac{-1 + 7}{2}, \frac{3 + 9}{2}\right) = (3, 6)$;

$L = \left(\frac{7 + 10}{2}, \frac{9 + 0}{2}\right) = \left(\frac{17}{2}, \frac{9}{2}\right)$; $M = \left(\frac{0 + 10}{2}, \frac{0 + 0}{2}\right) = (5, 0)$

b. a rhombus; $JK = \sqrt{\left(-\frac{1}{2} - 3\right)^2 + \left(\frac{3}{2} - 6\right)^2} = \sqrt{\left(-\frac{7}{2}\right)^2 + \left(-\frac{9}{2}\right)^2} =$

$\sqrt{\frac{49}{4} + \frac{81}{4}} = \frac{\sqrt{130}}{2}$; $KL = \sqrt{\left(\frac{17}{2} - 3\right)^2 + \left(\frac{9}{2} - 6\right)^2} = \sqrt{\left(\frac{11}{2}\right)^2 + \left(-\frac{3}{2}\right)^2} =$

$\sqrt{\frac{121}{4} + \frac{9}{4}} = \frac{\sqrt{130}}{2}$; $LM = \sqrt{\left(\frac{17}{2} - 5\right)^2 + \left(\frac{9}{2} - 0\right)^2} = \sqrt{\left(\frac{7}{2}\right)^2 + \left(\frac{9}{2}\right)^2} =$

$\sqrt{\frac{49}{4} + \frac{81}{4}} = \frac{\sqrt{130}}{2}$; $JM = \sqrt{\left(-\frac{1}{2} - 5\right)^2 + \left(\frac{3}{2} - 0\right)^2} = \sqrt{\left(-\frac{11}{2}\right)^2 + \left(\frac{3}{2}\right)^2} =$

$\sqrt{\frac{121}{4} + \frac{9}{4}} = \frac{\sqrt{130}}{2}$; $JK = KL = LM = JM$, so $JKLM$ is a rhombus.

22. Let M be the midpoint of $\overline{PQ}$. Then E is the midpoint of $\overline{PM}$.

$$M = \left(\frac{x_1 + x_2}{2}, \frac{y_1 + y_2}{2}\right) \text{ and } E = \left(\frac{x_1 + \frac{x_1 + x_2}{2}}{2}, \frac{y_1 + \frac{y_1 + y_2}{2}}{2}\right) =$$

$$\left(\frac{2x_1 + x_1 + x_2}{4}, \frac{2y_1 + y_1 + y_2}{4}\right) = \left(\frac{3}{4}x_1 + \frac{1}{4}x_2, \frac{3}{4}y_1 + \frac{1}{4}y_2\right).$$

C **23.** As in Ex. 22, let M be the midpt. of $\overline{PQ}$, E be the midpt. of $\overline{PM}$, and $E = \left(\frac{3}{4}x_1 + \frac{1}{4}x_2, \frac{3}{4}y_1 + \frac{1}{4}y_2\right)$. Then F is the midpt. of $\overline{EM}$. $F =$

$$\left(\frac{\frac{3}{4}x_1 + \frac{1}{4}x_2 + \frac{1}{2}x_1 + \frac{1}{2}x_2}{2}, \frac{\frac{3}{4}y_1 + \frac{1}{4}y_2 + \frac{1}{2}y_1 + \frac{1}{2}y_2}{2}\right) =$$

$$\left(\frac{3x_1 + x_2 + 2x_1 + 2x_2}{8}, \frac{3y_1 + y_2 + 2y_1 + 2y_2}{8}\right) = \left(\frac{5}{8}x_1 + \frac{3}{8}x_2, \frac{5}{8}y_1 + \frac{3}{8}y_2\right).$$

24. Draw $\overline{PQ}$ where $Q = (7, 1)$. The vertical lines through (3, 1), (4, 1), (5, 1), and (6, 1) divide $\overline{PQ}$, and thus $\overline{PD}$, into 5 ≅ segments. (If ∥ lines cut off ≅ segments on one transversal, then they cut off ≅ segments on every transversal.) The pt. T, where the vertical line through (4, 1) intersects $\overline{PD}$ is the required pt.; let T have coordinates $(4, t)$; since T lies on $\overline{PD}$, slope of $\overline{PT}$ = slope of $\overline{PD}$; $\frac{t-1}{4-2} = \frac{11-1}{7-2}$; $\frac{t-1}{2} = \frac{10}{5} = 2$; $t - 1 = 4$; $t = 5$; $T = (4, 5)$.

Page 547 • SELF-TEST 1

1. a. $|5 - 3| = |2| = 2$ **b.** $\left(\frac{5+3}{2}, \frac{1+1}{2}\right) = \left(\frac{8}{2}, \frac{2}{2}\right) = (4, 1)$

2. a. $\sqrt{(8-0)^2 + (-6-0)^2} = \sqrt{8^2 + (-6)^2} = \sqrt{64 + 36} = \sqrt{100} = 10$

b. $\left(\frac{8+0}{2}, \frac{-6+0}{2}\right) = \left(\frac{8}{2}, \frac{-6}{2}\right) = (4, -3)$

3. a. $\sqrt{(-2-8)^2 + (7-(-3))^2} = \sqrt{(-10)^2 + 10^2} = \sqrt{100 + 100} = \sqrt{200} = 10\sqrt{2}$ **b.** $\left(\frac{-2+8}{2}, \frac{7-3}{2}\right) = \left(\frac{6}{2}, \frac{4}{2}\right) = (3, 2)$

4. a. $\sqrt{(-3-(-5))^2 + (2-7)^2} = \sqrt{2^2 + (-5)^2} = \sqrt{4 + 25} = \sqrt{29}$

b. $\left(\frac{-3-5}{2}, \frac{2+7}{2}\right) = \left(\frac{-8}{2}, \frac{9}{2}\right) = \left(-4, \frac{9}{2}\right)$

5. $(x - 0)^2 + (y - 0)^2 = 9^2$; $x^2 + y^2 = 81$

6. $(x - (-1))^2 + (y - 2)^2 = 5^2$; $(x + 1)^2 + (y - 2)^2 = 25$

7. $(x - (-2))^2 + (y - 3)^2 = 6^2$; center, $(-2, 3)$; radius, 6 **8.** $\frac{4-0}{7-0} = \frac{4}{7}$

9. $\frac{-1-2}{1-(-4)} = \frac{-3}{5} = -\frac{3}{5}$ **10.** vertical

11. a. Slope of $\overleftrightarrow{PQ} = \frac{2-(-2)}{5-3} = \frac{4}{2} = 2$; slope of any line parallel to $\overleftrightarrow{PQ} = 2$

b. Slope of any line perpendicular to $\overleftrightarrow{PQ} = \frac{-1}{2} = -\frac{1}{2}$.

12. a. $\overrightarrow{AB} = (2 - (-4), 3 - 5) = (6, -2)$ **b.** $\overrightarrow{CD} = (-2 - 1, -1 - 2) = (-3, -3)$ **c.** $\overrightarrow{FE} = (3 - 3, 1 - (-3)) = (0, 4)$

13. a. $|\overrightarrow{AB}| = \sqrt{6^2 + (-2)^2} = \sqrt{36 + 4} = \sqrt{40} = 2\sqrt{10}$ **b.** $|\overrightarrow{CD}| = \sqrt{(-3)^2 + (-3)^2} = \sqrt{9 + 9} = \sqrt{18} = 3\sqrt{2}$ **c.** $|\overrightarrow{FE}| = \sqrt{0^2 + 4^2} = \sqrt{4^2} = 4$

14. **a.** $(-3 + 7, 2 - 11) = (4, -9)$ **b.** $(3(4), 3(-1)) + (-2(-5), -2(3)) = (12, -3) + (10, -6) = (12 + 10, -3 - 6) = (22, -9)$

15. Let $Q = (x, y)$; $\left(\frac{x + 9}{2}, \frac{y - 4}{2}\right) = (-3, 7)$; $\frac{x + 9}{2} = -3$ and $\frac{y - 4}{2} = 7$; $x + 9 = -6$ and $y - 4 = 14$, $x = -15$ and $y = 18$; $Q(-15, 18)$

Page 550 • CLASSROOM EXERCISES

1. **a.** $3 \cdot 0 - 2 \cdot 4 = -8 \neq 12$ **b.** $3 \cdot 2 - 2(-3) = 12$; $(2, -3)$ lies on the line.
c. $3 \cdot 3 - 2 \cdot \frac{3}{2} = 6 \neq 12$ **d.** $3 \cdot 0 - 2(-6) = 12$; $(0, -6)$ lies on the line.

2. **a.** $2(-2) + 3 \cdot 5 = 11 \neq 10$ **b.** $-4 + 2 \cdot 6 = 8$ and $2(-4) + 3 \cdot 6 = 10$, so $(-4, 6)$ is the int. pt. **c.** $2 \cdot 2 + 3 \cdot 3 = 13 \neq 10$ **d.** $-1 + 2 \cdot 4 = 7 \neq 8$

3. $2x + 3(0) = 6$; $x = 3$; $2 \cdot 0 + 3y = 6$; $y = 2$

4. $3x - 5 \cdot 0 = 15$; $x = 5$; $3 \cdot 0 - 5y = 15$; $y = -3$

5. $-4x + 3 \cdot 0 = 24$; $x = -6$; $-4 \cdot 0 + 3y = 24$; $y = 8$

6. $x + 3 \cdot 0 = 9$; $x = 9$; $0 + 3y = 9$; $y = 3$

7. $0 = 5x - 10$; $x = 2$; $y = 5 \cdot 0 - 10$; $y = -10$

8. $0 = 2x + 5$; $x = -\frac{5}{2}$; $y = 2 \cdot 0 + 5$; $y = 5$ **9.** $\frac{2}{5}$; -9

10. $y = -2x + 8$; slope, -2; y-intercept, 8

11. $-4y = -3x + 6$; $y = \frac{3}{4}x - \frac{3}{2}$; slope, $\frac{3}{4}$; y-intercept, $-\frac{3}{2}$

12. $y = 4$; $y = 0x + 4$; $m = 0$; answers will vary. Examples: $(1, 4)$, $(-1, 4)$, $(2, 4)$

13. Answers will vary. Examples: $(5, 1)$, $(5, -1)$, $(5, 4)$

Pages 550–552 • WRITTEN EXERCISES

A 1.

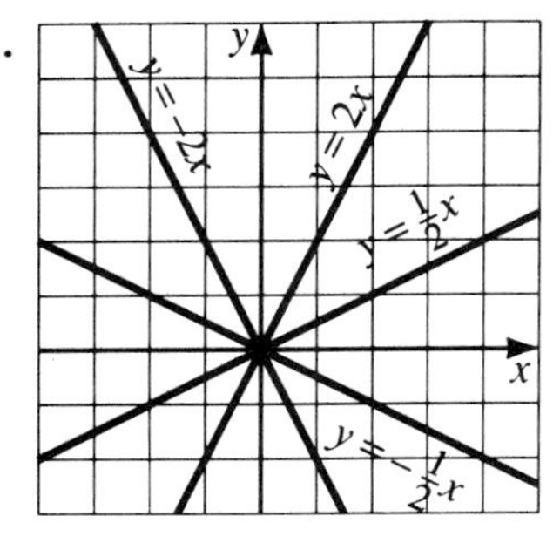

2.

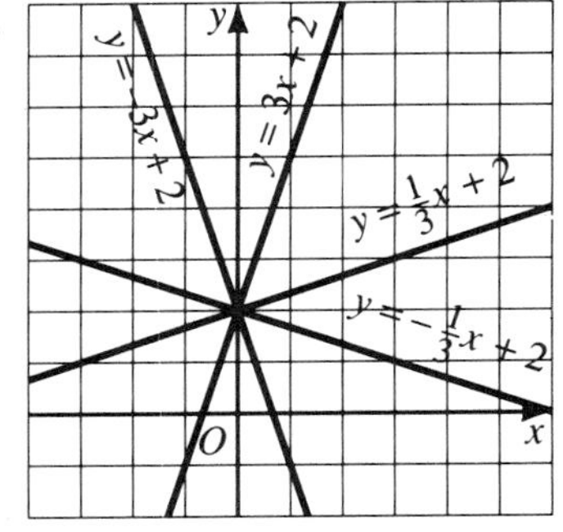

3.

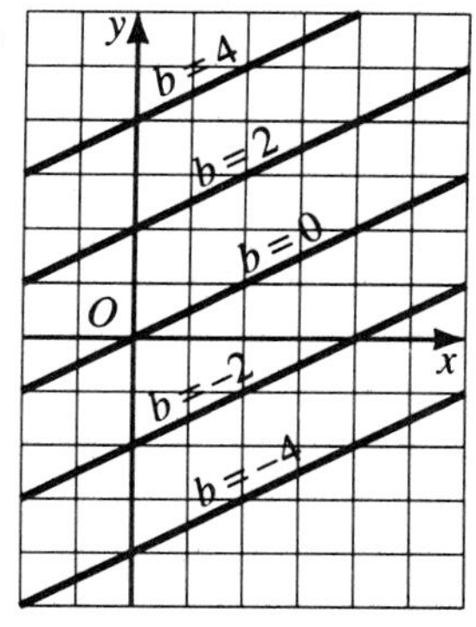

4.

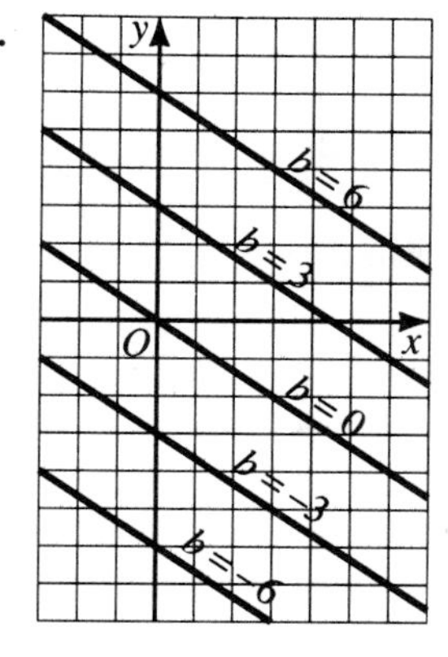

5.

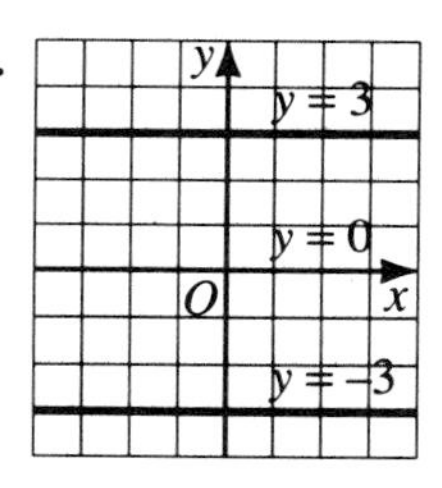

6.

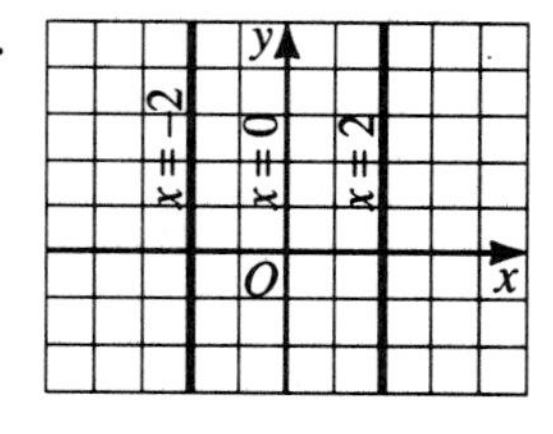

7. $3x + 0 = -21;\ x = -7;$
$3 \cdot 0 + y = -21;\ y = -21$

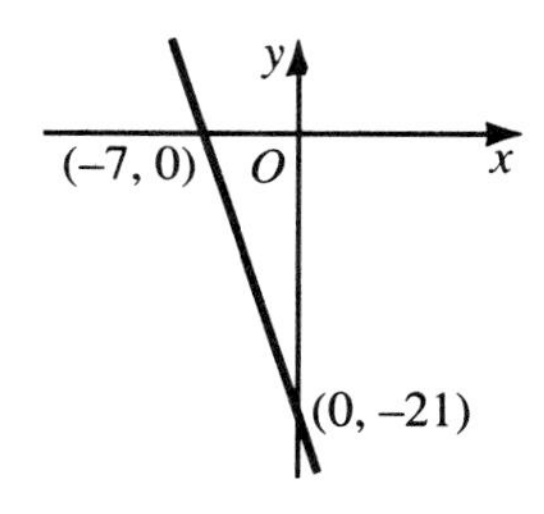

8. $4x - 5 \cdot 0 = 20;\ x = 5;$
$4 \cdot 0 - 5y = 20;\ y = -4$

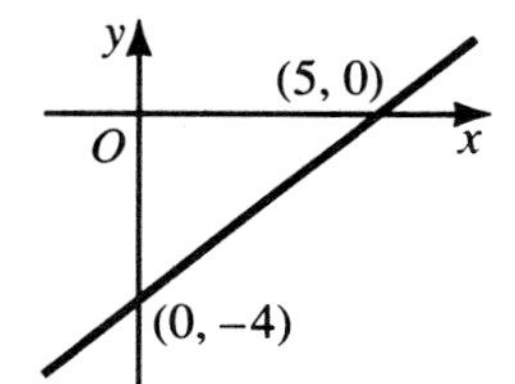

9. $3x + 2 \cdot 0 = 12;\ x = 4;$
$3 \cdot 0 + 2y = 12;\ y = 6$

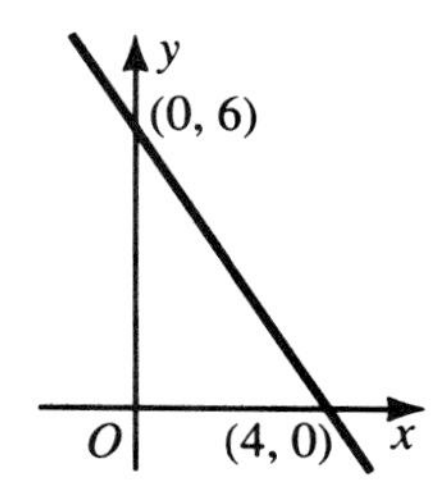

10. $3x - 2 \cdot 0 = 12;\ x = 4;$
$3 \cdot 0 - 2y = 12;\ y = -6$

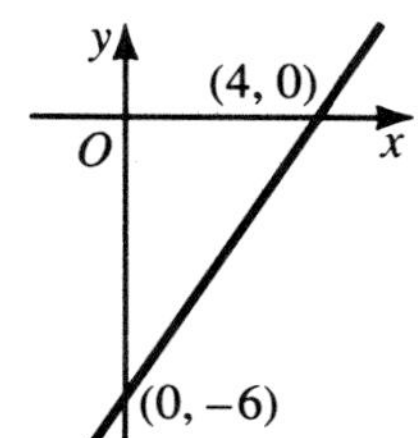

11. $5x + 8 \cdot 0 = 20$; $x = 4$;

$5 \cdot 0 + 8y = 20$; $y = \frac{5}{2}$

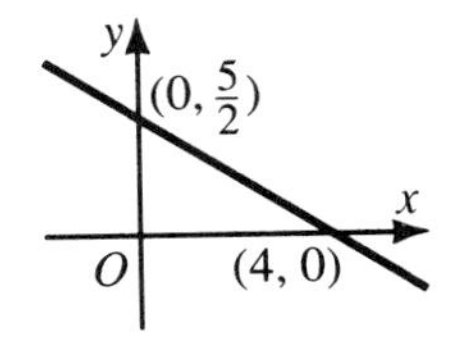

12. $3x + 4 \cdot 0 = -18$; $x = -6$;

$3 \cdot 0 + 4y = -18$; $y = -\frac{9}{2}$

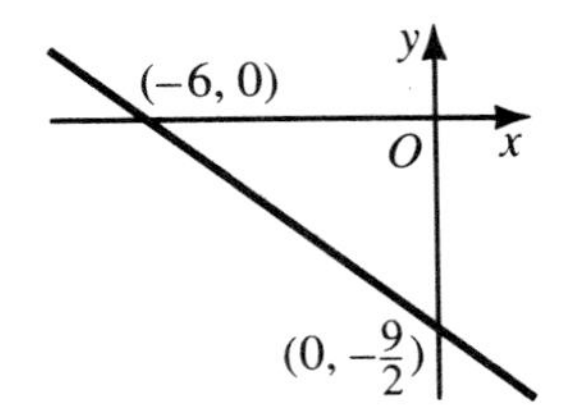

13. $m = 2$; $b = -3$

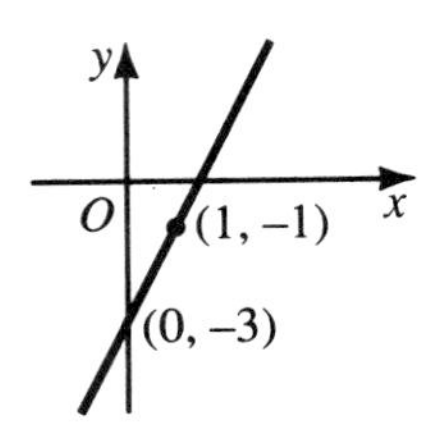

14. $m = 2$; $b = 3$

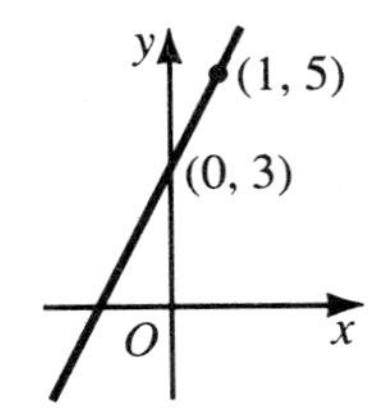

15. $m = -4$; $b = 0$

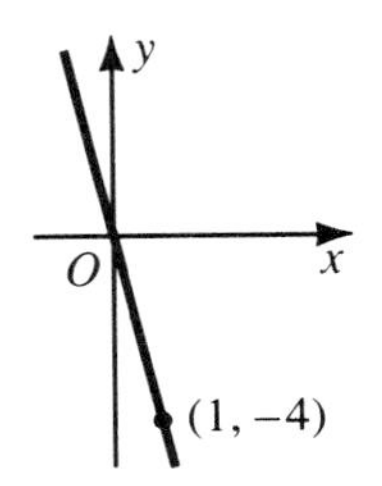

16. $m = \frac{3}{4}$; $b = 1$

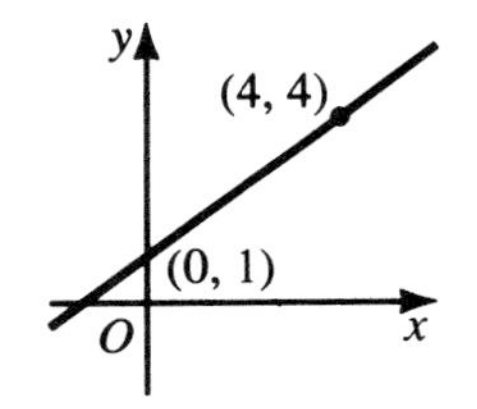

17. $m = -\frac{2}{3}$; $b = -4$

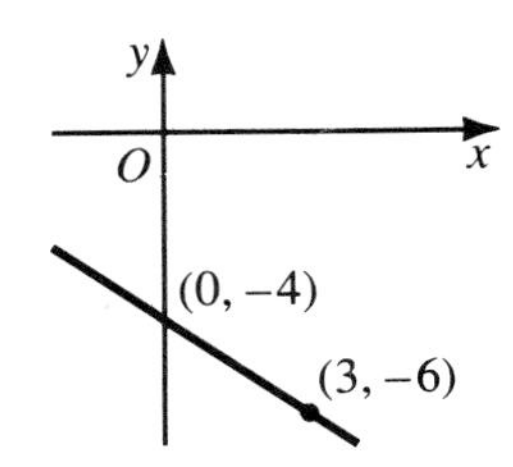

18. $m = \frac{5}{3}$; $b = -2$

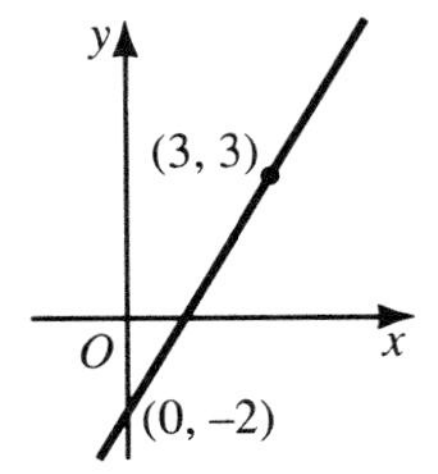

19. $4x + y = 10$; $y = -4x + 10$; $m = -4$; $b = 10$

20. $2x - y = 5$; $y = 2x - 5$; $m = 2$; $b = -5$

21. $5x - 2y = 10$; $-2y = -5x + 10$; $y = \frac{5}{2}x - 5$; $m = \frac{5}{2}$; $b = -5$

22. $3x + 4y = 12$; $4y = -3x + 12$; $y = -\frac{3}{4}x + 3$; $m = -\frac{3}{4}$; $b = 3$

23. $x - 4y = 6;\ -4y = -x + 6;\ y = \frac{1}{4}x - \frac{3}{2};\ m = \frac{1}{4};\ b = -\frac{3}{2}$

24. $4x + 3y = 8;\ 3y = -4x + 8;\ y = -\frac{4}{3}x + \frac{8}{3};\ m = -\frac{4}{3};\ b = \frac{8}{3}$

25.
$$\begin{array}{rcl} x + y &=& 3 \\ x - y &=& -1 \\ \hline 2x &=& 2 \end{array};\ x = 1;\ y = 2;\ (1, 2)$$

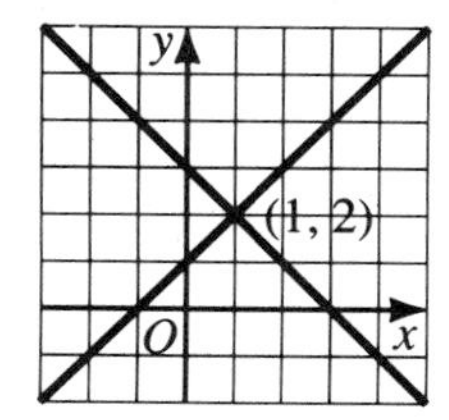

26.
$$\begin{array}{rcl} 2x + y &=& 7 \\ 3x + y &=& 9 \\ \hline -x &=& -2 \end{array};\ x = 2;\ y = 3;\ (2, 3)$$

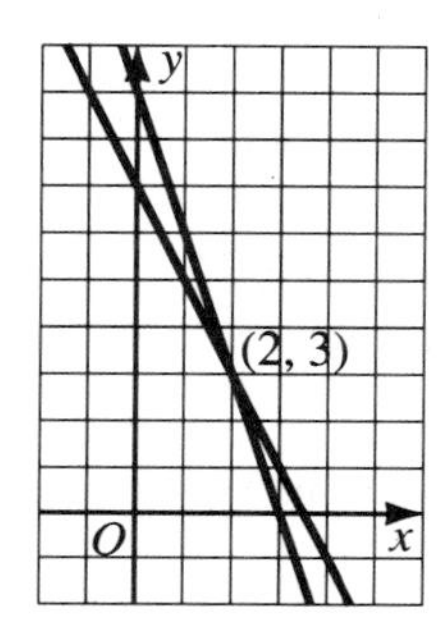

27.
$$\begin{array}{rcl} x + 2y &=& 10 \\ 3x - 2y &=& 6 \\ \hline 4x &=& 16 \end{array};\ x = 4;\ y = 3;\ (4, 3)$$

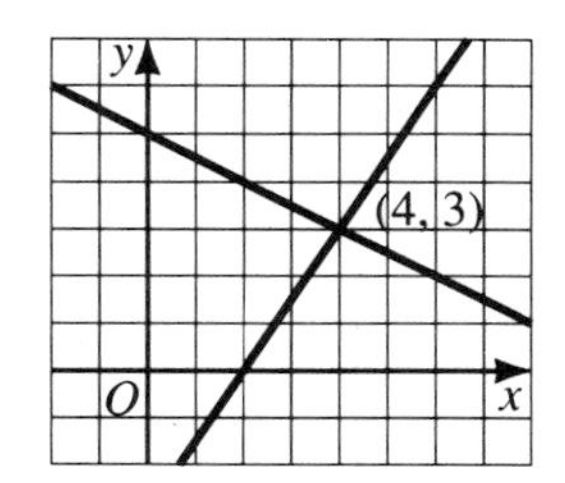

28. $3x + 2y = -30;\ y = x;\ 3x + 2x = -30;\ 5x = -30;\ x = -6;\ y = -6;\ (-6, -6)$

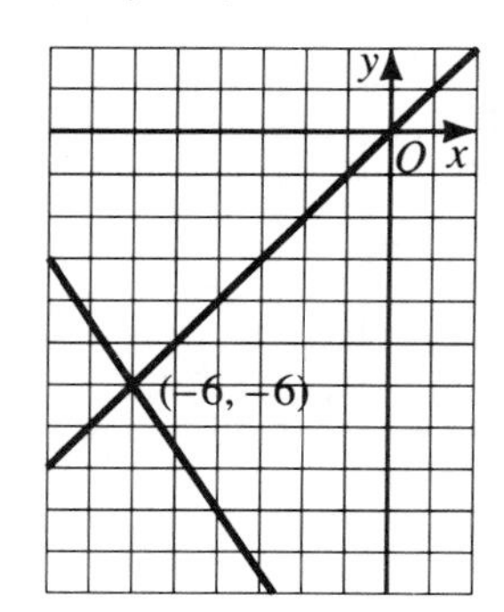

29.
$$\begin{array}{rcl} 4x + 5y &=& -7 \\ 2x - 3y &=& 13, \\ \hline \end{array} \quad \begin{array}{rcl} 4x + 5y &=& -7 \\ 4x - 6y &=& 26 \\ \hline 11y &=& -33 \end{array};\ y = -3;\ x = 2;\ (2, -3)$$

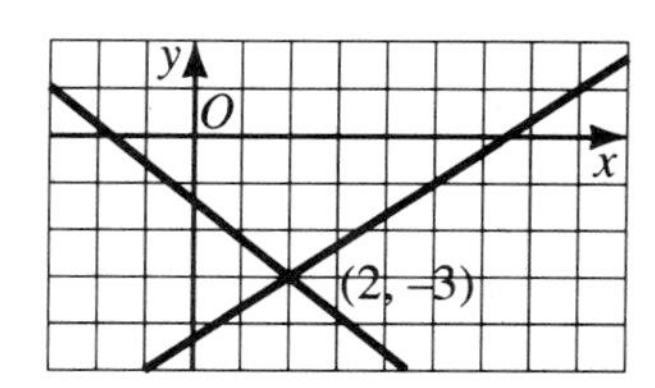

30. $\begin{array}{r} 3x + 2y = 8 \\ -x + 3y = 12, \\ \hline \end{array}$ $\begin{array}{r} 3x + 2y = 8 \\ -3x + 9y = 36 \\ \hline 11y = 44; \end{array}$ $y = 4;\ x = 0;\ (0, 4)$

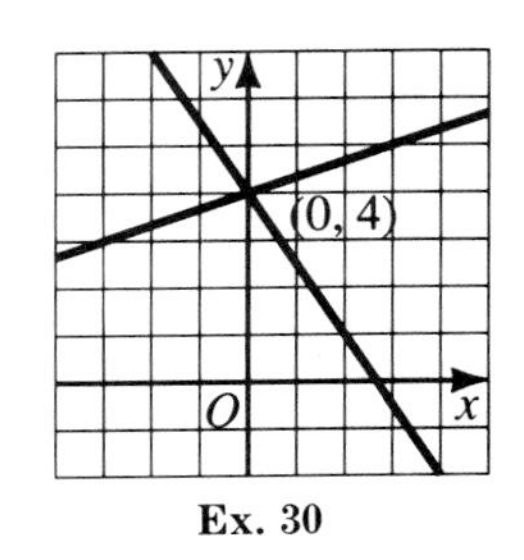

Ex. 30

B 31. a. $6x + 3y = 10;\ 3y = -6x + 10;\ y = -2x + \frac{10}{3};$

$y = -2x + \frac{10}{3}$ and $y = -2x + 5$ both have slope -2.

b. No; their y-intercepts are different.

c. $y = -2x + \frac{10}{3}$ and $y = -2x + 5;\ -2x + \frac{10}{3} = -2x + 5;\ \frac{10}{3} = 5;$ since this is false, there is no solution.

32. *Geometric*: The lines are parallel, and thus do not intersect.
Algebraic: If $y = 3x - 5$ and $y = 3x + 5$, then $3x - 5 = 3x + 5$; and $-5 = 5$; the pair of equations has no solution, so the lines do not intersect.

33. a. $2x - y = 7;\ 2x - 7 = y$ has slope 2; $x + 2y = 4;\ 2y = -x + 4;\ y = -\frac{1}{2}x + 2$ has slope $-\frac{1}{2}$. **b.** Since $2\left(-\frac{1}{2}\right) = -1$, the lines are $\perp$; two nonvertical lines are $\perp$ if and only if the product of their slopes is -1.

34. a.

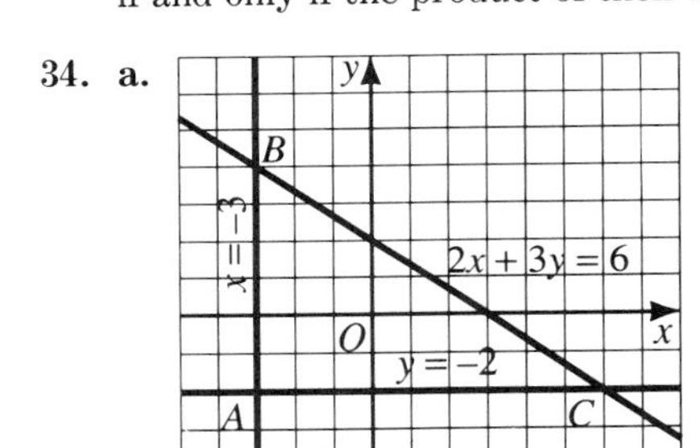

b. $A(-3, -2);\ 2(-3) + 3y = 6;\ 3y = 12;$
$y = 4;\ B(-3, 4);\ 2x + 3(-2) = 6;$
$2x = 12;\ x = 6;\ C(6, -2)$

c. area $= \frac{1}{2} \cdot AC \cdot AB = \frac{1}{2} \cdot 9 \cdot 6 = 27$

35. a.

b. $\frac{1}{2}x - 2 = -2x + 3;\ \frac{5}{2}x = 5;\ x = 2;$
$y = -2(2) + 3 = -1;\ A(2, -1);$
$-2x + 3 = 3x + 8;\ -5 = 5x;\ -1 = x;$
$y = 3(-1) + 8 = 5;\ B(-1, 5);$
$\frac{1}{2}x - 2 = 3x + 8;\ -10 = \frac{5}{2}x;$
$x = \frac{2}{5}(-10) = -4;\ y = 3(-4) + 8 = -4;$
$C(-4, -4)$

c. Slope of $\overline{AC} = \frac{1}{2}$, slope of $\overline{AB} = -2$; $\frac{1}{2}(-2) = -1$, so $\overline{AC} \perp \overline{AB}$;

area $= \frac{1}{2} \cdot AC \cdot AB$; $AC = \sqrt{(2-(-4))^2 + (-1-(-4))^2} =$

$\sqrt{6^2 + 3^2} = \sqrt{36 + 9} = \sqrt{45}$; $AB = \sqrt{(2-(-1))^2 + (-1-5)^2} =$

$\sqrt{3^2 + (-6)^2} = \sqrt{9 + 36} = \sqrt{45}$; Area $= \frac{1}{2}\sqrt{45} \cdot \sqrt{45} = \frac{45}{2} = 22\frac{1}{2}$

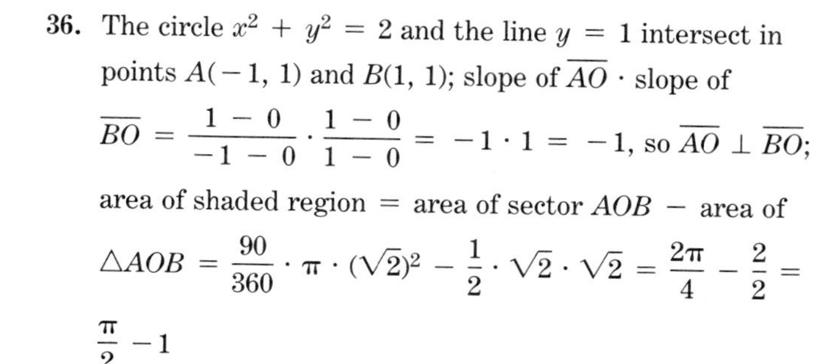

36. The circle $x^2 + y^2 = 2$ and the line $y = 1$ intersect in points $A(-1, 1)$ and $B(1, 1)$; slope of $\overline{AO}$ · slope of $\overline{BO} = \frac{1-0}{-1-0} \cdot \frac{1-0}{1-0} = -1 \cdot 1 = -1$, so $\overline{AO} \perp \overline{BO}$; area of shaded region = area of sector AOB − area of $\triangle AOB = \frac{90}{360} \cdot \pi \cdot (\sqrt{2})^2 - \frac{1}{2} \cdot \sqrt{2} \cdot \sqrt{2} = \frac{2\pi}{4} - \frac{2}{2} =$

$\frac{\pi}{2} - 1$

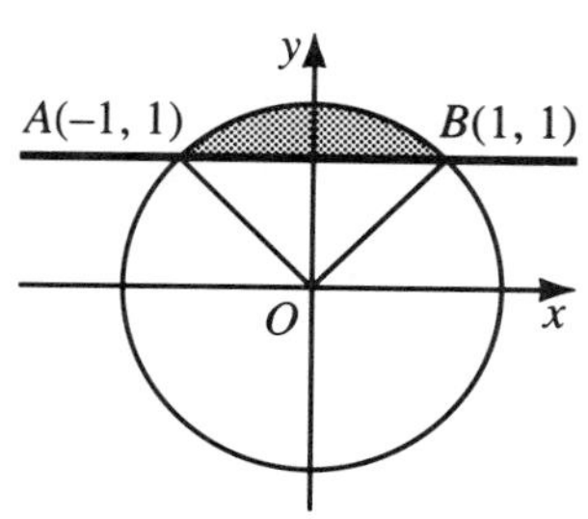

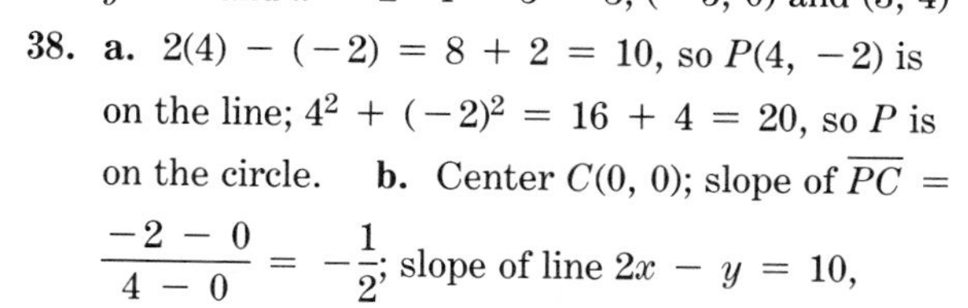

C 37. $x = 2y - 5$; $x^2 + y^2 = (2y-5)^2 + y^2 = 25$; $4y^2 - 20y + 25 + y^2 = 25$; $5y^2 - 20y = 0$; $5y(y-4) = 0$; $y = 0$ and $x = 2 \cdot 0 - 5 = -5$, or $y = 4$ and $x = 2 \cdot 4 - 5 = 3$; $(-5, 0)$ and $(3, 4)$

38. a. $2(4) - (-2) = 8 + 2 = 10$, so $P(4, -2)$ is on the line; $4^2 + (-2)^2 = 16 + 4 = 20$, so P is on the circle. **b.** Center $C(0, 0)$; slope of $\overline{PC} = \frac{-2-0}{4-0} = -\frac{1}{2}$; slope of line $2x - y = 10$, or $y = 2x - 10$, is 2; $-\frac{1}{2} \cdot 2 = -1$, so $\overline{PC}$ is $\perp$ to the line.

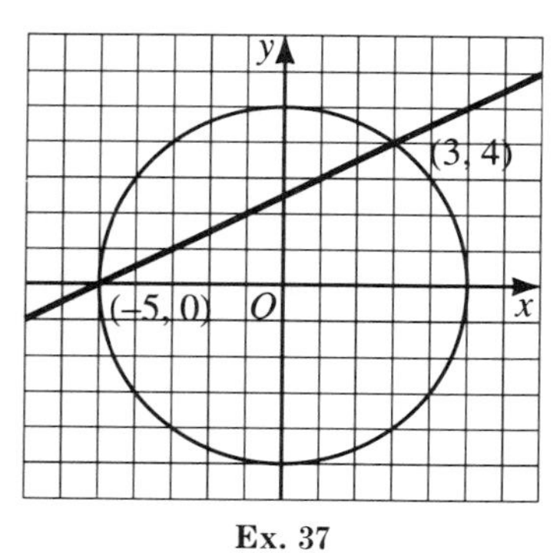

Ex. 37

c. Since the line is in the plane of the ⊙ and ⊥ to the radius $\overline{CP}$ at its outer endpoint, the line is tangent to the ⊙.

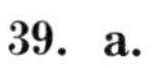

39. a.

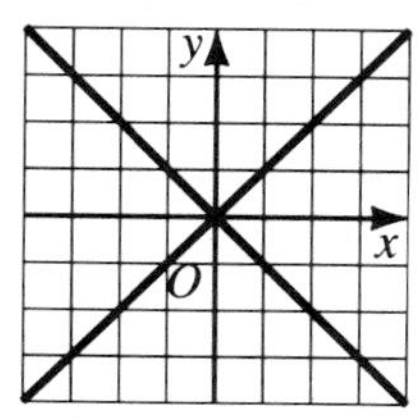

b.

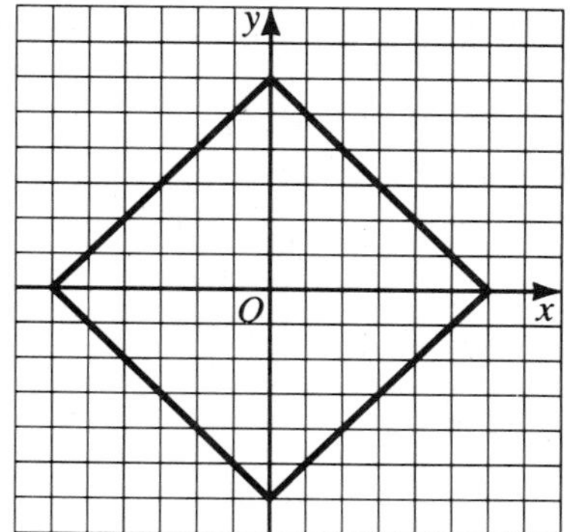

c.

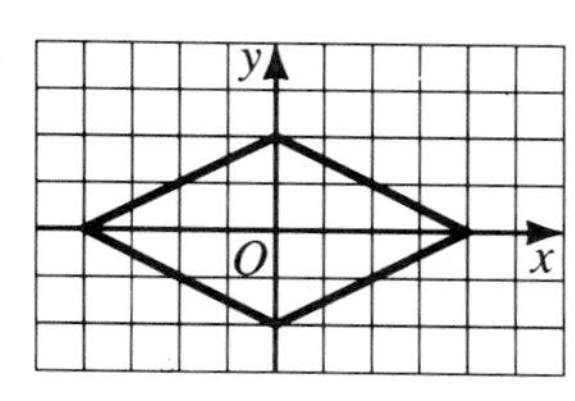

Page 552 • EXPLORATIONS

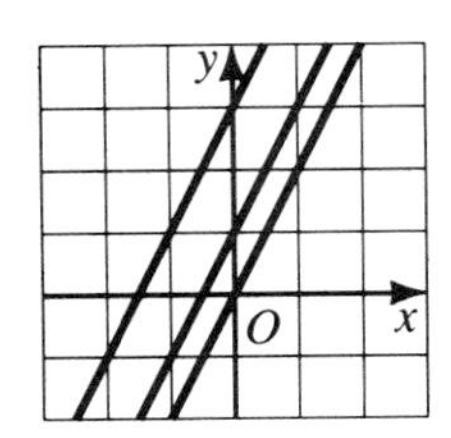

They are parallel; 2 nonvertical lines are $\parallel$ if and only if their slopes are $=$; $y = 2x + 7$

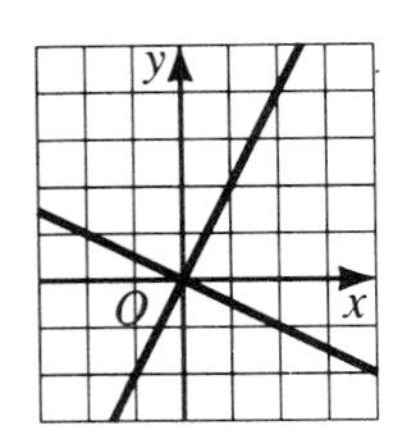

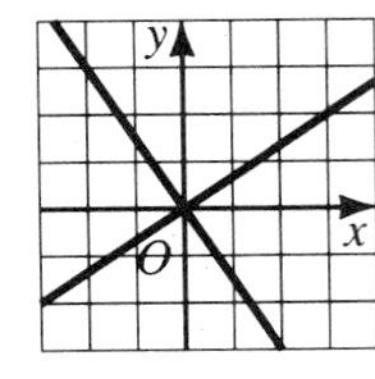

They are perpendicular; 2 nonvertical lines are $\perp$ if and only if the product of their slopes is -1; $y = -\frac{5}{4}x$

Page 552 • CHALLENGE

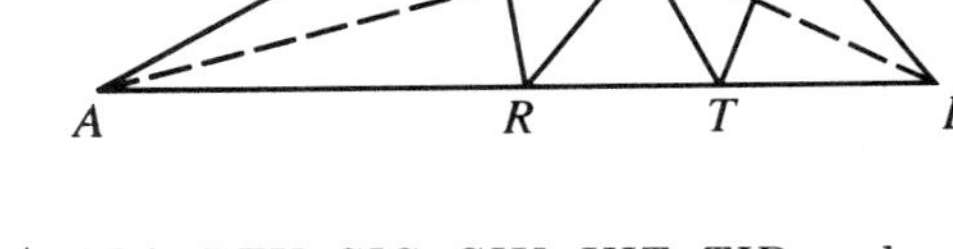

Let $\angle C$ be an obtuse $\angle$ and let $\angle A$ be the smaller acute $\angle$ of $\triangle CBA$. Let I be the pt. where the bisectors of $\angle C$, $\angle B$, and $\angle A$ intersect. At I draw 4 ∠s with measure $45 - \frac{1}{4}m\angle A$, two of the ∠s with $\overrightarrow{IA}$ as a side, the other two with $\overrightarrow{IB}$ as a side. Use R, S, T, and U to name the pts. where the 4 new rays intersect the sides of $\triangle CBA$ as shown. Draw $\overline{RS}$ and $\overline{TU}$. △s ARS, BTU, SIC, CIU, UIT, TIR, and RIS solve the problem. There are infinitely many solutions; you can replace $45 - \frac{1}{4}m\angle A$ above by $45 - k$, where k is any positive number less than $\frac{1}{2}m\angle A$.

Page 554 • CLASSROOM EXERCISES

1–10. Form of equations may vary.

1. $y = -\frac{1}{2}x + 5$ **2.** $y = \frac{3}{7}x + 8$

3. (2, 0) and (0, 4) are on the line; $y = mx + 4$; $m = \frac{4 - 0}{0 - 2} = -2$; $y = -2x + 4$

4. (2, 0) and (0, -6) are on the line; $y = mx - 6$; $m = \frac{-6 - 0}{0 - 2} = 3$; $y = 3x - 6$

5. $y = 0$ **6.** $x = 0$

7. $m = -\frac{4}{5}$; $y = -\frac{4}{5}x - 3$ **8.** $y = mx + 0 = mx$; $m = -1 \div \left(-\frac{7}{4}\right) = \frac{4}{7}$; $y = \frac{4}{7}x$

9. $y - 4 = \frac{5}{8}(x - 3)$ **10.** $y - 6 = -2(x - 8)$

11–13. Points may vary. **11.** $m = -1; (0, -7), (-7, 0)$ **12.** $m = \frac{1}{2}; (5, -2), (7, -1)$

13. $m = \frac{a}{b}; (d, c), (b + d, a + c)$

14. a. $OP = \sqrt{3^2 + 4^2} = 5$ **b.** $x^2 + y^2 = 25$ **c.** $l \perp \overline{OP}$ and $\overline{OP}$ has slope $\frac{4 - 0}{3 - 0} = \frac{4}{3}$, so l has slope $-\frac{3}{4}$. **d.** $y - 4 = -\frac{3}{4}(x - 3)$

Pages 555–556 • WRITTEN EXERCISES

A **1–32.** Form of equations may vary.

1. $y = 2x + 5$ **2.** $y = -3x + 6$ **3.** $y = \frac{1}{2}x - 8$ **4.** $y = \frac{3}{4}x - 9$

5. $y = -\frac{7}{5}x + 8$ **6.** $y = -\frac{3}{2}x - 7$

7. $y = mx + 2$; the line contains (8, 0) and (0, 2); $m = \frac{2 - 0}{0 - 8} = -\frac{1}{4}; y = -\frac{1}{4}x + 2$

8. $y = mx - 3$; the line contains (9, 0) and (0, −3); $m = \frac{-3 - 0}{0 - 9} = \frac{-3}{-9} = \frac{1}{3};$ $y = \frac{1}{3}x - 3$

9. $y = mx + 4$; the line contains (−8, 0) and (0, 4); $m = \frac{4 - 0}{0 - (-8)} = \frac{4}{8} = \frac{1}{2};$ $y = \frac{1}{2}x + 4$

10. $y = mx - 2$; the line contains (−5, 0) and (0, −2); $m = \frac{-2 - 0}{0 - (-5)} = -\frac{2}{5};$ $y = -\frac{2}{5}x - 2$

11. $y - 2 = 5(x - 1)$, or $y = 5x - 3$ **12.** $y - 8 = 4(x - 3)$, or $y = 4x - 4$

13. $y - 5 = \frac{1}{3}(x + 3)$, or $y = \frac{1}{3}x + 6$ **14.** $y + 6 = -\frac{2}{3}(x - 6)$, or $y = -\frac{2}{3}x - 2$

15. $y = -\frac{1}{2}(x + 4)$, or $y = -\frac{1}{2}x - 2$ **16.** $y - 3 = -\frac{2}{5}(x + 10)$, or $y = -\frac{2}{5}x - 1$

17. $m = \frac{7 - 1}{4 - 1} = \frac{6}{3} = 2; y - 1 = 2(x - 1)$, or $y = 2x - 1$

18. $m = \frac{-3 - 1}{-1 - 2} = \frac{4}{3}; y - 1 = \frac{4}{3}(x - 2)$, or $y = \frac{4}{3}x - \frac{5}{3}$

19. $m = \frac{3 - 1}{3 - (-3)} = \frac{2}{6} = \frac{1}{3}; y - 3 = \frac{1}{3}(x - 3)$, or $y = \frac{1}{3}x + 2$

20. $m = \frac{-5 - (-1)}{-6 - (-2)} = \frac{-4}{-4} = 1; y + 1 = 1(x + 2)$, or $y = x + 1$

21. The slope of the line is undefined; $x = 2$

22. The line has slope 0 and y-intercept 1; $y = 1$

23. The line is vertical and contains $(5, -3)$; $x = 5$

24. The line is vertical and contains $(-8, -2)$; $x = -8$

B 25. The line contains (5, 7) and has slope 3; $y - 7 = 3(x - 5)$, or $y = 3x - 8$

26. $3x + 5y = 15$, or $y = -\frac{3}{5}x + 3$, has slope $-\frac{3}{5}$; $y - 3 = -\frac{3}{5}(x + 1)$, or $3x + 5y = 12$

27. $8x - 5y = 0$, or $y = \frac{8}{5}x$, has slope $\frac{8}{5}$; a line $\perp$ to it has slope $-\frac{5}{8}$; $y + 2 = -\frac{5}{8}(x + 3)$, or $5x + 8y = -31$

28. $3x + 4y = 12$, or $y = -\frac{3}{4}x + 3$, has slope $-\frac{3}{4}$; a line $\perp$ to it has slope $\frac{4}{3}$; $y = \frac{4}{3}(x - 8)$, or $4x - 3y = 32$

29. Given $O(0, 0)$ and $P(10, 6)$, $\overline{OP}$ has slope $\frac{6 - 0}{10 - 0} = \frac{3}{5}$, so the $\perp$ bis. has slope $-\frac{5}{3}$; midpt. $M = \left(\frac{0 + 10}{2}, \frac{0 + 6}{2}\right) = (5, 3)$; $y - 3 = -\frac{5}{3}(x - 5)$, or $y = -\frac{5}{3}x + \frac{34}{3}$

30. Given $C(-3, 7)$ and $D(5, 1)$, $\overline{CD}$ has slope $\frac{1 - 7}{5 - (-3)} = \frac{-6}{8} = -\frac{3}{4}$, so the $\perp$ bis. of $\overline{CD}$ has slope $\frac{4}{3}$; midpt. of $\overline{CD} = \left(\frac{-3 + 5}{2}, \frac{7 + 1}{2}\right) = (1, 4)$; $y - 4 = \frac{4}{3}(x - 1)$, or $y = \frac{4}{3}x + \frac{8}{3}$

31. From Ex. 31, p. 534, the line has slope 1; $y - 5 = 1(x - 5)$, or $y = x$

32. Let j be the vertical line through $Q(-4, 0)$ and let P be the point where j intersects the given line. Since $m\angle POQ = 45$ and $\overline{PQ} \perp \overline{OQ}$, $\triangle POQ$ is a 45°–45°–90° $\triangle$; $OQ = 4 = PQ$, so $P = (-4, 4)$; $\overleftrightarrow{OP}$ has slope $\frac{4 - 0}{-4 - 0} = -1$ and y-int. 0; $y = -x$.

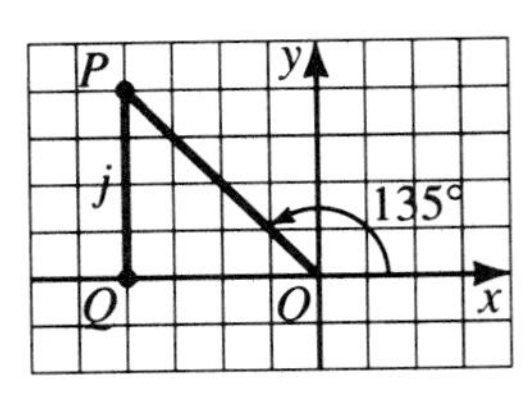

33. $y = 9kx - 1$ has slope $9k$; $kx + 4y = 12$, or $y = -\frac{k}{4}x + 3$, has slope $-\frac{1}{4}k$; if the lines are $\perp$, $9k\left(-\frac{1}{4}k\right) = -1$; $\frac{9}{4}k^2 = 1$; $k^2 = \frac{4}{9}$; $k = \frac{2}{3}$ or $-\frac{2}{3}$

34. **a.** Diag. $\overline{BC}$ has slope $\frac{-3 - 5}{7 - 3} = \frac{-8}{4} = -2$; since the diags. of a rhombus are $\perp$, diag. $\overline{EK}$ has slope $\frac{1}{2}$. **b.** $\overleftrightarrow{EK}$ contains the midpoint $\left(\frac{3 + 7}{2}, \frac{5 - 3}{2}\right) = (5, 1)$ of $\overline{BC}$; $y - 1 = \frac{1}{2}(x - 5)$, or $y = \frac{1}{2}x - \frac{3}{2}$

35. Given $A(2, 10)$, $B(10, 6)$, and $C(-6, -6)$, the ctr. of the $\odot$ is at the intersection of the $\perp$ bis. of $\overline{AB}$ and $\overline{BC}$. $\overline{AB}$ has slope $\frac{6 - 10}{10 - 2} = \frac{-4}{8} = -\frac{1}{2}$, so the $\perp$ bis. has slope 2; midpt. of $\overline{AB} = \left(\frac{2 + 10}{2}, \frac{10 + 6}{2}\right) = (6, 8)$; $y - 8 = 2(x - 6)$, or $y = 2x - 4$. $\overline{BC}$ has slope $\frac{-6 - 6}{-6 - 10} = \frac{-12}{-16} = \frac{3}{4}$, so the $\perp$ bis. has slope $-\frac{4}{3}$; midpt. of $\overline{BC} = \left(\frac{10 - 6}{2}, \frac{6 - 6}{2}\right) = (2, 0)$; $y = -\frac{4}{3}(x - 2)$. Solving the $\perp$ bis. equations simultaneously, $y = 2x - 4 = -\frac{4}{3}(x - 2)$; $6x - 12 = -4x + 8$; $10x = 20$; $x = 2$; $y = 2 \cdot 2 - 4 = 0$; center is $(2, 0)$.

C **36.** **a.** Midpt. N of $\overline{QR} = \left(\frac{-6 + 12}{2}, \frac{0 + 0}{2}\right) = (3, 0)$; median $\overline{SN}$ has slope $\frac{12 - 0}{0 - 3} = -4$ and equation $y = -4(x - 3)$, or $y = -4x + 12$, or $4x + y = 12$. Midpt. M of $\overline{RS} = \left(\frac{12 + 0}{2}, \frac{0 + 12}{2}\right) = (6, 6)$; median $\overline{QM}$ has slope $\frac{0 - 6}{-6 - 6} = \frac{1}{2}$ and equation $y = \frac{1}{2}(x + 6)$, or $x - 2y = -6$. Midpt. P of $\overline{QS} = \left(\frac{-6 + 0}{2}, \frac{0 + 12}{2}\right) = (-3, 6)$; median $\overline{PR}$ has slope $\frac{6 - 0}{-3 - 12} = -\frac{2}{5}$ and equation $y = -\frac{2}{5}(x - 12)$, or $2x + 5y = 24$. **b.** $\overline{SN}$ and $\overline{QM}$ intersect when $y = -4x + 12 = \frac{1}{2}(x + 6)$; $-8x + 24 = x + 6$; $18 = 9x$; $x = 2$; $y = -4 \cdot 2 + 12 = 4$; $G = (2, 4)$; since $-\frac{2}{5}(2 - 12) = -\frac{2}{5}(-10) = 4$, G lies on $\overline{PR}$ also. **c.** $QG = \sqrt{(-6 - 2)^2 + (0 - 4)^2} = \sqrt{(-8)^2 + (-4)^2} = \sqrt{64 + 16} = 4\sqrt{5}$; $QM = \sqrt{(-6 - 6)^2 + (0 - 6)^2} = \sqrt{(-12)^2 + (-6)^2} = \sqrt{144 + 36} = \sqrt{180} = 6\sqrt{5}$; $\frac{2}{3}QM = \frac{2}{3} \cdot 6\sqrt{5} = 4\sqrt{5} = QG$

37. a. From Ex. 36, midpt. N of $\overline{QR} = (3, 0)$, midpt. M of $\overline{RS} = (6, 6)$, and midpt. P of $\overline{QS} = (-3, 6)$. Slope of $\overline{QR} = \frac{0 - 0}{-6 - 12} = 0$, so the $\perp$ bis. of $\overline{QR}$ is a vertical line through (3, 0) and has equation $x = 3$. Slope of $\overline{RS} = \frac{12 - 0}{0 - 12} = -1$, so the $\perp$ bis. of $\overline{RS}$ has slope 1 and equation $y - 6 = 1(x - 6)$, or $y = x$. Slope of $\overline{QS} = \frac{12 - 0}{0 - (-6)} = 2$, so the $\perp$ bis. of $\overline{QS}$ has slope $-\frac{1}{2}$ and equation $y - 6 = -\frac{1}{2}(x + 3)$; $2y - 12 = -x - 3$, or $x + 2y = 9$ **b.** The $\perp$ bis. of $\overline{QR}$ and $\overline{RS}$ intersect when $x = 3$ and $y = x = 3$; $C = (3, 3)$; since $3 + 2(3) = 3 + 6 = 9$, C lies on the $\perp$ bis. of $\overline{QS}$ also. **c.** $CQ = \sqrt{(-6 - 3)^2 + (0 - 3)^2} = \sqrt{(-9)^2 + (-3)^2} = \sqrt{81 + 9} = 3\sqrt{10}$; $CR = \sqrt{(12 - 3)^2 + (0 - 3)^2} = \sqrt{9^2 + (-3)^2} = \sqrt{81 + 9} = 3\sqrt{10}$; $CS = \sqrt{(0 - 3)^2 + (12 - 3)^2} = \sqrt{(-3)^2 + 9^2} = \sqrt{9 + 81} = 3\sqrt{10}$; $CQ = CR = CS$, so C is equidistant from Q, R, and S. **d.** The circle has ctr. $C = (3, 3)$ and radius $3\sqrt{10}$; $(x - 3)^2 + (y - 3)^2 = (3\sqrt{10})^2$; $(x - 3)^2 + (y - 3)^2 = 90$

38. a. The altitude to each side has the same slope as the $\perp$ bis. to that side. From Ex. 37, the altitude to $\overline{QR}$ is a vertical line; since it contains $S(0, 12)$, the line containing the altitude to $\overline{QR}$ has equation $x = 0$. From Ex. 37, the altitudes to $\overline{RS}$ and $\overline{QS}$ have slopes 1 and $-\frac{1}{2}$, resp. The line containing the altitude to $\overline{RS}$ has equation $y = 1(x + 6) = x + 6$; the line containing the altitude to $\overline{QS}$ has equation $y = -\frac{1}{2}(x - 12)$; $2y = -x + 12$, or $x + 2y = 12$. **b.** The lines containing the altitudes to $\overline{QR}$ and $\overline{RS}$ intersect when $x = 0$ and $y = 0 + 6 = 6$; $H = (0, 6)$; since $0 + 2(6) = 12$, H lies on the altitude to $\overline{QS}$ also.

39. a. From Exs. 36–38, $G = (2, 4)$, $C = (3, 3)$, and $H = (0, 6)$; slope of $\overline{CG} = \frac{4 - 3}{2 - 3} = -1$ and slope of $\overline{GH} = \frac{6 - 4}{0 - 2} = \frac{2}{-2} = -1$, so G, C, and H are collinear. **b.** $GC = \sqrt{(3 - 2)^2 + (3 - 4)^2} = \sqrt{1^2 + (-1)^2} = \sqrt{1 + 1} = \sqrt{2}$; $GH = \sqrt{(0 - 2)^2 + (6 - 4)^2} = \sqrt{(-2)^2 + (2)^2} = \sqrt{4 + 4} = 2\sqrt{2} = 2GC$

Page 558 • CLASSROOM EXERCISES

1. $P(0, a)$; $T(a, a)$ 2. $M(2b, 0)$ 3. $J(0, d)$; $K(e, 0)$ 4. $M(i, 0)$; $E(h, g)$; $G(h + i, g)$
5. $G(j, 0)$; $D(j, k)$ 6. $P(b, 0)$

Pages 558–559 • WRITTEN EXERCISES

A 1. upper left: $(0, b)$; lower right: $(a, 0)$ 2. $(c + e, d)$

3. upper left: $(-f, 2f)$; upper right: $(f, 2f)$ 4. $\left(\frac{g + h}{2}, j\right)$

5. $(h + m, n)$ 6. $(n - p, q)$

B 7. Use 30°–60°–90° relationships; $\left(\frac{s}{2}, \frac{s\sqrt{3}}{2}\right)$

8. Let t be the x-coord. of the upper left vertex; $\sqrt{(t - 0)^2 + (4 - 0)^2} = \sqrt{(5 - 0)^2 + (0 - 0)^2}$; $\sqrt{t^2 + 4^2} = \sqrt{5^2}$; $t^2 + 16 = 25$; $t^2 = 9$; $t = 3$; upper left: $(3, 4)$. Since opp. sides are $\parallel$, the upper right vertex has y-coord. 4; since the top side has length 5, the x-coord. is $3 + 5 = 8$; upper right: $(8, 4)$.

9. Let t be the x-coord. of C; $\sqrt{(t - 0)^2 + (b - 0)^2} = \sqrt{(a - 0)^2 + (0 - 0)^2}$; $t^2 + b^2 = a^2$; $t = \sqrt{a^2 - b^2}$; $C = (\sqrt{a^2 - b^2}, b)$. Since $OABC$ is a ▱ and opp. sides are $\cong$, B has coord. $(\sqrt{a^2 - b^2} + a, b)$.

10. a. $\left(\frac{2w + 2t}{2}, \frac{2z + 0}{2}\right) = (w + t, z)$ b. $\left(\frac{2u + 0}{2}, \frac{2v + 0}{2}\right) = (u, v)$

c. $\left(\frac{w + t + u}{2}, \frac{z + v}{2}\right)$ d. $\left(\frac{2u + 2w}{2}, \frac{2v + 2z}{2}\right) = (u + w, v + z)$

e. $\left(\frac{2t + 0}{2}, \frac{0 + 0}{2}\right) = (t, 0)$ f. $\left(\frac{u + w + t}{2}, \frac{v + z}{2}\right) = \left(\frac{w + t + u}{2}, \frac{z + v}{2}\right)$

g. $\left(\frac{w + t + u}{2}, \frac{z + v}{2}\right)$

11–12. Answers may vary.

C 11. Let A have coord. $(a, 0)$. Then, since $\triangle OAB$ is a 30°–60°–90° $\triangle$, B has coordinates $(0, a\sqrt{3})$ and $AB = 2a$. F is pt. $(a + 2a, 0)$, or $(3a, 0)$. $\triangle OAB \cong \triangle PFE \cong \triangle QCB$ (AAS), so $OA = PF = a$; $OB = QB = a\sqrt{3}$. Thus, E is pt. $(4a, a\sqrt{3})$, C is pt. $(a, 2a\sqrt{3})$, and D is pt. $(3a, 2a\sqrt{3})$.

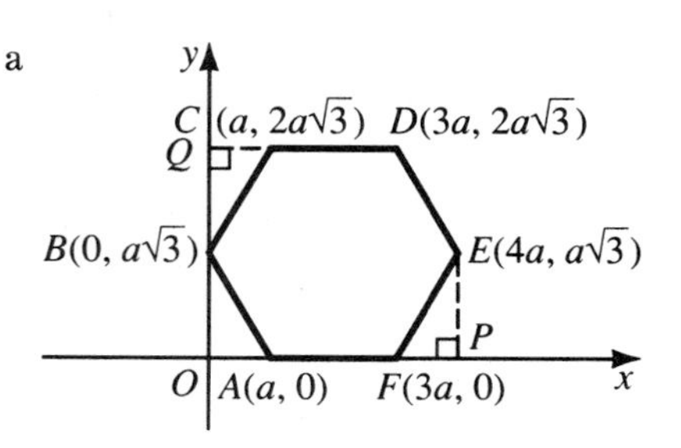

12. Let A have coord. $(a, 0)$. Then B has coord. $(0, a)$ and $AB = a\sqrt{2}$. Since $AH = a\sqrt{2}$, H has coord. $(a + a\sqrt{2}, 0)$. Similarly, C is pt. $(0, a + a\sqrt{2})$. $\triangle ABO \cong \triangle GHP$ (AAS Thm.), so $OA = PG = a$. Then G is pt. $(2a + a\sqrt{2}, a)$ and F is pt. $(2a + a\sqrt{2}, a + a\sqrt{2})$. $\triangle ABO \cong \triangle CDQ$ (AAS Thm.), so $OB = QD = a$. Then D and E are pts. $(a, 2a + a\sqrt{2})$ and $(a + a\sqrt{2}, 2a + a\sqrt{2})$, respectively.

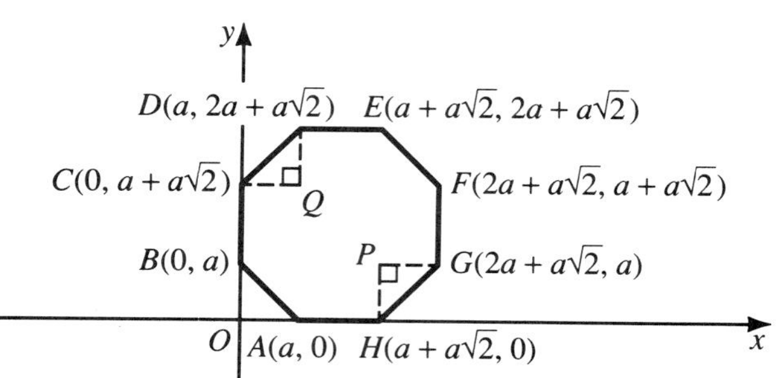

13. $HOJK$ is a trap., so $\overline{JK} \parallel \overline{OH}$, and slope of $\overline{JK}$ = slope of $\overline{OH}$. But the slope of $\overline{JK} = \frac{e - c}{d - b}$, and the slope of $\overline{OH} = \frac{0 - 0}{a - 0} = 0$, so $\frac{e - c}{d - b} = 0$ and $e = c$. Also, $HOJK$ is isos., so $JO = KH$. But $JO = \sqrt{b^2 + c^2}$, and $KH = \sqrt{(a - d)^2 + e^2}$, so $\sqrt{b^2 + c^2} = \sqrt{(a - d)^2 + e^2}$; $b^2 + c^2 = (a - d)^2 + e^2$. Then since $e = c$, $b^2 = (a - d)^2$; $b = a - d$; $d = a - b$ $(b > 0, a - d > 0)$.

Page 561 • CLASSROOM EXERCISES

1. Square; $OQ = QR = RS = OS = a$, and $m\angle QOS = 90$

2. $OR = \sqrt{(a - 0)^2 + (a - 0)^2} = \sqrt{a^2 + a^2} = a\sqrt{2}$;
$QS = \sqrt{(a - 0)^2 + (0 - a)^2} = \sqrt{a^2 + (-a)^2} = \sqrt{a^2 + a^2} = a\sqrt{2}$; $\overline{OR} \cong \overline{QS}$

3. slope of $\overline{OR} = \frac{a - 0}{a - 0} = 1$; slope of $\overline{QS} = \frac{a - 0}{0 - a} = -1$; $1(-1) = -1$, so $\overline{OR} \perp \overline{QS}$.

4. Midpt. of $\overline{OR} = \left(\frac{0 + a}{2}, \frac{0 + a}{2}\right) = \left(\frac{a}{2}, \frac{a}{2}\right) = \left(\frac{0 + a}{2}, \frac{a + 0}{2}\right)$ = midpt. of $\overline{QS}$, so $\overline{OR}$ and $\overline{QS}$ bisect each other.

5. a. k is a vert. line through (b, c); $x = b$ **b.** $-\frac{b - a}{c}$, or $\frac{a - b}{c}$ **c.** Line l has slope $\frac{a - b}{c}$ and y-intercept 0, so its equation is $y = \left(\frac{a - b}{c}\right)x + 0$, or $y = \left(\frac{a - b}{c}\right)x$. **d.** $x = b$ and $y = \left(\frac{a - b}{c}\right)x$ intersect at $x = b$ and $y = \left(\frac{a - b}{c}\right)b = \frac{ab - b^2}{c}$. **e.** $\frac{c - 0}{b - 0} = \frac{c}{b}$; $-\frac{b}{c}$ **f.** j has slope $-\frac{b}{c}$ and contains $(a, 0)$, so its equation is $y - 0 = -\frac{b}{c}(x - a)$, or $y = -\frac{b}{c}(x - a)$.

g. $x = b$ and $y = -\frac{b}{c}(x - a)$ intersect at $x = b$ and $y = -\frac{b}{c}(b - a) = \frac{ab - b^2}{c}$.

h. $\left(b, \frac{ab - b^2}{c}\right)$

Pages 562–563 • WRITTEN EXERCISES

A **1.** Let $OABC$ be a rectangle with coordinates as shown below. By the distance formula, $AC = \sqrt{(a - 0)^2 + (0 - b)^2} = \sqrt{a^2 + (-b)^2} = \sqrt{a^2 + b^2}$, and $OB = \sqrt{(a - 0)^2 + (b - 0)^2} = \sqrt{a^2 + b^2}$. Therefore, $\overline{AC} \cong \overline{OB}$.

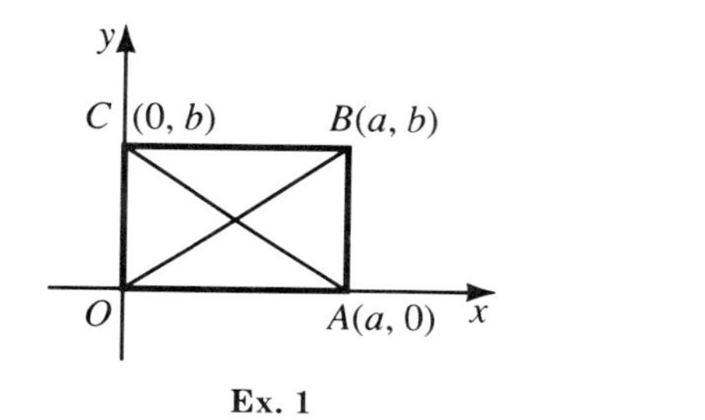

Ex. 1

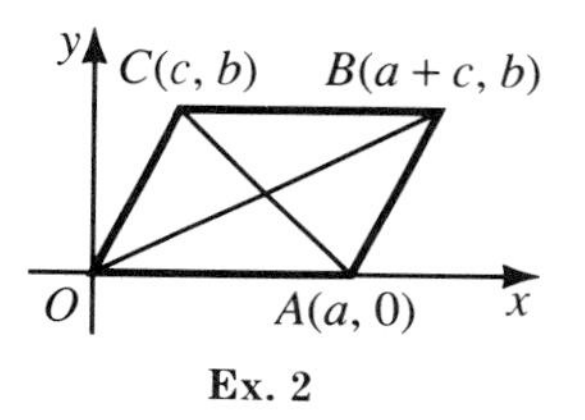

Ex. 2

2. Let $OABC$ be a ▱ with coordinates as shown above. By the midpt. formula, the midpt. of $\overline{AC}$ has coordinates $\left(\frac{a + c}{2}, \frac{b + 0}{2}\right) = \left(\frac{a + c}{2}, \frac{b}{2}\right)$, and the midpt. of $\overline{OB}$ has coordinates $\left(\frac{a + c + 0}{2}, \frac{b + 0}{2}\right) = \left(\frac{a + c}{2}, \frac{b}{2}\right)$. By the def. of seg. bis., diags. $\overline{AC}$ and $\overline{OB}$ bisect each other.

3. Let $OPQR$ be a rhombus with coordinates as shown. By the def. of a rhombus, $\overline{OP} \cong \overline{OR}$ or $OP = OR$. By the distance formula, $OP = \sqrt{(a - 0)^2 + (0 - 0)^2} = \sqrt{a^2 + 0^2} = \sqrt{a^2}$, and $OR = \sqrt{(b - 0)^2 + (c - 0)^2} = \sqrt{b^2 + c^2}$. Substituting, $\sqrt{a^2} = \sqrt{b^2 + c^2}$; $a^2 = b^2 + c^2$; $c^2 = a^2 - b^2$. Slope of $\overline{OQ} = \frac{c - 0}{(a + b) - 0} = \frac{c}{a + b}$; slope of $\overline{PR} = \frac{c - 0}{b - a} = \frac{c}{b - a}$. Slope of $\overline{OQ}$ · slope of $\overline{PR}$ = $\frac{c}{a + b} \cdot \frac{c}{b - a} = \frac{c^2}{b^2 - a^2} = \frac{a^2 - b^2}{b^2 - a^2} = -1$, by substitution. Therefore, $\overline{OQ} \perp \overline{PR}$ since two nonvertical lines are $\perp$ if and only if the product of their slopes is -1.

4. a. By the midpt. formula, the midpt. Q of $\overline{ON}$ is $\left(\frac{0 + b}{2}, \frac{0 + c}{2}\right) = \left(\frac{b}{2}, \frac{c}{2}\right)$; the midpt. R of $\overline{PM} = \left(\frac{a + d}{2}, \frac{0 + c}{2}\right) = \left(\frac{a + d}{2}, \frac{c}{2}\right)$. Slope of $\overline{OP} = \frac{0 - 0}{a - 0} = 0$, slope of $\overline{MN} = \frac{c - c}{d - b} = 0$, and slope of median $\overline{QR} = \frac{\frac{c}{2} - \frac{c}{2}}{\frac{a + d}{2} - \frac{b}{2}} = 0$. Therefore, $\overline{OP}$, $\overline{MN}$, and $\overline{QR}$ are $\parallel$ since their slopes are $=$.
b. $QR = \left|\frac{a + d}{2} - \frac{b}{2}\right| = \frac{a + d - b}{2}$; $OP = |a - 0| = a$; $MN = |d - b| = d - b$. The average of OP and MN is $\frac{1}{2}(a + (d - b)) = \frac{a + d - b}{2} = QR$.

5. By the midpt. formula, the midpt. G of $\overline{NP}$ is $\left(\frac{a + b}{2}, \frac{c + 0}{2}\right) = \left(\frac{a + b}{2}, \frac{c}{2}\right)$; the midpt. H of $\overline{OM}$ is $\left(\frac{d + 0}{2}, \frac{c + 0}{2}\right) = \left(\frac{d}{2}, \frac{c}{2}\right)$. Slope of $\overline{GH} = \frac{\frac{c}{2} - \frac{c}{2}}{\frac{a + b}{2} - \frac{d}{2}} = 0 =$ slope of $\overline{OP} =$ slope of $\overline{MN}$, from Ex. 4. Therefore, $\overline{GH} \parallel \overline{MN} \parallel \overline{OP}$ since nonvertical lines are $\parallel$ if and only if their slopes are $=$. $GH = \left|\frac{a + b}{2} - \frac{d}{2}\right| = \frac{a + b - d}{2}$. From Ex. 4, $OP = a$ and $MN = d - b$. $\frac{1}{2}(OP - MN) = \frac{1}{2}(a - (d - b)) = \frac{a + b - d}{2} = GH$.

6. a. Let $a = b + d$. By the distance formula, $NO = \sqrt{(b - 0)^2 + (c - 0)^2} = \sqrt{b^2 + c^2}$, and $MP = \sqrt{(d - a)^2 + (c - 0)^2} = \sqrt{(d - a)^2 + c^2}$. Substituting, $MP = \sqrt{(d - (b + d))^2 + c^2} = \sqrt{b^2 + c^2} = NO$. So $\overline{MP} \cong \overline{NO}$, and $MNOP$ is isos. by def. **b.** By the distance formula, $MO = \sqrt{(d - 0)^2 + (c - 0)^2} = \sqrt{d^2 + c^2}$, and $PN = \sqrt{(a - b)^2 + (0 - c)^2} = \sqrt{(a - b)^2 + c^2}$. Substituting, $PN = \sqrt{(b + d - b)^2 + c^2} = \sqrt{d^2 + c^2} = MO$. So diags. $\overline{PN}$ and $\overline{MO}$ are $\cong$.

B 7. Let M, N, P, and Q be the midpts. of $\overline{RO}$, $\overline{OS}$, $\overline{ST}$, and $\overline{RT}$, resp. By the midpt. formula, $M\left(\frac{2b + 0}{2}, \frac{2c + 0}{2}\right) = (b, c)$, $N\left(\frac{2a + 0}{2}, \frac{0 + 0}{2}\right) = (a, 0)$, $P\left(\frac{2d + 2a}{2}, \frac{2e + 0}{2}\right) = (d + a, e)$, and $Q\left(\frac{2b + 2d}{2}, \frac{2c + 2e}{2}\right) = (b + d, c + e)$. Slope of $\overline{MQ} = \frac{c + e - c}{b + d - b} = \frac{e}{d}$; slope of $\overline{NP} = \frac{e - 0}{d + a - a} = \frac{e}{d}$; slope of $\overline{MN} = \frac{0 - c}{a - b} = -\frac{c}{a - b}$; slope of $\overline{QP} = \frac{e - (c + e)}{d + a - (b + d)} = -\frac{c}{a - b}$. Thus, slope of $\overline{MQ}$ = slope of $\overline{NP}$, and slope of $\overline{MN}$ = slope of $\overline{QP}$. Therefore, $\overline{MQ} \parallel \overline{NP}$ and $\overline{MN} \parallel \overline{QP}$ since two nonvertical lines are $\parallel$ if and only if their slopes are $=$. $MNPQ$ is a $\square$ by def.

8. Let $DABC$ be an isos. trap. with $DC = AB$ and with coordinates as shown. Let M, N, P, and O be the midpts. of $\overline{AB}$, $\overline{BC}$, $\overline{CD}$, and $\overline{DA}$, resp. By the midpt. formula, $M\left(\frac{2a + 2b}{2}, \frac{0 + 2c}{2}\right) = (a + b, c)$, $N\left(\frac{2b - 2b}{2}, \frac{2c + 2c}{2}\right) = (0, 2c)$, $P\left(\frac{-2b - 2a}{2}, \frac{2c + 0}{2}\right) = (-b - a, c)$, and $O\left(\frac{-2a + 2a}{2}, \frac{0 + 0}{2}\right) = (0, 0)$. By the distance formula,
$MN = \sqrt{(a + b - 0)^2 + (c - 2c)^2} = \sqrt{(a + b)^2 + (-c)^2} = \sqrt{(a + b)^2 + c^2}$,
$NP = \sqrt{(0 - (-b - a))^2 + (2c - c)^2} = \sqrt{(b + a)^2 + (c)^2} = \sqrt{(a + b)^2 + c^2}$,
$PO = \sqrt{(0 - (-b - a))^2 + (0 - c)^2} = \sqrt{(b + a)^2 + (-c)^2} = \sqrt{(a + b)^2 + c^2}$,
and $OM = \sqrt{(a + b - 0)^2 + (c - 0)^2} = \sqrt{(a + b)^2 + c^2}$. Therefore, $\overline{MN} \cong \overline{NP} \cong \overline{PO} \cong \overline{OM}$, and $MNPO$ is a rhombus by def. Alternatively: By Ex. 7, $MNPO$ is a $\square$. Also, $MN = NP$ so consecutive sides $\overline{MN}$ and $\overline{NP}$ are $\cong$, and $MNPO$ is a rhombus.

9. Point C is on the circle with center O and radius a. The circle has equation $x^2 + y^2 = a^2$, so $b^2 + c^2 = a^2$. Slope of $\overline{CA} = \frac{c - 0}{b - (-a)} = \frac{c}{b + a}$; slope of $\overline{CB} = \frac{c - 0}{b - a} = \frac{c}{b - a}$. Slope of $\overline{CA}$ · slope of $\overline{CB} = \frac{c}{b + a} \cdot \frac{c}{b - a} = \frac{c^2}{b^2 - a^2}$, but $b^2 - a^2 = -c^2$. Substituting, slope of $\overline{CA}$ · slope of $\overline{CB} = \frac{c^2}{-c^2} = -1$. Thus, $\overline{CA}$ and $\overline{CB}$ are $\perp$ since two nonvertical lines are $\perp$ if and only if the prod. of their slopes is -1. Therefore, $\angle ACB$ is a right angle by the def. of $\perp$ lines.

10. Let $ORST$ be a $\square$ with coordinates as shown. By the distance formula, $(OR)^2 = (a - 0)^2 + (0 - 0)^2 = a^2 + 0^2 = a^2$; $(RS)^2 = (a + b - a)^2 + (c - 0)^2 = (b + 0)^2 + c^2 = b^2 + c^2$; $(ST)^2 = (a + b - b)^2 + (c - c)^2 = (a + 0)^2 + 0^2 = a^2$; $(TO)^2 = (b - 0)^2 + (c - 0)^2 = b^2 + c^2$; $(OS)^2 = (a + b - 0)^2 + (c - 0)^2 = (a + b)^2 + c^2$; $(RT)^2 = (a - b)^2 + (0 - c)^2 = (a - b)^2 + (-c)^2 = (a - b)^2 + c^2$. $(OS)^2 + (RT)^2 = (a + b)^2 + c^2 + (a - b)^2 + c^2 = a^2 + 2ab + b^2 + c^2 + a^2 - 2ab + b^2 + c^2 = 2a^2 + 2b^2 + 2c^2$; $(OR)^2 + (RS)^2 + (ST)^2 + (TO)^2 = a^2 + b^2 + c^2 + a^2 + b^2 + c^2 = 2a^2 + 2b^2 + 2c^2$. Therefore, $(OS)^2 + (RT)^2 = (OR)^2 + (RS)^2 + (ST)^2 + (TO)^2$.

y, T(b, c), S(a + b, c), O, R(a, 0), x

C **11.** M, N, and J are midpts. of $\overline{RS}$, $\overline{RO}$, and $\overline{OS}$; $\overline{OM}$ and $\overline{SN}$ intersect at P. By the midpt. formula, $M\left(\frac{6b + 6a}{2}, \frac{6c + 0}{2}\right) = (3b + 3a, 3c)$; $N\left(\frac{6b + 0}{2}, \frac{6c + 0}{2}\right) = (3b, 3c)$; $J\left(\frac{6a + 0}{2}, \frac{0 + 0}{2}\right) = (3a, 0)$. Slope of $\overline{OM} = \frac{3c - 0}{3b + 3a - 0} = \frac{3c}{3b + 3a} = \frac{c}{b + a}$; slope of $\overline{SN} = \frac{3c - 0}{3b - 6a} = \frac{c}{b - 2a}$; slope of $\overline{RJ} = \frac{0 - 6c}{3a - 6b} = \frac{2c}{2b - a}$.

Using the point-slope form, the equation of $\overleftrightarrow{OM}$ is $y - 0 = \frac{c}{b + a}(x - 0)$, or $y = \frac{c}{b + a}x$; the equation of $\overleftrightarrow{SN}$ is $y - 0 = \frac{c}{b - 2a}(x - 6a)$, or $y = \frac{c}{b - 2a}(x - 6a)$; the equation of $\overleftrightarrow{RJ}$ is $y - 0 = \frac{2c}{2b - a}(x - 3a)$, or $y = \frac{2c}{2b - a}(x - 3a)$. Solving the equations of $\overleftrightarrow{OM}$ and $\overleftrightarrow{SN}$ by the substitution method to find the coordinates of P, $\frac{c}{b + a}x = \frac{c}{b - 2a}(x - 6a)$; $cx(b - 2a) = c(x - 6a)(b + a)$; $xbc - 2acx = xbc + xac - 6abc - 6a^2c$; $-3acx = -6abc - 6a^2c$; $x = 2b + 2a$. Substituting to find y, $y = \frac{c}{b + a}(2b + 2a) = 2c$. Then P has coordinates $(2b + 2a, 2c)$. Substituting the x value for P into the equation for $\overleftrightarrow{RJ}$, $y = \frac{2c}{2b - a}(2b + 2a - 3a) = \frac{2c}{2b - a}(2b - a) = 2c$. Thus P is on $\overleftrightarrow{RJ}$. By the distance formula,

$OM = \sqrt{(3b + 3a - 0)^2 + (3c - 0)^2} =$
$\sqrt{(3(b + a))^2 + (3c)^2} = \sqrt{9((b + a)^2 + c^2)} = 3\sqrt{(b + a)^2 + c^2}$;
$OP = \sqrt{(2b + 2a - 0)^2 + (2c - 0)^2} = \sqrt{(2(b + a))^2 + (2c)^2} =$
$\sqrt{4((b + a)^2 + c^2)} = 2\sqrt{(b + a)^2 + c^2}$; $SN = \sqrt{(3b - 6a)^2 + (3c - 0)^2} =$
$\sqrt{(3(b - 2a))^2 + 9c^2} = \sqrt{9((b - 2a)^2 + c^2)} = 3\sqrt{(b - 2a)^2 + c^2}$;
$SP = \sqrt{(2b + 2a - 6a)^2 + (2c - 0)^2} = \sqrt{(2b - 4a)^2 + (2c)^2} =$
$\sqrt{(2(b - 2a))^2 + 4c^2} = \sqrt{4((b - 2a)^2 + c^2)} = 2\sqrt{(b - 2a)^2 + c^2}$;
$RJ = \sqrt{(3a - 6b)^2 + (0 - 6c)^2} = \sqrt{(3(a - 2b))^2 + (3(-2c))^2} =$
$\sqrt{9(a - 2b)^2 + 9(4c^2)} = \sqrt{9((a - 2b)^2 + 4c^2)} = 3\sqrt{(a - 2b)^2 + 4c^2}$;
$RP = \sqrt{(2b + 2a - 6b)^2 + (2c - 6c)^2} = \sqrt{(2a - 4b)^2 + (-4c)^2} =$
$\sqrt{(2(a - 2b))^2 + (2(-2c))^2} = \sqrt{4(a - 2b)^2 + 4(4c^2)} = \sqrt{4((a - 2b)^2 + 4c^2)} =$
$2\sqrt{(a - 2b)^2 + 4c^2}$. Multiplying by $\frac{2}{3}$, $\frac{2}{3}OM = 2\sqrt{(b + a)^2 + c^2}$; $\frac{2}{3}SN =$
$2\sqrt{(b - 2a)^2 + c^2}$; $\frac{2}{3}RJ = 2\sqrt{(a - 2b)^2 + 4c^2}$. Therefore, $OP = \frac{2}{3}OM$, $SP = \frac{2}{3}SN$,
and $RP = \frac{2}{3}RJ$.

12. Let C be the int. pt. of the $\perp$ bisectors of $\overline{OS}$ and $\overline{OR}$. From Ex. 11, the midpt. J of $\overline{OS} = (3a, 0)$, the midpt. M of $\overline{RS} = (3b + 3a, 3c)$, and the midpt. N of $\overline{OR} = (3b, 3c)$. Slope of $\overline{OS} = \frac{0 - 0}{6a - 0} = 0$, so the $\perp$ bis. of $\overline{OS}$ is a vert. line through J; its equation is $x = 3a$. Slope of $\overline{OR} = \frac{6c - 0}{6b - 0} = \frac{c}{b}$, so the $\perp$ bis. of $\overline{OR}$ has slope $-\frac{b}{c}$ and equation $y - 3c = -\frac{b}{c}(x - 3b)$. The $\perp$ bisectors of $\overline{OS}$ and $\overline{OR}$ intersect at $x = 3a$ and $y = -\frac{b}{c}(3a - 3b) + 3c = \frac{3b^2 + 3c^2 - 3ab}{c}$. C is pt. $\left(3a, \frac{3b^2 + 3c^2 - 3ab}{c}\right)$. $\overline{RS}$ has slope $\frac{6c - 0}{6b - 6a} = \frac{c}{b - a}$, so the $\perp$ bis. of $\overline{RS}$ has slope $\frac{a - b}{c}$ and equation $y - 3c = \frac{a - b}{c}(x - (3b + 3a))$; when $x = 3a$, $y = \frac{a - b}{c}(3a - 3b - 3a) + 3c = \frac{a - b}{c}(-3b) + 3c = \frac{3b^2 + 3c^2 - 3ab}{c}$, so C lies on the $\perp$ bis. of $\overline{RS}$ also.

13. From Ex. 12, the slopes of $\overline{OS}$, $\overline{OR}$, and $\overline{RS}$ are 0, $\frac{c}{b}$, and $\frac{c}{b - a}$, resp. Thus, the line containing the altitude from R is a vertical line and the lines containing the altitudes from S and O have slopes $-\frac{b}{c}$ and $\frac{a - b}{c}$, resp. The line through R has equation $x = 6b$, and the line through O has equation $y - 0 = \left(\frac{a - b}{c}\right)(x - 0)$ or $y = \left(\frac{a - b}{c}\right)x$; they int. at $x = 6b$ and $y = \left(\frac{a - b}{c}\right)6b = \frac{6ab - 6b^2}{c}$. Let H be pt. $\left(6b, \frac{6ab - 6b^2}{c}\right)$. The line containing the altitude from S has equation $y - 0 = -\frac{b}{c}(x - 6a)$, or $y = -\frac{b}{c}(x - 6a)$, and int. the line containing the altitude from R at $x = 6b$ and $y = -\frac{b}{c}(6b - 6a) = \frac{6ab - 6b^2}{c}$, that is, at pt. H. Thus, the lines containing the altitudes int. at $H\left(6b, \frac{6ab - 6b^2}{c}\right)$.

14. We have $C\left(3a, \frac{3b^2 + 3c^2 - 3ab}{c}\right)$, $G(2a + 2b, 2c)$, and $H\left(6b, \frac{6ab - 6b^2}{c}\right)$.

a. The slope of $\overline{CG} = \dfrac{2c - \frac{3b^2 + 3c^2 - 3ab}{c}}{2a + 2b - 3a} = \dfrac{2c^2 - 3b^2 - 3c^2 + 3ab}{c(2b - a)} = \dfrac{-3b^2 - c^2 + 3ab}{2bc - ac}$; slope of $\overline{GH} = \dfrac{\frac{6ab - 6b^2}{c} - 2c}{6b - (2a + 2b)} = \dfrac{6ab - 6b^2 - 2c^2}{c(4b - 2a)} = \dfrac{3ab - 3b^2 - c^2}{2bc - ac}$. Slope of $\overline{CG}$ = slope of $\overline{GH}$. Thus, C, G, and H are collinear, since if they were not collinear, $\overleftrightarrow{CG}$ and $\overleftrightarrow{GH}$ would be two parallel lines that intersect at G.

b. By the distance formula,

$$CG = \sqrt{(2a + 2b - 3a)^2 + \left(2c - \frac{3b^2 + 3c^2 - 3ab}{c}\right)^2} =$$

$$\sqrt{(2b - a)^2 + \left(\frac{2c^2 - 3b^2 - 3c^2 + 3ab}{c}\right)^2} =$$

$$\sqrt{(2b - a)^2 + \left(\frac{-3b^2 - c^2 + 3ab}{c}\right)^2};$$

$$CH = \sqrt{(6b - 3a)^2 + \left(\frac{6ab - 6b^2}{c} - \frac{3b^2 + 3c^2 - 3ab}{c}\right)^2} =$$

$$\sqrt{(3(2b - a))^2 + \left(\frac{9ab - 9b^2 - 3c^2}{c}\right)^2} = \sqrt{9(2b - a)^2 + \left(\frac{3(3ab - 3b^2 - c^2)}{c}\right)^2} =$$

$$\sqrt{9\left[(2b - a)^2 + \left(\frac{3ab - 3b^2 - c^2}{c}\right)^2\right]} = 3\sqrt{(2b - a)^2 + \left(\frac{-3b^2 - c^2 + 3ab}{c}\right)^2};$$

$$\frac{1}{3}CH = \frac{1}{3} \cdot 3\sqrt{(2b - a)^2 + \left(\frac{-3b^2 - c^2 + 3ab}{c}\right)^2} =$$

$$\sqrt{(2b - a)^2 + \left(\frac{-3b^2 - c^2 + 3ab}{c}\right)^2} = CG.$$

Page 563 • SELF-TEST 2

1. $2x - 5y = 20$; $-5y = -2x + 20$; $y = \frac{2}{5}x - 4$; slope, $\frac{2}{5}$; y-int., -4.

2.

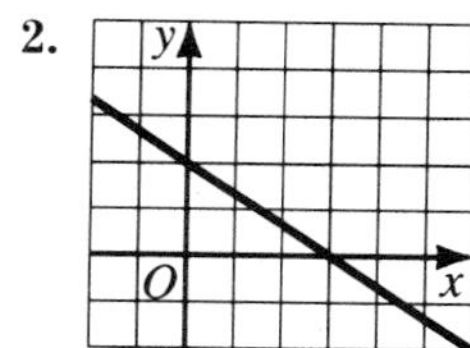

3. slope $= \frac{2 - 0}{1 - 5} = \frac{2}{-4} = -\frac{1}{2}$; $y - 0 = -\frac{1}{2}(x - 5)$; $y = -\frac{1}{2}x + \frac{5}{2}$ **4.** $y = 5$

5. Substituting $y = 3x - 4$ into $5x - 2y = 7$, $5x - 2(3x - 4) = 7$; $5x - 6x + 8 = 7$; $x = 1$; $y = 3(1) - 4 = -1$; $(1, -1)$

6. $(2e, 0)$ **7.** $(c + g, h)$ **8.** $(c - g, h)$

9. slope of $\overline{GO} = \frac{0 - (-1)}{0 - 4} = -\frac{1}{4}$; slope of $\overline{LD} = \frac{6 - 5}{2 - 6} = -\frac{1}{4}$; $\overline{GO} \parallel \overline{LD}$; slope of $\overline{OL} = \frac{6 - 0}{2 - 0} = 3$; slope of $\overline{DG} = \frac{5 - (-1)}{-6 - 4} = 3$; $\overline{OL} \parallel \overline{DG}$. Since both pairs of opp. sides are $\parallel$, $GOLD$ is a ▱ by def.

Page 564 • APPLICATION

1. 120 **2–4.** The measures of the $\angle$s where the soap films meet are 120.

Page 565 • EXTRA

1. y-axis **2.** z-axis **3.** x-axis **4.** all 3 axes **5.** xy-plane **6.** xz-plane
7. yz-plane **8.** all 3 planes

9.

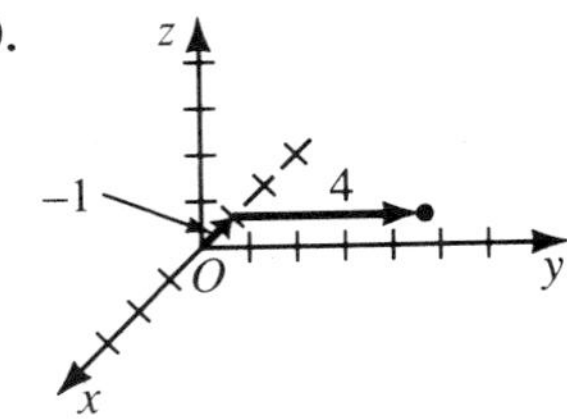

10.

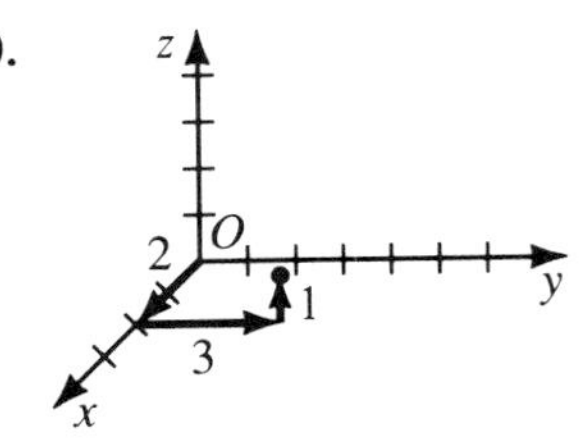

11.

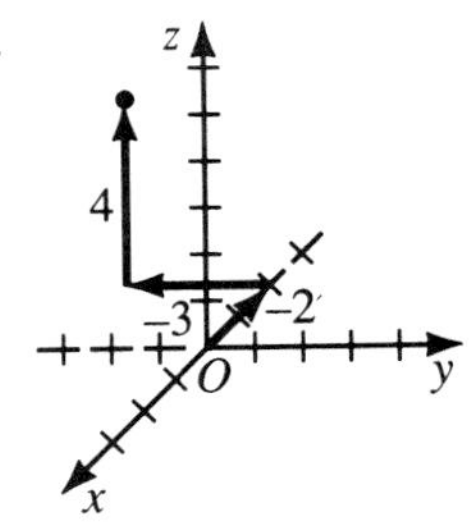

12.

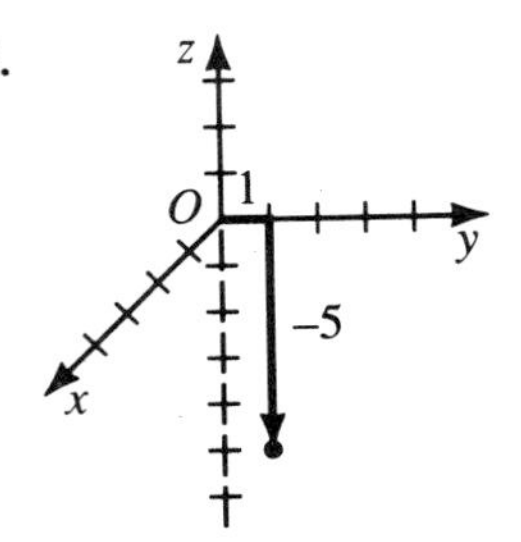

13.

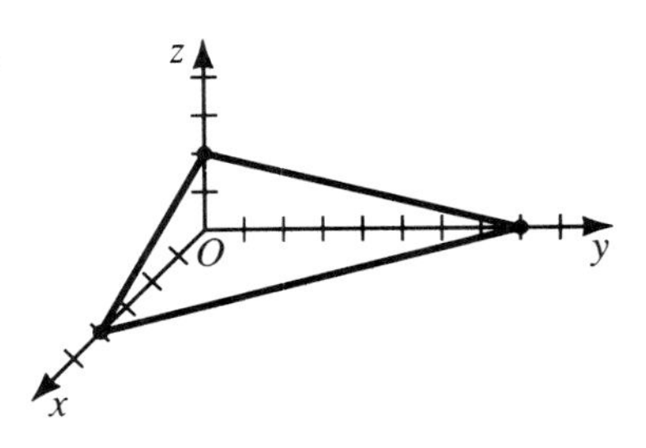

14.

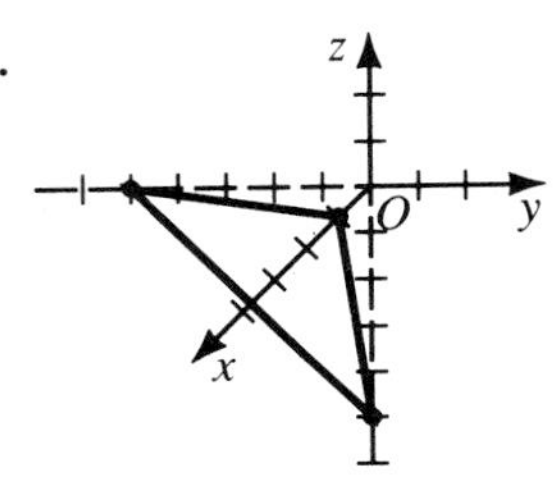

15.

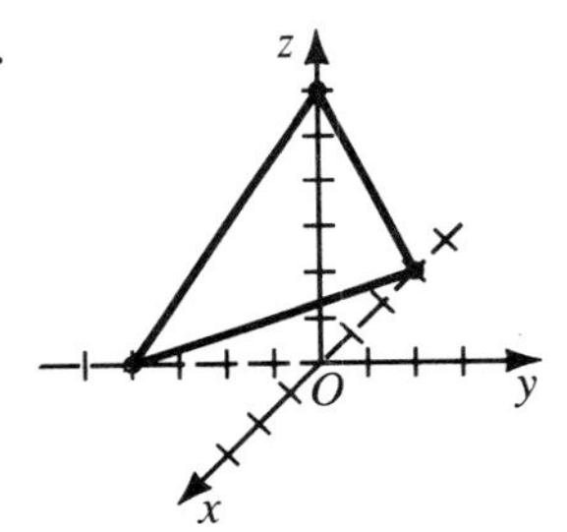

16.

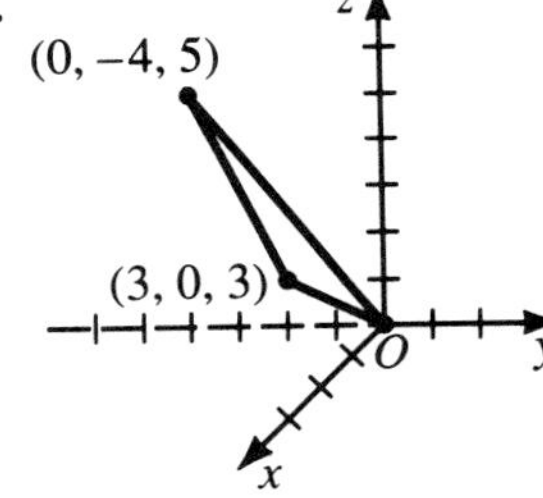

Page 567 • CHAPTER REVIEW

1.

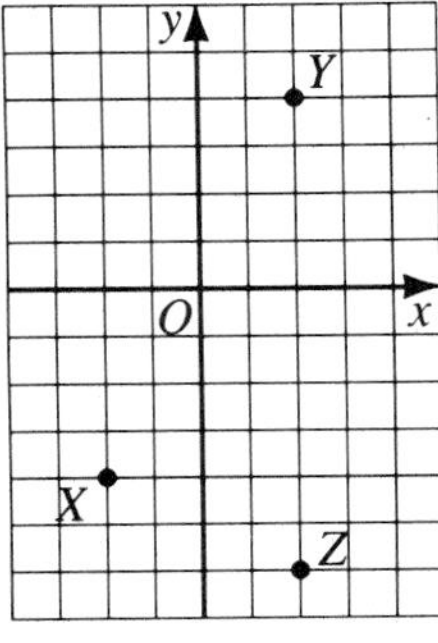

$XY = \sqrt{(2-(-2))^2 + (4-(-4))^2} = \sqrt{4^2 + 8^2} = \sqrt{16 + 64} = \sqrt{80} = 4\sqrt{5}$; $YZ = |4 - (-6)| = 10$; $XZ = \sqrt{(2-(-2))^2 + (-6-(-4))^2} = \sqrt{4^2 + (-2)^2} = \sqrt{16 + 4} = \sqrt{20} = 2\sqrt{5}$

2. $(XY)^2 + (XZ)^2 = 80 + 20 = 100 = (YZ)^2$, so $\triangle XYZ$ is a rt. $\triangle$. (If the square of one side of a $\triangle$ = the sum of the squares of the other 2 sides, then the $\triangle$ is a rt. $\triangle$.)

3. $(-3, 0)$; 10 **4.** $(5, -1)$; 7 **5.** $(x + 6)^2 + (y + 1)^2 = 9$

6. $\dfrac{-6 - (-1)}{15 - (-5)} = \dfrac{-5}{20} = -\dfrac{1}{4}$

7. $\frac{y-(-13)}{0-9} = \frac{2}{3}; \frac{y+13}{-9} = \frac{2}{3}; 3y + 39 = -18; 3y = -57; y = -19$

8. $5 = \frac{y-(-2)}{x-0}; 5 = \frac{y+2}{x}; y + 2 = 5x; y = 5x - 2$; answers may vary;

for example: $(1, 3), (2, 8), (-1, -7)$

9. 0

10. slope of $\overline{TQ} = \frac{-2-4}{-1-7} = \frac{-6}{-8} = \frac{3}{4}$; slope of $\overline{SR} = \frac{-2-1}{6-10} = \frac{-3}{-4} = \frac{3}{4}$; slope of $\overline{TS} = \frac{1-4}{10-7} = \frac{-3}{3} = -1$; slope of $\overline{QR} = \frac{-2-(-2)}{6-(-1)} = \frac{0}{7} = 0$; $\overline{TQ} \parallel \overline{SR}$; $\overline{TS}$ is not $\parallel$ to $\overline{QR}$; $QRST$ is a trap. by def.

11. $\frac{3}{4}; -\frac{4}{3}$

12. U is on $\overline{QT}$, so slope of $\overline{UT}$ = slope of $\overline{QT} = \frac{3}{4}; \frac{4-y}{7-x} = \frac{3}{4}; 16 - 4y = 21 - 3x$;

$3x - 4y = 5$; also, if $\overline{UR} \parallel \overline{ST}$, then $\frac{-2-y}{6-x} = -1$ = slope of $\overline{ST}$;

$-2 - y = -6 + x; y = 4 - x$; substituting, $3x - 4(4 - x) = 5$;

$3x - 16 + 4x = 5; 7x = 21; x = 3; y = 4 - 3 = 1; U(3, 1)$

13. **a.** $\overrightarrow{PQ} = (7 - 3, 1 - (-2)) = (4, 3)$ **b.** $|\overrightarrow{PQ}| = \sqrt{4^2 + 3^2} = \sqrt{16 + 9} = \sqrt{25} = 5$ **c.** $-2(4, 3) = (-2 \cdot 4, -2 \cdot 3) = (-8, -6)$

14. $(2, 6) + 3(1, -2) = (2, 6) + (3 \cdot 1, 3(-2)) = (2, 6) + (3, -6) = (2 + 3, 6 - 6) = (5, 0)$

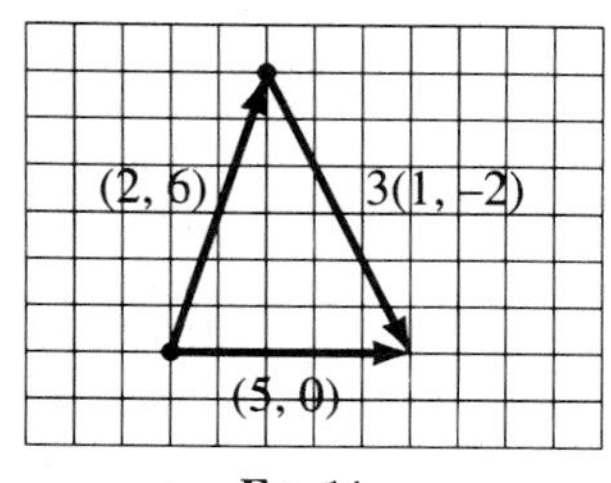

Ex. 14

15. $\left(\frac{7+1}{2}, \frac{-2-1}{2}\right) = \left(4, -\frac{3}{2}\right)$

16. $\left(\frac{-4+2}{2}, \frac{5-5}{2}\right) = (-1, 0)$

17. $\left(\frac{a-a}{2}, \frac{b+b}{2}\right) = (0, b)$

18. Let $R = (a, b)$; $\left(\frac{a+11}{2}, \frac{b-1}{2}\right) = (0, 5)$; $\frac{a+11}{2} = 0$ and $\frac{b-1}{2} = 5$;

$a + 11 = 0$ and $b - 1 = 10$; $a = -11$ and $b = 11$; $R(-11, 11)$.

19.

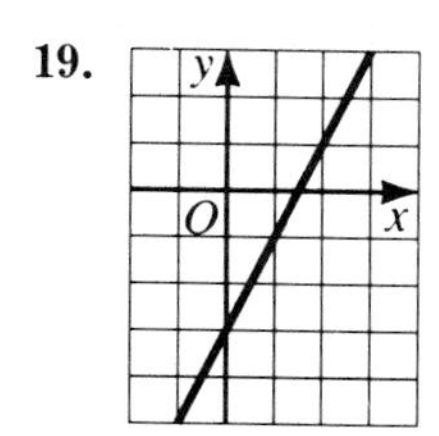

20.

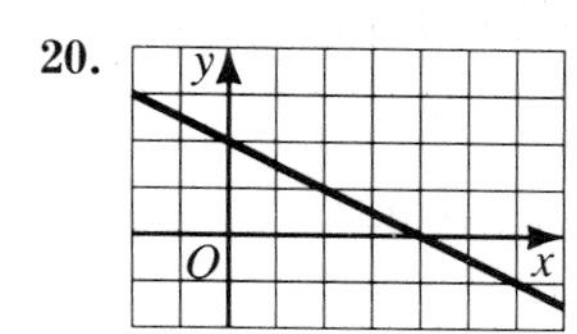

21. $y = 2x - 3$ and $x + 2y = 4$; $x + 2(2x - 3) = 4$; $x + 4x - 6 = 4$; $5x = 10$; $x = 2$; $y = 1$; $(2, 1)$

22. $y = 4x + 7$

23. $m = \frac{10 - 2}{3 - (-1)} = \frac{8}{4} = 2$; $y = 2x + b$; $2 = 2(-1) + b$; $b = 4$; $y = 2x + 4$

24. $(a + b, c)$

25. Let $OPQR$ be a ▱ with $Q(a + b, c)$ and the other coordinates as given. M is the midpt. of $\overline{RQ}$ and N is the midpt. of $\overline{OP}$. By the midpt. formula, $M\left(\frac{b + a + b}{2}, \frac{c + c}{2}\right) = \left(\frac{a}{2} + b, c\right)$; $N\left(\frac{0 + a}{2}, \frac{0 + 0}{2}\right) = \left(\frac{a}{2}, 0\right)$. Slope of $\overline{ON} = \frac{0 - 0}{\frac{a}{2} - 0} = 0$; slope of $\overline{MQ} = \frac{c - c}{a + b - \left(\frac{a}{2} + b\right)} = 0$; slope of $\overline{OM} = \frac{c - 0}{\frac{a}{2} + b - 0} = \frac{2c}{a + 2b}$; slope of $\overline{NQ} = \frac{c - 0}{a + b - \frac{a}{2}} = \frac{c}{\frac{2a + 2b - a}{2}} = \frac{2c}{a + 2b}$. Slope of $\overline{ON}$ = slope of $\overline{MQ}$; slope of $\overline{OM}$ = slope of $\overline{NQ}$. $\overline{ON} \parallel \overline{MQ}$ and $\overline{OM} \parallel \overline{NQ}$ since two nonvertical lines are $\parallel$ if and only if their slopes are =. $ONQM$ is a ▱ by def.

Page 568 • CHAPTER TEST

1. a. $MN = \sqrt{(2 - (-2))^2 + (4 - 1)^2} = \sqrt{4^2 + 3^2} = \sqrt{16 + 9} = \sqrt{25} = 5$

b. $m = \frac{4 - 1}{2 - (-2)} = \frac{3}{4}$ **c.** $\left(\frac{-2 + 2}{2}, \frac{1 + 4}{2}\right) = \left(\frac{0}{2}, \frac{5}{2}\right) = \left(0, \frac{5}{2}\right)$

2. From Ex. 1(b), $m = \frac{3}{4}$; $y - 1 = \frac{3}{4}(x + 2)$ or $y = \frac{3}{4}x + \frac{5}{2}$

3. From Ex. 1(a), $MN = 5$; $(x + 2)^2 + (y - 1)^2 = 25$

4. Let $Z = (x, y)$; $\left(\frac{x + 2}{2}, \frac{y + 4}{2}\right) = (-2, 1)$; $\frac{x + 2}{2} = -2$ and $\frac{y + 4}{2} = 1$; $x + 2 = -4$ and $y + 4 = 2$; $x = -6$ and $y = -2$; $Z(-6, -2)$

5–8. Form of equations may vary.

5. $y = -\frac{3}{2}x + 4$

6. $y = mx + 5$; the line contains $(3, 0)$, so $0 = 3m + 5$; $m = -\frac{5}{3}$; $y = -\frac{5}{3}x + 5$, or $5x + 3y = 15$

7. $y = -3x + 6$ has slope -3, so the required line has slope -3 also; $y - 5 = -3(x + 2)$

8. $y = -2x + 3$ has slope -2, so a line $\perp$ to it has slope $\frac{1}{2}$; $y = \frac{1}{2}x + 7$

9. **a.** $\overrightarrow{PQ} = (4 - (-2), 1 - 5) = (6, -4)$ **b.** $|\overrightarrow{PQ}| = \sqrt{6^2 + (-4)^2} = \sqrt{36 + 16} = \sqrt{52} = 2\sqrt{13}$

10. slope of (3, 6) $= \frac{6}{3} = 2$; $2 = \frac{k}{-2}$; $k = -4$

11. slope of (3, -5) is $\frac{-5}{3}$; slope of (c, 6) $= \frac{6}{c}$; $\frac{-5}{3}\left(\frac{6}{c}\right) = -1$; $\frac{-10}{c} = -1$; $c = 10$

12. $(5, -3) + 4(-2, 1) = (5, -3) + (4(-2), 4 \cdot 1) = (5, -3) + (-8, 4) = (5 - 8, -3 + 4) = (-3, 1)$

13. $x + 2y = 8$, so $x = 8 - 2y$; $3x - y = 3$; $3(8 - 2y) - y = 3$; $24 - 6y - y = 3$; $21 = 7y$; $y = 3$; $x = 8 - 2 \cdot 3 = 2$; (2, 3)

14.

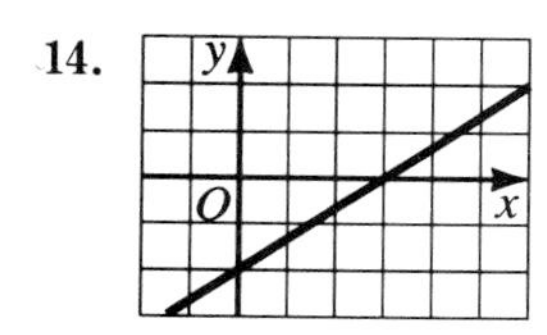

15.

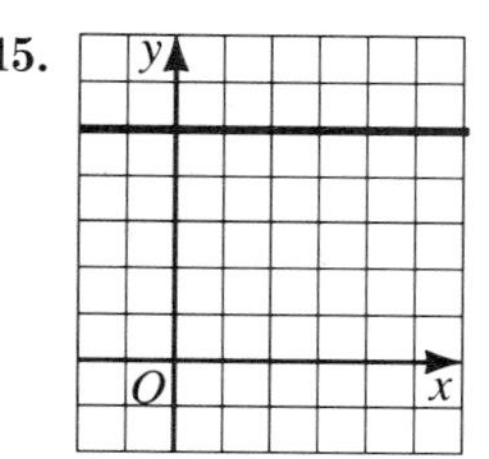

16. Points may vary. Examples: (5, 6), (8, 10), (-1, -2) **17.** $(f - g, h)$

18. $JL = \sqrt{(-3 - (-12))^2 + (-3 - 0)^2} = \sqrt{9^2 + (-3)^2} = \sqrt{81 + 9} = \sqrt{90} = 3\sqrt{10}$; $KL = \sqrt{(-3 - 0)^2 + (-3 - 6)^2} = \sqrt{(-3)^2 + (-9)^2} = \sqrt{9 + 81} = \sqrt{90} = 3\sqrt{10}$; $\overline{JL} \cong \overline{KL}$, so $\triangle JKL$ is isos.

19. slope of $\overline{JL} = \frac{-3 - 0}{-3 - (-12)} = \frac{-3}{9} = -\frac{1}{3}$; slope of $\overline{KL} = \frac{-3 - 6}{-3 - 0} = \frac{-9}{-3} = 3$;

slope of $\overline{JL}$ · slope of $\overline{KL} = -\frac{1}{3}(3) = -1$; $\overline{JL} \perp \overline{KL}$; $\triangle JKL$ is a rt. $\triangle$.

20. Let rectangle $OABC$ have coordinates as shown. By the midpt. formula, $\overline{CA}$ has midpt. $\left(\frac{0 + 2a}{2}, \frac{2b + 0}{2}\right) = (a, b)$; $\overline{OB}$ has midpt. $\left(\frac{2a + 0}{2}, \frac{2b + 0}{2}\right) = (a, b)$. $\overline{CA}$ and $\overline{OB}$ have the same midpt. By the def. of seg. bis., diags. $\overline{CA}$ and $\overline{OB}$ bisect each other.

y
C(0, 2b) B(2a, 2b)
O(0, 0) A(2a, 0) x

21. Let rectangle $ABCD$ have coordinates as shown with M, N, P, and Q the midpts. of $\overline{AB}$, $\overline{BC}$, $\overline{CD}$, and $\overline{DA}$, resp. By the midpt. formula, $M\left(\frac{-a+a}{2}, \frac{-b-b}{2}\right) = (0, -b)$; $N\left(\frac{a+a}{2}, \frac{b-b}{2}\right) = (a, 0)$;
$P\left(\frac{a-a}{2}, \frac{b+b}{2}\right) = (0, b)$; $Q\left(\frac{-a-a}{2}, \frac{b-b}{2}\right) = (-a, 0)$.

Slope of $\overline{QM} = \frac{-b-0}{0-(-a)} = -\frac{b}{a}$; slope of $\overline{PN} = \frac{b-0}{0-a} = -\frac{b}{a}$.
Slope of $\overline{QM}$ = slope of $\overline{PN}$, so $\overline{QM} \parallel \overline{PN}$. By the distance formula,
$QM = \sqrt{(-a-0)^2 + (0-(-b))^2} = \sqrt{(-a)^2 + b^2} = \sqrt{a^2+b^2}$;
$PN = \sqrt{(a-0)^2 + (0-b)^2} = \sqrt{a^2 + (-b)^2} = \sqrt{a^2+b^2}$; so $\overline{QM} \cong \overline{PN}$.
Therefore $QMNP$ is a ▱ since one pair of opp. sides are both $\cong$ and $\parallel$. By the distance formula, $MN = \sqrt{(a-0)^2 + (0-(-b))^2} = \sqrt{a^2+b^2}$. Thus, $\overline{MN} \cong \overline{QM}$ and $QMNP$ is a rhombus since it is a ▱ with two consec. sides $\cong$.

Page 569 • CUMULATIVE REVIEW: CHAPTERS 1–13

A **1.** $m\angle CBD = \frac{1}{2}m\angle ABC$; $\frac{3}{2}x + 21 = \frac{1}{2}(5x-4)$; $\frac{3}{2}x + 21 = \frac{5}{2}x - 2$; $23 = x$;
$m\angle ABC = 5 \cdot 23 - 4 = 111$; obtuse

2. Show that when the lines are cut by a trans.: (1) a pair of corr. ∠s are ≅; (2) a pair of alt. int. ∠s are ≅; (3) a pair of same-side int. ∠s are supp.; show that (4) in a plane, both lines are ⊥ to a third line; (5) both lines are ∥ to a third line.

3. No; no; draw a figure in which the diags. do not bisect each other. An example is given at the right.

Ex. 3

4. If $x = 1$, then $x \neq 0$. If $x = 0$, then $x \neq 1$; true.

5. a. Since $\angle AEB \cong \angle CED$ (Vert. ∠s are ≅.) and $\frac{AE}{CE} = \frac{BE}{DE}$, $\triangle AEB \sim \triangle CED$ (SAS ~ Thm.); therefore, $\angle B \cong \angle D$ (Corr. ∠s of ~ △s are ≅.).

b. $\frac{AE}{CE} = \frac{AB}{CD}$; $\frac{8}{12} = \frac{12}{x}$; $\frac{2}{3} = \frac{12}{x}$; $2x = 36$; $x = 18$ **c.** $2^2 : 3^2 = 4 : 9$

6. $12^2 + 35^2 = 144 + 1225 = 1369$; $37^2 = 1369$; right

7. **a.** $\cos A = \frac{AB}{AC} = \frac{1}{3}$ **b.** $\sin C = \frac{AB}{AC} = \frac{1}{3}$ **c.** $BC = \sqrt{3^2 - 1^2} = \sqrt{8} = 2\sqrt{2}$; $\tan A = \frac{BC}{AB} = \frac{2\sqrt{2}}{1} = 2\sqrt{2}$ **d.** $\cos C = \frac{BC}{AC} = \frac{2\sqrt{2}}{3}$

8. Using 30°–60°–90° △ relationships, $\frac{1}{2}s = \frac{\sqrt{3}}{\sqrt{3}} = 1$; $s = 2$;

$p = 6 \cdot 2 = 12$ cm; $A = \frac{1}{2}ap = \frac{1}{2} \cdot \sqrt{3} \cdot 12 = 6\sqrt{3}$ cm^2

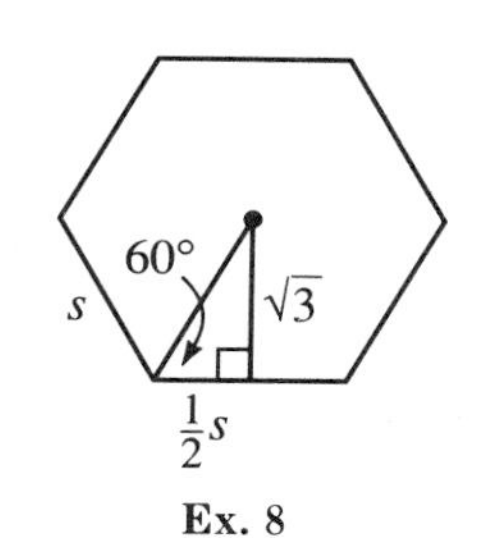

Ex. 8

9. T.A. = L.A. + $2B = 2\pi rh + 2\pi r^2 =$ $2\pi(10)(8.2) + 2\pi \cdot 10^2 = 164\pi + 200\pi =$ 364π; $V = \pi r^2 h = \pi(10^2)(8.2) = 820\pi$

10. The locus is a line ∥ to the two given lines and halfway between them.

11. $\frac{10}{18} = \frac{x}{12}$; $\frac{5}{9} = \frac{x}{12}$; $9x = 60$; $x = 6\frac{2}{3}$

12. $(x + y) + (x - y) = 180$; $2x = 180$; $x = 90$

13. $180 - x = \frac{1}{2}(104 + 78) = 91$; $x = 89$

B **14.** $9(9 + 11) = x^2$; $x^2 = 9 \cdot 20$; $x = \sqrt{180} = 6\sqrt{5}$; $9(9 + 11) = 10(10 + y)$; $180 = 10(10 + y)$; $10 + y = 18$; $y = 8$

15. Given: $\overrightarrow{BD}$ bis. $\angle ABC$; $\overrightarrow{BD} \perp \overline{AC}$

Prove: $\triangle ABC$ is isos.

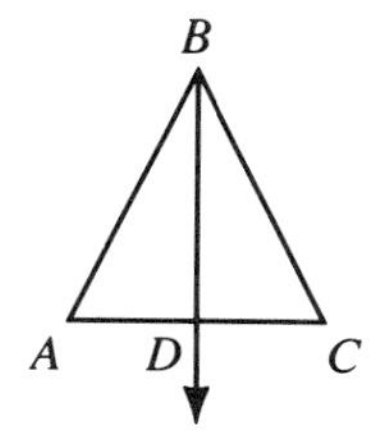

Statements	Reasons
1. $\overrightarrow{BD}$ bis. $\angle ABC$	1. Given
2. $\angle ABD \cong \angle CBD$	2. Def. of ∠ bis.
3. $\overrightarrow{BD} \perp \overline{AC}$	3. Given
4. $\angle BDA \cong \angle BDC$	4. If 2 lines are ⊥, then they form ≅ adj. ∠s.
5. $\overline{BD} \cong \overline{BD}$	5. Refl. Prop.
6. $\triangle ABD \cong \triangle CBD$	6. ASA Post.
7. $\overline{AB} \cong \overline{CB}$	7. Corr. parts of ≅ △s are ≅.
8. $\triangle ABC$ is isos.	8. Def. of isos. △

16. Draw an obtuse △. Use Const. 10.

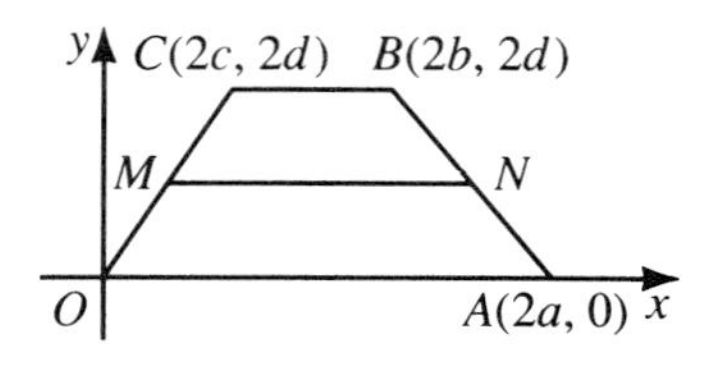

17. Let $OABC$ be a trap. with median $\overline{MN}$ and coordinates as shown. By the midpt. formula, $M\left(\frac{2c + 0}{2}, \frac{2d + 0}{2}\right) = (c, d)$; $N\left(\frac{2a + 2b}{2}, \frac{0 + 2d}{2}\right) = (a + b, d)$. Slope of

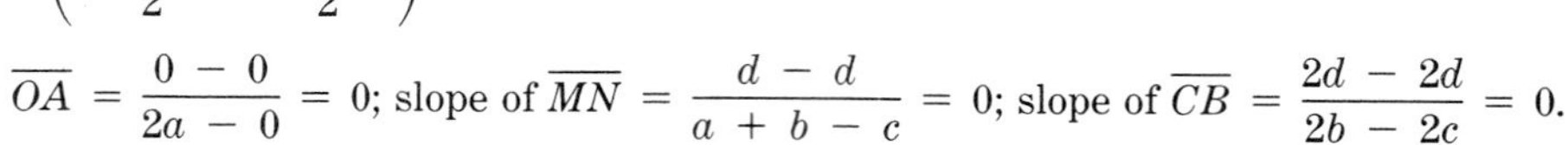

$\overline{OA} = \frac{0 - 0}{2a - 0} = 0$; slope of $\overline{MN} = \frac{d - d}{a + b - c} = 0$; slope of $\overline{CB} = \frac{2d - 2d}{2b - 2c} = 0$.

Slope of $\overline{OA}$ = slope of $\overline{MN}$ = slope of $\overline{CB}$. Therefore, $\overline{OA} \parallel \overline{MN} \parallel \overline{CB}$ since two nonvertical lines are $\parallel$ if and only if their slopes are $=$.

Page 574 • CLASSROOM EXERCISES

1. **a.** One pt. in A has 2 images in B. **b.** One pt. in B has no preimage in A. **c.** One pt. in B has 2 preimages in A. **d.** One pt. in A has no image in B.
2. **a.** 3; 6; 6 **b.** No; every positive number y has 2 preimages, y and $-y$.
3. **a.** $P'(2, 6)$; $Q'(8, 2)$ **b.** Yes **c.** No; $PQ \neq P'Q'$ **d.** $M\left(\frac{5}{2}, 2\right) \rightarrow M'(5, 4)$; yes
4. **a.** $g(8) = 2 \cdot 8 - 1 = 15$; $g(-8) = 2(-8) - 1 = -17$ **b.** $g(5) = 2 \cdot 5 - 1 = 9$ **c.** $7 = 2x - 1$; $2x = 8$; $x = 4$
5. **a.** $A'(1, 2)$; $B'(4, 6)$; $C'(6, 3)$; $D'(0, -1)$ **b.** $AB = \sqrt{3^2 + 4^2} = 5$; $A'B' = \sqrt{3^2 + 4^2} = 5$; $CD = \sqrt{6^2 + 4^2} = \sqrt{52} = 2\sqrt{13}$; $C'D' = \sqrt{6^2 + 4^2} = 2\sqrt{13}$ **c.** Yes **d.** $(x + 1, y + 2) = (0, 0)$, $x = -1$, $y = -2$, $(-1, -2)$; $(x + 1, y + 2) = (4, 5)$, $x = 3$, $y = 3$, $(3, 3)$
6. N **7.** No **8.** It distorts distances.
9. Let B be the vertex of a given $\angle$. Choose pts. A and C, one on each side of $\angle B$, and draw $\overline{AC}$. An isom. maps $\triangle ABC$ to $\triangle A'B'C'$ with $\triangle ABC \cong \triangle A'B'C'$ by Thm. 14-1. $\angle B$ and $\angle B'$ are corr. parts of $\cong$ ▲s, so $\angle B \cong \angle B'$. Thus the isom. maps an $\angle$ to a $\cong \angle$.
10. The diagonals drawn from one vertex separate the interior of a polygon into triangular regions. Under an isom., each of these ▲s maps to a $\cong \triangle$ (Thm. 14-1), and the areas of $\cong$ ▲s are $\cong$ by the Area $\cong$ Post. By the Area Add. Post., the area of the original polygon is the sum of the areas of its triangular regions. This sum = the sum of the areas of the image ▲s, which is the area of the image polygon.

Pages 574–576 • WRITTEN EXERCISES

A

1. $f: 8 \rightarrow 5(8) - 7 = 33$; $5x - 7 = 13$, $x = 4$
2. $g: 5 \rightarrow 8 - 3(5) = -7$; $8 - 3x = 0$, $x = \frac{8}{3}$
3. $f(3) = 3^2 + 1 = 10$; $f(-3) = (-3)^2 + 1 = 10$; no, since 10 has 2 preimages.
4. $h\left(\frac{1}{2}\right) = 6\left(\frac{1}{2}\right) + 1 = 4$; yes

5. a.

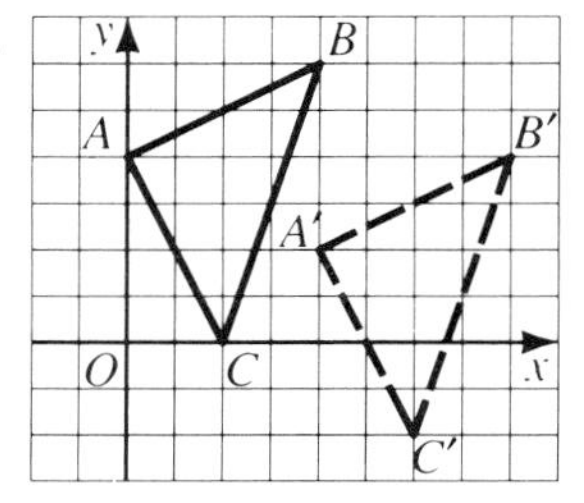

b. Yes **c.** $x + 4 = 12$, $x = 8$; $y - 2 = 6$, $y = 8$; (8, 8)

6. a.

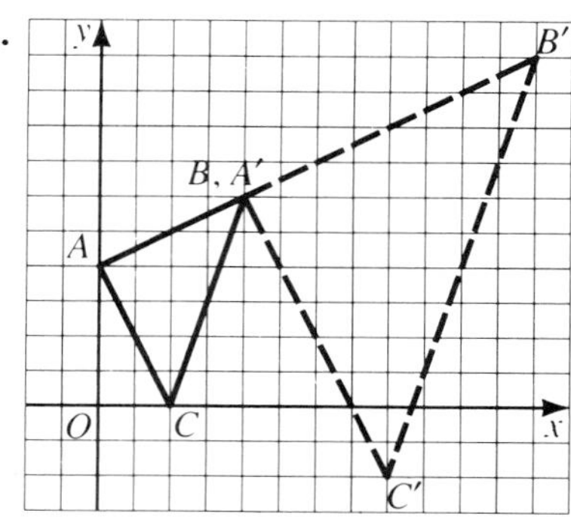

b. No **c.** $2x + 4 = 12$, $2x = 8$, $x = 4$; $2y - 2 = 6$, $2y = 8$, $y = 4$; (4, 4)

7. a.

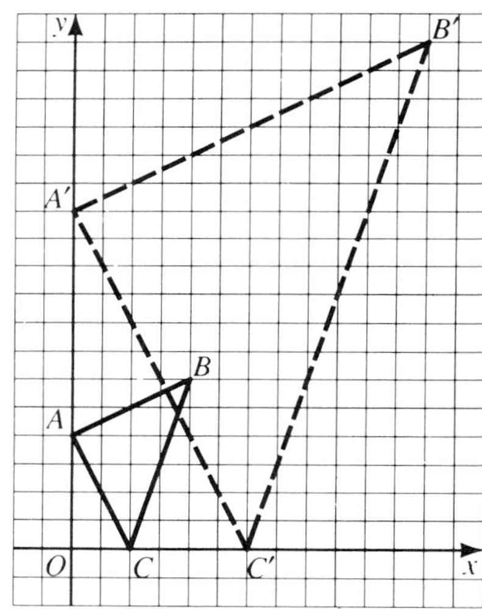

b. No **c.** $3x = 12$, $x = 4$; $3y = 6$, $y = 2$; (4, 2)

8. a.

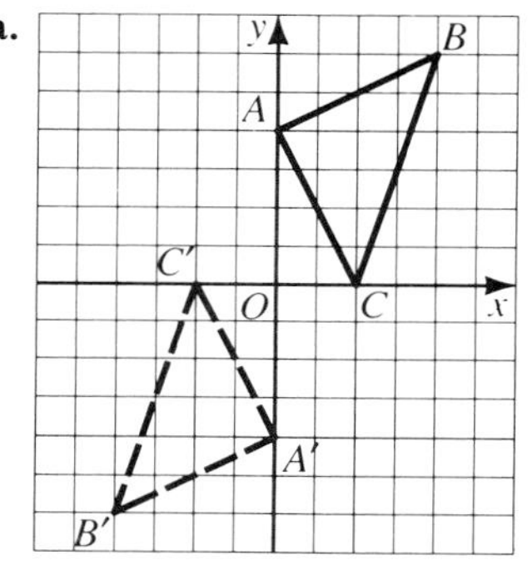

b. Yes **c.** $-x = 12$, $x = -12$; $-y = 6$, $y = -6$; $(-12, -6)$

9. a.

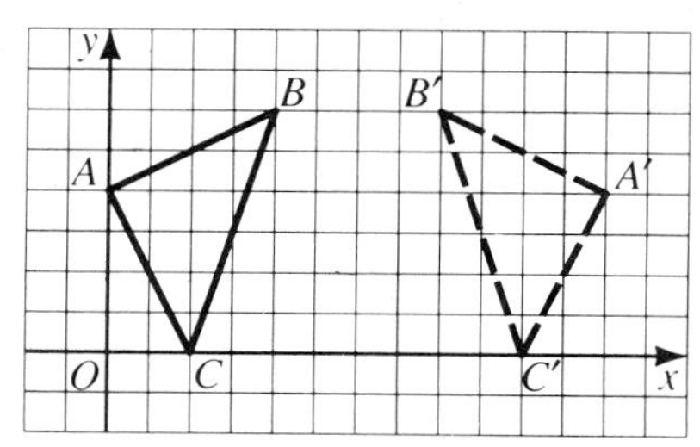

b. Yes **c.** $12 - x = 12$, $x = 0$; $y = 6$; (0, 6)

10. a.

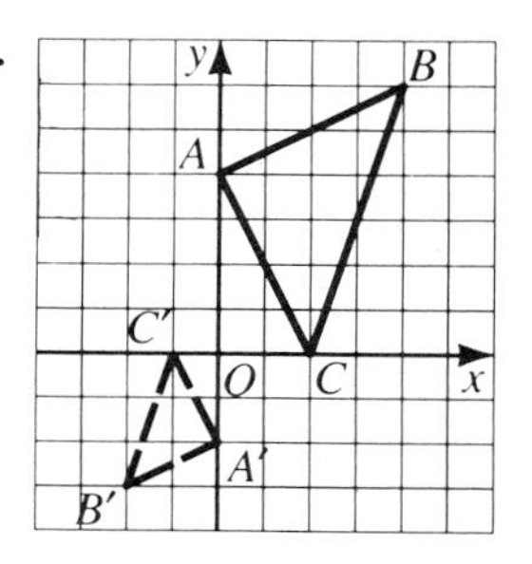

b. No **c.** $-\frac{1}{2}x = 12$, $x = -24$; $-\frac{1}{2}y = 6$, $y = -12$; $(-24, -12)$

11. **a.** Yes **b.** preserve **c.** No; $PQ \neq P'Q'$

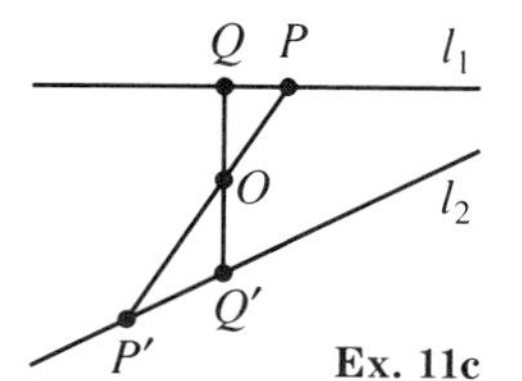

Ex. 11c

12. Map each pt. P on $\overline{XY}$ to the pt. P' where the line $\parallel$ to $\overline{YZ}$ through P intersects $\overline{XZ}$.

B 13. Let X be the intersection of diags. $\overline{AC}$ and $\overline{DB}$. For each pt. P on $\overline{DC}$, map P to the pt. P' where $\overrightarrow{PX}$ intersects $\overline{AB}$; no.

14. **a.** Yes **b, c.**

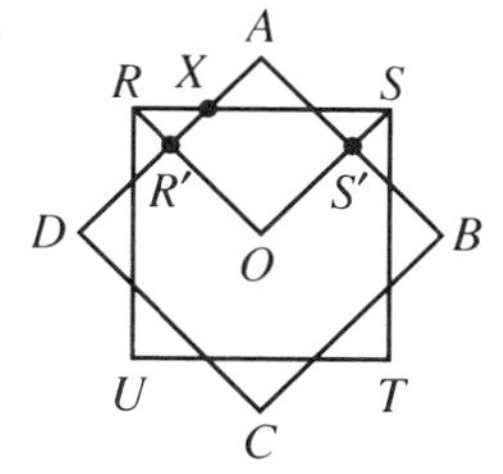

d. No **e.** Answers may vary. For example, rotate $\overrightarrow{OP}$ clockwise through an $\angle$ of 45°. Let P' be the intersection of the rotated ray and the blue square.

15. $ABCD$: $A = 4$, $p = 8$;
$A'B'C'D'$: $A = 4$, $p = 4 + 4\sqrt{2}$

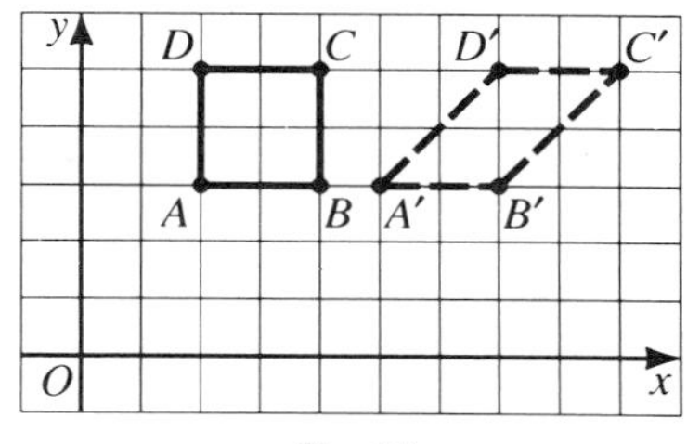

Ex. 15

16. The equator is mapped to itself.

17. Yes **18.** Less **19.** No

20. **a.** $P'(4, 0)$, $Q'(-3, 0)$, $R'(-3, 0)$; check students' graphs.
b. No; $PQ \neq P'Q'$
c. No; S is not one-to-one, since it maps the whole plane onto the x-axis; a transformation would map the plane onto itself.

21. $A'B' = 0$, but if A and B are different points, then $AB > 0$. $AB \neq A'B'$, so M does not preserve distance.

22. **a.** Yes **b.** P **c.** k is the $\perp$ bis. of $\overline{QQ'}$.

C 23. **a.** $A'(-6, 1)$, $B'(-3, 4)$, $C'(-1, -3)$; check students' graphs. **b.** For any 2 pts., $P(x_1, y_1)$ and $Q(x_2, y_2)$, $PQ = \sqrt{(x_2 - x_1)^2 + (y_2 - y_1)^2}$. The images of P and Q are $P'(-x_1, y_1)$ and $Q'(-x_2, y_2)$. Then $P'Q' = \sqrt{(-x_2 + x_1)^2 + (y_2 - y_1)^2} = \sqrt{[-1(x_2 - x_1)]^2 + (y_2 - y_1)^2} = \sqrt{(-1)^2(x_2 - x_1)^2 + (y_2 - y_1)^2} = \sqrt{(x_2 - x_1)^2 + (y_2 - y_1)^2} = PQ$. Since R preserves distance, it is an isometry.

Page 576 • EXPLORATIONS

$\triangle ABC \cong \triangle ABD$; $CE = DE$; the $\angle$s with vertex E are rt. $\angle$s.

Page 579 • CLASSROOM EXERCISES

1. reflection in line k **2.** D **3.** C **4.** $\overline{DC}$ **5.** B **6.** T **7.** $\overline{CB}$

8. $\angle UTS$ or $\angle STU$ **9.** W **10.** $\overline{WX}$ **11.** $\overline{TU}$ **12.** line k

13. $A'(-4, -5)$, $B'(2, -4)$, $C'(5, 1)$, $D'(x, -y)$

14. $E'(0, 5)$, $F'(-6, 3)$, $G'(3, 1)$, $H'(-x, y)$

15. **16.** **17.**

18.

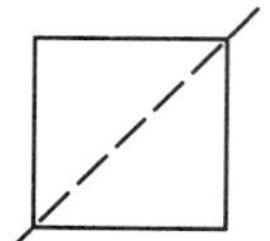

The square is its own image.

19. Yes; yes; by Thm. 14-2, a reflection is an isometry.

20. Answers may vary; examples are given. **a.** Under an isometry, the length of a seg. = the length of the image seg. **b.** Under a reflection, the area of a plane figure = the area of its image. **c.** Under a reflection, if the orientation of a figure, $A_1A_2 \cdots A_n$, is considered to be clockwise, then the orientation of the image, $A_1'A_2' \cdots A_n'$, is counterclockwise.

Pages 580–582 • WRITTEN EXERCISES

A **1.**

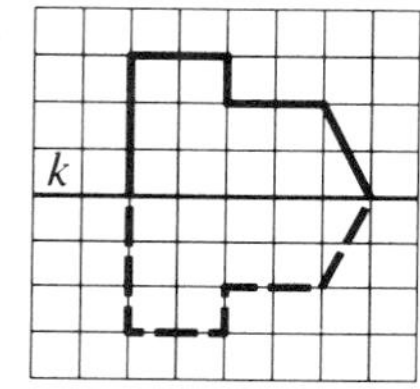

2.

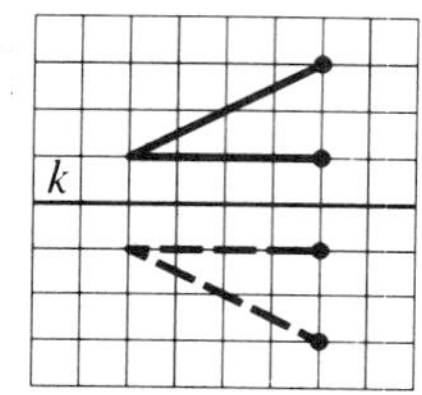

3.

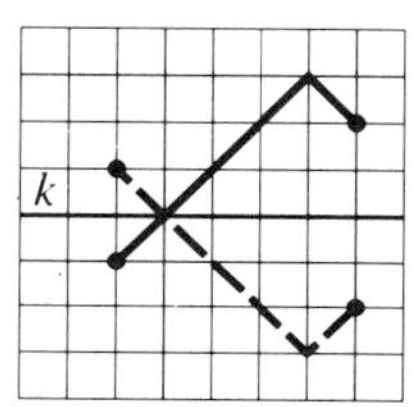

4.

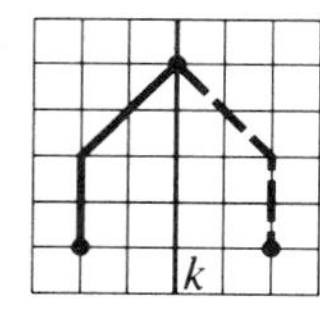

5.

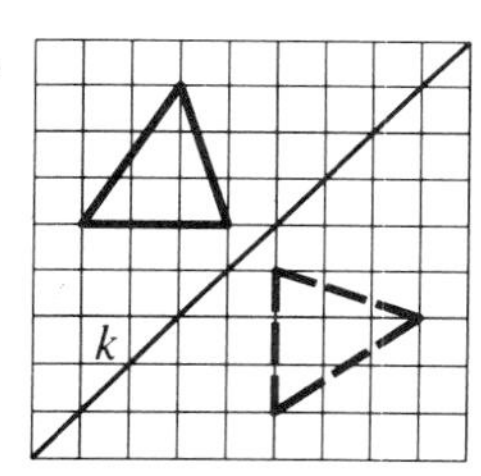

6.

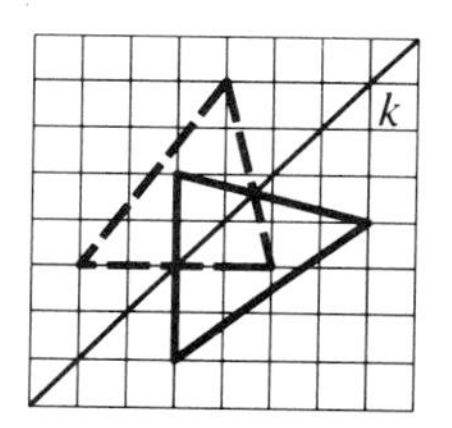

7. a. $(2, -4)$ **b.** $(-2, 4)$ **c.** $(4, 2)$ **8. a.** $(4, 0)$ **b.** $(-4, 0)$ **c.** $(0, 4)$

9. a. $(0, 2)$ **b.** $(0, -2)$ **c.** $(-2, 0)$ **10. a.** $(-2, -1)$ **b.** $(2, 1)$ **c.** $(1, -2)$

11. a. $(-3, 2)$ **b.** $(3, -2)$ **c.** $(-2, -3)$ **12. a.** $(0, 0)$ **b.** $(0, 0)$ **c.** $(0, 0)$

13. Answers may vary; for example: WOW, TOOT, AHA

14. Answers may vary; for example: ICEBOX, DECIDE, OBOE

B **15.**

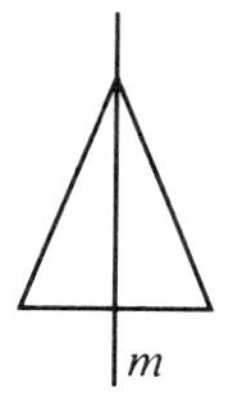

An isos. △ with m the ⊥ bis. of the base.

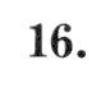

16.

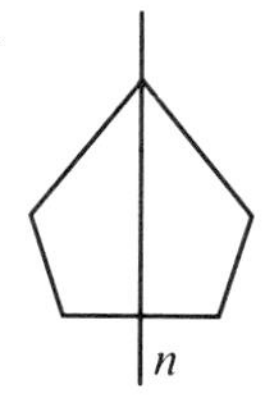

17. If P is not on plane X, then X is $\perp$ to and bisects $\overline{PP'}$. If P is on plane X, then $P' = P$.

18. $P = P'$ is on m, the $\perp$ bis. of $\overline{QQ'}$. So $PQ = P'Q'$. (If a pt. lies on the $\perp$ bis. of a seg., then it is equidistant from the endpts. of the seg.)

19. Let X and Y be the pts. where $\overline{PP'}$ and $\overline{QQ'}$ int. m, resp. $\triangle XYQ \cong \triangle XYQ'$, so $XQ = XQ'$ and $\angle QXY \cong \angle Q'XY$. Then $\angle PXQ \cong \angle P'XQ'$ and since $XP = XP'$, $\triangle XPQ \cong \triangle XP'Q'$ by SAS. Then $PQ = P'Q'$.

20. O is on m, the $\perp$ bis. of $\overline{PP'}$ and $\overline{QQ'}$. So $OP = OP'$ and $OQ = OQ'$. Then $OP + OQ = OP' + OQ'$, and $PQ = P'Q'$.

21. Const. k, the $\perp$ to t through A, int. t at P; const. $\overline{PA'}$ on k so that $AP = PA'$; $A' = R_t(A)$.

22. Const. the $\perp$ bis. of $\overline{BB'}$. **23.** Yes

24. a. $\angle 1 \cong \angle 3$ since $\angle 1$ and $\angle 3$ are vert. $\measuredangle$s. Let X be the pt. where $\overline{HH'}$ int. the wall; $\triangle PXH \cong \triangle PXH'$ (SAS), so $\angle 3 \cong \angle 2$. $\angle 1 \cong \angle 2$ by the Trans. Prop.

b. $BP + PH = BP + PH' = BH'$

25. Let P and Q be the pts. where $\overline{H'H''}$ and $\overline{HH'}$, resp., intersect the walls and let R and S be the pts. where $\overrightarrow{BH''}$ and $\overrightarrow{RH'}$, resp., intersect the walls. $\triangle RPH'' \cong \triangle RPH'$ and $\triangle SQH \cong \triangle SQH'$ (SAS) so $BR + RS + SH = BR + RS + SH' = BR + RH' = BR + RH'' = BH''$.

26. Find H' by reflecting H (the hole) in the line of the lower wall. Aim for H'.

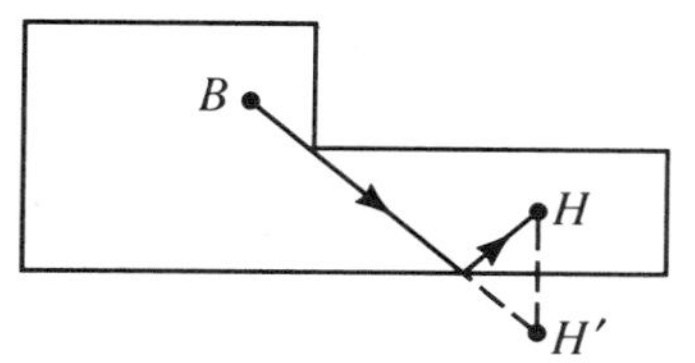

27. Find H' as in Ex. 26, above. Then find H'', the image of H' under reflection in the line of the left side wall. Aim for H''.

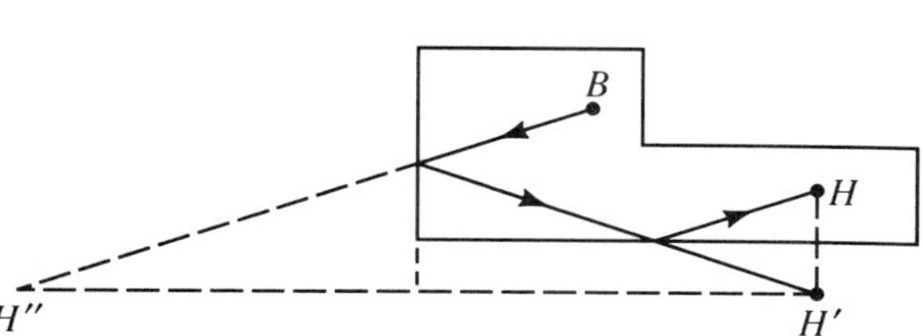

28. First find H', the image of H under reflection in the upper wall, then H'', the image of H' under reflection in the lower wall, then H''', the image of H'' under reflection in the line of the left side wall. Aim for H'''.

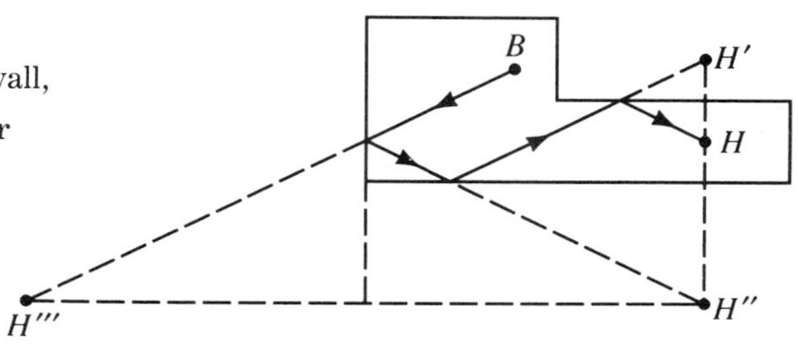

29. Yes

30. (0, 3) and (1, 5) are on the given line. Their images under reflection in the y-axis are (0, 3) and $(-1, 5)$; slope of image line $= \frac{5 - 3}{-1 - 0} = \frac{2}{-1} = -2$; equation: $y = -2x + 3$ (Form of equation may vary.)

31. (0, 5) and $(-5, 0)$ are on the given line. Their images under reflection in the x-axis are $(0, -5)$ and $(-5, 0)$; slope of image line $= \frac{0 - (-5)}{-5 - 0} = \frac{5}{-5} = -1$; y-intercept is -5; equation: $y = -x - 5$. (Form of equation may vary.)

32–37. Find M, the midpt. of $\overline{AA'}$; M is a pt. on k.

32. M has coordinates $\left(\frac{5 + 9}{2}, \frac{0 + 0}{2}\right) = \left(\frac{14}{2}, \frac{0}{2}\right) = (7, 0)$; $m = \frac{0 - 0}{9 - 5} = \frac{0}{4} = 0$; slope of k is undefined; equation: $x = 7$

33. M has coordinates $\left(\frac{1 + 3}{2}, \frac{4 + 4}{2}\right) = \left(\frac{4}{2}, \frac{8}{2}\right) = (2, 4)$; $m = \frac{4 - 4}{3 - 1} = \frac{0}{2} = 0$; slope of k is undefined; equation: $x = 2$

34. M has coordinates $\left(\frac{4 + 4}{2}, \frac{0 + 6}{2}\right) = \left(\frac{8}{2}, \frac{6}{2}\right) = (4, 3)$; slope of $\overline{AA'}$ is undefined; slope of $k = 0$; equation: $y = 3$

35. M has coordinates $\left(\frac{5 + 1}{2}, \frac{1 + 5}{2}\right) = \left(\frac{6}{2}, \frac{6}{2}\right) = (3, 3)$; $m = \frac{5 - 1}{1 - 5} = \frac{4}{-4} = -1$; slope of $k = 1$; equation: $y - 3 = 1(x - 3)$; $y = x$

36. M has coordinates $\left(\frac{0 + 4}{2}, \frac{2 + 6}{2}\right) = \left(\frac{4}{2}, \frac{8}{2}\right) = (2, 4)$; $m = \frac{6 - 2}{4 - 0} = \frac{4}{4} = 1$; slope of $k = -1$; equation: $y - 4 = -1(x - 2)$; $y - 4 = -x + 2$; $y = -x + 6$

37. M has coordinates $\left(\frac{-1 + 4}{2}, \frac{2 + 5}{2}\right) = \left(\frac{3}{2}, \frac{7}{2}\right)$; $m = \frac{5 - 2}{4 - (-1)} = \frac{3}{5}$; slope of $k = -\frac{5}{3}$; equation: $y - \frac{7}{2} = -\frac{5}{3}\left(x - \frac{3}{2}\right)$; $y = -\frac{5}{3}x + \frac{5}{2} + \frac{7}{2}$; $y = -\frac{5}{3}x + 6$

C 38. Let $T(t, -t)$ be the pt. where $\overline{AA'}$ intersects l if $A = (a, b)$. Since the slope of l is -1, the slope of $\overleftrightarrow{AT}$ is $\frac{b + t}{a - t} = 1$; by solving, $t = \frac{a - b}{2}$. Then $T\left(\frac{a - b}{2}, \frac{b - a}{2}\right)$ is the midpt. of $\overline{AA'}$. Use the midpt. formula to get $A'(-b, -a)$.

39. **a.** (6, 3) **b.** (10, −2) **c.** (13, 1) **d.** $(5 + (5 - x), y) = (10 - x, y)$

40. **a.** (4, 9) **b.** (0, 14) **c.** (−3, 11) **d.** $(x, 6 + (6 - y)) = (x, 12 - y)$

Page 583 • APPLICATION

1. 45 2. Yes

3. (Figure not drawn to scale.) Let B be highest pt. on wall that can be seen at D, highest pt. of mirror. $\angle BDC \cong \angle ADE$; $m\angle C = 90 = m\angle E$; $\triangle BDC \sim \triangle ADE$; $\frac{CD}{ED} = \frac{BC}{AE}$; $\frac{CD}{30} = \frac{300}{100}$; $CD = 90$; height of $B = 150 + 30 + 90 = 270$; 270 cm

4. Let $\overline{AB}$ be the person, E the eyes, and $\overline{CD}$ the mirror (and the person's image). Draw $\overline{CF} \perp \overline{AB}$ (and $\overline{CD}$) and $\overline{DH} \perp \overline{AB}$ (and $\overline{CD}$).

Prove: $CD = \frac{1}{2}AB$

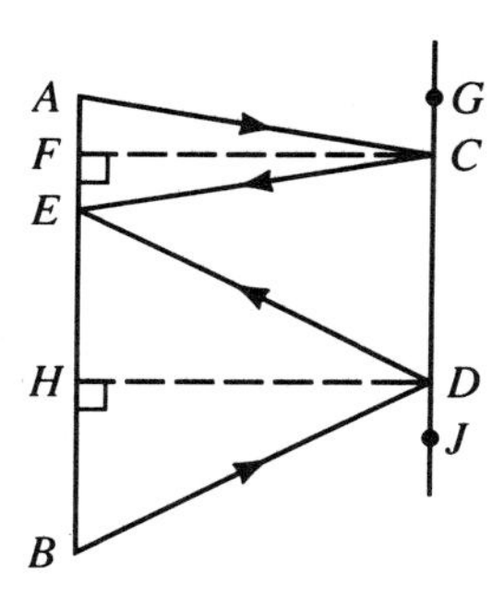

Proof:

Statements	Reasons
1. $\angle ACG \cong \angle ECD$; $\angle BDJ \cong \angle EDC$	1. The $\angle$ between a mirror and a reflected light ray $\cong$ the $\angle$ between the mirror and the initial light ray.
2. $\angle ACF \cong \angle ECF$; $\angle EDH \cong \angle BDH$	2. If 2 ∠s are comps. of ≅ ∠s, then the 2 ∠s are ≅.
3. $\triangle ACF \cong \triangle ECF$; $\triangle EDH \cong \triangle BDH$	3. ASA Post.
4. $AF = FE$; $EH = HB$	4. Corr. parts of ≅ △s are ≅.
5. $FH = FE + EH = \frac{1}{2}AE + \frac{1}{2}EB = \frac{1}{2}AB$	5. Seg. Add. Post., Substitution Prop.
6. $FH = CD$	6. Opp. sides of a ▱ are ≅.
7. $CD = \frac{1}{2}AB$	7. Substitution Prop.

5. Let M be the midpt. of $\overline{AD}$.

Statements	Reasons
1. $\overleftrightarrow{EM}$ is the $\perp$ bis. of $\overline{AD}$.	1. Given
2. $\triangle BAM \cong \triangle BDM$	2. SAS Post.
3. $m\angle DBM = m\angle ABM = x$	3. Corr. parts of $\cong$ $\triangle$s are $\cong$.
4. $m\angle EBD + x = 180$	4. $\angle$ Add. Post.
5. $m\angle CBE = x$	5. Given
6. $m\angle EBD + m\angle CBE = 180$	6. Substitution Prop.
7. $\angle CBD$ is a straight $\angle$, and D lies on $\overleftrightarrow{BC}$.	7. Protractor Post.

6. Let M be the midpt. of $\overline{AD}$.

Statements	Reasons
1. $\overleftrightarrow{EM}$ is the $\perp$ bis. of $\overline{AD}$.	1. Given
2. $AB = BD$; $AE = DE$	2. If a pt. lies on the $\perp$ bis. of a seg., then the pt. is equidistant from the endpts. of the seg.
3. $AE + EC = DE + EC > CD$	3. Substitution Prop., $\triangle$ Ineq. Thm.
4. $AB + BC = BD + BC = CD$	4. Substitution Prop., Seg. Add. Post.
5. $AE + EC > AB + BC$	5. Substitution Prop.

Page 585 • CLASSROOM EXERCISES

1. a. 3; down **b.** (7, 5) **c.** (−1, 4)

2. T glides pts. 5 units left and 4 units up; (−1, 10); (7, −1)

3. T glides pts. 1 unit right; (5, 6); (1, 3) **4.** $(x + 5, y + 3)$ **5.** $(x - 3, y + 3)$

6. $(x, y - 4)$ **7.** $(-5, -7)$

8. a. (1, 5) **b.** (5, 1) **c.** $(x + 2, y + 2)$; $(y + 2, x + 2)$

Pages 586–587 • WRITTEN EXERCISES

A 1.

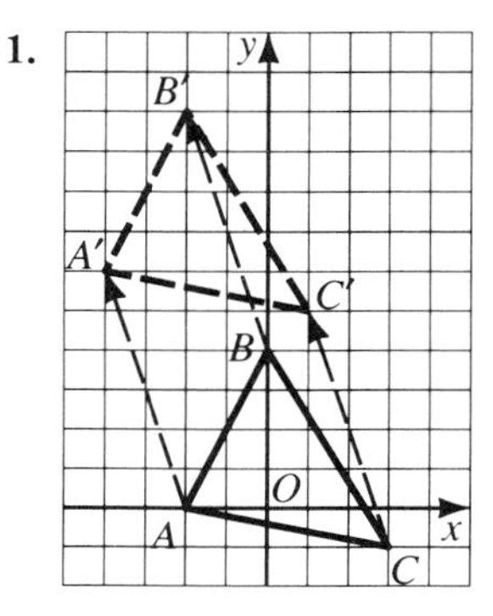

a. Yes **c.** Yes; yes

2.

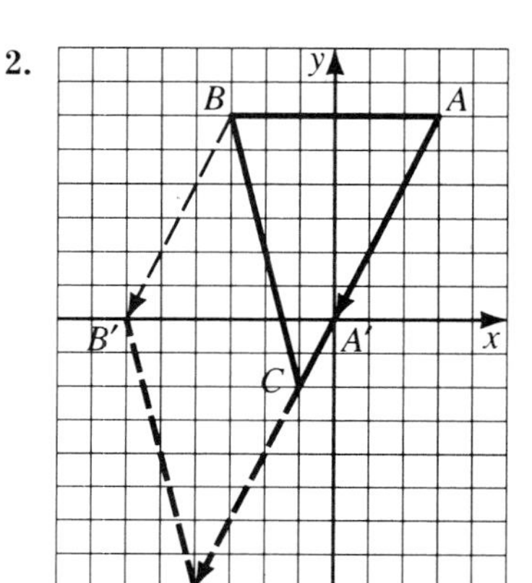

a. Yes **c.** Yes; yes

3. $0 + a = 5$; $a = 5$; $0 + b = 1$; $b = 1$; $T: (x, y) \rightarrow (x + 5, y + 1)$;
$T: (3, 3) \rightarrow (3 + 5, 3 + 1) = (8, 4)$

4. $1 + a = 3$; $a = 2$; $1 + b = 0$; $b = -1$; $T: (x, y) \rightarrow (x + 2, y - 1)$;
$T: (0, 0) \rightarrow (0 + 2, 0 - 1) = (2, -1)$

5. $-2 + a = 2$; $a = 4$; $3 + b = 6$; $b = 3$; $T: (x, y) \rightarrow (x + 4, y + 3)$; $x + 4 = 0$;
$x = -4$; $y + 3 = 0$; $y = -3$; $(-4, -3)$

6. $-1 + a = 5$; $a = 6$; $5 + b = 7$; $b = 2$; $T: (x, y) \rightarrow (x + 6, y + 2)$; $x + 6 = -1$;
$x = -7$; $y + 2 = 5$; $y = 3$; $(-7, 3)$

7.

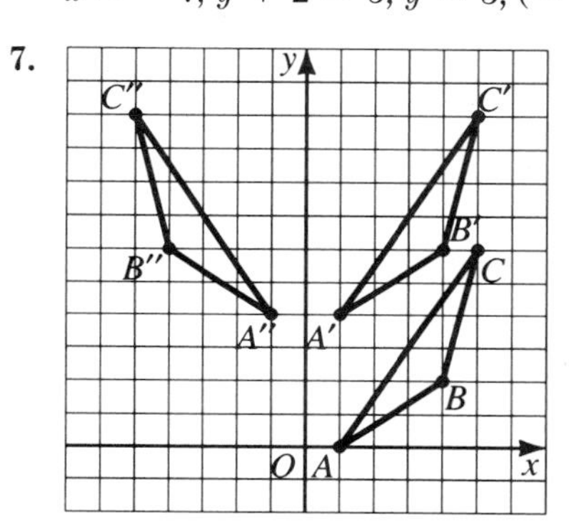

8.

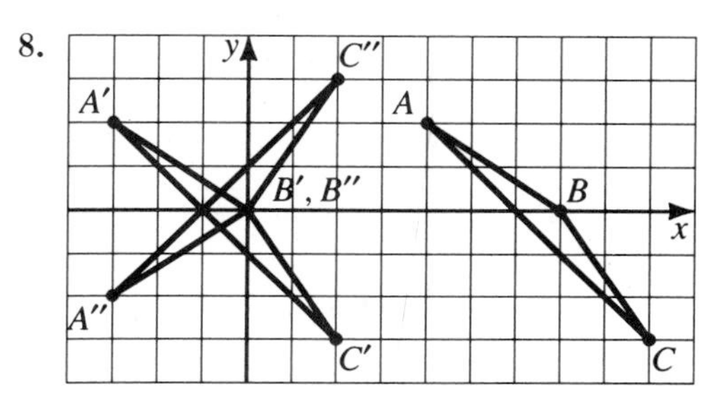

B **9.** The glide maps (x, y) to $(x, y + 4)$; the reflection maps $(x, y + 4)$ to $(-x, y + 4)$;
the glide reflection maps (x, y) to $(-x, y + 4)$.

10. The glide maps (x, y) to $(x - 7, y)$; the reflection maps $(x - 7, y)$ to $(x - 7, -y)$;
the glide reflection maps (x, y) to $(x - 7, -y)$.

11. a, b, c, d **12.** a, b, c

13. $R: (x, y) \rightarrow (x + 1, y + 2)$; $S: (x + 1, y + 2) \rightarrow ((x + 1) - 5, (y + 2) + 7) =$
$(x - 4, y + 9)$; $T: (x, y) \rightarrow (x - 4, y + 9)$

14. $R: (x, y) \rightarrow (x - 5, y - 3)$; $S: (x - 5, y - 3) \rightarrow ((x - 5) + 4, (y - 3) - 6) =$
$(x - 1, y - 9)$; $T: (x, y) \rightarrow (x - 1, y - 9)$

15. **a.**

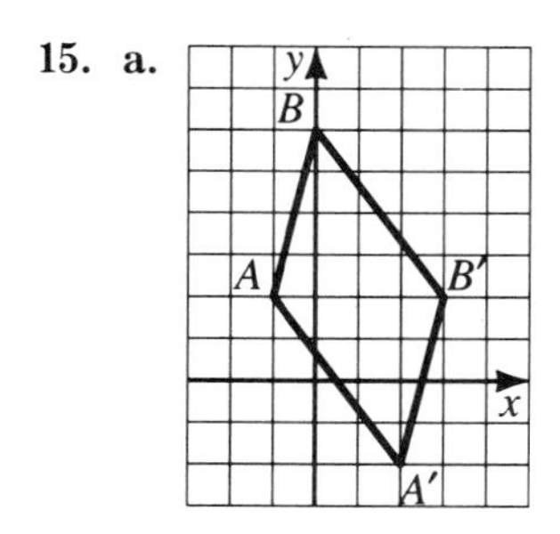

b. parallelogram; $AA' = BB' = \sqrt{3^2 + (-4)^2} = 5$;
$A'B' = AB = \sqrt{1^2 + 4^2} = \sqrt{17}$; perimeter $=$
$2 \cdot 5 + 2 \cdot \sqrt{17} = 10 + 2\sqrt{17}$

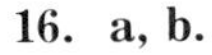
16. a, b.

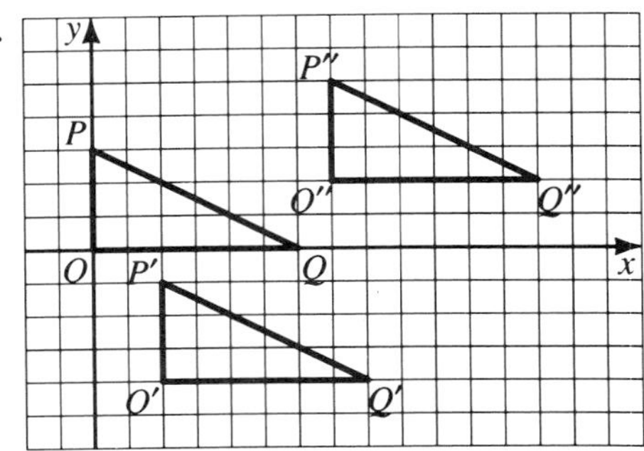

c. T_3: $(x, y) \rightarrow (x + 7, y + 2)$

d. $\overrightarrow{T_2} = (5, 6)$; $\overrightarrow{T_3} = (7, 2)$; $\overrightarrow{T_1} + \overrightarrow{T_2} = \overrightarrow{T_3}$

17. The midpts. of $\overline{AA'}$, $\overline{BB'}$, and $\overline{CC'}$ lie on the reflecting line. Proof: Let X be the image of A under the glide, k the reflecting line, and P the pt. where $\overline{XA'}$ intersects k. k is the $\perp$ bis. of $\overline{XA'}$ so P is the midpt. of $\overline{XA'}$ and $A'P = \frac{1}{2}A'X$.

Draw $\overline{AA'}$, intersecting k at Q. Since $k \parallel \overline{AX}$, $\triangle A'QP \sim \triangle A'AX$ and $\frac{A'Q}{A'A} = \frac{A'P}{A'X} = \frac{1}{2}$. Then Q is the midpt. of $\overline{AA'}$ and the midpt. of $\overline{AA'}$ is on k.

(The proofs for $\overline{BB'}$ and $\overline{CC'}$ are similar.)

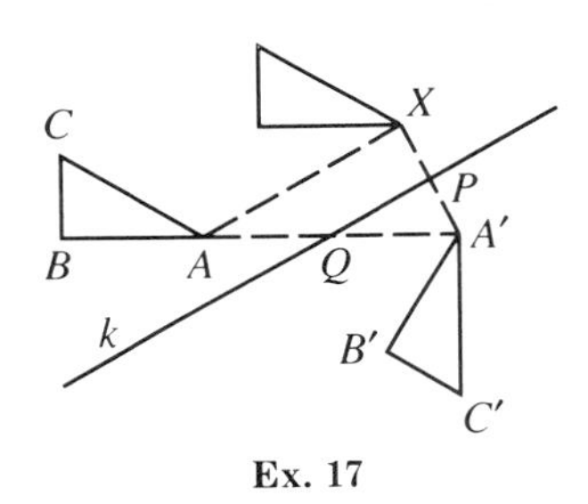

Ex. 17

Ex. 18

18. The reflecting line is the line containing the midpts. of $\overline{AA'}$, $\overline{BB'}$, and $\overline{CC'}$. Const. the $\perp$ bis. of any 2 of $\overline{AA'}$, $\overline{BB'}$, and $\overline{CC'}$, to locate midpts. M and P; $\overleftrightarrow{MP}$ is the line of reflection. To find the glide image of A, const. the $\perp$ to $\overleftrightarrow{MP}$ from pt. A', int. $\overleftrightarrow{MP}$ at X. Const. $\overline{XA''}$ on $\overrightarrow{A'X}$ such that $\overline{XA''} \cong \overline{XA'}$. Similarly, locate B'' and C''.

19. Let T be the translation and R_k the reflection. Let $T(P) = P'$, $T(Q) = Q'$, $R_k(P') = P''$, $R_k(Q') = Q''$. Since T and R_k are isometries, $PQ = P'Q'$ and $P'Q' = P''Q''$. Then $PQ = P''Q''$, so the glide reflection is an isometry.

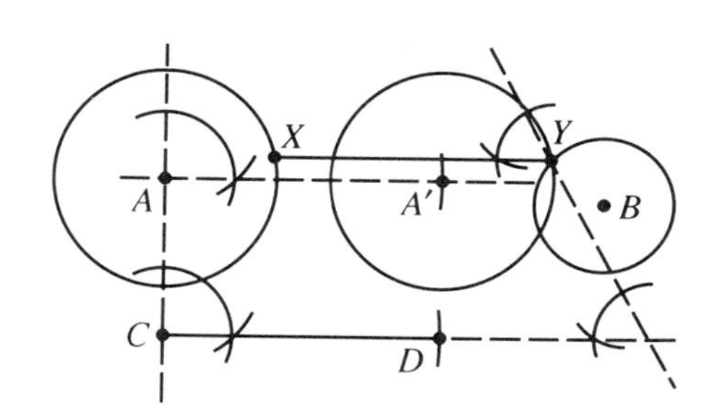

20. Const. a $\parallel$ to $\overline{CD}$ through A and const. $\overline{AA'}$ on the $\parallel$ such that $\overline{AA'} \cong \overline{CD}$. Draw $\odot A' \cong \odot A$. $\odot A'$ int. $\odot B$ at 2 pts.; name 1 pt. Y. Const. a $\parallel$ to $\overline{CD}$ through Y, int. $\odot A$ at X. Since Y is on $\odot A'$ and $\overline{XY} \parallel \overline{CD}$, Y is the image of X under the translation; $XY = CD$.

C 21. Let j and k be the upper and lower lines. Choose a pt. G on k and draw $\overline{GF}$. Const. a $\parallel$ to $\overline{EF}$ through G and const. $\overline{HG}$ on the $\parallel$ such that $\overline{HG} \cong \overline{EF}$. $EFGH$ is a ▱. Const. $\overline{HX} \parallel k$ with X on j, and $\overleftrightarrow{XY} \parallel \overleftrightarrow{GH}$ with Y on k. Two lines $\parallel$ to the same line are $\parallel$ to each other, so $\overline{XY} \parallel \overline{EF}$. Since $\overline{EF} \cong \overline{HG}$ and $\overline{XY} \cong \overline{HG}$, $\overline{XY} \cong \overline{EF}$.

22. T: $A \rightarrow A'$, R_l: $A' \rightarrow A''$; R_l: $A \rightarrow A_1$, T: $A_1 \rightarrow A_2$; $A'' \neq A_2$

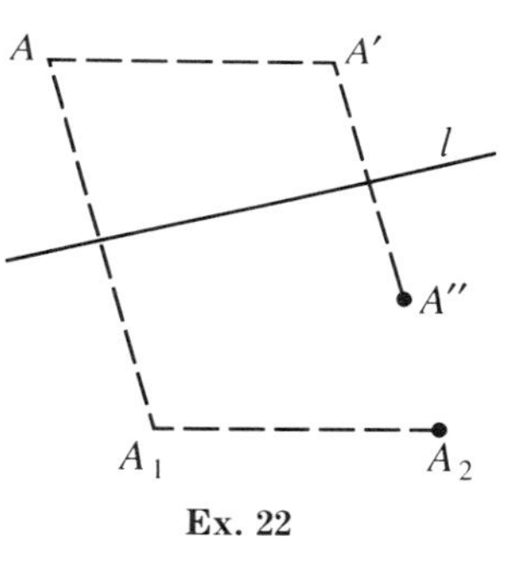

Ex. 22

23. Given: $P(x_1, y_1)$, $Q(x_2, y_2)$, T: $(x, y) \rightarrow (x + a, y + b)$
Prove: $PQ = P'Q'$, where $T(P) = P'$ and $T(Q) = Q'$
Proof: By the distance formula, $PQ = \sqrt{(x_2 - x_1)^2 + (y_2 - y_1)^2}$. $P' = T(P) = (x_1 + a, y_1 + b)$, and $Q' = T(Q) = (x_2 + a, y_2 + b)$, so $P'Q' = \sqrt{[(x_2 + a) - (x_1 + a)]^2 + [(y_2 + b) - (y_1 + b)]^2} = \sqrt{(x_2 - x_1)^2 + (y_2 - y_1)^2} = PQ$.

Page 589 • CLASSROOM EXERCISES

1–5. Answers may vary; examples are given.

1. $\mathcal{R}_{O,410}$ **2.** $\mathcal{R}_{O,320}$ **3.** $\mathcal{R}_{O,270}$ **4.** $\mathcal{R}_{O,40}$ **5.** $\mathcal{R}_{O,180}$ **6.** T, P, S **7.** S, T, P
8. P, S, T **9.** P **10.** S **11.** S **12. a.** $(-4, -1)$ **b.** $(-1, 4)$ **c.** $(1, -4)$
13. a. $(3, -5)$ **b.** $(-5, -3)$ **c.** $(5, 3)$
14. Yes; a half-turn is a rotation, and a rotation is an isometry.
15. Answers may vary; examples are given. **a.** Reflection in line k maps A to A'. **b.** A half-turn about the origin maps $(-2, 0)$ to $(2, 0)$. **c.** Translation T maps (x, y) to $(x - 1, y + 3)$. **d.** Rotation about pt. P through $10°$

Pages 590–592 • WRITTEN EXERCISES

A **1–5.** Answers may vary; examples are given.

1. $\mathcal{R}_{O,440}$ **2.** $\mathcal{R}_{O,345}$ **3.** $\mathcal{R}_{A,90}$ **4.** $\mathcal{R}_{B,0}$ **5.** $\mathcal{R}_{O,180}$ **6.** F **7.** C **8.** B **9.** E **10.** C **11.** D **12.** $D; B$ **13.** D **14.** $C; D$ **15.** rotation **16.** reflection **17.** half-turn **18.** translation **19.** rotation **20.** half-turn **21.** reflection **22.** translation

B **23.** 6 **24.** (3) and (8) **25.** a, b, c, d **26.** a, b, c, d

27.

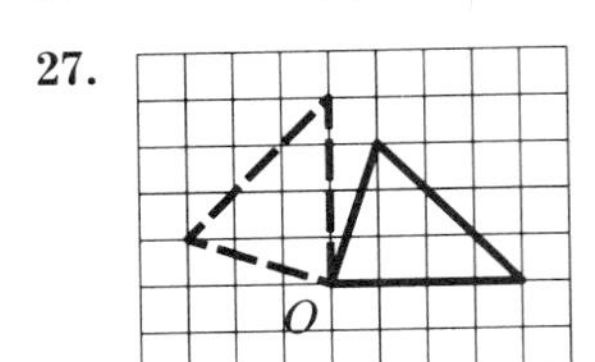

28.

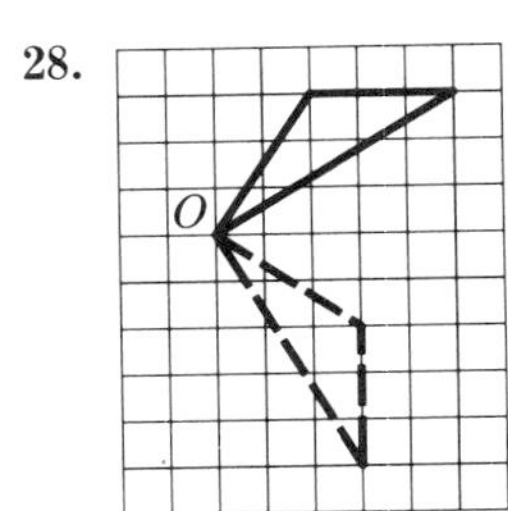

29.

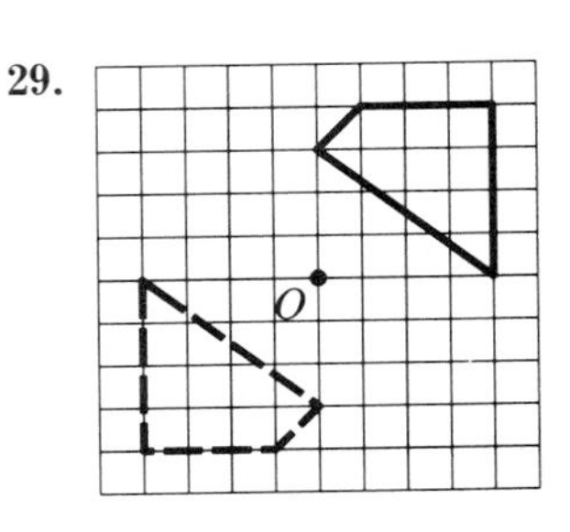

30. C is the midpt. of the seg. joining (1, 1) and (7, 3); $C = \left(\frac{1+7}{2}, \frac{1+3}{2}\right) = \left(\frac{8}{2}, \frac{4}{2}\right) =$ (4, 2).

31. Const. the $\perp$ bisectors of $\overline{AA'}$ and $\overline{BB'}$. Their intersection is O.

32. a, b.

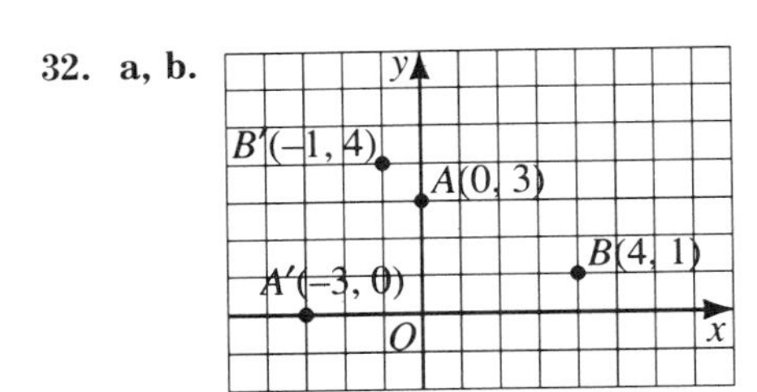

c. slope of $\overleftrightarrow{AB} = \frac{1-3}{4-0} = \frac{-2}{4} = -\frac{1}{2}$;

slope of $\overleftrightarrow{A'B'} = \frac{4-0}{-1-(-3)} = \frac{4}{2} = 2$;

$\overleftrightarrow{AB} \perp \overleftrightarrow{A'B'}$

d. A rotation is an isometry.

e. An isometry maps any $\triangle$ to a $\cong \triangle$.

f. $(-y, x)$

33. a, b.

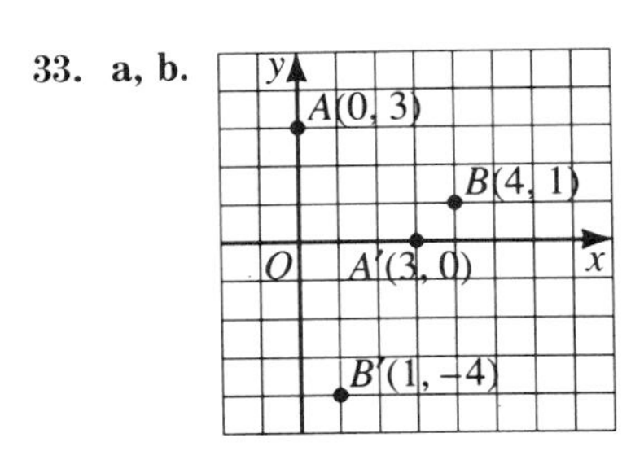

c. slope of $\overleftrightarrow{AB} = \frac{1-3}{4-0} = \frac{-2}{4} = -\frac{1}{2}$;

slope of $\overleftrightarrow{A'B'} = \frac{0-(-4)}{3-1} = \frac{4}{2} = 2$;

$\overleftrightarrow{AB} \perp \overleftrightarrow{A'B'}$

d. A rotation is an isometry.

e. An isometry maps any $\triangle$ to a $\cong \triangle$.

f. $(y, -x)$

34. In each case, (3, 2) is the midpoint of the preimage and image pts. Use the midpt. formula. **a.** P **b.** (6, 4) **c.** (3, 4) **d.** (5, 0) **e.** (8, 3) **f.** $(6 - x, 4 - y)$

35. Extend $\overrightarrow{OF}$ to int. l' at G, and let H be the int. of l and l'. $m\angle F'GO = 90 - x$, so $m\angle GHF = 90 - (90 - x) = x$.

36. **a.** $\mathcal{R}_{C,-60}$ **b.** $\overline{BE}$ is the image of $\overline{AD}$ under an isometry.

c. $\mathcal{R}_{C,-60}(\overleftrightarrow{AD}) = \overleftrightarrow{BE}$. By Ex. 35, if $\mathcal{R}_{O,x}(l) = l'$, then one of the ⦣s between l and l' has meas. $|x|$. Therefore, the meas. of an acute ∠ between $\overleftrightarrow{AD}$ and $\overleftrightarrow{BE} = 60$.

37. **a.** $\mathcal{R}_{C,90}$ **b.** $\overline{AD}$ is the image of $\overline{BE}$ under a rotation, which is an isometry.

c. If a rotation of 90° maps $\overline{BE}$ to $\overline{AD}$, then one of the ⦣s between $\overline{BE}$ and $\overline{AD}$ has measure 90.

C 38. **a.**

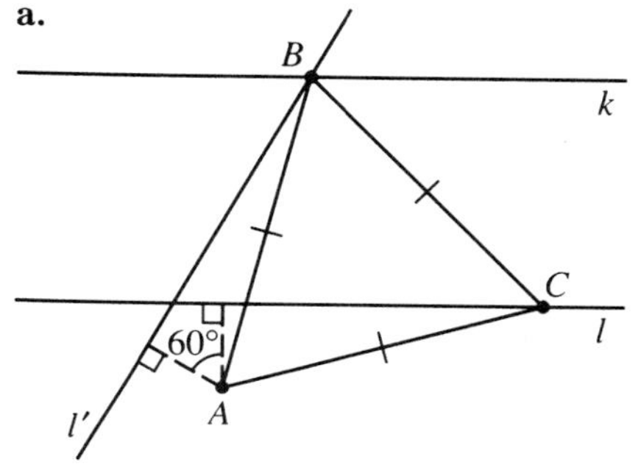

b. $\overline{AB}$ is the image of $\overline{AC}$ under $\mathcal{R}_{A,60}$ so by Ex. 35, $m\angle BAC = 60$; $AB = AC$ since a rotation is an isometry. Then $m\angle ABC = m\angle ACB = \frac{1}{2}(180 - 60) = 60$. $\triangle ABC$ is equiangular, so $\triangle ABC$ is equilateral.

c. Yes; construct a ⊥ to both l and k through A; reflect $\triangle ABC$ in the ⊥.

39. Locate X and Z as you did B and C in Ex. 38, using $\mathcal{R}_{A,90}$ instead of $\mathcal{R}_{A,60}$. With centers X and Z and radius AX, draw arcs int. at Y.

Page 592 • MIXED REVIEW EXERCISES

1. ODE, OFG **2.** 2 : 3

3. $\frac{2}{3} = \frac{3}{x}$, $x = \frac{9}{2}$; $\frac{2}{3} = \frac{y}{5}$, $y = \frac{10}{3}$; $\dfrac{\frac{10}{3}}{\frac{10}{3} + 5} = \frac{2}{z}$, $z = 5$; $\frac{3}{w} = \dfrac{\frac{10}{3}}{5}$, $w = \frac{9}{2}$

4. 3 : 5 **5.** $2^2 : 3^2 = 4 : 9$ **6.** $3^2 : 5^2 = 9 : 25$ **7.** $2^2 : 5^2 = 4 : 25$

Page 595 • CLASSROOM EXERCISES

1.

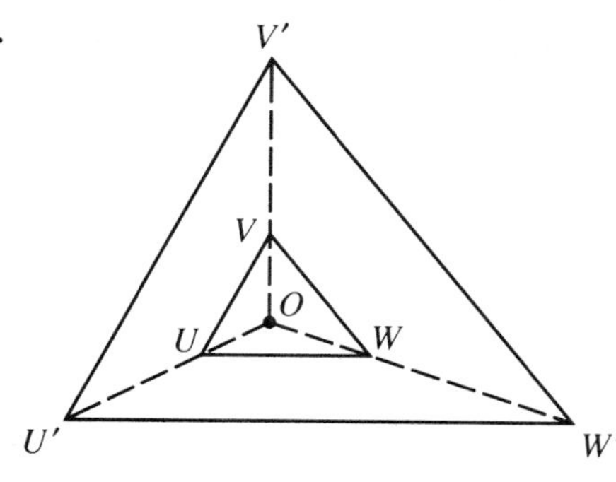

$\frac{OV'}{OV} = 3 = \frac{OU'}{OU} = \frac{OW'}{OW}$

2.

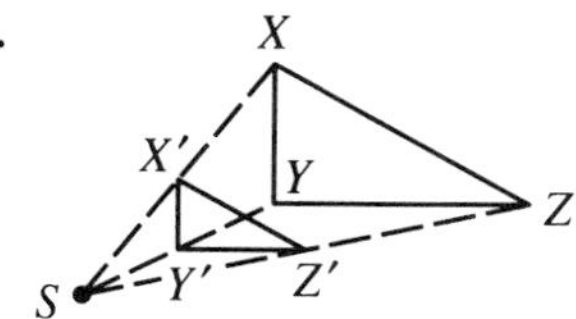

$\frac{SX'}{SX} = \frac{1}{2} = \frac{SY'}{SY} = \frac{SZ'}{SZ}$

3.

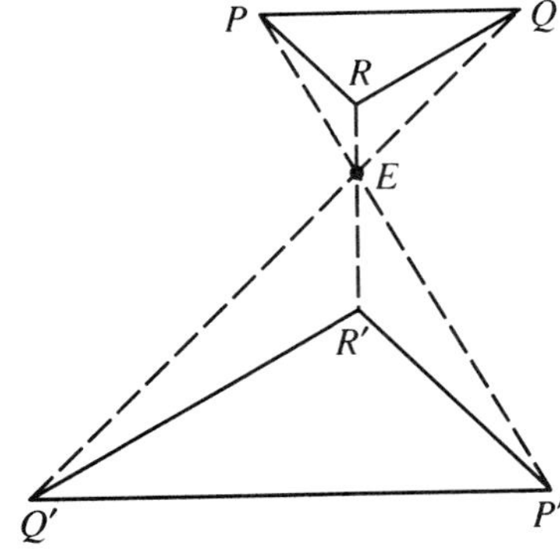

$$\frac{EP'}{EP} = 2 = \frac{EQ'}{EQ} = \frac{ER'}{ER}$$

4.

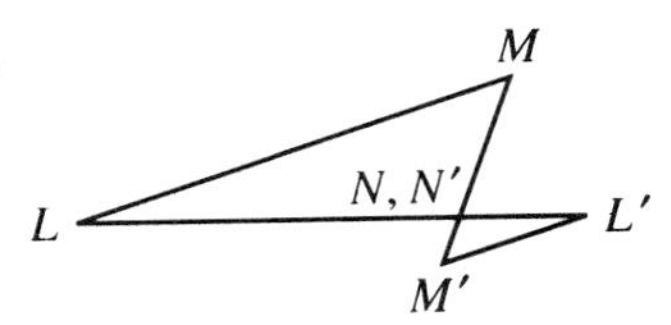

$$\frac{NL'}{NL} = \frac{1}{3} = \frac{NM'}{NM}; N = N'$$

5. $A'(8, 8)$; $B'(4, 4)$; $C'(6, 0)$ **6.** $(2x, 2y)$ **7.** $D_{O,\frac{1}{2}}$ **8.** $D_{O,-2}$ **9.** (6, 6)

10. (0, 6) **11.** $\frac{2}{5}$, contraction; -4, expansion; -1, half-turn

12. It maps every pt. onto itself. **13.** $D_{S,5}$: ⊙ with ctr. S, rad. 20; $D_{S,-1}$: ⊙S

14. line k

15. **a.** 6 **b.** 21 **c.** $OP' = 3(OP)$; $OP + PP' = 3(OP)$; $PP' = 2(OP) = 10$; $OP = 5$

16. A dilation maps any △ to a ~ △; corr. ∠s of ~ △s are ≅.

17. By Cor. 1, the image polygon has ∠s ≅ to those of the original polygon. By Cor. 2, the sides are in prop., so the polygons are ~ with scale factor k, and the ratio of the areas is $|k|^2 : 1 = k^2 : 1$, since $|k|^2 = k^2$ for all real k.

Pages 596–597 • WRITTEN EXERCISES

A **1.** $A'(12, 0)$, $B'(8, 4)$, $C'(4, -4)$ **2.** $A'(18, 0)$, $B'(12, 6)$, $C'(6, -6)$

3. $A'(3, 0)$, $B'(2, 1)$, $C'(1, -1)$ **4.** $A'(-3, 0)$, $B'(-2, -1)$, $C'(-1, 1)$

5. $A'(-12, 0)$, $B'(-8, -4)$, $C'(-4, 4)$ **6.** $A'(6, 0)$, $B'(4, 2)$, $C'(2, -2)$

7. $A'(6, 0)$, $B'(7, -1)$, $C'(8, 1)$ **8.** $A'(6, 0)$, $B'(2, 4)$, $C'(-2, -4)$

9. $k = \frac{8}{2} = 4$; expansion **10.** $k = \frac{4}{2} = 2$; expansion **11.** $k = \frac{1}{3}$; contraction

12. $k = \frac{-2}{4} = -\frac{1}{2}$; contraction **13.** $k = \frac{2}{3} \div \frac{1}{6} = \frac{2}{3} \cdot 6 = 4$; expansion

14. $k = \frac{18}{-6} = -3$; expansion

B **15.** b, d **16.** Yes **17.** a, b, c

18. 1 : 3; 1 : 9

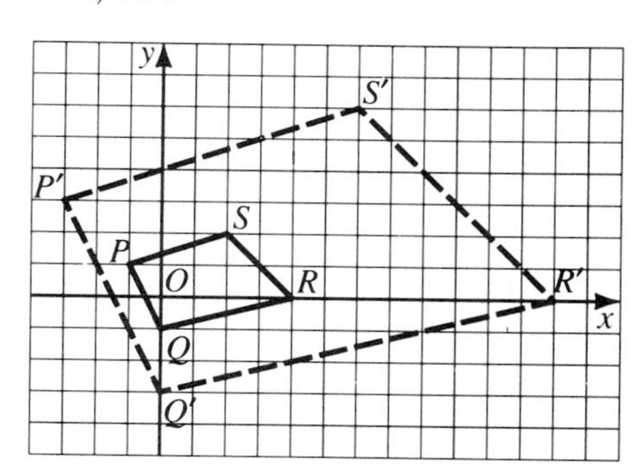

19. 3 : 2; 9 : 4

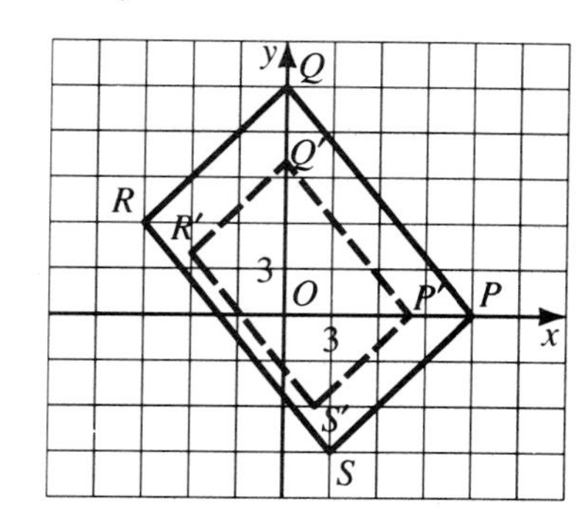

20. 1 : 2; 1 : 4

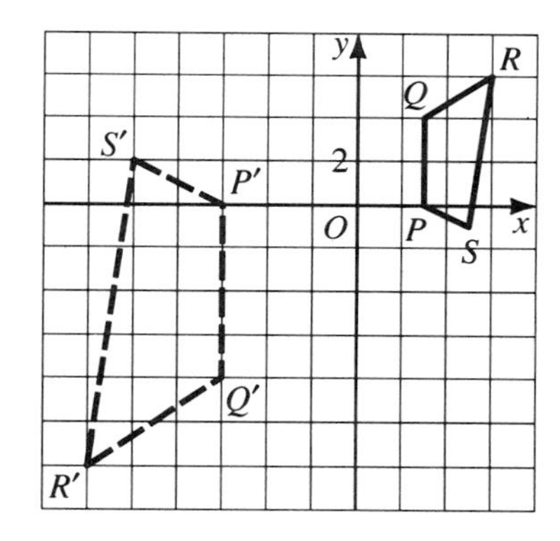

21. 2 : 1; 4 : 1

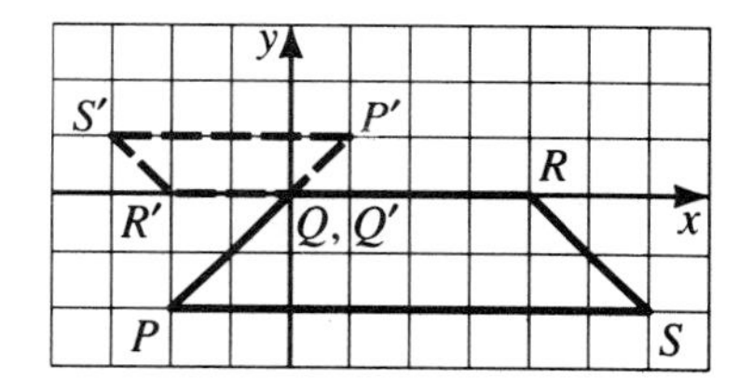

22. **a.** $1^2 : 2^2 = 1 : 4$ **b.** $1^3 : 2^3 = 1 : 8$

23. The scale factor of the original sphere to the image sphere is 4 : 3.

a. $4^2 : 3^2 = 16 : 9$ **b.** $4^3 : 3^3 = 64 : 27$

24. **a.** $\frac{2}{3}$ **b.** $\frac{1}{2}$ **c.** $D_{G,-\frac{1}{2}}$ **d.** $N; P$

25. **a.** slope of $\overline{PQ} = \frac{y_2 - y_1}{x_2 - x_1}$; slope of $\overline{P'Q'} = \frac{ky_2 - ky_1}{kx_2 - kx_1} = \frac{k(y_2 - y_1)}{k(x_2 - x_1)} = \frac{y_2 - y_1}{x_2 - x_1} =$ slope of $\overline{PQ}$ **b.** parallel

C **26.** $P'Q' = \sqrt{(kx_1 - kx_2)^2 + (ky_1 - ky_2)^2} = \sqrt{[k(x_1 - x_2)]^2 + [k(y_1 - y_2)]^2} =$
$\sqrt{k^2(x_1 - x_2)^2 + k^2(y_1 - y_2)^2} = \sqrt{k^2[(x_1 - x_2)^2 + (y_1 - y_2)^2]} =$
$|\,k\,|\sqrt{(x_1 - x_2)^2 + (y_1 - y_2)^2} = |\,k\,| \cdot PQ$

27. (a, b) is on $\overleftrightarrow{AA'}$ and $\overleftrightarrow{BB'}$; slope of $\overleftrightarrow{AA'} = \frac{8-4}{1-3} = \frac{4}{-2} = -2$; equation of $\overleftrightarrow{AA'}$: $y - 4 = -2(x - 3)$; $y = -2x + 10$; slope of $\overleftrightarrow{BB'} = 0$; equation of $\overleftrightarrow{BB'}$: $y = 2$; at (a, b): $2 = -2x + 10$; $2x = 8$; $x = 4$; $(a, b) = (4, 2)$; $A'B' = \sqrt{(1-1)^2 + (8-2)^2} = 6$; $AB = |4 - 2| = 2$; $k = \frac{6}{2} = 3$

28. Given: $D_{O,k} : \triangle ABC \rightarrow \triangle A'B'C'$
Prove: $\triangle ABC \sim \triangle A'B'C'$
Proof: $AB = \sqrt{(r-p)^2 + (s-q)^2}$; $A'B' = \sqrt{(kr - kp)^2 + (ks - kq)^2} = \sqrt{[k(r-p)]^2 + [k(s-q)]^2} = \sqrt{k^2(r-p)^2 + k^2(s-q)^2} = \sqrt{k^2[(r-p)^2 + (s-q)^2]} = |k| \cdot \sqrt{(r-p)^2 + (s-q)^2} = |k| \cdot AB$.
Similarly, $B'C' = |k| \cdot BC$ and $A'C' = |k| \cdot AC$. Since $\frac{A'B'}{AB} = \frac{B'C'}{BC} = \frac{A'C'}{AC} = |k|$, $\triangle ABC \sim \triangle A'B'C'$ by the SSS $\sim$ Thm.

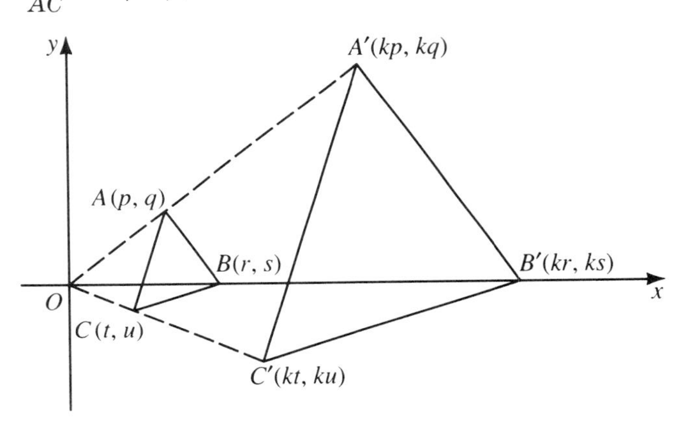

Page 597 • SELF-TEST 1

1. An isometry is a one-to-one mapping from the whole plane onto the whole plane that maps every segment to a $\cong$ segment.
2. $f(2) = 3 \cdot 2 - 7 = -1$; $3x - 7 = 2$, $x = 3$
3. $T(0, 0) = (1, -2)$; $x + 1 = 0$, $x = -1$; $y - 2 = 0$, $y = 2$; $T(-1, 2) = (0, 0)$
4. **a.** $(3, -5)$ **b.** $(-3, 5)$ **c.** $(5, 3)$ **5.** a, b
6. Examples: $\mathcal{R}_{O,330}$, $\mathcal{R}_{O,-390}$ **7.** B **8.** C **9.** $\overline{AB}$ **10.** $\overline{OB}$ **11.** M **12.** $\overline{AO}$

13. L **14.** NDO **15.** C **16.** Q **17.** C **18.** L

Pages 602–603 • CLASSROOM EXERCISES

1. **a.** 5 **b.** $g(f(4)) = g(5) = 15$ **c.** $g(f(x)) = g(x + 1) = 3(x + 1)$ **d.** 6 **e.** $f(g(2)) = f(6) = 7$ **f.** $f(g(x)) = f(3x) = 3x + 1$

2. **a.** 2 **b.** $g(f(4)) = g(2) = 9$ **c.** $g(f(x)) = g(\sqrt{x}) = \sqrt{x} + 7$ **d.** 9
e. $f(g(2)) = f(9) = 3$ **f.** $f(g(x)) = f(x + 7) = \sqrt{x + 7}$

3. C 4. B 5. C 6. C 7. A 8. C

9. $R_k \circ R_j$

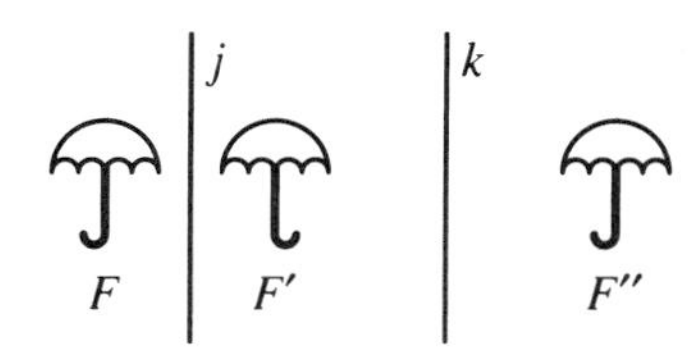

$R_j \circ R_k$

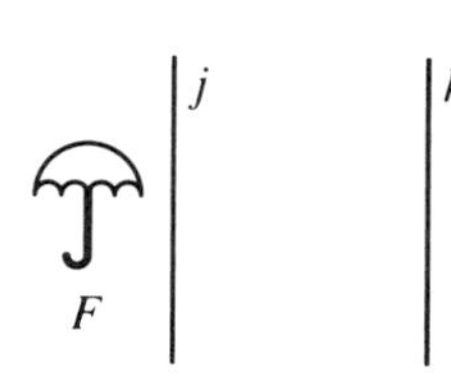

10. $R_k \circ R_j$

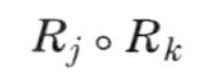

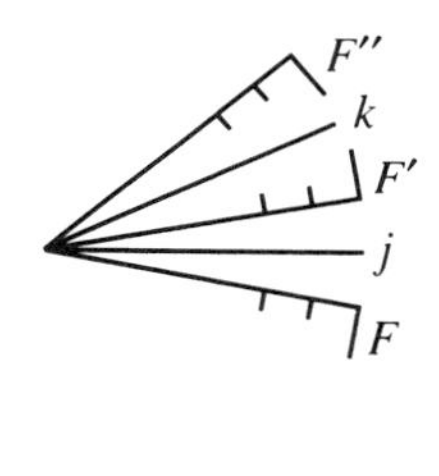

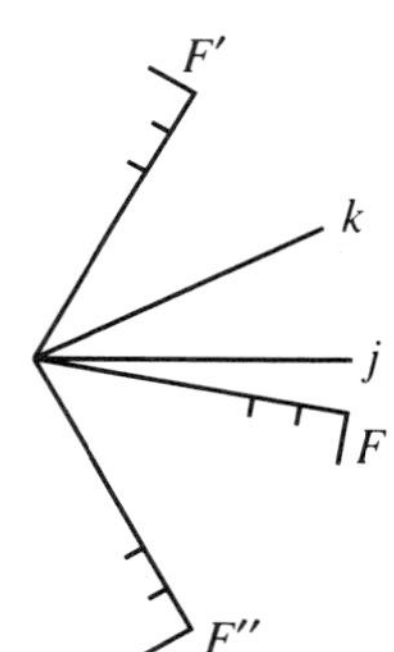

11. Let S and T be isometries, $T(P) = P'$, $T(Q) = Q'$, $S(P') = P''$, and $S(Q') = Q''$. Then $(S \circ T)(P) = S(T(P)) = S(P') = P''$, and $(S \circ T)(Q) = S(T(Q)) = S(Q') = Q''$. Since S and T are isometries, $PQ = P'Q'$ and $P'Q' = P''Q''$. Thus, $PQ = P''Q''$; $S \circ T$ preserves distance, so it is an isometry.

12. Since $\perp$ lines form rt. $\angle$s, the meas. of the $\angle$ of rotation is $2 \cdot 90 = 180$, by Thm. 14-8. A rotation of 180° about a point is a half-turn.

Pages 603–605 • WRITTEN EXERCISES

A 1. **a.** $g(f(2)) = g(4) = 1$ **b.** $g(f(x)) = g(x^2) = 2x^2 - 7$ **c.** $f(g(2)) = f(-3) = 9$
d. $f(g(x)) = f(2x - 7) = (2x - 7)^2$

2. **a.** $g(f(2)) = g(7) = -2$ **b.** $g(f(x)) = g(3x + 1) = 3x + 1 - 9 = 3x - 8$
c. $f(g(2)) = f(-7) = -20$ **d.** $f(g(x)) = f(x - 9) = 3(x - 9) + 1 = 3x - 26$

3. a. $k(h(3)) = k(2) = 8$ **b.** $k(h(5)) = k(3) = 27$ **c.** $k(h(x)) = k\left(\frac{x+1}{2}\right) = \left(\frac{x+1}{2}\right)^3$ **d.** $h(k(3)) = h(27) = 14$ **e.** $h(k(5)) = h(125) = 63$

f. $h(k(x)) = h(x^3) = \frac{x^3+1}{2}$

4. a. $k(h(3)) = k(8) = 23$ **b.** $k(h(5)) = k(24) = 55$ **c.** $k(h(x)) = k(x^2 - 1) = 2(x^2 - 1) + 7 = 2x^2 + 5$ **d.** $h(k(3)) = h(13) = 168$ **e.** $h(k(5)) = h(17) = 288$

f. $h(k(x)) = h(2x + 7) = (2x + 7)^2 - 1 = 4x^2 + 28x + 48$

5. $R_k \circ R_j$

F, j, F', k, F''

$R_j \circ R_k$

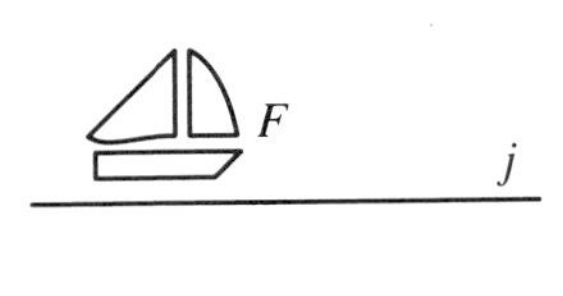

k

6. $R_k \circ R_j$

j, F, F', k, F''

$R_j \circ R_k$

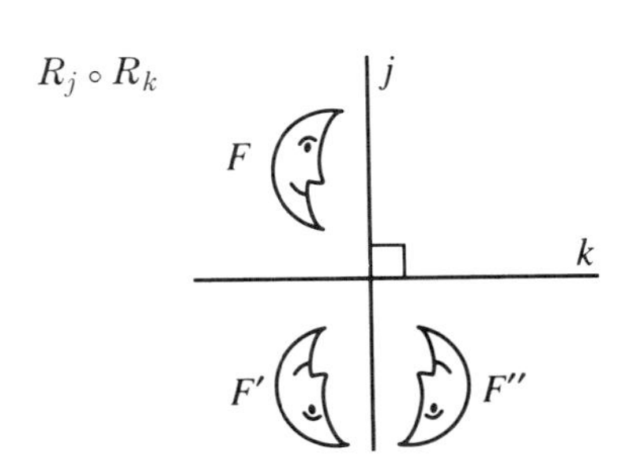

7. a. A B image

b. image A B

8. a.

j

image

b. image

C

9. a, b.

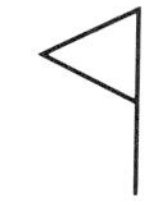

image E

10.

a. Yes **b.** Yes

c. $(x + 4, y + 3)$; $(x + 4, y + 3)$

11. a. Q **b.** S **c.** M **d.** Q **e.** Q

B 12. a. L **b.** L **c.** Q

13. b (also a and c if the absolute value of the scale factor of the dilation is 1)

14. a, b, c, d **15.** a, b, c, d

16. b, d (also a and c if the absolute values of the scale factors of the dilations are reciprocals)

17. $R_y : (3, 1) \rightarrow (-3, 1)$; $R_x : (-3, 1) \rightarrow (-3, -1)$

18. $H_O : (1, -2) \rightarrow (-1, 2)$; $R_y : (-1, 2) \rightarrow (1, 2)$

19. $H_O : (3, 0) \rightarrow (-3, 0)$; $H_A : (-3, 0) \rightarrow (9, 2)$

20. $H_A : (1, 1) \rightarrow (5, 1)$; $H_O : (5, 1) \rightarrow (-5, -1)$

21. $D_{O,2} : (2, 4) \rightarrow (4, 8)$; $R_x : (4, 8) \rightarrow (4, -8)$

22. $R_y : (-2, 1) \rightarrow (2, 1)$; $\mathcal{R}_{O,90} : (2, 1) \rightarrow (-1, 2)$

23. $\mathcal{R}_{O,-90} : (-1, -1) \rightarrow (-1, 1)$; $\mathcal{R}_{A,90} : (-1, 1) \rightarrow (3, -3)$

24. $D_{A,4} : (3, 0) \rightarrow (3, -3)$; $D_{O,-\frac{1}{3}} : (3, -3) \rightarrow (-1, 1)$

25. a. $R_l : (5, 2) \rightarrow (2, 5)$; $R_y : (2, 5) \rightarrow (-2, 5)$; $Q(-2, 5)$ **b.** Since the measure of an $\angle$ between $y = x$ and the y-axis is 45, $m\angle POQ = 2 \cdot 45 = 90$. **c.** slope of $\overline{OP} = \frac{2 - 0}{5 - 0} = \frac{2}{5}$; slope of $\overline{OQ} = \frac{5 - 0}{-2 - 0} = \frac{5}{-2} = -\frac{5}{2}$; two nonvertical lines are $\perp$ if and only if the product of their slopes is -1; $\frac{2}{5}\left(-\frac{5}{2}\right) = -1$, so $\overline{OP} \perp \overline{OQ}$.

d. $R_l : (x, y) \to (y, x)$; $R_y : (y, x) \to (-y, x)$; $R_y \circ R_l : (x, y) \to (-y, x)$.
$R_y : (x, y) \to (-x, y)$; $R_l : (-x, y) \to (y, -x)$; $R_l \circ R_y : (x, y) \to (y, -x)$.

26. **a.** $R_x : (-6, -2) \to (-6, 2)$; $R_k : (-6, 2) \to (-2, 6)$; $Q(-2, 6)$ **b.** slope of $\overline{OP} = \frac{-2-0}{-6-0} = \frac{1}{3}$; slope of $\overline{OQ} = \frac{6-0}{-2-0} = -3$; since $\frac{1}{3}(-3) = -1$, $\overline{OP} \perp \overline{OQ}$ and $m\angle POQ = 90$. (By Thm. 14-8, $m\angle POQ = 2$(meas. between $y = -x$ and x-axis) $= 2 \cdot 45 = 90$.) **c.** $R_x : (x, y) \to (x, -y)$; $R_k : (x, -y) \to (y, -x)$; $R_k \circ R_x : (x, y) \to (y, -x)$. $R_k : (x, y) \to (-y, -x)$; $R_x : (-y, -x) \to (-y, x)$; $R_x \circ R_k : (x, y) \to (-y, x)$.

C **27.** Const. B' so that k is the $\perp$ bis. of $\overline{BB'}$ ($R_k : B \to B'$). Const. line j, the $\perp$ bis. of $\overline{AB'}$.

28. **a.** The seg. that joins the midpts. of 2 sides of a $\triangle$ is half as long as the third side. **b.** Yes **c.** Yes; translation

29. translation; $\overline{P'Q'} \parallel \overline{PQ}$ and $\overline{P''Q''} \parallel \overline{P'Q'}$; so $\overline{PQ} \parallel \overline{P''Q''}$; $P'Q' = 2 \cdot PQ$ and $P''Q'' = \frac{1}{2} \cdot P'Q'$, so $P''Q'' = \frac{1}{2}(2 \cdot PQ) = PQ$; then $PP''Q''Q$ is a ▱, and $D_{B,\frac{1}{2}} \circ D_{A,2}$ is a translation.

30. **a.** $\mathcal{R}_{O,90}$: one (the origin); R_y : infinitely many (all pts. on the y-axis); $D_{O,3}$: one (the origin); T: none **b.** Let (x, y) be the fixed pt.; $D_{A,\frac{1}{4}}(x, y) = \left(1 + \frac{1}{4}(x - 1), \frac{y}{4}\right)$; $D_{O,2} \circ D_{A,\frac{1}{4}}(x, y) = \left(2 + \frac{1}{2}(x - 1), \frac{y}{2}\right) = (x, y)$; $2 + \frac{1}{2}(x - 1) = x$ and $\frac{y}{2} = y$; $\frac{3}{2} = \frac{1}{2}x$, so $x = 3$; $y = 0$; $(x, y) = (3, 0)$.

Page 607 • CLASSROOM EXERCISES

1. $\frac{1}{3}$ **2.** $\frac{1}{7}$ **3.** $\frac{5}{4}$ **4.** 2 **5.** 10 right **6.** 15 right **7.** 5 left **8.** 10 left **9.** identity **10.** T **11.** B **12.** C **13.** A **14.** A **15.** C **16.** B **17.** A **18.** C **19.** B **20.** 1 **21.** t **22.** T

23. **a.** R_l **b.** $\mathcal{R}_{O,-30}$ **c.** $T^{-1} : (x, y) \to (x + 4, y - 1)$ **d.** $D_{O,-1}$

24. For all real numbers, $ab = ba$, but for transformations in general, $T \circ S \neq S \circ T$. That is, composites of transformations are not generally commutative.

Pages 607–608 • WRITTEN EXERCISES

A **1.** $\frac{1}{4}$ **2.** $\frac{1}{9}$ **3.** $\frac{3}{2}$ **4.** 5 **5.** C **6.** D **7.** A **8.** D **9.** C **10.** B **11.** A **12.** B **13.** C **14.** P **15.** I **16.** H_O **17.** H_O **18.** $(x + 4, y)$ **19.** $(x + 6, y - 8)$ **20.** P

B **21.** $S^{-1}: (x, y) \rightarrow (x - 5, y - 2)$ **22.** $S^{-1}: (x, y) \rightarrow (x + 3, y + 1)$

23. $S^{-1}: (x, y) \rightarrow \left(\frac{1}{3}x, -2y\right)$ **24.** $S^{-1}: (x, y) \rightarrow (4x, 4y)$

25. $S^{-1}: (x, y) \rightarrow \left(x + 4, \frac{1}{4}y\right)$ **26.** $S^{-1}: (x, y) \rightarrow (y, x)$

27. $T: (x, y) \rightarrow \left(x + 2, y - \frac{1}{2}\right)$ **28.** $\mathcal{R}_{O,72}$

C **29.** **a.** $R_k \circ R_j$ moves each pt. 2 units right; $R_j \circ R_k$ moves each pt. 2 units left.
b. $R_k \circ R_k = I$ and $R_j \circ R_j = I$; $(R_k \circ R_j) \circ (R_j \circ R_k) = R_k \circ (R_j \circ R_j) \circ R_k = R_k \circ I \circ R_k = R_k \circ R_k = I$

30. **a.** S and T are inverses. **b.** $H_A \circ H_A = I$ and $H_B \circ H_B = I$; $(H_A \circ H_B) \circ (H_B \circ H_A) = H_A \circ (H_B \circ H_B) \circ H_A = H_A \circ I \circ H_A = H_A \circ H_A = I$

31. 1, 2. A composite of reflections in ⊥ lines is a half-turn about the pt. where the lines intersect. 3. Substitution Prop. 5. $R_2 \circ R_2 = I$; Substitution Prop. 6. Def. of identity mapping 7. A composite of reflections in 2 ∥ lines is a translation.

Pages 611–612 • CLASSROOM EXERCISES

1. 1 **2.** 3 **3.** 4 **4.** 0 **5.** 3 and 4 **6.** $\mathcal{R}_{O,120}$, $\mathcal{R}_{O,240}$, $\mathcal{R}_{O,360}$

7. Let P be the int. pt. of the lines of symmetry; $\mathcal{R}_{P,90}$, $\mathcal{R}_{P,180}$, $\mathcal{R}_{P,270}$, $\mathcal{R}_{P,360}$

8. R_l : line symmetry about the ⊥ bis. of the base; I

9. H_O : point symmetry about the int. pt. of the diags.; I

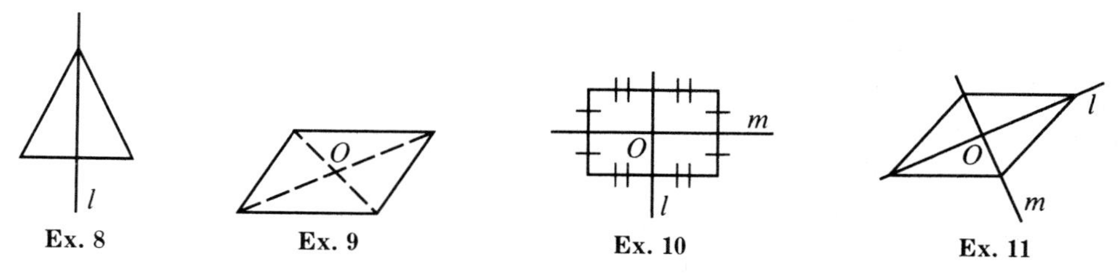

Ex. 8 Ex. 9 Ex. 10 Ex. 11

10. H_O : point symmetry about the int. pt. of the diags.; R_l and R_m : line symmetry about the lines joining the midpts. of the opp. sides; I

11. H_O : pt. symmetry about the int. pt. of the diags.; R_l and R_m : line symmetry about the diags.; I

12. **a.** Yes **b.** Yes **c.** No **d.** Yes **13.** **a.** 3 **b.** Infinitely many **c.** 4

14. Symmetry points for the tessellation of F's are shown at the right. Symmetry points for the tessellation of rhombuses are at every vertex and in the centers of the rhombuses.

15. Pt. symmetry; line symmetry about both fold lines

Ex. 14

Pages 612–614 • WRITTEN EXERCISES

A 1. **a.** 5 **b.** No **c.** Let O be the center of the circumscribed ⊙ of the starfish; $\mathscr{R}_{O,72}$; $\mathscr{R}_{O,144}$; $\mathscr{R}_{O,216}$; $\mathscr{R}_{O,288}$

2. **a.** 6 **b.** Yes **c.** Let O be the center of the circumscribed ⊙ of the snowflake; $\mathscr{R}_{O,60}$; $\mathscr{R}_{O,120}$; $\mathscr{R}_{O,180}$; $\mathscr{R}_{O,240}$; $\mathscr{R}_{O,300}$

3. **a.** 4 **b.** Yes **c.** Let O be the center of the circumscribed ⊙ of the flower; $\mathscr{R}_{O,90}$; $\mathscr{R}_{O,180}$; $\mathscr{R}_{O,270}$

4. **a.** 13 **b.** No **c.** Let O be the center of the circumscribed ⊙ of the cactus; $\mathscr{R}_{O,\frac{360n}{13}}$ for $n = 1, 2, 3, \ldots, 12$

5–7. Answers may vary depending on the way in which the letters are formed.

5. A, B, C, D, E, K, M, T, U, V, W, Y **6.** H, I, O X **7.** H, I, N, O, S, X, Z

8–11. Figures may vary; examples are given.

8.

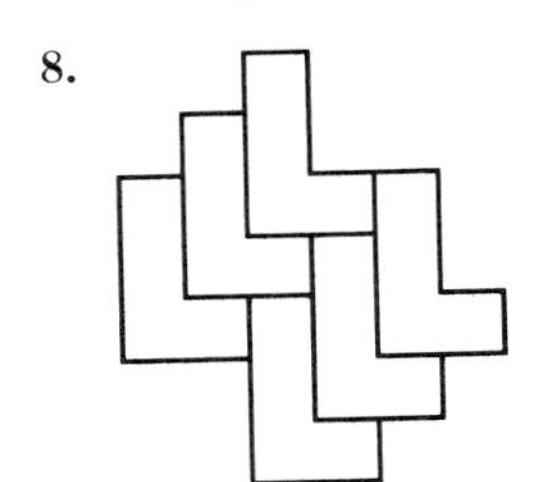

9.

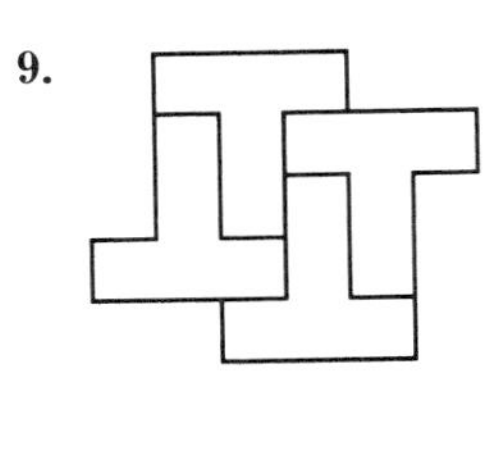

10.

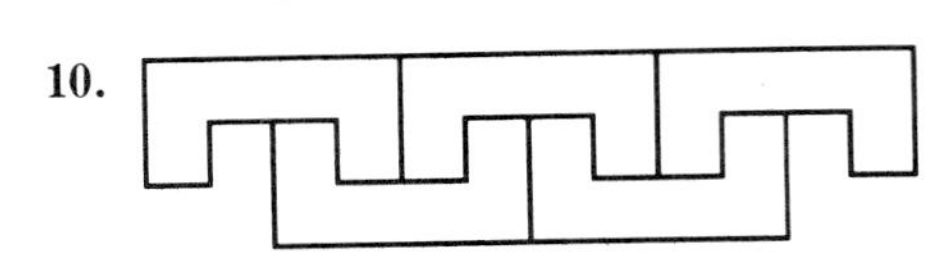

11.

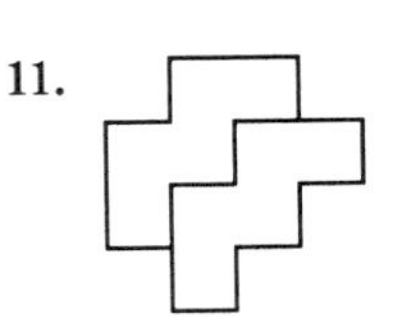

12.

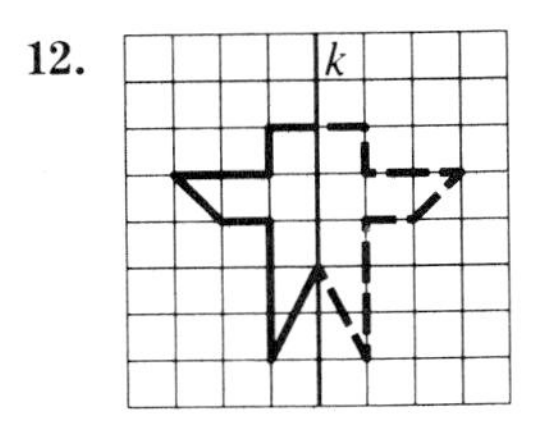

13.

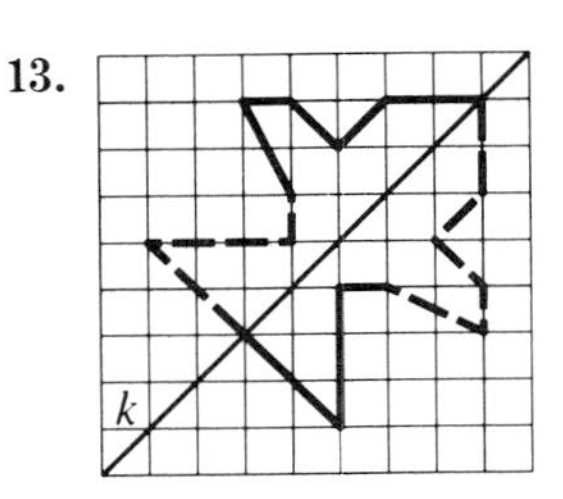

14.

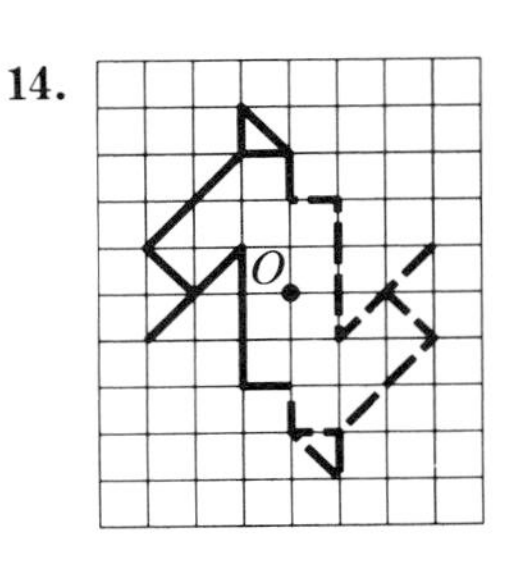

B **15.** 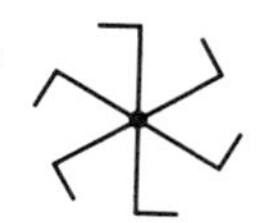 **16.** 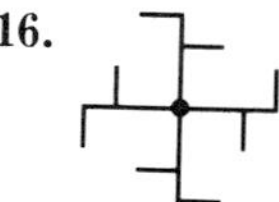 **17.**

18. **a.** bilateral symmetry **b.** 8 symmetry planes (4 through pairs of opp. legs, and 4 between pairs of adjacent legs); rotational symmetry of 45°, 90°, 135°, 180°, 225°, 270°, 315°, and 360°

19. **a.** The ellipse has line symmetry about each of its axes and pt. symmetry about the int. of the axes. **b.** If $a = b$, the original ellipse becomes a circle and the solid formed is a sphere with volume $\frac{4}{3}\pi a^3$. **c.** The ellipsoid shown has plane symmetry about the infinitely many planes that contain the major axis b of length $2b$; plane symmetry about the single plane that contains the minor axis a of length $2a$ and is $\perp$ to the major axis b; infinitely many rotational symmetries about the major axis b; 180° rotational symmetry about any line $\perp$ to the major axis and containing the intersection of the axes.

20. **a.** Yes **b.** Yes **c.** No **d.** Yes

21. **a.** non-isos. trap. **b.** isos. trap. **c.** not possible

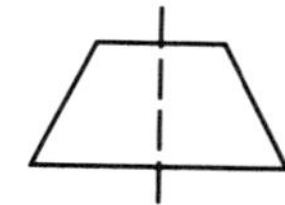

22. **a.** square **b.** rect. or rhombus **c.** not possible

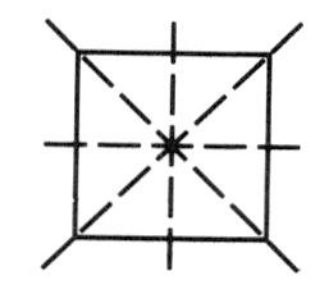

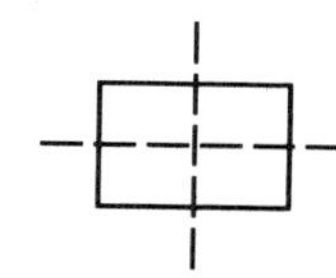

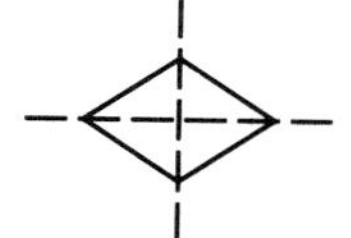

23. **a.** regular octagon **b.**  **c.** 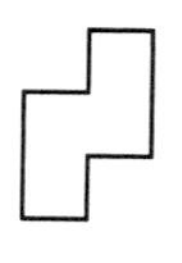

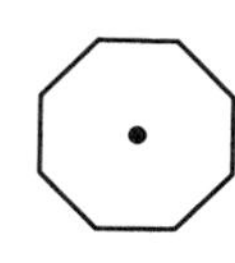

24. **a.** 90°, 180°, 270° rotational symmetries **b.** 60°: 6; 45°: 8; 30°: 12; in general, the number of coins visible when mirrors are arranged at $x°$ is $\frac{360}{x}$. Whenever this is an even integer, half of these coins will have words reading backwards.

25. **a.** at the midpts. of the sides **b.** translational

26. Let P be a pt. on the figure and P' be its image under $\mathscr{R}_{O,60}$. Let P'' be the image of P' under $\mathscr{R}_{O,60}$. P'' is also the image of P under $\mathscr{R}_{O,120}$. Thus the figure has 120° rotational symmetry. Similarly, it has 180°, 240°, 300°, and 360° rotational symmetries.

C **27.** Let $ABCDEF$ be a hexagon with symmetry pt. O. H_O maps $\overline{AB}$ to $\overline{DE}$. Since a rotation is an isometry, $AB = DE$. By def. of a rotation, $OA = OD$; $OB = OE$. Thus, $\triangle OAB \cong \triangle ODE$; $\angle OAB \cong \angle ODE$. Hence, $\overline{AB} \parallel \overline{DE}$.

28. $(50 \cdot n)°$ rotational symmetry for all integers n. Since the figure has 350° symmetry and $360° - 350° = 10°$, this answer is equivalent to $(10 \cdot n)°$ rotational symmetry for all integers n.

29. **a.** Planes: infinitely many (any plane containing $\overleftrightarrow{OA}$); axes of rotation: 1 ($\overleftrightarrow{OA}$).

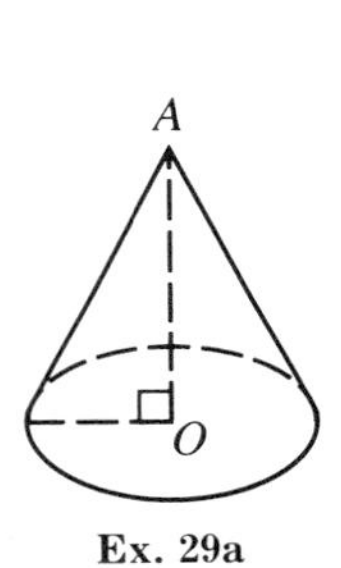

Ex. 29a

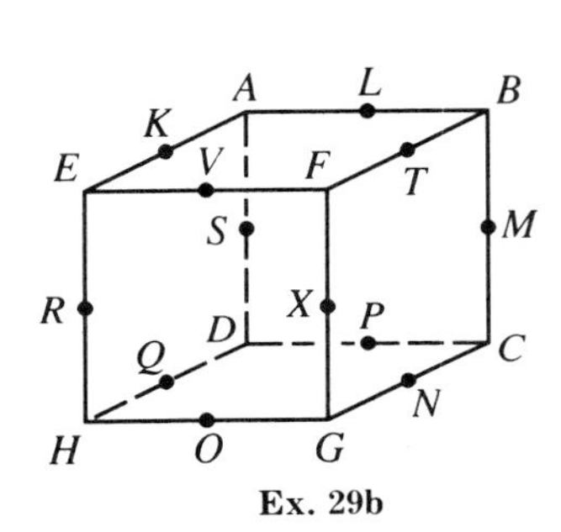

Ex. 29b

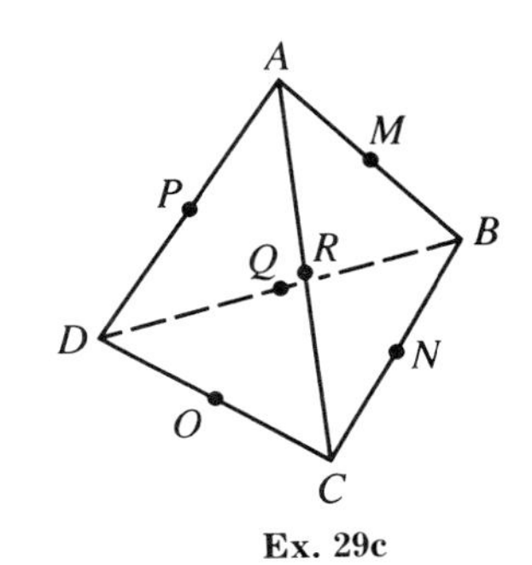

Ex. 29c

b. Vertices of the cube and midpts. of edges are labeled. Planes: 9 ($RSMX$, $KTNQ$, $VLPO$, $ADGF$, $EHCB$, $EACG$, $FBDH$, $FEDC$, $BAHG$); axes of rotation: 13 (the 3 lines joining the centers of opp. faces, $\overleftrightarrow{AG}$, $\overleftrightarrow{BH}$, $\overleftrightarrow{FD}$, $\overleftrightarrow{EC}$, $\overleftrightarrow{KN}$, $\overleftrightarrow{TQ}$, $\overleftrightarrow{LO}$, $\overleftrightarrow{VP}$, $\overleftrightarrow{RM}$, $\overleftrightarrow{SX}$) **c.** M, N, O, P, Q, and R are midpts. of edges. Planes: 6 (DNA, DRB, DMC, CPB, CQA, BOA); axes of rotation: 7 (the altitude from each of the 4 vertices, $\overleftrightarrow{MO}$, $\overleftrightarrow{PN}$, $\overleftrightarrow{RQ}$)

Page 614 • CHALLENGES

1. The locus of the cat is point C. Let $\overline{CD}$ be $\perp$ to $\overline{AJ}$, intersecting $\overline{AJ}$ at D. Then since $\overline{CD}$ is the median of trap. $AXYJ$, $CD = \frac{1}{2}(AX + JY) = \frac{1}{2}(AM + MJ) = \frac{1}{2} \cdot AJ$. Since $\overline{CD} \parallel \overline{AX} \parallel \overline{JY}$ and C is the midpt. of $\overline{XY}$, D is the midpt. of $\overline{AJ}$; $\overline{CD}$ bisects $\overline{AJ}$ for each $\overline{XY}$, so the cat stays on $\overline{CD}$ at distance CD from $\overline{AJ}$. That is, the cat doesn't change position.

2. Let $\overrightarrow{SR}$ int. the number line at pt. $T(a, 0)$. Since Q and S are the midpts. of 2 sides of $\triangle PAC$, $\overline{QS} \parallel \overline{AC}$ and $QS = \frac{1}{2}AC = \frac{1}{2}(26 - 8) = 9$. $\angle SQR \cong \angle TBR$ (alt. int. $\measuredangle$s) and $\angle SRQ \cong \angle TRB$ (vert. $\measuredangle$s); also, $\overline{QR} \cong \overline{BR}$ so $\triangle SRQ \cong \triangle TRB$ by ASA. Thus, $TB = QS = 9$, and since $OB = OT + TB = 12$, $OT = 3$; T has coord. 3.

Page 615 • SELF-TEST 2

1. B is mapped to R, then A. **2.** A is mapped to R, then A. **3.** B
4. P is mapped to Q, then P. **5.** y-axis **6.** No **7.** P **8.** T **9.** translation
10. a. $D_{O,\frac{1}{5}}$ **b.** $\mathcal{R}_{O,70}$ **c.** R_y **d.** $S^{-1}: (x, y) \to (x - 2, y + 3)$ **11.** 6

Pages 616–617 • EXTRA

1. Let k be the alt. to the base.

$\circ$	I	R_k
I	I	R_k
R_k	R_k	I

2. a. I, R_j, R_k, H_O **c.** yes

b.

$\circ$	I	R_j	R_k	H_O
I	I	R_j	R_k	H_O
R_j	R_j	I	H_O	R_k
R_k	R_k	H_O	I	R_j
H_O	H_O	R_k	R_j	I

3.

$\circ$	I	$\mathcal{R}_{O,120}$	$\mathcal{R}_{O,240}$
I	I	$\mathcal{R}_{O,120}$	$\mathcal{R}_{O,240}$
$\mathcal{R}_{O,120}$	$\mathcal{R}_{O,120}$	$\mathcal{R}_{O,240}$	I
$\mathcal{R}_{O,240}$	$\mathcal{R}_{O,240}$	I	$\mathcal{R}_{O,120}$

4.

$\circ$	I	$\mathcal{R}_{O,90}$	H_O	$\mathcal{R}_{O,270}$
I	I	$\mathcal{R}_{O,90}$	H_O	$\mathcal{R}_{O,270}$
$\mathcal{R}_{O,90}$	$\mathcal{R}_{O,90}$	H_O	$\mathcal{R}_{O,270}$	I
H_O	H_O	$\mathcal{R}_{O,270}$	I	$\mathcal{R}_{O,90}$
$\mathcal{R}_{O,270}$	$\mathcal{R}_{O,270}$	I	$\mathcal{R}_{O,90}$	H_O

5. a. 2 (H_O and I) **b.** all 4 **6.** Yes; yes

7. a.

$\circ$	I	$\mathcal{R}_{O,120}$	$\mathcal{R}_{O,240}$	R_j	R_k	R_l
I	I	$\mathcal{R}_{O,120}$	$\mathcal{R}_{O,240}$	R_j	R_k	R_l
$\mathcal{R}_{O,120}$	$\mathcal{R}_{O,120}$	$\mathcal{R}_{O,240}$	I	R_l	R_j	R_k
$\mathcal{R}_{O,240}$	$\mathcal{R}_{O,240}$	I	$\mathcal{R}_{O,120}$	R_k	R_l	R_j
R_j	R_j	R_k	R_l	I	$\mathcal{R}_{O,120}$	$\mathcal{R}_{O,240}$
R_k	R_k	R_l	R_j	$\mathcal{R}_{O,240}$	I	$\mathcal{R}_{O,120}$
R_l	R_l	R_j	R_k	$\mathcal{R}_{O,120}$	$\mathcal{R}_{O,240}$	I

b. Answers may vary; for example, $R_j \circ R_k = \mathcal{R}_{O,240}$ and $R_k \circ R_j = \mathcal{R}_{O,120}$.

8.

$\circ$	I	$\mathcal{R}_{O,90}$	H_O	$\mathcal{R}_{O,270}$	R_j	R_k	R_l	R_m
I	I	$\mathcal{R}_{O,90}$	H_O	$\mathcal{R}_{O,270}$	R_j	R_k	R_l	R_m
$\mathcal{R}_{O,90}$	$\mathcal{R}_{O,90}$	H_O	$\mathcal{R}_{O,270}$	I	R_l	R_m	R_k	R_j
H_O	H_O	$\mathcal{R}_{O,270}$	I	$\mathcal{R}_{O,90}$	R_k	R_j	R_m	R_l
$\mathcal{R}_{O,270}$	$\mathcal{R}_{O,270}$	I	$\mathcal{R}_{O,90}$	H_O	R_m	R_l	R_j	R_k
R_j	R_j	R_m	R_k	R_l	I	H_O	$\mathcal{R}_{O,270}$	$\mathcal{R}_{O,90}$
R_k	R_k	R_l	R_j	R_m	H_O	I	$\mathcal{R}_{O,90}$	$\mathcal{R}_{O,270}$
R_l	R_l	R_j	R_m	R_k	$\mathcal{R}_{O,90}$	$\mathcal{R}_{O,270}$	I	H_O
R_m	R_m	R_k	R_l	R_j	$\mathcal{R}_{O,270}$	$\mathcal{R}_{O,90}$	H_O	I

No. The symmetry group of the square is not a commutative group, since, for example, $R_j \circ \mathcal{R}_{O,90} = R_l$ and $\mathcal{R}_{O,90} \circ R_j = R_m$.

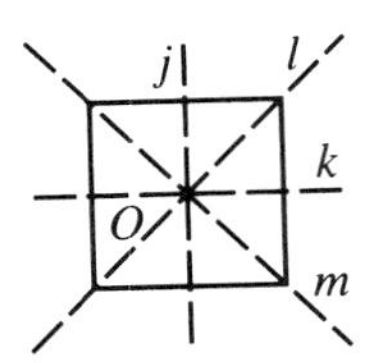

9. **a.** No; there is no identity, and the composite of 2 line symmetries is not a line symmetry. **b.** Yes; the rotational symmetries **c.** I and H_O

10. **a.** S^3 maps each fish to the third fish of the same color to its right; yes. **b.** T^{-1} maps each fish to the fish of the same color directly below; yes. **c.** $S \circ T$ maps each fish to the fish of the same color above and to the right; yes. **d.** infinitely many **e.** Yes

Page 619 • CHAPTER REVIEW

1. $\cong$ **2.** 18; 2 **3. a.** (6, 1); $\left(\frac{3}{2}, 5\right)$ **b.** No

4. **a.** $(-7, -5)$ **b.** $(7, 5)$ **c.** $(5, -7)$

5. Image: $y = -2x + 1$ **6. a.** $(x + 2, y - 4)$ **b.** Yes **c.** Yes **d.** Yes

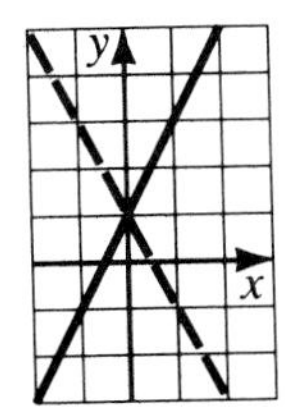

7. $(7, -2) \rightarrow (12, -2) \rightarrow (12, 2)$

8. **a.** $A'(-2, 3)$, $B'(-1, -1)$, $C'(3, 1)$ **b.** $A'(-3, -2)$, $B'(1, -1)$, $C'(-1, 3)$

9. a, c **10.** $(6, -4)$ **11.** $(1, 3)$

12. **a.** $(3, 1) \rightarrow (-3, 1) \rightarrow (-3, -1)$ **b.** $(3, 1) \rightarrow (-3, -1) \rightarrow (3, -1)$ **c.** $(3, 1) \rightarrow (1, -3) \rightarrow (1, 3)$

13. $(x + 1, y - 6)$ **14.** $D_{O,\frac{1}{4}}$ **15.** I **16.** -75 or 285 **17.** No **18.** Yes
19. Yes **20.** A regular pentagon

Page 620 • CHAPTER TEST

1. Translation **2.** Reflection **3.** Glide reflection **4.** Rotation **5.** 5; 2
6. (2, 4) **7.** (2, −4) **8.** (2, 1) **9.** $(4, 2) \rightarrow (4, -2) \rightarrow (-4, 2)$
10. $(4, 2) \rightarrow (-2, 4) \rightarrow (-4, -2)$ **11.** (2, 2) **12.** $(4, 2) \rightarrow (-5, 2) \rightarrow (2, -5)$
13. $(4, 2) \rightarrow (4, -2) \rightarrow (-4, -2) \rightarrow (-2, -4)$ **14.** H_O **15.** R_x **16.** $D_{O,-\frac{1}{2}}$
17. (2, 1) **18.** (6, 3) **19.** (−2, −1) **20.** True **21.** True **22.** True **23.** True
24. **a.** Yes; yes **b.** distance, ∠ meas., area, orientation
25. **a.** -2 **b.** $\frac{1}{2}$ **c.** 2

Page 621 • PREPARING FOR COLLEGE ENTRANCE EXAMS

1. B. $m = \frac{7 - (-1)}{-1 - 3} = -2$; slope of ⊥ bis. $= \frac{1}{2}$; midpt. $M =$
$\left(\frac{3 + (-1)}{2}, \frac{-1 + 7}{2}\right) = (1, 3)$; $y - 3 = \frac{1}{2}(x - 1)$; $2y - 6 = x - 1$;
$x - 2y = -5$
2. A. $d = \sqrt{(-6 - 0)^2 + (-16 - (-8))^2} = \sqrt{(-6)^2 + (-8)^2} = \sqrt{36 + 64} =$
$\sqrt{100} = 10$; $r = \frac{10}{2} = 5$; center $= \left(\frac{-6 + 0}{2}, \frac{-16 + (-8)}{2}\right) = (-3, -12)$;
equation: $(x + 3)^2 + (y + 12)^2 = 25$
3. E **4.** C **5.** C **6.** E **7.** A **8.** E **9.** B **10.** D

Pages 622–625 • CUMULATIVE REVIEW: CHAPTERS 1–14

True-False Exercises

A **1.** True **2.** True **3.** True **4.** False **5.** False **6.** True **7.** False **8.** True
9. False **10.** False **11.** True **12.** True
B **13.** False **14.** False **15.** True **16.** True **17.** True **18.** True **19.** False
20. True

Multiple-Choice Exercises

A **1.** d **2.** c **3.** b **4.** d **5.** d
B **6.** a **7.** c **8.** a **9.** c **10.** b **11.** b

Completion Exercises

A **1.** Add. Prop. of = **2.** corr.; alt. int.; s-s. int. **3.** 124 **4.** $\overline{BE}$ **5.** 33 **6.** $2\sqrt{13}$
7. $-\frac{3}{2}$ **8.** (−6, 1)

B 9. $(-5, 2)$ 10. 12.5 11. 22.5 12. $\frac{1}{2}; \frac{1}{16}$ 13. $\frac{15}{17}$ 14. 32° 15. $27\sqrt{7}$

16. 20π 17. 324π; 135π 18. 288π cm^3 19. $(4, -2)$

C 20. $36\sqrt{3}$ cm^2; $18\sqrt{2}$ cm^3 21. 1 : 7

Always-Sometimes-Never Exercises

A 1. N 2. N 3. S 4. S 5. A 6. A 7. N 8. A 9. S 10. A

B 11. A 12. S 13. N 14. S 15. A 16. A 17. S

Construction Exercises

A 1. Const. $\overleftrightarrow{AC} \perp \overleftrightarrow{BD}$ at E. Const. $\overrightarrow{EF}$, the bis. of $\angle DEC$; then const. $\overrightarrow{EG}$, the bis. of $\angle FEC$. $m\angle GEC = 22\frac{1}{2}$

2. Const. 8 3. Const. 11

4. On a line, mark off $AB = x$, $BC = y$. Construct the $\perp$ bis. of $\overline{AC}$ int. $\overline{AC}$ at D. $AD = \frac{1}{2}(x + y)$

B 5. Const. $AB = y$. Const. the $\perp$ to $\overline{AB}$ at A and the $\perp$ to $\overline{AB}$ at B. With ctr. B and rad. x, draw an arc int. the first $\perp$ at D. With ctr. A and rad. x, draw an arc int. the second $\perp$ at C. Draw $\overline{DC}$.

6. On a line, mark off $AB = x$, $BC = x$. Const. a $\perp$ to $\overline{AC}$ at C and const. $\overline{CD}$ on the $\perp$ with length y. Draw $\overline{AD}$.

7. On a line, mark off $AB = x$, $BC = x$, and $CD = x$. Use Const. 14 to const. the geom. mean of AD and y.

8. Const. the $\perp$ bis. of $\overline{AB}$, int. $\overline{AB}$ at C. Use Const. 12 to divide $\overline{AC}$ into 5 $\cong$ parts, $\overline{AW}$, $\overline{WX}$, $\overline{XY}$, $\overline{YZ}$, and $\overline{ZC}$. Const. $\overline{DE} \cong \overline{AY}$. Const. a $\perp$ to $\overline{DE}$ at D and one at E. Const. $\overline{DG}$ and $\overline{EF}$ both $\cong$ to $\overline{YC}$ on the $\perp$s. Draw $\overline{GF}$.

Proof Exercises

A 1.

Statements	Reasons
1. $\overline{PQ} \parallel \overline{RS}$	1. Given
2. $\angle OPQ \cong \angle ORS$; $\angle OQP \cong \angle OSR$	2. If 2 $\parallel$ lines are cut by a trans., then alt. int. $\angle$s are $\cong$.
3. $\triangle OQP \sim \triangle OSR$	3. AA $\sim$ Post.
4. $\frac{PO}{RO} = \frac{PQ}{RS}$	4. Corr. sides of $\sim$ $\triangle$s are in prop.

2.

Statements	Reasons
1. $\overline{PR} \perp \overline{QS}$	1. Given
2. $\angle POS$ and $\angle QOR$ are rt. $\angle$s.	2. Def. of $\perp$ lines
3. $\triangle POS$ and $\triangle QOR$ are rt. $\triangle$s.	3. Def. of rt. $\triangle$
4. $\overline{PS} \cong \overline{QR}$; $\overline{OS} \cong \overline{OR}$	4. Given
5. $\triangle POS \cong \triangle QOR$	5. HL Thm.
6. $\angle PSO \cong \angle QRO$	6. Corr. parts of $\cong$ $\triangle$s are $\cong$.

B **3.**

Statements	Reasons
1. $\angle OSR \cong \angle ORS$; $\angle OPQ \cong \angle OQP$	1. Given
2. $\overline{OS} \cong \overline{OR}$, or $OS = OR$; $\overline{OQ} \cong \overline{OP}$, or $OQ = OP$	2. If 2 $\angle$s of a $\triangle$ are $\cong$, then the sides opp. those $\angle$s are $\cong$.
3. $OS + OQ = OR + OP$	3. Add. Prop. of =
4. $OS + OQ = QS$; $OR + OP = PR$	4. Seg. Add. Post.
5. $QS = PR$, or $\overline{QS} \cong \overline{PR}$	5. Substitution Prop.
6. $\overline{SR} \cong \overline{SR}$	6. Refl. Prop.
7. $\triangle PSR \cong \triangle QRS$	7. SAS Post.

4. Given: Rectangle $ABCD$

Prove: $\overline{AX} \cong \overline{XC} \cong \overline{BX} \cong \overline{XD}$

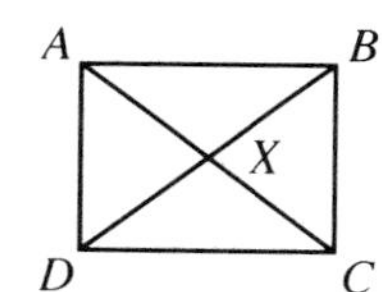

Statements	Reasons
1. $ABCD$ is a rect.	1. Given
2. $\overline{AC}$ and $\overline{BD}$ bis. each other.	2. The diags. of a $\square$ bis. each other.
3. X is the midpt. of $\overline{AC}$ and $\overline{BD}$.	3. Def. of bis.
4. $AX = XC = \frac{1}{2}AC$; $BX = XD = \frac{1}{2}BD$	4. Midpt. Thm.
5. $\overline{AC} \cong \overline{BD}$, or $AC = BD$	5. The diags. of a rect. are $\cong$.
6. $\frac{1}{2}AC = \frac{1}{2}BD$	6. Mult. Prop. of =
7. $AX = XC = BX = XD$, or $\overline{AX} \cong \overline{XC} \cong \overline{BX} \cong \overline{XD}$	7. Substitution Prop.

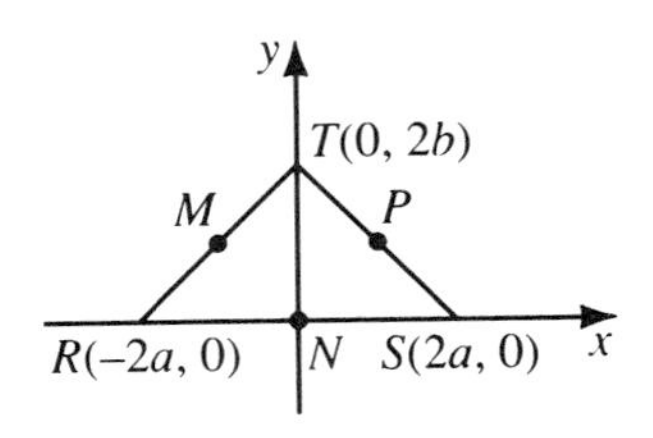

5. Proof: Let $R(-2a, 0)$, $S(2a, 0)$, and $T(0, 2b)$ be the vertices of an isos. △. Then $M(-a, b)$, $N(0, 0)$, and $P(a, b)$ are the midpts. of $\overline{TR}$, $\overline{RS}$, and $\overline{TS}$, resp.; $NM = \sqrt{(0 - (-a))^2 + (0 - b)^2} = \sqrt{a^2 + b^2}$; $NP = \sqrt{(a - 0)^2 + (b - 0)^2} = \sqrt{a^2 + b^2}$; then $\overline{NM} \cong \overline{NP}$ and $\triangle MNP$ is isos.

C 6. Let $ABCD$ be a trap. with bases $\overline{AB}$ and $\overline{CD}$ and assume temp. that $ABCD$ has 2 pairs of ≅ sides. *Case 1:* $\overline{AB} \cong \overline{DC}$ and $\overline{AD} \cong \overline{BC}$. Since $\overline{AB} \parallel \overline{DC}$ and $\overline{AB} \cong \overline{DC}$, $ABCD$ must be a ▱ and $\overline{AD} \parallel \overline{BC}$. This contradicts the def. of a trap. *Case 2:* $\overline{AB} \cong \overline{AD}$ and $\overline{BC} \cong \overline{DC}$. Draw $\overline{AC}$. $\triangle ADC \cong \triangle ABC$ (SSS) so $\angle D \cong \angle B$, $\angle DAC \cong \angle BAC$, and $\angle ACD \cong \angle ACB$. Since $\overline{AB} \parallel \overline{DC}$, $\angle ACD \cong \angle BAC \cong \angle DAC$. Then $m\angle BAD = m\angle DAC + m\angle BAC = m\angle ACB + m\angle ACD = m\angle BCD$. If both pairs of opp. ∠s of $ABCD$ are ≅, $ABCD$ is a ▱ and $\overline{AD} \parallel \overline{BC}$. This contradicts the def. of a trap. *Case 3:* ($\overline{AB} \cong \overline{BC}$ and $\overline{AD} \cong \overline{DC}$) is similar to Case 2. All three cases lead to a contradiction. Our temp. assumption must be false. It follows that a trap. cannot have 2 pairs of ≅ sides.

7. Given: $\overline{XY}$ tan. to ⊙O and ⊙P
Prove: $\overleftrightarrow{BZ}$ bis. $\overline{XY}$

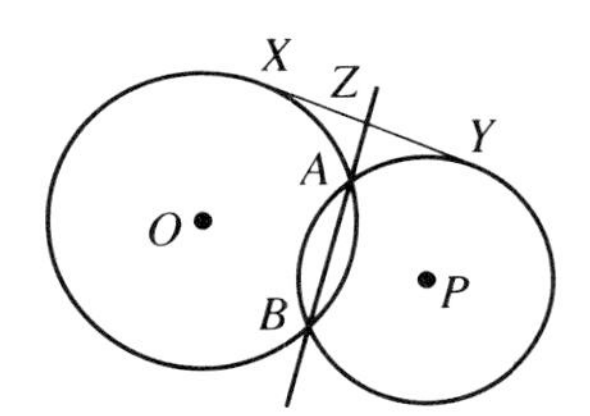

Statements	Reasons
1. $ZA \cdot ZB = (ZX)^2$; $ZA \cdot ZB = (ZY)^2$	1. When a secant seg. and a tan. seg. are drawn to a ⊙ from an ext. pt., the product of the secant seg. and its ext. seg. = the square of the tan. seg.
2. $(ZX)^2 = (ZY)^2$ or $ZX = ZY$	2. Substitution Prop.
3. Z is the midpt. of $\overline{XY}$.	3. Def. of midpt.
4. $\overrightarrow{BZ}$ bis. $\overline{XY}$	4. Def. of bis.

Examinations

Page 626 • CHAPTER 1

1. b **2.** d **3.** b **4.** a **5.** a **6.** b **7.** d **8.** c **9.** a **10.** b **11.** a

Page 627 • CHAPTER 2

1. c **2.** a **3.** b **4.** d **5.** d **6.** a **7.** b **8.** b **9.** a **10.** b **11.** a

Page 628 • CHAPTER 3

1. d **2.** a **3.** b **4.** a **5.** d **6.** b **7.** d **8.** c **9.** a **10.** c **11.** c
12. a **13.** d

Page 629 • CHAPTER 4

1. ASA or AAS **2.** AAS **3.** ASA or AAS **4.** HL **5.** SSS **6.** AAS
7. SAS **8.** HL **9.** d **10.** a **11.** a **12.** c **13.** b **14.** c

Page 630 • CHAPTER 5

1. a **2.** d **3.** a **4.** d **5.** b **6.** b **7.** a **8.** c **9.** b **10.** c **11.** c **12.** b

Page 631 • CHAPTER 6

1. b **2.** d **3.** a **4.** c **5.** b **6.** c **7.** a **8.** b **9.** a **10.** a **11.** d

Page 632 • CHAPTER 7

1. c **2.** d **3.** a **4.** d **5.** b **6.** b **7.** b **8.** a **9.** b **10.** d
11. c **12.** b

Page 633 • CHAPTER 8

1. a **2.** c **3.** c **4.** d **5.** b **6.** a **7.** c **8.** d **9.** c **10.** a **11.** c **12.** b

Page 634 • CHAPTER 9

1. c **2.** a **3.** c **4.** c **5.** c **6.** c **7.** a **8.** b **9.** b **10.** d **11.** a
12. d **13.** a **14.** d

Page 635 • CHAPTER 10

1. c **2.** d **3.** b **4.** b **5.** c **6.** d **7.** a **8.** d **9.** d **10.** d **11.** a **12.** c

Page 636 • CHAPTER 11

1. a **2.** d **3.** b **4.** c **5.** b **6.** a **7.** d **8.** b **9.** b **10.** c **11.** c
12. d **13.** b **14.** d **15.** d **16.** b

Page 637 • CHAPTER 12

1. d **2.** b **3.** a **4.** c **5.** b **6.** b **7.** a **8.** b **9.** a **10.** d **11.** c
12. a **13.** d **14.** b

Page 638 • CHAPTER 13

1. d **2.** d **3.** a **4.** c **5.** d **6.** b **7.** b **8.** c **9.** c **10.** a **11.** b **12.** b

Page 639 • CHAPTER 14

1. c **2.** c **3.** b **4.** c **5.** a **6.** d **7.** b **8.** c **9.** c **10.** d **11.** b
12. b **13.** d **14.** c **15.** b

Pages 645–646 • EXERCISES

1. I like the city and you like the country. 2. I don't like the city.
3. You don't like the country. 4. I like the city or you like the country.
5. I like the city or you don't like the country.
6. It is not true that I like the city and you like the country.
7. I don't like the city or you don't like the country.
8. I don't like the city and you like the country.
9. It is not true that I like the city or you like the country.
10. I don't like the city and you don't like the country.
11. $p \vee q$ 12. $\sim q$ 13. $\sim (p \vee q)$ 14. $\sim p \wedge \sim q$ 15. $\sim (p \wedge q)$ 16. $\sim p \vee \sim q$
17. Yes 18. Yes

19.

p	q	$\sim q$	$p \vee \sim q$
T	T	F	T
T	F	T	T
F	T	F	F
F	F	T	T

20.

p	q	$\sim p$	$\sim p \vee q$
T	T	F	T
T	F	F	F
F	T	T	T
F	F	T	T

21.

p	$\sim p$	$\sim (\sim p)$
T	F	T
F	T	F

22.

p	q	$p \wedge q$	$\sim (p \wedge q)$
T	T	T	F
T	F	F	T
F	T	F	T
F	F	F	T

23.

p	$\sim p$	$p \vee \sim p$
T	F	T
F	T	T

24.

p	$\sim p$	$p \wedge \sim p$
T	F	F
F	T	F

25.

p	q	r	$q \vee r$	$p \wedge (q \vee r)$
T	T	T	T	T
T	T	F	T	T
T	F	T	T	T
T	F	F	F	F
F	T	T	T	F
F	T	F	T	F
F	F	T	T	F
F	F	F	F	F

26.

p	q	r	$p \wedge q$	$p \wedge r$	$(p \wedge q) \vee (p \wedge r)$
T	T	T	T	T	T
T	T	F	T	F	T
T	F	T	F	T	T
T	F	F	F	F	F
F	T	T	F	F	F
F	T	F	F	F	F
F	F	T	F	F	F
F	F	F	F	F	F

Page 647 • EXERCISES

1. If you like to paint, then you are an artist.
2. If you are an artist, then you draw landscapes.
3. If you aren't an artist, then you don't draw landscapes.
4. It is not true that if you like to paint, then you are an artist.
5. If you like to paint and you are an artist, then you draw landscapes.
6. You like to paint, and if you are an artist, then you draw landscapes.
7. If you draw landscapes or you are an artist, then you like to paint.
8. You draw landscapes, or if you are an artist, then you like to paint.

9. $b \rightarrow k$ **10.** $k \rightarrow \sim s$ **11.** $(\sim b \vee \sim k) \rightarrow s$ **12.** $s \wedge (b \rightarrow k)$
13. $\sim (b \rightarrow s)$ **14.** $\sim b \rightarrow (k \wedge s)$

15. a.

p	q	$\sim p$	$\sim q$	$\sim p \rightarrow \sim q$
T	T	F	F	T
T	F	F	T	T
F	T	T	F	F
F	F	T	T	T

Yes; no **b.** Yes; yes

16.

p	q	$\sim q$	$p \rightarrow \sim q$
T	T	F	F
T	F	T	T
F	T	F	T
F	F	T	T

17.

p	q	$p \rightarrow q$	$\sim (p \rightarrow q)$
T	T	T	F
T	F	F	T
F	T	T	F
F	F	T	F

18.

p	q	$\sim q$	$p \wedge \sim q$
T	T	F	F
T	F	T	T
F	T	F	F
F	F	T	F

19. $\sim (p \rightarrow q)$ and $p \wedge \sim q$ are logically equivalent.

20.

p	q	$p \rightarrow q$	$q \rightarrow p$	$(p \rightarrow q) \wedge (q \rightarrow p)$
T	T	T	T	T
T	F	F	T	F
F	T	T	F	F
F	F	T	T	T

Page 649 • EXERCISES

1. 1. Given 2. Step 1 and Simplification 3. Given 4. Steps 2 and 3 and Modus Ponens

2. 1. Given 2. Given 3. Steps 1 and 2 and Modus Ponens 4. Given 5. Steps 3 and 4 and Modus Ponens

3.

Statements	Reasons
1. $p \vee q$	1. Given
2. $\sim p$	2. Given
3. q	3. Steps 1 and 2 and Disjunctive Syllogism
4. $q \rightarrow s$	4. Given
5. s	5. Steps 3 and 4 and Modus Ponens

4.

Statements	Reasons
1. $a \rightarrow b$	1. Given
2. $\sim b$	2. Given
3. $\sim a$	3. Steps 1 and 2 and Modus Tollens
4. $a \vee c$	4. Given
5. c	5. Steps 3 and 4 and Disjunctive Syllogism

5.

Statements	Reasons
1. $a \wedge b$	1. Given
2. a	2. Step 1 and Simplification
3. $a \rightarrow \sim c$	3. Given
4. $\sim c$	4. Steps 2 and 3 and Modus Ponens
5. $c \vee d$	5. Given
6. d	6. Steps 4 and 5 and Disjunctive Syllogism

6.

Statements	Reasons
1. $p \wedge q$	1. Given
2. p	2. Step 1 and Simplification
3. $p \rightarrow \sim s$	3. Given
4. $\sim s$	4. Steps 2 and 3 and Modus Ponens
5. $r \rightarrow s$	5. Given
6. $\sim r$	6. Steps 4 and 5 and Modus Tollens

7. Given: $w \rightarrow g$; $g \rightarrow p$; $w \wedge y$

Prove: p

Statements	Reasons
1. $w \wedge y$	1. Given
2. w	2. Step 1 and Simplification
3. $w \rightarrow g$	3. Given
4. g	4. Steps 2 and 3 and Modus Ponens
5. $g \rightarrow p$	5. Given
6. p	6. Steps 4 and 5 and Modus Ponens

8. Given: $\sim l \wedge p$; $s \vee r$; $\sim l \rightarrow \sim s$

Prove: r

Statements	Reasons
1. $\sim l \wedge p$	1. Given
2. $\sim l$	2. Step 1 and Simplification
3. $\sim l \rightarrow \sim s$	3. Given
4. $\sim s$	4. Steps 2 and 3 and Modus Ponens
5. $s \vee r$	5. Given
6. r	6. Steps 4 and 5 and Disjunctive Syllogism

Pages 650–651 • EXERCISES

1. Part (c) is a tautology.

a.

p	q	$p \vee q$	$(p \vee q) \rightarrow p$
T	T	T	T
T	F	T	T
F	T	T	F
F	F	F	T

b.

p	q	$p \rightarrow q$	$(p \rightarrow q) \rightarrow p$
T	T	T	T
T	F	F	T
F	T	T	F
F	F	T	F

c.

p	q	$p \wedge q$	$(p \wedge q) \rightarrow p$
T	T	T	T
T	F	F	T
F	T	F	T
F	F	F	T

d.

p	q	$p \rightarrow q$	$(p \rightarrow q) \rightarrow q$
T	T	T	T
T	F	F	T
F	T	T	T
F	F	T	F

2.

p	q	r	$p \vee q \vee r$	$p \wedge q \wedge r$	$\sim (p \wedge q \wedge r)$	$(p \vee q \vee r) \vee \sim (p \wedge q \wedge r)$
T	T	T	T	T	F	T
T	T	F	T	F	T	T
T	F	T	T	F	T	T
T	F	F	T	F	T	T
F	T	T	T	F	T	T
F	T	F	T	F	T	T
F	F	T	T	F	T	T
F	F	F	F	F	T	T

3.

p	q	r	$p \wedge q \wedge r$	$p \vee q \vee r$	$(p \wedge q \wedge r) \rightarrow (p \vee q \vee r)$
T	T	T	T	T	T
T	T	F	F	T	T
T	F	T	F	T	T
T	F	F	F	T	T
F	T	T	F	T	T
F	T	F	F	T	T
F	F	T	F	T	T
F	F	F	F	F	T

4.

p	q	r	$(p \to q)$	$(q \to r)$	$p \wedge (p \to q) \wedge (q \to r)$	$[p \wedge (p \to q) \wedge (q \to r)] \to r$
T	T	T	T	T	T	T
T	T	F	T	F	F	T
T	F	T	F	T	F	T
T	F	F	F	T	F	T
F	T	T	T	T	F	T
F	T	F	T	F	F	T
F	F	T	T	T	F	T
F	F	F	T	T	F	T

5. **a.** "The sandwich costs $3.50." **b.** Perhaps it's not true that if I have enough money I'll buy milk. Maybe I'm allergic to milk. Or maybe milk costs more than a dollar.

6. Step 5 involves dividing by zero, since $x = 1$ and $x - 1 = 0$.

Page 652 • EXERCISES

1. 1. Given 2. Step 1 and Double Negation 3. Given
4. Steps 2 and 3 and Modus Tollens

2. 1. Given 2. Step 1 and Distributive Rule 3. Step 2 and Simplification
4. Step 3 and Commutative Rule 5. Given 6. Steps 4 and 5 and Disjunctive Syllogism

3.

Statements	Reasons
1. $a \wedge (b \wedge c)$	1. Given
2. $(a \wedge b) \wedge c$	2. Step 1 and Associative Rule
3. $c \wedge (a \wedge b)$	3. Step 2 and Commutative Rule
4. c	4. Step 3 and Simplification

4.

Statements	Reasons
1. $(p \wedge q) \rightarrow s$	1. Given
2. $\sim s$	2. Given
3. $\sim (p \wedge q)$	3. Steps 1 and 2 and Modus Tollens
4. $\sim p \vee \sim q$	4. Step 3 and DeMorgan's Rule

5.

Statements	Reasons
1. $p \vee \sim q$	1. Given
2. $\sim q \vee p$	2. Step 1 and Commutative Rule
3. q	3. Given
4. $\sim (\sim q)$	4. Step 3 and Double Negation
5. p	5. Steps 2 and 4 and Disjunctive Syllogism

6.

Statements	Reasons
1. $\sim q \rightarrow \sim p$	1. Given
2. $p \rightarrow q$	2. Step 1 and Contrapositive Rule
3. p	3. Given
4. q	4. Steps 2 and 3 and Modus Ponens
5. $q \rightarrow r$	5. Given
6. r	6. Steps 4 and 5 and Modus Ponens

7.

Statements	Reasons
1. $p \vee (q \wedge s)$	1. Given
2. $(p \vee q) \wedge (p \vee s)$	2. Step 1 and Distributive Rule
3. $(p \vee s) \wedge (p \vee q)$	3. Step 2 and Commutative Rule
4. $p \vee s$	4. Step 3 and Simplification

8.

Statements	Reasons
1. $t \vee (r \vee s)$	1. Given
2. $(r \vee s) \vee t$	2. Step 1 and Commutative Rule
3. $\sim r \wedge \sim s$	3. Given
4. $\sim (r \vee s)$	4. Step 3 and DeMorgan's Rule
5. t	5. Steps 2 and 4 and Disjunctive Syllogism

9. Given: $c \rightarrow t$; $\sim c \rightarrow \sim s$; s

Prove: t

Statements	Reasons
1. $\sim c \rightarrow \sim s$	1. Given
2. s	2. Given
3. $\sim (\sim s)$	3. Step 2 and Double Negation
4. $\sim (\sim c)$	4. Steps 1 and 3 and Modus Tollens
5. c	5. Step 4 and Double Negation
6. $c \rightarrow t$	6. Given
7. t	7. Steps 5 and 6 and Modus Ponens

10. Given: $p \vee j$; $v \vee k$; $v \rightarrow r$; $k \rightarrow \sim p$; $\sim r$

Prove: j

Statements	Reasons
1. $v \rightarrow r$	1. Given
2. $\sim r$	2. Given
3. $\sim v$	3. Steps 1 and 2 and Modus Tollens
4. $v \vee k$	4. Given
5. k	5. Steps 3 and 4 and Disjunctive Syllogism
6. $k \rightarrow \sim p$	6. Given
7. $\sim p$	7. Steps 5 and 6 and Modus Ponens
8. $p \vee j$	8. Given
9. j	9. Steps 7 and 8 and Disjunctive Syllogism

Page 654 • EXERCISES

1. $p \wedge r$ **2.** $r \vee s$ **3.** $s \wedge (t \vee p)$ **4.** $(r \wedge p) \vee (\sim r \wedge q)$

5. $(t \vee s) \wedge (\sim t \vee s)$ **6.** $(r \vee s) \vee \sim r$

7. Electricity can always pass through the circuit $p \vee \sim p$ but can never pass through the circuit $p \wedge \sim p$.

p —— $\sim p$

$p \wedge \sim p$

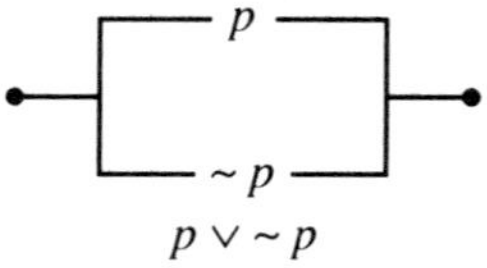

p

$\sim p$

$p \vee \sim p$

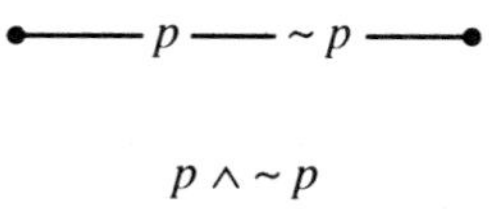

8.

9.

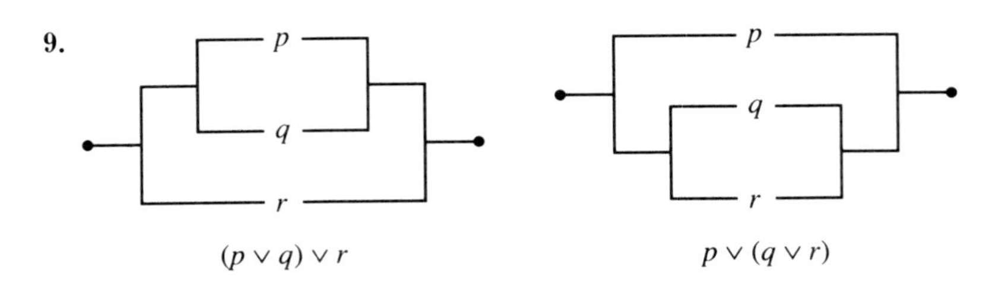

10.

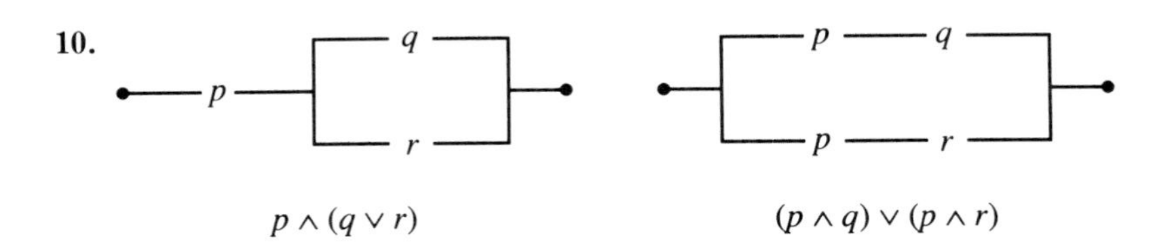

11.

p	q	$\sim q$	$p \vee q$	$(p \vee q) \vee \sim q$
T	T	F	T	T
T	F	T	T	T
F	T	F	T	T
F	F	T	F	T

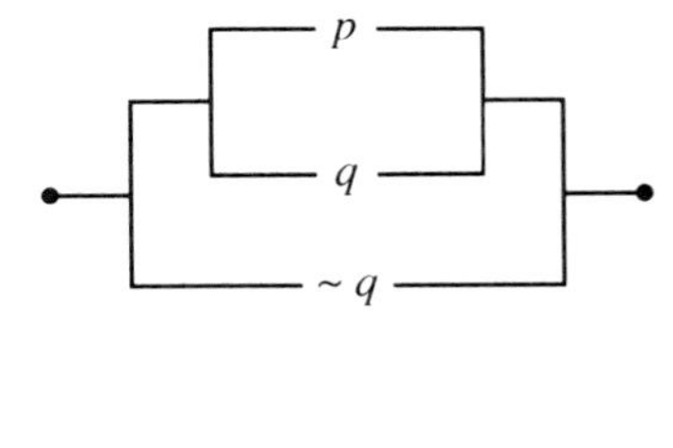

12.

p	q	$\sim q$	$p \vee q$	$p \vee \sim q$	$(p \vee q) \wedge (p \vee \sim q)$
T	T	F	T	T	T
T	F	T	T	T	T
F	T	F	T	F	F
F	F	T	F	T	F

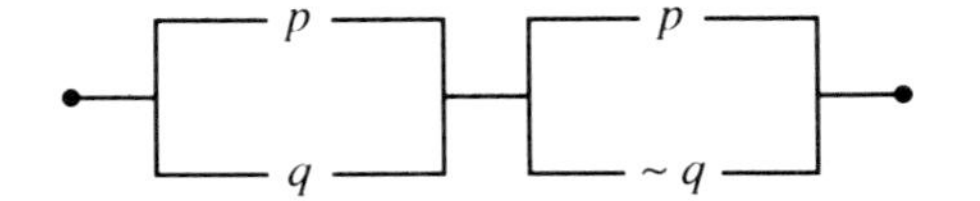

The circuit is equivalent to one that contains just switch p.

Flow Proofs

Page 656 • EXERCISES

1. Flow Proof:

$$\left.\begin{array}{l} 1.\ \overline{WO} \cong \overline{ZO} \\ 2.\ \overline{XO} \cong \overline{YO} \\ 3.\ \angle WOX \cong \angle ZOY \end{array}\right\} \rightarrow 4.\ \triangle WOX \cong \triangle ZOY \rightarrow 5.\ \angle W \cong \angle Z$$

Reasons

1. Given
2. Given
3. Vert. $\angle$s are $\cong$.
4. SAS Post.
5. Corr. parts of $\cong$ $\triangle$s are $\cong$.

2. Flow Proof:

$$1.\ \overline{AB} \parallel \overline{DC} \rightarrow \left.\begin{array}{l} 2.\ \angle BAC \cong \angle DCA \\ 3.\ \overline{AB} \cong \overline{DC} \\ 4.\ \overline{AC} \cong \overline{AC} \end{array}\right\} \rightarrow 5.\ \triangle ABC \cong \triangle CDA$$

Reasons

1. Given
2. If 2 $\parallel$ lines are cut by a trans., then alt. int. $\angle$s are $\cong$.
3. Given
4. Refl. Prop.
5. SAS Post.

3. Flow Proof:

$$1.\ \angle 1 \cong \angle 2 \rightarrow \left.\begin{array}{l} 2.\ \overline{ME} \cong \overline{MD} \\ 3.\ \angle 1 \cong \angle 3 \\ 4.\ \overline{EN} \cong \overline{DG} \end{array}\right\} \rightarrow 5.\ \triangle MEN \cong \triangle MDG \rightarrow 6.\ \angle 4 \cong \angle 5$$

Reasons

1. Given
2. If 2 $\angle$s of a $\triangle$ are $\cong$, then the sides opp. those $\angle$s are $\cong$.
3. Given
4. Given
5. SAS Post.
6. Corr. parts of $\cong$ $\triangle$s are $\cong$.

4. Flow Proof:

1. $t \perp l \rightarrow$ 2. $m\angle 1 = 90$
3. $l \parallel n \rightarrow$ 4. $m\angle 2 = m\angle 1$

} $\rightarrow$ 5. $m\angle 2 = 90 \rightarrow$ 6. $t \perp n$

Reasons

1. Given
2. Def. of $\perp$ lines
3. Given
4. If 2 $\parallel$ lines are cut by a trans., then corr. $\angle$s are $\cong$.
5. Substitution Prop.
6. Def. of $\perp$ lines

5. Flow Proof:

1. $\overline{RS} \perp \overline{ST}$; $\overline{TU} \perp \overline{ST} \rightarrow$ 2. $m\angle S = 90$; $m\angle T = 90 \rightarrow$ 3. $\angle S \cong \angle T$
4. V is the midpt. of $\overline{ST}$. $\rightarrow$ 5. $\overline{SV} \cong \overline{VT}$
6. $\angle RVS \cong \angle UVT$

} $\rightarrow$ 7. $\triangle RSV \cong \triangle UTV$

Reasons

1. Given
2. Def. of $\perp$ lines
3. Def. of $\cong$ $\angle$s
4. Given
5. Def. of midpt.
6. Vert. $\angle$s are $\cong$.
7. ASA Post.

6. Flow Proof:

1. $\overline{KL} \perp \overline{LA}$; $\overline{KJ} \perp \overline{JA} \rightarrow$ 2. $m\angle L = 90$; $m\angle J = 90 \rightarrow$ 3. $\angle L \cong \angle J$
4. $\overrightarrow{AK}$ bisects $\angle LAJ$. $\rightarrow$ 5. $\angle 1 \cong \angle 2$
6. $\overline{KA} \cong \overline{KA}$

} $\rightarrow$ 7. $\triangle LKA \cong \triangle JKA \rightarrow$ 8. $\overline{LK} \cong \overline{JK}$

Reasons

1. Given
2. Def. of $\perp$ lines
3. Def. of $\cong$ $\angle$s
4. Given
5. Def. of $\angle$ bis.
6. Refl. Prop.
7. AAS Thm.
8. Corr. parts of $\cong$ $\triangle$s are $\cong$.

7. Flow Proof:

1. $\overline{LF} \cong \overline{KF}$, 2. $\overline{LA} \cong \overline{KA}$, 3. $\overline{FA} \cong \overline{FA}$ $\rightarrow$ 4. $\triangle FLA \cong \triangle FKA \rightarrow$ 5. $\angle 1 \cong \angle 2$, 6. $\overline{LF} \cong \overline{KF}$, 7. $\overline{FJ} \cong \overline{FJ}$ $\rightarrow$ 8. $\triangle FLJ \cong \triangle FKJ \rightarrow$ 9. $\overline{LJ} \cong \overline{KJ}$

Reasons

1. Given
2. Given
3. Refl. Prop.
4. SSS Post.
5. Corr. parts of $\cong$ $\triangle$s are $\cong$.
6. See Step 1.
7. Refl. Prop.
8. SAS Post.
9. Corr. parts of $\cong$ $\triangle$s are $\cong$.

8. Flow Proof:

1. $\overline{AS} \parallel \overline{BT} \rightarrow$ {2. $m\angle 1 = m\angle 4$, 3. $m\angle 2 = m\angle 5$}, 4. $m\angle 4 = m\angle 5$ $\rightarrow$ 5. $m\angle 1 = m\angle 2 \rightarrow$ 6. $\overrightarrow{SA}$ bis. $\angle BSR$.

Reasons

1. Given
2. If 2 $\parallel$ lines are cut by a trans., then corr. $\angle$s are $\cong$.
3. If 2 $\parallel$ lines are cut by a trans., then alt. int. $\angle$s are $\cong$.
4. Given
5. Substitution Prop.
6. Def. of $\angle$ bis.

9. Flow Proof:

1. $\overline{SF} \cong \overline{SK} \rightarrow$ 2. $\angle F \cong \angle K$

3. $FL = AK$, 4. $LA = LA$ $\rightarrow$ 5. $FL + LA = AK + LA$

5. and 6. $FA = FL + LA; KL = AK + LA$ $\rightarrow$ 7. $FA = KL$

8. M is the midpt. of $\overline{SF}$. 9. N is the midpt. of $\overline{SK}$. $\rightarrow$ 10. $FM = \frac{1}{2}SF; KN = \frac{1}{2}SK$

11. $SF = SK \rightarrow$ 12. $\frac{1}{2}SF = \frac{1}{2}SK$

10. and 12. $\rightarrow$ 13. $FM = KN$

2., 7. and 13. $\rightarrow$ 14. $\triangle FAM \cong \triangle KLN \rightarrow$ 15. $\overline{AM} \cong \overline{LN}$

Reasons

1. Given
2. Isos. $\triangle$ Thm.
3. Given
4. Refl. Prop.
5. Add. Prop. of =
6. Seg. Add. Post.
7. Substitution Prop.
8. Given
9. Given
10. Midpt. Thm.
11. See Step 1.
12. Mult. Prop. of =
13. Substitution Prop.
14. SAS Post.
15. Corr. parts of $\cong$ ⚠s are $\cong$.

Pages 658–659 • EXERCISES

1. The sum is 360. In the tiling, the 4 $\angle$s about each vertex of one of the quads. are the 4 $\angle$s of the basic quad. Since the sum of the $\angle$ measures about a pt. is 360, the sum of the $\angle$ measures of the quad. is 360.
2. Yes; the upper left, upper right, and center quads.
3. **a.** All corr. $\angle$s of the transversals are $\cong$.

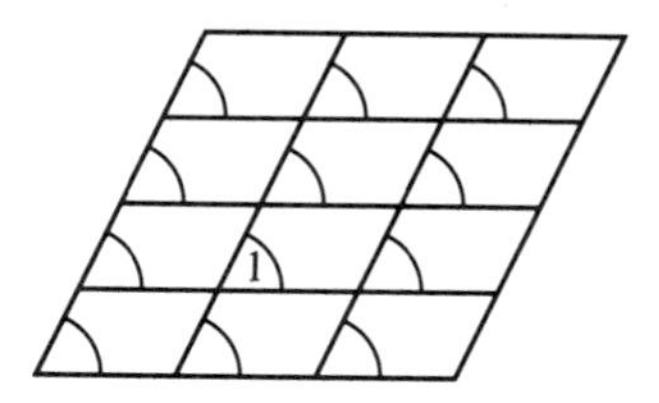

b. Vert. $\angle$s are $\cong$; Postulate 10

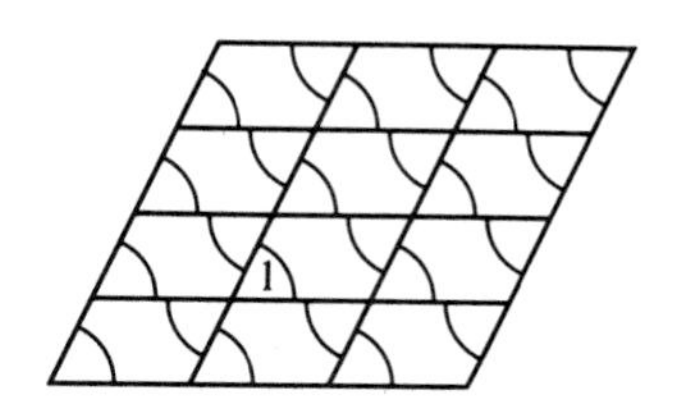

c. Opp. $\angle$s of a $\square$ are $\cong$.

4. **a.** 360; the pencil rotated through $\angle$s which are the supps. of $\angle$s A, B, and C. Since $m\angle A + m\angle B + m\angle C = 180$, $(180 - m\angle A) + (180 - m\angle B) + (180 - m\angle C) = 3 \cdot 180 - (m\angle A + m\angle B + m\angle C) = 540 - 180 = 360$.

   b. The sum of the measures of the ext. $\angle$s of a $\triangle$, one $\angle$ at each vertex, is 360.

5. **a.**

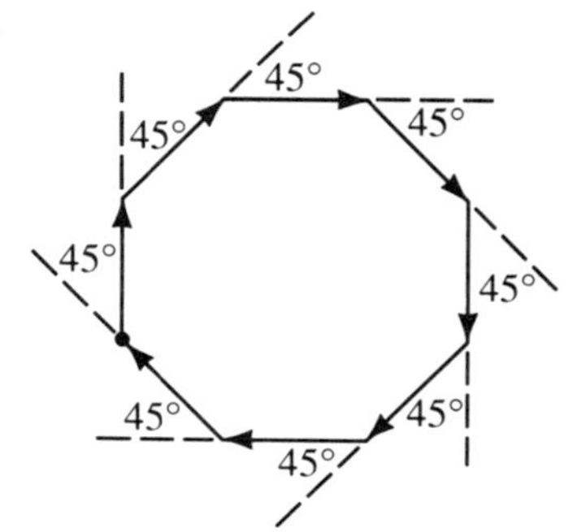

b. 360

6. 360; paths may vary.

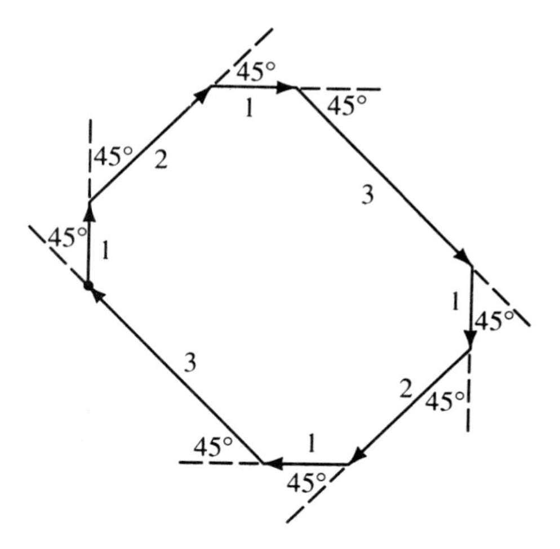

7. 360; paths may vary. The sum of the measures of the ext. $\angle$s of a convex polygon, one $\angle$ at each vertex, is 360.

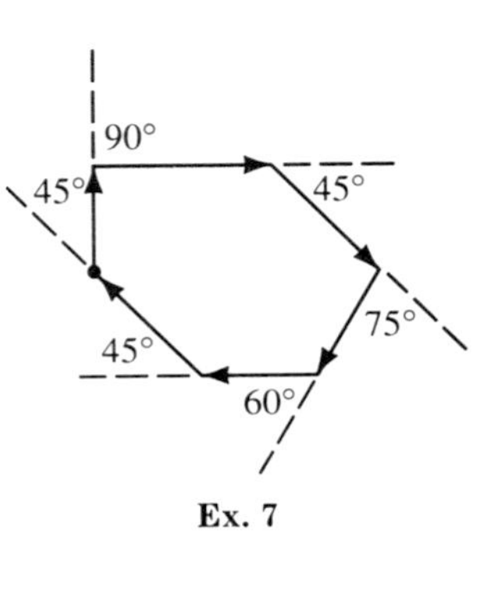

Ex. 7

Pages 659–660 • EXERCISES

1–4. Check students' cut-outs.

1.

2.

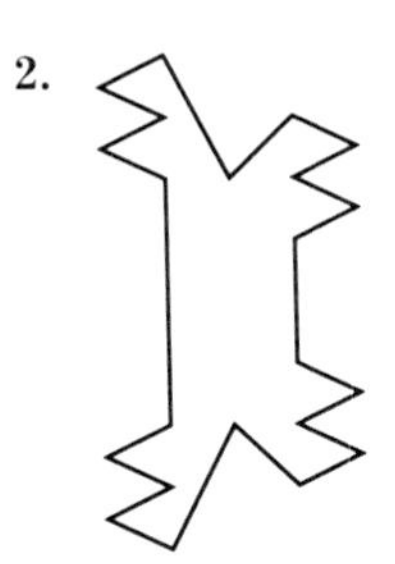

3.

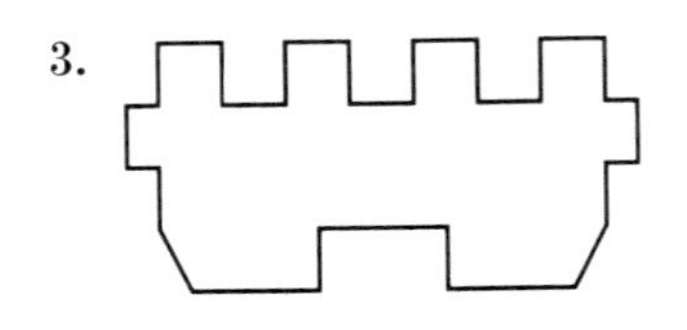

4.

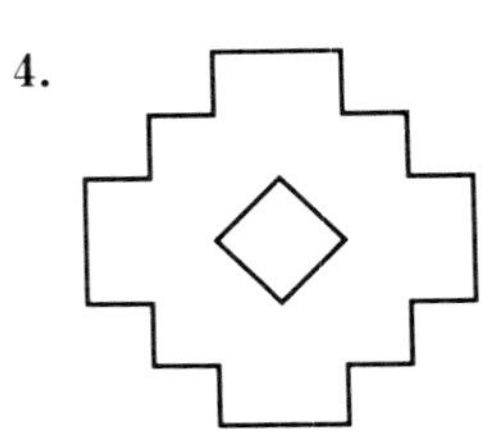

5. **a–b.** Check students' cut-outs. **c.** $\angle A \cong \angle B$; the base $\angle$s of an isos. $\triangle$ are $\cong$; the two $\angle$s coincided when the cut was made.

6. **a.** Rotation about O **b.** O is the midpt. of the seg. joining the eyes.

7. **a.** A translation maps fig. I to fig. III.

b. The distance between corr. pts. in figs. I and III is twice the distance between l and m.

8. Half of the figures will face left. Check students' cut-outs.

9. Cut a symmetric figure, or fan fold the paper from top to bottom.

Pages 661–662 • EXERCISES

1. *Method 1*: $AB = \sqrt{(0 - 5)^2 + (3 - 7)^2} = \sqrt{(-5)^2 + (-4)^2} = \sqrt{41}$;
$DC = \sqrt{(1 - 6)^2 + (-3 - 1)^2} = \sqrt{(-5)^2 + (-4)^2} = \sqrt{41}$;
$AD = \sqrt{(6 - 5)^2 + (1 - 7)^2} = \sqrt{1^2 + (-6)^2} = \sqrt{37}$;
$BC = \sqrt{(1 - 0)^2 + (-3 - 3)^2} = \sqrt{1^2 + (-6)^2} = \sqrt{37}$. So $\overline{AB} \cong \overline{DC}$ and $\overline{AD} \cong \overline{BC}$; therefore, $ABCD$ is a $\square$.

Method 2: Slope of $\overline{AB} = \frac{3 - 7}{0 - 5} = \frac{-4}{-5} = \frac{4}{5}$; slope of $\overline{DC} = \frac{-3 - 1}{1 - 6} = \frac{-4}{-5} = \frac{4}{5}$. Slope of $\overline{AD} = \frac{1 - 7}{6 - 5} = \frac{-6}{1} = -6$; slope of $\overline{BC} = \frac{-3 - 3}{1 - 0} = \frac{-6}{1} = -6$. So $\overline{AB} \parallel \overline{DC}$ and $\overline{AD} \parallel \overline{BC}$; therefore, $ABCD$ is a ▱.

Method 3: Midpt. of $\overline{AC} = \left(\frac{5 + 1}{2}, \frac{7 + (-3)}{2}\right) = (3, 2)$; midpt. of $\overline{BD} = \left(\frac{0 + 6}{2}, \frac{3 + 1}{2}\right) = (3, 2)$. So diags. $\overline{AC}$ and $\overline{BD}$ bisect each other; therefore, $ABCD$ is a ▱.

2. *Method 1*: $AB = \sqrt{(-3 - (-2))^2 + (2 - 6)^2} = \sqrt{(-1)^2 + (-4)^2} = \sqrt{17}$;
$DC = \sqrt{(2 - 3)^2 + (-4 - 0)^2} = \sqrt{(-1)^2 + (-4)^2} = \sqrt{17}$;
$AD = \sqrt{(3 - (-2))^2 + (0 - 6)^2} = \sqrt{5^2 + (-6)^2} = \sqrt{61}$;
$BC = \sqrt{(2 - (-3))^2 + (-4 - 2)^2} = \sqrt{5^2 + (-6)^2} = \sqrt{61}$. So $\overline{AB} \cong \overline{DC}$ and $\overline{AD} \cong \overline{BC}$; therefore, $ABCD$ is a ▱.

Method 2: Slope of $\overline{AB} = \frac{2 - 6}{-3 - (-2)} = \frac{-4}{-1} = 4$; slope of $\overline{DC} = \frac{-4 - 0}{2 - 3} = \frac{-4}{-1} = 4$. Slope of $\overline{AD} = \frac{0 - 6}{3 - (-2)} = \frac{-6}{5} = -\frac{6}{5}$; slope of $\overline{BC} = \frac{-4 - 2}{2 - (-3)} = \frac{-6}{5} = -\frac{6}{5}$. So $\overline{AB} \parallel \overline{DC}$ and $\overline{AD} \parallel \overline{BC}$; therefore, $ABCD$ is a ▱.

Method 3: Midpt. of $\overline{AC} = \left(\frac{-2 + 2}{2}, \frac{6 + (-4)}{2}\right) = (0, 1)$; midpt. of $\overline{BD} = \left(\frac{-3 + 3}{2}, \frac{2 + 0}{2}\right) = (0, 1)$. So diags. $\overline{AC}$ and $\overline{BD}$ bisect each other; therefore, $ABCD$ is a ▱.

3. Midpt. of $\overline{DF} = \left(\frac{3 + 3}{2}, \frac{5 + 4}{2}\right) = (3, 4.5)$; midpt. of $\overline{EG} = \left(\frac{5 + 0}{2}, \frac{7 + 1}{2}\right) = (2.5, 4)$. No; the diags. do not bisect each other, so $DEFG$ is not a ▱. (Method 1 or Method 2 could also be used.)

4. Midpt. of $\overline{DF} = \left(\frac{3 + 5}{2}, \frac{-2 + 6}{2}\right) = (4, 2)$; midpt. of $\overline{EG} = \left(\frac{-2 + 10}{2}, \frac{5 + (-1)}{2}\right) = (4, 2)$. Yes; the diags. bisect each other, so $DEFG$ is a ▱.

(Method 1 or Method 2 could also be used.)

5. $\overline{PS} \parallel \overline{QR}$; slope of $\overline{QR} = \frac{4 - 2}{8 - 5} = \frac{2}{3} = \frac{\text{change in } y}{\text{change in } x}$. So from Q to R, the change in x is $+3$ and the change in y is $+2$. Since $\overline{PS}$ and $\overline{QR}$ have the same slope and direction, from P to S, the change in x is $+3$ and the change in y is $+2$. $P = (0, 0)$, so $S = (0 + 3, 0 + 2) = (3, 2)$.

6. $\overline{PS} \parallel \overline{QR}$; slope of $\overline{PS} = \frac{5 - 0}{0 - (-2)} = \frac{5}{2} = \frac{\text{change in } y}{\text{change in } x}$. So from P to S, the change in x is $+2$ and the change in y is $+5$. Since $\overline{PS}$ and $\overline{QR}$ have the same slope and direction, from Q to R, the change in x is $+2$ and the change in y is $+5$. $Q = (2, 1)$, so $R = (2 + 2, 1 + 5) = (4, 6)$.

7. a. Check students' drawings. **b.** $M = \left(\frac{0 + 2}{2}, \frac{0 + 6}{2}\right) = (1, 3)$; $N = \left(\frac{4 + 2}{2}, \frac{2 + 6}{2}\right) = (3, 4)$. **c.** Slope of $\overline{MN} = \frac{4 - 3}{3 - 1} = \frac{1}{2}$; slope of $\overline{OI} = \frac{2 - 0}{4 - 0} = \frac{2}{4} = \frac{1}{2}$; $\overline{MN} \parallel \overline{OI}$; $OMNI$ is a trap.

8. a. Check students' drawings. **b.** $M = \left(\frac{0 + b}{2}, \frac{0 + c}{2}\right) = \left(\frac{b}{2}, \frac{c}{2}\right)$; $N = \left(\frac{a + b}{2}, \frac{0 + c}{2}\right) = \left(\frac{a + b}{2}, \frac{c}{2}\right)$ **c.** Slope of $\overline{MN} = \frac{\frac{c}{2} - \frac{c}{2}}{\frac{a + b}{2} - \frac{b}{2}} = \frac{0}{\frac{a}{2}} = 0$; slope of $\overline{OI} = \frac{0 - 0}{a - 0} = \frac{0}{a} = 0$; $\overline{MN} \parallel \overline{OI}$; $OMNI$ is a trap.

9. a. Slope of $\overline{AB} = \frac{-1 - 5}{4 - (-4)} = \frac{-6}{8} = -\frac{3}{4}$; slope of $\overline{DC} = \frac{3 - 9}{7 - (-1)} = \frac{-6}{8} = -\frac{3}{4}$; since their slopes are $=$, $\overline{AB} \parallel \overline{DC}$. Slope of $\overline{AD} = \frac{9 - 5}{-1 - (-4)} = \frac{4}{3}$; slope of $\overline{BC} = \frac{3 - (-1)}{7 - 4} = \frac{4}{3}$; since their slopes are $=$, $\overline{AD} \parallel \overline{BC}$. Slope of $\overline{AD}$ · slope of $\overline{AB} = \left(\frac{4}{3}\right)\left(-\frac{3}{4}\right) = -1$; slope of $\overline{AB}$ · slope of $\overline{BC} = \left(-\frac{3}{4}\right)\left(\frac{4}{3}\right) = -1$; slope of $\overline{BC}$ · slope of $\overline{DC} = \left(\frac{4}{3}\right)\left(-\frac{3}{4}\right) = -1$; slope of $\overline{DC}$ · slope of $\overline{AD} = \left(-\frac{3}{4}\right)\left(\frac{4}{3}\right) = -1$. Therefore, adjacent sides are $\perp$. **b.** Rectangle

10. a. slope of $\overline{RS} = \frac{0 - (-3)}{9 - 5} = \frac{3}{4}$; slope of $\overline{UT} = \frac{8 - 5}{3 - (-1)} = \frac{3}{4}$; slope of $\overline{ST} = \frac{8 - 0}{3 - 9} = \frac{8}{-6} = -\frac{4}{3}$; slope of $\overline{RU} = \frac{5 - (-3)}{-1 - 5} = \frac{8}{-6} = -\frac{4}{3}$. Thus $RSTU$ is a ▱. Also, slope of $\overline{RS}$ · slope of $\overline{RU} = \left(\frac{3}{4}\right)\left(-\frac{4}{3}\right) = -1$. Hence $\overline{RS} \perp \overline{RU}$ and $\angle R$ is a rt. $\angle$, so $RSTU$ is a rect. because a ▱ with one rt. $\angle$ is a rect.
b. $RT = \sqrt{(3 - 5)^2 + (8 - (-3))^2} = \sqrt{(-2)^2 + 11^2} = \sqrt{125} = 5\sqrt{5}$;
$SU = \sqrt{(-1 - 9)^2 + (5 - 0)^2} = \sqrt{(-10)^2 + 5^2} = \sqrt{125} = 5\sqrt{5}$

11. **a.** $DE = \sqrt{(2 - (-4))^2 + (3 - 1)^2} = \sqrt{6^2 + 2^2} = \sqrt{40} = 2\sqrt{10}$;
$EF = \sqrt{(4 - 2)^2 + (9 - 3)^2} = \sqrt{2^2 + 6^2} = \sqrt{40} = 2\sqrt{10}$;
$FG = \sqrt{(-2 - 4)^2 + (7 - 9)^2} = \sqrt{(-6)^2 + (-2)^2} = \sqrt{40} = 2\sqrt{10}$;
$GD = \sqrt{(-4 - (-2))^2 + (1 - 7)^2} = \sqrt{(-2)^2 + (-6)^2} = \sqrt{40} = 2\sqrt{10}$. So $\overline{DE} \cong \overline{EF} \cong \overline{FG} \cong \overline{GD}$, and $DEFG$ is a rhombus by def. **b.** Slope of $\overline{DF} = \dfrac{9 - 1}{4 - (-4)} = \dfrac{8}{8} = 1$; slope of $\overline{EG} = \dfrac{7 - 3}{-2 - 2} = \dfrac{4}{-4} = -1$; since $(1)(-1) = -1$, the diags. are $\perp$.

12. The resulting figure is a ▱ with base length and med. length = the sum of the base lengths of the trap., say $a + b$. This is twice the length of the med. of the trap., so the length of the med. of the trap. is $\dfrac{a + b}{2}$.

13–34. See solutions given earlier.

Pages 663–664 • EXERCISES

1.

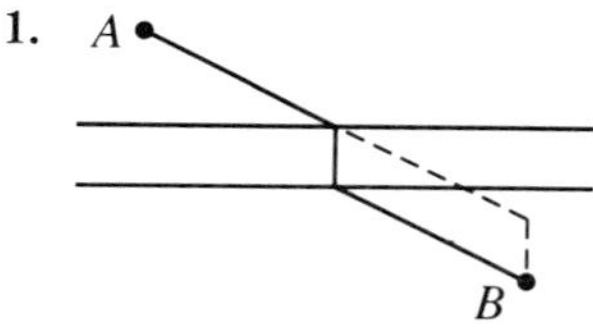

2.

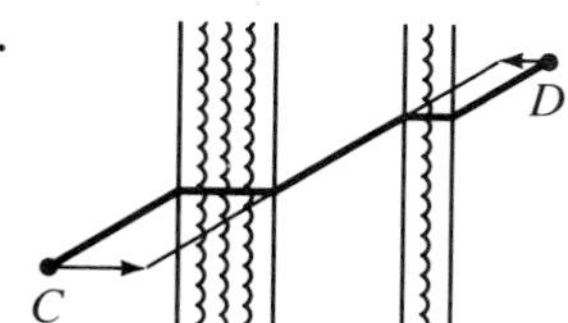

3. Construct the bridge between (3, 3) and (4, 3).

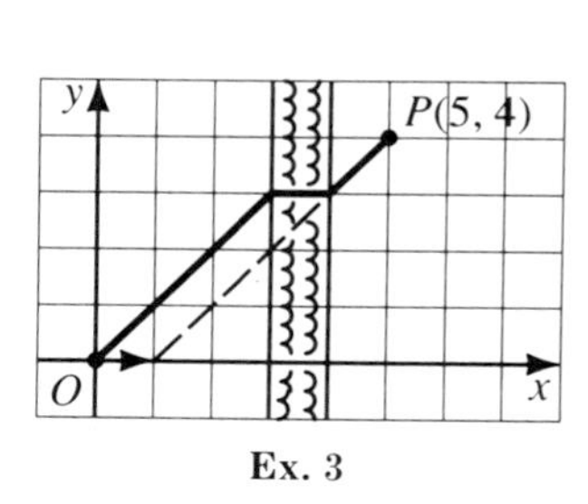

Ex. 3

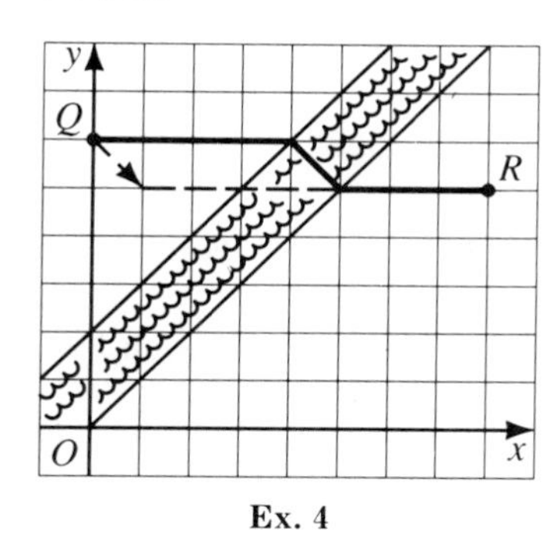

Ex. 4

4. Construct the bridge between (4, 6) and (5, 5).

5.

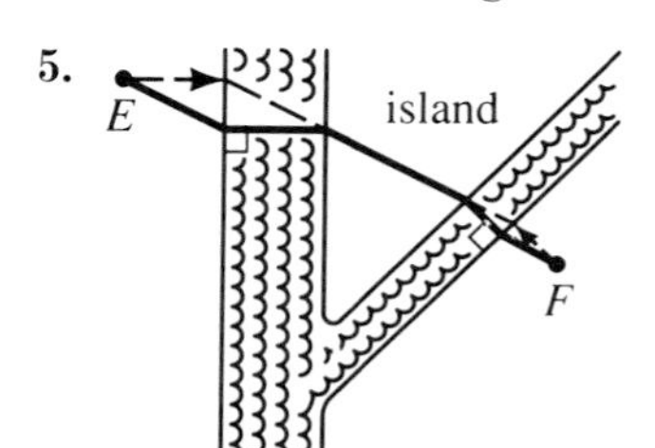

The non-bridge portions of the path are $\parallel$.

6. Let NT be the shortest distance, with T on the given line. Slope of $\overline{NT} = -3$, and the equation of $\overleftrightarrow{NT}$ is $y - 10 = -3(x - 0)$ or $y = -3x + 10$. Solving the equations of the 2 lines simultaneously gives $\frac{1}{3}x = -3x + 10$ or $x = -9x + 30$, so $x = 3$ and $y = \frac{1}{3}x = 1$. Thus $T = (3, 1)$ and $NT = \sqrt{(3 - 0)^2 + (1 - 10)^2} = \sqrt{3^2 + (-9)^2} = \sqrt{90} = 3\sqrt{10}$.
7. Let OR be the shortest distance, with R on the given line. Slope of $\overline{OR} = -\frac{1}{2}$, and the equation of $\overleftrightarrow{OR}$ is $y = -\frac{1}{2}x$. Solving the 2 equations simultaneously gives $-\frac{1}{2}x = 2x + 5$ or $-x = 4x + 10$, so $x = -2$ and $y = -\frac{1}{2}x = 1$. Thus $R = (-2, 1)$ and $OR = \sqrt{(-2 - 0)^2 + (1 - 0)^2} = \sqrt{(-2)^2 + 1^2} = \sqrt{5}$.
8. Let PQ be the shortest distance, with Q on the given line. Slope of $\overline{PQ} = -\frac{3}{2}$, and the equation of $\overleftrightarrow{PQ}$ is $y - (-1) = -\frac{3}{2}(x - 2)$ or $y + 1 = -\frac{3}{2}x + 3$ or $y = -\frac{3}{2}x + 2$. Solving the 2 equations simultaneously gives $-\frac{3}{2}x + 2 = \frac{2}{3}x + 2$ or $-\frac{3}{2}x = \frac{2}{3}x$ or $-9x = 4x$ or $13x = 0$, so $x = 0$ and $y = -\frac{3}{2}x + 2 = 0 + 2 = 2$. Thus $Q = (0, 2)$ and $PQ = \sqrt{(0 - 2)^2 + (2 - (-1))^2} = \sqrt{(-2)^2 + 3^2} = \sqrt{13}$.

Pages 664–665 • EXERCISES

1. Scale factor: about $\frac{2}{3}$; rotation through 90°
2. Scale factor: $-\frac{1}{2}$; rotation through 180°
3. $\overrightarrow{RS} \parallel \overrightarrow{R'S'}$ because a dilation maps any ray onto a ray $\parallel$ to itself; since the scale factor is positive, the 2 rays have the same direction.
4. $\overrightarrow{JK} \parallel \overrightarrow{J'K'}$ because a dilation maps any ray onto a ray $\parallel$ to itself; since the scale factor is negative, the 2 rays are oppositely directed.
5. Under a dilation of scale factor k, areas are related by a scale factor of k^2. Thus $D_1 : \text{I} \rightarrow \text{II}$ has a scale factor $= \sqrt{\frac{100}{25}} = \sqrt{4} = 2$, and $D_2 : \text{II} \rightarrow \text{III}$ has a scale factor $= \sqrt{\frac{900}{100}} = \sqrt{9} = 3$.

6. Dilate $WXYZ$ with center B until Z is on $\overline{AC}$. Since W' is on $\overrightarrow{BW}$, X' is on $\overrightarrow{BX}$ and Y' is on $\overrightarrow{BY}$, all 4 vertices of $W'X'Y'Z'$ will be on $\triangle ABC$.

7. $D_{0,3}$ maps l onto a line $\parallel$ to itself, so the slope of l' is 2. Also, the y-int. of l, (0, 1), is mapped onto (0, 3), the y-int. of l'. Thus the equation of l' is $y = 2x + 3$.

8. a. Check students' drawings. **b.** The scale factor is 4 since $A''B'' = 2 \cdot A'B' = 2 \cdot (2 \cdot AB) = 4 \cdot AB$. **c.** Check students' drawings. The center is on $\overline{PQ}$, about $\frac{1}{3}$ of the way from P to Q.

9, 10. See solutions given earlier.

Pages 666–667 • EXERCISES

1. $OP = \sqrt{(8 - 0)^2 + (4 - 0)^2} = \sqrt{64 + 16} = \sqrt{80}$, so $(OP)^2 = 80$;
$PQ = \sqrt{(6 - 8)^2 + (8 - 4)^2} = \sqrt{4 + 16} = \sqrt{20}$, so $(PQ)^2 = 20$;
$OQ = \sqrt{(6 - 0)^2 + (8 - 0)^2} = \sqrt{36 + 64} = \sqrt{100}$, so $(OQ)^2 = 100$.
Since $(OQ)^2 = (OP)^2 + (PQ)^2$, $\triangle OPQ$ is a rt. $\triangle$.

2. $OR = \sqrt{(-1 - 0)^2 + (5 - 0)^2} = \sqrt{1 + 25} = \sqrt{26}$, so $(OR)^2 = 26$;
$RS = \sqrt{(-6 - (-1))^2 + (0 - 5)^2} = \sqrt{25 + 25} = \sqrt{50}$, so $(RS)^2 = 50$;
$OS = \sqrt{(-6 - 0)^2 + (0 - 0)^2} = \sqrt{36}$, so $(OS)^2 = 36$. Since $(RS)^2 < (OR)^2 + (OS)^2$, $\triangle ORS$ is acute.

3. $AB = \sqrt{(3 - 1)^2 + (-2 - 1)^2} = \sqrt{4 + 9} = \sqrt{13}$, so $(AB)^2 = 13$;
$BC = \sqrt{(8 - 3)^2 + (1 - (-2))^2} = \sqrt{25 + 9} = \sqrt{34}$, so $(BC)^2 = 34$;
$AC = \sqrt{(8 - 1)^2 + (1 - 1)^2} = \sqrt{49 + 0} = \sqrt{49}$, so $(AC)^2 = 49$. Since $(AC)^2 > (AB)^2 + (BC)^2$, $\triangle ABC$ is obtuse.

4. $DE = \sqrt{(0 - (-5))^2 + (7 - 0)^2} = \sqrt{25 + 49} = \sqrt{74}$, so $(DE)^2 = 74$;
$EF = \sqrt{(3 - 0)^2 + (-3 - 7)^2} = \sqrt{9 + 100} = \sqrt{109}$, so $(EF)^2 = 109$;
$DF = \sqrt{(3 - (-5))^2 + (-3 - 0)^2} = \sqrt{64 + 9} = \sqrt{73}$, so $(DF)^2 = 73$.
Since $(EF)^2 < (DE)^2 + (DF)^2$, $\triangle DEF$ is acute.

5. a. slope of $\overline{AB} = \frac{3 - 0}{5 - 3} = \frac{3}{2}$; slope of $\overline{BC} = \frac{7 - 3}{-1 - 5} = \frac{4}{-6} = -\frac{2}{3}$. Since slope of $\overline{AB} \cdot$ slope of $\overline{BC} = \left(\frac{3}{2}\right)\left(-\frac{2}{3}\right) = -1$, $\overline{AB} \perp \overline{BC}$, so $\angle B$ is a rt. $\angle$ and $\triangle ABC$ is a rt. $\triangle$. **b.** $AB = \sqrt{(5 - 3)^2 + (3 - 0)^2} = \sqrt{4 + 9} = \sqrt{13}$, so $(AB)^2 = 13$;
$BC = \sqrt{(-1 - 5)^2 + (7 - 3)^2} = \sqrt{36 + 16} = \sqrt{52}$, so $(BC)^2 = 52$;
$AC = \sqrt{(-1 - 3)^2 + (7 - 0)^2} = \sqrt{16 + 49} = \sqrt{65}$, so $(AC)^2 = 65$.
Since $(AC)^2 = (AB)^2 + (BC)^2$, $\triangle ABC$ is a rt. $\triangle$.

6. a. The hypotenuse is $\overline{AC}$. $M = \left(\frac{3 + (-1)}{2}, \frac{0 + 7}{2}\right) = \left(1, \frac{7}{2}\right)$

b. $MB = \sqrt{(5 - 1)^2 + \left(3 - \frac{7}{2}\right)^2} = \sqrt{16 + \frac{1}{4}} = \sqrt{\frac{65}{4}} = \frac{1}{2}\sqrt{65}$; from Ex. 5(b), $AC = \sqrt{65}$ so $\frac{1}{2}AC = \frac{1}{2}\sqrt{65}$. Thus $MB = \frac{1}{2}AC$. **c.** Thm. 5-15: The midpt. of the hyp. of a rt. $\triangle$ is equidistant from the three vertices.

7. Use $\mathscr{R}_{N,\,-90}$ followed by $D_{N,2}$; the scale factor is 2.

8. a. Check students' drawings. **b.** $\overrightarrow{OQ} = \overrightarrow{OP} + \overrightarrow{PQ} = (4, 3) + (-2, 11) = (2, 14)$

c. $|\overrightarrow{OP}| = \sqrt{(4 - 0)^2 + (3 - 0)^2} = \sqrt{16 + 9} = \sqrt{25} = 5$;

$|\overrightarrow{PQ}| = \sqrt{(2 - 4)^2 + (14 - 3)^2} = \sqrt{4 + 121} = \sqrt{125} = 5\sqrt{5}$;

$|\overrightarrow{OQ}| = \sqrt{(2 - 0)^2 + (14 - 0)^2} = \sqrt{4 + 196} = \sqrt{200} = 10\sqrt{2}$

d. No; $(OQ)^2 > (OP)^2 + (PQ)^2$, so $\triangle OPQ$ is obtuse.

9. The area of the large square is c^2. The figure at the right shows that the dashed line divides the area into two squares of area b^2 and a^2. Hence $c^2 = a^2 + b^2$, and the Pythagorean Theorem is suggested.

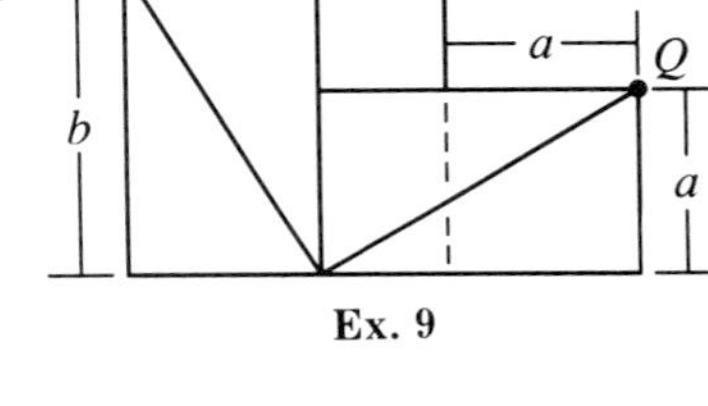

Ex. 9

10. a. The common center of the 2 squares; 45° or any odd multiple of 45°. **b.** The common center of the 2 squares; let a be the length of a side of the outer square, then the scale factor is

$$\frac{\text{half the diag. of the inner sq.}}{\text{half the diag. of the outer sq.}} = \frac{\frac{a}{2}}{\frac{a\sqrt{2}}{2}} = \frac{\sqrt{2}}{2}.$$

11–21. See solutions given earlier.

Pages 668–669 • EXERCISES

1. a. $(x - 6)^2 + y^2 = 25$ **b.** Yes, since $(2 - 6)^2 + 3^2 = 25$.

c. No; check students' drawings.

2–4. Check students' sketches.

2. a. $x^2 + y^2 = 25$; $0^2 + 5^2 = 25$; $4^2 + 3^2 = 16 + 9 = 25$

b. $M = \left(\frac{0 + 4}{2}, \frac{5 + 3}{2}\right) = (2, 4)$ **c.** Slope of $\overline{OM} = \frac{4 - 0}{2 - 0} = 2$; slope of $\overline{AB} = \frac{3 - 5}{4 - 0} = \frac{-2}{4} = -\frac{1}{2}$. Since slope of $\overline{OM}$ · slope of $\overline{AB} = 2\left(-\frac{1}{2}\right) = -1$, $\overline{OM} \perp \overline{AB}$.

3. a. $x^2 + y^2 = 100$; $6^2 + 8^2 = 36 + 64 = 100$; $(-8)^2 + 6^2 = 64 + 36 = 100$

b. $M = \left(\frac{6 + (-8)}{2}, \frac{8 + 6}{2}\right) = (-1, 7)$ **c.** slope of $\overline{OM} = \frac{7 - 0}{-1 - 0} = -7$;

slope of $\overline{AB} = \frac{6 - 8}{-8 - 6} = \frac{-2}{-14} = \frac{1}{7}$. Since slope of $\overline{OM}$ · slope of $\overline{AB}$ =

$(-7)\left(\frac{1}{7}\right) = -1$, $\overline{OM} \perp \overline{AB}$.

4. a. $x^2 + y^2 = (5\sqrt{2})^2 = 50$; $5^2 + 5^2 = 25 + 25 = 50$; $(-7)^2 + 1^2 = 49 + 1 = 50$

b. $M = \left(\frac{5 + (-7)}{2}, \frac{5 + 1}{2}\right) = (-1, 3)$ **c.** slope of $\overline{OM} = \frac{3 - 0}{-1 - 0} = -3$;

slope of $\overline{AB} = \frac{1 - 5}{-7 - 5} = \frac{-4}{-12} = \frac{1}{3}$. Since slope of $\overline{OM}$ · slope of $\overline{AB}$ =

$(-3)\left(\frac{1}{3}\right) = -1$, $\overline{OM} \perp \overline{AB}$.

5. Check students' sketches. $r_1 = \sqrt{225} = 15$; $r_2 = \sqrt{25} = 5$. $C_1 = (0, 0)$; $C_2 = (6, 8)$; $C_1C_2 = \sqrt{(6 - 0)^2 + (8 - 0)^2} = \sqrt{36 + 64} = \sqrt{100} = 10$. Let $\overrightarrow{C_1C_2}$ intersect the first ⊙ at P. Then $C_1P = 15$ and $C_1C_2 = 10$, so $C_2P = 15 - 10 = 5$. Thus C_2P is a radius of the second ⊙, so P is on the second ⊙ and the ⊙s are internally tangent at P.

6. a, c. Check students' sketches. **b.** $x^2 + (2x - 5)^2 = 25$; $x^2 + 4x^2 - 20x + 25 = 25$; $5x^2 - 20x = 0$; $5x(x - 4) = 0$; $x = 0$ or $x = 4$. If $x = 0$, $y = 2x - 5 = 0 - 5 = -5$. If $x = 4$, $y = 2x - 5 = 2 \cdot 4 - 5 = 3$. The solutions are $(0, -5)$ and $(4, 3)$.

7. a. Circle O; PA', the other tan. to ⊙O from P. **b.** $\overline{PA} \cong \overline{PA'}$ **c.** Tangents to a circle from a pt. are ≅.

8. a. B **b.** $\overline{AB} \perp \overline{PQ}$ because the line of reflection is the ⊥ bis. of every seg. joining a pt. and its image. **c.** Check students' sketches. $\overline{X'Y'}$ is the other common ext. tan. **d.** Common external tangents of two circles are ≅.

9. Rotate C about A through $-60°$ to find B. Then B and C are images of each other under the two rotations about A; since C is on the image of ⊙P under the first rotation, B is on ⊙P. Also $AC = AB$ and $m\angle ABC = m\angle ACB$. But $m\angle CAB = 60$; thus each ∠ of △ABC measures 60.

10–18. See solutions given earlier.

Page 670 • EXERCISES

1–4. Check students' constructions. The medians intersect at G, the altitudes at H, and the $\perp$ bisectors of the sides at C.

1. Yes **2.** $HG : GC = 2 : 1$ **3.** $1 : 4$

5. $|x| + |y| = 10$ or $|y| = -|x| + 10$

Quad. I: $y = -x + 10$; Quad. II: $y = x + 10$;

Quad. III: $-y = x + 10$ or $y = -x - 10$;

Quad. IV: $-y = -x + 10$ or $y = x - 10$

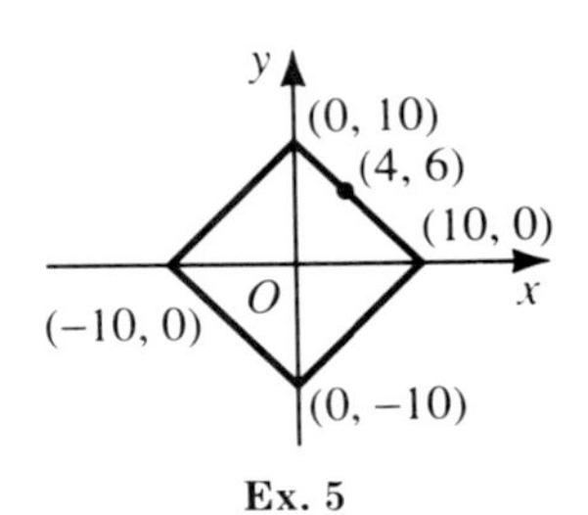

Ex. 5

6. The image of $\odot P$ under a half-turn about M intersects l at 2 pts., Y_1 and Y_2. Their preimages are the required X_1, X_2.

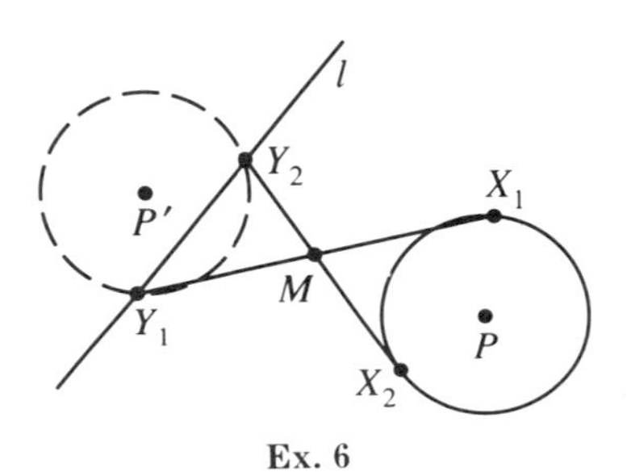

Ex. 6

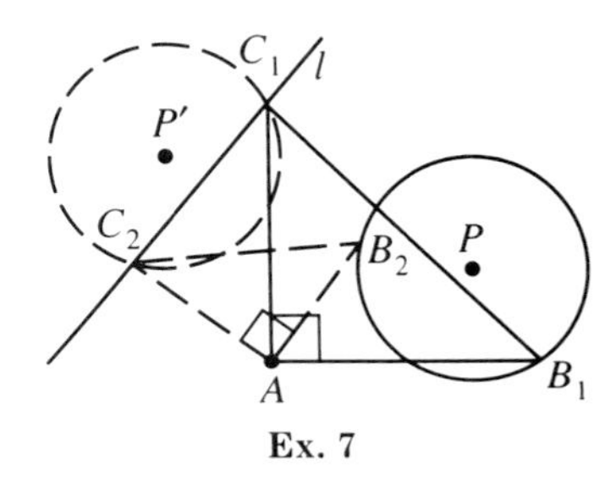

Ex. 7

7. Rotate $\odot P$ 90° about A to locate 2 pts., C_1 and C_2; then rotate each $-90°$ about A to get B_1 and B_2.

8–16. See solutions given earlier.

Pages 671–672 • EXERCISES

1. $A = 32$; rotate the shaded area through 90° about P. Use a similar rotation on the left side. Then the area of the rect. formed is $4 \cdot 8 = 32$.

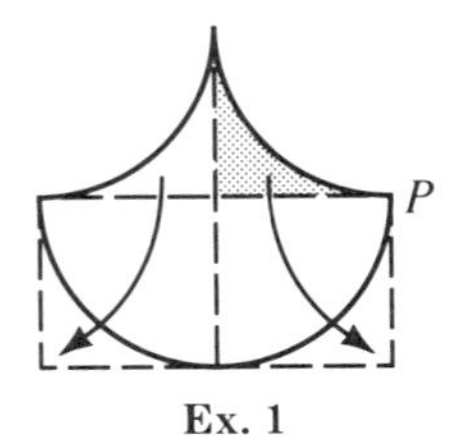

Ex. 1

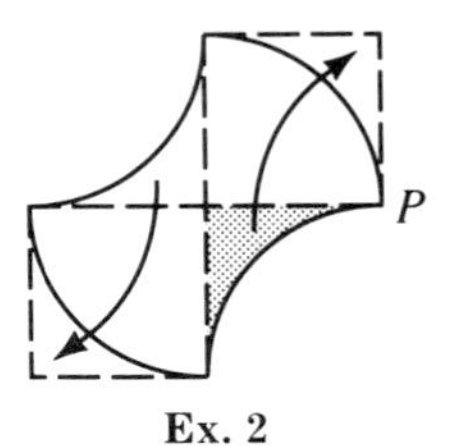

Ex. 2

2. $A = 32$; rotate the shaded area $-90°$ about P. Use a similar rotation on the other side of the figure. Then the figure formed consists of 2 squares, each 4 units on a side; their combined area is $2 \cdot 4 \cdot 4 = 32$.

3. $A = 64$; rotate the shaded area 90° about P; use a similar rotation on the left side of the figure. The figure formed is a square 8 units on a side with area $= 8 \cdot 8 = 64$.

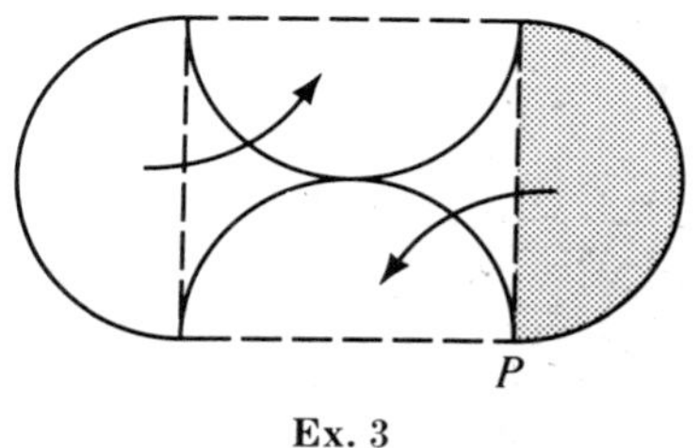

Ex. 3

4. **a.** -45; $\frac{\sqrt{2}}{2}$ **b.** III, IV

c. Region I: $\frac{1}{2} \cdot 4 \cdot 4 = 8$; since the ratio of the dilation is $\frac{\sqrt{2}}{2}$, the ratio of the areas is $\left(\frac{\sqrt{2}}{2}\right)^2 = \frac{2}{4} = \frac{1}{2}$. Thus the areas for regions II–V are 4, 2, 1, and $\frac{1}{2}$, resp.

5. **a.** -60; $\frac{1}{2}$ **b.** II, III, IV **c.** The ratio of the areas is $\left(\frac{1}{2}\right)^2 = \frac{1}{4}$. So the areas for regions II–IV are $\frac{1}{4}$, $\frac{1}{16}$, and $\frac{1}{64}$, resp.

6. Area $= 5$

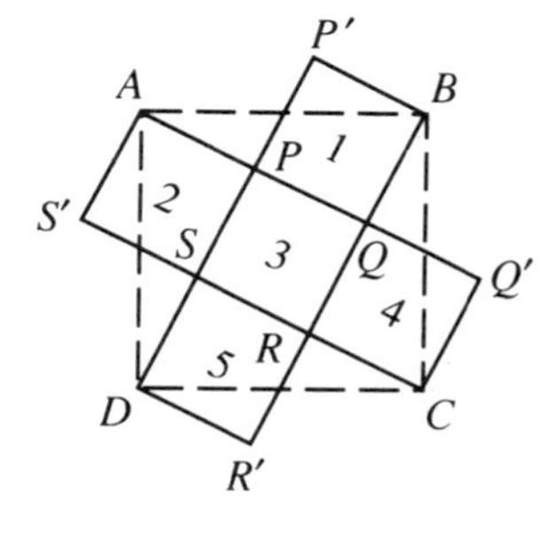

7. Area of $\triangle ABC = 7$ (area of $\triangle XYZ$)

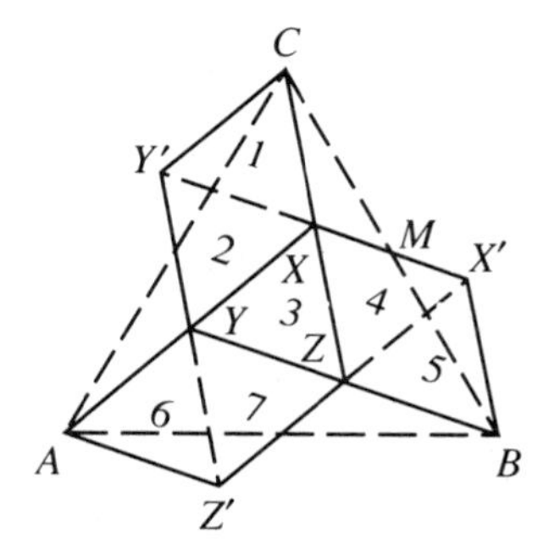

8–10. See solutions given earlier.

Pages 674–675 • EXERCISES

1. **a.**

Statements	Reasons
1. $\overline{AC}$ is the $\perp$ bis. of $\overline{BD}$.	1. Given
2. $AD = AB$ and $CD = CB$	2. If a pt. lies on the $\perp$ bis. of a seg., then the pt. is equidistant from the endpts. of the seg.

b. Reflect $\triangle ADC$ in $\overline{AC}$. Since $\overline{AC}$ is the $\perp$ bis. of $\overline{BD}$, D is mapped to B; A and C are mapped to themselves. Then the image of $\triangle ADC$ is $\triangle ABC$. The isometry preserves distance, so $AD = AB$ and $CD = CB$. **c.** Assign coordinates $A(0, a)$, $B(b, 0)$, $C(0, -c)$, and $D(-b, 0)$. Using the distance formula, $AD = \sqrt{b^2 + a^2}$, $AB = \sqrt{(-b)^2 + a^2}$, $CD = \sqrt{b^2 + (-c)^2}$, and $BC = \sqrt{b^2 + c^2}$. So $AD = AB$ and $CD = CB$.

2. a.

Statements	Reasons
1. O is the midpt. of $\overline{AA'}$, $\overline{BB'}$, and $\overline{CC'}$.	1. Given
2. $\overline{OC} \cong \overline{OC'}$; $\overline{OB} \cong \overline{OB'}$; $\overline{OA} \cong \overline{OA'}$	2. Def. of midpt.
3. $\angle AOC \cong \angle A'OC'$; $\angle AOB \cong \angle A'OB'$; $\angle COB \cong \angle C'OB'$	3. Vert. $\angle$s are $\cong$.
4. $\triangle AOC \cong \triangle A'OC'$; $\triangle AOB \cong \triangle A'OB'$; $\triangle COB \cong \triangle C'OB'$	4. SAS Post.
5. $\overline{AC} \cong \overline{A'C'}$; $\overline{AB} \cong \overline{A'B'}$; $\overline{BC} \cong \overline{B'C'}$	5. Corr. parts of $\cong$ $\triangle$s are $\cong$.
6. $\triangle ABC \cong \triangle A'B'C'$	6. SSS Post.

b. Rotate $\triangle ABC$ 180° about O. Since O is the midpt. of $\overline{BB'}$, B is mapped to B'. Similarly, A is mapped to A' and C is mapped to C'. Then the image of $\triangle ABC$ is $\triangle A'B'C'$. A rotation is an isometry, and an isometry maps a $\triangle$ to a $\cong$ $\triangle$, so $\triangle ABC \cong \triangle A'B'C'$.

3. a.

Statements	Reasons
1. $\triangle ABX$ and $\triangle BCY$ are equilateral.	1. Given
2. $\overline{XB} \cong \overline{AB}$; $\overline{BC} \cong \overline{BY}$	2. Def. of equilateral $\triangle$
3. $m\angle XBA = m\angle CBY = 60$	3. An equilateral $\triangle$ has 3 60° $\angle$s.
4. $m\angle ABC = m\angle ABC$	4. Refl. Prop.
5. $m\angle XBA + m\angle ABC = m\angle CBY + m\angle ABC$	5. Add. Prop. of =
6. $m\angle XBC = m\angle XBA + m\angle ABC$; $m\angle ABY = m\angle CBY + m\angle ABC$	6. $\angle$ Add. Post.
7. $m\angle XBC = m\angle ABY$ or $\angle XBC \cong \angle ABY$	7. Substitution Prop.
8. $\triangle XBC \cong \triangle ABY$	8. SAS Post.
9. $\overline{XC} \cong \overline{AY}$, or $AY = XC$	9. Corr. parts of $\cong$ $\triangle$s are $\cong$.

b. $\triangle ABX$ and $\triangle BCY$ are equilateral, so each $\triangle$ has 3 60° $\angle$s. $\mathscr{R}_{B,-60}$ maps Y to C since $m\angle CBY = 60$, and $\mathscr{R}_{B,-60}$ maps A to X since $m\angle XBA = 60$. Then the image of $\overline{AY}$ under $\mathscr{R}_{B,-60}$ is $\overline{XC}$, and $AY = XC$ since a rotation is an isometry.

4. a. A', B', and C' are the midpts. of the sides of $\triangle ABC$, so $A'B' = \frac{1}{2}AB$, $A'C' = \frac{1}{2}AC$, and $B'C' = \frac{1}{2}BC$, since the seg. joining the midpts. of 2 sides of a $\triangle$ is half as long as the third side. Then $\frac{AB}{A'B'} = \frac{2}{1}$, $\frac{AC}{A'C'} = \frac{2}{1}$, and $\frac{BC}{B'C'} = \frac{2}{1}$, and by the SSS $\sim$ Thm., $\triangle ABC \sim \triangle A'B'C'$ with scale factor $2 : 1$. The ratio of areas is $2^2 : 1^2$, or $4 : 1$.

b. The center of the dilation is the intersection of medians $\overline{AA'}$, $\overline{BB'}$, and $\overline{CC'}$. Let the center be pt. M. Since A' lies on the ray opposite $\overrightarrow{MA}$, the dilation must be negative. The medians of a $\triangle$ intersect in a pt. that is two thirds of the distance from each vertex to the midpt. of the opp. side (Thm. 10-4), so $AM : MA' = BM : MB' = CM : MC' = 2 : 1$. Then the scale factor of the dilation is $-\frac{1}{2}$, and the ratio of areas is $1 : \left(-\frac{1}{2}\right)^2 = 4 : 1$.

5–20. Methods of proof may vary.

5. Square. Proof (coordinate approach): $WX = \sqrt{(-a)^2 + a^2} = a\sqrt{2}$; $XY = \sqrt{a^2 + a^2} = a\sqrt{2}$; $YZ = \sqrt{a^2 + (-a)^2} = a\sqrt{2}$; $ZW = \sqrt{(-a)^2 + (-a)^2} = a\sqrt{2}$; so $WXYZ$ has 4 $\cong$ sides. Slope of $\overline{WX} = \frac{a - 0}{0 - a} = \frac{a}{-a} = -1$; slope of $\overline{XY} = \frac{0 - (-a)}{a - 0} = \frac{a}{a} = 1$; slope of $\overline{YZ} = \frac{-a - 0}{0 - (-a)} = \frac{-a}{a} = -1$; slope of $\overline{ZW} = \frac{0 - a}{-a - 0} = \frac{-a}{-a} = 1$; so $\overline{WX} \perp \overline{XY}$, $\overline{XY} \perp \overline{YZ}$, $\overline{YZ} \perp \overline{ZW}$, $\overline{ZW} \perp \overline{WX}$, and $WXYZ$ has 4 rt. ∠s. Therefore, $WXYZ$ is a square by def.

Alternate proof (transformational approach): Use a rotation and a dilation to map $ABCD$ to $A'B'C'D'$. The rotation is 45° about P, the pt. of intersection of the diags. of $ABCD$. The center of the dilation is also P. To find the scale factor of the dilation, consider $\triangle A'AB'$. Since A' and B' are midpts. of $\cong$ sides, $\overline{A'A} \cong \overline{AB'}$, and $\triangle A'AB'$ is a 45°–45°–90° $\triangle$. Let $AB' = x$; then $A'B' = x\sqrt{2}$ and $AB = 2x$. So $\frac{A'B'}{AB} = \frac{x\sqrt{2}}{2x} = \frac{\sqrt{2}}{2}$, and the scale factor of the dilation is $\frac{\sqrt{2}}{2}$.

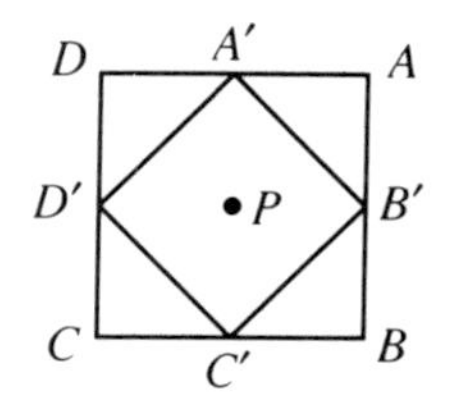

Therefore, $D_{P,\frac{\sqrt{2}}{2}} \circ \mathscr{R}_{P,45} : ABCD \rightarrow A'B'C'D'$, and $A'B'C'D'$ is a square.

6. Rectangle. Proof (coordinate approach): Assign coordinates as shown. Slope of $\overline{ZW} = 0 =$ slope of $\overline{YX}$, so both $\overline{ZW}$ and $\overline{YX}$ are horizontal. Slope of $\overline{ZY}$ and slope of $\overline{WX}$ are undefined, so both $\overline{ZY}$ and $\overline{WX}$ are vertical. $\overline{ZW} \perp \overline{WX}$, $\overline{WX} \perp \overline{XY}$, $\overline{XY} \perp \overline{YZ}$, and $\overline{YZ} \perp \overline{ZW}$, so $WXYZ$ has 4 rt. ∠s and is a rect. by def.

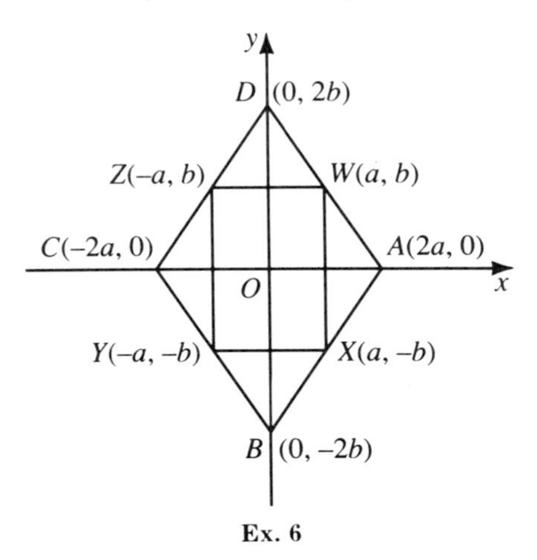

Ex. 6

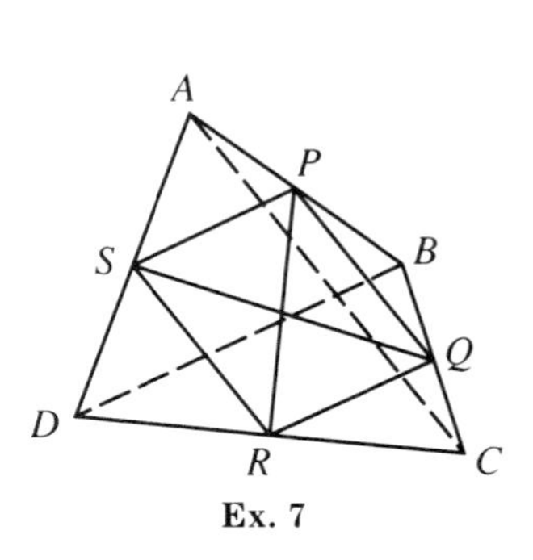

Ex. 7

7. Given: Quad. $ABCD$; P, Q, R, and S are midpts. of $\overline{AB}$, $\overline{BC}$, $\overline{CD}$, and $\overline{DA}$, resp.
Prove: $\overline{PR}$ and $\overline{QS}$ bis. each other.

Statements	Reasons
1. P, Q, R, and S are midpts. of $\overline{AB}$, $\overline{BC}$, $\overline{CD}$, and $\overline{DA}$, resp.	1. Given
2. $\overline{SP} \parallel \overline{DB}$ and $SP = \frac{1}{2}DB$; $\overline{RQ} \parallel \overline{DB}$ and $RQ = \frac{1}{2}DB$	2. The seg. that joins the midpts. of 2 sides of a △ is ∥ to the third side and is half as long as the third side.
3. $\overline{SP} \parallel \overline{RQ}$	3. 2 lines ∥ to a third line are ∥ to each other.
4. $SP = RQ$, or $\overline{SP} \cong \overline{RQ}$	4. Substitution Prop.
5. $SPQR$ is a ▱.	5. If one pair of opp. sides of a quad. are both ≅ and ∥, then the quad. is a ▱.
6. $\overline{PR}$ and $\overline{QS}$ bis. each other.	6. Diags. of a ▱ bis. each other.

Alternate proof (coordinate approach): Use the midpt. formula to find coords. $P(b + d, c + e)$, $Q(a + d, e)$, $R(a, 0)$, and $S(b, c)$. Then, the midpt. of $\overline{PR}$ is $\left(\frac{a + b + d}{2}, \frac{c + e}{2}\right)$, and the midpt. of $\overline{QS}$ is $\left(\frac{a + b + d}{2}, \frac{c + e}{2}\right)$. Since $\overline{PR}$ and $\overline{QS}$ have the same midpt., $\overline{PR}$ and $\overline{QS}$ bisect each other.

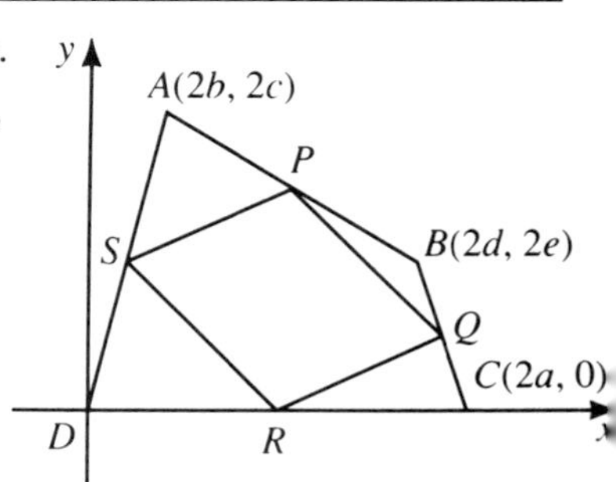

8. **a.** Slope of $\overline{AB} = \frac{3-1}{2-(-4)} = \frac{2}{6} = \frac{1}{3}$; slope of $\overline{BC} = \frac{9-3}{4-2} = \frac{6}{2} = 3$; slope of $\overline{CD} = \frac{7-9}{-2-4} = \frac{-2}{-6} = \frac{1}{3}$; slope of $\overline{DA} = \frac{7-1}{-2-(-4)} = \frac{6}{2} = 3$; $\overline{AB} \parallel \overline{CD}$ and $\overline{BC} \parallel \overline{DA}$, so $ABCD$ is a ▱ by def. Also, slope of $\overline{AC} = \frac{9-1}{4-(-4)} = \frac{8}{8} = 1$, and slope of $\overline{DB} = \frac{3-7}{2-(-2)} = \frac{-4}{4} = -1$, so $\overline{AC} \perp \overline{DB}$. **b.** $ABCD$ is a rhombus.

9. Proof (synthetic approach): Draw $\overline{XY}$ through P to form rectangles $AXYD$ and $XBCY$. Then $AX = DY$ and $CY = BX$. Using the Pythagorean Thm., $(PA)^2 = (AX)^2 + (XP)^2$, $(PB)^2 = (BX)^2 + (XP)^2$, $(PC)^2 = (CY)^2 + (YP)^2 = (BX)^2 + (YP)^2$, and $(PD)^2 = (DY)^2 + (YP)^2 = (AX)^2 + (YP)^2$. Then $(PA)^2 + (PC)^2 = (AX)^2 + (XP)^2 + (BX)^2 + (YP)^2 = (PB)^2 + (PD)^2$.

Alternate proof (coordinate approach): Assign coords. $A(0, a)$, $B(c, a)$, $C(c, 0)$, $D(0, 0)$, and $P(x, y)$. Use the distance formula to prove $(PA)^2 + (PC)^2 = (PB)^2 + (PD)^2$.

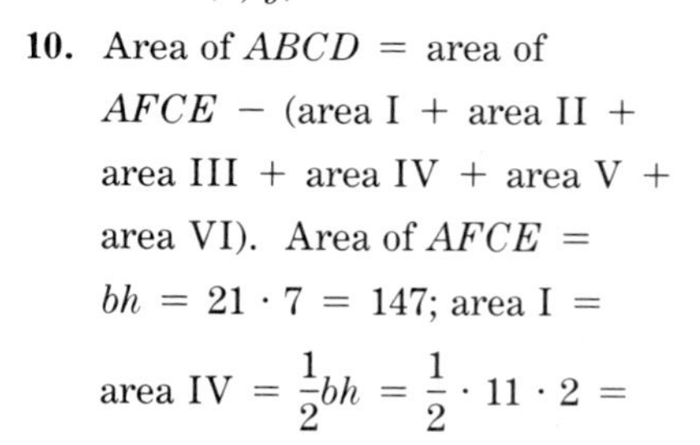

10. Area of $ABCD$ = area of $AFCE$ − (area I + area II + area III + area IV + area V + area VI). Area of $AFCE = bh = 21 \cdot 7 = 147$; area I = area IV $= \frac{1}{2}bh = \frac{1}{2} \cdot 11 \cdot 2 = 11$; area II = area V $= bh = 10 \cdot 2 = 20$; area III = area VI $= \frac{1}{2}bh = \frac{1}{2} \cdot 10 \cdot 5 = 25$. So, area of $ABCD = 147 - (11 + 20 + 25 + 11 + 20 + 25) = 147 - 112 = 35$.

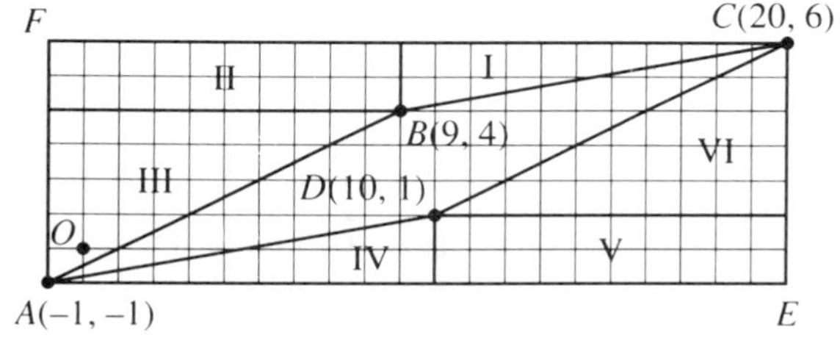

Alternate method (coordinate approach): Slope of $\overline{AB} = \frac{1}{2}$; slope of $\overline{CD} = \frac{1}{2}$; slope of $\overline{BC} = \frac{2}{11}$; slope of $\overline{AD} = \frac{2}{11}$. Therefore, $\overline{AB} \parallel \overline{CD}$ and $\overline{BC} \parallel \overline{AD}$, so $ABCD$ is a ▱. The line through A and $\perp$ to $\overleftrightarrow{CD}$ has equation $y + 1 = -2(x + 1)$ or $y = -2x - 3$. $\overline{CD}$ has equation $y - 1 = \frac{1}{2}(x - 10)$ or $y = \frac{1}{2}x - 4$. Solving these equations simultaneously gives $P(0.4, -3.8)$ as the intersection pt. Then area of $ABCD = bh = CD \cdot AP = \sqrt{10^2 + 5^2} \cdot \sqrt{(1.4)^2 + (-2.8)^2} = 35$.

11. Proof (synthetic approach): See figure below. Extend $\overrightarrow{AB}$ and $\overrightarrow{DC}$ to intersect at X. $\angle X$ is a rt. $\angle$. Draw diags. $\overline{AC}$ and $\overline{BD}$. Using the Pythagorean Thm., $(AC)^2 = (AX)^2 + (XC)^2$, $(BD)^2 = (BX)^2 + (XD)^2$, $(AD)^2 = (AX)^2 + (XD)^2$, and $(BC)^2 = (BX)^2 + (XC)^2$. Then $(AC)^2 + (BD)^2 = (AX)^2 + (XC)^2 + (BX)^2 + (XD)^2 = (AX)^2 + (XD)^2 + (BX)^2 + (XC)^2 = (AD)^2 + (BC)^2$.

Alternate proof (coordinate approach): Place the x-axis along $\overline{AB}$ and the y-axis along $\overline{CD}$. Assign coordinates as follows: $A(a, 0)$, $B(b, 0)$, $C(0, c)$, and $D(0, d)$. Then use the distance formula to show that $(AC)^2 = a^2 + c^2$, $(BD)^2 = b^2 + d^2$, $(AD)^2 = a^2 + d^2$, and $(BC)^2 = b^2 + c^2$. So $(AC)^2 + (BD)^2 = (AD)^2 + (BC)^2$.

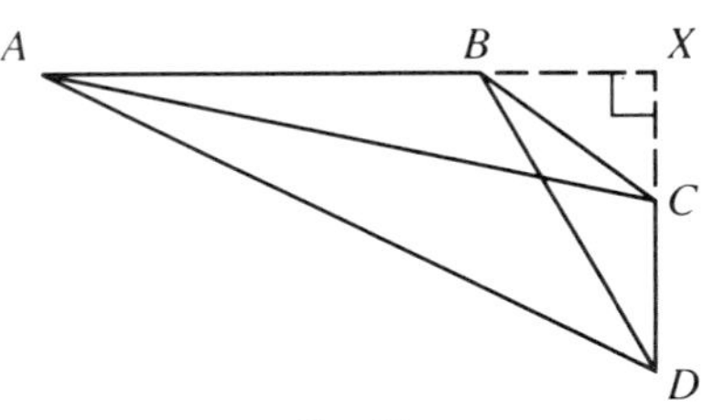

Ex. 11

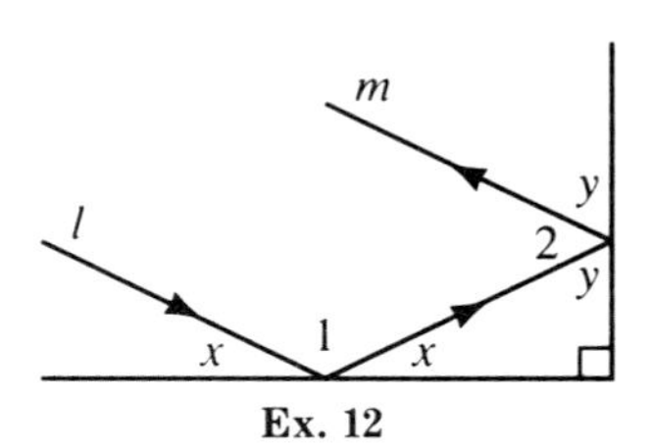

Ex. 12

12. $y = 90 - x$, so $m\angle 2 = 180 - 2y = 180 - 2(90 - x) = 2x$; $m\angle 1 = 180 - 2x$; $m\angle 1 + m\angle 2 = (180 - 2x) + 2x = 180$, so $\angle 1$ and $\angle 2$ are supp. int. $\angle$s, and rays l and m are $\parallel$.

13. Translate line l toward line m a distance AB. Then l' intersects m at Y. The line through Y and $\parallel$ to $\overline{AB}$ intersects l at X. So $\overline{XY} \parallel \overline{AB}$ and $\overline{XY} \cong \overline{AB}$.

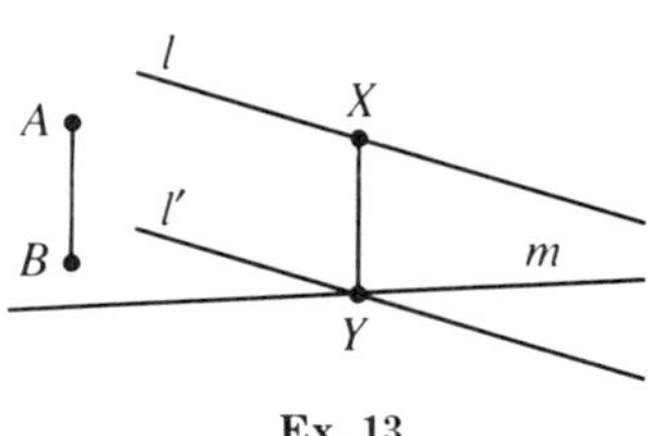

Ex. 13

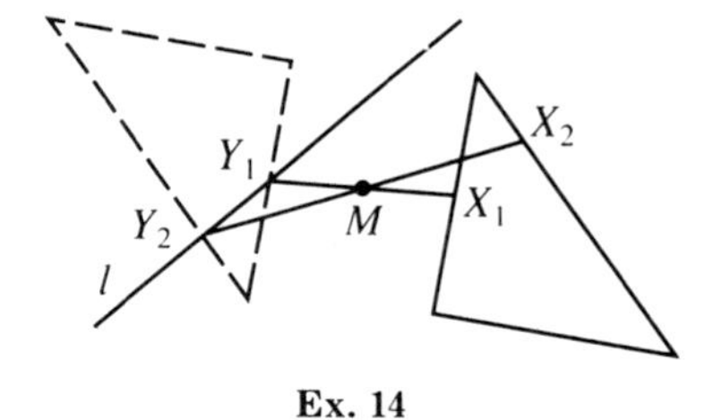

Ex. 14

14. Rotate the $\triangle$ 180° about M. The image under the half-turn intersects l at 2 pts., Y_1 and Y_2. Draw $\overrightarrow{Y_1M}$ to intersect the $\triangle$ at X_1, the preimage of Y_1. Draw $\overrightarrow{Y_2M}$ to intersect the $\triangle$ at X_2, the preimage of Y_2. M is the midpt. of $\overline{X_1Y_1}$ and $\overline{X_2Y_2}$ (2 solutions).

15. Const. $\parallel$ lines j, k, and l. Choose A on k. Rotate l 60° about A to get B on j. Rotate B $-60°$ about A to get C on l. Draw equilateral $\triangle ABC$.

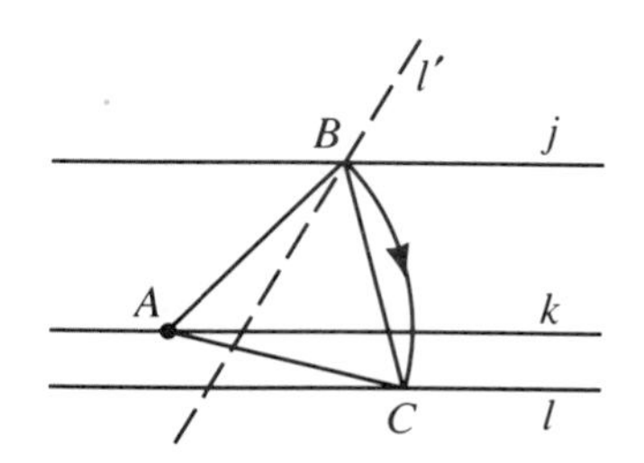

16. Let B be the pt. of tangency, $A = (9, 13)$, and $C = (0, 0)$. $AC = \sqrt{9^2 + 13^2} = \sqrt{250}$, so $(AC)^2 = 250$; $BC = \sqrt{25}$, so $(BC)^2 = 25$. In rt. $\triangle ABC$, $(AB)^2 = (AC)^2 - (BC)^2 = 250 - 25 = 225$. So $AB = \sqrt{225} = 15$.

17.

Statements	Reasons
1. $\overline{MN}$ is the median of trap. $ABCD$.	1. Given
2. N is the midpt. of $\overline{CB}$.	2. Def. of median of a trap.
3. $\overline{MN} \parallel \overline{AB}$ and $\overline{MN} \parallel \overline{DC}$	3. The median of a trap. is $\parallel$ to the bases.
4. X is the midpt. of $\overline{AC}$; Y is the midpt. of $\overline{DB}$.	4. A line that contains the midpt. of one side of a $\triangle$ and is $\parallel$ to another side passes through the midpt. of the third side.
5. $XN = \frac{1}{2}AB$; $YN = \frac{1}{2}DC$	5. The seg. that joins the midpts. of 2 sides of a $\triangle$ is half as long as the third side.
6. $XN = XY + YN$	6. Seg. Add. Post.
7. $\frac{1}{2}AB = XY + \frac{1}{2}DC$	7. Substitution Prop.
8. $XY = \frac{1}{2}(AB - DC)$	8. Algebra

18.

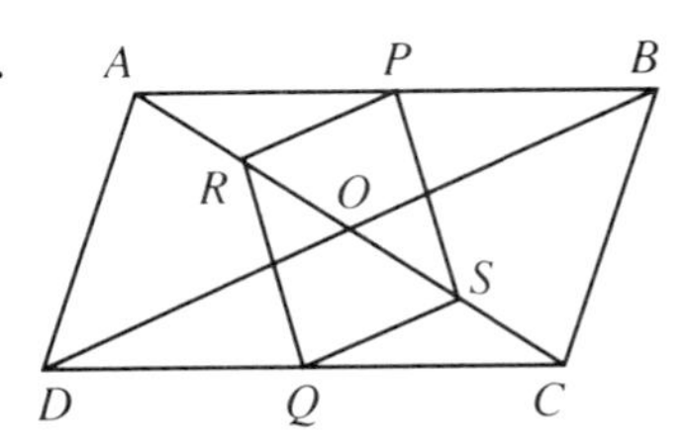

Statements	Reasons
1. P is the midpt. of $\overline{AB}$ and R is the midpt. of $\overline{AO}$; Q is the midpt. of $\overline{CD}$ and S is the midpt. of $\overline{CO}$.	1. Given
2. $\overline{RP} \parallel \overline{OB}$ and $RP = \frac{1}{2}OB$; $\overline{QS} \parallel \overline{DO}$ and $QS = \frac{1}{2}DO$	2. The seg. that joins the midpts. of 2 sides of a $\triangle$ is $\parallel$ to the third side and half as long as the third side.
3. $\overline{RP} \parallel \overline{QS}$	3. 2 lines $\parallel$ to a third line are $\parallel$ to each other.
4. $ABCD$ is a ▱.	4. Given
5. $DO = OB$	5. Diags. of a ▱ bis. each other.
6. $\frac{1}{2}DO = \frac{1}{2}OB$	6. Mult. Prop. of =
7. $RP = QS$, or $\overline{RP} \cong \overline{QS}$	7. Substitution Prop.
8. $PRQS$ is a ▱.	8. If one pair of opp. sides of a quad. are both $\cong$ and $\parallel$, then the quad. is a ▱.

19. Reflect the circle $(x - 2)^2 + (y - 4)^2 = 25$ in the y-axis. Its image is the circle with equation $(x + 2)^2 + (y - 4)^2 = 25$, which intersects the other circle in 2 pts., $Y_1(-5, 8)$ and $Y_2(-6, 1)$. Let X_1 and X_2 be the preimages of Y_1 and Y_2. Then $X_1 = (5, 8)$ and $X_2 = (6, 1)$. The y-axis is the $\perp$ bis. of $\overline{X_1Y_1}$ and $\overline{X_2Y_2}$ (2 solutions).

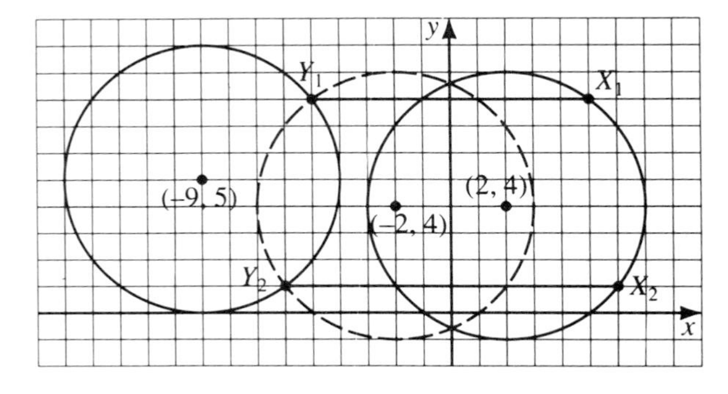

20. Place the x-axis along $\overline{AB}$ with $A(0, 0)$; then $B(12, 0)$, $C(a, 0)$, and $P(a, b)$. Draw the $\perp$ from X to $\overline{AB}$ int. $\overline{AB}$ at R. $\angle XAR \cong \angle APC$ since they are both comps. of $\angle PAR$. By AAS, $\triangle XAR \cong \triangle APC$, so $AR = PC = b$ and $XR = AC = a$. Thus, $X(b, -a)$ may be assigned. Draw the $\perp$ from Y to $\overline{AB}$ int. $\overline{AB}$ at S. $\angle YBS \cong \angle BPC$ since they are both comps. of $\angle PBC$. By AAS, $\triangle YBS \cong \triangle BPC$, so $SB = PC = b$ and $SY = CB = 12 - a$. Thus, $Y(12 - b, a - 12)$ may be assigned. Then the midpt. of $\overline{XY}$ is $\left(\frac{b + (12 - b)}{2}, \frac{-a + (a - 12)}{2}\right) = (6, -6)$. Since $M(6, -6)$ does not involve a or b, the position of M does not depend on the position of P.

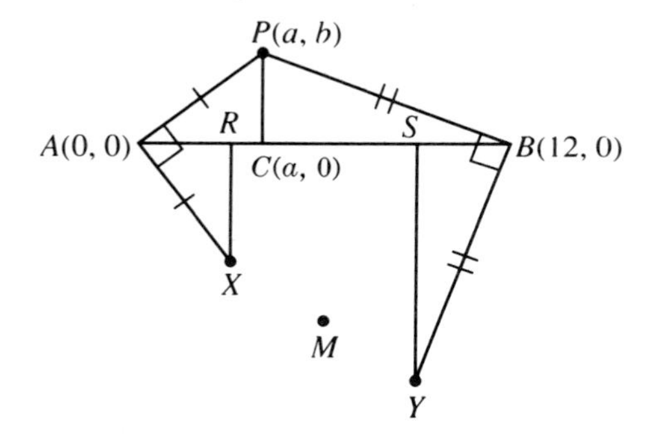

Pages 677–678 • EXERCISES

1. No Euler circuit is possible since vertices B and C have odd valences.

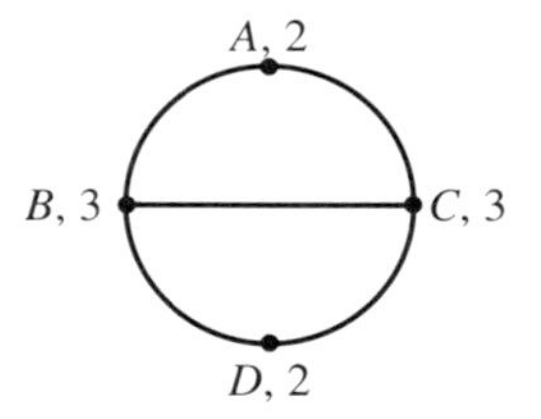

2. An Euler circuit is possible since every vertex has an even valence. A possible circuit is $AEDECDABCA$.

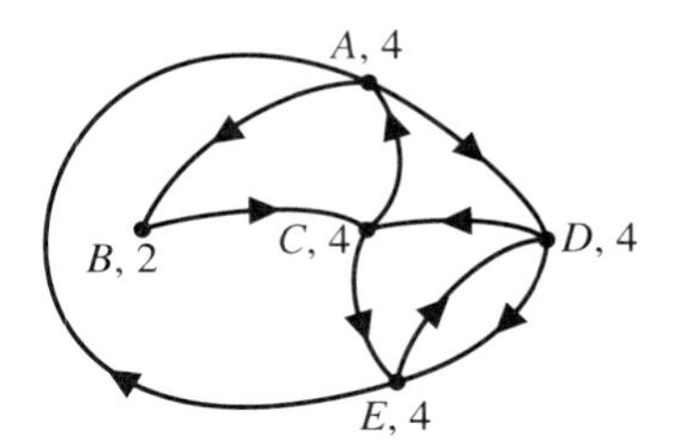

3. A possible path is $CBDCAB$, shown by the graph.

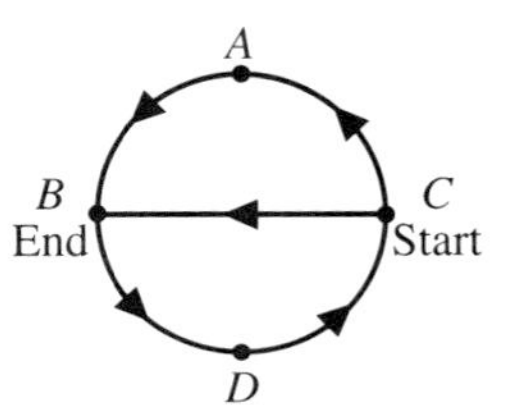

4.

a. No, since vertices D and H each have an odd valence (5).

b. Yes, a possible path is $HABCDEFGHBDFHD$.

5.

a. No, since there are an odd number of doorways to/from the outdoors and to/from the utility room.

b. Yes; a possible path starts outdoors and ends in the utility room: outdoors, foyer, den, kitchen, foyer, living room, dining room, kitchen, utility room, outdoors, utility room.

c. It is possible to travel a path starting at one vertex with an odd valence and ending at another vertex with an odd valence, only if the graph contains exactly two vertices with odd valences.

6. a.

Vertex	Invalence	Outvalence
A	2	2
B	1	1
C	1	3
D	1	1
E	2	2
F	1	1
G	2	2
H	1	1
K	3	1

b. No; the invalence must equal the outvalence for every vertex, and this is not true for C or K.

c. Yes; change the direction of the arrow between K and C. A possible order is $KGHABCAKCDEGFEK$.

7. a.

Team	Wins	Losses
Bears	2	0
Lions	1	1
Tigers	2	1
Eagles	0	2
Panthers	1	2

b.

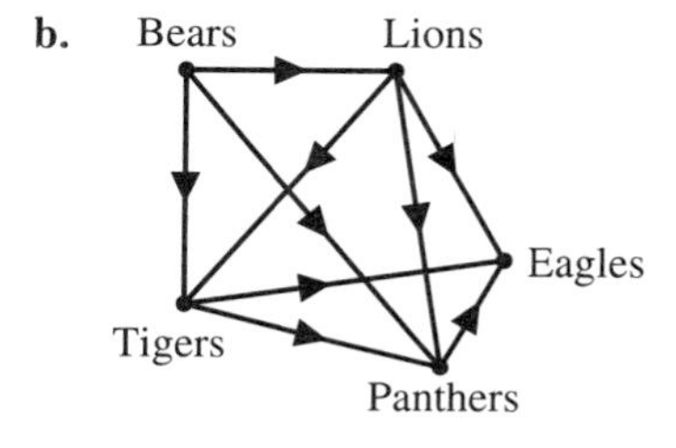

Bears and Tigers have the most wins.

c. Bears should be ranked first since they have three wins, no losses, and a victory over the only other team with three wins (Lions). Eagles should be ranked last since they have no wins, three losses, and a loss to the only other team with three losses (Panthers).

Pages 679–681 • EXERCISES

1. No Hamilton circuit is possible.
2. Yes, a Hamilton circuit is possible: $LDGFXYHPEZVL$.
3. **a.** Yes

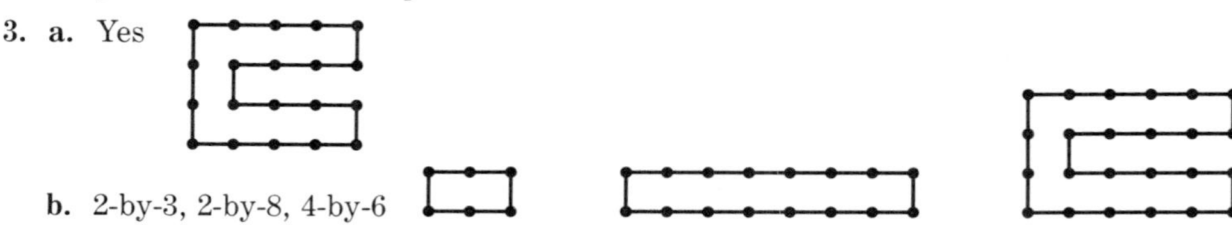

b. 2-by-3, 2-by-8, 4-by-6

c. At least one of x or y must be even for a Hamilton circuit to be possible.

4.

Circuit	Total Distance
FEHGF	140 + 80 + 150 + 100 = 470
FEGHF	140 + 200 + 150 + 130 = 620
FHEGF	130 + 80 + 200 + 100 = 510
FHGEF	130 + 150 + 200 + 140 = 620
FGEHF	100 + 200 + 80 + 130 = 510
FGHEF	100 + 150 + 80 + 140 = 470

The shortest circuit is *FGHEF*, or its reverse, *FEHGF*.

5. Yes; the circuit using the nearest neighbor algorithm is *FGHEF*, the shortest circuit.

6. **a.** $14! = 87{,}178{,}291{,}200$

   b. $14 \times 14! \div 10^9 \approx 1220.5$ s; $1220.5 \div 60 = 20.34$ min $\approx$ 20 min 20 s

7. **a.** $20! = 2{,}432{,}902{,}008{,}176{,}640{,}000$

   b. 20

   c. $20 \times 20! = 48{,}658{,}040{,}163{,}532{,}800{,}000$

   d. $20 \times 20! \div 10^9 \approx 4.8658 \times 10^{10}$ s; $4.8658 \times 10^{10} \div 60 \div 60 \div 24 \div 365 \approx$ 1543 years

8. **a.** $25! \approx 1.551121004 \times 10^{25}$

   b. 25

   c. $25 \times 25!$

   d. $25 \times 25! \div 10^9 \div 60 \div 60 \div 24 \div 365 \approx 12{,}296{,}431{,}097$ years

9. **a.**

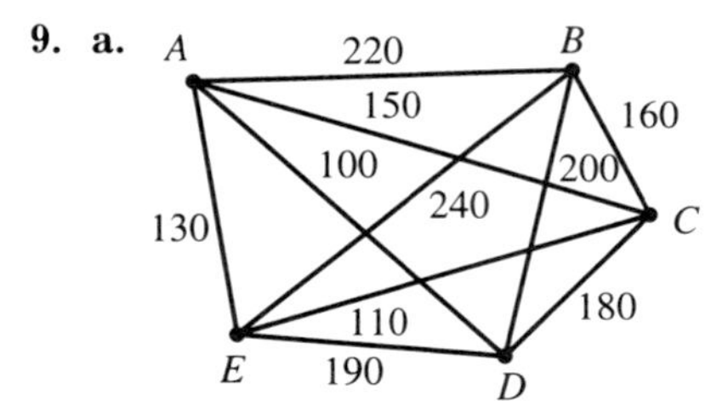

   b. Using the nearest neighbor algorithm, the least expensive circuit is *ADCEBA* = \$100 + \$180 + \$110 + \$240 + \$220 = \$850.

   c. A less expensive circuit is *AECBDA* = \$130 + \$110 + \$160 + \$200 + \$100 = \$700.

10. Yes

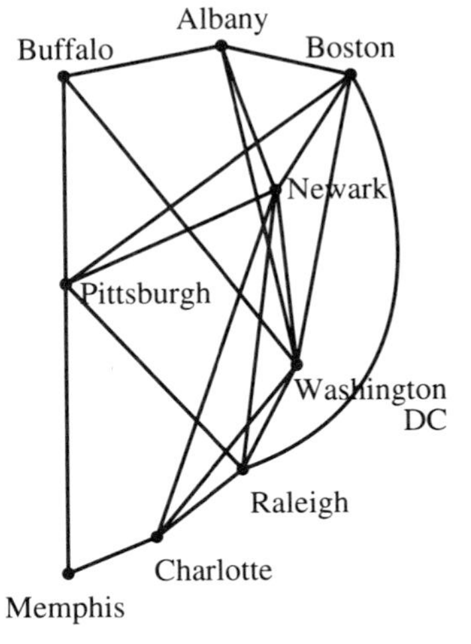

11. Example: The postal service collecting mail from mailboxes or delivering packages to specific destinations.

Pages 682–683 • EXERCISES

1.

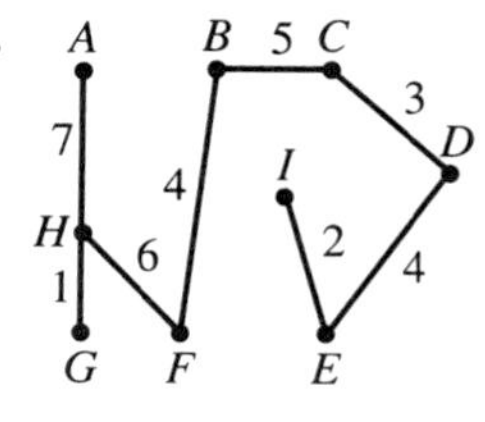

$1 + 2 + 3 + 4 + 4 + 5 + 6 + 7 = 32$

2.

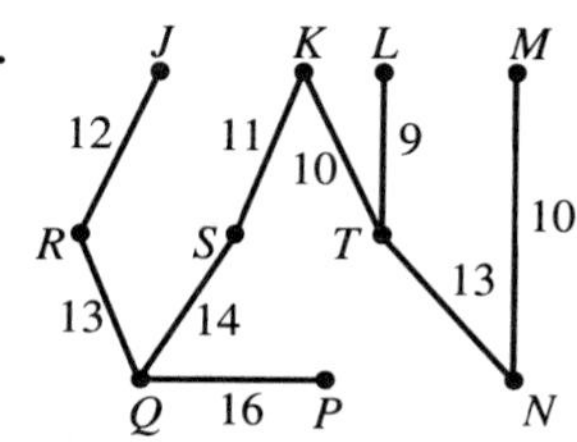

$9 + 10 + 10 + 11 + 12 + 13 + 13 + 14 + 16 = 108$

3. If a graph has n vertices, its minimal spanning tree has $n - 1$ edges, as in Exercises 1 and 2 above.

4. a. Person: to save the greatest amount of money using coupons. Company: to make the greatest profit on options.

b.

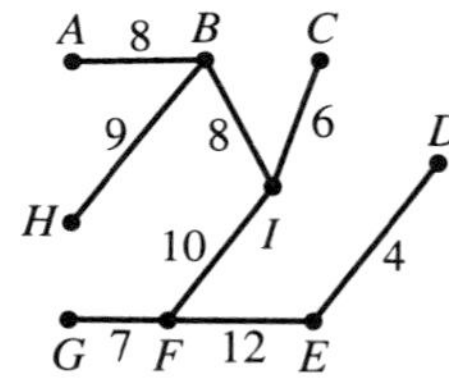

$12 + 10 + 9 + 8 + 8 + 7 + 6 + 4 = 64$

c. Cost of the network formed by the maximal spanning tree is \$6400, from Exercise 4(b). Cost of the network formed by the minimal spanning tree is \$3200, from Exercise 1. Difference in costs is \$6400 − \$3200 = \$3200.

5.

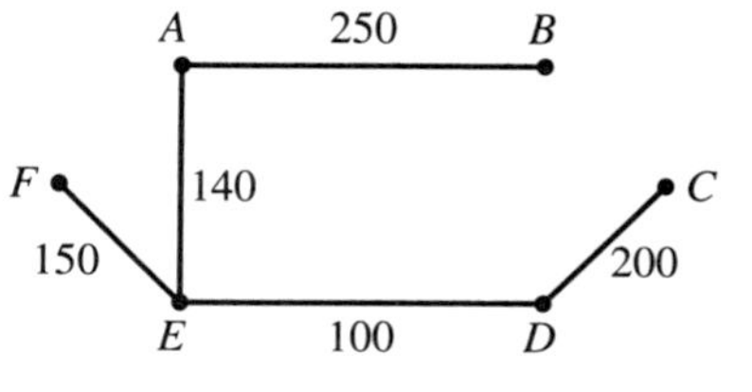

\$100 + \$140 + \$150 + \$200 + \$250 = \$840

6.

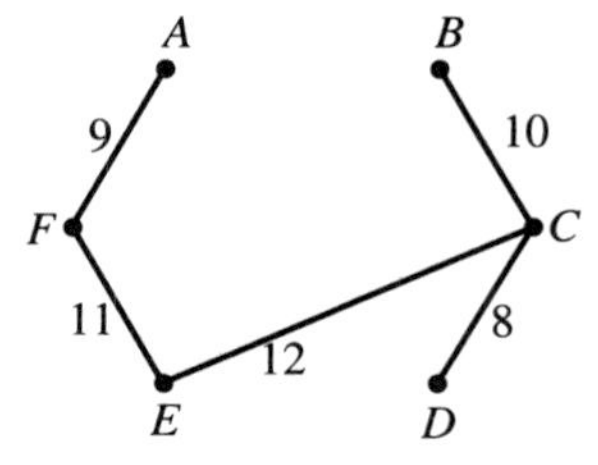

$8 + 9 + 10 + 11 + 12 = 50$; \$50,000

7. a.

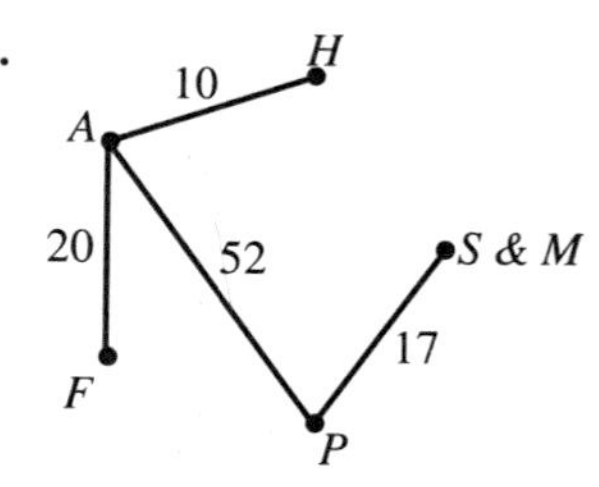

$52 + 10 + 17 + 20 = 99$; \$99,000

b.

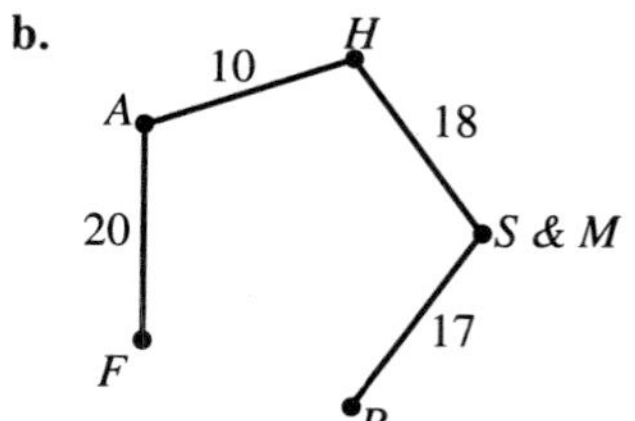

$10 + 17 + 18 + 20 = 65$;
\$99,000 − \$65,000 = \$34,000

Fractal Geometry

Pages 685–686 • EXERCISES

1–3. Answers may vary slightly, depending on whether students swing the compass inside or outside the coastline, and depending on their accuracy and estimation skills. Examples are given.

1.

Ruler Length	Number of Sides	Perimeter
1 in.	7	7 in.
$\frac{1}{2}$ in.	15	$7\frac{1}{4}$ in.
$\frac{1}{4}$ in.	30	$7\frac{1}{2}$ in.

2.

Ruler Length	Number of Segments	Length
1 in.	4	$3\frac{1}{2}$ in.
$\frac{1}{2}$ in.	12	$5\frac{1}{2}$ in.
$\frac{1}{4}$ in.	30	$7\frac{1}{2}$ in.

3.

Ruler Length	Number of Sides	Perimeter
4 in.	5	17 in.
2 in.	10	$18\frac{1}{4}$ in.
1 in.	19	$18\frac{3}{4}$ in.

a. Yes **b.** The approximate length of the circumference will always be less than 6π, or 18.85, in. **c.** The circumference of a circle has a measurable length; a jagged coastline does not.

Pages 688–690 • EXERCISES

1. a.

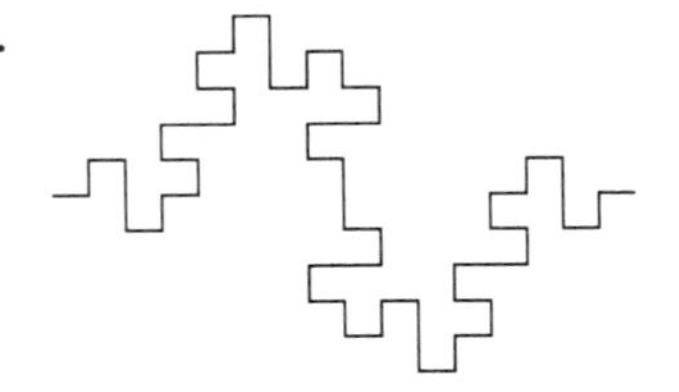

b.

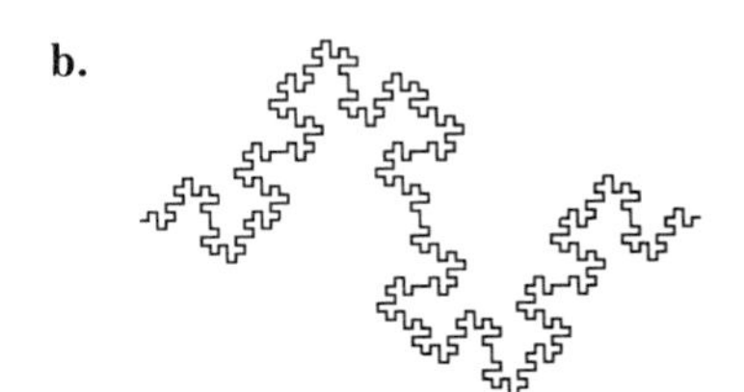

c.

Level Number	Edge Length	Number of Edges	Total Length
0	1	1	1
1	$\frac{1}{4}$	8	2
2	$\frac{1}{16}$	64	4
3	$\frac{1}{64}$	512	8

2. a.

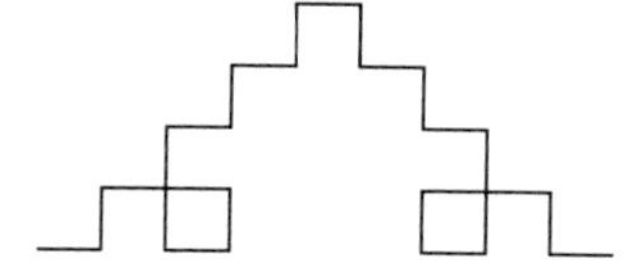

b.

c.

Level Number	Edge Length	Number of Edges	Total Length
0	1	1	1
1	$\frac{1}{3}$	5	$\frac{5}{3} = 1\frac{2}{3}$
2	$\frac{1}{9}$	25	$\frac{25}{9} = 2\frac{7}{9}$
3	$\frac{1}{27}$	125	$\frac{125}{27} = 4\frac{17}{27}$

3. a.

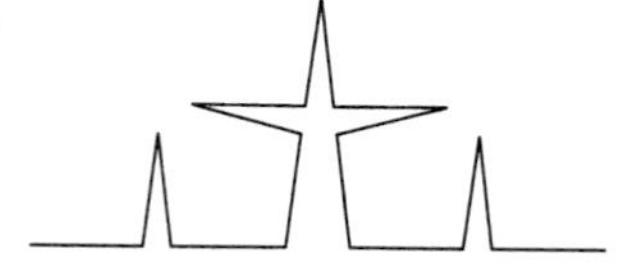

b.

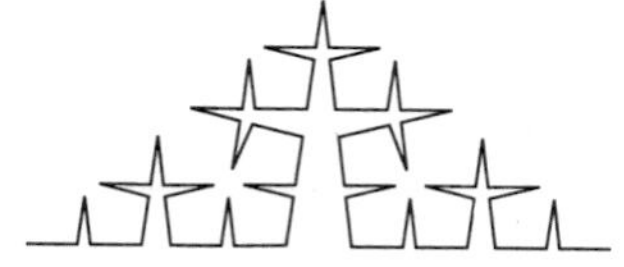

c.

Level Number	Edge Length	Number of Edges	Total Length
0	1	1	1
1	$\frac{4}{9}$	4	$\frac{16}{9} = 1\frac{7}{9}$
2	$\frac{16}{81}$	16	$\frac{256}{81} = 3\frac{13}{81}$
3	$\frac{64}{729}$	64	$\frac{4096}{729} = 5\frac{451}{729}$

4. a.

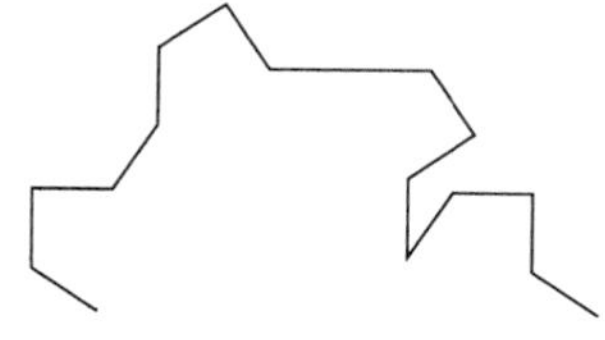

b.

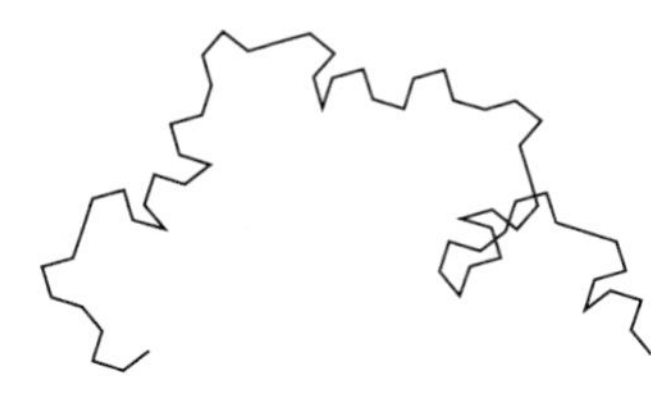

c.

Level Number	Edge Length	Number of Edges	Total Length
0	1	1	1
1	$\frac{2}{5}$	4	$\frac{8}{5} = 1\frac{3}{5}$
2	$\frac{4}{25}$	16	$\frac{64}{25} = 2\frac{14}{25}$
3	$\frac{8}{125}$	64	$\frac{512}{125} = 4\frac{12}{125}$

5. a.

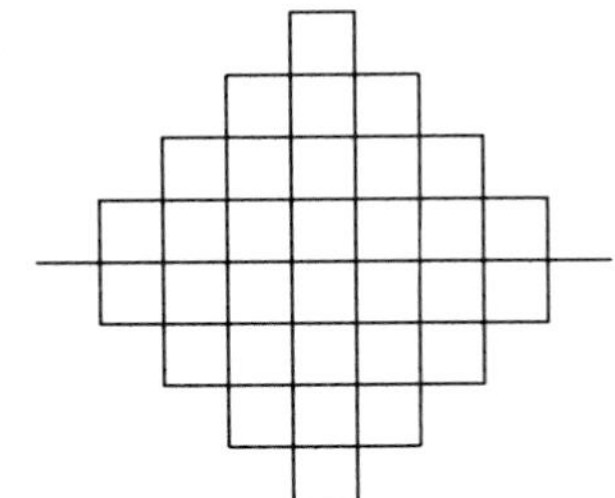

b.

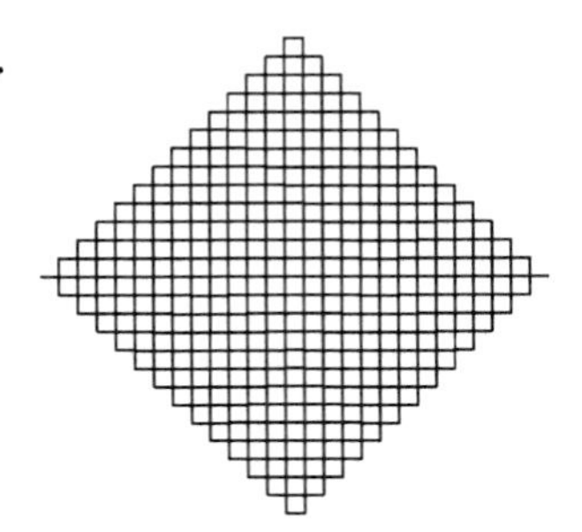

c.

Level Number	Edge Length	Number of Edges	Total Length
0	1	1	1
1	$\frac{1}{3}$	9	3
2	$\frac{1}{9}$	81	9
3	$\frac{1}{27}$	729	27

6. a. As the pre-fractal level increases, the edge length for the Sierpiński gasket decreases by a factor of $\frac{1}{2}$. **b.** The number of edges increases by a factor of 3. **c.** The sum of the lengths increases by a factor of $\frac{3}{2}$.

7. a. As the pre-fractal level increases, the edge length for the Cantor set decreases by a factor of $\frac{1}{3}$. **b.** The number of edges increases by a factor of 2. **c.** The sum of the lengths decreases by a factor of $\frac{2}{3}$.

8. a.

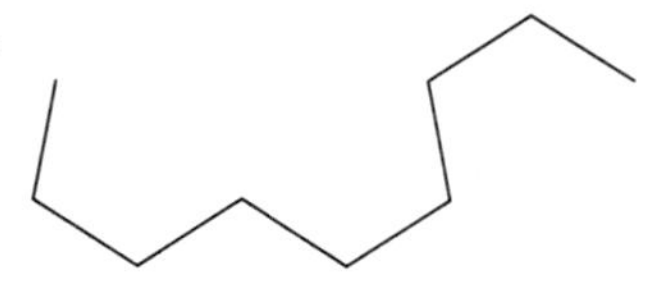

b.

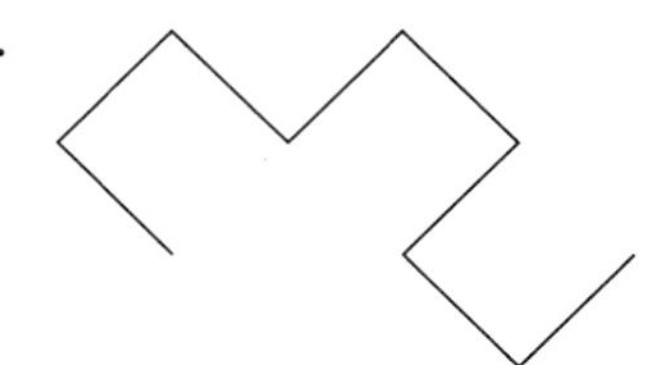

9. The level 3 pre-fractal for the Cantor set is defined by 16 endpoints. The boldface points were given as endpoints of the level 1 pre-fractal:
$(\mathbf{0}, \mathbf{0})$, $\left(\frac{1}{27}, 0\right)$, $\left(\frac{2}{27}, 0\right)$, $\left(\frac{1}{9}, 0\right)$, $\left(\frac{2}{9}, 0\right)$, $\left(\frac{7}{27}, 0\right)$, $\left(\frac{8}{27}, 0\right)$, $\left(\mathbf{\frac{1}{3}}, \mathbf{0}\right)$, $\left(\mathbf{\frac{2}{3}}, \mathbf{0}\right)$, $\left(\frac{19}{27}, 0\right)$, $\left(\frac{20}{27}, 0\right)$, $\left(\frac{7}{9}, 0\right)$, $\left(\frac{8}{9}, 0\right)$, $\left(\frac{25}{27}, 0\right)$, $\left(\frac{26}{27}, 0\right)$, $(\mathbf{1}, \mathbf{0})$.

10. Since starting with a solid triangle and deleting its midtriangle is equivalent to replacing the triangle with three similar triangles with scale factor $\frac{1}{2}$, the resulting form is also a Sierpiński gasket. Assume the area of the first triangle is 1 square unit. The table below shows data for the sequence of pre-fractals. The area approaches zero.

Level Number	Triangle Area	Number of Triangles	Sum of Areas
0	1	1	1
1	$\frac{1}{4}$	3	$\frac{3}{4}$
2	$\frac{1}{16}$	9	$\frac{9}{16}$
3	$\frac{1}{64}$	27	$\frac{27}{64}$

11. The area of the snowflake fractal is 1.6 times the area of the original equilateral triangle. To create each pre-fractal after level 0, we add a new triangle to every edge of the previous pre-fractal. The number of edges increases by a factor of 4 for each new pre-fractal, so after level 1, the number of new triangles to add also increases by a factor of 4. Each of the new triangles is $\frac{1}{9}$ the size of a new triangle added to the previous pre-fractal. Thus, after level 1, the new area added in creating the next pre-fractal is $\frac{4}{9}$ the area added in the previous step. The table below shows the areas added to each level, and the total area as it grows.

Level Number	Number of Edges	Number of New Triangles	Area of Each New Triangle	New Area Added	Total Area
0	3	1	1	1	1
1	12	3	$\frac{1}{9}$	$\frac{3}{9} = \frac{1}{3}$	$1 + \frac{1}{3} \approx 1.333$
2	48	12	$\frac{1}{81}$	$\frac{12}{81} = \frac{4}{27}$	$1 + \frac{1}{3} + \frac{4}{27} \approx 1.481$
3	192	48	$\frac{1}{729}$	$\frac{48}{729} = \frac{16}{243}$	$1 + \frac{1}{3} + \frac{4}{27} + \frac{16}{243} \approx$ 1.547

(Continued)

To prove that the total area approaches 1.6, you must use the formula for the sum of an infinite geometric series, generally a topic of advanced algebra. The sum $\frac{1}{3} + \frac{4}{27} + \frac{16}{243} + \ldots$ is an infinite geometric series with first term $\frac{1}{3}$ and common ratio $\frac{4}{9}$. Its sum is $\dfrac{\frac{1}{3}}{1 - \frac{4}{9}} = \frac{1}{3} \cdot \frac{9}{5} = \frac{3}{5}$. Therefore, the sum $1 + \frac{1}{3} + \frac{4}{27} + \frac{16}{243} + \ldots = 1 + \frac{3}{5} = 1.6$.

The snowflake fractal is an example of a closed curve of infinite length that bounds a finite area.

Page 691 • EXERCISES

1–8. Approximations should match the value derived by exact calculation: $D = \frac{\log N}{\log R}$.

1. $D = \frac{\log N}{\log R} = \frac{\log 8}{\log 4} = 1.5$

2. $D = \frac{\log N}{\log R} = \frac{\log 5}{\log 3} \approx 1.46$

3. $D = \frac{\log N}{\log R} = \dfrac{\log 4}{\log \frac{9}{4}} \approx 1.71$

4. $D = \frac{\log N}{\log R} = \dfrac{\log 4}{\log \frac{5}{2}} \approx 1.51$

5. $D = \frac{\log N}{\log R} = \frac{\log 9}{\log 3} = 2$

6. For the Sierpiński gasket, $N = 3$, $R = 2$, and $D = \frac{\log N}{\log R} = \frac{\log 3}{\log 2} \approx 1.58$.

7. For the Cantor set, $N = 2$, $R = 3$, and $D = \frac{\log N}{\log R} = \frac{\log 2}{\log 3} \approx 0.63$.

8. Students' fractals will vary.

Portfolio Projects

Assessing Student Performance on Portfolio Projects

The Portfolio Projects are designed to encourage independent or group exploration of mathematical concepts within new applications. The projects are intended to be idea-starters. Students should use their creativity both in forming the problem and in shaping the solution. Because of this, you should expect a variety of responses from students. Similarly, you can evaluate student performance in a number of ways.

Some teachers prefer an analytic scoring system, assigning a specific number of points to each of a list of predetermined steps (for example, awarding 2 points for understanding the problem, recognizing what information is given and what must be found; 2 points for making a plan and choosing an appropriate problem-solving strategy; 2 points for carrying out the plan appropriately; and 2 points for the presentation of results). Other teachers prefer a holistic approach, assessing each student's work in its entirety rather than as a collection of discrete parts.

Holistic Scoring and Scoring Rubrics

Many teachers who use holistic scoring start with written descriptions, or scoring rubrics, that distinguish among different levels of performance. Rather than assigning so many points per part of an answer, scoring rubrics tend to focus on qualitative distinctions. As you read the student responses the first time through, the rubrics should help you sort the works into four categories. The first group includes responses that show complete understanding and in which the ideas are clearly expressed. The second group includes responses that show some understanding of the basic ideas and concepts but which are incomplete or unclear. The third group includes responses that show little or no understanding of the problem or its solution and suggest the need for some remediation. The fourth group includes responses that show no effort. After the initial sorting of all student responses, the responses within each category can be arranged into subcategories as necessary in a second reading.

General Rubrics and Specific Rubrics

Scoring rubrics help clarify the key assessment criteria for yourself and for your students. General scoring rubrics serve as a guideline. This broad framework helps establish a dependable system, familiar to students, colleagues, administrators, and parents. One such general rubric follows:

Complete Understanding

4 points — Complete responses that demonstrate competence

Some Understanding

3 points — Responses show competence, though they may not be complete or entirely accurate

2 points — Responses show limited competence and contain significant errors

Little or No Understanding

1 point — Responses that show effort but little evidence of competence

No Attempt

0 points — Responses that show no effort

This is an example of a 4-point scoring system, with the primary emphasis being the distinction among the strongest responses, the weakest responses, and those in the middle. Which general rubric you use will depend somewhat on your own needs and preferences. Many educators with experience using rubrics, however, recommend using the general rubrics only as a framework from which specific rubrics for each problem or activity are written.

Specific rubrics allow you to highlight the project's distinctive requirements. The process of creating specific scoring rubrics for a project helps define the project's essential elements. Sharing these rubrics with students helps focus their work within the project. Many teachers, in fact, involve the students in writing the project's specific scoring rubrics.

Writing Scoring Rubrics

Scoring rubrics for a specific project should be general enough to allow for the unexpected (in order to recognize what each student might bring to the problem). At the same time, they need to be specific enough to make qualitative distinctions among the levels of performance. The elements you may want to consider in writing specific rubrics for a project are mathematical content, connections, the problem-solving process, reasoning, and communication.

The project-specific rubrics you develop should reflect your instructional goals and the needs of your students. Consider the elements as you identify the qualities that make up a complete response for a given project, as well as the qualitative differences between top-level and lower-level performance. Each of the Portfolio Projects will call upon students to use their abilities in all five of these areas, but some projects rely more heavily on one of these areas than on another. Your scoring rubrics for a particular project should focus primarily on the key elements for that project.

Suggestions for Specific Scoring Rubrics

Drawing a Star (Chapter 1)
Complete Understanding
Drawings are clean and accurate, labeled with the directions used and the name of the resulting figures. Questions posed in the project are answered. Paragraph is clear and understandable, using sentences as well as math equations.
Some Understanding
Drawings are done with some degree of accuracy and some labeling. Questions may have been missed or misunderstood. Paragraph may be fragmented or misses some cases.
Little or No Understanding
Some drawings are done. No show of understanding past following directions is given.

Tessellations (Chapter 3)
This project has a wide scope for students with both artistic and mathematical talents.
Complete understanding
Drawings are clean, accurate, and include labeling. A variety of polygons is used.
Complexity may be present in the patterns; color may be used to reinforce patterns.
Some Understanding
Drawings are done with some degree of accuracy. At least the minimum number of examples are given with some labeling. Several different polygons are used.
Little or No Understanding
Drawings are imprecise. Few examples are given. Patterns are limited in their complexity.

Using Rubrics to Encourage Excellence

Assessment should not be considered the end of the instructional sequence. Many teachers find holistic scoring rubrics useful in encouraging students toward higher levels of performance. Scoring rubrics can be particularly helpful in a "review-and-revise" cycle within the process of working on the projects. The review and preliminary rating can be done by the teacher, or it can be done within a conference among students. In either case, because holistic scoring rubrics focus on qualitative distinctions among varying levels of performance, this preliminary review often helps the student see more clearly how to move his or her work from mid-range to top-level performance.